Table for Evaluating $\int_0^L m\, m'\, dx$

$\int_0^L m\, m'\, dx$	rectangle m', L	triangle m', L	trapezoid m'_1, m'_2, L	parabola m', L
rectangle m, L	$mm'L$	$\dfrac{1}{2}mm'L$	$\dfrac{1}{2}m(m'_1+m'_2)L$	$\dfrac{2}{3}mm'L$
triangle m, L	$\dfrac{1}{2}mm'L$	$\dfrac{1}{3}mm'L$	$\dfrac{1}{6}m(m'_1+2m'_2)L$	$\dfrac{5}{12}mm'L$
trapezoid m_1, m_2, L	$\dfrac{1}{2}m'(m_1+m_2)L$	$\dfrac{1}{6}m'(m_1+2m_2)L$	$\dfrac{1}{6}[m'_1(2m_1+m_2)+m'_2(m_1+2m_2)]L$	$\dfrac{1}{12}[m'(3m_1+5m_2)]L$
triangle peak (a, b), m	$\dfrac{1}{2}mm'L$	$\dfrac{1}{6}mm'(L+a)$	$\dfrac{1}{6}m[m'_1(L+b)+m_2(L+a)]$	$\dfrac{1}{12}mm'\left(3+\dfrac{3a}{L}-\dfrac{a^2}{L^2}\right)L$
triangle m, L	$\dfrac{1}{2}mm'L$	$\dfrac{1}{6}mm'L$	$\dfrac{1}{6}m(2m'_1+m'_2)L$	$\dfrac{1}{4}mm'L$

Beam Deflections and Slopes

Loading	$v +$ ↑	$\theta +$ ↖	Equation $+$ ↑ $+$ ↖
cantilever with P at free end, x, L	$v_{max}=-\dfrac{PL^3}{3EI}$ at $x=L$	$\theta_{max}=-\dfrac{PL^2}{2EI}$ at $x=L$	$v=\dfrac{P}{6EI}(x^3-3Lx^2)$
cantilever with M_O at free end, x, L	$v_{max}=\dfrac{M_OL^2}{2EI}$ at $x=L$	$\theta_{max}=\dfrac{M_OL}{EI}$ at $x=L$	$v=\dfrac{M_O}{2EI}x^2$

Beam Deflections and Slopes (continued)

	$v_{max} = -\dfrac{wL^4}{8EI}$ at $x = L$	$\theta_{max} = -\dfrac{wL^3}{6EI}$ at $x = L$	$v = -\dfrac{w}{24EI}(x^4 - 4Lx^3 + 6L^2x^2)$
	$v_{max} = -\dfrac{PL^3}{48EI}$ at $x = L/2$	$\theta_{max} = \pm\dfrac{PL^2}{16EI}$ at $x = 0$ or $x = L$	$v = \dfrac{P}{48EI}(4x^3 - 3L^2x)$, $0 \le x \le L/2$
		$\theta_L = -\dfrac{Pab(L + b)}{6LEI}$ $\theta_R = \dfrac{Pab(L + a)}{6LEI}$	$v = -\dfrac{Pbx}{6LEI}(L^2 - b^2 - x^2)$ $0 \le x \le a$
	$v_{max} = -\dfrac{5wL^4}{384EI}$ at $x = \dfrac{L}{2}$	$\theta_{max} = \pm\dfrac{wL^3}{24EI}$	$v = -\dfrac{wx}{24EI}(x^3 - 2Lx^2 + L^3)$
		$\theta_L = -\dfrac{3wL^3}{128EI}$ $\theta_R = \dfrac{7wL^3}{384EI}$	$v = -\dfrac{wx}{384EI}(16x^3 - 24Lx^2 + 9L^3)$ $0 \le x \le L/2$ $v = -\dfrac{wL}{384EI}(8x^3 - 24Lx^2 + 17L^2x - L^3)$ $L/2 \le x \le L$
	$v_{max} = -\dfrac{M_OL^2}{9\sqrt{3}EI}$	$\theta_L = -\dfrac{M_OL}{6EI}$ $\theta_R = \dfrac{M_OL}{3EI}$	$v = -\dfrac{M_Ox}{6EIL}(L^2 - x^2)$

STRUCTURAL ANALYSIS
EIGHTH EDITION IN SI UNITS

Thank you for purchasing a new copy of *Structural Analysis*, Eighth Edition in SI Units, by R. C. Hibbeler. The one-time access code below provides access to the Video Solutions found on the Companion Website.

For students only:
The Video Solutions provide a complete step-by-step solution walkthroughs of representative homework problems from each chapter. These videos offer:

- **Fully-worked Solutions** — Showing every step of representative homework problems, to help students make vital connections between concepts.
- **Self-paced Instruction** — Students can navigate each problem and select, play, rewind, fast-forward, stop, and jump-tosections within each problem's solution.
- **24/7Access** — Help whenever you need it.

To access the Video Solutions found on the companion website for the first time:

1. Go to http://www.pearsoned-asia.com/hibbeler
2. Click on the title *Structural Analysis*, Eighth Edition in SI Units
3. Click on "Video Solutions".
4. Enter the one-time access code given below.

5. Follow the instructions on screen from then on.   HSAS-TREAD-GAMIC-SCOUT-ANETO-VIRES

For instructors only:
To access the suite of Instructor's Resources available for your use on the Companion Website, please follow steps 1 and 2 above and click on "Instructor's Resources". You will be prompted to enter a User ID and Password. Please proceed if you are already registered and have an existing instructor's access.

If you do not have an instructor's access, please contact your Pearson Education Representative. He/She will then register for you and provide you with a User ID and Password.

The Video Solutions can be found on the Instructor's Resources webpage on the Companion Website.

IMPORTANT: The access code on this page can only be used once to establish a subscription to the Video Solutions on the Structural Analysis, Eighth Edition in SI Units Companion Website. Each copy of this title sold through your local bookstore is shrink-wrapped to protect the one-time access code. If you have bought a copy of the book that is not shrink-wrapped, please contact your local bookstore. Alternatively, you can purchase a subscription to the Video Solutions by following steps 1 and 2 above and click on "Get Access".

PRENTICE HALL   www.pearsoned-asia.com

STRUCTURAL ANALYSIS

EIGHTH EDITION IN SI UNITS

STRUCTURAL ANALYSIS

EIGHTH EDITION IN SI UNITS

R. C. Hibbeler

SI Conversion by
Tan Kiang Hwee
National University of Singapore

PRENTICE HALL

Singapore   London   New York   Toronto   Sydney   Tokyo   Madrid
Mexico City   Munich   Paris   Capetown   Hong Kong   Montreal

Published in 2011 by
Pearson Education South Asia Pte Ltd
23/25 First Lok Yang Road, Jurong
Singapore 629733

Pearson Asia Pacific offices: *Bangkok, Beijing, Hong Kong, Jakarta, Kuala Lumpur, Manila, New Delhi, Seoul, Singapore, Taipei, Tokyo*

Original edition *Structural Analysis*, Eighth Edition, by R. C. Hibbeler, published by Prentice-Hall, Inc.
(ISBN 0-13-257053-4)
Copyright © 2012 by R. C. Hibbeler

Previous editions copyright © 2009, 2006, 2002, 1999, 1995, 1985 by R. C. Hibbeler.

Printed in Singapore

4   3   2   1
14   13   12   11

ISBN   978-981-06-8713-7

The author and publisher of this book have used their best efforts in preparing this book. These efforts include the development, research, and testing of theories and programs to determine their effectiveness. The author and publisher make no warranty of any kind, expressed or implied, with regard to these programs or the documentation contained in this book. The author and publisher shall not be liable in any event for incidental or consequential damages in connection with, or arising out of, the furnishing, performance, or use of these programs.

PRENTICE HALL   www.pearsoned-asia.com

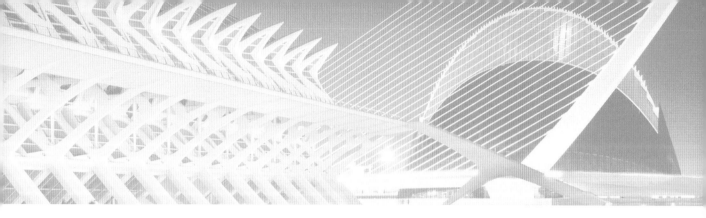

TO THE STUDENT

With the hope that this work will stimulate an interest in Structural Analysis and provide an acceptable guide to its understanding.

Structural Analysis, Eighth Edition in SI Units is developed for the effective study of the theory and application of structural analysis as it applies to trusses, beams, and frames. Emphasis is placed on developing the student's ability to both model and analyze a structure and to provide realistic applications encountered in professional practice.

Although technology has reduced the often lengthy calculations required when large or complicated structures are analyzed, it is Hibbeler's opinion that students taking a first course in this subject should also be well versed in the classical methods. Practice in applying these methods will develop a deeper understanding of the basic engineering sciences of statics and mechanics of materials.

This book goes beyond theory and steps into the information age by specially developing animations that breathe new life into classical concepts. Newly developed animations complement the video solutions as part of an enhanced visual learning kit that aids understanding. Anchored by the classic Hibbeler's methods of problem-solving, this is the definitive authority on structural theory gracing lecture halls today.

The book contains these New and Enhanced Elements

1. **Animations help students visualize the invisible forces governing structures**
2. **24/7 video solution walkthroughs offer independent revision**
3. **Diversity of problems and probing analysis build critical thinking**
4. **Realistic diagrams and photos illustrate theories in practice**
5. **Comprehensive and concise organization pave the way for systematic study**
6. **Wide-ranging, time-prudent resources for instructors and students**

Animations Help Students Visualize the Invisible Forces Governing Structures

Animations available on the Companion Website

On the Companion Website, students have access to specially created animations. These animations cover concepts that were identified by the adaptor as important teaching concepts. Students can access the Companion Website with the Access Code provided. The animations help students visualize the relation between mathematical explanation and real structure, breaking down complicated sequences and building up fuller analysis. They lend a graphic component in tutorial and lecture, assisting lecturers in demonstrating the teaching of concepts with greater ease and clarity.

List of Animations

1. Method of Joints
   Page 96
   Example 3.2

2. Methods of Sections
   Page 108
   Example 3.6

3. Internal Loadings M & V
   Page 150
   Section 4.3

4. Methods of Superpositions
   Page 169
   Figure 4-21

5. Influence Lines
   Page 207
   Example 6.1

6. Qualitative Influence Lines
   Pages 219, 220, and 221
   Examples 6.9(a), 6.10(b), and 6.11(b)

7. Moment-Area Theorems
   Page 321
   Example 8.8

8. Conjugate-Beam Method
   Page 330
   Example 8.14

9. Principle of Virtual Work
   Page 351
   Example 9.1

10. Force Method of Analysis
    Pages 398 and 399
    Figures 10-3 and 10-4

11. Force Method of Analysis
    Page 400
    Figure 10-5

12. Moment Distribution for Beams
    Page 491
    Section 12.2

Lecturers can demonstrate
the different methods of
analysis step-by-step.

Maximize the use of
class contact time.

Students 'see' how the
variables they apply in the
mathematical equations affect
the analysis of the structure.

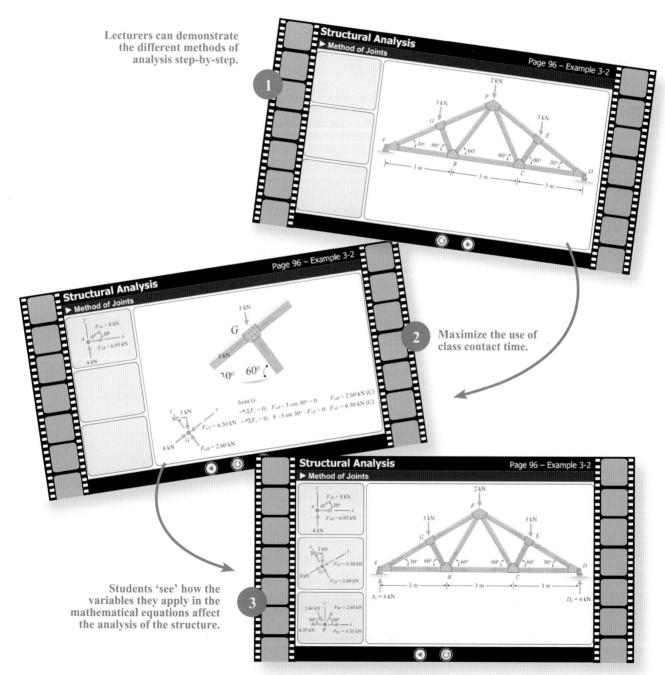

2 24/7 Video Solution Walkthroughs Offer Independent Revision

Video Solutions

An invaluable resource in and out of the classroom, these complete, solution walkthroughs of representative homework problems from each chapter, offer fully-worked solutions, self-paced instruction and 24/7 accessibility. Lecturers and students can harness this resource to gain independent exposure to a wide range of examples applying formulae to actual structures. Each Video Solution is flagged by a monitor icon. Icons with "US" inside the monitor indicate that the videos are in imperial units. Unmarked ones are in SI units.

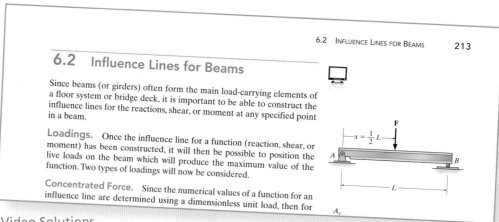

6.2 Influence Lines for Beams

Since beams (or girders) often form the main load-carrying elements of a floor system or bridge deck, it is important to be able to construct the influence lines for the reactions, shear, or moment at any specified point in a beam.

Loadings. Once the influence line for a function (reaction, shear, or moment) has been constructed, it will then be possible to position the live loads on the beam which will produce the maximum value of the function. Two types of loadings will now be considered.

Concentrated Force. Since the numerical values of a function for an influence line are determined using a dimensionless unit load, then for

$x = \frac{1}{2}L$

List of Video Solutions

1. Chapter 1: Section 1.3-1 (US)
    Floor Load

2. Chapter 1: Section 1.3-2 (US)
    Wind Load

3. Chapter 2: Section 2.1 (US)
    Idealized Floor Loads

4. Chapter 2: Section 2.4
    Determinacy and Stability

5. Chapter 2: Section 2.5 (US)
    Analysis of a Determinate Frame

6. Chapter 3: Section 3.2
    Classification of Trusses

7. Chapter 3: Section 3.5
    Coplanar Truss Analysis

8. Chapter 3: Section 3.8 (US)
    Space Truss Analysis

9. Chapter 4: Section 4.1 (US)
    Internal Loadings at a Point

10. Chapter 4: Section 4.3 (US)
    Beam Internal Loading Diagrams

11. Chapter 4: Section 4.4
    Frame Internal Loading Diagrams

12. Chapter 5: Section 5.2
    Cables with Concentrated Loads

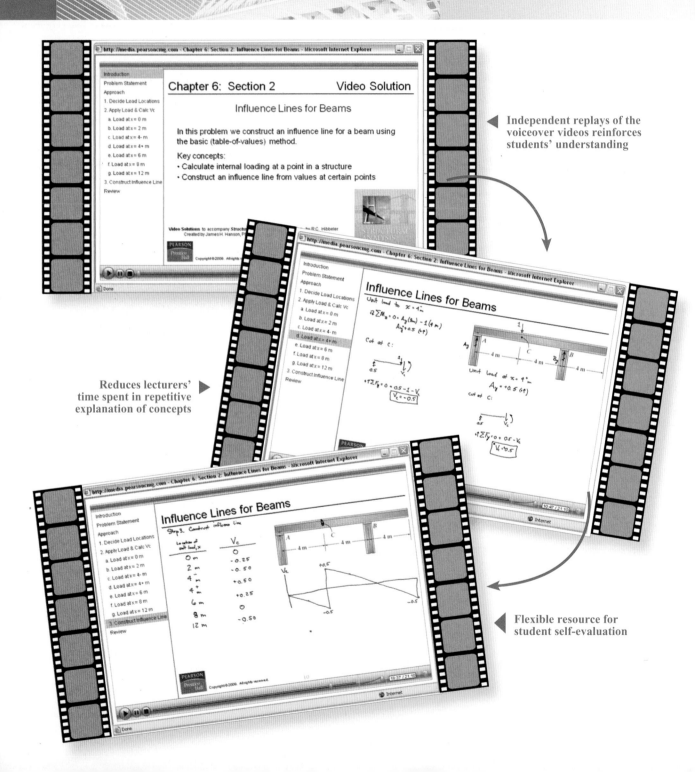

Independent replays of the voiceover videos reinforces students' understanding

Reduces lecturers' time spent in repetitive explanation of concepts

Flexible resource for student self-evaluation

Diversity of Problems and Probing Analysis Build Critical Thinking

24   CHAPTER 1   TYPES OF STRUCTURES AND LOADS

1 | EXAMPLE 1.4

The unheated storage facility shown in Fig. 1–14 is located on flat open terrain in southern Illinois, where the specified ground snow load is 0.72 kN/m^2. Determine the design snow load on the roof which has a slope of 4%.

SOLUTION
Since the roof slope is $< 5\%$, we will use Eq. 1–5. Here, $C_e = 0.8$ due to the open area, $C_t = 1.2$ and $I_s = 0.8$. Thus,

$$p_f = 0.7 C_e C_t I_s p_g$$
$$= 0.7(0.8)(1.2)(0.8)(0.72 \text{ kN/m}^2) = 0.39 \text{ kN/m}^2$$

Since $p_g = 0.72 \text{ kN/m}^2 < 0.96 \text{ kN/m}^2$, then also

$$p_f = I p_g = 0.8(0.72 \text{ kN/m}^2) = 0.58 \text{ kN/m}^2$$

By comparison, choose

$$p_f = 0.58 \text{ kN/m}^2 \qquad Ans.$$

Fig. 1–14

◀ **Example Problems**

Students are able to exercise their problem-solving skills through these problems which have a range of possible solutions. Concluding notes have also been incorporated in the examples enabling the student to extend the analysis of the example with detailed solutions.

Earthquake Loads. Earthquakes produce loadings on a structure through its interaction with the ground and its response characteristics. These loadings result from the structure's distortion caused by the ground's motion and the lateral resistance of the structure. Their magnitude depends on the amount and type of ground accelerations and the mass and stiffness of the structure. In order to provide some insight as to the nature of earthquake loads, consider the simple structural model shown in Fig. 1–15. This model may represent a single-story building, where the top block is the "lumped" mass of the roof, and the middle block is the lumped stiffness of all the building's columns. During an earthquake the ground vibrates both horizontally and vertically. The horizontal accelerations create shear forces in the column that put the block in sequential motion with the ground. If the column is *stiff* and the block has a *small* mass, the period of vibration of the block will be *short* and the block will accelerate with the same motion as the ground and undergo only slight relative displacements. For an actual structure which is designed to have large amounts of bracing and stiff connections this can be beneficial, since less stress is developed in the members. On the other hand, if the column in Fig 1–15 is very flexible and the block has a large mass, then earthquake-induced motion will cause small accelerations of the block and large relative displacements.

In practice the effects of a structure's acceleration, velocity, and displacement can be determined and represented as an *earthquake*

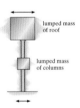

lumped mass of roof

lumped mass of columns

Fig. 1–15

Project Problems ▶

Project Problems that involve real structural systems are included at the end of selected chapters. They provide the student with insight as to how loads are determined and transmitted through the structure.

PROJECT PROBLEM

2–1P. The railroad trestle bridge shown in the photo is supported by reinforced concrete bents. Assume the two simply supported side girders, track bed, and two rails have a weight of 7.5 kN/m and the load imposed by a train is 100 kN/m (see Fig. 1–11). Each girder is 6 m long. Apply the load over the entire bridge and determine the compressive force in the columns of each bent. For the analysis assume all joints are pin connected and neglect the weight of the bent. Are these realistic assumptions?

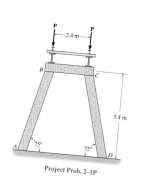

Project Prob. 2–1P

CHAPTER 4 INTERNAL LOADINGS DEVELOPED IN STRUCTURAL MEMBERS

146

PROBLEMS

4–1. Determine the internal normal force, shear force, and bending moment in the beam at points C and D. Assume the support at A is a pin and B is a roller.

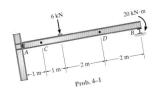

Prob. 4–1

4–2. Determine the internal normal force, shear force, and bending moment in the beam at points C and D. Assume

4–3. The boom DF of the jib crane and the column DE have a uniform weight of 750 N/m. If the hoist and load weigh 1350 N, determine the internal normal force, shear force, and bending moment in the crane at points A, B, and C.

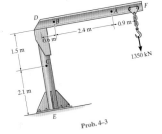

Prob. 4–3

***4–4.** Determine the internal normal force, shear force, and bending moment at point D. Take $w = 150$ N/m.

◀ Problems

These problems depict realistic situations encountered in practice. They are developed to test student's ability to apply the concepts. A wider range of questions is given for student's practice and application. Lecturers have more questions to select, modify and add as new questions to their resources.

Realistic Diagrams and Photos Illustrate Theories in Practice

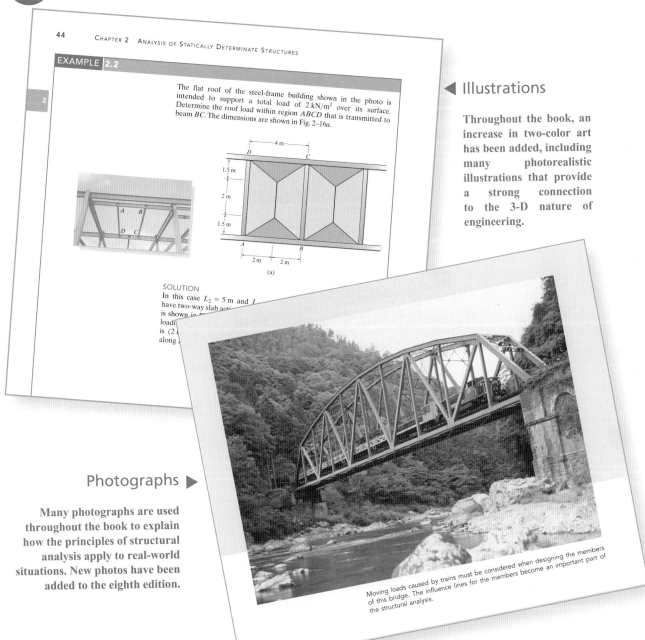

44   CHAPTER 2   ANALYSIS OF STATICALLY DETERMINATE STRUCTURES

EXAMPLE 2.2

The flat roof of the steel-frame building shown in the photo is intended to support a total load of $2\ kN/m^2$ over its surface. Determine the roof load within region $ABCD$ that is transmitted to beam BC. The dimensions are shown in Fig. 2–16a.

(a)

SOLUTION
In this case $L_2 = 5\ m$ and L
have two-way slab act
is shown in
loadi
is (2
along

◄ Illustrations

Throughout the book, an increase in two-color art has been added, including many photorealistic illustrations that provide a strong connection to the 3-D nature of engineering.

Photographs ▶

Many photographs are used throughout the book to explain how the principles of structural analysis apply to real-world situations. New photos have been added to the eighth edition.

Moving loads caused by trains must be considered when designing the members of this bridge. The influence lines for the members become an important part of the structural analysis.

Comprehensive and Concise Organization Pave the Way for Systematic Learning

Framework

The contents of each chapter are arranged into sections with specific topics categorized by title headings, providing clear cues that allows students to progressively follow topics.

382

CHAPTER 9 DEFLECTIONS USING ENERGY METHODS

$$U_s = K \int_0^L \frac{V^2\,dx}{2AG} \qquad \frac{\partial U_s}{\partial P} = \int_0^L \frac{V}{AG}\left(\frac{\partial V}{\partial P}\right)dx$$

$$U_t = \int_0^L \frac{T^2\,dx}{2JG} \qquad \frac{\partial U_t}{\partial P} = \int_0^L \frac{T}{JG}\left(\frac{\partial T}{\partial P}\right)dx$$

These effects, however, will not be included in the analysis of the problems in this text since beam and frame deflections are caused mainly by bending strain energy. Larger frames, or those with unusual geometry, can be analyzed by computer, where these effects can readily be incorporated into the analysis.

Procedure for Analysis

The following procedure provides a method that may be used to determine the deflection and/or slope at a point in a beam or frame using Castigliano's theorem.

External Force P or Couple Moment M'

- Place a force **P** on the beam or frame at the point and in the direction of the desired displacement.
- If the slope is to be determined, place a couple moment **M'** at the point.
- It is assumed that both P and M' have a *variable magnitude* in order to obtain the changes $\partial M/\partial P$ or $\partial M/\partial M'$.

Internal Moments M

- Establish appropriate x coordinates that are valid within regions of the beam or frame where there is no discontinuity of force, distributed load, or couple moment.
- Calculate the internal moment M as a function of P or M' and each x coordinate. Also, compute the partial derivative $\partial M/\partial P$ or $\partial M/\partial M'$ for each coordinate x.
- After M and $\partial M/\partial P$ or $\partial M/\partial M'$ have been determined, assign P or M' its numerical value if it has replaced a real force or couple moment. Otherwise, set P or M' equal to zero.

Castigliano's Theorem

- Apply Eq. 9–28 or 9–29 to determine the desired displacement Δ or slope θ. It is important to retain the algebraic signs for corresponding values of M and $\partial M/\partial P$ or $\partial M/\partial M'$.
- If the resultant sum of all the definite integrals is positive, Δ or θ is in the same direction as **P** or **M'**.

◀ **Procedure for Analysis**

Discussions relevant to a particular theory are succinct, yet thorough. In most cases, this is followed by a "Procedure for Analysis" guide, which provides students with a summary of the important concepts and a systematic approach for applying the concepts. The example problems are solved using this outlined method in order to clarify its numerical application. Problems are given at the end of each chapter and are arranged to cover the material in sequential order; moreover, for any topic they are arranged in order of increasing difficulty.

Wide-ranging, Time-prudent Resources for Instructors and Students

Resources for Instructors

Instructor's Solutions Manual.
An instructor's solutions manual was prepared by the author to aid lecturers in tutorial and exam preparation.

Presentation Resources
Powerpoint slide files are available for download from the Instructor Resource Center at www.pearsoned-asia. com/Hibbeler. If you are in need of a login and password for this site, please contact your local Pearson Prentice Hall representative.

Video Solutions
Make efficient use of class time by showing students the complete and concise problem-solving approaches that they can access anytime and view at their own pace.

Resources for Students

STRAN
A program you can use to solve two and three dimensional trusses and beams, and two dimensional frames. Instructions for downloading and how to use the program are available on the Companion Website.

Animations
These specially created animations cover concepts that were identified by the adaptor as important teaching concepts. The animations help students visualize the relation between mathematical explanation and real structure, breaking down complicated sequences and building up fuller analysis.

E-Text
Students can also view the book as an e-text, making it convenient and accessible anywhere.

Video Solutions
The Companion Website provides Video Solutions coverage that offer

- **Fully-worked Solutions** — Showing every step of representative homework problems, to help students make vital connections between concepts.
- **Self-paced Instruction** — Students can navigate each problem and select, play, rewind, fast-forward, stop, and jump-to sections within each problem's solution.
- **24/7 Access** — Help whenever students need it with over 20 hours of helpful review.

ACKNOWLEDGEMENTS

Over one hundred of my colleagues in the teaching profession and many of my students have made valuable suggestions that have helped in the development of this book, and I would like to hereby acknowledge all of their comments. I personally would like to thank the reviewers contracted by my editor for this new edition, namely:

Thomas H. Miller, *Oregon State University*
Hayder A. Rasheed, *Kansas State University*
Jeffrey A. Laman, *Penn State University*
Jerry R. Bayless, *University of Missouri—Rolla*
Paolo Gardoni, *Texas A&M University*
Timothy Ross, *University of New Mexico*
F. Wayne Klaiber, *Iowa State University*
Husam S. Najm, *Rutgers University*

Also, the constructive comments from Kai Beng Yap, and Barry Nolan, both practicing engineers are greatly appreciated. Finally, I would like to acknowledge the support I received from my wife Conny, who has always been very helpful in preparing the manuscript for publication.

I would greatly appreciate hearing from you if at any time you have any comments or suggestions regarding the contents of this edition.

RUSSELL CHARLES HIBBELER
hibbeler@bellsouth.net

About the Adaptor

Dr. Kiang Hwee TAN is Professor of Civil Engineering at the National University of Singapore. He obtained his Dr.Eng.(1985) degree from the University of Tokyo in Japan. He has more than 30 years of experience in teaching and research, and has published more than 200 refereed papers in the field of structural engineering, and a book titled *Concrete Beams with Openings: Analysis and Design* (1999).

CREDITS

CONTENTS

13
Beams and Frames Having Nonprismatic Members 523

14
Truss Analysis Using the Stiffness Method 539

15
Beam Analysis Using the Stiffness Method 575

16
Plane Frame Analysis Using the Stiffness Method 595

Appendices

STRUCTURAL ANALYSIS

The diamond pattern framework (cross bracing) of these high-rise buildings is used to resist loadings due to wind.

Types of Structures and Loads

1

This chapter provides a discussion of some of the preliminary aspects of structural analysis. The phases of activity necessary to produce a structure are presented first, followed by an introduction to the basic types of structures, their components, and supports. Finally, a brief explanation is given of the various types of loads that must be considered for an appropriate analysis and design.

1.1   Introduction

A *structure* refers to a system of connected parts used to support a load. Important examples related to civil engineering include buildings, bridges, and towers; and in other branches of engineering, ship and aircraft frames, tanks, pressure vessels, mechanical systems, and electrical supporting structures are important.

When designing a structure to serve a specified function for public use, the engineer must account for its safety, esthetics, and serviceability, while taking into consideration economic and environmental constraints. Often this requires several independent studies of different solutions before final judgment can be made as to which structural form is most appropriate. This design process is both creative and technical and requires a fundamental knowledge of material properties and the laws of mechanics which govern material response. Once a preliminary design of a structure is proposed, the structure must then be *analyzed* to ensure that it has its required stiffness and strength. To analyze a structure properly, certain idealizations must be made as to how the members are supported and connected together. The loadings are determined from codes and local specifications, and the forces in the members and their displacements are found using the theory of structural analysis, which is the subject matter of this text. The results of this analysis then can be used to

1

redesign the structure, accounting for a more accurate determination of the weight of the members and their size. Structural design, therefore, follows a series of successive approximations in which every cycle requires a structural analysis. In this book, the structural analysis is applied to civil engineering structures; however, the method of analysis described can also be used for structures related to other fields of engineering.

1.2   Classification of Structures

It is important for a structural engineer to recognize the various types of elements composing a structure and to be able to classify structures as to their form and function. We will introduce some of these aspects now and expand on them at appropriate points throughout the text.

Structural Elements.   Some of the more common elements from which structures are composed are as follows.

Tie Rods.   Structural members subjected to a *tensile force* are often referred to as *tie rods* or *bracing struts*. Due to the nature of this load, these members are rather slender, and are often chosen from rods, bars, angles, or channels, Fig. 1–1.

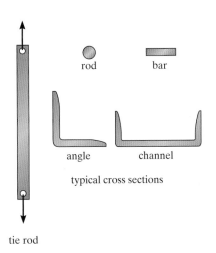

rod   bar

angle   channel

typical cross sections

tie rod

Fig. 1–1

Beams.   Beams are usually straight horizontal members used primarily to carry vertical loads. Quite often they are classified according to the way they are supported, as indicated in Fig. 1–2. In particular, when the cross section varies the beam is referred to as tapered or haunched. Beam cross sections may also be "built up" by adding plates to their top and bottom.

Beams are primarily designed to resist bending moment; however, if they are short and carry large loads, the internal shear force may become quite large and this force may govern their design. When the material used for a beam is a metal such as steel or aluminum, the cross section is most efficient when it is shaped as shown in Fig. 1–3. Here the forces developed in the top and bottom *flanges* of the beam form the necessary couple used to resist the applied moment **M**, whereas the *web* is effective in resisting the applied shear **V**. This cross section is commonly referred to as a "wide flange," and it is normally formed as a single unit in a rolling mill in lengths up to 23 m. If shorter lengths are needed, a cross section having tapered flanges is sometimes selected. When the beam is required to have a very large span and the loads applied are rather large, the cross section may take the form of a *plate girder*. This member is fabricated by using a large plate for the web and welding or bolting plates to its ends for flanges. The girder is often transported to the field in segments, and

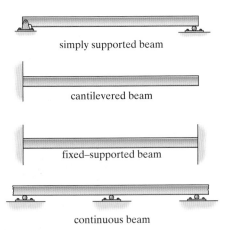

simply supported beam

cantilevered beam

fixed–supported beam

continuous beam

Fig. 1–2

Fig. 1–3

The prestressed concrete girders are simply supported and are used for this highway bridge.

the segments are designed to be spliced or joined together at points where the girder carries a small internal moment. (See the photo below.)

Concrete beams generally have rectangular cross sections, since it is easy to construct this form directly in the field. Because concrete is rather weak in resisting tension, steel "reinforcing rods" are cast into the beam within regions of the cross section subjected to tension. Precast concrete beams or girders are fabricated at a shop or yard in the same manner and then transported to the job site.

Beams made from timber may be sawn from a solid piece of wood or laminated. *Laminated* beams are constructed from solid sections of wood, which are fastened together using high-strength glues.

Shown are typical splice plate joints used to connect the steel girders of a highway bridge.

The steel reinforcement cage shown on the right and left is used to resist any tension that may develop in the concrete beams which will be formed around it.

Wide-flange members are often used for columns. Here is an example of a beam column.

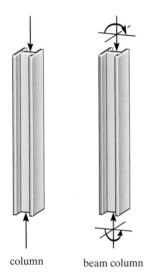

column   beam column

Fig. 1–4

Columns. Members that are generally vertical and resist axial compressive loads are referred to as *columns,* Fig. 1–4. Tubes and wide-flange cross sections are often used for metal columns, and circular and square cross sections with reinforcing rods are used for those made of concrete. Occasionally, columns are subjected to both an axial load and a bending moment as shown in the figure. These members are referred to as *beam columns.*

Types of Structures. The combination of structural elements and the materials from which they are composed is referred to as a *structural system.* Each system is constructed of one or more of four basic types of structures. Ranked in order of complexity of their force analysis, they are as follows.

Trusses. When the span of a structure is required to be large and its depth is not an important criterion for design, a truss may be selected. *Trusses* consist of slender elements, usually arranged in triangular fashion. *Planar trusses* are composed of members that lie in the same plane and are frequently used for bridge and roof support, whereas *space trusses* have members extending in three dimensions and are suitable for derricks and towers.

Due to the geometric arrangement of its members, loads that cause the entire truss to bend are converted into tensile or compressive forces in the members. Because of this, one of the primary advantages of a truss, compared to a beam, is that it uses less material to support a given load, Fig. 1–5. Also, a truss is constructed from *long and slender elements,* which can be arranged in various ways to support a load. Most often it is

Loading causes bending of truss, which develops compression in top members, tension in bottom members.

Fig. 1–5

economically feasible to use a truss to cover spans ranging from 9 m to 122 m, although trusses have been used on occasion for spans of greater lengths.

Cables and Arches. Two other forms of structures used to span long distances are the cable and the arch. *Cables* are usually flexible and carry their loads in tension. They are commonly used to support bridges, Fig. 1–6a, and building roofs. When used for these purposes, the cable has an advantage over the beam and the truss, especially for spans that are greater than 46 m. Because they are always in tension, cables will not become unstable and suddenly collapse, as may happen with beams or trusses. Furthermore, the truss will require added costs for construction and increased depth as the span increases. Use of cables, on the other hand, is limited only by their sag, weight, and methods of anchorage.

The *arch* achieves its strength in compression, since it has a reverse curvature to that of the cable. The arch must be rigid, however, in order to maintain its shape, and this results in secondary loadings involving shear and moment, which must be considered in its design. Arches are frequently used in bridge structures, Fig. 1–6b, dome roofs, and for openings in masonry walls.

Cables support their loads in tension.
(a)

Arches support their loads in compression.
(b)

Fig. 1–6

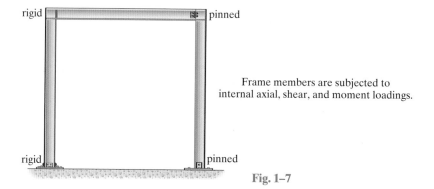

rigid   pinned

rigid   pinned

Frame members are subjected to
internal axial, shear, and moment loadings.

Fig. 1–7

Here is an example of a steel frame that is
used to support a crane rail. The frame can
be assumed fixed connected at its top joints
and pinned at the supports.

Frames.   Frames are often used in buildings and are composed of beams
and columns that are either pin or fixed connected, Fig. 1–7. Like trusses,
frames extend in two or three dimensions. The loading on a frame causes
bending of its members, and if it has rigid joint connections, this structure
is generally "indeterminate" from a standpoint of analysis. The strength of
such a frame is derived from the moment interactions between the beams
and the columns at the rigid joints.

Surface Structures.   A *surface structure* is made from a material having
a very small thickness compared to its other dimensions. Sometimes this
material is very flexible and can take the form of a tent or air-inflated
structure. In both cases the material acts as a membrane that is subjected
to pure tension.

Surface structures may also be made of rigid material such as reinforced
concrete. As such they may be shaped as folded plates, cylinders, or
hyperbolic paraboloids, and are referred to as *thin plates* or *shells*.
These structures act like cables or arches since they support loads
primarily in tension or compression, with very little bending. In spite of
this, plate or shell structures are generally very difficult to analyze, due
to the three-dimensional geometry of their surface. Such an analysis is
beyond the scope of this text and is instead covered in texts devoted
entirely to this subject.

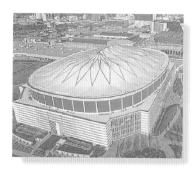

The roof of the "Georgia Dome" in Atlanta,
Georgia can be considered as a thin membrane.

1.3   Loads

Once the dimensional requirements for a structure have been defined, it becomes necessary to determine the loads the structure must support. Often, it is the anticipation of the various loads that will be imposed on the structure that provides the basic type of structure that will be chosen for design. For example, high-rise structures must endure large lateral loadings caused by wind, and so shear walls and tubular frame systems are selected, whereas buildings located in areas prone to earthquakes must be designed having ductile frames and connections.

Once the structural form has been determined, the actual design begins with those elements that are subjected to the primary loads the structure is intended to carry, and proceeds in sequence to the various supporting members until the foundation is reached. Thus, a building floor slab would be designed first, followed by the supporting beams, columns, and last, the foundation footings. In order to design a structure, it is therefore necessary to first specify the loads that act on it.

The design loading for a structure is often specified in codes. In general, the structural engineer works with two types of codes: general building codes and design codes. *General building codes* specify the requirements of governmental bodies for minimum design loads on structures and minimum standards for construction. *Design codes* provide detailed technical standards and are used to establish the requirements for the actual structural design. Table 1–1 lists some of the important codes used in practice. It should be realized, however, that codes provide only a general guide for design. *The ultimate responsibility for the design lies with the structural engineer.*

TABLE 1–1   Codes
General Building Codes
Minimum Design Loads for Buildings and Other Structures, ASCE/SEI 7-10, American Society of Civil Engineers *International Building Code*
Design Codes
Building Code Requirements for Reinforced Concrete, Am. Conc. Inst. (ACI) *Manual of Steel Construction,* American Institute of Steel Construction (AISC) *Standard Specifications for Highway Bridges,* American Association of State Highway and Transportation Officials (AASHTO) *National Design Specification for Wood Construction,* American Forest and Paper Association (AFPA) *Manual for Railway Engineering,* American Railway Engineering Association (AREA)

Since a structure is generally subjected to several types of loads, a brief discussion of these loadings will now be presented to illustrate how one must consider their effects in practice.

Dead Loads. *Dead loads* consist of the weights of the various structural members and the weights of any objects that are permanently attached to the structure. Hence, for a building, the dead loads include the weights of the columns, beams, and girders, the floor slab, roofing, walls, windows, plumbing, electrical fixtures, and other miscellaneous attachments.

In some cases, a structural dead load can be estimated satisfactorily from simple formulas based on the weights and sizes of similar structures. Through experience one can also derive a "feeling" for the magnitude of these loadings. For example, the average weight for timber buildings is 1.9–2.4 kN/m^2, for steel framed buildings it is 2.9–3.6 kN/m^2, and for reinforced concrete buildings it is 5.3–6.2 kN/m^2. Ordinarily, though, once the materials and sizes of the various components of the structure are determined, their weights can be found from tables that list their densities.

The densities of typical materials used in construction are listed in Table 1–2, and a portion of a table listing the weights of typical building

TABLE 1–2 Minimum Densities for Design Loads from Materials*	
	kN/m^3
Aluminum	26.7
Concrete, plain cinder	17.0
Concrete, plain stone	22.6
Concrete, reinforced cinder	17.4
Concrete, reinforced stone	23.6
Clay, dry	9.9
Clay, damp	17.3
Sand and gravel, dry, loose	15.7
Sand and gravel, wet	18.9
Masonry, lightweight solid concrete	16.5
Masonry, normal weight	21.2
Plywood	5.7
Steel, cold-drawn	77.3
Wood, Douglas Fir	5.3
Wood, Southern Pine	5.8
Wood, spruce	4.5

*Reproduced with permission from American Society of Civil Engineers *Minimum Design Loads for Buildings and Other Structures*, ASCE/SEI 7-10. Copies of this standard may be purchased from ASCE at www.pubs.asce.org.

TABLE 1–3   Minimum Design Dead Loads*

Walls	kN/m²
100 mm clay brick	1.87
200 mm clay brick	3.78
300 mm clay brick	5.51
Frame Partitions and Walls	
Exterior stud walls with brick veneer	2.30
Windows, glass, frame and sash	0.38
Wood studs 50 × 100 mm unplastered	0.19
Wood studs 50 × 100 mm plastered one side	0.57
Wood studs 50 × 100 mm plastered two sides	0.96
Floor Fill	
Cinder concrete, per mm	0.017
Lightweight concrete, plain, per mm	0.015
Stone concrete, per mm	0.023
Ceilings	
Acoustical fiberboard	0.05
Plaster on tile or concrete	0.24
Suspended metal lath and gypsum plaster	0.48
Asphalt shingles	0.10
Fiberboard, 13 mm	0.04

*Reproduced with permission from American Society of Civil Engineers *Minimum Design Loads for Buildings and Other Structures*, ASCE/SEI 7-10.

components is given in Table 1–3. Although calculation of dead loads based on the use of tabulated data is rather straightforward, it should be realized that in many respects these loads will have to be estimated in the initial phase of design. These estimates include nonstructural materials such as prefabricated facade panels, electrical and plumbing systems, etc. Furthermore, even if the material is specified, the unit weights of elements reported in codes may vary from those given by manufacturers, and later use of the building may include some changes in dead loading. As a result, estimates of dead loadings can be in error by 15% to 20% or more.

Normally, the dead load is not large compared to the design load for simple structures such as a beam or a single-story frame; however, for multistory buildings it is important to have an accurate accounting of all the dead loads in order to properly design the columns, especially for the lower floors.

EXAMPLE 1.1

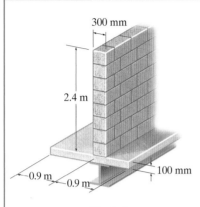

300 mm

2.4 m

0.9 m   0.9 m

100 mm

Fig. 1–8

The floor beam in Fig. 1–8 is used to support the 1.8 m width of a lightweight plain concrete slab having a thickness of 100 mm. The slab serves as a portion of the ceiling for the floor below, and therefore its bottom is coated with plaster. Furthermore, an 2.4-m-high, 300-mm-thick lightweight solid concrete block wall is directly over the top flange of the beam. Determine the loading on the beam measured per meter of length of the beam.

SOLUTION

Using the data in Tables 1–2 and 1–3, we have

Concrete slab: $[0.015 \text{ kN}/(\text{m}^2 \cdot \text{mm})](100 \text{ mm})(1.8 \text{ m}) = 2.70 \text{ kN/m}$
Plaster ceiling: $(0.24 \text{ kN/m}^2)(1.8 \text{ m}) = 0.43 \text{ kN/m}$
Block wall: $(16.5 \text{ kN/m}^3)(2.4 \text{ m})(0.3 \text{ m}) = \underline{11.88 \text{ kN/m}}$
Total load 15.01 kN/m

Ans.

Live Loads.

Live Loads can vary both in their magnitude and location. They may be caused by the weights of objects temporarily placed on a structure, moving vehicles, or natural forces. The minimum live loads specified in codes are determined from studying the history of their effects on existing structures. Usually, these loads include additional protection against excessive deflection or sudden overload. In Chapter 6 we will develop techniques for specifying the proper location of live loads on the structure so that they cause the greatest stress or deflection of the members. Various types of live loads will now be discussed.

Building Loads.

The floors of buildings are assumed to be subjected to *uniform live loads,* which depend on the purpose for which the building is designed. These loadings are generally tabulated in local, state, or national codes. A representative sample of such *minimum live loadings,* taken from the ASCE 7-10 Standard, is shown in Table 1–4. The values are determined from a history of loading various buildings. They include some protection against the possibility of overload due to emergency situations, construction loads, and serviceability requirements due to vibration. In addition to uniform loads, some codes specify *minimum concentrated live loads,* caused by hand carts, automobiles, etc., which must also be applied anywhere to the floor system. For example, both uniform and concentrated live loads must be considered in the design of an automobile parking deck.

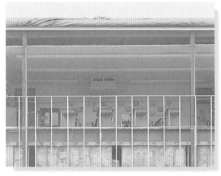

The live floor loading in this classroom consists of desks, chairs and laboratory equipment. For design the ASCE 7-10 Standard specifies a loading of 1.92 kN/m².

TABLE 1–4   Minimum Live Loads*

Occupancy or Use	Live Load kN/m²	Occupancy or Use	Live Load kN/m²
Assembly areas and theaters		Residential	
Fixed seats	2.87	Dwellings (one- and two-family)	1.92
Movable seats	4.79	Hotels and multifamily houses	
Garages (passenger cars only)	2.40	Private rooms and corridors	1.92
Office buildings		Public rooms and corridors	4.79
Lobbies	4.79	Schools	
Offices	2.40	Classrooms	1.92
Storage warehouse		Corridors above first floor	3.83
Light	6.00		
Heavy	11.97		

*Reproduced with permission from *Minimum Design Loads for Buildings and Other Structures*, ASCE/SEI 7-10.

For some types of buildings having very large floor areas, many codes will allow a *reduction* in the uniform live load for a *floor*, since it is unlikely that the prescribed live load will occur simultaneously throughout the entire structure at any one time. For example, ASCE 7-10 allows a reduction of live load on a member having an *influence area* ($K_{LL} A_T$) of 37.2 m² or more. This reduced live load is calculated using the following equation:

$$L = L_o\left(0.25 + \frac{4.57}{\sqrt{K_{LL} A_T}}\right) \quad (1\text{–}1)$$

where

L = reduced design live load per square meter of area supported by the member.

L_o = unreduced design live load per square meter of area supported by the member (see Table 1–4).

K_{LL} = live load element factor. For interior columns $K_{LL} = 4$.

A_T = tributary area in square meters.*

The reduced live load defined by Eq. 1–1 is limited to not less than 50% of L_o for members supporting one floor, or not less than 40% of L_o for members supporting more than one floor. No reduction is allowed for loads exceeding 4.79 kN/m², or for structures used for public assembly, garages, or roofs. Example 1–2 illustrates Eq. 1–1's application.

*Specific examples of the determination of tributary areas for beams and columns are given in Sec. 2–1.

EXAMPLE 1.2

A two-story office building shown in the photo has interior columns that are spaced 6.6 m apart in two perpendicular directions. If the (flat) roof loading is 0.96 kN/m², determine the reduced live load supported by a typical interior column located at ground level.

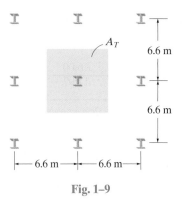

Fig. 1–9

SOLUTION

As shown in Fig. 1–9, each interior column has a tributary area or effective loaded area of $A_T = (6.6 \text{ m})(6.6 \text{ m}) = 43.56 \text{ m}^2$. A ground-floor column therefore supports a roof live load of

$$F_R = (0.96 \text{ kN/m}^2)(43.56 \text{ m}^2) = 41.8 \text{ kN}$$

This load cannot be reduced, since it is not a floor load. For the second floor, the live load is taken from Table 1–4: $L_o = 2.4 \text{ kN/m}^2$. Since $K_{LL} = 4$, then $4A_T = 4(43.56 \text{ m}^2) = 174.2 \text{ m}^2$ and $174.2 \text{ m}^2 > 37.2 \text{ m}^2$, the live load can be reduced using Eq. 1–1. Thus,

$$L = 2.4\left(0.25 + \frac{4.57}{\sqrt{174.2}}\right) = 1.43 \text{ kN/m}^2$$

The load reduction here is $(1.43/2.4)100\% = 59.6\% > 50\%$. O.K. Therefore,

$$F_F = (1.43 \text{ kN/m}^2)(43.56 \text{ m}^2) = 62.3 \text{ kN}$$

The total live load supported by the ground-floor column is thus

$$F = F_R + F_F = 41.8 \text{ kN} + 62.3 \text{ kN} = 104.1 \text{ kN} \qquad Ans.$$

Highway Bridge Loads.

The primary live loads on bridge spans are those due to traffic, and the heaviest vehicle loading encountered is that caused by a series of trucks. Specifications for truck loadings on highway bridges are reported in the *LRFD Bridge Design Specifications* of the American Association of State and Highway Transportation Officials (AASHTO). For two-axle trucks, these loads are designated with an H, followed by the weight of the truck in tons and another number which gives the year of the specifications in which the load was reported. H-series truck weights vary from 10 to 20 tons. However, bridges located on major highways, which carry a great deal of traffic, are often designed for two-axle trucks plus a one-axle semitrailer as in Fig. 1–10. These are designated as HS loadings. In general, a truck loading selected for design depends upon the type of bridge, its location, and the type of traffic anticipated.

Fig. 1–10

The size of the "standard truck" and the distribution of its weight is also reported in the specifications. Although trucks are assumed to be on the road, all lanes on the bridge need not be fully loaded with a row of trucks to obtain the critical load, since such a loading would be highly improbable. The details are discussed in Chapter 6.

Railroad Bridge Loads.

The loadings on railroad bridges, as in Fig. 1–11, are specified in the *Specifications for Steel Railway Bridges* published by the American Railroad Engineers Association (AREA). Normally, E loads, as originally devised by Theodore Cooper in 1894, were used for design. B. Steinmann has since updated Cooper's load distribution and has devised a series of M loadings, which are currently acceptable for design. Since train loadings involve a complicated series of concentrated forces, to simplify hand calculations, tables and graphs are sometimes used in conjunction with influence lines to obtain the critical load. Also, computer programs are used for this purpose.

Fig. 1–11

1

Impact Loads.   Moving vehicles may bounce or sidesway as they move over a bridge, and therefore they impart an *impact* to the deck. The percentage increase of the live loads due to impact is called the *impact factor, I*. This factor is generally obtained from formulas developed from experimental evidence. For example, for highway bridges the AASHTO specifications require that

$$I = \frac{15.24}{L + 38.1} \qquad \text{but not larger than 0.3}$$

where L is the length of the span in meters that is subjected to the live load.

In some cases provisions for impact loading on the structure of a building must also be taken into account. For example, the ASCE 7-10 Standard requires the weight of elevator machinery to be increased by 100%, and the loads on any hangers used to support floors and balconies to be increased by 33%.

Wind Loads.   When structures block the flow of wind, the wind's kinetic energy is converted into potential energy of pressure, which causes a wind loading. The effect of wind on a structure depends upon the density and velocity of the air, the angle of incidence of the wind, the shape and stiffness of the structure, and the roughness of its surface. For design purposes, wind loadings can be treated using either a static or a dynamic approach.

For the static approach, the fluctuating pressure caused by a constantly blowing wind is approximated by a mean velocity pressure that acts on the structure. This pressure q is defined by its kinetic energy, $q = \frac{1}{2}\rho V^2$, where ρ is the density of the air and V is its velocity. According to the ASCE 7-10 Standard, this equation is modified to account for the importance of the structure, its height, and the terrain in which it is located. It is represented as

$$q_z = 0.613 K_z K_{zt} K_d V^2 \ (\text{N/m}^2) \qquad (1\text{--}2)$$

where

$V =$ the velocity in m/s of a 3-second gust of wind measured 10 m above the ground. Specific values depend upon the "category" of the structure obtained from a wind map. For example, the interior portion of the continental United States reports a wind

Hurricane winds caused this damage to a condominium in Miami, Florida.

speed of 47 m/s if the structure is an agricultural or storage building, since it is of low risk to human life in the event of a failure. The wind speed is 54 m/s for cases where the structure is a hospital, since its failure would cause substantial loss of human life.

K_z = the velocity pressure exposure coefficient, which is a function of height and depends upon the ground terrain. Table 1–5 lists values for a structure which is located in open terrain with scattered low-lying obstructions.

K_{zt} = a factor that accounts for wind speed increases due to hills and escarpments. For flat ground $K_{zt} = 1.0$.

K_d = a factor that accounts for the direction of the wind. It is used only when the structure is subjected to combinations of loads (see Sec. 1–4). For wind acting alone, $K_d = 1.0$.

TABLE 1–5 Velocity Pressure Exposure Coefficient for Terrain with Low-Lying Obstructions

z (m)	K_z
0–4.6	0.85
6.1	0.90
7.6	0.94
9.1	0.98
12.2	1.04
15.2	1.09

Design Wind Pressure for Enclosed Buildings. Once the value for q_z is obtained, the design pressure can be determined from a list of relevant equations listed in the ASCE 7-10 Standard. The choice depends upon the flexibility and height of the structure, and whether the design is for the main wind-force resisting system, or for the building's components and cladding. For example, using a "directional procedure" the *wind-pressure* on an enclosed building of any height is determined using a two-termed equation resulting from both external and internal pressures, namely,

$$p = qGC_p - q_h(GC_{pi}) \qquad (1\text{--}3)$$

Here

$q = \quad q_z$ for the windward wall at height z above the ground (Eq. 1–2), and $q = q_h$ for the leeward walls, side walls, and roof, where $z = h$, the mean height of the roof.

$G = \quad$ a wind-gust effect factor, which depends upon the exposure. For example, for a rigid structure, $G = 0.85$.

$C_p = \quad$ a wall or roof pressure coefficient determined from a table. These tabular values for the walls and a roof pitch of $\theta = 10°$ are given in Fig. 1–12. Note in the elevation view that the pressure will vary with height on the windward side of the building, whereas on the remaining sides and on the roof the pressure is assumed to be constant. Negative values indicate pressures acting away from the surface.

$(GC_{pi}) = \quad$ the internal pressure coefficient, which depends upon the type of openings in the building. For fully enclosed buildings $(GC_{pi}) = \pm0.18$. Here the signs indicate that either positive or negative (suction) pressure can occur within the building.

Application of Eq. 1–3 will involve calculations of wind pressures from each side of the building, with due considerations for the possibility of either positive or negative pressures acting on the building's interior.

Wind blowing on a wall will tend to tip a building or cause it to sidesway. To prevent this engineers often use cross bracing to provide stability. Also, see p. 46.

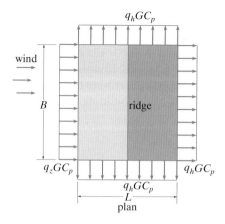

Surface	L/B	C_p	Use with
Windward wall	All values	0.8	q_z
Leeward wall	0–1	−0.5	q_h
	2	−0.3	
	≥4	−0.2	
Side walls	All values	−0.7	q_h

Wall pressure coefficients, C_p

(a)

		Windward angle θ		Leeward angle
Wind direction	h/L	10°		$\theta = 10°$
Normal to ridge	≤0.25	−0.7		−0.3
	0.5	−0.9		−0.5
	>1.0	−1.3		−0.7

Maximum negative roof pressure coefficients, C_p, for use with q_h

(b)

Fig. 1–12

For high-rise buildings or those having a shape or location that makes them wind sensitive, it is recommended that a *dynamic approach* be used to determine the wind loadings. The methodology for doing this is also outlined in the ASCE 7-10 Standard. It requires wind-tunnel tests to be performed on a scale model of the building and those surrounding it, in order to simulate the natural environment. The pressure effects of the wind on the building can be determined from pressure transducers attached to the model. Also, if the model has stiffness characteristics that are in proper scale to the building, then the dynamic deflections of the building can be determined.

EXAMPLE | 1.3

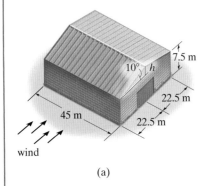

(a)

Fig. 1–13

The enclosed building shown in the photo and in Fig. 1–13a is used for storage purposes and is located outside of Chicago, Illinois on open flat terrain. When the wind is directed as shown, determine the design wind pressure acting on the roof and sides of the building using the ASCE 7-10 Specifications.

SOLUTION

First the wind pressure will be determined using Eq. 1–2. The basic wind speed is $V = 47$ m/s, since the building is used for storage. Also, for flat terrain, $K_{zt} = 1.0$. Since only wind loading is being considered, $K_d = 1.0$. Therefore,

$$q_z = 0.613\, K_z K_{zt} K_d V^2$$
$$= 0.613\, K_z (1.0)(1.0)(47)^2$$
$$= 1354\, K_z$$

From Fig. 1–13a, $h' = 22.5 \tan 10° = 3.97$ m so that the mean height of the roof is $h = 7.5 + 3.97/2 = 9.48$ m. Using the values of K_z in Table 1–5, calculated values of the pressure profile are listed in the table in Fig. 1–13b. Note the value of K_z was determined by linear interpolation for $z = h$, i.e., $(1.04 - 0.98)/(12.2 - 9.1) = (1.04 - K_z)/(12.2 - 9.48)$, $K_z = 0.987$, and so $q_h = 1354(0.987) = 1337$ N/m^2.

In order to apply Eq. 1–3 the gust factor is $G = 0.85$, and $(GC_{pi}) = \pm 0.18$. Thus,

$$p = qGC_p - q_h(GC_{pi})$$
$$= q(0.85)C_p - 1337(\pm 0.18)$$
$$= 0.85qC_p \mp 241 \tag{1}$$

The pressure loadings are obtained from this equation using the calculated values for q_z listed in Fig. 1–13b in accordance with the wind-pressure profile in Fig. 1–12.

z (m)	K_z	q_z (N/m^2)
0–4.6	0.85	1151
6.1	0.90	1219
7.6	0.94	1273
h = 9.48	0.987	1337

(b)

Windward Wall.   Here the pressure varies with height z since $q_z G C_p$ must be used. For all values of L/B, $C_p = 0.8$, so that from Eq. (1),

$$p_{0-4.6} = 542 \text{ N/m}^2 \quad \text{or} \quad 1024 \text{ N/m}^2$$
$$p_{6.1} = 588 \text{ N/m}^2 \quad \text{or} \quad 1070 \text{ N/m}^2$$
$$p_{7.6} = 625 \text{ N/m}^2 \quad \text{or} \quad 1107 \text{ N/m}^2$$

Leeward Wall.   Here $L/B = 2(22.5)/45 = 1$, so that $C_p = -0.5$. Also, $q = q_h$ and so from Eq. (1),

$$p = -809 \text{ N/m}^2 \quad \text{or} \quad -327 \text{ N/m}^2$$

Side Walls.   For all values of L/B, $C_p = -0.7$, and therefore since we must use $q = q_h$ in Eq. (1), we have

$$p = -1037 \text{ N/m}^2 \quad \text{or} \quad -555 \text{ N/m}^2$$

Windward Roof.   Here $h/L = 9.48/2(22.5) = 0.211 < 0.25$, so that $C_p = -0.7$ and $q = q_h$. Thus,

$$p = -1037 \text{ N/m}^2 \quad \text{or} \quad -555 \text{ N/m}^2$$

Leeward Roof.   In this case $C_p = -0.3$; therefore with $q = q_h$, we get

$$p = -582 \text{ N/m}^2 \quad \text{or} \quad -100 \text{ N/m}^2$$

These two sets of loadings are shown on the elevation of the building, representing either positive or negative (suction) internal building pressure, Fig. 1–13c. The main framing structure of the building must resist these loadings as well as for separate loadings calculated from wind blowing on the front or rear of the building.

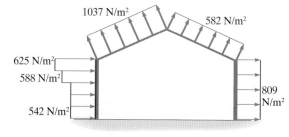
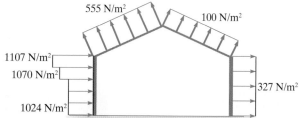

(c)

Design Wind Pressure for Signs. If the structure represents a sign, the wind will produce a *resultant force* acting on the face of the sign which is determined from

$$F = q_h G C_f A_s \tag{1-4}$$

Here

q_h = the wind pressure evaluated at the height h, measured from the ground to the top of the sign.

G = the wind-gust coefficient factor defined previously.

C_f = a force coefficient which depends upon the aspect ratio (width B of the sign to height s of the sign), and the clear area ratio (sign height s to the elevation h, measured from the ground to the top of the sign). For cases of wind directed normal to the sign and through its center, for $B/s = 4$, values are listed in Table 1–6.

A_s = the area of the face of the sign in m².

TABLE 1–6 Force Coefficients for Above-Ground Solid Signs, C_f	
s/h	C_f
1	1.35
0.9	1.45
0.5	1.70
0.2	1.80
≤0.16	1.85

To allow for normal and oblique wind directions, the calculated resultant force is assumed to act either through the geometric center of the face of the sign or at other specified locations on the face of the sign which depend upon the ratios s/h and B/s.

Hurricane winds acting on the face of this sign were strong enough to noticeably bend the two supporting arms causing the material to yield. Proper design would have prevented this.

Snow Loads. In some parts of the country, roof loading due to snow can be quite severe, and therefore protection against possible failure is of primary concern. Design loadings typically depend on the building's general shape and roof geometry, wind exposure, location, its importance, and whether or not it is heated. Like wind, snow loads in the ASCE 7-10 Standard are generally determined from a zone map reporting 50-year recurrence intervals of an extreme snow depth. For example, on the relatively flat elevation throughout the mid-section of Illinois and Indiana, the ground snow loading is 0.96 kN/m^2. However, for areas of Montana, specific case studies of ground snow loadings are needed due to the variable elevations throughout the state. Specifications for snow loads are covered in the ASCE 7-10 Standard, although no single code can cover all the implications of this type of loading.

Excessive snow and ice loadings act on this roof.

If a roof is flat, defined as having a slope of less than 5%, then the pressure loading on the roof can be obtained by modifying the ground snow loading, p_g, by the following empirical formula

$$p_f = 0.7 C_e C_t I_s p_g \qquad (1\text{–}5)$$

Here

C_e = an exposure factor which depends upon the terrain. For example, for a fully exposed roof in an unobstructed area, $C_e = 0.8$, whereas if the roof is sheltered and located in the center of a large city, then $C_e = 1.2$.

C_t = a thermal factor which refers to the average temperature within the building. For unheated structures kept below freezing $C_t = 1.2$, whereas if the roof is supporting a normally heated structure, then $C_t = 1.0$.

I_s = the importance factor as it relates to occupancy. For example, $I_s = 0.80$ for agriculture and storage facilities, and $I_s = 1.20$ for schools and hospitals.

If $p_g \leq 0.96 \text{ kN/m}^2$, then use the *largest value* for p_f, either computed from the above equation or from $p_f = I_s p_g$. If $p_g > 0.96 \text{ kN/m}^2$, then use $p_f = I_s(0.96 \text{ kN/m}^2)$.

EXAMPLE 1.4

The unheated storage facility shown in Fig. 1–14 is located on flat open terrain in southern Illinois, where the specified ground snow load is 0.72 kN/m². Determine the design snow load on the roof which has a slope of 4%.

Fig. 1–14

SOLUTION

Since the roof slope is < 5%, we will use Eq. 1–5. Here, $C_e = 0.8$ due to the open area, $C_t = 1.2$ and $I_s = 0.8$. Thus,

$$p_f = 0.7 C_e C_t I_s p_g$$
$$= 0.7(0.8)(1.2)(0.8)(0.72 \text{ kN/m}^2) = 0.39 \text{ kN/m}^2$$

Since $p_g = 0.72 \text{ kN/m}^2 < 0.96 \text{ kN/m}^2$, then also

$$p_f = I p_g = 0.8(0.72 \text{ kN/m}^2) = 0.58 \text{ kN/m}^2$$

By comparison, choose

$$p_f = 0.58 \text{ kN/m}^2 \qquad Ans.$$

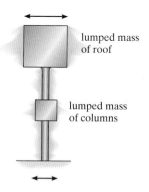

lumped mass of roof

lumped mass of columns

Fig. 1–15

Earthquake Loads. Earthquakes produce loadings on a structure through its interaction with the ground and its response characteristics. These loadings result from the structure's distortion caused by the ground's motion and the lateral resistance of the structure. Their magnitude depends on the amount and type of ground accelerations and the mass and stiffness of the structure. In order to provide some insight as to the nature of earthquake loads, consider the simple structural model shown in Fig. 1–15. This model may represent a single-story building, where the top block is the "lumped" mass of the roof, and the middle block is the lumped stiffness of all the building's columns. During an earthquake the ground vibrates both horizontally and vertically. The horizontal accelerations create shear forces in the column that put the block in sequential motion with the ground. If the column is *stiff* and the block has a *small* mass, the period of vibration of the block will be *short* and the block will accelerate with the same motion as the ground and undergo only slight relative displacements. For an actual structure which is designed to have large amounts of bracing and stiff connections this can be beneficial, since less stress is developed in the members. On the other hand, if the column in Fig 1–15 is very flexible and the block has a large mass, then earthquake-induced motion will cause small accelerations of the block and large relative displacements.

In practice the effects of a structure's acceleration, velocity, and displacement can be determined and represented as an *earthquake response spectrum.* Once this graph is established, the earthquake loadings can be calculated using a *dynamic analysis* based on the theory of structural dynamics. This type of analysis is gaining popularity, although it is often quite elaborate and requires the use of a computer. Even so, such an analysis becomes mandatory if the structure is large.

Some codes require that specific attention be given to earthquake design, especially in areas of the country where strong earthquakes predominate. Also, these loads should be seriously considered when designing high-rise buildings or nuclear power plants. In order to assess the importance of earthquake design consideration, one can check the seismic ground-acceleration maps published in the ASCE 7-10 Standard. These maps provide the peak ground accelerations caused by an earthquake along with risk coefficients. Regions vary from low risk, such as parts of Texas, to very high risk, such as along the west coast of California.

For small structures, a *static analysis* for earthquake design may be satisfactory. This case approximates the dynamic loads by a set of externally applied *static forces* that are applied laterally to the structure. One such method for doing this is reported in the ASCE 7-10 Standard. It is based upon finding a seismic response coefficient, C_s, determined from the soil properties, the ground accelerations, and the vibrational response of the structure. For most structures, this coefficient is then multiplied by the structure's total dead load W to obtain the "base shear" in the structure. The value of C_s is actually determined from

$$C_s = \frac{S_{DS}}{R/I_e}$$

where

S_{DS} = the spectral response acceleration for short periods of vibration.

R = a response modification factor that depends upon the ductility of the structure. Steel frame members which are highly ductile can have a value as high as 8, whereas reinforced concrete frames can have a value as low as 3.

I_e = the importance factor that depends upon the use of the building. For example, $I_e = 1$ for agriculture and storage facilities, and $I_e = 1.5$ for hospitals and other essential facilities.

With each new publication of the Standard, values of these coefficients are updated as more accurate data about earthquake response become available.

Hydrostatic and Soil Pressure.

When structures are used to retain water, soil, or granular materials, the pressure developed by these loadings becomes an important criterion for their design. Examples of such types of structures include tanks, dams, ships, bulkheads, and retaining walls. Here the laws of hydrostatics and soil mechanics are applied to define the intensity of the loadings on the structure.

The design of this retaining wall requires estimating the soil pressure acting on it. Also, the gate of the lock will be subjected to hydrostatic pressure that must be considered for its design.

Other Natural Loads. Several other types of live loads may also have to be considered in the design of a structure, depending on its location or use. These include the effect of blast, temperature changes, and differential settlement of the foundation.

1.4   Structural Design

Whenever a structure is designed, it is important to give consideration to both material and load uncertainties. These uncertainties include a possible variability in material properties, residual stress in materials, intended measurements being different from fabricated sizes, loadings due to vibration or impact, and material corrosion or decay.

ASD. Allowable-stress design (ASD) methods include *both* the material and load uncertainties into a single factor of safety. The many types of loads discussed previously can occur simultaneously on a structure, but it is very unlikely that the maximum of all these loads will occur at the same time. For example, both maximum wind and earthquake loads normally do not act simultaneously on a structure. For *allowable-stress design* the computed elastic stress in the material must not exceed the allowable stress for each of various load combinations. Typical load combinations as specified by the ASCE 7-10 Standard include

- dead load
- 0.6 (dead load) + 0.6 (wind load)
- 0.6 (dead load) + 0.7 (earthquake load)

LRFD. Since uncertainty can be considered using probability theory, there has been an increasing trend to separate material uncertainty from load uncertainty. This method is called *strength design* or LRFD (load and resistance factor design). For example, to account for the uncertainty of loads, this method uses load factors applied to the loads or combinations of loads. According to the ASCE 7-10 Standard, some of the load factors and combinations are

- 1.4 (dead load)
- 1.2 (dead load) + 1.6 (live load) + 0.5 (snow load)
- 0.9 (dead load) + 1.0 (wind load)
- 0.9 (dead load) + 1.0 (earthquake load)

In all these cases, the combination of loads is thought to provide a maximum, yet realistic loading on the structure.

PROBLEMS

1–1. The floor of a heavy storage warehouse building is made of 150-mm-thick stone concrete. If the floor is a slab having a length of 4.5 m and width of 3 m, determine the resultant force caused by the dead load and the live load.

1–2. The floor of the office building is made of 100-mm-thick lightweight concrete. If the office floor is a slab having a length of 6 m and width of 4.5 m, determine the resultant force caused by the dead load and the live load.

Prob. 1–2

1–3. The T-beam is made from concrete having a specific weight of 24 kN/m³. Determine the dead load per meter length of beam. Neglect the weight of the steel reinforcement.

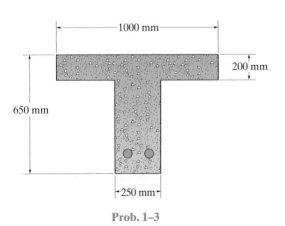

*1–4. The "New Jersey" barrier is commonly used during highway construction. Determine its weight per meter of length if it is made from plain stone concrete.

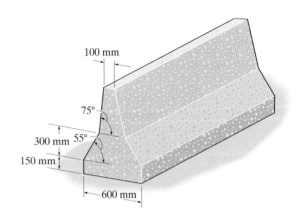

Prob. 1–4

1–5. The floor of a light storage warehouse is made of 150-mm-thick lightweight plain concrete. If the floor is a slab having a length of 7 m and width of 3 m, determine the resultant force caused by the dead load and the live load.

1–6. The prestressed concrete girder is made from plain stone concrete and four 19-mm cold-form steel reinforcing rods. Determine the dead weight of the girder per meter of its length.

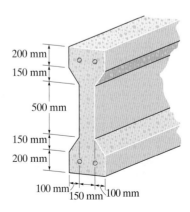

Prob. 1–6

Prob. 1–3

1–7. The wall is 2.5 m high and consists of 51 mm × 102 mm studs plastered on one side. On the other side is 13 mm fiberboard, and 102 mm clay brick. Determine the average load in kN/m of length of wall that the wall exerts on the floor.

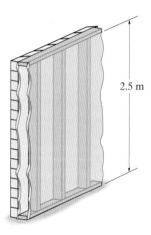

2.5 m

Prob. 1–7

1–8. A building wall consists of exterior stud walls with brick veneer and 13 mm fiberboard on one side. If the wall is 4 m high, determine the load in kN/m that it exerts on the floor.

1–9. The interior wall of a building is made from 50 × 100 mm wood studs, plastered on two sides. If the wall is 3.6 m high, determine the load in kN/m of length of wall that it exerts on the floor.

1–10. The second floor of a light manufacturing building is constructed from a 125-mm-thick stone concrete slab with an added 100-mm cinder concrete fill as shown. If the suspended ceiling of the first floor consists of metal lath and gypsum plaster, determine the dead load for design in kN per square meter of floor area.

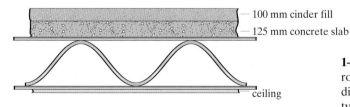

100 mm cinder fill
125 mm concrete slab

ceiling

Prob. 1–10

1–11. A four-story office building has interior columns spaced 9 m apart in two perpendicular directions. If the flat-roof live loading is estimated to be 1.5 kN/m^2, determine the reduced live load supported by a typical interior column located at ground level.

1–12. A two-story light storage warehouse has interior columns that are spaced 3.6 m apart in two perpendicular directions. If the live loading on the roof is estimated to be 1.25 kN/m^2, determine the reduced live load supported by a typical interior column at (a) the ground-floor level, and (b) the second-floor level.

1–13. The office building has interior columns spaced 5 m apart in perpendicular directions. Determine the reduced live load supported by a typical interior column located on the first floor under the offices.

Prob. 1–13

1–14. A two-story hotel has interior columns for the rooms that are spaced 6 m apart in two perpendicular directions. Determine the reduced live load supported by a typical interior column on the first floor under the public rooms.

1–15. Wind blows on the side of a fully enclosed hospital located on open flat terrain in Arizona. Determine the external pressure acting over the windward wall, which has a height of 9 m. The roof is flat.

1–17. A closed storage building is located on open flat terrain in central Ohio. If the side wall of the building is 6 m high, determine the external wind pressure acting on the windward and leeward walls. Each wall is 18 m long. Assume the roof is essentially flat.

Prob. 1–17

Prob. 1–15

*1–16.** Wind blows on the side of the fully enclosed hospital located on open flat terrain in Arizona. Determine the external pressure acting on the leeward wall, which has a length of 60 m and a height of 9 m.

1–18. The light metal storage building is on open flat terrain in central Oklahoma. If the side wall of the building is 4.2 m high, what are the two values of the external wind pressure acting on this wall when the wind blows on the back of the building? The roof is essentially flat and the building is fully enclosed.

Prob. 1–16

Prob. 1–18

1–19. Determine the resultant force acting perpendicular to the face of the billboard and through its center if it is located in Michigan on open flat terrain. The sign is rigid and has a width of 12 m and a height of 3 m. Its top side is 15 m from the ground.

1–21. The school building has a flat roof. It is located in an open area where the ground snow load is 0.68 kN/m². Determine the snow load that is required to design the roof.

Prob. 1–21

1–22. The hospital is located in an open area and has a flat roof and the ground snow load is 1.5 kN/m². Determine the design snow load for the roof.

Prob. 1–19

Prob. 1–22

*1–20. A hospital located in central Illinois has a flat roof. Determine the snow load in kN/m² that is required to design the roof.

CHAPTER REVIEW

The basic structural elements are:

Tie Rods—Slender members subjected to tension. Often used for bracing.

Beams—Members designed to resist bending moment. They are often fixed or pin supported and can be in the form of a steel plate girder, reinforced concrete, or laminated wood.

Columns—Members that resist axial compressive force. If the column also resists bending, it is called a *beam column*.

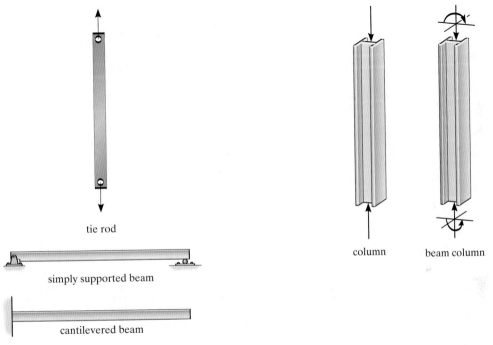

tie rod

simply supported beam

cantilevered beam

column   beam column

The types of structures considered in this book consist of *trusses* made from slender pin-connected members forming a series of triangles, *cables and arches,* which carry tensile and compressive loads, respectively, and *frames* composed of pin- or fixed-connected beams and columns.

(Ref: Section 1.2)

Loads are specified in codes such as the ASCE 7-10 code. *Dead loads* are fixed and refer to the weights of members and materials. *Live loads* are movable and consist of uniform building floor loads, traffic and train loads on bridges, impact loads due to vehicles and machines, wind loads, snow loads, earthquake loads, and hydrostatic and soil pressure.

(Ref: Section 1.3)

Oftentimes the elements of a structure, like the beams and girders of this building frame, are connected together in a manner whereby the analysis can be considered statically determinate.

Analysis of Statically Determinate Structures

2

In this chapter we will direct our attention to the most common form of structure that the engineer will have to analyze, and that is one that lies in a plane and is subjected to a force system that lies in the same plane. We begin by discussing the importance of choosing an appropriate analytical model for a structure so that the forces in the structure may be determined with reasonable accuracy. Then the criteria necessary for structural stability are discussed. Finally, the analysis of statically determinate, planar, pin-connected structures is presented.

2.1   Idealized Structure

An exact analysis of a structure can never be carried out, since estimates always have to be made of the loadings and the strength of the materials composing the structure. Furthermore, points of application for the loadings must also be estimated. It is important, therefore, that the structural engineer develop the ability to model or idealize a structure so that he or she can perform a practical force analysis of the members. In this section we will develop the basic techniques necessary to do this.

Notice that the deck of this concrete bridge is made so that one section can be considered roller supported on the other section.

Support Connections.   Structural members are joined together in various ways depending on the intent of the designer. The three types of joints most often specified are the pin connection, the roller support, and the fixed joint. A pin-connected joint and a roller support allow some freedom for slight rotation, whereas a fixed joint allows no relative rotation between the connected members and is consequently more expensive to fabricate. Examples of these joints, fashioned in metal and concrete, are shown in Figs. 2–1 and 2–2, respectively. For most timber structures, the members are assumed to be pin connected, since bolting or nailing them will not sufficiently restrain them from rotating with respect to each other.

Idealized models used in structural analysis that represent pinned and fixed supports and pin-connected and fixed-connected joints are shown in Figs. 2–3a and 2–3b. In reality, however, all connections exhibit some stiffness toward joint rotations, owing to friction and material behavior. In this case a more appropriate model for a support or joint might be that shown in Fig. 2–3c. If the torsional spring constant $k = 0$, the joint is a pin, and if $k \rightarrow \infty$, the joint is fixed.

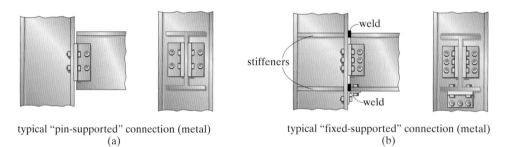

typical "pin-supported" connection (metal)
(a)

typical "fixed-supported" connection (metal)
(b)

Fig. 2–1

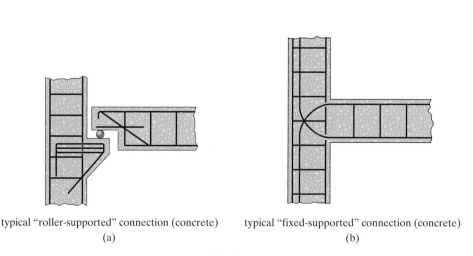

typical "roller-supported" connection (concrete)
(a)

typical "fixed-supported" connection (concrete)
(b)

Fig. 2–2

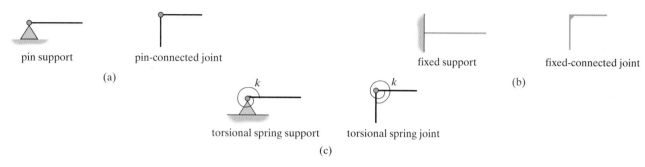

pin support   pin-connected joint   fixed support   fixed-connected joint

(a)   (b)

torsional spring support   torsional spring joint

(c)

Fig. 2–3

When selecting a particular model for each support or joint, the engineer must be aware of how the assumptions will affect the actual performance of the member and whether the assumptions are reasonable for the structural design. For example, consider the beam shown in Fig. 2–4a, which is used to support a concentrated load **P**. The angle connection at support A is like that in Fig. 2–1a and can therefore be idealized as a typical pin support. Furthermore, the support at B provides an approximate point of smooth contact and so it can be idealized as a roller. The beam's thickness can be neglected since it is small in comparison to the beam's length, and therefore the idealized model of the beam is as shown in Fig. 2–4b. The analysis of the loadings in this beam should give results that closely approximate the loadings in the actual beam. To show that the model is appropriate, consider a specific case of a beam made of steel with $P = 35.6$ kN and $L = 6.1$ m. One of the major simplifications made here was assuming the support at A to be a pin. Design of the beam using standard code procedures* indicates that a W10 × 19 would be adequate for supporting the load. Using one of the deflection methods of Chapter 8, the rotation at the "pin" support can be calculated as $\theta = 0.0103$ rad $= 0.59°$. From Fig. 2–4c, such a rotation only moves the top or bottom flange a distance of $\Delta = \theta r = (0.0103 \text{ rad})(130 \text{ mm}) = 1.34$ mm! This *small amount* would certainly be accommodated by the connection fabricated as shown in Fig. 2–1a, and therefore the pin serves as an appropriate model.

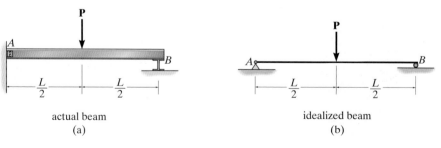

actual beam   idealized beam
(a)   (b)

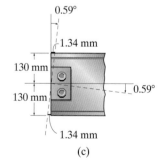

0.59°
1.34 mm
130 mm
130 mm
0.59°
1.34 mm
(c)

Fig. 2–4

*Codes such as the *Manual of Steel Construction,* American Institute of Steel Construction.

A typical rocker support used for a bridge girder.

Rollers and associated bearing pads are used to support the prestressed concrete girders of a highway bridge.

Other types of connections most commonly encountered on coplanar structures are given in Table 2–1. It is important to be able to recognize the symbols for these connections and the kinds of reactions they exert on their attached members. This can easily be done by noting how the connection *prevents* any degree of freedom or displacement of the member. In particular, the support will develop a *force* on the member if it *prevents translation* of the member, and it will develop a *moment* if it *prevents rotation* of the member. For example, a member in contact with a smooth surface (3) is prevented from translating only in one direction, which is perpendicular or normal to the surface. Hence, the surface exerts only a *normal force* $\mathbf{F}$ on the member in this direction. The magnitude of this force represents *one unknown*. Also note that the member is free to rotate on the surface, so that a moment cannot be developed by the surface on the member. As another example, the fixed support (7) prevents *both* translation and rotation of a member at the point of connection. Therefore, this type of support exerts two force components and a moment on the member. The "curl" of the moment lies in the plane of the page, since rotation is prevented in that plane. Hence, there are *three unknowns* at a fixed support.

In reality, all supports actually exert *distributed surface loads* on their contacting members. The concentrated forces and moments shown in Table 2–1 represent the *resultants* of these load distributions. This representation is, of course, an idealization; however, it is used here since the surface area over which the distributed load acts is considerably *smaller* than the *total* surface area of the connecting members.

The short link is used to connect the two girders of the highway bridge and allow for thermal expansion of the deck.

Typical pin used to support the steel girder of a railroad bridge.

TABLE 2–1 Supports for Coplanar Structures

Type of Connection	Idealized Symbol	Reaction	Number of Unknowns
(1) light cable weightless link			One unknown. The reaction is a force that acts in the direction of the cable or link.
(2) rollers rocker			One unknown. The reaction is a force that acts perpendicular to the surface at the point of contact.
(3) smooth contacting surface			One unknown. The reaction is a force that acts perpendicular to the surface at the point of contact.
(4) smooth pin-connected collar			One unknown. The reaction is a force that acts perpendicular to the surface at the point of contact.
(5) smooth pin or hinge			Two unknowns. The reactions are two force components.
(6) slider fixed-connected collar			Two unknowns. The reactions are a force and a moment.
(7) fixed support			Three unknowns. The reactions are the moment and the two force components.

2

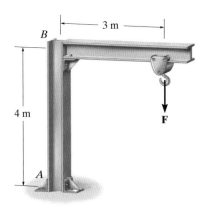

B

3 m

4 m

F

A

actual structure

(a)

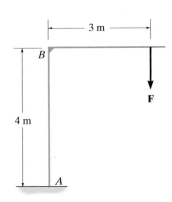

3 m

B

F

4 m

A

idealized structure

(b)

Fig. 2–5

Idealized Structure. Having stated the various ways in which the connections on a structure can be idealized, we are now ready to discuss some of the techniques used to represent various structural systems by idealized models.

As a first example, consider the jib crane and trolley in Fig. 2–5a. For the structural analysis we can neglect the thickness of the two main members and will assume that the joint at *B* is fabricated to be rigid. Furthermore, the support connection at *A* can be modeled as a fixed support and the details of the trolley excluded. Thus, the members of the idealized structure are represented by two connected lines, and the load on the hook is represented by a single concentrated force **F**, Fig. 2–5b. This idealized structure shown here as a *line drawing* can now be used for applying the principles of structural analysis, which will eventually lead to the design of its two main members.

Beams and girders are often used to support building floors. In particular, a *girder* is the main load-carrying element of the floor, whereas the smaller elements having a shorter span and connected to the girders are called *beams*. Often the loads that are applied to a beam or girder are transmitted to it by the floor that is supported by the beam or girder. Again, it is important to be able to appropriately idealize the system as a series of models, which can be used to determine, to a close approximation, the forces acting in the members. Consider, for example, the framing used to support a typical floor slab in a building, Fig. 2–6a. Here the slab is supported by *floor joists* located at even intervals, and these in turn are supported by the two side girders *AB* and *CD*. For analysis it is reasonable to assume that the joints are pin and/or roller connected to the girders and that the girders are pin and/or roller connected to the columns. The top view of the structural framing plan for this system is shown in Fig. 2–6b. In this "graphic" scheme, notice that the "lines" representing the joists do not touch the girders and the lines for the girders do not touch the columns. This symbolizes pin- and/ or roller-supported connections. On the other hand, if the framing plan is intended to represent fixed-connected members, such as those that are welded

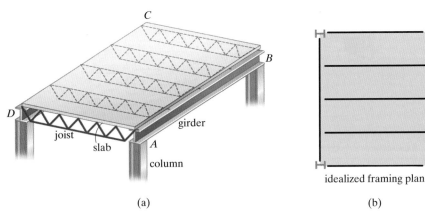

C

B

D

joist

girder

slab

A

column

idealized framing plan

(a)

(b)

Fig. 2–6

instead of simple bolted connections, then the lines for the beams or girders would touch the columns as in Fig. 2–7. Similarly, a fixed-connected overhanging beam would be represented in top view as shown in Fig. 2–8. If reinforced concrete construction is used, the beams and girders are represented by double lines. These systems are generally all fixed connected and therefore the members are drawn to the supports. For example, the structural graphic for the cast-in-place reinforced concrete system in Fig. 2–9a is shown in top view in Fig. 2–9b. The lines for the beams are dashed because they are below the slab.

Structural graphics and idealizations for timber structures are similar to those made of metal. For example, the structural system shown in Fig. 2–10a represents beam-wall construction, whereby the roof deck is supported by wood joists, which deliver the load to a masonry wall. The joists can be assumed to be simply supported on the wall, so that the idealized framing plan would be like that shown in Fig. 2–10b.

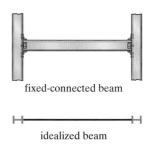

fixed-connected beam

idealized beam

Fig. 2–7

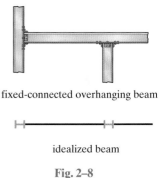

fixed-connected overhanging beam

idealized beam

Fig. 2–8

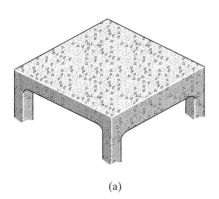

(a)

idealized framing plan

(b)

Fig. 2–9

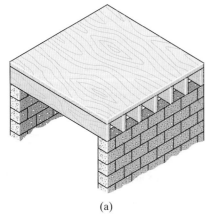

(a)

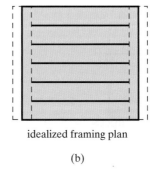

idealized framing plan

(b)

Fig. 2–10

The structural framework of this building consists of concrete floor joists, which were formed on site using metal pans. These joists are simply supported on the girders, which in turn are simply supported on the columns.

Tributary Loadings.   When flat surfaces such as walls, floors, or roofs are supported by a structural frame, it is necessary to determine how the load on these surfaces is transmitted to the various structural elements used for their support. There are generally two ways in which this can be done. The choice depends on the geometry of the structural system, the material from which it is made, and the method of its construction.

One-Way System.   A slab or deck that is supported such that it delivers its load to the supporting members by one-way action, is often referred to as a *one-way slab*. To illustrate the method of load transmission, consider the framing system shown in Fig. 2–11a where the beams AB, CD, and EF rest on the girders AE and BF. If a uniform load of 4.8 kN/m² is placed on the slab, then the center beam CD is assumed to support the load acting on the *tributary area* shown dark shaded on the structural framing plan in Fig. 2–11b. Member CD is therefore subjected to a *linear* distribution of load of $(4.8\ \text{kN/m}^2)(1.5\ \text{m}) = 7.2\ \text{kN/m}$, shown on the idealized beam in Fig. 2–11c. The reactions on this beam (10.8 kN) would then be applied to the center of the girders AE (and BF), shown idealized in Fig. 2–11d. Using this same concept, do you see how the remaining portion of the slab loading is transmitted to the ends of the girder as 5.4 kN?

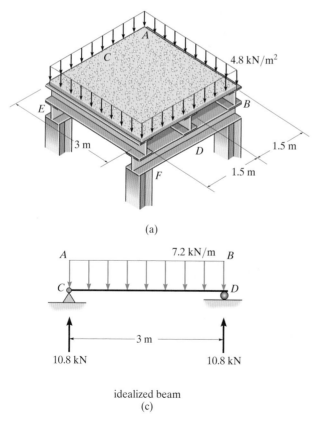

(a)

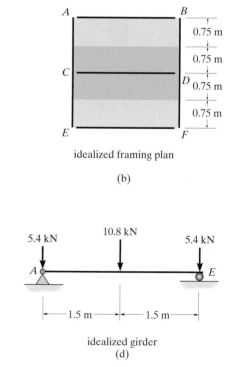

idealized framing plan

(b)

idealized beam
(c)

idealized girder
(d)

Fig. 2–11

An example of one-way slab construction of a steel frame building having a poured concrete floor on a corrugated metal deck. The load on the floor is considered to be transmitted to the beams, not the girders.

For some floor systems the beams and girders are connected to the columns at the *same elevation,* as in Fig. 2–12a. If this is the case, the slab can in some cases also be considered a "one-way slab." For example, if the slab is reinforced concrete with reinforcement in *only one direction,* or the concrete is poured on a corrugated metal deck, as in the above photo, then one-way action of load transmission can be assumed. On the other hand, if the slab is flat on top and bottom and is reinforced in *two directions,* then consideration must be given to the *possibility* of the load being transmitted to the supporting members from either one or two directions. For example, consider the slab and framing plan in Fig. 2–12b. According to the American Concrete Institute, ACI 318 code, if $L_2 > L_1$ and if the span ratio $(L_2/L_1) > 2$, *the slab will behave as a one-way slab,* since as L_1 becomes smaller, the beams AB, CD, and EF provide the greater stiffness to carry the load.

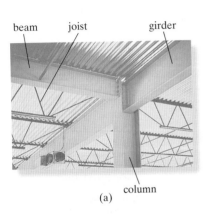

beam   joist   girder

column

(a)

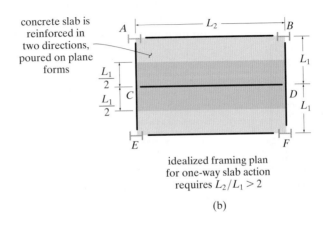

concrete slab is reinforced in two directions, poured on plane forms

idealized framing plan for one-way slab action requires $L_2/L_1 > 2$

(b)

Fig. 2–12

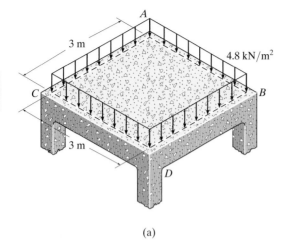

(a)

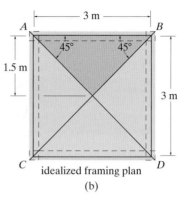

idealized framing plan

(b)

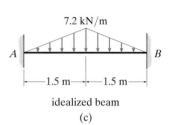

idealized beam

(c)

Fig. 2–13

Two-Way System.

If, according to the ACI 318 concrete code the support ratio in Fig. 2–12b is $(L_2/L_1) \leq 2$, the load is assumed to be delivered to the supporting beams and girders in two directions. When this is the case the slab is referred to as a *two-way slab*. To show one method of treating this case, consider the square reinforced concrete slab in Fig. 2–13a, which is supported by four 3-m-long edge beams, AB, BD, DC, and CA. Here $L_2/L_1 = 1$. Due to two-way slab action, the assumed *tributary area* for beam AB is shown dark shaded in Fig. 2–13b. This area is determined by constructing diagonal 45° lines as shown. Hence if a uniform load of 4.8 kN/m² is applied to the slab, a peak intensity of $(4.8 \text{ kN/m}^2)(1.5 \text{ m}) = 7.2 \text{ kN/m}$ will be applied to the center of beam AB, resulting in a *triangular* load distribution shown in Fig. 2–13c. For other geometries that cause two-way action, a similar procedure can be used. For example, if $L_2/L_1 = 1.5$ it is then necessary to construct 45° lines that intersect as shown in Fig. 2–14a. A 4.8-kN/m² loading placed on the slab will then produce *trapezoidal* and *triangular* distributed loads on members AB and AC, Fig. 2–14b and 2–14c, respectively.

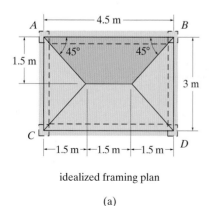

idealized framing plan

(a)

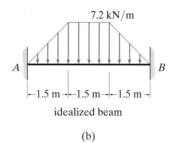

idealized beam

(b)

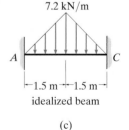

idealized beam

(c)

Fig. 2–14

The ability to reduce an actual structure to an idealized form, as shown by these examples, can only be gained by experience. To provide practice at doing this, the example problems and the problems for solution throughout this book are presented in somewhat realistic form, and the associated problem statements aid in explaining how the connections and supports can be modeled by those listed in Table 2–1. In engineering practice, if it becomes doubtful as to how to model a structure or transfer the loads to the members, it is best to consider *several* idealized structures and loadings and then design the actual structure so that it can resist the loadings in all the idealized models.

EXAMPLE 2.1

The floor of a classroom is to be supported by the bar joists shown in Fig. 2–15a. Each joist is 4.5 m long and they are spaced 0.75 m on centers. The floor itself is to be made from lightweight concrete that is 100 mm thick. Neglect the weight of the joists and the corrugated metal deck, and determine the load that acts along each joist.

(a)

SOLUTION

The dead load on the floor is due to the weight of the concrete slab. From Table 1–3 for 100 mm of lightweight concrete it is $(100)(0.015 \text{ kN/m}^2) = 1.50 \text{ kN/m}^2$. From Table 1–4, the live load for a classroom is 1.92 kN/m². Thus the total floor load is $1.50 \text{ kN/m}^2 + 1.92 \text{ kN/m}^2 = 3.42 \text{ kN/m}^2$. For the floor system, $L_1 = 0.75 \text{ m}$ and $L_2 = 4.5 \text{ m}$. Since $L_2/L_1 > 2$ the concrete slab is treated as a one-way slab. The tributary area for each joist is shown in Fig. 2–15b. Therefore the uniform load along its length is

$$w = 3.42 \text{ kN/m}^2(0.75 \text{ m}) = 2.57 \text{ kN/m}^2$$

This loading and the end reactions on each joist are shown in Fig. 2–15c.

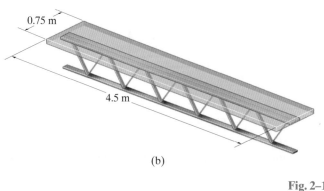

0.75 m

4.5 m

(b)

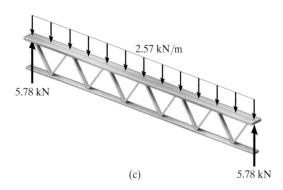

2.57 kN/m

5.78 kN

(c)

5.78 kN

Fig. 2–15

EXAMPLE 2.2

The flat roof of the steel-frame building shown in the photo is intended to support a total load of 2 kN/m^2 over its surface. Determine the roof load within region $ABCD$ that is transmitted to beam BC. The dimensions are shown in Fig. 2–16a.

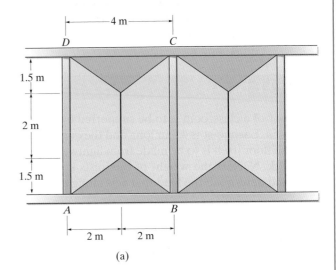

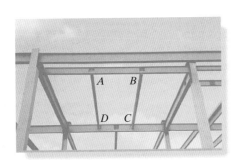

(a)

SOLUTION

In this case $L_2 = 5 \text{ m}$ and $L_1 = 4 \text{ m}$. Since $L_2/L_1 = 1.25 < 2$, we have two-way slab action. The tributary loading along each edge beam is shown in Fig. 2–16a, where the lighter shaded trapezoidal area of loading is transmitted to member BC. The peak intensity of this loading is $(2 \text{ kN/m}^2)(2 \text{ m}) = 4 \text{ kN/m}$. As a result, the distribution of load along BC is shown in Fig. 2–16b.

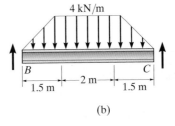

(b)

Fig. 2–16

This process of tributary load transmission should *also* be calculated for the region to the right of BC shown in the photo, and this load should *also* be placed on BC. See the next example.

EXAMPLE 2.3

The concrete girders shown in the photo of the passenger car parking garage span 10 m and are 5 m on center. If the floor slab is 125 mm thick and made of reinforced stone concrete, and the specified live load is 2.4 kN/m² (see Table 1–4), determine the distributed load the floor system transmits to each interior girder.

SOLUTION

Here, $L_2 = 10$ m and $L_1 = 5$ m, so that $L_2/L_1 = 2$. We have a two-way slab. From Table 1–2, for reinforced stone concrete, the specific weight of the concrete is 23.6 kN/m³. Thus the design floor loading is

$$p = 23.6 \text{ kN/m}^3\left(\frac{125}{1000}\text{m}\right) + 2.4 \text{ kN/m}^2 = 5.35 \text{ kN/m}^2$$

A trapezoidal distributed loading is transmitted to each interior girder AB from each of its sides. The maximum intensity of each of these distributed loadings is $(5.35 \text{ kN/m}^2)(2.5 \text{ m}) = 13.38 \text{ kN/m}$, so that on the girder this intensity becomes $2(13.38 \text{ kN/m}) = 26.75 \text{ kN/m}$, Fig. 2–17b. *Note:* For design, consideration should also be given to the weight of the girder.

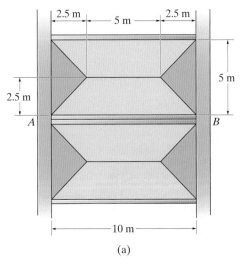

(a)

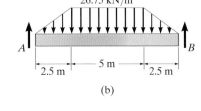

(b)

Fig. 2–17

2.2   Principle of Superposition

The principle of superposition forms the basis for much of the theory of structural analysis. It may be stated as follows: *The total displacement or internal loadings (stress) at a point in a structure subjected to several external loadings can be determined by adding together the displacements or internal loadings (stress) caused by each of the external loads acting separately.* For this statement to be valid it is necessary that a *linear relationship* exist among the loads, stresses, and displacements.

Two requirements must be imposed for the principle of superposition to apply:

1. The material must behave in a linear-elastic manner, so that Hooke's law is valid, and therefore the load will be proportional to displacement.

2. The geometry of the structure must not undergo significant change when the loads are applied, i.e., small displacement theory applies. Large displacements will significantly change the position and orientation of the loads. An example would be a cantilevered thin rod subjected to a force at its end.

Throughout this text, these two requirements will be satisfied. Here only linear-elastic material behavior occurs; and the displacements produced by the loads will not significantly change the directions of applied loadings nor the dimensions used to compute the moments of forces.

The walls on the sides of this building are used to strengthen its structure when the building is subjected to large hurricane wind loadings applied to its front or back. These walls are called "shear walls."

2.3   Equations of Equilibrium

It may be recalled from statics that a structure or one of its members is in equilibrium when it maintains a balance of force and moment. In general this requires that the force and moment equations of equilibrium be satisfied along three independent axes, namely,

$$\Sigma F_x = 0 \quad \Sigma F_y = 0 \quad \Sigma F_z = 0$$
$$\Sigma M_x = 0 \quad \Sigma M_y = 0 \quad \Sigma M_z = 0 \tag{2–1}$$

The principal load-carrying portions of most structures, however, lie in a single plane, and since the loads are also coplanar, the above requirements for equilibrium reduce to

$$\Sigma F_x = 0$$
$$\Sigma F_y = 0 \tag{2–2}$$
$$\Sigma M_O = 0$$

Here ΣF_x and ΣF_y represent, respectively, the algebraic sums of the x and y components of all the forces acting on the structure or one of its members, and ΣM_O represents the algebraic sum of the moments of these force components about an axis perpendicular to the x–y plane (the z axis) and passing through point O.

Whenever these equations are applied, *it is first necessary to draw a free-body diagram of the structure or its members.* If a member is selected, it must be *isolated* from its supports and surroundings and its outlined shape drawn. All the forces and couple moments must be shown that act *on the member.* In this regard, the types of reactions at the supports can be determined using Table 2–1. Also, recall that forces common to two members act with equal magnitudes but opposite directions on the respective free-body diagrams of the members.

If the *internal loadings* at a specified point in a member are to be determined, the *method of sections* must be used. This requires that a "cut" or section be made perpendicular to the axis of the member at the point where the internal loading is to be determined. A free-body diagram of either segment of the "cut" member is isolated and the internal loads are then determined from the equations of equilibrium applied to the segment. In general, the internal loadings acting at the section will consist of a normal force **N**, shear force **V**, and bending moment **M**, as shown in Fig. 2–18.

We will cover the principles of statics that are used to determine the external reactions on structures in Sec. 2–5. Internal loadings in structural members will be discussed in Chapter 4.

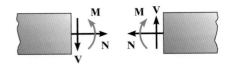

internal loadings

Fig. 2–18

2.4   Determinacy and Stability

Before starting the force analysis of a structure, it is necessary to establish the determinacy and stability of the structure.

Determinacy.   The equilibrium equations provide both the *necessary and sufficient* conditions for equilibrium. When all the forces in a structure can be determined strictly from these equations, the structure is referred to as *statically determinate*. Structures having more unknown forces than available equilibrium equations are called *statically indeterminate*. As a general rule, a structure can be identified as being either statically determinate or statically indeterminate by drawing free-body diagrams of all its members, or selective parts of its members, and then comparing the total number of unknown reactive force and moment components with the total number of available equilibrium equations.* For a coplanar structure there are at most *three* equilibrium equations for each part, so that if there is a total of n parts and r force and moment reaction components, we have

$$\begin{aligned} r = 3n, \text{ statically determinate} \\ r > 3n, \text{ statically indeterminate} \end{aligned} \qquad (2\text{--}3)$$

In particular, if a structure is *statically indeterminate,* the additional equations needed to solve for the unknown reactions are obtained by relating the applied loads and reactions to the displacement or slope at different points on the structure. These equations, which are referred to as *compatibility equations,* must be equal in number to the *degree of indeterminacy* of the structure. Compatibility equations involve the geometric and physical properties of the structure and will be discussed further in Chapter 10.

We will now consider some examples to show how to classify the determinacy of a structure. The first example considers beams; the second example, pin-connected structures; and in the third we will discuss frame structures. Classification of trusses will be considered in Chapter 3.

---

*Drawing the free-body diagrams is not strictly necessary, since a "mental count" of the number of unknowns can also be made and compared with the number of equilibrium equations.

EXAMPLE 2.4

Classify each of the beams shown in Figs. 2–19a through 2–19d as statically determinate or statically indeterminate. If statically indeterminate, report the number of degrees of indeterminacy. The beams are subjected to external loadings that are assumed to be known and can act anywhere on the beams.

SOLUTION

Compound beams, i.e., those in Fig. 2–19c and 2–19d, which are composed of pin-connected members must be disassembled. Note that in these cases, the unknown reactive forces acting between each member must be shown in equal but opposite pairs. The free-body diagrams of each member are shown. Applying $r = 3n$ or $r > 3n$, the resulting classifications are indicated.

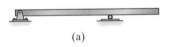

(a)

$r = 3, n = 1, 3 = 3(1)$

Statically determinate   *Ans.*

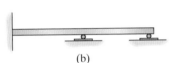

(b)

$r = 5, n = 1, 5 > 3(1)$

Statically indeterminate to the second degree   *Ans.*

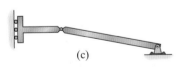

(c)

$r = 6, n = 2, 6 = 3(2)$

Statically determinate   *Ans.*

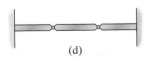

(d)

$r = 10, n = 3, 10 > 3(3)$

Statically indeterminate to the first degree   *Ans.*

Fig. 2–19

EXAMPLE 2.5

Classify each of the pin-connected structures shown in Figs. 2–20a through 2–20d as statically determinate or statically indeterminate. If statically indeterminate, report the number of degrees of indeterminacy. The structures are subjected to arbitrary external loadings that are assumed to be known and can act anywhere on the structures.

SOLUTION

Classification of pin-connected structures is similar to that of beams. The free-body diagrams of the members are shown. Applying $r = 3n$ or $r > 3n$, the resulting classifications are indicated.

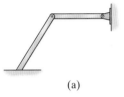

(a)

$r = 7, n = 2, 7 > 6$
Statically indeterminate to the first degree   *Ans.*

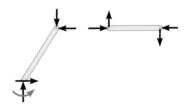

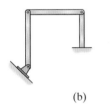

(b)

$r = 9, n = 3, 9 = 9,$
Statically determinate   *Ans.*

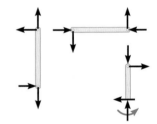

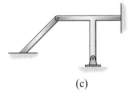

(c)

$r = 10, n = 2, 10 > 6,$
Statically indeterminate to the fourth degree   *Ans.*

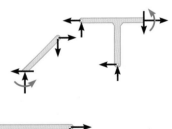

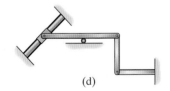

(d)

$r = 9, n = 3, 9 = 9,$
Statically determinate   *Ans.*

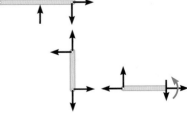

Fig. 2–20

EXAMPLE 2.6

Classify each of the frames shown in Figs. 2–21a through 2–21c as statically determinate or statically indeterminate. If statically indeterminate, report the number of degrees of indeterminacy. The frames are subjected to external loadings that are assumed to be known and can act anywhere on the frames.

SOLUTION

Unlike the beams and pin-connected structures of the previous examples, frame structures consist of members that are connected together by rigid joints. Sometimes the members form internal loops as in Fig. 2–21a. Here *ABCD* forms a closed loop. In order to classify these structures, it is necessary to use the method of sections and "cut" the loop apart. The free-body diagrams of the sectioned parts are drawn and the frame can then be classified. Notice that only *one section* through the loop is required, since once the unknowns at the section are determined, the internal forces at any point in the members can then be found using the method of sections and the equations of equilibrium. A second example of this is shown in Fig. 2–21b. Although the frame in Fig. 2–21c has no closed loops we can use this same method, using vertical sections, to classify it. For this case we can *also* just draw its complete free-body diagram. The resulting classifications are indicated in each figure.

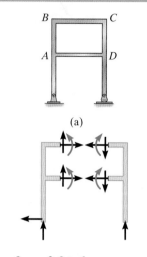

(a)

$r = 9, n = 2, 9 > 6,$
Statically indeterminate to the third degree *Ans.*

(a)

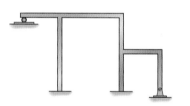

$r = 9, n = 1, 9 > 3,$
Statically indeterminate to the sixth degree *Ans.*

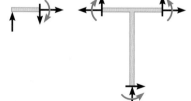

(This frame has no closed loops.)

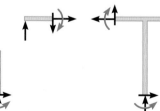

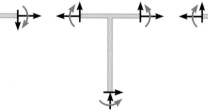

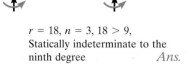

$r = 18, n = 3, 18 > 9,$
Statically indeterminate to the ninth degree *Ans.*

(b)

(c)

$r = 18, n = 4, 18 > 12,$
Statically indeterminate to the sixth degree *Ans.*

Fig. 2–21

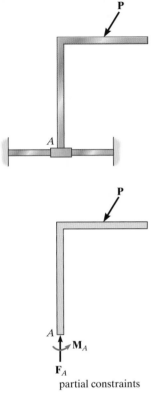

partial constraints

Fig. 2–22

Stability.   To ensure the equilibrium of a structure or its members, it is not only necessary to satisfy the equations of equilibrium, but the members must also be properly held or constrained by their supports. Two situations may occur where the conditions for proper constraint have not been met.

Partial Constraints.   In some cases a structure or one of its members may have *fewer* reactive forces than equations of equilibrium that must be satisfied. The structure then becomes only *partially constrained*. For example, consider the member shown in Fig. 2–22 with its corresponding free-body diagram. Here the equation $\Sigma F_x = 0$ will not be satisfied for the loading conditions and therefore the member will be unstable.

Improper Constraints.   In some cases there may be as many unknown forces as there are equations of equilibrium; however, *instability* or movement of a structure or its members can develop because of *improper constraining* by the supports. This can occur if all the *support reactions are concurrent* at a point. An example of this is shown in Fig. 2–23. From the free-body diagram of the beam it is seen that the summation of moments about point O will *not* be equal to zero $(Pd \neq 0)$; thus rotation about point O will take place.

Another way in which improper constraining leads to instability occurs when the *reactive forces* are all *parallel*. An example of this case is shown in Fig. 2–24. Here when an inclined force $\mathbf{P}$ is applied, the summation of forces in the horizontal direction will not equal zero.

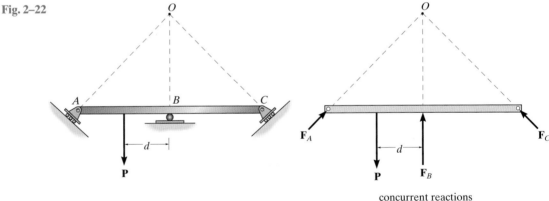

concurrent reactions

Fig. 2–23

parallel reactions

Fig. 2–24

In general, then, a structure will be geometrically unstable—that is, it will move slightly or collapse—if there are fewer reactive forces than equations of equilibrium; or if there are enough reactions, instability will occur if the lines of action of the reactive forces intersect at a common point or are parallel to one another. If the structure consists of several members or components, local instability of one or several of these members can generally be determined *by inspection*. If the members form a collapsible mechanism, the structure will be unstable. We will now formalize these statements for a *coplanar structure* having n members or components with r unknown reactions. Since three equilibrium equations are available for each member or component, we have

$$
\begin{array}{ll}
r < 3n & \text{unstable} \\
r \geq 3n & \text{unstable if member reactions are} \\
 & \text{concurrent or parallel or some of the} \\
 & \text{components form a collapsible mechanism}
\end{array}
\qquad (2\text{--}4)
$$

If the structure is unstable, *it does not matter* if it is statically determinate or indeterminate. In all cases such types of structures must be avoided in practice.

The following examples illustrate how structures or their members can be classified as stable or unstable. Structures in the form of a truss will be discussed in Chapter 3.

The K-bracing on this frame provides lateral support from wind and vertical support of the floor girders. Notice the use of concrete grout, which is applied to insulate the steel to keep it from losing its stiffness in the event of a fire.

EXAMPLE 2.7

Classify each of the structures in Figs. 2–25a through 2–25d as stable or unstable. The structures are subjected to arbitrary external loads that are assumed to be known.

SOLUTION

The structures are classified as indicated.

(a)

Fig. 2–25

The member is *stable* since the reactions are nonconcurrent and nonparallel. It is also statically determinate.   *Ans.*

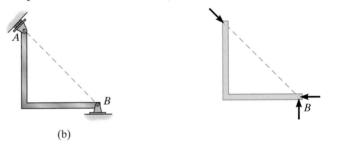

(b)

The member is *unstable* since the three reactions are concurrent at *B*.   *Ans.*

(c)

The beam is *unstable* since the three reactions are all parallel.   *Ans.*

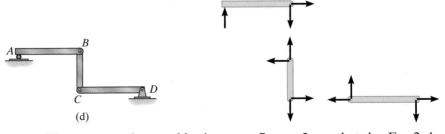

(d)

The structure is *unstable* since $r = 7$, $n = 3$, so that, by Eq. 2–4, $r < 3n$, $7 < 9$. Also, this can be seen by inspection, since *AB* can move horizontally without restraint.   *Ans.*

PROBLEMS

2–1. The steel framework is used to support the reinforced stone concrete slab that is used for an office. The slab is 200 mm thick. Sketch the loading that acts along members *BE* and *FED*. Take $a = 2$ m, $b = 5$ m. *Hint:* See Tables 1–2 and 1–4.

2–2. Solve Prob. 2–1 with $a = 3$ m, $b = 4$ m.

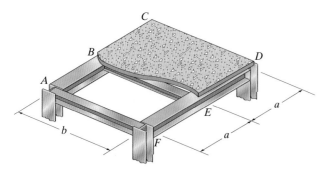

Probs. 2–1/2–2

2–3. The floor system used in a school classroom consists of a 100-mm reinforced stone concrete slab. Sketch the loading that acts along the joist *BF* and side girder *ABCDE*. Set $a = 3$ m, $b = 9$ m. *Hint:* See Tables 1–2 and 1–4.

*2–4. Solve Prob. 2–3 with $a = 3$ m, $b = 4.5$ m.

2–5. Solve Prob. 2–3 with $a = 2.25$ m, $b = 6$ m.

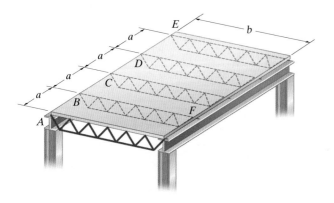

Probs. 2–3/2–4/2–5

2–6. The frame is used to support a 50-mm-thick plywood floor of a residential dwelling. Sketch the loading that acts along members *BG* and *ABCD*. Set $a = 1.5$ m, $b = 4.5$ m. *Hint:* See Tables 1–2 and 1–4.

2–7. Solve Prob. 2–6, with $a = 2.4$ m, $b = 2.4$ m.

*2–8. Solve Prob. 2–6, with $a = 2.7$ m, $b = 4.5$ m.

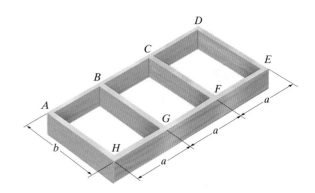

Probs. 2–6/2–7/2–8

2–9. The steel framework is used to support the 100-mm reinforced stone concrete slab that carries a uniform live loading of 24 kN/m². Sketch the loading that acts along members *BE* and *FED*. Set $b = 3$ m, $a = 2.25$ m. *Hint:* See Table 1–2.

2–10. Solve Prob. 2–9, with $b = 3.6$ m, $a = 1.2$ m.

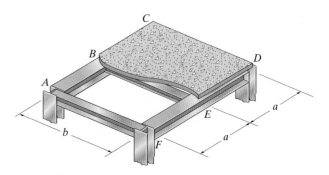

Probs. 2–9/2–10

2–11. Classify each of the structures as statically determinate, statically indeterminate, or unstable. If indeterminate, specify the degree of indeterminacy. The supports or connections are to be assumed as stated.

*2–12. Classify each of the frames as statically determinate or indeterminate. If indeterminate, specify the degree of indeterminacy. All internal joints are fixed connected.

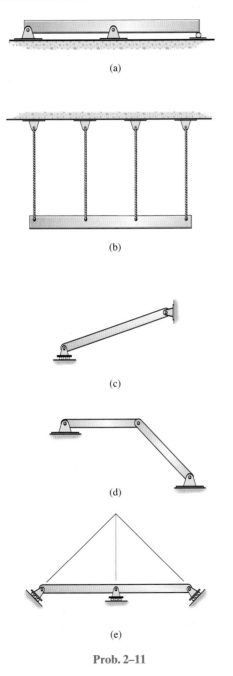

(a)

(b)

(c)

(d)

(e)

Prob. 2–11

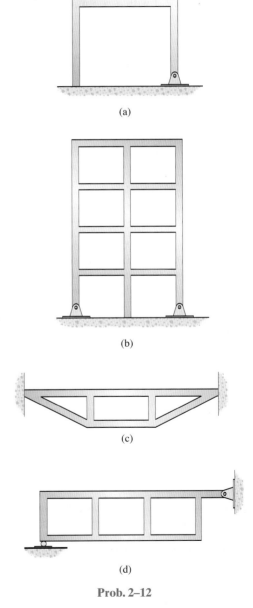

(a)

(b)

(c)

(d)

Prob. 2–12

2–13. Classify each of the structures as statically determinate, statically indeterminate, stable, or unstable. If indeterminate, specify the degree of indeterminacy. The supports or connections are to be assumed as stated.

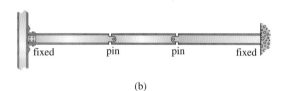

(a)

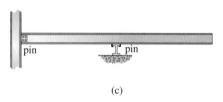

(b)

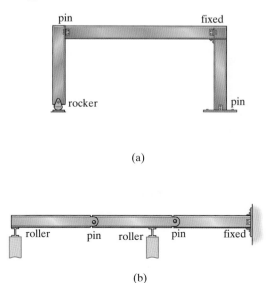

(c)

Prob. 2–13

2–14. Classify each of the structures as statically determinate, statically indeterminate, stable, or unstable. If indeterminate, specify the degree of indeterminacy. The supports or connections are to be assumed as stated.

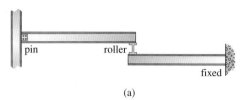

(a)

(b)

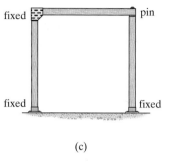

(c)

Prob. 2–14

2–15. Classify each of the structures as statically determinate, statically indeterminate, or unstable. If indeterminate, specify the degree of indeterminacy.

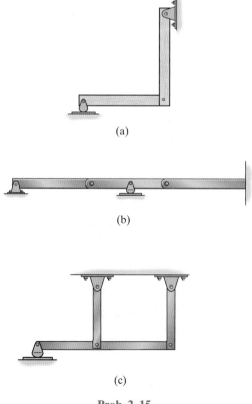

(a)

(b)

(c)

Prob. 2–15

*2–16. Classify each of the structures as statically determinate, statically indeterminate, or unstable. If indeterminate, specify the degree of indeterminacy.

2–17. Classify each of the structures as statically determinate, statically indeterminate, stable, or unstable. If indeterminate, specify the degree of indeterminacy.

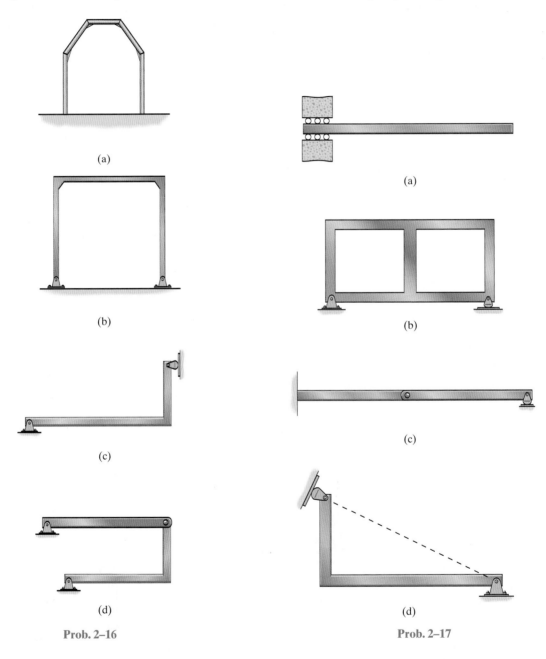

(a)

(b)

(c)

(d)

Prob. 2–16

(a)

(b)

(c)

(d)

Prob. 2–17

2.5   Application of the Equations of Equilibrium

Occasionally, the members of a structure are connected together in such a way that the joints can be assumed as pins. Building frames and trusses are typical examples that are often constructed in this manner. Provided a pin-connected coplanar structure is properly constrained and contains no more supports or members than are necessary to prevent collapse, the forces acting at the joints and supports can be determined by applying the three equations of equilibrium ($\Sigma F_x = 0$, $\Sigma F_y = 0$, $\Sigma M_O = 0$) to each member. Understandably, once the forces at the joints are obtained, the size of the members, connections, and supports can then be determined on the basis of design code specifications.

To illustrate the method of force analysis, consider the three-member frame shown in Fig. 2–26a, which is subjected to loads $\mathbf{P}_1$ and $\mathbf{P}_2$. The free-body diagrams of each member are shown in Fig. 2–26b. In total there are nine unknowns; however, nine equations of equilibrium can be written, three for each member, so the problem is *statically determinate*. For the actual solution it is *also* possible, and sometimes convenient, to consider a portion of the frame or its entirety when applying some of these nine equations. For example, a free-body diagram of the entire frame is shown in Fig. 2–26c. One could determine the three reactions $\mathbf{A}_x$, $\mathbf{A}_y$, and $\mathbf{C}_x$ on this "rigid" pin-connected system, then analyze *any two* of its members, Fig. 2–26b, to obtain the other six unknowns. Furthermore, the answers can be checked in part by applying the three equations of equilibrium to the remaining "third" member. To summarize, this problem can be solved by writing *at most* nine equilibrium equations using free-body diagrams of any members and/or combinations of connected members. Any more than nine equations written would *not* be unique from the original nine and would only serve to check the results.

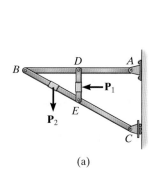

(a)

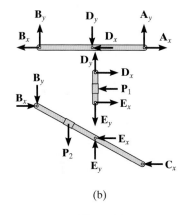

(b)

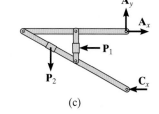

(c)

Fig. 2–26

Consider now the two-member frame shown in Fig. 2–27a. Here the free-body diagrams of the members reveal six unknowns, Fig. 2–27b; however, six equilibrium equations, three for each member, can be written, so again the problem is statically determinate. As in the previous case, a free-body diagram of the entire frame can also be used for part of the analysis, Fig. 2–27c. Although, as shown, the frame has a tendency to collapse without its supports, by rotating about the pin at B, this will not happen since the force system acting on it must still hold it in equilibrium. Hence, if so desired, all six unknowns can be determined by applying the three equilibrium equations to the entire frame, Fig. 2–27c, and also to either one of its members.

The above two examples illustrate that if a structure is properly supported and contains no more supports or members than are necessary to prevent collapse, the frame becomes statically determinate, and so the unknown forces at the supports and connections can be determined from the equations of equilibrium applied to each member. Also, if the structure remains *rigid* (noncollapsible) when the supports are removed (Fig. 2–26c), all three support reactions can be determined by applying the three equilibrium equations to the entire structure. However, if the structure appears to be nonrigid (collapsible) after removing the supports (Fig. 2–27c), it must be dismembered and equilibrium of the individual members must be considered in order to obtain enough equations to determine *all* the support reactions.

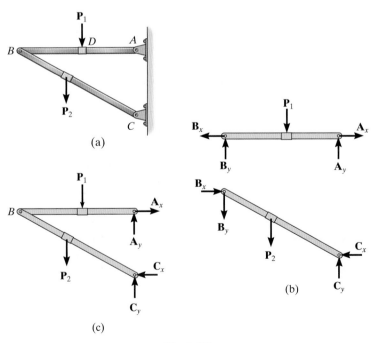

Fig. 2–27

Procedure for Analysis

The following procedure provides a method for determining the *joint reactions* for structures composed of pin-connected members.

Free-Body Diagrams

- Disassemble the structure and draw a free-body diagram of each member. Also, it may be convenient to supplement a member free-body diagram with a free-body diagram of the *entire structure*. Some or all of the support reactions can then be determined using this diagram.

- Recall that reactive forces common to two members act with equal magnitudes but opposite directions on the respective free-body diagrams of the members.

- All two-force members should be identified. These members, regardless of their shape, have no external loads on them, and therefore their free-body diagrams are represented with equal but opposite collinear forces acting on their ends.

- In many cases it is possible to tell by inspection the proper arrowhead sense of direction of an unknown force or couple moment; however, if this seems difficult, the directional sense can be assumed.

Equations of Equilibrium

- Count the total number of unknowns to make sure that an equivalent number of equilibrium equations can be written for solution. Except for two-force members, recall that in general three equilibrium equations can be written for each member.

- Many times, the solution for the unknowns will be straightforward if the moment equation $\Sigma M_O = 0$ is applied about a point (O) that lies at the intersection of the lines of action of as many unknown forces as possible.

- When applying the force equations $\Sigma F_x = 0$ and $\Sigma F_y = 0$, orient the x and y axes along lines that will provide the simplest reduction of the forces into their x and y components.

- If the solution of the equilibrium equations yields a *negative* magnitude for an unknown force or couple moment, it indicates that its arrowhead sense of direction is *opposite* to that which was assumed on the free-body diagram.

EXAMPLE 2.8

Determine the reactions on the beam shown in Fig. 2–28a.

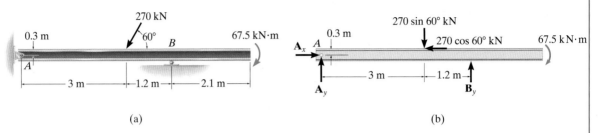

(a)   (b)

Fig. 2–28

SOLUTION

Free-Body Diagram. As shown in Fig. 2–28b, the 270-kN force is resolved into x and y components. Furthermore, the 2.1-m dimension line is not needed since a couple moment is a *free vector* and can therefore act anywhere on the beam for the purpose of computing the external reactions.

Equations of Equilibrium. Applying Eqs. 2–2 in a sequence, using previously calculated results, we have

$$\xrightarrow{+} \Sigma F_x = 0; \quad A_x - 270 \cos 60° = 0 \quad A_x = 135 \text{ kN} \quad \textit{Ans}$$

$$\curvearrowleft + \Sigma M_A = 0; \quad -270 \sin 60°(3) + 270 \cos 60°(0.3) + B_y(4.2) - 67.5 = 0 \quad B_y = 173.4 \text{ kN} \textit{ Ans.}$$

$$+ \uparrow \Sigma F_y = 0; \quad -270 \sin 60° + 173.4 + A_y = 0 \quad A_y = 60.4 \text{ kN} \quad \textit{Ans.}$$

EXAMPLE 2.9

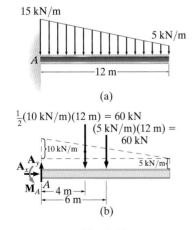

(a)

(b)

Fig. 2–29

Determine the reactions on the beam in Fig. 2–29a.

SOLUTION

Free-Body Diagram. As shown in Fig. 2–29b, the trapezoidal distributed loading is segmented into a triangular and a uniform load. The *areas* under the triangle and rectangle represent the *resultant* forces. These forces act through the centroid of their corresponding areas.

Equations of Equilibrium

$$\xrightarrow{+} \Sigma F_x = 0; \quad A_x = 0 \quad \textit{Ans}$$

$$+ \uparrow \Sigma F_y = 0; \quad A_y - 60 - 60 = 0 \quad A_y = 120 \text{ kN} \quad \textit{Ans.}$$

$$\curvearrowleft + \Sigma M_A = 0; \quad -60(4) - 60(6) + M_A = 0 \quad M_A = 600 \text{ kN} \cdot \text{m} \quad \textit{Ans.}$$

EXAMPLE 2.10

Determine the reactions on the beam in Fig. 2–30a. Assume A is a pin and the support at B is a roller (smooth surface).

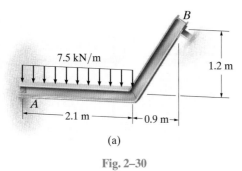

(a)

Fig. 2–30

SOLUTION

Free-Body Diagram.   As shown in Fig. 2–30b, the support ("roller") at B exerts a *normal force* on the beam at its point of contact. The line of action of this force is defined by the 3–4–5 triangle.

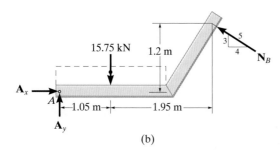

(b)

Equations of Equilibrium.   Resolving $\mathbf{N}_B$ into x and y components and summing moments about A yields a direct solution for N_B. Why? Using this result, we can then obtain A_x and A_y.

$$\zeta+\Sigma M_A = 0; \qquad -15.75(1.05) + \left(\tfrac{4}{5}\right)N_B(1.2) + \left(\tfrac{3}{5}\right)N_B(3) = 0 \quad \textit{Ans.}$$

$$N_B = 5.99 \text{ kN}$$

$$\xrightarrow{+}\Sigma F_x = 0; \qquad A_x - \tfrac{4}{5}(5.99) = 0 \qquad\qquad A_x = 4.79 \text{ kN} \quad \textit{Ans.}$$

$$+\uparrow\Sigma F_y = 0; \qquad A_y - 15.75 + \tfrac{3}{5}(5.99) = 0 \qquad A_y = 12.15 \text{ kN} \quad \textit{Ans.}$$

EXAMPLE 2.11

The compound beam in Fig. 2–31a is fixed at A. Determine the reactions at A, B, and C. Assume that the connection at B is a pin and C is a roller.

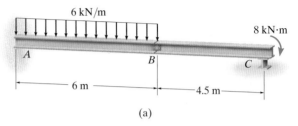

(a)

Fig. 2–31

SOLUTION

Free-Body Diagrams.   The free-body diagram of each segment is shown in Fig. 2–31b. Why is this problem statically determinate?

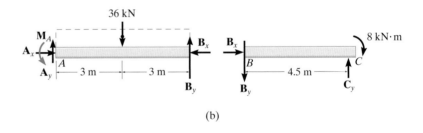

(b)

Equations of Equilibrium.   There are six unknowns. Applying the six equations of equilibrium, using previously calculated results, we have

Segment BC:

$\zeta + \Sigma M_C = 0$;	$-8 + B_y(4.5) = 0$	$B_y = 1.78$ kN	*Ans.*
$+\uparrow \Sigma F_y = 0$;	$-1.78 + C_y = 0$	$C_y = 1.78$ kN	*Ans.*
$\xrightarrow{+} \Sigma F_x = 0$;	$B_x = 0$		*Ans.*

Segment AB:

$\zeta + \Sigma M_A = 0$;	$M_A - 36(3) + 1.78(6) = 0$		
	$M_A = 97.3$ kN $\cdot$ m		*Ans.*
$+\uparrow \Sigma F_y = 0$;	$A_y - 36 + 1.78 = 0$	$A_y = 34.2$ kN	*Ans.*
$\xrightarrow{+} \Sigma F_x = 0$;	$A_x - 0 = 0$	$A_x = 0$	*Ans.*

EXAMPLE 2.12

Determine the horizontal and vertical components of reaction at the pins A, B, and C of the two-member frame shown in Fig. 2–32a.

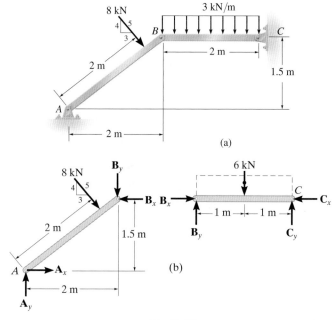

Fig. 2–32

SOLUTION

Free-Body Diagrams. The free-body diagram of each member is shown in Fig. 2–32b.

Equations of Equilibrium. Applying the six equations of equilibrium in the following sequence allows a direct solution for each of the six unknowns.

Member BC:

$$\curvearrowleft + \Sigma M_C = 0; \quad -B_y(2) + 6(1) = 0 \qquad\qquad B_y = 3 \text{ kN} \qquad Ans.$$

Member AB:

$$\curvearrowleft + \Sigma M_A = 0; \quad -8(2) - 3(2) + B_x(1.5) = 0 \quad B_x = 14.7 \text{ kN} \qquad Ans.$$

$$\xrightarrow{+} \Sigma F_x = 0; \quad A_x + \tfrac{3}{5}(8) - 14.7 = 0 \qquad A_x = 9.87 \text{ kN} \qquad Ans.$$

$$+\uparrow \Sigma F_y = 0; \quad A_y - \tfrac{4}{5}(8) - 3 = 0 \qquad A_y = 9.40 \text{ kN} \qquad Ans.$$

Member BC:

$$\xrightarrow{+} \Sigma F_x = 0; \quad 14.7 - C_x = 0 \qquad\qquad C_x = 14.7 \text{ kN} \qquad Ans.$$

$$+\uparrow \Sigma F_y = 0; \quad 3 - 6 + C_y = 0 \qquad\qquad C_y = 3 \text{ kN} \qquad Ans.$$

EXAMPLE 2.13

The side of the building in Fig. 2–33a is subjected to a wind loading that creates a uniform *normal* pressure of 15 kPa on the windward side and a suction pressure of 5 kPa on the leeward side. Determine the horizontal and vertical components of reaction at the pin connections A, B, and C of the supporting gable arch.

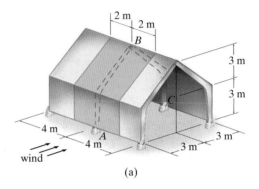

(a)

Fig. 2–33

SOLUTION

Since the loading is evenly distributed, the central gable arch supports a loading acting on the walls and roof of the dark-shaded tributary area. This represents a uniform distributed load of (15 kN/m^2) $(4 \text{ m}) = 60 \text{ kN/m}$ on the windward side and $(5 \text{ kN/m}^2)(4 \text{ m}) = 20 \text{ kN/m}$ on the leeward side, Fig. 2–33b.

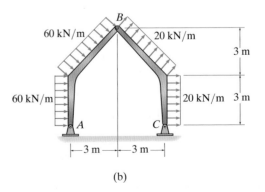

(b)

Free-Body Diagrams. Simplifying the distributed loadings, the free-body diagrams of the entire frame and each of its parts are shown in Fig. 2–33c.

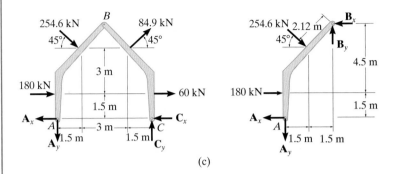
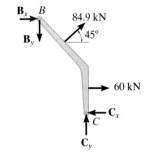

(c)

Equations of Equilibrium. Simultaneous solution of equations is avoided by applying the equilibrium equations in the following sequence using previously computed results.*

Entire Frame:

$$\zeta+\Sigma M_A = 0; \quad -(180 + 60)(1.5) - (254.6 + 84.9)\cos 45°(4.5)$$
$$- (254.6 \sin 45°)(1.5) + (84.9 \sin 45°)(4.5) + C_y(6) = 0$$
$$C_y = 240.0 \text{ kN} \qquad \qquad Ans.$$

$$+\uparrow\Sigma F_y = 0; \quad -A_y - 254.6 \sin 45° + 84.9 \sin 45° + 240.0 = 0$$
$$A_y = 120.0 \text{ kN} \qquad \qquad Ans.$$

Member AB:

$$\zeta+\Sigma M_B = 0; \quad -A_x(6) + 120.0(3) + 180(4.5) + 254.6(2.12) = 0$$
$$A_x = 285.0 \text{ kN} \qquad \qquad Ans.$$

$$\xrightarrow{+}\Sigma F_x = 0; \quad -285.0 + 180 + 254.6 \cos 45° - B_x = 0$$
$$B_x = 75.0 \text{ kN} \qquad \qquad Ans.$$

$$+\uparrow\Sigma F_y = 0; \quad -120.0 - 254.6 \sin 45° + B_y = 0$$
$$B_y = 300.0 \text{ kN} \qquad \qquad Ans.$$

Member CB:

$$\xrightarrow{+}\Sigma F_x = 0; \quad -C_x + 60 + 84.9 \cos 45° + 75.0 = 0$$
$$C_x = 195.0 \text{ kN} \qquad \qquad Ans.$$

*The problem can also be solved by applying the six equations of equilibrium only to the two members. If this is done, it is best to first sum moments about point A on member AB, then point C on member CB. By doing this, one obtains two equations to be solved simultaneously for B_x and B_y.

2

CHAPTER REVIEW

Supports—Structural members are often assumed to be pin connected if slight relative rotation can occur between them, and fixed connected if no rotation is possible.

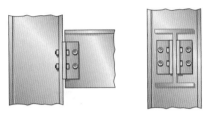

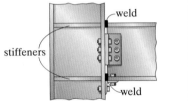

weld

stiffeners

weld

typical "pin-supported" connection (metal)   typical "fixed-supported" connection (metal)

(Ref: Section 2.1)

Idealized Structures—By making assumptions about the supports and connections as being either roller supported, pinned, or fixed, the members can then be represented as lines, so that we can establish an idealized model that can be used for analysis.

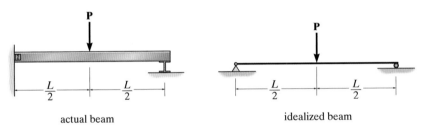

P   P

$\frac{L}{2}$ $\frac{L}{2}$ $\frac{L}{2}$ $\frac{L}{2}$

actual beam   idealized beam

(Ref: Section 2.1)

The tributary loadings on slabs can be determined by first classifying the slab as a one-way or two-way slab. As a general rule, if L_2 is the largest dimension, and $L_2/L_1 > 2$, the slab will behave as a one-way slab. If $L_2/L_1 \leq 2$, the slab will behave as a two-way slab.

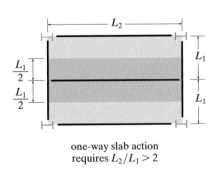

L_2

$\frac{L_1}{2}$
$\frac{L_1}{2}$

L_1
L_1

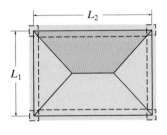

L_2

L_1

one-way slab action
requires $L_2/L_1 > 2$

two-way slab action
requires $L_2/L_1 \leq 2$

(Ref: Section 2.1)

Principle of Superposition—Either the loads or displacements can be added together provided the material is linear elastic and only small displacements of the structure occur.

(Ref: Section 2.2)

Equilibrium—Statically determinate structures can be analyzed by disassembling them and applying the equations of equilibrium to each member. The analysis of a statically determinate structure requires first drawing the free-body diagrams of all the members, and then applying the equations of equilibrium to each member.

$$\Sigma F_x = 0$$
$$\Sigma F_y = 0$$
$$\Sigma M_O = 0$$

The number of equations of equilibrium for all n members of a structure is $3n$. If the structure has r reactions, then the structure is *statically determinate* if

$$r = 3n$$

and *statically indeterminate* if

$$r > 3n$$

The additional number of equations required for the solution refers to the degree of indeterminacy.

(Ref: Sections 2.3 & 2.4)

Stability—If there are fewer reactions than equations of equilibrium, then the structure will be unstable because it is partially constrained. Instability due to improper constraints can also occur if the lines of action of the reactions are concurrent at a point or parallel to one another.

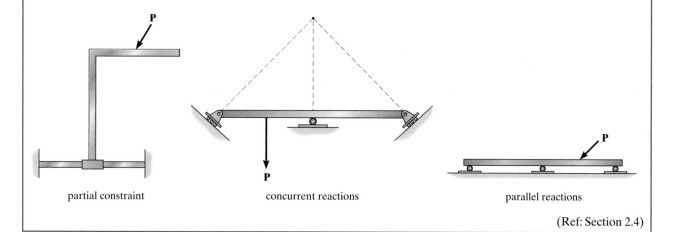

partial constraint   concurrent reactions   parallel reactions

(Ref: Section 2.4)

FUNDAMENTAL PROBLEMS

F2–1. Determine the horizontal and vertical components of reaction at the pins A, B, and C.

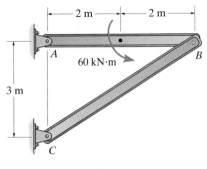

F2–1

F2–2. Determine the horizontal and vertical components of reaction at the pins A, B, and C.

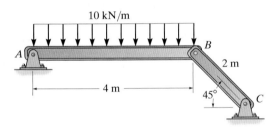

F2–2

F2–3. Determine the horizontal and vertical components of reaction at the pins A, B, and C.

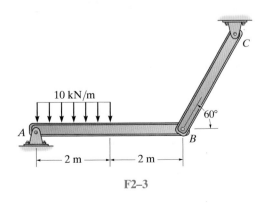

F2–3

F2–4. Determine the horizontal and vertical components of reaction at the roller support A, and fixed support B.

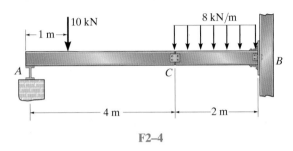

F2–4

F2–5. Determine the horizontal and vertical components of reaction at pins A, B, and C of the two-member frame.

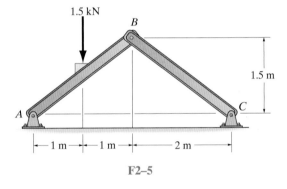

F2–5

F2–6. Determine the components of reaction at the roller support A and pin support C. Joint B is fixed connected.

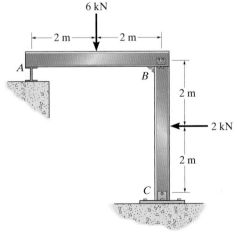

F2–6

F2–7. Determine the horizontal and vertical components of reaction at the pins A, B, and D of the three-member frame. The joint at C is fixed connected.

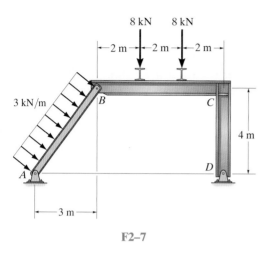

F2–7

F2–8. Determine the components of reaction at the fixed support D and the pins A, B, and C of the three-member frame. Neglect the thickness of the members.

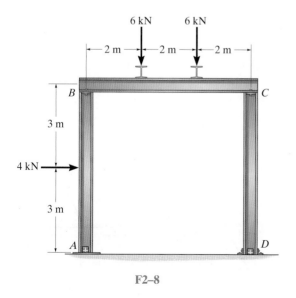

F2–8

F2–9. Determine the components of reaction at the fixed support D and the pins A, B, and C of the three-member frame. Neglect the thickness of the members.

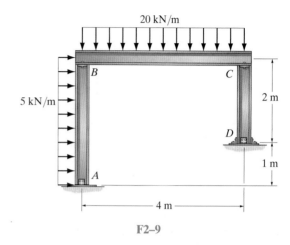

F2–9

F2–10. Determine the components of reaction at the fixed support D and the pins A, B, and C of the three-member frame. Neglect the thickness of the members.

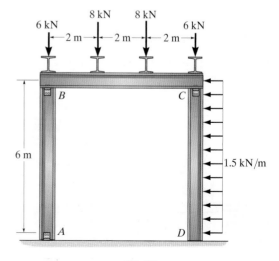

F2–10

2

PROBLEMS

2–18. Determine the reactions on the beam. Neglect the thickness of the beam.

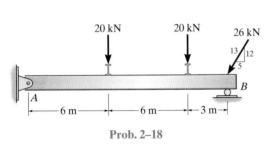

Prob. 2–18

2–19. Determine the reactions on the beam.

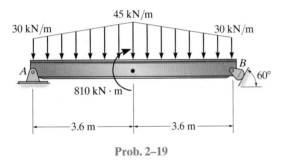

Prob. 2–19

*2–20. Determine the reactions on the beam.

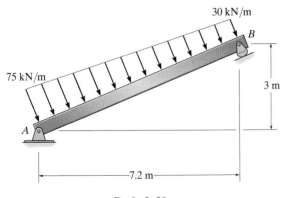

Prob. 2–20

2–21. Determine the reactions at the supports A and B of the compound beam. Assume there is a pin at C.

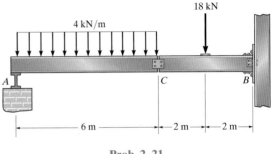

Prob. 2–21

2–22. Determine the reactions at the supports A, B, D, and F.

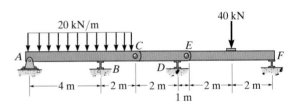

Prob. 2–22

2–23. The compound beam is pin supported at C and supported by a roller at A and B. There is a hinge (pin) at D. Determine the reactions at the supports. Neglect the thickness of the beam.

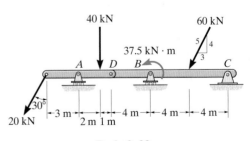

Prob. 2–23

*2–24. Determine the reactions on the beam. The support at *B* can be assumed to be a roller.

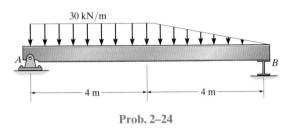

30 kN/m

A   B

4 m   4 m

Prob. 2–24

2–27. The compound beam is fixed at *A* and supported by a rocker at *B* and *C*. There are hinges (pins) at *D* and *E*. Determine the reactions at the supports.

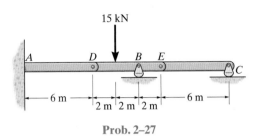

15 kN

A   D   B   E   C

6 m   2 m 2 m 2 m   6 m

Prob. 2–27

2–25. Determine the reactions at the smooth support *C* and pinned support *A*. Assume the connection at *B* is fixed connected.

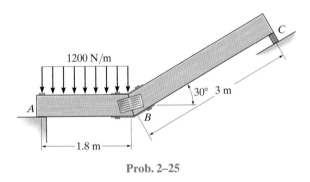

1200 N/m

C

30° 3 m

A   B

1.8 m

Prob. 2–25

*2–28. Determine the reactions at the supports *A* and *B*. The floor decks *CD*, *DE*, *EF*, and *FG* transmit their loads to the girder on smooth supports. Assume *A* is a roller and *B* is a pin.

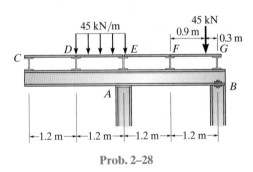

45 kN

45 kN/m   0.9 m

0.3 m

C   D   E   F G

A   B

1.2 m 1.2 m 1.2 m 1.2 m

Prob. 2–28

2–26. Determine the reactions at the truss supports *A* and *B*. The distributed loading is caused by wind.

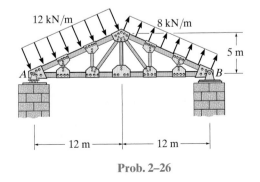

12 kN/m   8 kN/m

5 m

A   B

12 m   12 m

Prob. 2–26

2–29. Determine the reactions at the supports *A* and *B* of the compound beam. There is a pin at *C*.

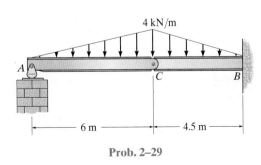

4 kN/m

A   C   B

6 m   4.5 m

Prob. 2–29

2–30. Determine the reactions at the supports A and B of the compound beam. There is a pin at C.

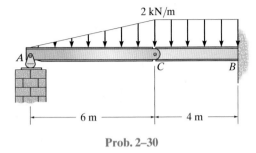

Prob. 2–30

2–31. The beam is subjected to the two concentrated loads as shown. Assuming that the foundation exerts a linearly varying load distribution on its bottom, determine the load intensities w_1 and w_2 for equilibrium (a) in terms of the parameters shown; (b) set $P = 2$ kN, $L = 3$ m.

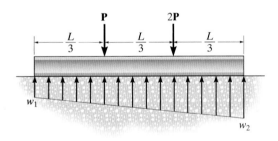

Prob. 2–31

*2–32. The cantilever footing is used to support a wall near its edge A so that it causes a uniform soil pressure under the footing. Determine the uniform distribution loads, w_A and w_B, measured in kN/m at pads A and B, necessary to support the wall forces of 40 kN and 100 kN.

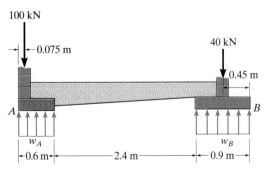

Prob. 2–32

2–33. Determine the horizontal and vertical components of reaction acting at the supports A and C.

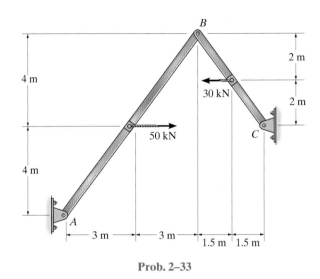

Prob. 2–33

2–34. Determine the reactions at the smooth support A and the pin support B. The joint at C is fixed connected.

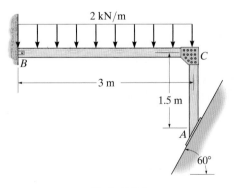

Prob. 2–34

2–35. Determine the reactions at the supports *A* and *B*.

2–37. Determine the horizontal and vertical components force at pins *A* and *C* of the two-member frame.

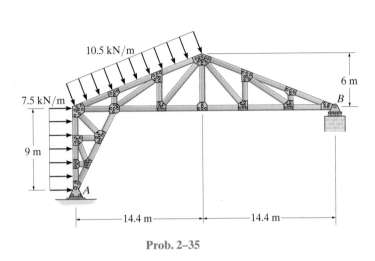

Prob. 2–35

Prob. 2–37

*2–36. Determine the horizontal and vertical components of reaction at the supports *A* and *B*. Assume the joints at *C* and *D* are fixed connections.

2–38. The wall crane supports a load of 7 kN. Determine the horizontal and vertical components of reaction at the pins *A* and *D*. Also, what is the force in the cable at the winch *W*?

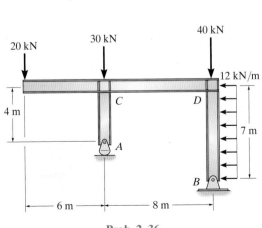

Prob. 2–36

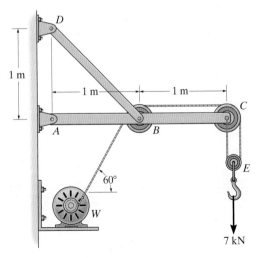

Prob. 2–38

2–39.   Determine the resultant forces at pins B and C on member ABC of the four-member frame.

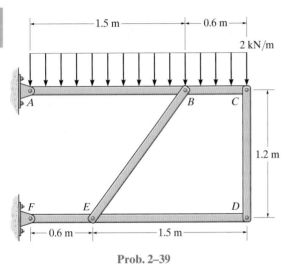

2 kN/m

Prob. 2–39

2–41.   Determine the horizontal and vertical reactions at the connections A and C of the gable frame. Assume that A, B, and C are pin connections. The purlin loads such as D and E are applied perpendicular to the center line of each girder.

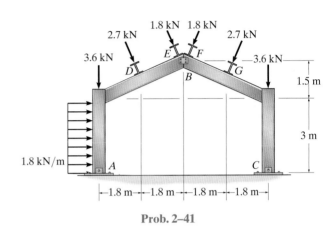

Prob. 2–41

*2–40.   Determine the reactions at the supports A and D. Assume A is fixed and B and C and D are pins.

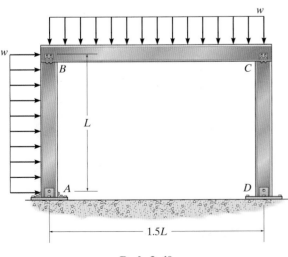

Prob. 2–40

2–42.   Determine the horizontal and vertical components of reaction at A, C, and D. Assume the frame is pin connected at A, C, and D, and there is a fixed-connected joint at B.

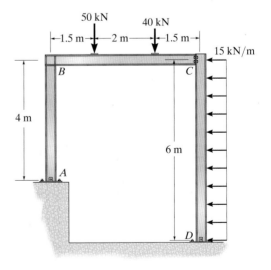

Prob. 2–42

2–43. Determine the horizontal and vertical components at A, B, and C. Assume the frame is pin connected at these points. The joints at D and E are fixed connected.

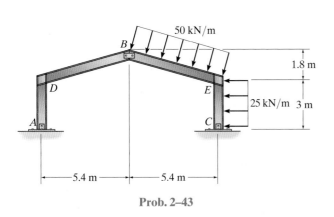

Prob. 2–43

***2–44.** Determine the reactions at the supports A and B. The joints C and D are fixed connected.

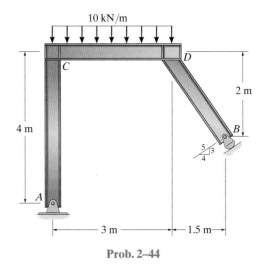

Prob. 2–44

PROJECT PROBLEM

2–1P. The railroad trestle bridge shown in the photo is supported by reinforced concrete bents. Assume the two simply supported side girders, track bed, and two rails have a weight of 7.5 kN/m and the load imposed by a train is 100 kN/m (see Fig. 1–11). Each girder is 6 m long. Apply the load over the entire bridge and determine the compressive force in the columns of each bent. For the analysis assume all joints are pin connected and neglect the weight of the bent. Are these realistic assumptions?

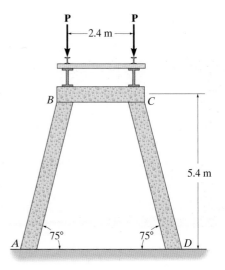

Project Prob. 2–1P

The forces in the members of this bridge can be analyzed using either the method of joints or the method of sections.

Analysis of Statically Determinate Trusses

3

In this chapter we will develop the procedures for analyzing statically determinate trusses using the method of joints and the method of sections. First, however, the determinacy and stability of a truss will be discussed. Then the analysis of three forms of planar trusses will be considered: simple, compound, and complex. Finally, at the end of the chapter we will consider the analysis of a space truss.

3.1 Common Types of Trusses

A *truss* is a structure composed of slender members joined together at their end points. The members commonly used in construction consist of wooden struts, metal bars, angles, or channels. The joint connections are usually formed by bolting or welding the ends of the members to a common plate, called a *gusset plate*, as shown in Fig. 3–1, or by simply passing a large bolt or pin through each of the members. Planar trusses lie in a single plane and are often used to support roofs and bridges.

The gusset plate is used to connect eight members of the truss supporting structure for a water tank.

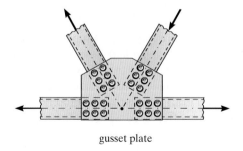

gusset plate

Fig. 3–1

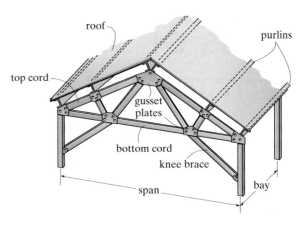

Fig. 3–2

Roof Trusses.

Roof trusses are often used as part of an industrial building frame, such as the one shown in Fig. 3–2. Here, the roof load is transmitted to the truss at the joints by means of a series of *purlins*. The roof truss along with its supporting columns is termed a *bent*. Ordinarily, roof trusses are supported either by columns of wood, steel, or reinforced concrete, or by masonry walls. To keep the bent rigid, and thereby capable of resisting horizontal wind forces, knee braces are sometimes used at the supporting columns. The space between adjacent bents is called a *bay*. Bays are economically spaced at about 4.5 m for spans around 18 m and about 6 m for spans of 30 m. Bays are often tied together using diagonal bracing in order to maintain rigidity of the building's structure.

Trusses used to support roofs are selected on the basis of the span, the slope, and the roof material. Some of the more common types of trusses used are shown in Fig. 3–3. In particular, the scissors truss, Fig. 3–3a, can be used for short spans that require overhead clearance. The Howe and Pratt trusses, Figs. 3–3b and 3–3c, are used for roofs of moderate span, about 18 m to 30 m. If larger spans are required to support the roof, the fan truss or Fink truss may be used, Figs. 3–3d and 3–3e. These trusses may be built with a cambered bottom cord such as that shown in Fig. 3–3f. If a flat roof or nearly flat roof is to be selected, the Warren truss, Fig. 3–3g, is often used. Also, the Howe and Pratt trusses may be modified for flat roofs. Sawtooth trusses, Fig. 3–3h, are often used where column spacing is not objectionable and uniform lighting is important. A textile mill would be an example. The bowstring truss, Fig. 3–3i, is sometimes selected for garages and small airplane hangars; and the arched truss, Fig. 3–3j, although relatively expensive, can be used for high rises and long spans such as field houses, gymnasiums, and so on.

Although more decorative than structural, these simple Pratt trusses are used for the entrance of a building.

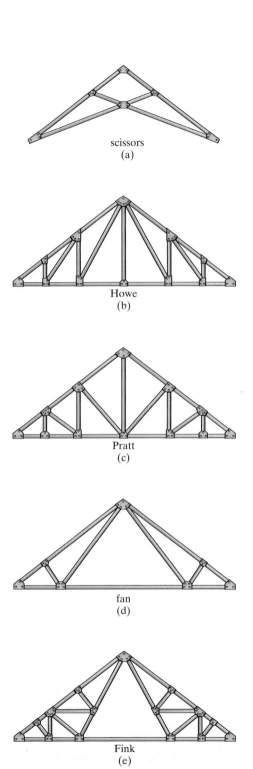

scissors
(a)

Howe
(b)

Pratt
(c)

fan
(d)

Fink
(e)

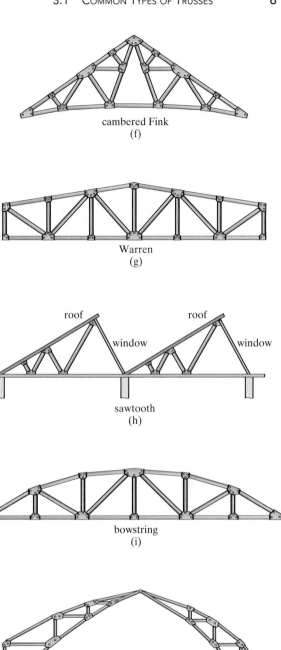

cambered Fink
(f)

Warren
(g)

roof   roof

window   window

sawtooth
(h)

bowstring
(i)

three-hinged arch
(j)

Fig. 3–3

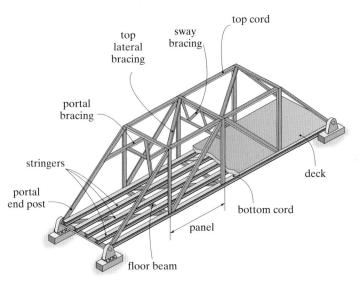

Fig. 3–4

Parker trusses are used to support this bridge.

Bridge Trusses. The main structural elements of a typical bridge truss are shown in Fig. 3–4. Here it is seen that a load on the *deck* is first transmitted to *stringers,* then to *floor beams,* and finally to the joints of the two supporting side trusses. The top and bottom cords of these side trusses are connected by top and bottom *lateral bracing,* which serves to resist the lateral forces caused by wind and the sidesway caused by moving vehicles on the bridge. Additional stability is provided by the *portal* and *sway bracing.* As in the case of many long-span trusses, a roller is provided at one end of a bridge truss to allow for thermal expansion.

A few of the typical forms of bridge trusses currently used for single spans are shown in Fig. 3–5. In particular, the Pratt, Howe, and Warren trusses are normally used for spans up to 60 m in length. The most common form is the Warren truss with verticals, Fig. 3–5c. For larger spans, a truss with a polygonal upper cord, such as the Parker truss, Fig. 3–5d, is used for some savings in material. The Warren truss with verticals can also be fabricated in this manner for spans up to 90 m. The greatest economy of material is obtained if the diagonals have a slope between 45° and 60° with the horizontal. If this rule is maintained, then for spans greater than 90 m, the depth of the truss must increase and consequently the panels will get longer. This results in a heavy deck system and, to keep the weight of the deck within tolerable limits, *subdivided* trusses have been developed. Typical examples include the Baltimore and subdivided Warren trusses, Figs. 3–5e and 3–5f. Finally, the K-truss shown in Fig. 3–5g can also be used in place of a subdivided truss, since it accomplishes the same purpose.

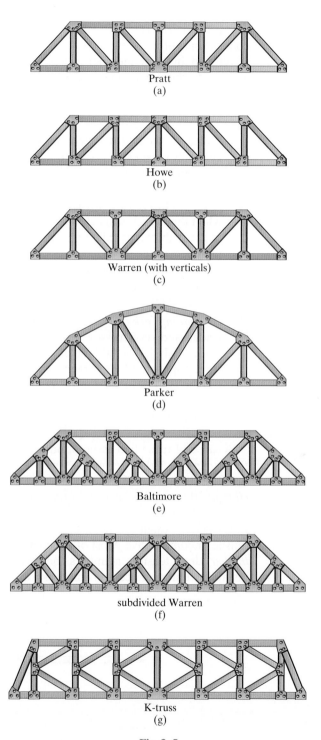

Pratt
(a)

Howe
(b)

Warren (with verticals)
(c)

Parker
(d)

Baltimore
(e)

subdivided Warren
(f)

K-truss
(g)

Fig. 3–5

Assumptions for Design.

To design both the members and the connections of a truss, it is first necessary to determine the *force* developed in each member when the truss is subjected to a given loading. In this regard, two important assumptions will be made in order to idealize the truss.

1. *The members are joined together by smooth pins.* In cases where bolted or welded joint connections are used, this assumption is generally satisfactory provided the center lines of the joining members are concurrent at a point, as in Fig. 3–1. It should be realized, however, that the actual connections do give some *rigidity* to the joint and this in turn introduces bending of the connected members when the truss is subjected to a load. The bending stress developed in the members is called *secondary stress,* whereas the stress in the members of the idealized truss, having pin-connected joints, is called *primary stress.* A secondary stress analysis of a truss can be performed using a computer, as discussed in Chapter 16. For some types of truss geometries these stresses may be large.

2. *All loadings are applied at the joints.* In most situations, such as for bridge and roof trusses, this assumption is true. Frequently in the force analysis, the weight of the members is neglected, since the force supported by the members is large in comparison with their weight. If the weight is to be included in the analysis, it is generally satisfactory to apply it as a vertical force, half of its magnitude applied at each end of the member.

Because of these two assumptions, *each truss member acts as an axial force member,* and therefore the forces acting at the ends of the member must be directed along the axis of the member. If the force tends to *elongate* the member, it is a *tensile force (T),* Fig. 3–6a; whereas if the force tends to *shorten* the member, it is a *compressive force (C),* Fig. 3–6b. In the actual design of a truss it is important to state whether the force is tensile or compressive. Most often, compression members must be made *thicker* than tension members, because of the buckling or sudden instability that may occur in compression members.

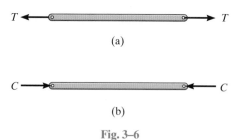

T T

(a)

C C

(b)

Fig. 3–6

3.2   Classification of Coplanar Trusses

Before beginning the force analysis of a truss, it is important to classify the truss as simple, compound, or complex, and then to be able to specify its determinacy and stability.

Simple Truss.   To prevent collapse, the framework of a truss must be rigid. Obviously, the four-bar frame *ABCD* in Fig. 3–7 will collapse unless a diagonal, such as *AC*, is added for support. The simplest framework that is rigid or stable is a *triangle*. Consequently, a *simple truss* is constructed by starting with a basic triangular element, such as *ABC* in Fig. 3–8, and connecting two members (*AD* and *BD*) to form an additional element. Thus it is seen that as each additional element of two members is placed on the truss, the number of joints is increased by one.

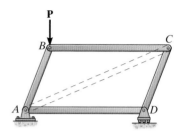

Fig. 3–7

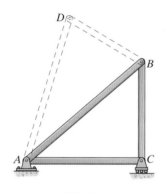

Fig. 3–8

An example of a simple truss is shown in Fig. 3–9, where the basic "stable" triangular element is *ABC*, from which the remainder of the joints, *D*, *E*, and *F*, are established in alphabetical sequence. For this method of construction, however, it is important to realize that simple trusses *do not* have to consist entirely of triangles. An example is shown in Fig. 3–10, where starting with triangle *ABC*, bars *CD* and *AD* are added to form joint *D*. Finally, bars *BE* and *DE* are added to form joint *E*.

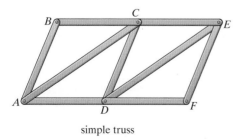

simple truss

Fig. 3–9

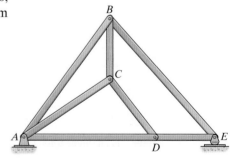

simple truss

Fig. 3–10

Compound Truss.   A *compound truss* is formed by connecting two or more simple trusses together. Quite often this type of truss is used to support loads acting over a *large span,* since it is cheaper to construct a somewhat lighter compound truss than to use a heavier single simple truss.

There are three ways in which simple trusses are joined together to form a compound truss. The trusses may be connected by a common joint and bar. An example is given in Fig. 3–11a, where the shaded truss *ABC* is connected to the shaded truss *CDE* in this manner. The trusses may be joined by three bars, as in the case of the shaded truss *ABC* connected to the larger truss *DEF,* Fig. 3–11b. And finally, the trusses may be joined where bars of a large simple truss, called the *main truss,* have been *substituted* by simple trusses, called *secondary trusses.* An example is shown in Fig. 3–11c, where dashed members of the main truss *ABCDE* have been *replaced* by the secondary shaded trusses. If this truss carried roof loads, the use of the secondary trusses might be more economical, since the dashed members may be subjected to excessive bending, whereas the secondary trusses can better transfer the load.

Complex Truss.   A *complex truss* is one that cannot be classified as being either simple or compound. The truss in Fig. 3–12 is an example.

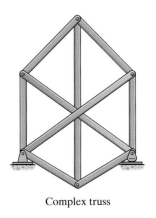

Complex truss

Fig. 3–12

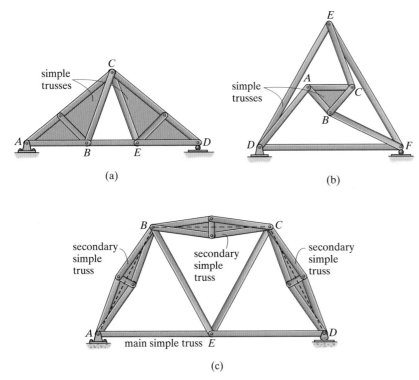

(a)

(b)

(c)

Various types of compound trusses

Fig. 3–11

Determinacy. For any problem in truss analysis, it should be realized that the total number of *unknowns* includes the forces in b number of bars of the truss and the total number of external support reactions r. Since the truss members are all straight axial force members lying in the same plane, the force system acting at each joint is *coplanar and concurrent*. Consequently, rotational or moment equilibrium is automatically satisfied at the joint (or pin), and it is only necessary to satisfy $\Sigma F_x = 0$ and $\Sigma F_y = 0$ to ensure translational or force equilibrium. Therefore, only two equations of equilibrium can be written for each joint, and if there are j number of joints, the total number of equations available for solution is $2j$. By simply comparing the total number of unknowns $(b + r)$ with the total number of available equilibrium equations, it is therefore possible to specify the determinacy for either a simple, compound, or complex truss. We have

$$
\begin{array}{ll}
b + r = 2j & \text{statically determinate} \\
b + r > 2j & \text{statically indeterminate}
\end{array}
\qquad (3\text{–}1)
$$

In particular, the *degree of indeterminacy* is specified by the difference in the numbers $(b + r) - 2j$.

Stability. If $b + r < 2j$, a truss will be *unstable*, that is, it will collapse, since there will be an insufficient number of bars or reactions to constrain all the joints. Also, a truss can be unstable if it is statically determinate or statically indeterminate. In this case the stability will have to be determined either by inspection or by a force analysis.

External Stability. As stated in Sec. 2–4, a *structure (or truss) is externally unstable if all of its reactions are concurrent or parallel*. For example, the two trusses in Fig. 3–13 are externally unstable since the support reactions have lines of action that are either concurrent or parallel.

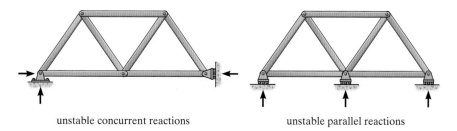

unstable concurrent reactions   unstable parallel reactions

Fig. 3–13

Internal Stability. The internal stability of a truss can often be checked by careful inspection of the arrangement of its members. If it can be determined that each joint is held fixed so that it cannot move in a "rigid body" sense with respect to the other joints, then the truss will be stable. Notice that *a simple truss will always be internally stable,* since by the nature of its construction it requires starting from a basic triangular element and adding successive "rigid elements," each containing two additional members and a joint. The truss in Fig. 3–14 exemplifies this construction, where, starting with the shaded triangle element ABC, the successive joints D, E, F, G, H have been added.

If a truss is constructed so that it does not hold its joints in a fixed position, it will be unstable or have a "critical form." An obvious example of this is shown in Fig. 3–15, where it can be seen that no restraint or fixity is provided between joints C and F or B and E, and so the truss will collapse under load.

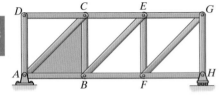

Fig. 3–14

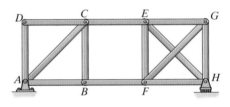

Fig. 3–15

To determine the internal stability of a *compound truss*, it is necessary to identify the way in which the simple trusses are connected together. For example, the compound truss in Fig. 3–16 is unstable since the inner simple truss ABC is connected to the outer simple truss DEF using three bars, AD, BE, and CF, which are *concurrent* at point O. Thus an external load can be applied to joint A, B, or C and cause the truss ABC to rotate slightly.

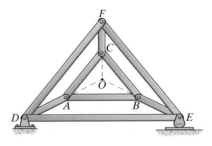

Fig. 3–16

If a truss is identified as *complex,* it may not be possible to tell by inspection if it is stable. For example, it can be shown by the analysis discussed in Sec. 3–7 that the complex truss in Fig. 3–17 is unstable or has a "critical form" only if the dimension $d = d'$. If $d \neq d'$ it is stable.

The instability of any form of truss, be it simple, compound, or complex, may also be noticed by using a computer to solve the $2j$ simultaneous equations written for all the joints of the truss. If inconsistent results are obtained, the truss will be unstable or have a critical form.

If a computer analysis is not performed, the methods discussed previously can be used to check the stability of the truss. To summarize, if the truss has b bars, r external reactions, and j joints, then if

$b + r < 2j$	unstable	
$b + r \geq 2j$	unstable if truss support reactions are concurrent or parallel or if some of the components of the truss form a collapsible mechanism	(3–2)

Bear in mind, however, that *if a truss is unstable, it does not matter whether it is statically determinate or indeterminate.* Obviously, the use of an unstable truss is to be avoided in practice.

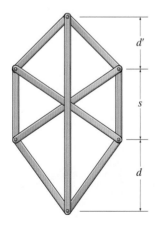

Fig. 3–17

EXAMPLE 3.1

Classify each of the trusses in Fig. 3–18 as stable, unstable, statically determinate, or statically indeterminate. The trusses are subjected to arbitrary external loadings that are assumed to be known and can act anywhere on the trusses.

SOLUTION

Fig. 3–18a. *Externally stable,* since the reactions are not concurrent or parallel. Since $b = 19$, $r = 3$, $j = 11$, then $b + r = 2j$ or $22 = 22$. Therefore, the truss is *statically determinate*. By inspection the truss is *internally stable*.

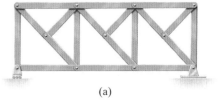

(a)

Fig. 3–18

Fig. 3–18b. *Externally stable.* Since $b = 15$, $r = 4$, $j = 9$, then $b + r > 2j$ or $19 > 18$. The truss is *statically indeterminate* to the first degree. By inspection the truss is *internally stable*.

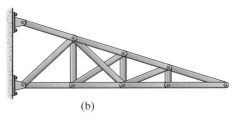

(b)

Fig. 3–18c. *Externally stable.* Since $b = 9$, $r = 3$, $j = 6$, then $b + r = 2j$ or $12 = 12$. The truss is *statically determinate*. By inspection the truss is *internally stable*.

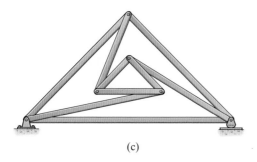

(c)

Fig. 3–18d. *Externally stable.* Since $b = 12$, $r = 3$, $j = 8$, then $b + r < 2j$ or $15 < 16$. The truss is *internally unstable*.

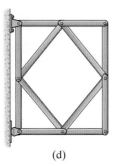

(d)

PROBLEMS

3–1. Classify each of the following trusses as statically determinate, statically indeterminate, or unstable. If indeterminate, state its degree.

3–2. Classify each of the following trusses as stable, unstable, statically determinate, or statically indeterminate. If indeterminate state its degree.

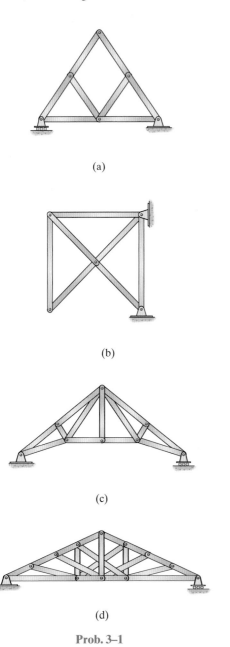

(a)

(b)

(c)

(d)

Prob. 3–1

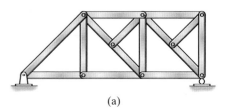

(a)

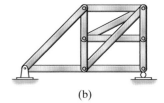

(b)

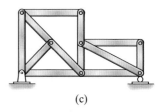

(c)

Prob. 3–2

3–3. Classify each of the following trusses as statically determinate, indeterminate, or unstable. If indeterminate, state its degree.

*3–4. Classify each of the following trusses as statically determinate, statically indeterminate, or unstable. If indeterminate, state its degree.

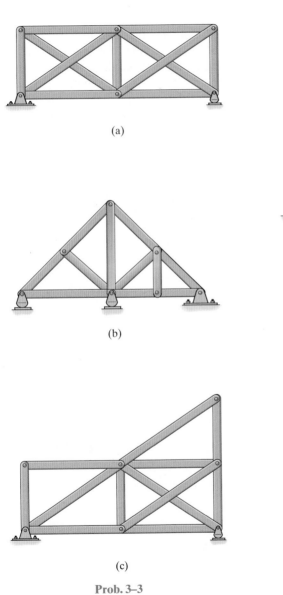

Prob. 3–3

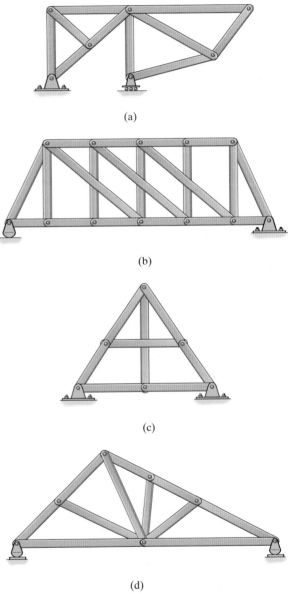

Prob. 3–4

3.3   The Method of Joints

If a truss is in equilibrium, then each of its joints must also be in equilibrium. Hence, the method of joints consists of satisfying the equilibrium conditions $\Sigma F_x = 0$ and $\Sigma F_y = 0$ for the forces exerted *on the pin* at each joint of the truss.

When using the method of joints, it is necessary to draw each joint's free-body diagram before applying the equilibrium equations. Recall that the *line of action* of each member force acting on the joint is *specified* from the geometry of the truss, since the force in a member passes along the axis of the member. As an example, consider joint B of the truss in Fig. 3–19a. From the free-body diagram, Fig. 3–19b, the only unknowns are the *magnitudes* of the forces in members BA and BC. As shown, $\mathbf{F}_{BA}$ is "pulling" on the pin, which indicates that member BA is in *tension,* whereas $\mathbf{F}_{BC}$ is "pushing" on the pin, and consequently member BC is in *compression.* These effects are clearly demonstrated by using the method of sections and isolating the joint with small segments of the member connected to the pin, Fig. 3–19c. Notice that pushing or pulling on these small segments indicates the effect of the member being either in compression or tension.

In all cases, the joint analysis should start at a joint having at least one known force and at most two unknown forces, as in Fig. 3–19b. In this way, application of $\Sigma F_x = 0$ and $\Sigma F_y = 0$ yields two algebraic equations that can be solved for the two unknowns. When applying these equations, the correct sense of an unknown member force can be determined using one of two possible methods.

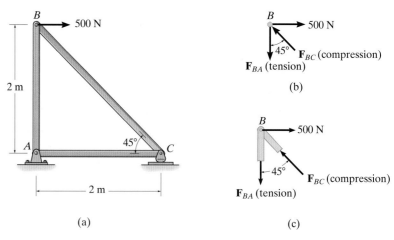

(a)

(b)

(c)

Fig. 3–19

1. *Always assume the unknown member forces acting on the joint's free-body diagram to be in tension, i.e., "pulling" on the pin.* If this is done, then numerical solution of the equilibrium equations will yield *positive scalars for members in tension and negative scalars for members in compression.* Once an unknown member force is found, use its *correct* magnitude and sense (T or C) on subsequent joint free-body diagrams.

2. *The correct sense of direction of an unknown member force can, in many cases, be determined "by inspection."* For example, $\mathbf{F}_{BC}$ in Fig. 3–19b must push on the pin (compression) since its horizontal component, $F_{BC} \sin 45°$, must balance the 500-N force ($\Sigma F_x = 0$). Likewise, $\mathbf{F}_{BA}$ is a tensile force since it balances the vertical component, $F_{BC} \cos 45°$ ($\Sigma F_y = 0$). In more complicated cases, the sense of an unknown member force can be *assumed;* then, after applying the equilibrium equations, the assumed sense can be verified from the numerical results. A *positive* answer indicates that the sense is *correct,* whereas a *negative* answer indicates that the sense shown on the free-body diagram must be *reversed.* This is the method we will use in the example problems which follow.

Procedure for Analysis

The following procedure provides a means for analyzing a truss using the method of joints.

- Draw the free-body diagram of a joint having at least one known force and at most two unknown forces. (If this joint is at one of the supports, it may be necessary to calculate the external reactions at the supports by drawing a free-body diagram of the entire truss.)

- Use one of the two methods previously described for establishing the sense of an unknown force.

- The x and y axes should be oriented such that the forces on the free-body diagram can be easily resolved into their x and y components. Apply the two force equilibrium equations $\Sigma F_x = 0$ and $\Sigma F_y = 0$, solve for the two unknown member forces, and verify their correct directional sense.

- Continue to analyze each of the other joints, where again it is necessary to choose a joint having at most two unknowns and at least one known force.

- Once the force in a member is found from the analysis of a joint at one of its ends, the result can be used to analyze the forces acting on the joint at its other end. Remember, a member in *compression* "pushes" on the joint and a member in *tension* "pulls" on the joint.

EXAMPLE 3.2

Determine the force in each member of the roof truss shown in the photo. The dimensions and loadings are shown in Fig. 3–20a. State whether the members are in tension or compression.

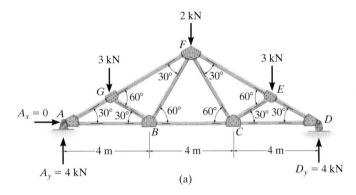

(a)

(b)

(c)

(d)

Fig. 3–20

SOLUTION

Only the forces in half the members have to be determined, since the truss is symmetric with respect to *both* loading and geometry.

Joint A, Fig. 3–20b.   We can start the analysis at joint A. Why? The free-body diagram is shown in Fig. 3–20b.

$$+\uparrow \Sigma F_y = 0; \quad 4 - F_{AG} \sin 30° = 0 \qquad F_{AG} = 8 \text{ kN (C)} \qquad \textit{Ans.}$$

$$\xrightarrow{+} \Sigma F_x = 0; \quad F_{AB} - 8 \cos 30° = 0 \qquad F_{AB} = 6.928 \text{ kN (T)} \quad \textit{Ans.}$$

Joint G, Fig. 3–20c.   In this case note how the orientation of the x, y axes avoids simultaneous solution of equations.

$$+\nwarrow \Sigma F_y = 0; \quad F_{GB} \sin 60° - 3 \cos 30° = 0$$

$$F_{GB} = 3.00 \text{ kN (C)} \qquad \textit{Ans.}$$

$$+\nearrow \Sigma F_x = 0; \quad 8 - 3 \sin 30° - 3.00 \cos 60° - F_{GF} = 0$$

$$F_{GF} = 5.00 \text{ kN (C)} \qquad \textit{Ans.}$$

Joint B, Fig. 3–20d.

$$+\uparrow \Sigma F_y = 0; \quad F_{BF} \sin 60° - 3.00 \sin 30° = 0$$

$$F_{BF} = 1.73 \text{ kN (T)} \qquad \textit{Ans.}$$

$$\xrightarrow{+} \Sigma F_x = 0; \quad F_{BC} + 1.73 \cos 60° + 3.00 \cos 30° - 6.928 = 0$$

$$F_{BC} = 3.46 \text{ kN (T)} \qquad \textit{Ans.}$$

EXAMPLE 3.3

Determine the force in each member of the scissors truss shown in Fig. 3–21a. State whether the members are in tension or compression. The reactions at the supports are given.

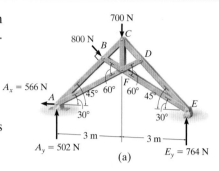

(a)

SOLUTION

The truss will be analyzed in the following sequence:

Joint E, Fig. 3–21b. Note that simultaneous solution of equations is avoided by the x, y axes orientation.

$+\nearrow \Sigma F_y = 0;$ $764 \cos 30° - F_{ED} \sin 15° = 0$

$\qquad F_{ED} = 2557 \text{ N (C)}$ *Ans.*

$+\searrow \Sigma F_x = 0;$ $2557 \cos 15° - F_{EF} - 764 \sin 30° = 0$

$\qquad F_{EF} = 2088 \text{ N (T)}$ *Ans.*

Joint D, Fig. 3–21c.

$+\swarrow \Sigma F_x = 0;$ $-F_{DF} \sin 75° = 0$ $F_{DF} = 0$ *Ans.*

$+\nwarrow \Sigma F_y = 0;$ $-F_{DC} + 2557 = 0$ $F_{DC} = 2557 \text{ N (C)}$ *Ans.*

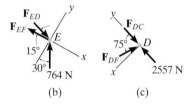

(b)   (c)

Joint C, Fig. 3–21d.

$\xrightarrow{+} \Sigma F_x = 0;$ $F_{CB} \sin 45° - 2557 \sin 45° = 0$

$\qquad F_{CB} = 2557 \text{ N (C)}$ *Ans.*

$+\uparrow \Sigma F_y = 0;$ $-F_{CF} - 700 + 2(2557) \cos 45° = 0$

$\qquad F_{CF} = 2916 \text{ N (T)}$ *Ans.*

Joint B, Fig. 3–21e.

$+\nwarrow \Sigma F_y = 0;$ $F_{BF} \sin 75° - 800 = 0$ $F_{BF} = 828 \text{ N (C)}$ *Ans.*

$+\swarrow \Sigma F_x = 0;$ $2557 + 828 \cos 75° - F_{BA} = 0$

$\qquad F_{BA} = 2771 \text{ N (C)}$ *Ans.*

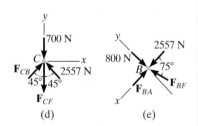

(d)   (e)

Joint A, Fig. 3–21f.

$\xrightarrow{+} \Sigma F_x = 0;$ $F_{AF} \cos 30° - 2771 \cos 45° - 566 = 0$

$\qquad F_{AF} = 2916 \text{ N (T)}$ *Ans.*

$+\uparrow \Sigma F_y = 0;$ $502 - 2771 \sin 45° + 2916 \sin 30° = 0$ check

Notice that since the reactions have been calculated, a further check of the calculations can be made by analyzing the last joint F. Try it and find out.

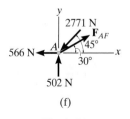

(f)

Fig. 3–21

3.4   Zero-Force Members

Truss analysis using the method of joints is greatly simplified if one is able to first determine those members that support *no loading*. These *zero-force members* may be necessary for the stability of the truss during construction and to provide support if the applied loading is changed. The zero-force members of a truss can generally be determined by inspection of the joints, and they occur in two cases.

Case 1.   Consider the truss in Fig. 3–22a. The two members at joint C are connected together at a right angle *and* there is no external load on the joint. The free-body diagram of joint C, Fig. 3–22b, indicates that the force in each member must be zero in order to maintain equilibrium. Furthermore, as in the case of joint A, Fig. 3–22c, this must be true regardless of the angle, say θ, between the members.

Case 2.   Zero-force members also occur at joints having a geometry as joint D in Fig. 3–23a. Here *no external load acts on the joint,* so that a force summation in the y direction, Fig. 3–23b, which is perpendicular to the two collinear members, requires that $F_{DF} = 0$. Using this result, FC is also a zero-force member, as indicated by the force analysis of joint F, Fig. 3–23c.

In summary, then, if only two non-collinear members form a truss joint and no external load or support reaction is applied to the joint, the members must be zero-force members, Case 1. Also, if three members form a truss joint for which two of the members are collinear, the third member is a zero-force member, provided no external force or support reaction is applied to the joint, Case 2. Particular attention should be directed to these conditions of joint geometry and loading, since the analysis of a truss can be considerably simplified by *first* spotting the zero-force members.

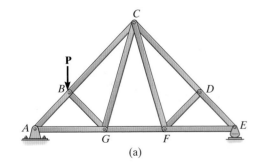

(a)

$$\xleftarrow{+} \Sigma F_x = 0;\ F_{CB} = 0$$

$$+\downarrow \Sigma F_y = 0;\ F_{CD} = 0$$

(b)

$$+\uparrow \Sigma F_y = 0;\ F_{AB}\sin\theta = 0$$
$$F_{AB} = 0 \ \text{(since } \sin\theta \neq 0\text{)}$$

$$\xrightarrow{+} \Sigma F_x = 0;\ -F_{AE} + 0 = 0$$
$$F_{AE} = 0$$

(c)

Fig. 3–22

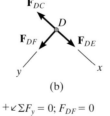

$$+\swarrow \Sigma F_y = 0;\ F_{DF} = 0$$

(b)

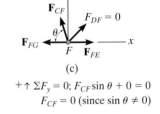

(c)

$$+\uparrow \Sigma F_y = 0;\ F_{CF}\sin\theta + 0 = 0$$
$$F_{CF} = 0 \ \text{(since } \sin\theta \neq 0\text{)}$$

Fig. 3–23

EXAMPLE 3.4

Using the method of joints, indicate all the members of the truss shown in Fig. 3–24a that have zero force.

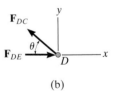

(b)

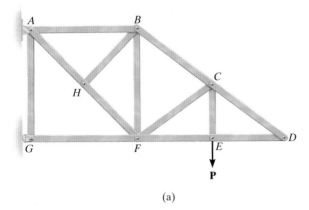

(a)

Fig. 3–24

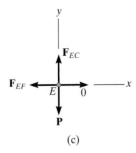

(c)

SOLUTION

Looking for joints similar to those discussed in Figs. 3–22 and 3–23, we have

Joint D, Fig. 3–24b.

$+\uparrow \Sigma F_y = 0;$ $F_{DC} \sin \theta = 0$ $F_{DC} = 0$ *Ans.*

$\xrightarrow{+} \Sigma F_x = 0;$ $F_{DE} + 0 = 0$ $F_{DE} = 0$ *Ans.*

Joint E, Fig. 3–24c.

$\xleftarrow{+} \Sigma F_x = 0;$ $F_{EF} = 0$ *Ans.*

(Note that $F_{EC} = P$ and an analysis of joint C would yield a force in member CF.)

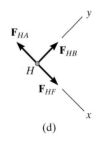

(d)

Joint H, Fig. 3–24d.

$+\nearrow \Sigma F_y = 0;$ $F_{HB} = 0$ *Ans.*

Joint G, Fig. 3–24e. The rocker support at G can only exert an x component of force on the joint, i.e., $\mathbf{G}_x$. Hence,

$+\uparrow \Sigma F_y = 0;$ $F_{GA} = 0$ *Ans.*

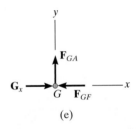

(e)

FUNDAMENTAL PROBLEMS

F3–1. Determine the force in each member of the truss and state whether it is in tension or compression.

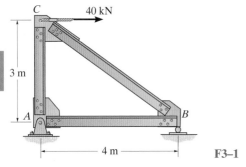

F3–1

F3–2. Determine the force in each member of the truss and state whether it is in tension or compression.

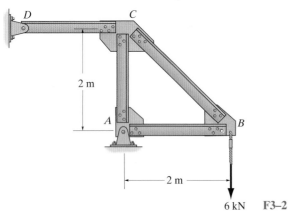

6 kN   F3–2

F3–3. Determine the force in each member of the truss and state whether it is in tension or compression.

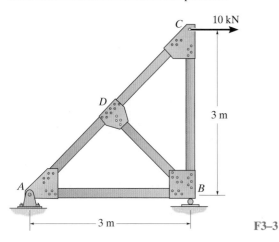

F3–3

F3–4. Determine the force in each member of the truss and state whether it is in tension or compression.

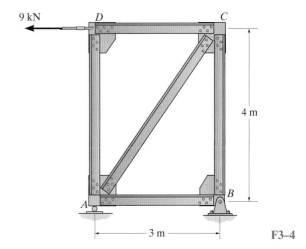

F3–4

F3–5. Determine the force in each member of the truss and state whether it is in tension or compression.

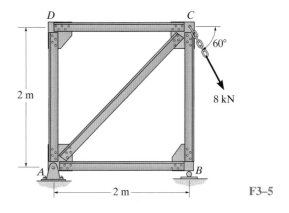

F3–5

F3–6. Determine the force in each member of the truss and state whether it is in tension or compression.

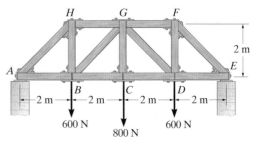

F3–6

PROBLEMS

3–5. A sign is subjected to a wind loading that exerts horizontal forces of 1.5 kN on joints B and C of one of the side supporting trusses. Determine the force in each member of the truss and state if the members are in tension or compression.

3–7. Determine the force in each member of the truss. State whether the members are in tension or compression. Set $P = 8$ kN.

***3–8.** If the maximum force that any member can support is 8 kN in tension and 6 kN in compression, determine the maximum force P that can be supported at joint D.

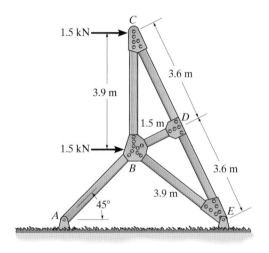

Prob. 3–5

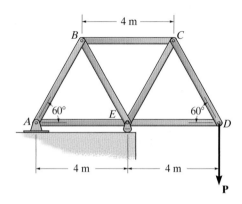

Probs. 3–7/3–8

3–6. Determine the force in each member of the truss. Indicate if the members are in tension or compression. Assume all members are pin connected.

3–9. Determine the force in each member of the truss. State if the members are in tension or compression.

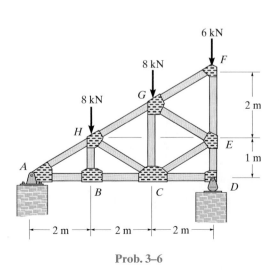

Prob. 3–6

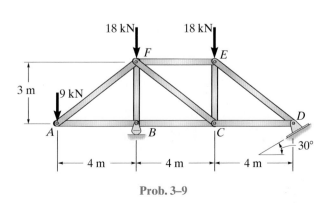

Prob. 3–9

3–10. Determine the force in each member of the truss. State if the members are in tension or compression.

*3–12. Determine the force in each member of the truss. State if the members are in tension or compression. Assume all members are pin connected. $AG = GF = FE = ED$.

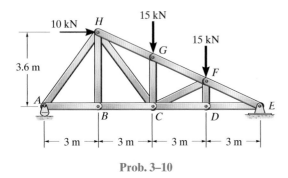

Prob. 3–10

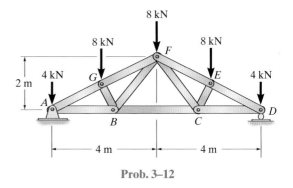

Prob. 3–12

3–11. Determine the force in each member of the truss. State if the members are in tension or compression. Assume all members are pin connected.

3–13. Determine the force in each member of the truss and state if the members are in tension or compression.

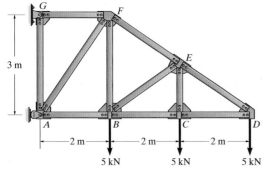

Prob. 3–11

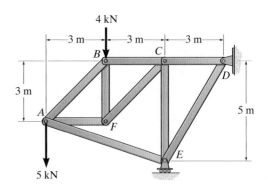

Prob. 3–13

3–14. Determine the force in each member of the roof russ. State if the members are in tension or compression.

*3–16. Determine the force in each member of the truss. State if the members are in tension or compression.

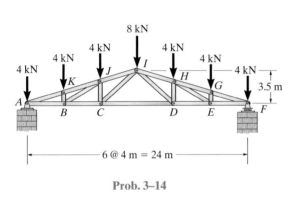

Prob. 3–14

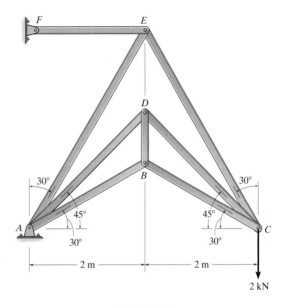

Prob. 3–16

3–15. Determine the force in each member of the roof truss. State if the members are in tension or compression. Assume all members are pin connected.

3–17. Determine the force in each member of the roof truss. State if the members are in tension or compression. Assume B is a pin and C is a roller support.

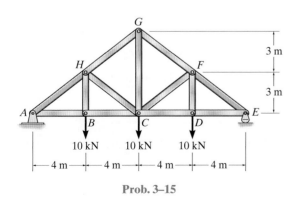

Prob. 3–15

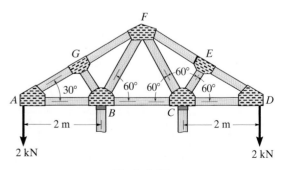

Prob. 3–17

3.5   The Method of Sections

If the forces in only a few members of a truss are to be found, the method of sections generally provides the most direct means of obtaining these forces. The *method of sections* consists of passing an *imaginary section* through the truss, thus cutting it into two parts. Provided the entire truss is in equilibrium, each of the two parts must also be in equilibrium; and as a result, the three equations of equilibrium may be applied to either one of these two parts to determine the member forces at the "cut section."

When the method of sections is used to determine the force in a particular member, a decision must be made as to how to "cut" or section the truss. Since only *three* independent equilibrium equations ($\Sigma F_x = 0$, $\Sigma F_y = 0$, $\Sigma M_O = 0$) can be applied to the isolated portion of the truss, try to select a section that, in general, passes through not more than *three* members in which the forces are unknown. For example, consider the truss in Fig. 3–25a. If the force in member *GC* is to be determined, section *aa* will be appropriate. The free-body diagrams of the two parts are shown in Figs. 3–25b and 3–25c. In particular, note that the line of action of each force in a sectioned member is specified from the *geometry* of the truss, since the force in a member passes along the axis of the member. Also, the member forces acting on one part of the truss are equal but opposite to those acting on the other part—Newton's third law. As shown, members assumed to be in *tension* (*BC* and *GC*) are subjected to a "pull," whereas the member in *compression* (*GF*) is subjected to a "push."

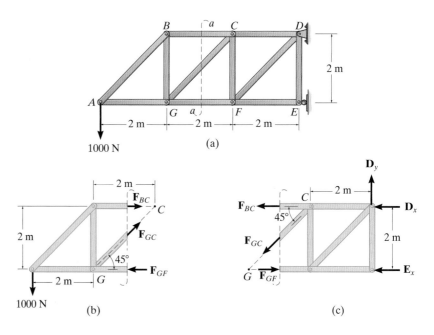

The three unknown member forces $\mathbf{F}_{BC}$, $\mathbf{F}_{GC}$, and $\mathbf{F}_{GF}$ can be obtained by applying the three equilibrium equations to the free-body diagram in Fig. 3–25b. If, however, the free-body diagram in Fig. 3–25c is considered, the three support reactions $\mathbf{D}_x$, $\mathbf{D}_y$, and $\mathbf{E}_x$ will have to be determined *first*. Why? (This, of course, is done in the usual manner by considering a free-body diagram of the *entire truss*.) When applying the equilibrium equations, consider ways of writing the equations so as to yield a *direct solution* for each of the unknowns, rather than having to solve simultaneous equations. For example, summing moments about C in Fig. 3–25b would yield a direct solution for $\mathbf{F}_{GF}$ since $\mathbf{F}_{BC}$ and $\mathbf{F}_{GC}$ create zero moment about C. Likewise, $\mathbf{F}_{BC}$ can be obtained directly by summing moments about G. Finally, $\mathbf{F}_{GC}$ can be found directly from a force summation in the vertical direction, since $\mathbf{F}_{GF}$ and $\mathbf{F}_{BC}$ have no vertical components.

As in the method of joints, there are two ways in which one can determine the correct sense of an unknown member force.

1. *Always assume that the unknown member forces at the cut section are in tension, i.e., "pulling" on the member.* By doing this, the numerical solution of the equilibrium equations will yield *positive scalars for members in tension and negative scalars for members in compression.*

2. *The correct sense of an unknown member force can in many cases be determined "by inspection."* For example, $\mathbf{F}_{BC}$ is a tensile force as represented in Fig. 3–25b, since moment equilibrium about G requires that $\mathbf{F}_{BC}$ create a moment opposite to that of the 1000-N force. Also, $\mathbf{F}_{GC}$ is tensile since its vertical component must balance the 1000-N force. In more complicated cases, the sense of an unknown member force may be *assumed*. If the solution yields a *negative* scalar, it indicates that the force's sense is *opposite* to that shown on the free-body diagram. This is the method we will use in the example problems which follow.

A truss bridge being constructed over Lake Shasta in northern California.

Procedure for Analysis

The following procedure provides a means for applying the method of sections to determine the forces in the members of a truss.

Free-Body Diagram

- Make a decision as to how to "cut" or section the truss through the members where forces are to be determined.
- Before isolating the appropriate section, it may first be necessary to determine the truss's *external* reactions, so that the three equilibrium equations are used *only* to solve for member forces at the cut section.
- Draw the free-body diagram of that part of the sectioned truss which has the least number of forces on it.
- Use one of the two methods described above for establishing the sense of an unknown force.

Equations of Equilibrium

- Moments should be summed about a point that lies at the intersection of the lines of action of two unknown forces; in this way, the third unknown force is determined directly from the equation.
- If two of the unknown forces are *parallel*, forces may be summed *perpendicular* to the direction of these unknowns to determine *directly* the third unknown force.

An example of a Warren truss (with verticals)

EXAMPLE 3.5

Determine the force in members *GJ* and *CO* of the roof truss shown in the photo. The dimensions and loadings are shown in Fig. 3–26a. State whether the members are in tension or compression. The reactions at the supports have been calculated.

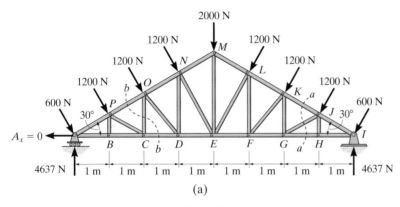

(a)

Fig. 3–26

SOLUTION

Member CF.

Free-Body Diagram. The force in member *GJ* can be obtained by considering the section *aa* in Fig. 3–26a. The free-body diagram of the right part of this section is shown in Fig. 3–26b.

Equations of Equilibrium. A direct solution for F_{GJ} can be obtained by applying $\Sigma M_I = 0$. Why? For simplicity, slide F_{GJ} to point *G* (principle of transmissibility), Fig. 3–26b. Thus,

$$\zeta + \Sigma M_I = 0; \quad -F_{GJ}\sin 30°(2) + 1200(1.155) = 0$$

$$F_{GJ} = 1386 \text{ N (C)} \qquad Ans.$$

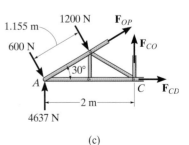

(b)

Member GC.

Free-Body Diagram. The force in *CO* can be obtained by using section *bb* in Fig. 3–26a. The free-body diagram of the left portion of the section is shown in Fig. 3–26c.

Equations of Equilibrium. Moments will be summed about point *A* in order to eliminate the unknowns F_{OP} and F_{CD}.

$$\zeta + \Sigma M_A = 0; \quad -1200(1.155) + F_{CO}(2) = 0$$

$$F_{CO} = 693 \text{ N (T)} \qquad Ans.$$

(c)

EXAMPLE 3.6

Determine the force in members *GF* and *GD* of the truss shown in Fig. 3–27a. State whether the members are in tension or compression. The reactions at the supports have been calculated.

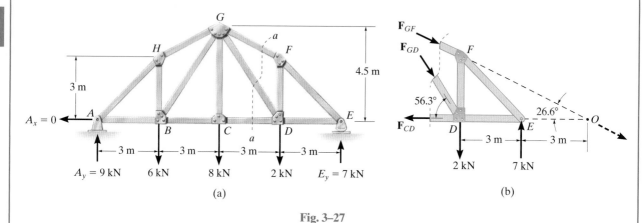

(a)   (b)

Fig. 3–27

SOLUTION

Free-Body Diagram.   Section *aa* in Fig. 3–27a will be considered. Why? The free-body diagram to the right of this section is shown in Fig. 3–27b. The distance *EO* can be determined by proportional triangles or realizing that member *GF* drops vertically $4.5 - 3 = 1.5$ m in 3 m, Fig. 3–27a. Hence to drop 4.5 m from G the distance from *C* to *O* must be 9 m. Also, the angles that $\mathbf{F}_{GD}$ and $\mathbf{F}_{GF}$ make with the horizontal are $\tan^{-1}(4.5/3) = 56.3°$ and $\tan^{-1}(4.5/9) = 26.6°$, respectively.

Equations of Equilibrium.   The force in *GF* can be determined directly by applying $\Sigma M_D = 0$. Why? For the calculation use the principle of transmissibility and slide $\mathbf{F}_{GF}$ to point *O*. Thus,

$$\zeta + \Sigma M_D = 0; \qquad -F_{GF} \sin 26.6°(6) + 7(3) = 0$$
$$F_{GF} = 7.83 \text{ kN (C)} \qquad\qquad Ans.$$

The force in *GD* is determined directly by applying $\Sigma M_O = 0$. For simplicity use the principle of transmissibility and slide $\mathbf{F}_{GD}$ to *D*. Hence,

$$\zeta + \Sigma M_O = 0; \qquad -7(3) + 2(6) + F_{GD} \sin 56.3°(6) = 0$$
$$F_{GD} = 1.80 \text{ kN (C)} \qquad\qquad Ans.$$

EXAMPLE 3.7

Determine the force in members BC and MC of the K-truss shown in Fig. 3–28a. State whether the members are in tension or compression. The reactions at the supports have been calculated.

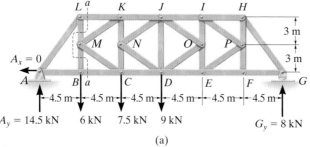

$A_x = 0$

$A_y = 14.5$ kN 6 kN 7.5 kN 9 kN

$G_y = 8$ kN

(a)

SOLUTION

Free-Body Diagram. Although section aa shown in Fig. 3–28a cuts through four members, it is possible to solve for the force in member BC using this section. The free-body diagram of the left portion of the truss is shown in Fig. 3–28b.

Equations of Equilibrium. Summing moments about point L eliminates *three* of the unknowns, so that

$\zeta+\Sigma M_L = 0;\quad -14.5(4.5) + F_{BC}(6) = 0$

$F_{BC} = 10.9$ kN (T)   *Ans.*

Free-Body Diagrams. The force in MC can be obtained indirectly by first obtaining the force in MB from vertical force equilibrium of joint B, Fig. 3–28c, i.e., $F_{MB} = 6$ kN (T). Then from the free-body diagram in Fig. 3–28b.

$+\uparrow\Sigma F_y = 0;\quad 14.5 - 6 + 6 - F_{ML} = 0$

$F_{ML} = 14.5$ kN (T)

Using these results, the free-body diagram of joint M is shown in Fig. 3–28d.

Equations of Equilibrium.

$\xrightarrow{+}\Sigma F_x = 0;\quad \left(\frac{3}{\sqrt{13}}\right)F_{MC} - \left(\frac{3}{\sqrt{13}}\right)F_{MK} = 0$

$+\uparrow\Sigma F_y = 0;\quad 14.5 - 6 - \left(\frac{2}{\sqrt{13}}\right)F_{MC} - \left(\frac{2}{\sqrt{13}}\right)F_{MK} = 0$

$F_{MK} = 7.66$ kN (C) $F_{MC} = 7.66$ kN (T)   *Ans.*

Sometimes, as in this example, application of both the method of sections and the method of joints leads to the most direct solution to the problem.

It is also possible to solve for the force in MC by using the result for $\mathbf{F}_{BC}$. In this case, pass a vertical section through $LK, MK, MC,$ and BC, Fig. 3–28a. Isolate the left section and apply $\Sigma M_K = 0$.

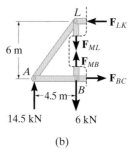

(b)

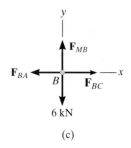

(c)

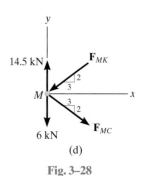

(d)

Fig. 3–28

3.6   Compound Trusses

In Sec. 3–2 it was stated that compound trusses are formed by connecting two or more simple trusses together either by bars or by joints. Occasionally this type of truss is best analyzed by applying *both* the method of joints and the method of sections. It is often convenient to first recognize the type of construction as listed in Sec. 3–2 and then perform the analysis using the following procedure.

EXAMPLE 3.8

Indicate how to analyze the compound truss shown in Fig. 3–29a. The reactions at the supports have been calculated.

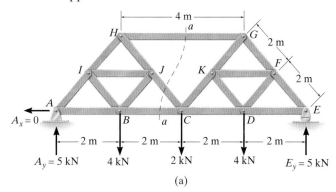

(a)

SOLUTION

The truss is a compound truss since the simple trusses *ACH* and *CEG* are connected by the pin at *C* and the bar *HG*.

Section *aa* in Fig. 3–29a cuts through bar *HG* and two other members having unknown forces. A free-body diagram for the left part is shown in Fig. 3–29b. The force in *HG* is determined as follows:

$$\zeta + \Sigma M_C = 0; \quad -5(4) + 4(2) + F_{HG}(4 \sin 60°) = 0$$

$$F_{HG} = 3.46 \text{ kN (C)}$$

We can now proceed to determine the force in each member of the simple trusses using the method of joints. For example, the free-body diagram of *ACH* is shown in Fig. 3–29c. The joints of this truss can be analyzed in the following sequence:

Joint A: Determine the force in *AB* and *AI*.
Joint H: Determine the force in *HI* and *HJ*.
Joint I: Determine the force in *IJ* and *IB*.
Joint B: Determine the force in *BC* and *BJ*.
Joint J: Determine the force in *JC*.

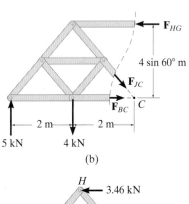

(b)

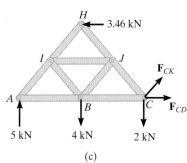

(c)

Fig. 3–29

EXAMPLE 3.9

Compound roof trusses are used in a garden center, as shown in the photo. They have the dimensions and loading shown in Fig. 3–30a. Indicate how to analyze this truss.

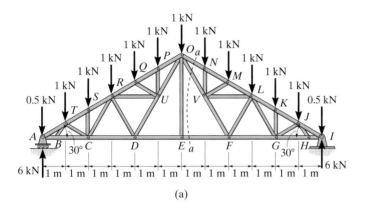

(a)

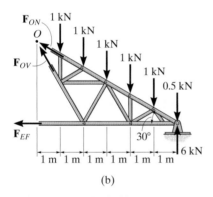

(b)

Fig. 3–30

SOLUTION

We can obtain the force in *EF* by using section *aa* in Fig. 3–30a. The free-body diagram of the right segment is shown in Fig. 3–30b

$$\zeta+\Sigma M_O = 0; \quad -1(1) - 1(2) - 1(3) - 1(4) - 1(5) - 0.5(6) + 6(6) - F_{EF}(6\tan 30°) = 0$$

$$F_{EF} = 5.20 \text{ kN (T)} \qquad \qquad Ans.$$

By inspection notice that *BT*, *EO*, and *HJ* are zero-force members since $+\uparrow\Sigma F_y = 0$ at joints *B*, *E*, and *H*, respectively. Also, by applying $+\nwarrow\Sigma F_y = 0$ (perpendicular to *AO*) at joints *P*, *Q*, *S*, and *T*, we can directly determine the force in members *PU*, *QU*, *SC*, and *TC*, respectively.

3

EXAMPLE 3.10

Indicate how to analyze the compound truss shown in Fig. 3–31a. The reactions at the supports have been calculated.

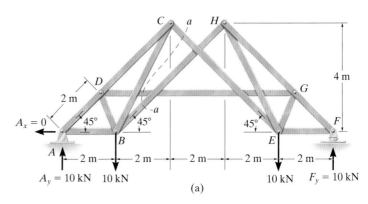

Fig. 3–31

SOLUTION

The truss may be classified as a type 2 compound truss since the simple trusses *ABCD* and *FEHG* are connected by three nonparallel or nonconcurrent bars, namely, *CE*, *BH*, and *DG*.

Using section *aa* in Fig. 3–31a we can determine the force in each connecting bar. The free-body diagram of the left part of this section is shown in Fig. 3–31b. Hence,

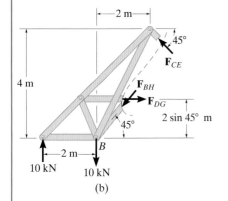

$$\zeta+\Sigma M_B = 0; \quad -10(2) - F_{DG}(2\sin 45°) + F_{CE}\cos 45°(4)$$
$$+ F_{CE}\sin 45°(2) = 0 \tag{1}$$

$$+\uparrow\Sigma F_y = 0; \quad 10 - 10 - F_{BH}\sin 45° + F_{CE}\sin 45° = 0 \tag{2}$$

$$\xrightarrow{+}\Sigma F_x = 0; \quad -F_{BH}\cos 45° + F_{DG} - F_{CE}\cos 45° = 0 \tag{3}$$

From Eq. (2), $F_{BH} = F_{CE}$; then solving Eqs. (1) and (3) simultaneously yields

$$F_{BH} = F_{CE} = 8.9 \text{ kN (C)} \qquad F_{DG} = 12.6 \text{ kN (T)}$$

Analysis of each connected simple truss can now be performed using the method of joints. For example, from Fig. 3–31c, this can be done in the following sequence.

Joint A: Determine the force in *AB* and *AD*.
Joint D: Determine the force in *DC* and *DB*.
Joint C: Determine the force in *CB*.

FUNDAMENTAL PROBLEMS

F3–7. Determine the force in members *HG*, *BG*, and *BC* and state whether they are in tension or compression.

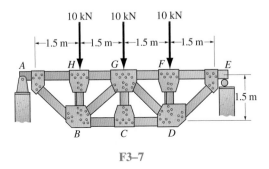

F3–7

F3–8. Determine the force in members *HG*, *HC*, and *BC* and state whether they are in tension or compression.

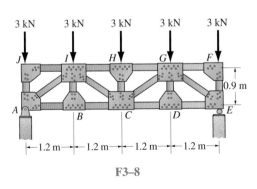

F3–8

F3–9. Determine the force in members *ED*, *BD*, and *BC* and state whether they are in tension or compression.

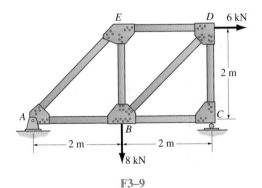

F3–9

F3–10. Determine the force in members *GF*, *CF*, and *CD* and state whether they are in tension or compression.

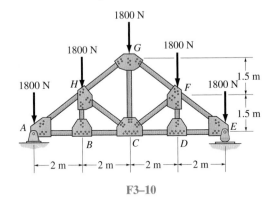

F3–10

F3–11. Determine the force in members *FE*, *FC*, and *BC* and state whether they are in tension or compression.

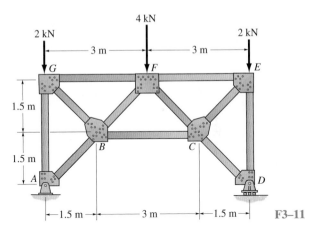

F3–11

F3–12. Determine the force in members *GF*, *CF*, and *CD* and state whether they are in tension or compression.

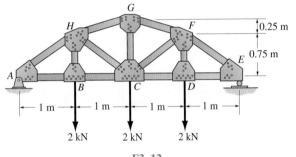

F3–12

PROBLEMS

3–18. Determine the force in members GF, FC, and CD of the bridge truss. State if the members are in tension or compression. Assume all members are pin connected.

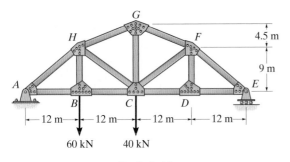

Prob. 3–18

3–19. Determine the force in members JK, JN, and CD. State if the members are in tension or compression. Identify all the zero-force members.

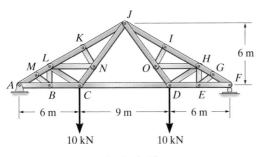

Prob. 3–19

*3–20.** Determine the force in members GF, FC, and CD of the cantilever truss. State if the members are in tension or compression. Assume all members are pin connected.

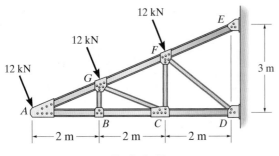

Prob. 3–20

3–21. The *Howe* truss is subjected to the loading shown. Determine the forces in members GF, CD, and GC. State if the members are in tension or compression. Assume all members are pin connected.

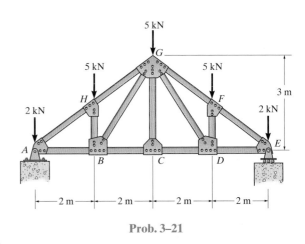

Prob. 3–21

3–22. Determine the force in members BG, HG, and BC of the truss and state if the members are in tension or compression.

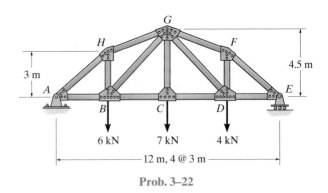

Prob. 3–22

3–23. Determine the force in members *GF*, *CF*, and *CD* of the roof truss and indicate if the members are in tension or compression.

3–25. Determine the force in members *IH*, *ID*, and *CD* of the truss. State if the members are in tension or compression. Assume all members are pin connected.

3–26. Determine the force in members *JI*, *IC*, and *CD* of the truss. State if the members are in tension or compression. Assume all members are pin connected.

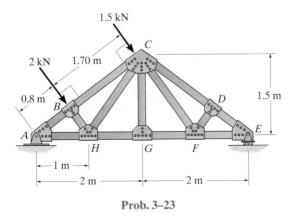

Prob. 3–23

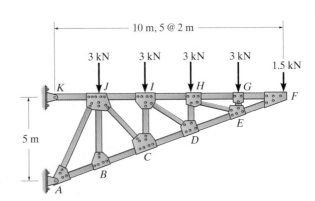

Probs. 3–25/3–26

*3–24. Determine the force in members *GF*, *FB*, and *BC* of the *Fink* truss and state if the members are in tension or compression.

3–27. Determine the forces in members *KJ*, *CD*, and *CJ* of the truss. State if the members are in tension or compression.

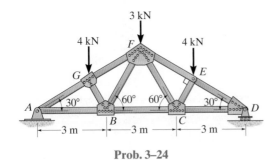

Prob. 3–24

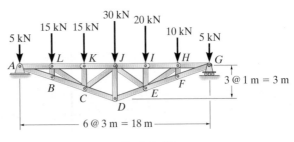

Prob. 3–27

3.7   Complex Trusses

The member forces in a complex truss can be determined using the method of joints; however, the solution will require writing the two equilibrium equations for each of the j joints of the truss and then solving the complete set of $2j$ equations *simultaneously*.* This approach may be impractical for hand calculations, especially in the case of large trusses. Therefore, a more direct method for analyzing a complex truss, referred to as the *method of substitute members*, will be presented here.

Procedure for Analysis

With reference to the truss in Fig. 3–32a, the following steps are necessary to solve for the member forces using the substitute-member method.

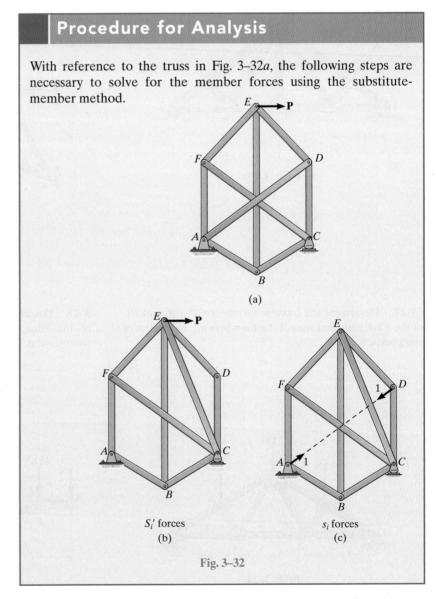

S_i' forces
(b)

s_i forces
(c)

Fig. 3–32

*This can be readily accomplished using a computer as will be shown in Chapter 14.

Reduction to Stable Simple Truss
Determine the reactions at the supports and begin by imagining how to analyze the truss by the method of joints, i.e., progressing from joint to joint and solving for each member force. If a joint is reached where there are *three unknowns*, remove one of the members at the joint and replace it by an *imaginary* member elsewhere in the truss. By doing this, reconstruct the truss to be a stable simple truss.

For example, in Fig. 3–32a it is observed that each joint will have three *unknown* member forces acting on it. Hence we will remove member AD and replace it with the imaginary member EC, Fig. 3–32b. This truss can now be analyzed by the method of joints for the two types of loading that follow.

External Loading on Simple Truss
Load the simple truss with the actual loading **P**, then determine the force S_i' in each member i. In Fig. 3–32b, provided the reactions have been determined, one could start at joint A to determine the forces in AB and AF, then joint F to determine the forces in FE and FC, then joint D to determine the forces in DE and DC (both of which are zero), then joint E to determine EB and EC, and finally joint B to determine the force in BC.

Remove External Loading from Simple Truss
Consider the simple truss without the external load **P**. Place equal but opposite collinear *unit loads* on the truss at the two joints from which the member was removed. If these forces develop a force s_i in the ith truss member, then by proportion an unknown force x in the removed member would exert a force xs_i in the ith member.

From Fig. 3–32c the equal but opposite unit loads will create *no reactions* at A and C when the equations of equilibrium are applied to the entire truss. The s_i forces can be determined by analyzing the joints in the same sequence as before, namely, joint A, then joints F, D, E, and finally B.

Superposition
If the effects of the above two loadings are combined, the force in the ith member of the truss will be

$$S_i = S_i' + xs_i \tag{1}$$

In particular, for the substituted member EC in Fig. 3–32b the force $S_{EC} = S_{EC}' + xs_{EC}$. Since member EC does not actually exist on the original truss, we will choose x to have a magnitude such that it yields *zero force* in EC. Hence,

$$S_{EC}' + xs_{EC} = 0 \tag{2}$$

or $x = -S_{EC}'/s_{EC}$. Once the value of x has been determined, the force in the other members i of the complex truss can be determined from Eq. (1).

EXAMPLE 3.11

Determine the force in each member of the complex truss shown in Fig. 3–33a. Assume joints B, F, and D are on the same horizontal line. State whether the members are in tension or compression.

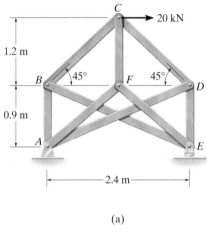

(a)

Fig. 3–33

SOLUTION

Reduction to Stable Simple Truss. By inspection, each joint has three unknown member forces. A joint analysis can be performed by hand if, for example, member CF is removed and member DB substituted, Fig. 3–33b. The resulting truss is stable and will not collapse.

External Loading on Simple Truss. As shown in Fig. 3–33b, the support reactions on the truss have been determined. Using the method of joints, we can first analyze joint C to find the forces in members CB and CD; then joint F, where it is seen that FA and FE are zero-force members; then joint E to determine the forces in members EB and ED; then joint D to determine the forces in DA and DB; then finally joint B to determine the force in BA. Considering tension as positive and compression as negative, these S_i' forces are recorded in column 2 of Table 1.

20 kN

C

B 45° F 45° D

20 kN A   E

17.5 kN   17.5 kN

(b)

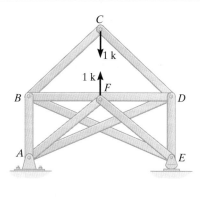

(c)

Remove External Loading from Simple Truss. The unit load acting on the truss is shown in Fig. 3–33c. These equal but opposite forces create no external reactions on the truss. The joint analysis follows the same sequence as discussed previously, namely, joints $C, F,$ $E, D,$ and B. The results of the s_i force analysis are recorded in column 3 of Table 1.

Superposition. We require

$$S_{DB} = S'_{DB} + xs_{DB} = 0$$

Substituting the data for S'_{DB} and s_{DB}, where S'_{DB} is negative since the force is compressive, we have

$$-10.0 + x(1.167) = 0 \qquad x = 8.567$$

The values of xs_i are recorded in column 4 of Table 1, and the actual member forces $S_i = S'_i + xs_i$ are listed in column 5.

TABLE 1				
Member	S'_i	s_i	xs_i	S_i
CB	14.1	−0.707	−6.06	8.08 (T)
CD	−14.1	−0.707	−6.06	20.2 (C)
FA	0	0.833	7.14	7.14 (T)
FE	0	0.833	7.14	7.14 (T)
EB	0	−0.712	−6.10	6.10 (C)
ED	−17.5	−0.250	−2.14	19.6 (C)
DA	21.4	−0.712	−6.10	15.3 (T)
DB	−10.0	1.167	10.0	0
BA	10.0	−0.250	−2.14	7.86 (T)

3

Fig. 3–34

The roof of this pavilion is supported using a system of space trusses.

3.8   Space Trusses

A *space truss* consists of members joined together at their ends to form a stable three-dimensional structure. In Sec. 3–2 it was shown that the simplest form of a stable two-dimensional truss consists of the members arranged in the form of a triangle. We then built up the simple plane truss from this basic triangular element by adding two members at a time to form further elements. In a similar manner, the simplest element of a stable space truss is a *tetrahedron*, formed by connecting six members together with four joints as shown in Fig. 3–34. Any additional members added to this basic element would be redundant in supporting the force **P**. A simple space truss can be built from this basic tetrahedral element by adding three additional members and another joint forming multiconnected tetrahedrons.

Determinacy and Stability.   Realizing that in three dimensions there are three equations of equilibrium available for each joint ($\Sigma F_x = 0$, $\Sigma F_y = 0$, $\Sigma F_z = 0$), then for a space truss with j number of joints, $3j$ equations are available. If the truss has b number of bars and r number of reactions, then like the case of a planar truss (Eqs. 3–1 and 3–2) we can write

$b + r < 3j$	unstable truss	
$b + r = 3j$	statically determinate—check stability	(3–3)
$b + r > 3j$	statically indeterminate—check stability	

The *external stability* of the space truss requires that the support reactions keep the truss in force and moment equilibrium about any and all axes. This can sometimes be checked by inspection, although if the truss is unstable a solution of the equilibrium equations will give inconsistent results. *Internal stability* can sometimes be checked by careful inspection of the member arrangement. Provided each joint is held fixed by its supports or connecting members, so that it cannot move with respect to the other joints, the truss can be classified as internally stable. Also, if we do a force analysis of the truss and obtain inconsistent results, then the truss configuration will be unstable or have a "critical form."

Assumptions for Design.   The members of a space truss may be treated as axial-force members provided the external loading is applied at the joints and the joints consist of ball-and-socket connections. This assumption is justified provided the joined members at a connection intersect at a common point and the weight of the members can be neglected. In cases where the weight of a member is to be included in the analysis, it is generally satisfactory to apply it as a vertical force, half of its magnitude applied to each end of the member.

For the force analysis the supports of a space truss are generally modeled as a short link, plane roller joint, slotted roller joint, or a ball-and-socket joint. Each of these supports and their reactive force components are shown in Table 3–1.

TABLE 3–1 Supports and Their Reactive Force Components

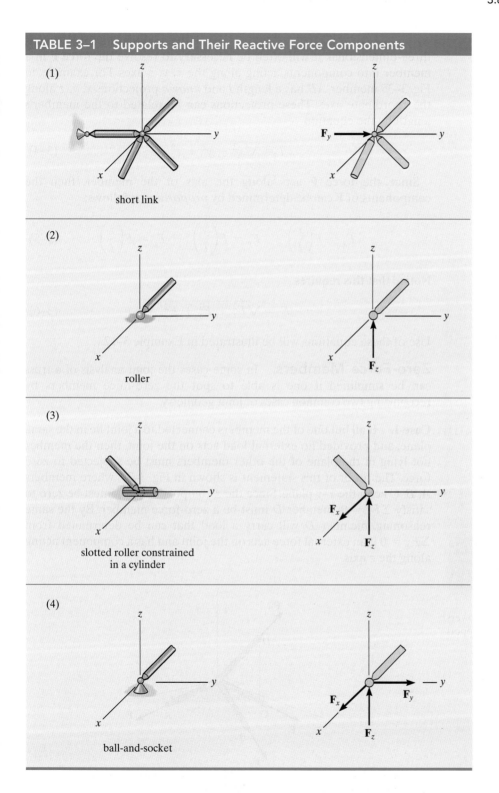

(1) short link

(2) roller

(3) slotted roller constrained in a cylinder

(4) ball-and-socket

3

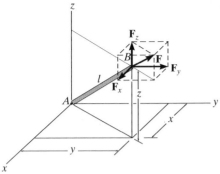

Fig. 3–35

x, y, z, Force Components.   Since the analysis of a space truss is three-dimensional, it will often be necessary to resolve the force **F** in a member into components acting along the x, y, z axes. For example, in Fig. 3–35 member AB has a length l and *known* projections x, y, z along the coordinate axes. These projections can be related to the member's length by the equation

$$ l = \sqrt{x^2 + y^2 + z^2} \tag{3–4} $$

Since the force **F** acts along the axis of the member, then the components of **F** can be determined by *proportion* as follows:

$$ F_x = F\left(\frac{x}{l}\right) \qquad F_y = F\left(\frac{y}{l}\right) \qquad F_z = F\left(\frac{z}{l}\right) \tag{3–5} $$

Notice that this requires

$$ F = \sqrt{F_x^2 + F_y^2 + F_z^2} \tag{3–6} $$

Use of these equations will be illustrated in Example 3–12.

Zero-Force Members.   In some cases the joint analysis of a truss can be simplified if one is able to spot the zero-force members by recognizing two common cases of joint geometry.

Case 1.   If all but one of the members connected to a joint lie in the same plane, and provided no external load acts on the joint, then the member not lying in the plane of the other members must be subjected to zero force. The proof of this statement is shown in Fig. 3–36, where members A, B, C lie in the x–y plane. Since the z component of $\mathbf{F}_D$ must be zero to satisfy $\Sigma F_z = 0$, member D must be a zero-force member. By the same reasoning, member D will carry a load that can be determined from $\Sigma F_z = 0$ if an external force acts on the joint and has a component acting along the z axis.

Because of their cost effectiveness, towers such as these are often used to support multiple electric transmission lines.

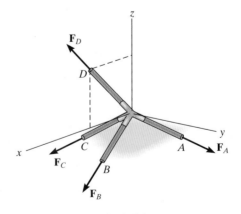

Fig. 3–36

Case 2. If it has been determined that all but two of several members connected at a joint support zero force, then the two remaining members must also support zero force, provided they do not lie along the same line. This situation is illustrated in Fig. 3–37, where it is known that A and C are zero-force members. Since $\mathbf{F}_D$ is collinear with the y axis, then application of $\Sigma F_x = 0$ or $\Sigma F_z = 0$ requires the x or z component of $\mathbf{F}_B$ to be zero. Consequently, $F_B = 0$. This being the case, $F_D = 0$ since $\Sigma F_y = 0$.

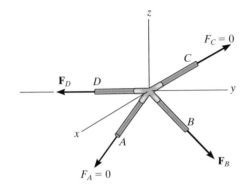

Fig. 3–37

Particular attention should be directed to the foregoing two cases of joint geometry and loading, since the analysis of a space truss can be considerably simplified by first spotting the zero-force members.

Procedure for Analysis

Either the method of sections or the method of joints can be used to determine the forces developed in the members of a space truss.

Method of Sections
If only a *few* member forces are to be determined, the method of sections may be used. When an imaginary section is passed through a truss and the truss is separated into two parts, the force system acting on either one of the parts must satisfy the six scalar equilibrium equations: $\Sigma F_x = 0$, $\Sigma F_y = 0$, $\Sigma F_z = 0$, $\Sigma M_x = 0$, $\Sigma M_y = 0$, $\Sigma M_z = 0$. By proper choice of the section and axes for summing forces and moments, many of the unknown member forces in a space truss can be computed *directly*, using a single equilibrium equation. In this regard, recall that the *moment* of a force about an axis is *zero* provided *the force is parallel to the axis or its line of action passes through a point on the axis.*

Method of Joints
Generally, if the forces in *all* the members of the truss must be determined, the method of joints is most suitable for the analysis. When using the method of joints, it is necessary to solve the three scalar equilibrium equations $\Sigma F_x = 0$, $\Sigma F_y = 0$, $\Sigma F_z = 0$ at each joint. Since it is relatively easy to draw the free-body diagrams and apply the equations of equilibrium, the method of joints is very consistent in its application.

EXAMPLE 3.12

Determine the force in each member of the space truss shown in Fig. 3–38a. The truss is supported by a ball-and-socket joint at A, a slotted roller joint at B, and a cable at C.

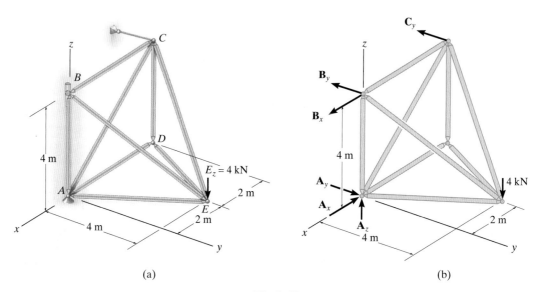

(a)   (b)

Fig. 3–38

SOLUTION

The truss is statically determinate since $b + r = 3j$ or $9 + 6 = 3(5)$, Fig. 3–38b.

Support Reactions.   We can obtain the support reactions from the free-body diagram of the entire truss, Fig. 3–38b, as follows:

$$\Sigma M_y = 0; \qquad -4(2) + B_x(4) = 0 \qquad B_x = 2 \text{ kN}$$
$$\Sigma M_z = 0; \qquad\qquad C_y = 0$$
$$\Sigma M_x = 0; \qquad B_y(4) - 4(4) = 0 \qquad B_y = 4 \text{ kN}$$
$$\Sigma F_x = 0; \qquad\quad 2 - A_x = 0 \qquad A_x = 2 \text{ kN}$$
$$\Sigma F_y = 0; \qquad\quad A_y - 4 = 0 \qquad A_y = 4 \text{ kN}$$
$$\Sigma F_z = 0; \qquad\quad A_z - 4 = 0 \qquad A_z = 4 \text{ kN}$$

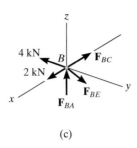

(c)

Joint B. We can begin the method of joints at B since there are three unknown member forces at this joint, Fig. 3–38c. The components of $\mathbf{F}_{BE}$ can be determined by proportion to the length of member BE, as indicated by Eqs. 3–5. We have

$$\Sigma F_y = 0; \qquad -4 + F_{BE}\left(\tfrac{4}{6}\right) = 0 \qquad F_{BE} = 6 \text{ kN (T)} \qquad \text{Ans.}$$

$$\Sigma F_x = 0; \qquad 2 - F_{BC} - 6\left(\tfrac{2}{6}\right) = 0 \qquad F_{BC} = 0 \qquad \text{Ans.}$$

$$\Sigma F_z = 0; \qquad F_{BA} - 6\left(\tfrac{4}{6}\right) = 0 \qquad F_{BA} = 4 \text{ kN (C)} \qquad \text{Ans.}$$

Joint A. Using the result for $F_{BA} = 4 \text{ kN (C)}$, the free-body diagram of joint A is shown in Fig. 3–38d. We have

$$\Sigma F_z = 0; \qquad 4 - 4 + F_{AC}\sin 45° = 0$$
$$F_{AC} = 0 \qquad \text{Ans.}$$

$$\Sigma F_y = 0; \qquad -F_{AE}\left(\tfrac{2}{\sqrt{5}}\right) + 4 = 0$$
$$F_{AE} = 4.47 \text{ kN (C)} \qquad \text{Ans.}$$

$$\Sigma F_x = 0; \qquad -2 + F_{AD} + 4.47\left(\tfrac{1}{\sqrt{5}}\right) = 0$$
$$F_{AD} = 0 \qquad \text{Ans.}$$

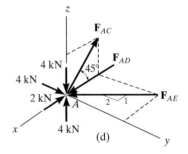

(d)

Joint D. By inspection the members at joint D, Fig. 3–38a, support zero force, since the arrangement of the members is similar to either of the two cases discussed in reference to Figs. 3–36 and 3–37. Also, from Fig. 3–38e,

$$\Sigma F_x = 0; \qquad\qquad F_{DE} = 0 \qquad \text{Ans.}$$

$$\Sigma F_z = 0; \qquad\qquad F_{DC} = 0 \qquad \text{Ans.}$$

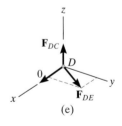

(e)

Joint C. By observation of the free-body diagram, Fig. 3–38f,

$$F_{CE} = 0 \qquad \text{Ans.}$$

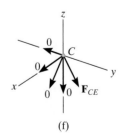

(f)

EXAMPLE 3.13

Determine the zero-force members of the truss shown in Fig. 3–39a. The supports exert components of reaction on the truss as shown.

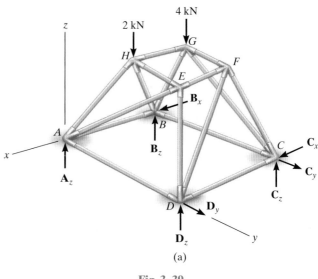

(a)

Fig. 3–39

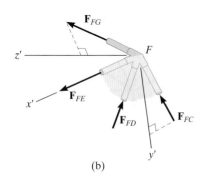

(b)

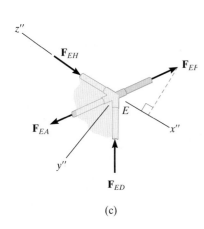

(c)

SOLUTION

The free-body diagram, Fig. 3–39a, indicates there are eight unknown reactions for which only six equations of equilibrium are available for solution. Although this is the case, the reactions can be determined, since $b + r = 3j$ or $16 + 8 = 3(8)$.

To spot the zero-force members, we must compare the conditions of joint geometry and loading to those of Figs. 3–36 and 3–37. Consider joint F, Fig. 3–39b. Since members FC, FD, FE lie in the x'–y' plane and FG is not in this plane, FG is a zero-force member. ($\Sigma F_{z'} = 0$ must be satisfied.) In the same manner, from joint E, Fig. 3–39c, EF is *a zero-force member*, since it does not lie in the y''–z'' plane. ($\Sigma F_{x''} = 0$ must be satisfied.) Returning to joint F, Fig. 3–39b, it can be seen that $F_{FD} = F_{FC} = 0$, since $F_{FE} = F_{FG} = 0$, and there are no external forces acting on the joint. Use this procedure to show that AB is a zero force member.

The numerical force analysis of the joints can now proceed by analyzing joint G ($F_{GF} = 0$) to determine the forces in GH, GB, GC. Then analyze joint H to determine the forces in HE, HB, HA; joint E to determine the forces in EA, ED; joint A to determine the forces in AB, AD, and A_z; joint B to determine the force in BC and B_x, B_z; joint D to determine the force in DC and D_y, D_z; and finally, joint C to determine C_x, C_y, C_z.

PROBLEMS

*3–28. Determine the forces in all the members of the complex truss. State if the members are in tension or compression. *Hint:* Substitute member AD with one placed between E and C.

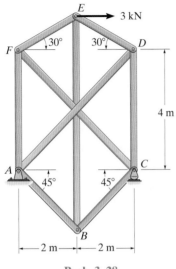

Prob. 3–28

3–29. Determine the forces in all the members of the lattice (complex) truss. State if the members are in tension or compression. *Hint:* Substitute member JE by one placed between K and F.

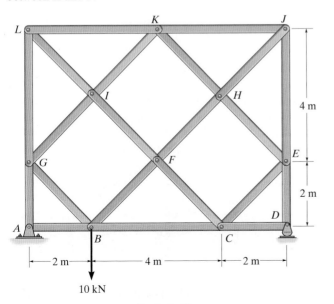

Prob. 3–29

3–30. Determine the force in each member and state if the members are in tension or compression.

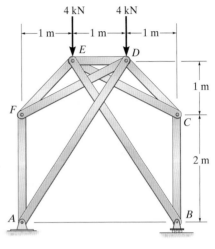

Prob. 3–30

3–31. Determine the force in all the members of the complex truss. State if the members are in tension or compression.

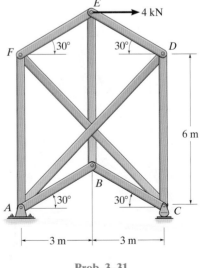

Prob. 3–31

*3–32. Determine the force developed in each member of the space truss and state if the members are in tension or compression. The crate has a weight of 5 kN.

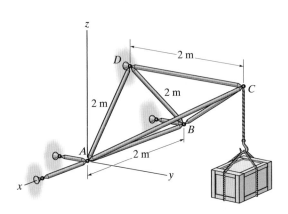

Prob. 3–32

3–33. Determine the force in each member of the space truss and state if the members are in tension or compression. *Hint:* The support reaction at *E* acts along member *EB*. Why?

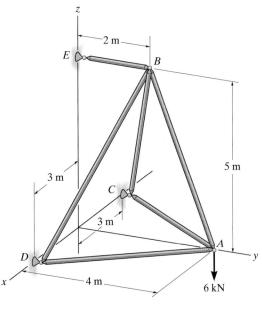

Prob. 3–33

3–34. Determine the force in each member of the space truss and state if the members are in tension or compression. The truss is supported by ball-and socket joints at *C, D, E,* and *G. Note:* Although this truss is indeterminate to the first degree, a solution is possible due to symmetry of geometry and loading.

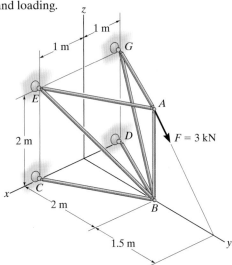

Prob. 3–34

3–35. Determine the force in members *FE* and *ED* of the space truss and state if the members are in tension or compression. The truss is supported by a ball-and-socket joint at *C* and short links at *A* and *B*.

*3–36. Determine the force in members *GD, GE,* and *FD* of the space truss and state if the members are in tension or compression.

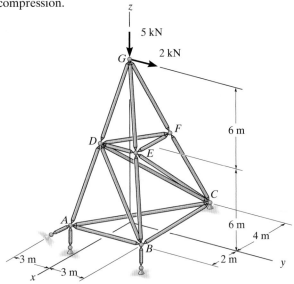

Probs. 3–35/3–36

3–37. Determine the force in each member of the space truss. Indicate if the members are in tension or compression.

3–38. Determine the force in members BE, DF, and BC of the space truss and state if the members are in tension or compression.

3–39. Determine the force in members CD, ED, and CF of the space truss and state if the members are in tension or compression.

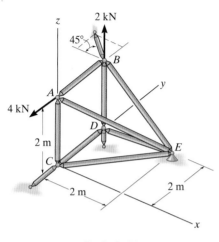

Prob. 3–37

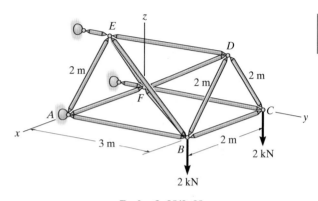

Probs. 3–38/3–39

PROJECT PROBLEM

3–1P. The Pratt roof trusses are uniformly spaced every 5 m. The deck, roofing material, and the purlins have an average weight of 0.28 kN/m^2. The building is located in New York where the anticipated snow load is 1.0 kN/m^2 and the anticipated ice load is 0.4 kN/m^2. These loadings occur over the horizontal projected area of the roof. Determine the force in each member due to dead load, snow, and ice loads. Neglect the weight of the truss members and assume A is pinned and E is a roller.

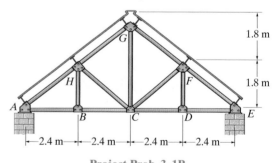

Project Prob. 3–1P

CHAPTER REVIEW

Trusses are composed of slender members joined together at their end points to form a series of triangles.

(Ref: Section 3.1)

For analysis we assume the members are pin connected, and the loads are applied at the joints. Thus, the members will either be in tension or compression.	
	(Ref: Section 3.1)

Trusses can be classified in three ways:

Simple trusses are formed by starting with an initial triangular element and connecting to it two other members and a joint to form a second triangle, etc.

Compound trusses are formed by connecting together two or more simple trusses using a common joint and/or additional member.

Complex trusses are those that cannot be classified as either simple or compound.

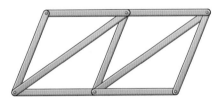

simple truss

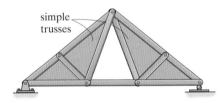

simple trusses

compound truss

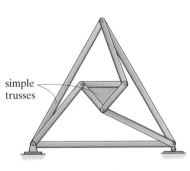

simple trusses

compound truss

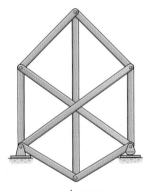

complex truss

(Ref: Section 3.2)

If the number of bars or members of a truss is b, and there are r reactions and j joints, then if	
$b + r = 2j$ the truss will be statically determinate	
$b + r > 2j$ the truss will be statically indeterminate	
(Ref: Section 3.2)	

The truss will be externally unstable if the reactions are concurrent or parallel.

Internal stability can be checked by counting the number of bars b, reactions r, and joints j.

If $b + r < 2j$ the truss is unstable.

If $b + r \geq 2j$ it may still be unstable, so it becomes necessary to inspect the truss and look for bar arrangements that form a parallel mechanism, without forming a triangular element.

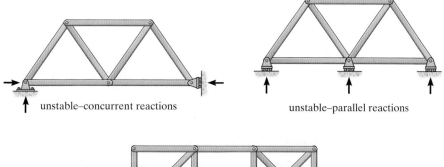

unstable–concurrent reactions   unstable–parallel reactions

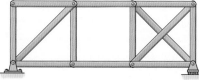

unstable internally

(Ref: Section 3.2)

Planar trusses can be analyzed by the *method of joints*. This is done by selecting each joint in sequence, having at most one known force and at least two unknowns. The free-body diagram of each joint is constructed and two force equations of equilibrium, $\Sigma F_x = 0$, $\Sigma F_y = 0$, are written and solved for the unknown member forces.

The method of sections requires passing a section through the truss and then drawing a free-body diagram of one of its sectioned parts. The member forces cut by the section are then found from the three equations of equilibrium. Normally a single unknown can be found if one sums moments about a point that eliminates the two other forces.

Compound and complex trusses can also be analyzed by the method of joints and the method of sections. The "method of substitute members" can be used to obtain a direct solution for the force in a particular member of a complex truss.

(Ref: Section 3.3)

The simply supported beams and girders of this building frame were designed to resist the internal shear and moment acting throughout their lengths.

Internal Loadings Developed in Structural Members

4

Before a structural member can be proportioned, it is necessary to determine the force and moment that act within it. In this chapter we will develop the methods for finding these loadings at specified points along a member's axis and for showing the variation graphically using the shear and moment diagrams. Applications are given for both beams and frames.

4.1 Internal Loadings at a Specified Point

As discussed in Sec. 2–3, the internal load at a specified point in a member can be determined by using the *method of sections*. In general, this loading for a coplanar structure will consist of a normal force $\mathbf{N}$, shear force $\mathbf{V}$, and bending moment $\mathbf{M}$.* It should be realized, however, that these loadings actually represent the *resultants* of the *stress distribution* acting over the member's cross-sectional area at the cut section. Once the resultant internal loadings are known, the magnitude of the stress can be determined provided an assumed distribution of stress over the cross-sectional area is specified.

*Three-dimensional frameworks can also be subjected to a *torsional moment*, which tends to twist the member about its axis.

Sign Convention. Before presenting a method for finding the internal normal force, shear force, and bending moment, we will need to establish a sign convention to define their "positive" and "negative" values.* Although the choice is arbitrary, the sign convention to be adopted here has been widely accepted in structural engineering practice, and is illustrated in Fig. 4–1a. On the *left-hand face* of the cut member the normal force **N** acts to the right, the internal shear force **V** acts downward, and the moment **M** acts counterclockwise. In accordance with Newton's third law, an equal but opposite normal force, shear force, and bending moment must act on the right-hand face of the member at the section. Perhaps an easy way to remember this sign convention is to isolate a small segment of the member and note that *positive normal force tends to elongate the segment,* Fig. 4–1b; *positive shear tends to rotate the segment clockwise,* Fig. 4–1c; and *positive bending moment tends to bend the segment concave upward,* so as to "hold water," Fig. 4–1d.

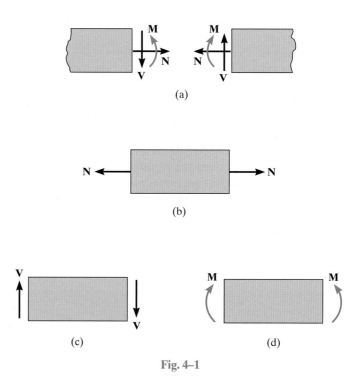

(a)

(b)

(c)   (d)

Fig. 4–1

*This will be convenient later in Secs. 4–2 and 4–3 where we will express V and M as functions of x and then *plot* these functions. Having a sign convention is similar to assigning coordinate directions x positive to the right and y positive upward when plotting a function $y = f(x)$.

Procedure for Analysis

The following procedure provides a means for applying the method of sections to determine the internal normal force, shear force, and bending moment at a specific location in a structural member.

Support Reactions

- Before the member is "cut" or sectioned, it may be necessary to determine the member's support reactions so that the equilibrium equations are used only to solve for the internal loadings when the member is sectioned.

- If the member is part of a pin-connected structure, the pin reactions can be determined using the methods of Sec. 2–5.

Free-Body Diagram

- Keep all distributed loadings, couple moments, and forces acting on the member in their *exact location,* then pass an imaginary section through the member, perpendicular to its axis at the point where the internal loading is to be determined.

- After the section is made, draw a free-body diagram of the segment that has the least number of loads on it. At the section indicate the unknown resultants **N**, **V**, and **M** acting in their *positive* directions (Fig. 4–1a).

Equations of Equilibrium

- Moments should be summed at the section about axes that pass through the *centroid* of the member's cross-sectional area, in order to eliminate the unknowns **N** and **V** and thereby obtain a direct solution for **M**.

- If the solution of the equilibrium equations yields a quantity having a negative magnitude, the assumed directional sense of the quantity is opposite to that shown on the free-body diagram.

EXAMPLE 4.1

The building roof shown in the photo has a weight of 1.8 kN/m² and is supported on 8-m long simply supported beams that are spaced 1 m apart. Each beam, shown in Fig. 4–2b transmits its loading to two girders, located at the front and back of the building. Determine the internal shear and moment in the front girder at point C, Fig. 4–2a. Neglect the weight of the members.

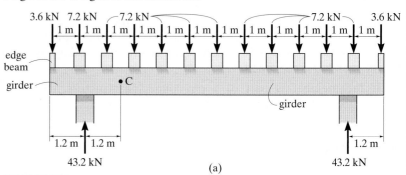

(a)

SOLUTION

Support Reactions. The roof loading is transmitted to each beam as a one-way slab ($L_2/L_1 = 8$ m/1 m $= 8 > 2$). The tributary loading on each interior beam is therefore $(1.8 \text{ kN/m}^2)(1 \text{ m}) = 1.8 \text{ kN/m}$. (The two edge beams support 0.9 kN/m.) From Fig. 4–2b, the reaction of each interior beam on the girder is $(1.8 \text{ kN/m})(8 \text{ m})/2 = 7.2 \text{ kN}$.

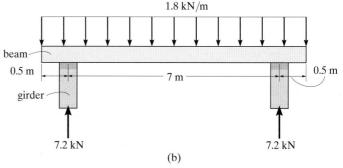

(b)

Free-Body Diagram. The free-body diagram of the girder is shown in Fig. 4–2a. Notice that each column reaction is

$$[(2(3.6 \text{ kN}) + 11(7.2 \text{ kN})]/2 = 43.2 \text{ kN}$$

The free-body diagram of the left girder segment is shown in Fig. 4–2c. Here the internal loadings are assumed to act in their positive directions.

Equations of Equilibrium

$$+\uparrow \Sigma F_y = 0; \qquad 43.2 - 3.6 - 2(7.2) - V_C = 0 \qquad V_C = 25.2 \text{ kN} \qquad \textit{Ans.}$$

$$\zeta + \Sigma M_C = 0; \qquad M_C + 7.2(0.4) + 7.2(1.4) + 3.6(2.4) - 43.2(1.2) = 0 \qquad M_C = 30.2 \text{ kN} \cdot \text{m} \qquad \textit{Ans.}$$

Fig. 4–2

(c)

EXAMPLE 4.2

Determine the internal shear and moment acting at a section passing through point C in the beam shown in Fig. 4–3a.

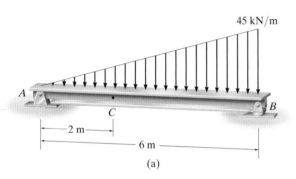

(a)

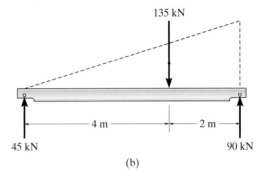

(b)

Fig. 4–3

SOLUTION

Support Reactions.   Replacing the distributed load by its resultant force and computing the reactions yields the results shown in Fig. 4–3b.

Free-Body Diagram.   Segment AC will be considered since it yields the simplest solution, Fig. 4–3c. The distributed load intensity at C is computed by proportion, that is,

$$w_C = (2 \text{ m}/6 \text{ m})(45 \text{ kN/m}) = 15 \text{ kN/m}$$

Equations of Equilibrium.

$$+\uparrow \Sigma F_y = 0; \qquad 45 - 15 - V_C = 0 \qquad\qquad V_C = 30 \text{ kN} \qquad Ans.$$

$$\zeta + \Sigma M_C = 0; \qquad -45(2) + 15(0.667) + M_C = 0$$

$$M_C = 80 \text{ kN} \cdot \text{m} \quad Ans.$$

This problem illustrates the importance of *keeping* the distributed loading on the beam until *after* the beam is sectioned. If the beam in Fig. 4–3b were sectioned at C, the effect of the distributed load on segment AC would not be recognized, and the result $V_C = 45$ kN and $M_C = 90$ kN · m would be wrong.

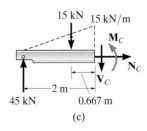

(c)

EXAMPLE 4.3

The 45-kN force in Fig. 4–4a is supported by the floor panel *DE*, which in turn is simply supported at its ends by floor beams. These beams transmit their loads to the simply supported girder *AB*. Determine the internal shear and moment acting at point *C* in the girder.

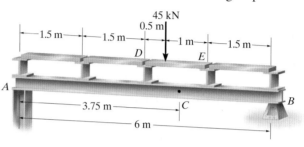

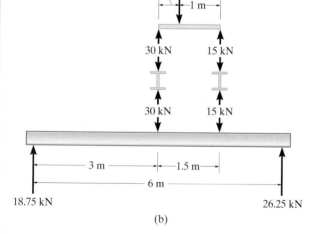

(b)

Fig. 4–4

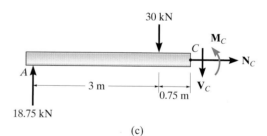

(c)

SOLUTION

Support Reactions.   Equilibrium of the floor panel, floor beams, and girder is shown in Fig. 4–4b. It is advisable to check these results.

Free-Body Diagram.   The free-body diagram of segment *AC* of the girder will be used since it leads to the simplest solution, Fig. 4–4c. Note that there are *no loads* on the floor beams supported by *AC*.

Equations of Equilibrium.

$$+\uparrow \Sigma F_y = 0; \qquad\qquad 18.75 - 30 - V_C = 0 \qquad V_C = -11.25 \text{ kN} \quad Ans.$$

$$\zeta + \Sigma M_C = 0; \quad -18.75(3.75) + 30(0.75) + M_C = 0$$

$$M_C = 47.8 \text{ kN} \cdot \text{m} \quad Ans.$$

4.2   Shear and Moment Functions

The design of a beam requires a detailed knowledge of the *variations* of the internal shear force V and moment M acting at each point along the axis of the beam. The internal normal force is generally not considered for two reasons: (1) in most cases the loads applied to a beam act perpendicular to the beam's axis and hence produce only an internal shear force and bending moment, and (2) for design purposes the beam's resistance to shear, and particularly to bending, is more important than its ability to resist normal force. An important exception to this occurs, however, when beams are subjected to compressive axial forces, since the buckling or instability that may occur has to be investigated.

The variations of V and M as a function of the position x of an arbitrary point along the beam's axis can be obtained by using the method of sections discussed in Sec. 4–1. Here, however, it is necessay to locate the imaginary section or cut at an arbitrary distance x from one end of the beam rather than at a specific point.

In general, the internal shear and moment functions will be discontinuous, or their slope will be discontinuous, at points where the type or magnitude of the distributed load changes or where concentrated forces or couple moments are applied. Because of this, shear and moment functions must be determined for each region of the beam located *between* any two discontinuities of loading. For example, coordinates x_1, x_2, and x_3 will have to be used to describe the variation of V and M throughout the length of the beam in Fig. 4–5a. These coordinates will be valid only within regions from A to B for x_1, from B to C for x_2, and from C to D for x_3. Although each of these coordinates has the same origin, as noted here, this does not have to be the case. Indeed, it may be easier to develop the shear and moment functions using coordinates x_1, x_2, x_3 having origins at A, B, and D as shown in Fig. 4–5b. Here x_1 and x_2 are positive to the right and x_3 is positive to the left.

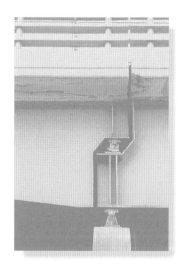

Additional reinforcement, provided by vertical plates called *stiffeners,* is used over the pin and rocker supports of these bridge girders. Here the reactions will cause large internal shear in the girders and the stiffeners will prevent localized buckling of the girder flanges or web. Also, note the tipping of the rocker support caused by the thermal expansion of the bridge deck.

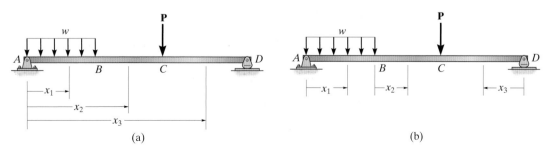

(a)   (b)

Fig. 4–5

4

Procedure for Analysis

The following procedure provides a method for determining the variation of shear and moment in a beam as a function of position x.

Support Reactions

- Determine the support reactions on the beam and resolve all the external forces into components acting perpendicular and parallel to the beam's axis.

Shear and Moment Functions

- Specify separate coordinates x and associated origins, extending into regions of the beam between concentrated forces and/or couple moments, or where there is a discontinuity of distributed loading.
- Section the beam perpendicular to its axis at each distance x, and from the free-body diagram of one of the segments determine the unknowns V and M at the cut section as functions of x. On the free-body diagram, V and M should be shown acting in their *positive directions,* in accordance with the sign convention given in Fig. 4–1.
- V is obtained from $\Sigma F_y = 0$ and M is obtained by summing moments about the point S located at the cut section, $\Sigma M_S = 0$.
- The results can be checked by noting that $dM/dx = V$ and $dV/dx = w$, where w is positive when it acts upward, away from the beam. These relationships are developed in Sec. 4–3.

The joists, beams, and girders used to support this floor can be designed once the internal shear and moment are known throughout their lengths.

EXAMPLE | 4.4

Determine the shear and moment in the beam shown in Fig. 4–6a as a function of x.

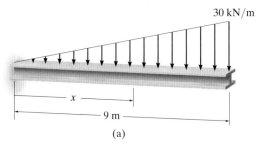

30 kN/m

x

9 m

(a)

Fig. 4–6

SOLUTION

Support Reactions. For the purpose of computing the support reactions, the distributed load is replaced by its resultant force of 135 kN, Fig. 4–6b. It is important to remember, however, that this resultant is not the actual load on the beam.

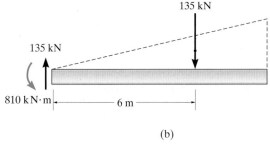

135 kN

135 kN

810 kN·m

6 m

(b)

Shear and Moment Functions. A free-body diagram of the beam segment of length x is shown in Fig. 4–6c. Note that the intensity of the triangular load at the section is found by proportion; that is, $w/x = 30/9$ or $w = 10x/3$. With the load intensity known, the result-ant of the distributed loading is found in the usual manner as shown in the figure. Thus,

$$+\uparrow \Sigma F_y = 0; \qquad 135 - \frac{1}{2}\left(\frac{10x}{3}\right)x - V = 0$$

$$V = 135 - 1.667x^2 \qquad Ans.$$

$$\zeta+\Sigma M_S = 0; \qquad 810 - 135x + \left[\frac{1}{2}\left(\frac{10x}{3}\right)x\right]\frac{x}{3} + M = 0$$

$$M = -810 + 135x - 0.556x^3 \qquad Ans.$$

Note that $dM/dx = V$ and $dV/dx = -10x/3 = w$, which serves as a check of the results.

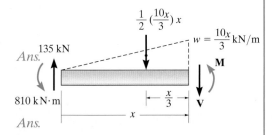

$\frac{1}{2}(\frac{10x}{3})x$

$w = \frac{10x}{3}$ kN/m

135 kN

M

810 kN·m

$\frac{x}{3}$

x

V

(c)

EXAMPLE 4.5

Determine the shear and moment in the beam shown in Fig. 4–7a as a function of x.

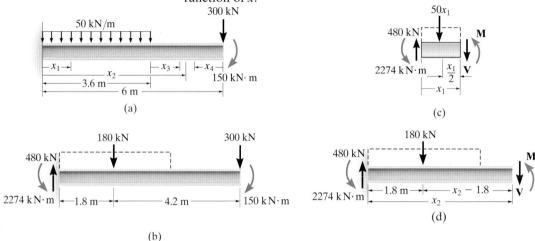

Fig. 4–7

SOLUTION

Support Reactions. The reactions at the fixed support are $V = 480$ kN and $M = 2274$ kN $\cdot$ m, Fig. 4–7b.

Shear and Moment Functions. Since there is a discontinuity of distributed load at $x = 3.6$ m, two regions of x must be considered in order to describe the shear and moment functions for the entire beam. Here x_1 is appropriate for the left 3.6 m and x_2 can be used for the remaining segment.

$0 \le x_1 \le 3.6$ m. Notice that V and M are shown in the positive directions, Fig. 4–7c.

$$+\uparrow \Sigma F_y = 0; \qquad 480 - 50x_1 - V = 0, \quad V = 480 - 50x_1 \qquad \textit{Ans.}$$

$$\zeta + \Sigma M_S = 0; \qquad 2274 - 480x_1 + 50x_1\left(\frac{x_1}{2}\right) + M = 0$$

$$M = -2274 + 480x_1 - 25x_1^2 \qquad \textit{Ans.}$$

3.6 m $\le x_2 \le 6$ m, Fig. 4–7d.

$$+\uparrow \Sigma F_y = 0; \qquad 480 - 180 - V = 0, \quad V = 300 \qquad \textit{Ans.}$$

$$\zeta + \Sigma M_S = 0; \qquad 2274 - 480x_2 + 180(x_2 - 1.8) + M = 0$$

$$M = 300x_2 - 1950 \qquad \textit{Ans.}$$

These results can be partially checked by noting that when $x_2 = 6$ m, then $V = 300$ kN and $M = -150$ kN $\cdot$ m. Also, note that $dM/dx = V$ and $dV/dx = w$.

EXAMPLE 4.6

Determine the shear and moment in the beam shown in Fig. 4–8a as a function of x.

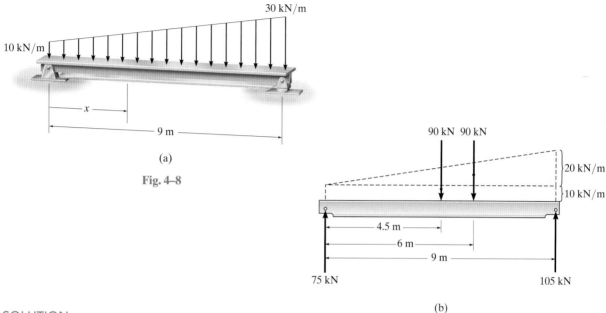

(a)

Fig. 4–8

(b)

SOLUTION

Support Reactions. To determine the support reactions, the distributed load is divided into a triangular and rectangular loading, and these loadings are then replaced by their resultant forces. These reactions have been computed and are shown on the beam's free-body diagram, Fig. 4–8b.

Shear and Moment Functions. A free-body diagram of the cut section is shown in Fig. 4–8c. As above, the trapezoidal loading is replaced by rectangular and triangular distributions. Note that the intensity of the triangular load at the cut is found by proportion. The resultant force of each distributed loading and its location are indicated. Applying the equilibrium equations, we have

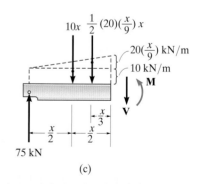

(c)

$$+\uparrow \Sigma F_y = 0; \qquad 75 - 10x - \left[\frac{1}{2}(20)\left(\frac{x}{9}\right)x\right] - V = 0$$

$$V = 75 - 10x - 1.11x^2 \qquad \textit{Ans.}$$

$$\zeta + \Sigma M_S = 0; \qquad -75x + (10x)\left(\frac{x}{2}\right) + \left[\frac{1}{2}(20)\left(\frac{x}{9}\right)x\right]\frac{x}{3} + M = 0$$

$$M = 75x - 5x^2 - 0.370x^3 \qquad \textit{Ans.}$$

FUNDAMENTAL PROBLEMS

F4–1. Determine the internal normal force, shear force, and bending moment acting at point C in the beam.

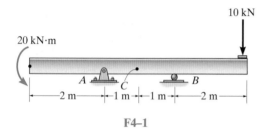

F4–1

F4–2. Determine the internal normal force, shear force, and bending moment acting at point C in the beam.

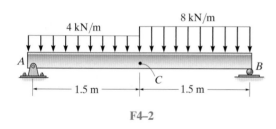

F4–2

F4–3. Determine the internal normal force, shear force, and bending moment acting at point C in the beam.

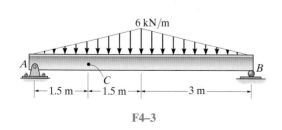

F4–3

F4–4. Determine the internal normal force, shear force, and bending moment acting at point C in the beam.

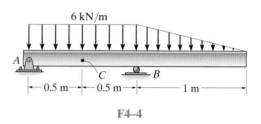

F4–4

F4–5. Determine the internal normal force, shear force, and bending moment acting at point C in the beam.

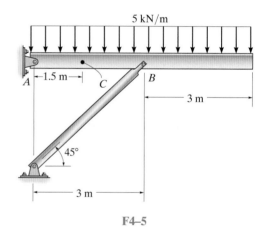

F4–5

F4–6. Determine the internal normal force, shear force, and bending moment acting at point C in the beam.

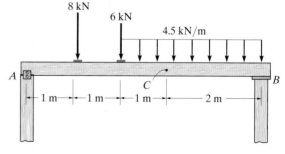

F4–6

F4–7. Determine the internal shear and moment in the beam as a function of x.

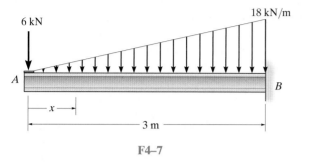

F4–7

F4–8. Determine the internal shear and moment in the beam as a function of x.

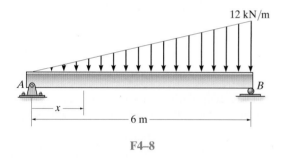

F4–8

F4–9. Determine the internal shear and moment in the beam as a function of x throughout the beam.

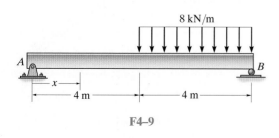

F4–9

F4–10. Determine the internal shear and moment in the beam as a function of x throughout the beam.

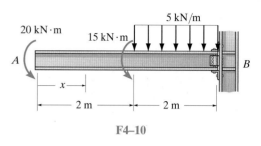

F4–10

F4–11. Determine the internal shear and moment in the beam as a function of x throughout the beam.

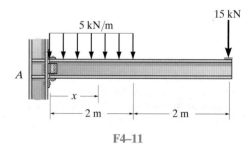

F4–11

F4–12. Determine the internal shear and moment in the beam as a function of x throughout the beam.

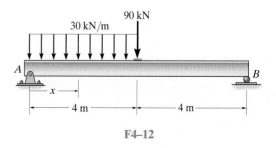

F4–12

PROBLEMS

4–1. Determine the internal normal force, shear force, and bending moment in the beam at points C and D. Assume the support at A is a pin and B is a roller.

4–3. The boom DF of the jib crane and the column DE have a uniform weight of 750 N/m. If the hoist and load weigh 1350 N, determine the internal normal force, shear force, and bending moment in the crane at points A, B, and C.

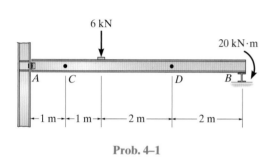

Prob. 4–1

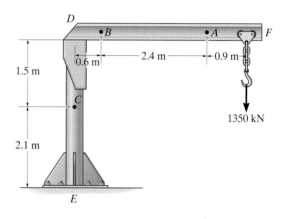

Prob. 4–3

4–2. Determine the internal normal force, shear force, and bending moment in the beam at points C and D. Assume the support at B is a roller. Point D is located just to the right of the 50-kN load.

*4–4. Determine the internal normal force, shear force, and bending moment at point D. Take $w = 150$ N/m.

4–5. The beam AB will fail if the maximum internal moment at D reaches 800 N $\cdot$ m or the normal force in member BC becomes 1500 N. Determine the largest load w it can support.

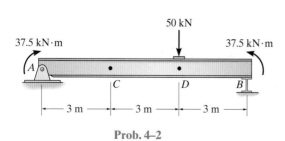

Prob. 4–2

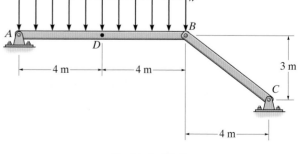

Probs. 4–4/4–5

4–6. Determine the internal normal force, shear force, and bending moment in the beam at points C and D. Assume the support at A is a roller and B is a pin.

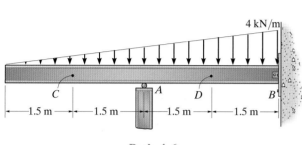

4 kN/m

—1.5 m—— 1.5 m—— 1.5 m—— 1.5 m—

Prob. 4–6

4–9. Determine the internal normal force, shear force, and bending moment in the beam at point C. The support at A is a roller and B is pinned.

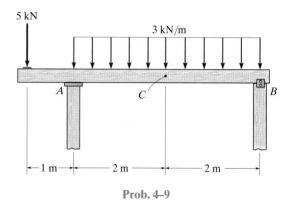

5 kN

3 kN/m

—1 m—— 2 m —— 2 m —

Prob. 4–9

4–7. Determine the internal normal force, shear force, and bending moment at point C. Assume the reactions at the supports A and B are vertical.

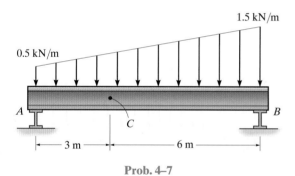

1.5 kN/m

0.5 kN/m

— 3 m —— 6 m —

Prob. 4–7

4–10. Determine the internal normal force, shear force, and bending moment at point C. Assume the reactions at the supports A and B are vertical.

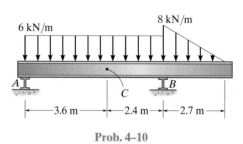

6 kN/m   8 kN/m

—3.6 m—— 2.4 m——2.7 m—

Prob. 4–10

*4–8. Determine the internal normal force, shear force, and bending moment at point D. Assume the reactions at the supports A and B are vertical.

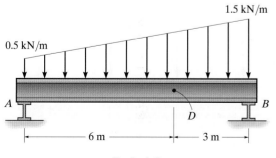

1.5 kN/m

0.5 kN/m

— 6 m ——— 3 m —

Prob. 4–8

4–11. Determine the internal normal force, shear force, and bending moment at points D and E. Assume the reactions at the supports A and B are vertical.

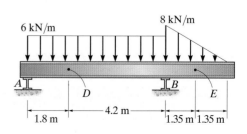

6 kN/m   8 kN/m

—1.8 m—— 4.2 m ——1.35 m 1.35 m—

Prob. 4–11

*4–12. Determine the shear and moment throughout the beam as a function of x.

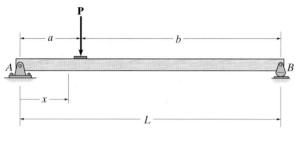

Prob. 4–12

4–15. Determine the shear and moment throughout the beam as a function of x.

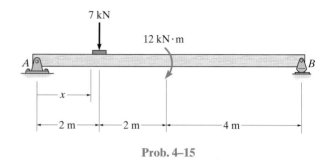

Prob. 4–15

4–13. Determine the shear and moment in the floor girder as a function of x. Assume the support at A is a pin and B is a roller.

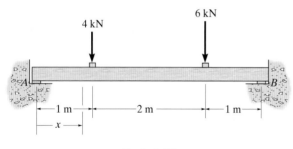

Prob. 4–13

*4–16. Determine the shear and moment throughout the beam as a function of x.

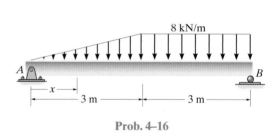

Prob. 4–16

4–14. Determine the shear and moment throughout the beam as a function of x.

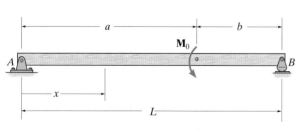

Prob. 4–14

4–17. Determine the shear and moment throughout the beam as a function of x.

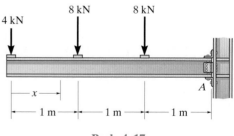

Prob. 4–17

4–18.   Determine the shear and moment throughout the beam as functions of x.

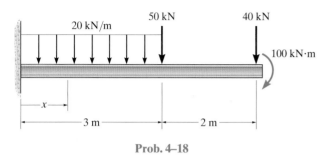

Prob. 4–18

4–19.   Determine the shear and moment throughout the beam as functions of x.

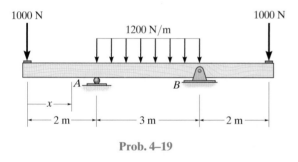

Prob. 4–19

*4–20.   Determine the shear and moment in the beam as functions of x.

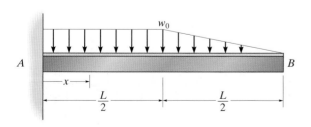

Prob. 4–20

4–21.   Determine the shear and moment in the beam as a function of x.

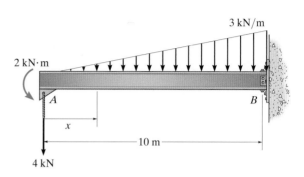

Prob. 4–21

4–22.   Determine the shear and moment throughout the tapered beam as a function of x.

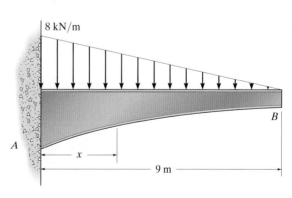

Prob. 4–22

4.3 Shear and Moment Diagrams for a Beam

If the variations of V and M as functions of x obtained in Sec. 4–2 are plotted, the graphs are termed the *shear diagram* and *moment diagram,* respectively. In cases where a beam is subjected to *several* concentrated forces, couples, and distributed loads, plotting V and M versus x can become quite tedious since several functions must be plotted. In this section a simpler method for constructing these diagrams is discussed—a method based on differential relations that exist between the load, shear, and moment.

To derive these relations, consider the beam AD in Fig. 4–9a, which is subjected to an arbitrary distributed loading $w = w(x)$ and a series of concentrated forces and couples. In the following discussion, *the distributed load will be considered positive when the loading acts upward* as shown. We will consider the free-body diagram for a small segment of the beam having a length Δx, Fig. 4–9b. Since this segment has been chosen at a point x along the beam that is *not* subjected to a concentrated force or couple, any results obtained will not apply at points of concentrated loading. The internal shear force and bending moment shown on the free-body diagram are assumed to act in the *positive direction* according to the established sign convention, Fig. 4–1. Note that both the shear force and moment acting on the right face must be increased by a small, finite amount in order to keep the segment in equilibrium. The distributed loading has been replaced by a concentrated force $w(x)\Delta x$ that acts at a fractional distance $\epsilon(\Delta x)$ from the right end, where $0 < \epsilon < 1$. (For example, if $w(x)$ is uniform or constant, then $w(x)\Delta x$ will act at $\frac{1}{2}\Delta x$, so $\epsilon = \frac{1}{2}$.) Applying the equations of equilibrium, we have

$$+\uparrow \Sigma F_y = 0; \quad V + w(x)\,\Delta x - (V + \Delta V) = 0$$
$$\Delta V = w(x)\,\Delta x$$

$$\zeta + \Sigma M_O = 0; \quad -V\Delta x - M - w(x)\,\Delta x\,\epsilon(\Delta x) + (M + \Delta M) = 0$$
$$\Delta M = V\Delta x + w(x)\,\epsilon(\Delta x)^2$$

The many concentrated loadings acting on this reinforced concrete beam create a variation of the internal loading in the beam. For this reason, the shear and moment diagrams must be drawn in order to properly design the beam.

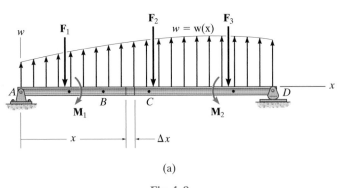

(a)

Fig. 4–9

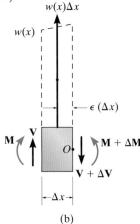

(b)

Dividing by Δx and taking the limit as $\Delta x \rightarrow 0$, these equations become

$$\frac{dV}{dx} = w(x)$$

$$\left.\begin{array}{c}\text{Slope of}\\\text{Shear Diagram}\end{array}\right\} = \left\{\begin{array}{c}\text{Intensity of}\\\text{Distributed Load}\end{array}\right.$$

(4–1)

$$\frac{dM}{dx} = V$$

$$\left.\begin{array}{c}\text{Slope of}\\\text{Moment Diagram}\end{array}\right\} = \{\text{Shear}$$

(4–2)

As noted, Eq. 4–1 states that *the slope of the shear diagram at a point (dV/dx) is equal to the intensity of the distributed load w(x) at the point.* Likewise, Eq. 4–2 states that *the slope of the moment diagram (dM/dx) is equal to the intensity of the shear at the point.*

Equations 4–1 and 4–2 can be "integrated" from one point to another between concentrated forces or couples (such as from B to C in Fig. 4–9a), in which case

$$\Delta V = \int w(x)\, dx$$

$$\left.\begin{array}{c}\text{Change in}\\\text{Shear}\end{array}\right\} = \left\{\begin{array}{c}\text{Area under}\\\text{Distributed Loading}\\\text{Diagram}\end{array}\right.$$

(4–3)

and

$$\Delta M = \int V(x)\, dx$$

$$\left.\begin{array}{c}\text{Change in}\\\text{Moment}\end{array}\right\} = \left\{\begin{array}{c}\text{Area under}\\\text{Shear Diagram}\end{array}\right.$$

(4–4)

As noted, Eq. 4–3 states that *the change in the shear between any two points on a beam equals the area under the distributed loading diagram between the points.* Likewise, Eq. 4–4 states that *the change in the moment between the two points equals the area under the shear diagram between the points.* If the areas under the load and shear diagrams are easy to compute, Eqs. 4–3 and 4–4 provide a method for determining the numerical values of the shear and moment at various points along a beam.

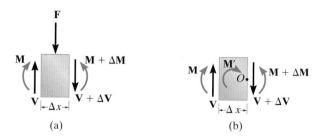

(a)   (b)

Fig. 4–10

From the derivation it should be noted that Eqs. 4–1 and 4–3 cannot be used at points where a concentrated force acts, since these equations do not account for the sudden change in shear at these points. Similarly, because of a discontinuity of moment, Eqs. 4–2 and 4–4 cannot be used at points where a couple moment is applied. In order to account for these two cases, we must consider the free-body diagrams of differential elements of the beam in Fig. 4–9a which are located at concentrated force and couple moments. Examples of these elements are shown in Figs. 4–10a and 4–10b, respectively. From Fig. 4–10a it is seen that force equilibrium requires the change in shear to be

$$+\uparrow \Sigma F_y = 0; \qquad\qquad \Delta V = -F \qquad\qquad (4\text{–}5)$$

Thus, when **F** acts *downward* on the beam, ΔV is negative so that the shear diagram shows a "jump" *downward*. Likewise, if **F** acts *upward*, the jump (ΔV) is *upward*. From Fig. 4–10b, letting $\Delta x \to 0$, moment equilibrium requires the change in moment to be

$$\zeta + \Sigma M_O = 0; \qquad\qquad \Delta M = M' \qquad\qquad (4\text{–}6)$$

In this case, if an external couple moment **M'** is applied *clockwise*, ΔM is positive, so that the moment diagram jumps *upward*, and when **M** acts *counterclockwise*, the jump (ΔM) must be *downward*.

Procedure for Analysis

The following procedure provides a method for constructing the shear and moment diagrams for a beam using Eqs. 4–1 through 4–6.

Support Reactions

- Determine the support reactions and resolve the forces acting on the beam into components which are perpendicular and parallel to the beam's axis.

Shear Diagram

- Establish the V and x axes and plot the values of the shear at the two *ends* of the beam.
- Since $dV/dx = w$, the *slope* of the *shear diagram* at any point is equal to the intensity of the *distributed loading* at the point. (Note that w is positive when it acts upward.)
- If a numerical value of the shear is to be determined at the point, one can find this value either by using the method of sections as discussed in Sec. 4–1 or by using Eq. 4–3, which states that the *change in the shear force* is equal to the *area under the distributed loading diagram.*
- Since $w(x)$ is *integrated* to obtain V, if $w(x)$ is a curve of degree n, then $V(x)$ will be a curve of degree $n + 1$. For example, if $w(x)$ is uniform, $V(x)$ will be linear.

Moment Diagram

- Establish the M and x axes and plot the values of the moment at the ends of the beam.
- Since $dM/dx = V$, the *slope* of the *moment diagram* at any point is equal to the intensity of the *shear* at the point.
- At the point where the shear is zero, $dM/dx = 0$, and therefore this may be a point of maximum or minimum moment.
- If the numerical value of the moment is to be determined at a point, one can find this value either by using the method of sections as discussed in Sec. 4–1 or by using Eq. 4–4, which states that the *change in the moment* is equal to the *area under the shear diagram.*
- Since $V(x)$ is *integrated* to obtain M, if $V(x)$ is a curve of degree n, then $M(x)$ will be a curve of degree $n + 1$. For example, if $V(x)$ is linear, $M(x)$ will be parabolic.

4

EXAMPLE 4.7

The two horizontal members of the power line support frame are subjected to the cable loadings shown in Fig. 4–11a. Draw the shear and moment diagrams for each member.

SOLUTION

Support Reactions. Each pole exerts a force of 6 kN on each member as shown on the free-body diagram.

Shear Diagram. The end points $x = 0$, $V = -4$ kN and $x = 6$ m, $V = 4$ kN are plotted first, Fig. 4–11b. As indicated, the shear between each concentrated force is *constant* since $w = dV/dx = 0$. The shear just to the right of point B (or C and D) can be determined by the method of sections, Fig. 4–11d. The shear diagram can also be established by "following the load" on the free-body diagram. Beginning at A the 4 kN load acts downward so $V_A = -4$ kN. No load acts between A and B so the shear is constant. At B the 6 kN force acts upward, so the shear jumps up 6 kN, from -4 kN to $+2$ kN, etc.

Moment Diagram. The moment at the end points $x = 0$, $M = 0$ and $x = 6$ m, $M = 0$ is plotted first, Fig. 4–11c. The slope of the moment diagram within each 1.5-m-long region is constant because V is constant. Specific values of the moment, such as at C, can be determined by the method of sections, Fig. 4–11d, or by finding the change in moment by the area under the shear diagram. For example, since $M_A = 0$ at A, then at C, $M_C = M_A + \Delta M_{AC} = 0 + (-4)(1.5) + 2(1.5) = -3$ kN·m.

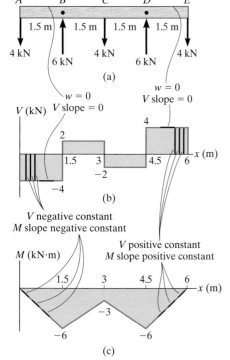

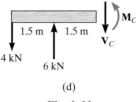

Fig. 4–11

EXAMPLE 4.8

Draw the shear and moment diagrams for the beam in Fig. 4–12a.

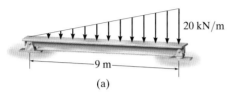

(a)

Fig. 4–12

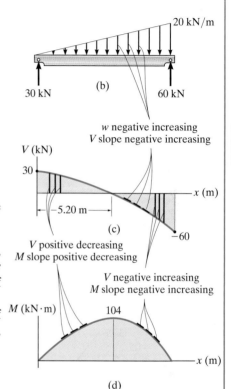

(b)

30 kN   60 kN

w negative increasing
V slope negative increasing

V (kN)

30

−5.20 m

(c)

−60

V positive decreasing
M slope positive decreasing

V negative increasing
M slope negative increasing

M (kN·m)   104

(d)

SOLUTION

Support Reactions.   The reactions have been calculated and are shown on the free-body diagram of the beam, Fig. 4–12b.

Shear Diagram.   The end points $x = 0$, $V = +30$ kN and $x = 9$ m, $V = -60$ kN are first plotted. Note that the shear diagram *starts* with zero slope since $w = 0$ at $x = 0$, and ends with a slope of $w = -20$ kN/m.

The point of zero shear can be found by using the method of sections from a beam segment of length x, Fig. 4–12e. We require $V = 0$, so that

$$+\uparrow \Sigma F_y = 0; \qquad 30 - \frac{1}{2}\left[20\left(\frac{x}{9}\right)\right]x = 0 \qquad x = 5.20 \text{ m}$$

Moment Diagram.   For $0 < x < 5.20$ m the value of shear is positive but decreasing and so the slope of the moment diagram is also positive and decreasing $(dM/dx = V)$. At $x = 5.20$ m, $dM/dx = 0$. Likewise for 5.20 m $< x < 9$ m, the shear and so the slope of the moment diagram are negative increasing as indicated.

The maximum value of moment is at $x = 5.20$ m since $dM/dx = V = 0$ at this point, Fig. 4–12d. From the free-body diagram in Fig. 4–12e we have

$$\zeta + \Sigma M_S = 0; \qquad -30(5.20) + \frac{1}{2}\left[20\left(\frac{5.20}{9}\right)\right](5.20)\left(\frac{5.20}{3}\right) + M = 0$$

$$M = 104 \text{ kN} \cdot \text{m}$$

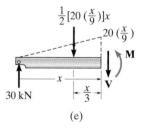

(e)

30 kN

EXAMPLE 4.9

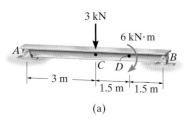

(a)

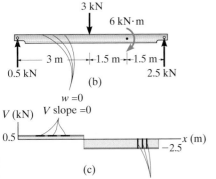

(b)

(c)

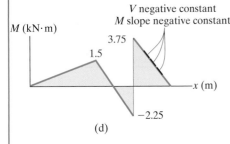

(d)

Draw the shear and moment diagrams for the beam shown in Fig. 4–13a.

SOLUTION

Support Reactions. The reactions are calculated and indicated on the free-body diagram.

Shear Diagram. The values of the shear at the end points A ($V_A = +0.5$ kN) and B ($V_B = -2.5$ kN) are plotted. At C the shear is *discontinuous* since there is a *concentrated force* of 3 kN there. The value of the shear just to the right of C can be found by sectioning the beam at this point. This yields the free-body diagram shown in equilibrium in Fig. 4–13e. This point ($V = -2.5$ kN) is plotted on the shear diagram. Notice that no jump or discontinuity in shear occurs at D, the point where the 6 kN·m couple moment is applied, Fig. 4–13b.

Moment Diagram. The moment at each end of the beam is zero, Fig. 4–13d. The value of the moment at C can be determined by the method of sections, Fig. 4–13e, or by finding the area under the shear diagram between A and C. Since $M_A = 0$,

$$M_C = M_A + \Delta M_{AC} = 0 + (0.5 \text{ kN})(3 \text{ m})$$
$$M_C = 1.5 \text{ kN} \cdot \text{m}$$

Also, since $M_C = 1.5$ kN·m, the moment at D is

$$M_D = M_C + \Delta M_{CD} = 1.5 \text{ kN} \cdot \text{m} + (-2.5 \text{ kN})(1.5 \text{ m})$$
$$M_D = -2.25 \text{ kN} \cdot \text{m}$$

A jump occurs at point D due to the couple moment of 6 kN·m. The method of sections, Fig. 4–13f, gives a value of $+3.75$ kN·m just to the right of D.

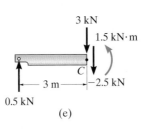

(e)

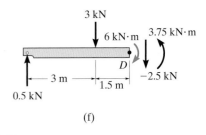

(f)

Fig. 4–13

EXAMPLE 4.10

Draw the shear and moment diagrams for each of the beams shown in Fig. 4–14.

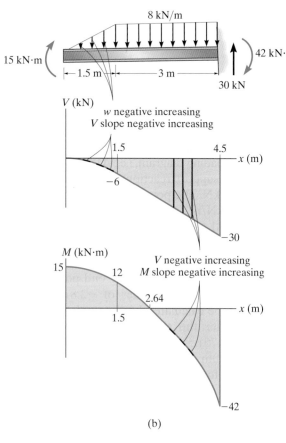

(b)

Fig. 4–14

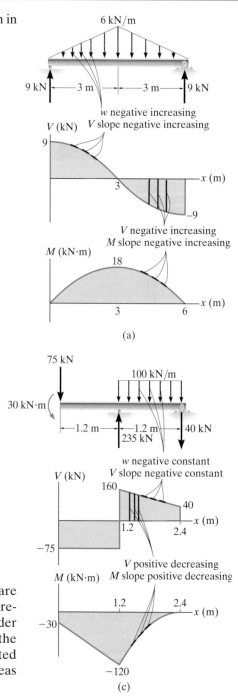

SOLUTION

In each case the support reactions have been calculated and are shown in the top figures. Following the techniques outlined in the previous examples, the shear and moment diagrams are shown under each beam. Carefully notice how they were established, based on the slope and moment, where $dV/dx = w$ and $dM/dx = V$. Calculated values are found using the method of sections or finding the areas under the load or shear diagrams.

EXAMPLE 4.11

The beam shown in the photo is used to support a portion of the overhang for the entranceway of the building. The idealized model for the beam with the load acting on it is shown in Fig. 4–15a. Assume B is a roller and C is pinned. Draw the shear and moment diagrams for the beam.

SOLUTION

Support Reactions. The reactions are calculated in the usual manner. The results are shown in Fig. 4–15b.

Shear Diagram. The shear at the ends of the beam is plotted first, i.e., $V_A = 0$ and $V_C = -2.19$ kN, Fig. 4–15c. To find the shear to the left of B use the method of sections for segment AB, or calculate the area under the distributed loading diagram, i.e., $\Delta V = V_B - 0 = -10(0.75)$, $V_{B^-} = -7.50$ kN. The support reaction causes the shear to jump up $-7.50 + 15.31 = 7.81$ kN. The point of zero shear can be determined from the slope -10 kN/m, or by proportional triangles, $7.81/x = 2.19/(1 - x)$, $x = 0.781$ m. Notice how the V diagram follows the negative slope, defined by the constant negative distributed loading.

Moment Diagram. The moment at the end points is plotted first, $M_A = M_C = 0$, Fig. 4–15d. The values of -2.81 and 0.239 on the moment diagram can be calculated by the method of sections, or by finding the areas under the shear diagram. For example, $\Delta M = M_B - 0 = \frac{1}{2}(-7.50)(0.75) = -2.81$, $M_B = -2.81$ kN·m. Likewise, show that the maximum positive moment is 0.239 kN·m. Notice how the M diagram is formed, by following the slope, defined by the V diagram.

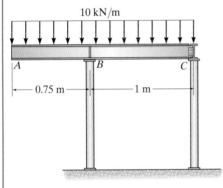

(a)

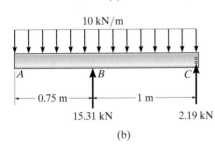

(b)

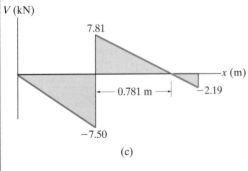

(c)

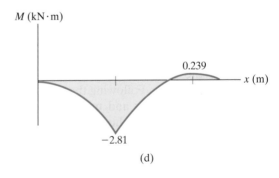

(d)

Fig. 4–15

EXAMPLE 4.12

Draw the shear and moment diagrams for the compound beam shown in Fig. 4–16a. Assume the supports at A and C are rollers and B and E are pin connections.

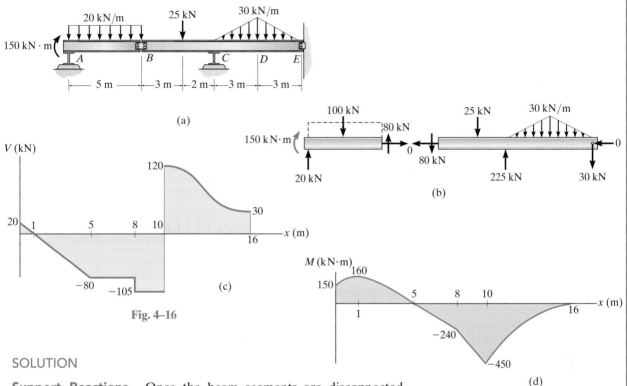

Fig. 4–16

SOLUTION

Support Reactions. Once the beam segments are disconnected from the pin at B, the support reactions can be calculated as shown in Fig. 4–16b.

Shear Diagram.   As usual, we start by plotting the end shear at A and E, Fig. 4–16c. The shape of the V diagram is formed by following its slope, defined by the loading. Try to establish the values of shear using the appropriate areas under the load diagram (w curve) to find the change in shear. The zero value for shear at $x = 1$ m can either be found by proportional triangles, or by using statics, as was done in Fig. 4–12e of Example 4–8.

Moment Diagram.   The end moments $M_A = 150$ kN · m and $M_E = 0$ are plotted first, Fig. 4–16d. Study the diagram and note how the various curves are established using $dM/dx = V$. Verify the numerical values for the peaks using statics or by calculating the appropriate areas under the shear diagram to find the change in moment.

FUNDAMENTAL PROBLEMS

F4–13. Draw the shear and moment diagrams for the beam. Indicate values at the supports and at the points where a change in load occurs.

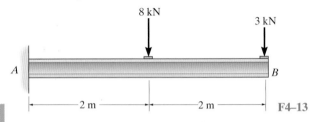

F4–13

F4–14. Draw the shear and moment diagrams for the beam. Indicate values at the supports and at the points where a change in load occurs.

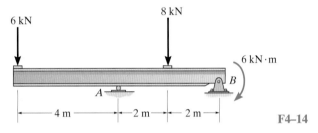

F4–14

F4–15. Draw the shear and moment diagrams for the beam. Indicate values at the supports and at the points where a change in load occurs.

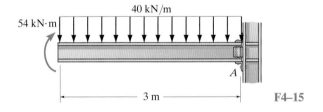

F4–15

F4–16. Draw the shear and moment diagrams for the beam. Indicate values at the supports and at the points where a change in load occurs.

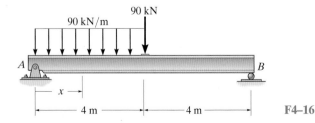

F4–16

F4–17. Draw the shear and moment diagrams for the beam. Indicate values at the supports and at the points where a change in load occurs.

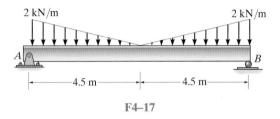

F4–17

F4–18. Draw the shear and moment diagrams for the beam. Indicate values at the supports and at the points where a change in load occurs.

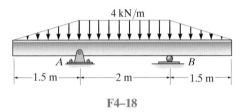

F4–18

F4–19. Draw the shear and moment diagrams for the beam. Indicate values at the supports and at the points where a change in load occurs.

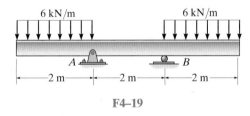

F4–19

F4–20. Draw the shear and moment diagrams for the beam. Indicate values at the supports and at the points where a change in load occurs.

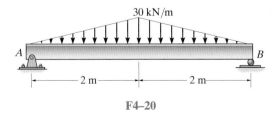

F4–20

PROBLEMS

4–23. Draw the shear and moment diagrams for the beam.

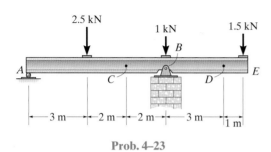

Prob. 4–23

*4–24. Draw the shear and moment diagrams for the beam.

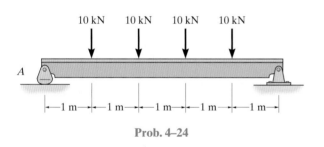

Prob. 4–24

4–25. Draw the shear and moment diagrams for the beam.

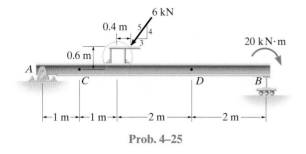

Prob. 4–25

4–26. Draw the shear and moment diagrams for the beam.

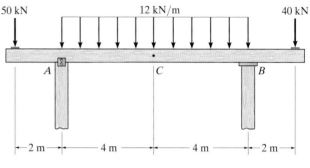

Prob. 4–26

4–27. Draw the shear and moment diagrams for the beam.

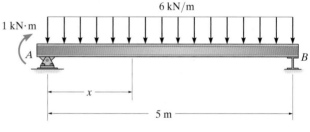

Prob. 4–27

*4–28. Draw the shear and moment diagrams for the beam (a) in terms of the parameters shown; (b) set $M_0 = 500 \text{ N} \cdot \text{m}$, $L = 8 \text{ m}$.

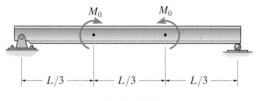

Prob. 4–28

4–29. Draw the shear and moment diagrams for the beam.

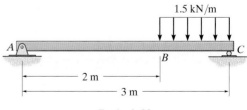

Prob. 4–29

4–30. Draw the shear and bending-moment diagrams for the beam.

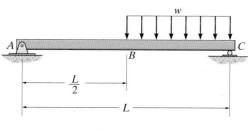

Prob. 4–30

4–31. Draw the shear and moment diagrams for the beam.

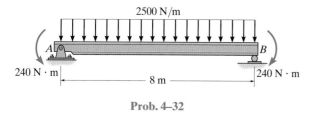

Prob. 4–31

*4–32. Draw the shear and moment diagrams for the beam.

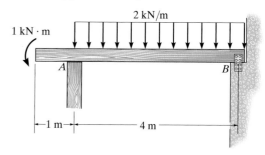

Prob. 4–32

4–33. Draw the shear and moment diagrams for the beam.

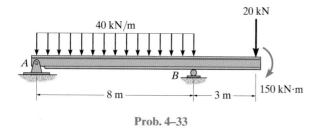

Prob. 4–33

4–34. Draw the shear and moment diagrams for the beam.

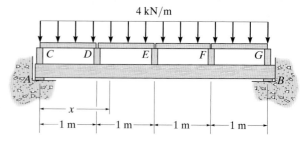

Prob. 4–34

4–35. Draw the shear and moment diagrams for the beam.

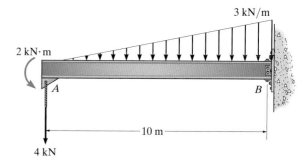

Prob. 4–35

*4–36. Draw the shear and moment diagrams of the beam. Assume the support at B is a pin and A is a roller.

Prob. 4–36

4–37. Draw the shear and moment diagrams for the beam. Assume the support at B is a pin.

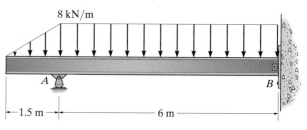

Prob. 4–37

4.4   Shear and Moment Diagrams for a Frame

Recall that a *frame* is composed of several connected members that are either fixed or pin connected at their ends. The design of these structures often requires drawing the shear and moment diagrams for each of the members. To analyze any problem, we can use the procedure for analysis outlined in Sec. 4–3. This requires first determining the reactions at the frame supports. Then, using the method of sections, we find the axial force, shear force, and moment acting at the ends of each member. Provided all loadings are resolved into components acting parallel and perpendicular to the member's axis, the shear and moment diagrams for each member can then be drawn as described previously.

When drawing the moment diagram, one of two sign conventions is used in practice. In particular, if the frame is made of *reinforced concrete,* designers often draw the moment diagram positive on the tension side of the frame. In other words, if the moment produces tension on the outer surface of the frame, the moment diagram is drawn positive on this side. Since concrete has a low tensile strength, it will then be possible to tell at a glance on which side of the frame the reinforcement steel must be placed. In this text, however, we will use the opposite sign convention and *always draw the moment diagram positive on the compression side of the member.* This convention follows that used for beams discussed in Sec. 4–1.

The following examples illustrate this procedure numerically.

The simply supported girder of this concrete building frame was designed by first drawing its shear and moment diagrams.

EXAMPLE 4.13

Draw the moment diagram for the tapered frame shown in Fig. 4–17a. Assume the support at A is a roller and B is a pin.

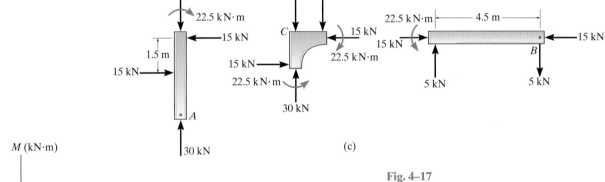

(a)

(b)

Fig. 4–17

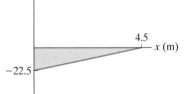

member CB

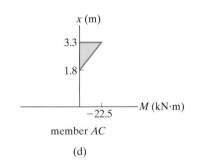

member AC

(c)

(d)

SOLUTION

Support Reactions. The support reactions are shown on the free-body diagram of the entire frame, Fig. 4–17b. Using these results, the frame is then sectioned into two members, and the internal reactions at the joint ends of the members are determined, Fig. 4–17c. Note that the external 25-kN load is shown only on the free-body diagram of the joint at C.

Moment Diagram. In accordance with our positive sign convention, and using the techniques discussed in Sec. 4–3, the moment diagrams for the frame members are shown in Fig. 4–17d.

EXAMPLE | 4.14

Draw the shear and moment diagrams for the frame shown in Fig. 4–18a. Assume A is a pin, C is a roller, and B is a fixed joint. Neglect the thickness of the members.

SOLUTION

Notice that the distributed load acts over a length of $3 \text{ m} \sqrt{2} = 4.243 \text{ m}$. The reactions on the entire frame are calculated and shown on its free-body diagram, Fig. 4–18b. From this diagram the free-body diagrams of each member are drawn, Fig. 4–18c. The distributed loading on BC has components along BC and perpendicular to its axis of $(2.121 \text{ kN/m}) \cos 45° = (2.121 \text{ kN/m}) \sin 45° = 1.5 \text{ kN/m}$ as shown. Using these results, the shear and moment diagrams are also shown in Fig. 4–18c.

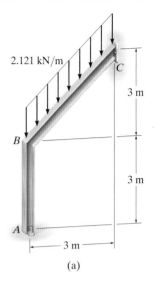

(a)

Fig. 4–18

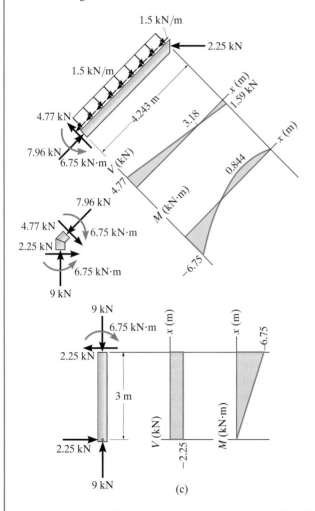

(c)

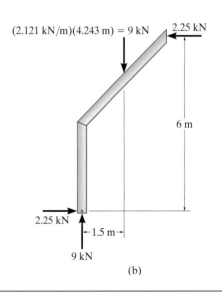

(b)

EXAMPLE | 4.15

Draw the shear and moment diagrams for the frame shown in Fig. 4–19a. Assume A is a pin, C is a roller, and B is a fixed joint.

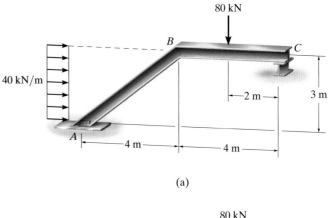

(a)

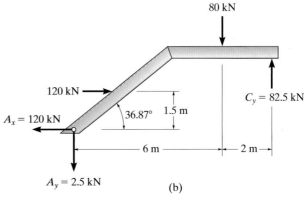

(b)

Fig. 4–19

SOLUTION

Support Reactions. The free-body diagram of the entire frame is shown in Fig. 4–19b. Here the distributed load, which represents wind loading, has been replaced by its resultant, and the reactions have been computed. The frame is then sectioned at joint B and the internal loadings at B are determined, Fig. 4–19c. As a check, equilibrium is satisfied at joint B, which is also shown in the figure.

Shear and Moment Diagrams. The components of the distributed load, $(72 \text{ kN})/(5 \text{ m}) = 14.4 \text{ kN/m}$ and $(96 \text{ kN})/(5 \text{ m}) = 19.2 \text{ kN/m}$, are shown on member AB, Fig. 4–19d. The associated shear and moment diagrams are drawn for each member as shown in Figs. 4–19d and 4–19e.

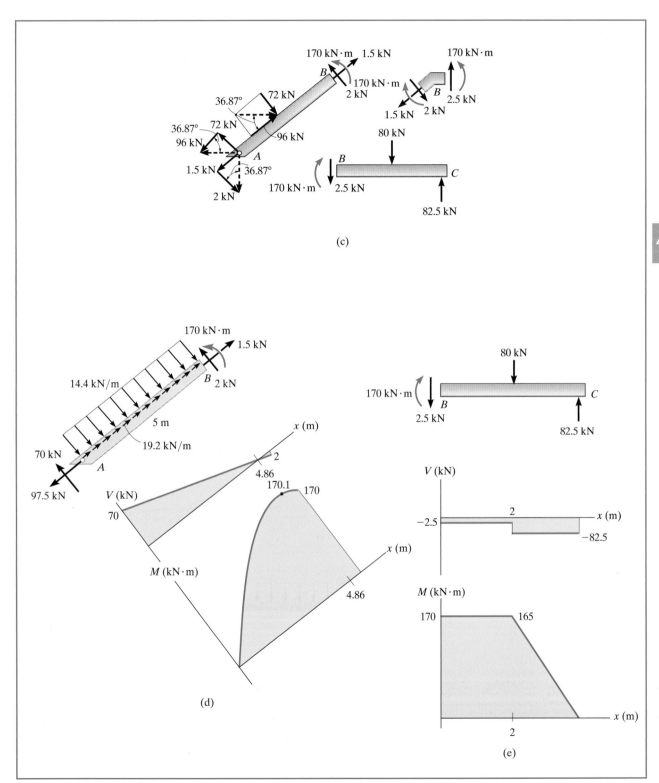

(c)

(d)

(e)

4.5   Moment Diagrams Constructed by the Method of Superposition

Since beams are used primarily to resist bending stress, it is important that the moment diagram accompany the solution for their design. In Sec. 4–3 the moment diagram was constructed by *first* drawing the shear diagram. If we use the principle of superposition, however, each of the loads on the beam can be treated separately and the moment diagram can then be constructed in a series of parts rather than a single and sometimes complicated shape. It will be shown later in the text that this can be particularly advantageous when applying geometric deflection methods to determine both the deflection of a beam and the reactions on statically indeterminate beams.

Most loadings on beams in structural analysis will be a combination of the loadings shown in Fig. 4–20. Construction of the associated moment diagrams has been discussed in Example 4–8. To understand how to use

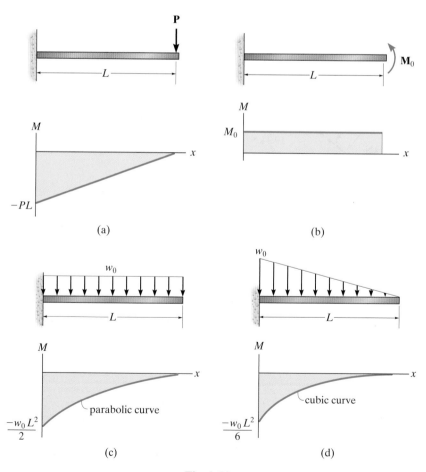

Fig. 4–20

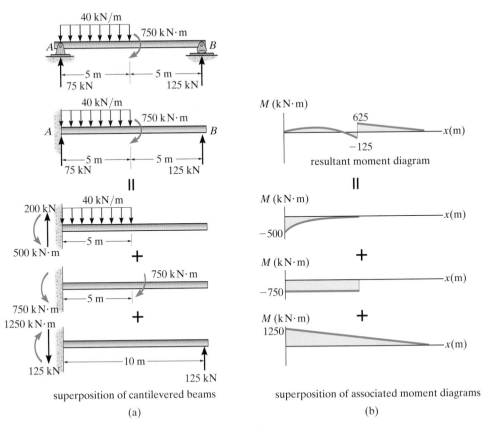

superposition of cantilevered beams   superposition of associated moment diagrams

(a)   (b)

Fig. 4–21

the method of superposition to construct the moment diagram consider the simply supported beam at the top of Fig. 4–21a. Here the reactions have been calculated and so the force system on the beam produces a zero force and moment resultant. The moment diagram for this case is shown at the top of Fig. 4–21b. Note that this *same* moment diagram is produced for the *cantilevered beam* when it is subjected to the same statically equivalent system of loads as the simply supported beam. Rather than considering *all the loads* on this beam *simultaneously* when drawing the moment diagram, we can instead *superimpose* the results of the loads acting separately on the three cantilevered beams shown in Fig. 4–21a. Thus, if the moment diagram for each cantilevered beam is drawn, Fig. 4–21b, the superposition of these diagrams yields the resultant moment diagram for the simply supported beam. For example, from each of the separate moment diagrams, the moment at end A is $M_A = -500 - 750 + 1250 = 0$, as verified by the top moment diagram in Fig. 4–21b. In some cases it is often *easier* to construct and use a separate series of statically equivalent moment diagrams for a beam, *rather than* construct the beam's more complicated "resultant" moment diagram.

In a similar manner, we can also simplify construction of the "resultant" moment diagram for a beam by using a superposition of "simply supported" beams. For example, the loading on the beam shown at the top of Fig. 4–22a is equivalent to the beam loadings shown below it. Consequently, the separate moment diagrams for each of these three beams can be used *rather than* drawing the resultant moment diagram shown in Fig. 4–22b.

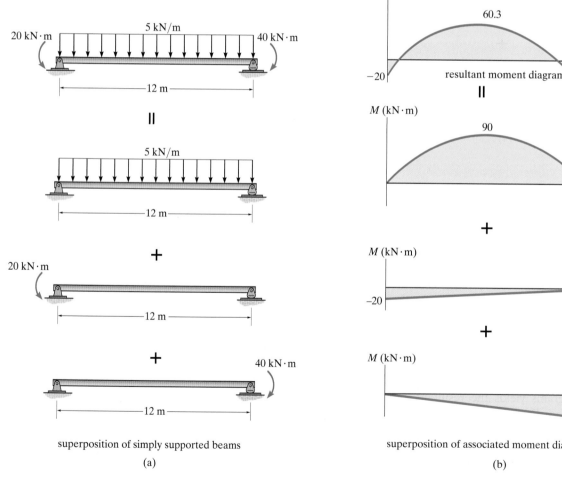

superposition of simply supported beams

(a)

superposition of associated moment diagrams

(b)

Fig. 4–22

EXAMPLE 4.16

Draw the moment diagrams for the beam shown at the top of Fig. 4–23a using the method of superposition. Consider the beam to be cantilevered from the support at B.

SOLUTION

If the beam were supported as a cantilever from B, it would be subjected to the statically equivalent loadings shown in Fig. 4–23a. The superimposed three cantilevered beams are shown below it together with their associated moment diagrams in Fig. 4–23b. (As an aid to their construction, refer to Fig. 4–20.) Although *not needed here,* the sum of these diagrams will yield the resultant moment diagram for the beam. For practice, try drawing this diagram and check the results.

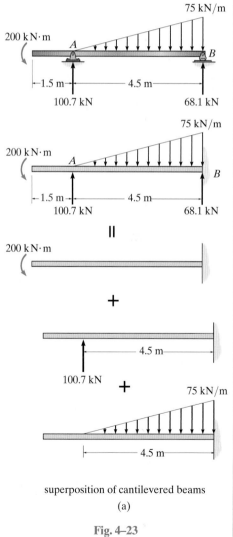

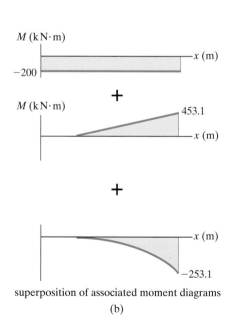

superposition of associated moment diagrams

(b)

superposition of cantilevered beams

(a)

Fig. 4–23

EXAMPLE 4.17

Draw the moment diagrams for the beam shown at the top of Fig. 4–24a using the method of superposition. Consider the beam to be cantilevered from the pin at A.

SOLUTION
The superimposed cantilevered beams are shown in Fig. 4–24a together with their associated moment diagrams, Fig. 4–24b. Notice that the reaction at the pin (100.7 kN) is not considered since it produces no moment diagram. As an exercise verify that the resultant moment diagram is given at the top of Fig. 4–24b.

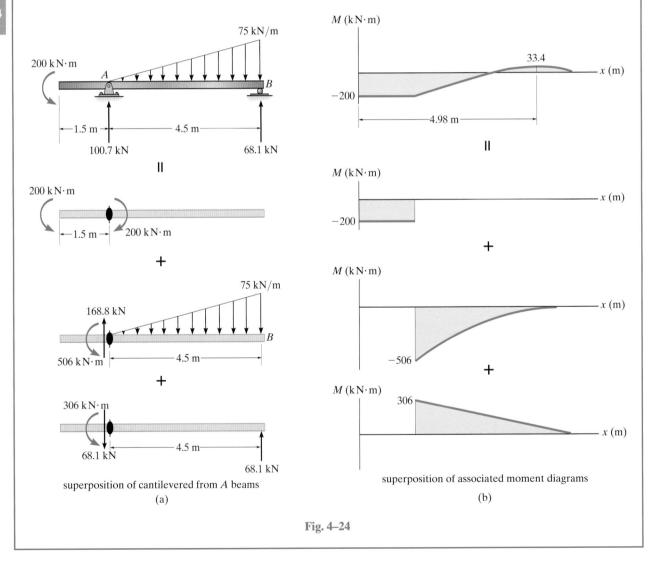

superposition of cantilevered from A beams
(a)

superposition of associated moment diagrams
(b)

Fig. 4–24

PROBLEMS

4–38. Draw the shear and moment diagrams for each of the three members of the frame. Assume the frame is pin connected at A, C, and D and there is a fixed joint at B.

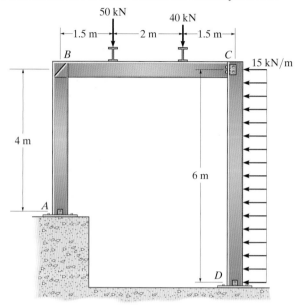

Prob. 4–38

4–39. Draw the shear and moment diagrams for each member of the frame. Assume the support at A is a pin and D is a roller.

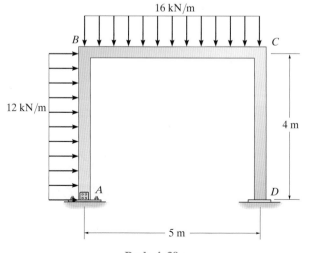

Prob. 4–39

*4–40. Draw the shear and moment diagrams for each member of the frame. Assume A is a rocker, and D is pinned.

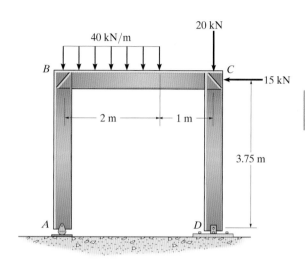

Prob. 4–40

4–41. Draw the shear and moment diagrams for each member of the frame. Assume the frame is pin connected at B, C, and D and A is fixed.

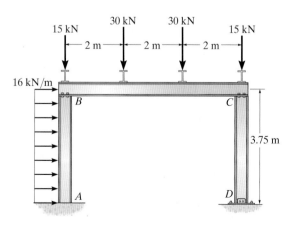

Prob. 4–41

4–42. Draw the shear and moment diagrams for each member of the frame. Assume A is fixed, the joint at B is a pin, and support C is a roller.

*4–44. Draw the shear and moment diagrams for each member of the frame. Assume the frame is roller supported at A and pin supported at C.

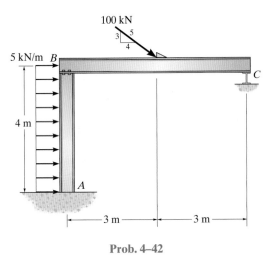

Prob. 4–42

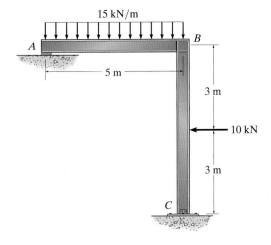

Prob. 4–44

4–43. Draw the shear and moment diagrams for each member of the frame. Assume the frame is pin connected at A, and C is a roller.

4–45. Draw the shear and moment diagrams for each member of the frame. The members are pin connected at A, B, and C.

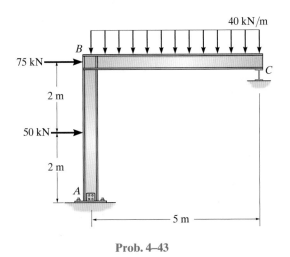

Prob. 4–43

Prob. 4–45

4–46. Draw the shear and moment diagrams for each member of the frame.

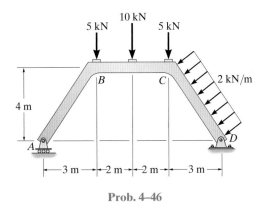

Prob. 4–46

*4–48. Draw the shear and moment diagrams for each member of the frame. The joints at A, B, and C are pin connected.

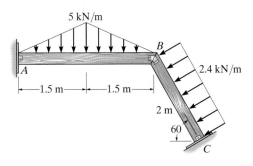

Prob. 4–48

4–47. Draw the shear and moment diagrams for each member of the frame. Assume the joint at A is a pin and support C is a roller. The joint at B is fixed. The wind load is transferred to the members at the girts and purlins from the simply supported wall and roof segments.

4–49. Draw the shear and moment diagrams for each of the three members of the frame. Assume the frame is pin connected at B, C, and D and A is fixed.

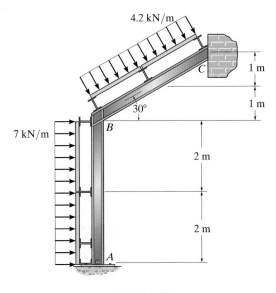

Prob. 4–47

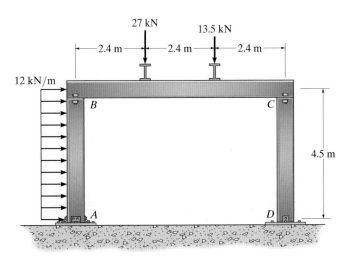

Prob. 4–49

4–50. Draw the moment diagrams for the beam using the method of superposition. The beam is cantilevered from *A*.

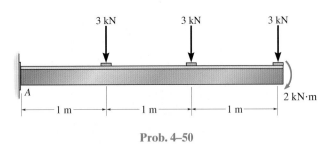

Prob. 4–50

4–51. Draw the moment diagrams for the beam using the method of superposition.

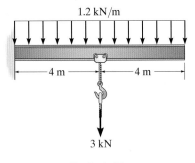

Prob. 4–51

*4–52. Draw the moment diagrams for the beam using the method of superposition. Consider the beam to be cantilevered from end *A*.

4–53. Draw the moment diagrams for the beam using the method of superposition. Consider the beam to be simply supported at *A* and *B* as shown.

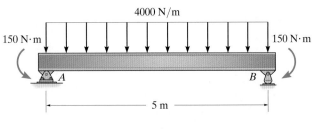

Probs. 4–52/4–53

4–54. Draw the moment diagrams for the beam using the method of superposition. Consider the beam to be cantilevered from the pin support at *A*.

4–55. Draw the moment diagrams for the beam using the method of superposition. Consider the beam to be cantilevered from the rocker at *B*.

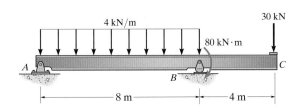

Probs. 4–54/4–55

*4–56. Draw the moment diagrams for the beam using the method of superposition. Consider the beam to be cantilevered from end *C*.

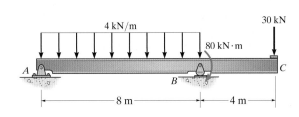

Prob. 4–56

4–57. Draw the moment diagrams for the beam using the method of superposition. Consider the beam to be simply supported at *A* and *B* as shown.

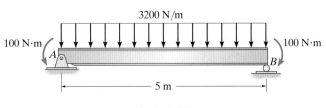

Prob. 4–57

PROJECT PROBLEMS

4–1P. The balcony located on the third floor of a motel is shown in the photo. It is constructed using a 100-mm-thick concrete (plain stone) slab which rests on the four simply supported floor beams, two cantilevered side girders AB and HG, and the front and rear girders. The idealized framing plan with average dimensions is shown in the adjacent figure. According to local code, the balcony live load is 2 kN/m². Draw the shear and moment diagrams for the front girder BG and a side girder AB. Assume the front girder is a channel that has a weight of 0.36 kN/m and the side girders are wide flange sections that have a weight of 0.66 kN/m. Neglect the weight of the floor beams and front railing. For this solution treat each of the five slabs as two-way slabs.

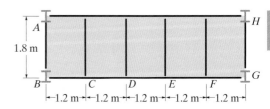

Prob. 4–1P

4–2P. The canopy shown in the photo provides shelter for the entrance of a building. Consider all members to be simply supported. The bar joists at C, D, E, F each have a weight of 0.6 kN and are 6 m long. The roof is 100 mm thick and is to be plain lightweight concrete having a density of 16 kN/m³. Live load caused by drifting snow is assumed to be trapezoidal, with 3 kN/m² at the right (against the wall) and 1 kN/m² at the left (overhang). Assume the concrete slab is simply supported between the joists. Draw the shear and moment diagrams for the side girder AB. Neglect its weight.

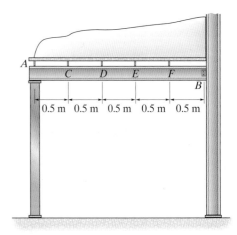

Prob. 4–2P

4–3P. The idealized framing plan for a floor system located in the lobby of an office building is shown in the figure. The floor is made using 100-mm-thick reinforced stone concrete. If the walls of the elevator shaft are made from 100-mm-thick lightweight solid concrete masonry, having a height of 3 m, determine the maximum moment in beam AB. Neglect the weight of the members.

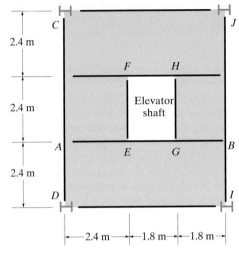

Prob. 4–3P

CHAPTER REVIEW

Structural members subjected to planar loads support an internal normal force **N**, shear force **V**, and bending moment **M**. To find these values at a specific point in a member, the method of sections must be used. This requires drawing a free-body diagram of a segment of the member, and then applying the three equations of equilibrium. Always show the three internal loadings on the section in their positive directions.	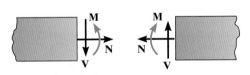 positive sign convention (Ref: Section 4.1)
The internal shear and moment can be expressed as a function of x along the member by establishing the origin at a fixed point (normally at the left end of the member, and then using the method of sections, where the section is made a distance x from the origin). For members subjected to several loads, different x coordinates must extend between the loads	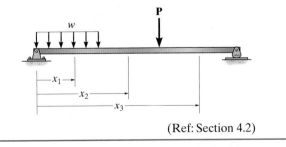 (Ref: Section 4.2)

Shear and moment diagrams for structural members can be drawn by plotting the shear and moment functions. They also can be plotted using the two graphical relationships.

$$\frac{dV}{dx} = w(x)$$

$$\left.\begin{array}{c}\text{Slope of}\\\text{Shear Diagram}\end{array}\right\} = \left\{\begin{array}{c}\text{Intensity of}\\\text{Distributed Load}\end{array}\right.$$

$$\frac{dM}{dx} = V$$

$$\left.\begin{array}{c}\text{Slope of}\\\text{Moment Diagram}\end{array}\right\} = \{\text{Shear}$$

Note that a point of zero shear locates the point of maximum moment since $V = dM/dx = 0$.

$$\Delta V = \int w(x)\,dx$$

$$\left.\begin{array}{c}\text{Change in}\\\text{Shear}\end{array}\right\} = \left\{\begin{array}{c}\text{Area under}\\\text{Distributed Loading}\\\text{Diagram}\end{array}\right.$$

$$\Delta M = \int V(x)\,dx$$

$$\left.\begin{array}{c}\text{Change in}\\\text{Moment}\end{array}\right\} = \left\{\begin{array}{c}\text{Area under}\\\text{Shear Diagram}\end{array}\right.$$

(Ref: Section 4.3)

A force acting downward on the beam will cause the shear diagram to jump downwards, and a counterclockwise couple moment will cause the moment diagram to jump downwards.

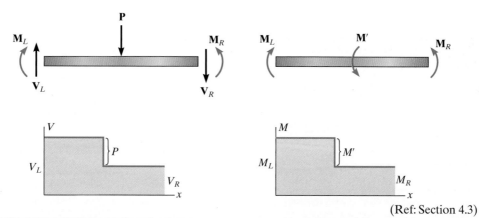

(Ref: Section 4.3)

Using the method of superposition, the moment diagrams for a member can be represented by a series of simpler shapes. The shapes represent the moment diagram for each of the separate loadings. The resultant moment diagram is then the algebraic addition of the separate diagrams.

(Ref: Section 4.5)

This parabolic arch bridge supports the deck above it.

Cables and Arches

5

Cables and arches often form the main load-carrying element in many types of structures, and in this chapter we will discuss some of the important aspects related to their structural analysis. The chapter begins with a general discussion of cables, followed by an analysis of cables subjected to a concentrated load and to a uniform distributed load. Since most arches are statically indeterminate, only the special case of a three-hinged arch will be considered. The analysis of this structure will provide some insight regarding the fundamental behavior of all arched structures.

5.1   Cables

Cables are often used in engineering structures for support and to transmit loads from one member to another. When used to support suspension roofs, bridges, and trolley wheels, cables form the main load-carrying element in the structure. In the force analysis of such systems, the weight of the cable itself may be neglected; however, when cables are used as guys for radio antennas, electrical transmission lines, and derricks, the cable weight may become important and must be included in the structural analysis. Two cases will be considered in the sections that follow: a cable subjected to concentrated loads and a cable subjected to a distributed load. Provided these loadings are coplanar with the cable, the requirements for equilibrium are formulated in an identical manner.

When deriving the necessary relations between the force in the cable and its slope, we will make the assumption that the cable is *perfectly flexible* and *inextensible*. Due to its flexibility, the cable offers no resistance to shear or bending and, therefore, the force acting in the cable is always tangent to the cable at points along its length. Being inextensible, the cable has a constant length both before and after the load is applied. As a result, once the load is applied, the geometry of the cable remains fixed, and the cable or a segment of it can be treated as a rigid body.

Fig. 5–1

The deck of a cable-stayed bridge is supported by a series of cables attached at various points along the deck and pylons.

5.2 Cable Subjected to Concentrated Loads

When a cable of negligible weight supports several concentrated loads, the cable takes the form of several straight-line segments, each of which is subjected to a constant tensile force. Consider, for example, the cable shown in Fig. 5–1. Here θ specifies the angle of the cable's *cord AB*, and L is the cable's span. If the distances L_1, L_2, and L_3 and the loads $\mathbf{P}_1$ and $\mathbf{P}_2$ are known, then the problem is to determine the *nine unknowns consisting of the tension in each of the three* segments, the *four* components of reaction at A and B, and the sags y_C and y_D at the *two* points C and D. For the solution we can write *two* equations of force equilibrium at each of points A, B, C, and D. This results in a total of *eight equations*. To complete the solution, it will be necessary to know something about the geometry of the cable in order to obtain the necessary ninth equation. For example, if the cable's total *length l* is specified, then the Pythagorean theorem can be used to relate *l* to each of the three segmental lengths, written in terms of θ, y_C, y_D, L_1, L_2, and L_3. Unfortunately, this type of problem cannot be solved easily by hand. Another possibility, however, is to specify one of the sags, either y_C or y_D, instead of the cable length. By doing this, the equilibrium equations are then sufficient for obtaining the unknown forces and the remaining sag. Once the sag at each point of loading is obtained, *l* can then be determined by trigonometry.

When performing an equilibrium analysis for a problem of this type, the forces in the cable can also be obtained by writing the equations of equilibrium for the entire cable or any portion thereof. The following example numerically illustrates these concepts.

EXAMPLE 5.1

Determine the tension in each segment of the cable shown in Fig. 5–2a. Also, what is the dimension h?

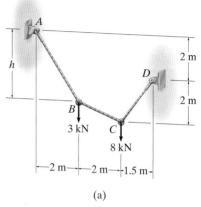

(a)

SOLUTION
By inspection, there are four unknown external reactions (A_x, A_y, D_x, and D_y) and three unknown cable tensions, one in each cable segment. These seven unknowns along with the sag h can be determined from the eight available equilibrium equations ($\Sigma F_x = 0$, $\Sigma F_y = 0$) applied to points A through D.

A more direct approach to the solution is to recognize that the slope of cable CD is specified, and so a free-body diagram of the entire cable is shown in Fig. 5–2b. We can obtain the tension in segment CD as follows:

$\zeta + \Sigma M_A = 0$;

$$T_{CD}(3/5)(2 \text{ m}) + T_{CD}(4/5)(5.5 \text{ m}) - 3 \text{ kN}(2 \text{ m}) - 8 \text{ kN}(4 \text{ m}) = 0$$

$$T_{CD} = 6.79 \text{ kN} \qquad\qquad Ans.$$

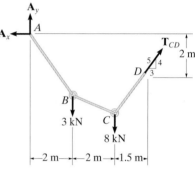

(b)

Now we can analyze the equilibrium of points C and B in sequence.
Point C (Fig. 5–2c);

$\xrightarrow{+} \Sigma F_x = 0$; $6.79 \text{ kN}(3/5) - T_{BC} \cos \theta_{BC} = 0$

$+\uparrow \Sigma F_y = 0$; $6.79 \text{ kN}(4/5) - 8 \text{ kN} + T_{BC} \sin \theta_{BC} = 0$

$\qquad\qquad \theta_{BC} = 32.3° \qquad T_{BC} = 4.82 \text{ kN} \qquad Ans.$

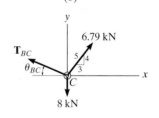

(c)

Point B (Fig. 5–2d);

$\xrightarrow{+} \Sigma F_x = 0$; $-T_{BA} \cos \theta_{BA} + 4.82 \text{ kN} \cos 32.3° = 0$

$+\uparrow \Sigma F_y = 0$; $T_{BA} \sin \theta_{BA} - 4.82 \text{ kN} \sin 32.3° - 3 \text{ kN} = 0$

$\qquad\qquad \theta_{BA} = 53.8° \qquad T_{BA} = 6.90 \text{ kN} \qquad Ans.$

Hence, from Fig. 5–2a,

$$h = (2 \text{ m}) \tan 53.8° = 2.74 \text{ m} \qquad\qquad Ans.$$

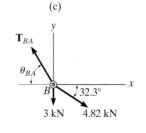

(d)

Fig. 5–2

5.3 Cable Subjected to a Uniform Distributed Load

Cables provide a very effective means of supporting the dead weight of girders or bridge decks having very long spans. A suspension bridge is a typical example, in which the deck is suspended from the cable using a series of close and equally spaced hangers.

In order to analyze this problem, we will first determine the shape of a cable subjected to a uniform *horizontally* distributed vertical load w_0, Fig. 5–3a. Here the x,y axes have their origin located at the lowest point on the cable, such that the slope is zero at this point. The free-body diagram of a small segment of the cable having a length Δs is shown in Fig. 5–3b. Since the tensile force in the cable changes continuously in both magnitude and direction along the cable's length, this change is denoted on the free-body diagram by ΔT. The distributed load is represented by its resultant force $w_0\Delta x$, which acts at $\Delta x/2$ from point O. Applying the equations of equilibrium yields

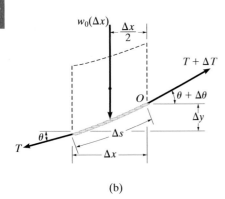

(a)

(b)

Fig. 5–3

$$\xrightarrow{+} \Sigma F_x = 0; \qquad -T\cos\theta + (T + \Delta T)\cos(\theta + \Delta\theta) = 0$$

$$+\uparrow \Sigma F_y = 0; \qquad -T\sin\theta - w_0(\Delta x) + (T + \Delta T)\sin(\theta + \Delta\theta) = 0$$

$$\zeta + \Sigma M_O = 0; \qquad w_0(\Delta x)(\Delta x/2) - T\cos\theta\,\Delta y + T\sin\theta\,\Delta x = 0$$

Dividing each of these equations by Δx and taking the limit as $\Delta x \to 0$, and hence $\Delta y \to 0$, $\Delta\theta \to 0$, and $\Delta T \to 0$, we obtain

$$\frac{d(T\cos\theta)}{dx} = 0 \tag{5–1}$$

$$\frac{d(T\sin\theta)}{dx} = w_0 \tag{5–2}$$

$$\frac{dy}{dx} = \tan\theta \tag{5–3}$$

Integrating Eq. 5–1, where $T = F_H$ at $x = 0$, we have:

$$T\cos\theta = F_H \tag{5–4}$$

which indicates the horizontal component of force at *any point* along the cable remains *constant*.

Integrating Eq. 5–2, realizing that $T\sin\theta = 0$ at $x = 0$, gives

$$T\sin\theta = w_0 x \tag{5–5}$$

Dividing Eq. 5–5 by Eq. 5–4 eliminates T. Then using Eq. 5–3, we can obtain the slope at any point,

$$\tan\theta = \frac{dy}{dx} = \frac{w_0 x}{F_H} \tag{5–6}$$

Performing a second integration with $y = 0$ at $x = 0$ yields

$$y = \frac{w_0}{2F_H} x^2 \qquad (5\text{–}7)$$

This is the equation of a *parabola*. The constant F_H may be obtained by using the boundary condition $y = h$ at $x = L$. Thus,

$$F_H = \frac{w_0 L^2}{2h} \qquad (5\text{–}8)$$

Finally, substituting into Eq. 5–7 yields

$$y = \frac{h}{L^2} x^2 \qquad (5\text{–}9)$$

From Eq. 5–4, the maximum tension in the cable occurs when θ is maximum; i.e., at $x = L$. Hence, from Eqs. 5–4 and 5–5,

$$T_{\max} = \sqrt{F_H^2 + (w_0 L)^2} \qquad (5\text{–}10)$$

Or, using Eq. 5–8, we can express $T_{\max}$ in terms of w_0, i.e.,

$$T_{\max} = w_0 L \sqrt{1 + (L/2h)^2} \qquad (5\text{–}11)$$

Realize that we have neglected the weight of the cable, which is *uniform* along the *length* of the cable, and not along its horizontal projection. Actually, a cable subjected to its own weight and free of any other loads will take the form of a *catenary* curve. However, if the sag-to-span ratio is small, which is the case for most structural applications, this curve closely approximates a parabolic shape, as determined here.

From the results of this analysis, it follows that a cable will *maintain a parabolic shape,* provided the dead load of the deck for a suspension bridge or a suspended girder will be *uniformly distributed* over the horizontal projected length of the cable. Hence, if the girder in Fig. 5–4a is supported by a series of hangers, which are close and uniformly spaced, the load in each hanger must be the *same* so as to ensure that the cable has a parabolic shape.

Using this assumption, we can perform the structural analysis of the girder or any other framework which is freely suspended from the cable. In particular, if the girder is simply supported as well as supported by the cable, the analysis will be statically indeterminate to the first degree, Fig. 5–4b. However, if the girder has an internal pin at some intermediate point along its length, Fig. 5–4c, then this would provide a condition of zero moment, and so a determinate structural analysis of the girder can be performed.

The Verrazano-Narrows Bridge at the entrance to New York Harbor has a main span of 4260 ft (1.30 km).

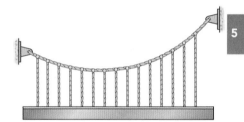

(a)

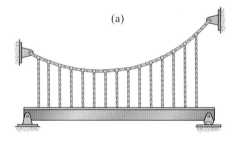

(b)

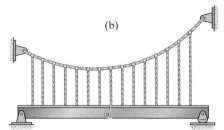

(c)

Fig. 5–4

EXAMPLE 5.2

The cable in Fig. 5–5a supports a girder which weighs 12 kN/m. Determine the tension in the cable at points A, B, and C.

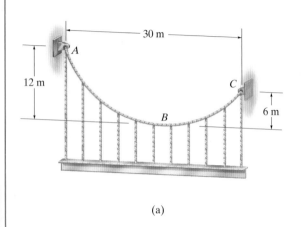

(a)

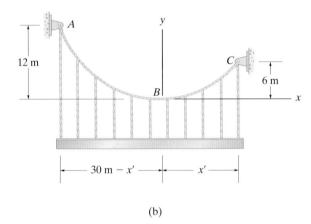

(b)

Fig. 5–5

SOLUTION

The origin of the coordinate axes is established at point B, the lowest point on the cable, where the slope is zero, Fig. 5–5b. From Eq. 5–7, the parabolic equation for the cable is:

$$y = \frac{w_0}{2F_H} x^2 = \frac{12 \text{ kN/m}}{2F_H} x^2 = \frac{6}{F_H} x^2 \qquad (1)$$

Assuming point C is located x' from B, we have

$$6 = \frac{6}{F_H} x'^2$$

$$F_H = 1.0x'^2 \qquad (2)$$

Also, for point A,

$$12 = \frac{6}{F_H} [-(30 - x')]^2$$

$$12 = \frac{6}{1.0x'^2} [-(30 - x')]^2$$

$$x'^2 + 60x' - 900 = 0$$

$$x' = 12.43 \text{ m}$$

Thus, from Eqs. 2 and 1 (or Eq. 5–6) we have

$$F_H = 1.0(12)^2 = 154.4 \text{ kN}$$

$$\frac{dy}{dx} = \frac{12}{154.4} x = 0.07772x \qquad (3)$$

At point A,

$$x = -(30 - 12.43) = -17.57 \text{ m}$$

$$\tan \theta_A = \frac{dy}{dx}\Big|_{x=-17.57} = 0.07772(-17.57) = -1.366$$

$$\theta_A = -53.79°$$

Using Eq. 5–4,

$$T_A = \frac{F_H}{\cos \theta_A} = \frac{154.4}{\cos(-53.79°)} = 261.4 \text{ kN} \qquad Ans.$$

At point B, $x = 0$,

$$\tan \theta_B = \frac{dy}{dx}\Big|_{x=0} = 0, \qquad \theta_B = 0°$$

$$T_B = \frac{F_H}{\cos \theta_B} = \frac{154.4}{\cos 0°} = 154.4 \text{ kN} \qquad Ans.$$

At point C,

$$x = 12.43 \text{ m}$$

$$\tan \theta_C = \frac{dy}{dx}\Big|_{x=12.43} = 0.07772(12.43) = 0.9660$$

$$\theta_C = 44.0°$$

$$T_C = \frac{F_H}{\cos \theta_C} = \frac{154.4}{\cos 44.0°} = 214.6 \text{ kN} \qquad Ans.$$

EXAMPLE 5.3

The suspension bridge in Fig. 5–6a is constructed using the two stiffening trusses that are pin connected at their ends C and supported by a pin at A and a rocker at B. Determine the maximum tension in the cable IH. The cable has a parabolic shape and the bridge is subjected to the single load of 50 kN.

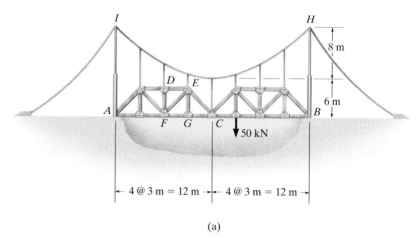

(a)

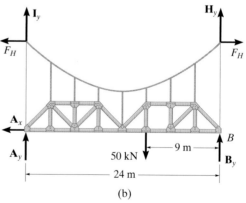

(b)

Fig. 5–6

SOLUTION
The free-body diagram of the cable-truss system is shown in Fig. 5–6b. According to Eq. 5–4 ($T \cos \theta = F_H$), the horizontal component of cable tension at I and H must be constant, F_H. Taking moments about B, we have

$$\zeta + \Sigma M_B = 0; \qquad -I_y(24\text{ m}) - A_y(24\text{ m}) + 50\text{ kN}(9\text{ m}) = 0$$

$$I_y + A_y = 18.75$$

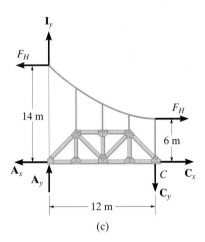

(c)

If only half the suspended structure is considered, Fig. 5–6c, then summing moments about the pin at C, we have

$$\zeta + \Sigma M_C = 0; \quad F_H(14\ \text{m}) - F_H(6\ \text{m}) - I_y(12\ \text{m}) - A_y(12\ \text{m}) = 0$$

$$I_y + A_y = 0.667F_H$$

From these two equations,

$$18.75 = 0.667F_H$$

$$F_H = 28.125\ \text{kN}$$

To obtain the maximum tension in the cable, we will use Eq. 5–11, but first it is necessary to determine the value of an assumed uniform distributed loading w_0 from Eq. 5–8:

$$w_0 = \frac{2F_H h}{L^2} = \frac{2(28.125\ \text{kN})(8\ \text{m})}{(12\ \text{m})^2} = 3.125\ \text{kN/m}$$

Thus, using Eq. 5–11, we have

$$T_{max} = w_0 L \sqrt{1 + (L/2h)^2}$$

$$= 3.125(12\ \text{m})\sqrt{1 + (12\ \text{m}/2(8\ \text{m}))^2}$$

$$= 46.9\ \text{kN} \qquad\qquad\qquad Ans.$$

PROBLEMS

5–1. Determine the tension in each segment of the cable and the cable's total length.

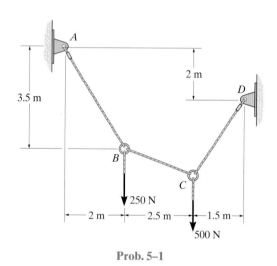

Prob. 5–1

5–2. Cable $ABCD$ supports the loading shown. Determine the maximum tension in the cable and the sag of point B.

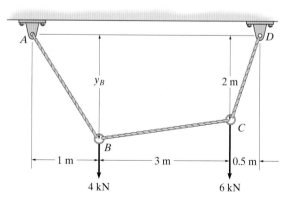

Prob. 5–2

5–3. Determine the tension in each cable segment and the distance y_D.

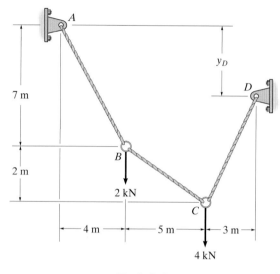

Prob. 5–3

*5–4. The cable supports the loading shown. Determine the distance x_B the force at point B acts from A. Set $P = 200$ N.

5–5. The cable supports the loading shown. Determine the magnitude of the horizontal force **P** so that $x_B = 3$ m.

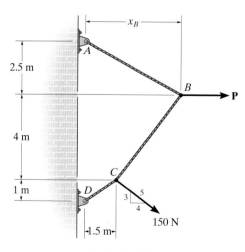

Probs. 5–4/5–5

5–6. Determine the forces P_1 and P_2 needed to hold the cable in the position shown, i.e., so segment CD remains horizontal. Also find the maximum loading in the cable.

*5–8. The cable supports the uniform load of $w_0 = 12$ kN/m. Determine the tension in the cable at each support A and B.

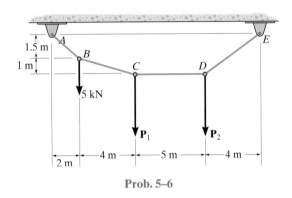

Prob. 5–6

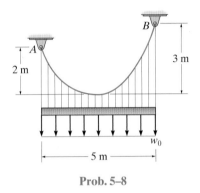

Prob. 5–8

5–7. The cable is subjected to the uniform loading. If the slope of the cable at point O is zero, determine the equation of the curve and the force in the cable at O and B.

5–9. Determine the maximum and minimum tension in the cable.

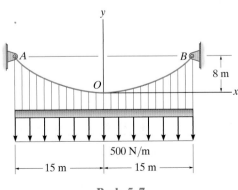

Prob. 5–7

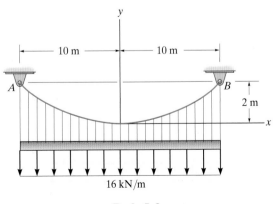

Prob. 5–9

5–10. Determine the maximum uniform loading w, measured in kN/m, that the cable can support if it is capable of sustaining a maximum tension of 15 kN before it will break.

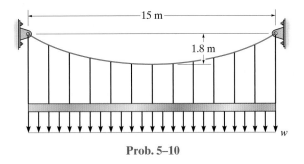

Prob. 5–10

5–11. The cable is subjected to a uniform loading of $w = 3.75$ kN/m. Determine the maximum and minimum tension in the cable.

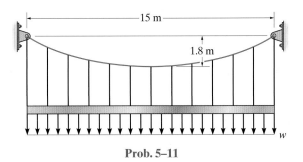

Prob. 5–11

*5–12. The cable shown is subjected to the uniform load w_0. Determine the ratio between the rise h and the span L that will result in using the minimum amount of material for the cable.

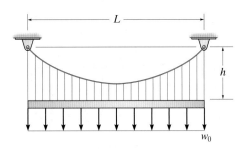

Prob. 5–12

5–13. The trusses are pin connected and suspended from the parabolic cable. Determine the maximum force in the cable when the structure is subjected to the loading shown.

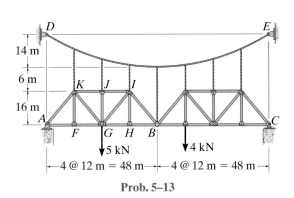

Prob. 5–13

5–14. Determine the maximum and minimum tension in the parabolic cable and the force in each of the hangers. The girder is subjected to the uniform load and is pin connected at B.

5–15. Draw the shear and moment diagrams for the pin-connected girders AB and BC. The cable has a parabolic shape.

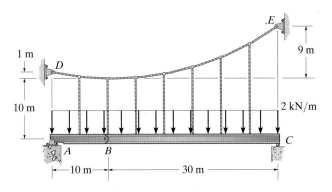

Probs. 5–14/5–15

*5–16. The cable will break when the maximum tension reaches $T_{max} = 5000$ kN. Determine the maximum uniform distributed load w required to develop this maximum tension.

5–17. The cable is subjected to a uniform loading of $w = 60$ kN/m. Determine the maximum and minimum tension in cable.

5–19. The beams AB and BC are supported by the cable that has a parabolic shape. Determine the tension in the cable at points D, F, and E, and the force in each of the equally spaced hangers.

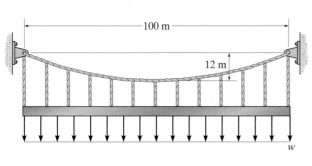

Probs. 5–16/5–17

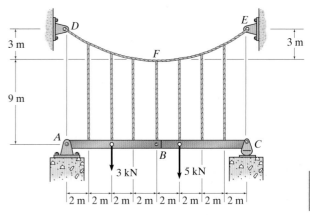

Prob. 5–19

5–18. The cable AB is subjected to a uniform loading of 200 N/m. If the weight of the cable is neglected and the slope angles at points A and B are 30° and 60°, respectively, determine the curve that defines the cable shape and the maximum tension developed in the cable.

*5–20. Draw the shear and moment diagrams for beams AB and BC. The cable has a parabolic shape.

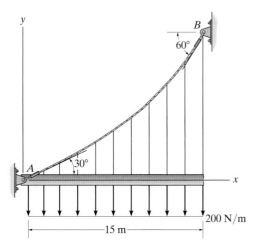

Prob. 5–18

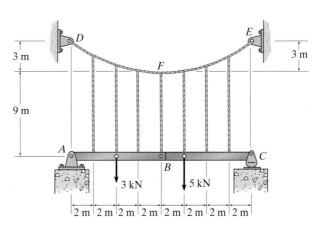

Prob. 5–20

5.4   Arches

Like cables, arches can be used to reduce the bending moments in long-span structures. Essentially, an arch acts as an inverted cable, so it receives its load mainly in compression although, because of its rigidity, it must also resist some bending and shear depending upon how it is loaded and shaped. In particular, if the arch has a *parabolic shape* and it is subjected to a *uniform* horizontally distributed vertical load, then from the analysis of cables it follows that *only compressive forces* will be resisted by the arch. Under these conditions the arch shape is called a *funicular arch* because no bending or shear forces occur within the arch.

A typical arch is shown in Fig. 5–7, which specifies some of the nomenclature used to define its geometry. Depending upon the application, several types of arches can be selected to support a loading. A *fixed arch,* Fig. 5–8a, is often made from reinforced concrete. Although it may require less material to construct than other types of arches, it must have solid foundation abutments since it is indeterminate to the third degree and, consequently, additional stresses can be introduced into the arch due to relative settlement of its supports. A *two-hinged arch,* Fig. 5–8b, is commonly made from metal or timber. It is indeterminate to the first degree, and although it is not as rigid as a fixed arch, it is somewhat insensitive to settlement. We could make this structure statically determinate by replacing one of the hinges with a roller. Doing so, however, would remove the capacity of the structure to resist bending along its span, and as a result it would serve as a curved beam, and *not* as an arch. A *three-hinged arch,* Fig. 5–8c, which is also made from metal or timber, is statically determinate. Unlike statically indeterminate arches, it is not affected by settlement or temperature changes. Finally, if two- and three-hinged arches are to be constructed without the need for larger foundation abutments and if clearance is not a problem, then the supports can be connected with a tie rod, Fig. 5–8d. A *tied arch* allows the structure to behave as a rigid unit, since the tie rod carries the horizontal component of thrust at the supports. It is also unaffected by relative settlement of the supports.

extrados (or back)

crown

springline

intrados (or soffit)

haunch

centerline rise

abutment

Fig. 5–7

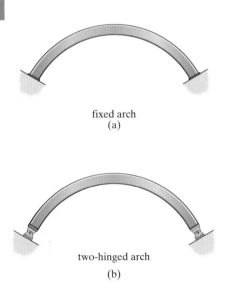

fixed arch
(a)

two-hinged arch
(b)

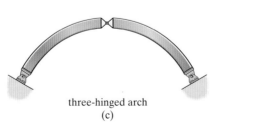

three-hinged arch
(c)

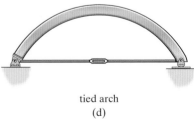

tied arch
(d)

Fig. 5–8

5.5 Three-Hinged Arch

To provide some insight as to how arches transmit loads, we will now consider the analysis of a three-hinged arch such as the one shown in Fig. 5–9a. In this case, the third hinge is located at the crown and the supports are located at different elevations. In order to determine the reactions at the supports, the arch is disassembled and the free-body diagram of each member is shown in Fig. 5–9b. Here there are six unknowns for which six equations of equilibrium are available. One method of solving this problem is to apply the moment equilibrium equations about points A and B. Simultaneous solution will yield the reactions C_x and C_y. The support reactions are then determined from the force equations of equilibrium. Once obtained, the internal normal force, shear, and moment loadings at any point along the arch can be found using the method of sections. Here, of course, the section should be taken perpendicular to the axis of the arch at the point considered. For example, the free-body diagram for segment AD is shown in Fig. 5–9c.

Three-hinged arches can also take the form of two pin-connected trusses, each of which would replace the arch ribs AC and CB in Fig. 5–9a. The analysis of this form follows the same procedure outlined above. The following examples numerically illustrate these concepts.

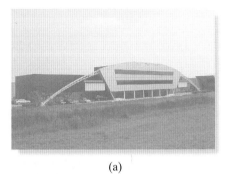

(a)

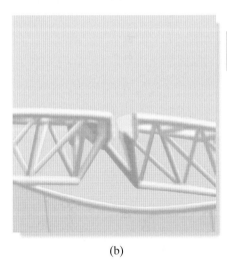

(b)

The three-hinge truss arch is used to support a portion of the roof loading of this building (a). The close-up photo shows the arch is pinned at its top (b).

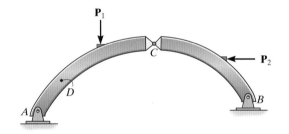

(a)

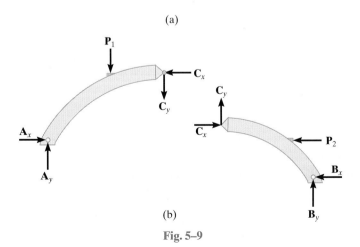

(b)

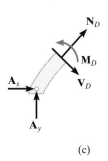

(c)

Fig. 5–9

EXAMPLE 5.4

The three-hinged open-spandrel arch bridge like the one shown in the photo has a parabolic shape. If this arch were to support a uniform load and have the dimensions shown in Fig. 5–10*a*, show that the arch is subjected *only to axial compression* at any intermediate point such as point *D*. Assume the load is uniformly transmitted to the arch ribs.

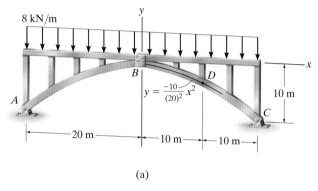

8 kN/m

$y = \dfrac{-10}{(20)^2} x^2$

10 m

20 m   10 m   10 m

(a)

Fig. 5–10

SOLUTION

Here the supports are at the same elevation. The free-body diagrams of the entire arch and part *BC* are shown in Fig. 5–10*b* and Fig. 5–10*c*. Applying the equations of equilibrium, we have:

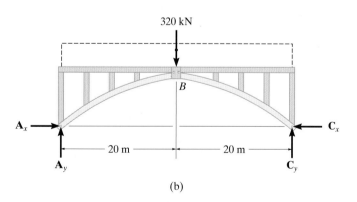

320 kN

A_x C_x

20 m   20 m

A_y C_y

(b)

Entire arch:

$$\zeta + \Sigma M_A = 0; \qquad C_y(40\text{ m}) - 320\text{ kN}(20\text{ m}) = 0$$

$$C_y = 160\text{ kN}$$

Arch segment *BC*:

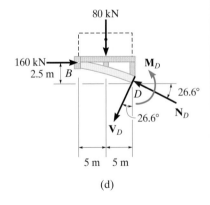

(c)

$$\curvearrowleft + \Sigma M_B = 0; \quad -160 \text{ kN}(10 \text{ m}) + 160 \text{ kN}(20 \text{ m}) - C_x(10 \text{ m}) = 0$$

$$C_x = 160 \text{ kN}$$

$$\xrightarrow{+} \Sigma F_x = 0; \qquad B_x = 160 \text{ kN}$$

$$+\uparrow \Sigma F_y = 0; \qquad B_y - 160 \text{ kN} + 160 \text{ kN} = 0$$

$$B_y = 0$$

A section of the arch taken through point D, $x = 10 \text{ m}$, $y = -10(10)^2/(20)^2 = -2.5 \text{ m}$, is shown in Fig. 5–10*d*. The slope of the segment at D is

$$\tan \theta = \frac{dy}{dx} = \frac{-20}{(20)^2} x \bigg|_{x=10 \text{ m}} = -0.5$$

$$\theta = -26.6°$$

Applying the equations of equilibrium, Fig. 5–10*d* we have

$$\xrightarrow{+} \Sigma F_x = 0; \quad 160 \text{ kN} - N_D \cos 26.6° - V_D \sin 26.6° = 0$$

$$+\uparrow \Sigma F_y = 0; \quad -80 \text{ kN} + N_D \sin 26.6° - V_D \cos 26.6° = 0$$

$$\curvearrowleft + \Sigma M_D = 0; \quad M_D + 80 \text{ kN}(5 \text{ m}) - 160 \text{ kN}(2.5 \text{ m}) = 0$$

$$N_D = 178.9 \text{ kN} \qquad \textit{Ans.}$$

$$V_D = 0 \qquad \textit{Ans.}$$

$$M_D = 0 \qquad \textit{Ans.}$$

(d)

Note: If the arch had a different shape or if the load were nonuniform, then the internal shear and moment would be nonzero. Also, if a simply supported beam were used to support the distributed loading, it would have to resist a maximum bending moment of $M = 1600 \text{ kN} \cdot \text{m}$. By comparison, it is more efficient to structurally resist the load in direct compression (although one must consider the possibility of buckling) than to resist the load by a bending moment.

EXAMPLE 5.5

The three-hinged tied arch is subjected to the loading shown in Fig. 5–11a. Determine the force in members CH and CB. The dashed member GF of the truss is intended to carry no force.

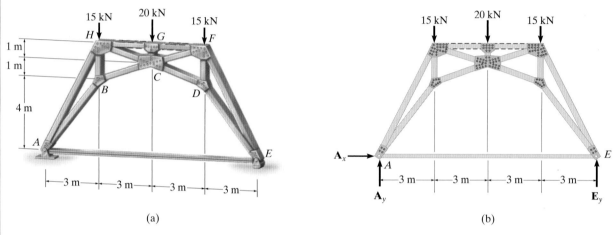

Fig. 5–11

SOLUTION

The support reactions can be obtained from a free-body diagram of the entire arch, Fig. 5–11b:

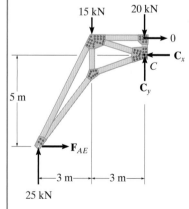

$$\zeta + \Sigma M_A = 0; \quad E_y(12\text{ m}) - 15\text{ kN}(3\text{ m}) - 20\text{ kN}(6\text{ m}) - 15\text{ kN}(9\text{ m}) = 0$$

$$E_y = 25\text{ kN}$$

$$\xrightarrow{+} \Sigma F_x = 0; \qquad A_x = 0$$

$$+\uparrow \Sigma F_y = 0; \quad A_y - 15\text{ kN} - 20\text{ kN} - 15\text{ kN} + 25\text{ kN} = 0$$

$$A_y = 25\text{ kN}$$

The force components acting at joint C can be determined by considering the free-body diagram of the left part of the arch, Fig. 5–11c. First, we determine the force:

$$\zeta + \Sigma M_C = 0; \quad F_{AE}(5\text{ m}) - 25\text{ kN}(6\text{ m}) + 15\text{ kN}(3\text{ m}) = 0$$

$$F_{AE} = 21.0\text{ kN}$$

Then,

$\xrightarrow{+} \Sigma F_x = 0;$ $-C_x + 21.0 \text{ kN} = 0,$ $C_x = 21.0 \text{ kN}$

$+\uparrow \Sigma F_y = 0;$ $25 \text{ kN} - 15 \text{ kN} - 20 \text{ kN} + C_y = 0,$ $C_y = 10 \text{ kN}$

To obtain the forces in CH and CB, we can use the method of joints as follows:

Joint G; Fig. 5–11d,

$+\uparrow \Sigma F_y = 0;$ $F_{GC} - 20 \text{ kN} = 0$

$F_{GC} = 20 \text{ kN (C)}$

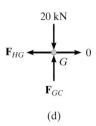

(d)

Joint C; Fig. 5–11e,

$\xrightarrow{+} \Sigma F_x = 0;$ $F_{CB}\left(\frac{3}{\sqrt{10}}\right) - 21.0 \text{ kN} - F_{CH}\left(\frac{3}{\sqrt{10}}\right) = 0$

$+\uparrow \Sigma F_y = 0;$ $F_{CB}\left(\frac{1}{\sqrt{10}}\right) + F_{CH}\left(\frac{1}{\sqrt{10}}\right) - 20 \text{ kN} + 10 \text{ kN} = 0$

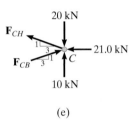

(e)

Thus,

$$F_{CB} = 26.9 \text{ kN (C)} \qquad Ans.$$
$$F_{CH} = 4.74 \text{ kN (T)} \qquad Ans.$$

Note: Tied arches are sometimes used for bridges. Here the deck is supported by suspender bars that transmit their load to the arch. The deck is in tension so that it supports the actual thrust or horizontal force at the ends of the arch.

EXAMPLE 5.6

The three-hinged trussed arch shown in Fig. 5–12a supports the symmetric loading. Determine the required height h_1 of the joints B and D, so that the arch takes a funicular shape. Member HG is intended to carry no force.

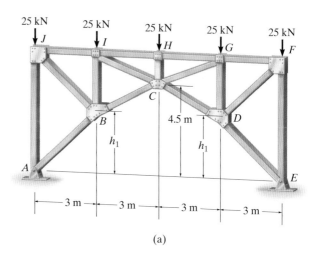

(a)

SOLUTION

For a symmetric loading, the funicular shape for the arch must be *parabolic* as indicated by the dashed line (Fig. 5–12b). Here we must find the equation which fits this shape. With the x, y axes having an origin at C, the equation is of the form $y = -cx^2$. To obtain the constant c, we require

$$-(4.5 \text{ m}) = -c(6 \text{ m})^2$$

$$c = 0.125/\text{m}$$

Therefore,

$$y_D = -(0.125/\text{m})(3 \text{ m})^2 = -1.125 \text{ m}$$

So that from Fig. 5–12a,

$$h_1 = 4.5 \text{ m} - 1.125 \text{ m} = 3.375 \text{ m} \qquad \textit{Ans.}$$

Using this value, if the method of joints is now applied to the truss, the results will show that the top cord and diagonal members will all be zero-force members, and the symmetric loading will be supported *only by the bottom cord* members AB, BC, CD, and DE of the truss.

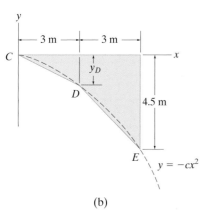

(b)

Fig. 5–12

PROBLEMS

5–21. The tied three-hinged arch is subjected to the loading shown. Determine the components of reaction at A and C and the tension in the cable.

5–23. The three-hinged spandrel arch is subjected to the loading shown. Determine the internal moment in the arch at point D.

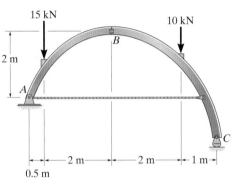

Prob. 5–21

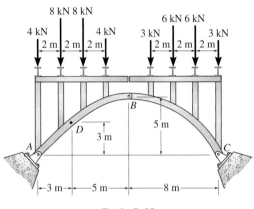

Prob. 5–23

5–22. Determine the resultant forces at the pins A, B, and C of the three-hinged arched roof truss.

*5–24. The tied three-hinged arch is subjected to the loading shown. Determine the components of reaction A and C, and the tension in the rod.

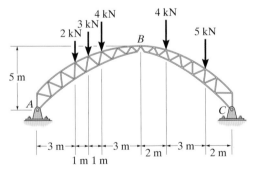

Prob. 5–22

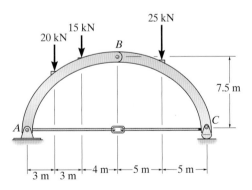

Prob. 5–24

5–25. The bridge is constructed as a *three-hinged trussed arch*. Determine the horizontal and vertical components of reaction at the hinges (pins) at *A*, *B*, and *C*. The dashed member *DE* is intended to carry *no* force.

5–26. Determine the design heights h_1, h_2, and h_3 of the bottom cord of the truss so the three-hinged trussed arch responds as a funicular arch.

*5–28.** The three-hinged spandrel arch is subjected to the uniform load of 20 kN/m. Determine the internal moment in the arch at point *D*.

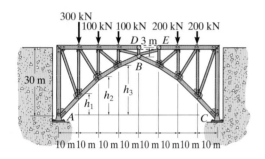

Probs. 5–25/5–26

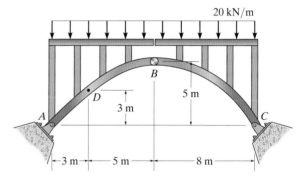

Prob. 5–28

5–27. Determine the horizontal and vertical components of reaction at *A*, *B*, and *C* of the three-hinged arch. Assume *A*, *B*, and *C* are pin connected.

5–29. The arch structure is subjected to the loading shown. Determine the horizontal and vertical components of reaction at *A* and *D*, and the tension in the rod *AD*.

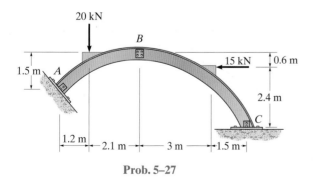

Prob. 5–27

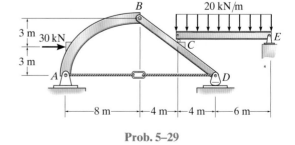

Prob. 5–29

CHAPTER REVIEW

Cables support their loads in tension if we consider them perfectly flexible.

(Ref: Section 5.1)

If the cable is subjected to concentrated loads then the force acting in each cable segment is determined by applying the equations of equilibrium to the free-body diagram of groups of segments of the cable or to the joints where the forces are applied.

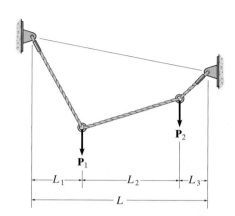

(Ref: Section 5.2)

If the cable supports a uniform load over a projected horizontal distance, then the shape of the cable takes the form of a parabola.

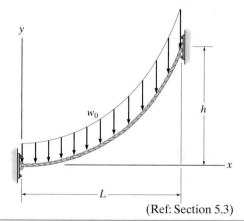

(Ref: Section 5.3)

Arches are designed primarily to carry a compressive force. A parabolic shape is required to support a uniform loading distributed over its horizontal projection.

(Ref: Section 5.4)

Three-hinged arches are statically determinate and can be analyzed by separating the two members and applying the equations of equilibrium to each member.

three-hinged arch

(Ref: Section 5.5)

Moving loads caused by trains must be considered when designing the members of this bridge. The influence lines for the members become an important part of the structural analysis.

Influence Lines for Statically Determinate Structures

6

Influence lines have important application for the design of structures that resist large live loads. In this chapter we will discuss how to draw the influence line for a statically determinate structure. The theory is applied to structures subjected to a distributed load or a series of concentrated forces, and specific applications to floor girders and bridge trusses are given. The determination of the absolute maximum live shear and moment in a member is discussed at the end of the chapter.

6.1 Influence Lines

In the previous chapters we developed techniques for analyzing the forces in structural members due to *dead* or *fixed loads*. It was shown that the *shear* and *moment diagrams* represent the most descriptive methods for displaying the variation of these loads in a member. If a structure is subjected to a *live* or *moving load*, however, the variation of the shear and bending moment in the member is best described using the *influence line*. An influence line represents the variation of either the reaction, shear, moment, or deflection at a *specific point* in a member as a concentrated force moves over the member. Once this line is constructed, one can tell at a glance where the moving load should be placed on the structure so that it creates the greatest influence at the specified point. Furthermore, the magnitude of the associated reaction, shear, moment, or deflection at the point can then be calculated from the ordinates of the influence-line diagram. For these reasons, influence lines play an important part in the design of bridges, industrial crane rails, conveyors, and other structures where loads move across their span.

Although the procedure for constructing an influence line is rather basic, one should clearly be aware of the *difference* between constructing an influence line and constructing a shear or moment diagram. Influence lines represent the effect of a *moving load* only at a *specified point* on a member, whereas shear and moment diagrams represent the effect of *fixed loads* at *all points* along the axis of the member.

Procedure for Analysis

Either of the following two procedures can be used to construct the influence line at a specific point P in a member for any function (reaction, shear, or moment). For both of these procedures we will choose the moving force to have a *dimensionless magnitude of unity.**

Tabulate Values

- Place a unit load at various locations, x, along the member, and at *each* location use statics to determine the value of the function (reaction, shear, or moment) at the specified point.

- If the influence line for a vertical force *reaction* at a point on a beam is to be constructed, consider the reaction to be *positive* at the point when it acts *upward* on the beam.

- If a shear or moment influence line is to be drawn for a point, take the shear or moment at the point as positive according to the same sign convention used for drawing shear and moment diagrams. (See Fig. 4–1.)

- All statically determinate beams will have influence lines that consist of straight line segments. After some practice one should be able to minimize computations and locate the unit load *only* at points representing the *end points* of each line segment.

- To avoid errors, it is recommended that one first construct a table, listing "unit load at x" versus the corresponding value of the function calculated at the specific point; that is, "reaction R," "shear V," or "moment M." Once the load has been placed at various points along the span of the member, the tabulated values can be plotted and the influence-line segments constructed.

Influence-Line Equations

- The influence line can also be constructed by placing the unit load at a *variable* position x on the member and then computing the value of R, V, or M at the point as a function of x. In this manner, the equations of the various line segments composing the influence line can be determined and plotted.

*The reason for this choice will be explained in Sec. 6–2.

EXAMPLE 6.1

Construct the influence line for the vertical reaction at A of the beam in Fig. 6–1a.

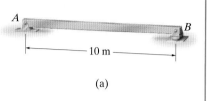

(a)

Fig. 6–1

SOLUTION

Tabulate Values.   A unit load is placed on the beam at each selected point x and the value of A_y is calculated by summing moments about B. For example, when $x = 2.5$ m and $x = 5$ m, see Figs. 6–1b and 6–1c, respectively. The results for A_y are entered in the table, Fig. 6–1d. A plot of these values yields the influence line for the reaction at A, Fig. 6–1e.

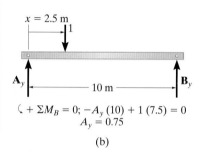

$\zeta + \Sigma M_B = 0;\ -A_y(10) + 1(7.5) = 0$
$A_y = 0.75$

(b)

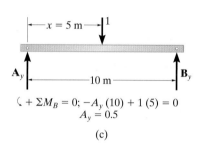

$\zeta + \Sigma M_B = 0;\ -A_y(10) + 1(5) = 0$
$A_y = 0.5$

(c)

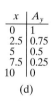

x	A_y
0	1
2.5	0.75
5	0.5
7.5	0.25
10	0

(d)

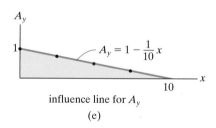

$A_y = 1 - \dfrac{1}{10}x$

influence line for A_y

(e)

Influence-Line Equation.   When the unit load is placed a variable distance x from A, Fig. 6–1f, the reaction A_y as a function of x can be determined from

$$\zeta + \Sigma M_B = 0; \qquad -A_y(10) + (10 - x)(1) = 0$$
$$A_y = 1 - \tfrac{1}{10}x$$

This line is plotted in Fig. 6–1e.

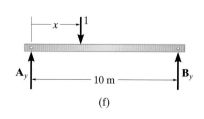

(f)

EXAMPLE 6.2

Construct the influence line for the vertical reaction at B of the beam in Fig. 6–2a.

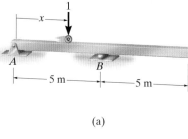

(a)

Fig. 6–2

SOLUTION

Tabulate Values. Using statics, verify that the values for the reaction B_y listed in the table, Fig. 6–2b, are correctly computed for each position x of the unit load. A plot of the values yields the influence line in Fig. 6–2c.

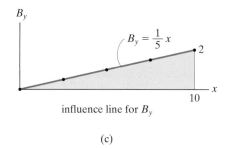

$$B_y = \frac{1}{5}x$$

influence line for B_y

(c)

x	B_y
0	0
2.5	0.5
5	1
7.5	1.5
10	2

(b)

Influence-Line Equation. Applying the moment equation about A, in Fig. 6–2d,

$$\zeta+\Sigma M_A = 0; \quad B_y(5) - 1(x) = 0$$
$$B_y = \tfrac{1}{5}x$$

This is plotted in Fig. 6–2c.

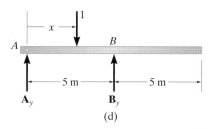

(d)

EXAMPLE 6.3

Construct the influence line for the shear at point C of the beam in Fig. 6–3a.

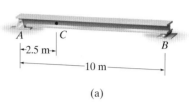

(a)

Fig. 6–3

SOLUTION

Tabulate Values. At each selected position x of the unit load, the method of sections is used to calculate the value of V_C. Note in particular that the unit load must be placed just to the left $(x = 2.5^-)$ and just to the right $(x = 2.5^+)$ of point C since the shear is discontinuous at C, Figs. 6–3b and 6–3c. A plot of the values in Fig. 6–3d yields the influence line for the shear at C, Fig. 6–3e.

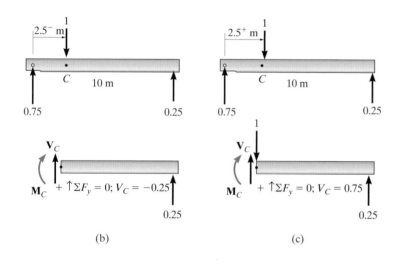

(b)   (c)

x	V_C
0	0
2.5^-	-0.25
2.5^+	0.75
5	0.5
7.5	0.25
10	0

(d)

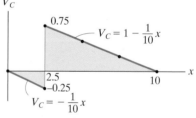

influence line for V_C

(e)

Influence-Line Equations. Here two equations have to be determined since there are two segments for the influence line due to the discontinuity of shear at C, Fig. 6–3f. These equations are plotted in Fig. 6–3e.

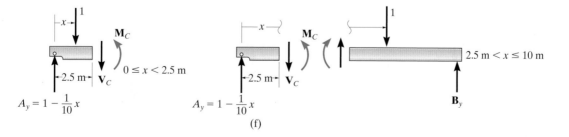

(f)

EXAMPLE 6.4

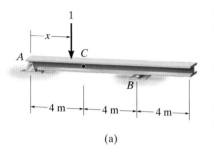

(a)

Fig. 6–4

Construct the influence line for the shear at point C of the beam in Fig. 6–4a.

SOLUTION

Tabulate Values. Using statics and the method of sections, verify that the values of the shear V_C at point C in Fig. 6–4b correspond to each position x of the unit load on the beam. A plot of the values in Fig. 6–4b yields the influence line in Fig. 6–4c.

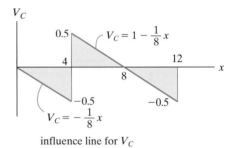

influence line for V_C

(c)

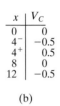

x	V_C
0	0
4^-	-0.5
4^+	0.5
8	0
12	-0.5

(b)

Influence-Line Equations.   From Fig. 6–4d, verify that

$$V_C = -\tfrac{1}{8}x \qquad 0 \le x < 4\text{ m}$$

$$V_C = 1 - \tfrac{1}{8}x \qquad 4\text{ m} < x \le 12\text{ m}$$

These equations are plotted in Fig. 6–4c.

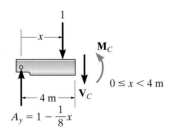
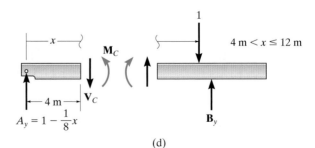

(d)

EXAMPLE 6.5

Construct the influence line for the moment at point C of the beam in Fig. 6–5a.

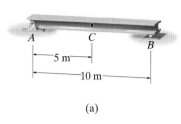

(a)

Fig. 6–5

SOLUTION

Tabulate Values.   At each selected position of the unit load, the value of M_C is calculated using the method of sections. For example, see Fig. 6–5b for $x = 2.5$ m. A plot of the values in Fig. 6–5c yields the influence line for the moment at C, Fig. 6–5d.

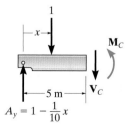

(b)

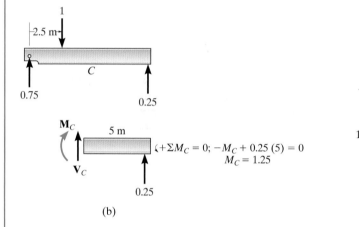

x	M_C
0	0
2.5	1.25
5	2.5
7.5	1.25
10	0

(c)

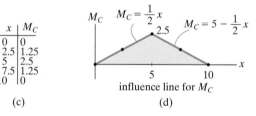

influence line for M_C

(d)

Influence-Line Equations.   The two line segments for the influence line can be determined using $\Sigma M_C = 0$ along with the method of sections shown in Fig. 6–5e. These equations when plotted yield the influence line shown in Fig. 6–5d.

$\downarrow + \Sigma M_C = 0;$ $M_C + 1(5 - x) - \left(1 - \frac{1}{10}x\right)5 = 0$ $\downarrow + \Sigma M_C = 0;$ $M_C - \left(1 - \frac{1}{10}x\right)5 = 0$

$M_C = \frac{1}{2}x$ $0 \le x < 5$ m $M_C = 5 - \frac{1}{2}x$ $5\text{ m} < x \le 10$ m

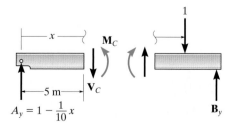

(e)

EXAMPLE 6.6

Construct the influence line for the moment at point C of the beam in Fig. 6–6a.

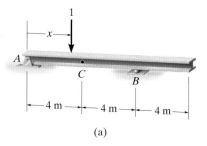

(a)

Fig. 6–6

SOLUTION

Tabulate Values. Using statics and the method of sections, verify that the values of the moment M_C at point C in Fig. 6–6b correspond to each position x of the unit load. A plot of the values in Fig. 6–6b yields the influence line in Fig. 6–6c.

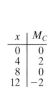

x	M_C
0	0
4	2
8	0
12	−2

(b)

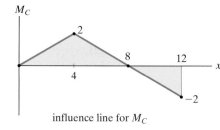

influence line for M_C

(c)

Influence-Line Equations. From Fig. 6–6d verify that

$$M_C = \tfrac{1}{2}x \qquad\qquad 0 \le x < 4 \text{ m}$$
$$M_C = 4 - \tfrac{1}{2}x \qquad 4 \text{ m} < x \le 12 \text{ m}$$

These equations are plotted in Fig. 6–6c.

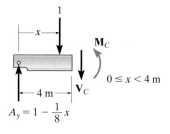
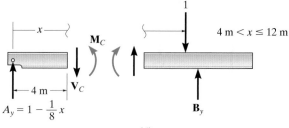

(d)

6.2 Influence Lines for Beams

Since beams (or girders) often form the main load-carrying elements of a floor system or bridge deck, it is important to be able to construct the influence lines for the reactions, shear, or moment at any specified point in a beam.

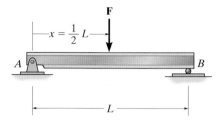

Loadings. Once the influence line for a function (reaction, shear, or moment) has been constructed, it will then be possible to position the live loads on the beam which will produce the maximum value of the function. Two types of loadings will now be considered.

Concentrated Force. Since the numerical values of a function for an influence line are determined using a dimensionless unit load, then for any concentrated force $\mathbf{F}$ acting on the beam at any position x, *the value of the function can be found by multiplying the ordinate of the influence line at the position x by the magnitude of* $\mathbf{F}$. For example, consider the influence line for the reaction at A for the beam AB, Fig. 6–7. If the *unit load* is at $x = \frac{1}{2}L$, the reaction at A is $A_y = \frac{1}{2}$ as indicated from the influence line. Hence, if the force F kN is at this same point, the reaction is $A_y = \left(\frac{1}{2}\right)(F)$ kN. Of course, this same value can also be determined by statics. Obviously, the *maximum influence* caused by $\mathbf{F}$ occurs when it is placed on the beam at the same location as the *peak* of the influence line— in this case at $x = 0$, where the reaction would be $A_y = (1)(F)$ kN.

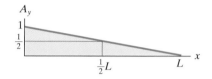

influence line for A_y

Fig. 6–7

Uniform Load. Consider a portion of a beam subjected to a uniform load w_0, Fig. 6–8. As shown, each dx segment of this load creates a concentrated force of $dF = w_0\,dx$ on the beam. If $d\mathbf{F}$ is located at x, where the beam's influence-line ordinate for some function (reaction, shear, moment) is y, then the value of the function is $(dF)(y) = (w_0\,dx)y$. The effect of all the concentrated forces $d\mathbf{F}$ is determined by integrating over the entire length of the beam, that is, $\int w_0 y\,dx = w_0\int y\,dx$. Also, since $\int y\,dx$ is equivalent to the *area* under the influence line, then, in general, *the value of a function caused by a uniform distributed load is simply the area under the influence line for the function multiplied by the intensity of the uniform load*. For example, in the case of a uniformly loaded beam shown in Fig. 6–9, the reaction $\mathbf{A}_y$ can be determined from the influence line as $A_y = (\text{area})(w_0) = \left[\frac{1}{2}(1)(L)\right]w_0 = \frac{1}{2}w_0 L$. This value can of course also be determined from statics.

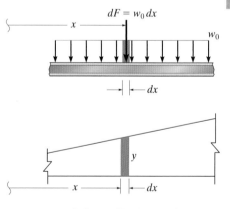

influence line for function

Fig. 6–8

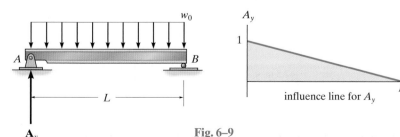

Fig. 6–9

influence line for A_y

EXAMPLE 6.7

Determine the maximum *positive* shear that can be developed at point C in the beam shown in Fig. 6–10a due to a concentrated moving load of 4 kN and a uniform moving load of 2 kN/m.

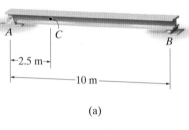

(a)

Fig. 6–10

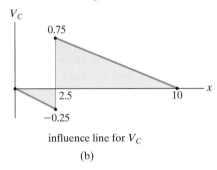

influence line for V_C

(b)

SOLUTION

The influence line for the shear at C has been established in Example 6–3 and is shown in Fig. 6–10b.

Concentrated Force.   The maximum positive shear at C will occur when the 4-kN force is located at $x = 2.5^+$ m, since this is the positive peak of the influence line. The ordinate of this peak is +0.75; so that

$$V_C = 0.75(4 \text{ kN}) = 3 \text{ kN}$$

Uniform Load.   The uniform moving load creates the maximum positive influence for V_C when the load acts on the beam between $x = 2.5^+$ m and $x = 10$ m, since within this region the influence line has a positive area. The magnitude of $\mathbf{V}_C$ due to this loading is

$$V_C = \left[\tfrac{1}{2}(10 \text{ m} - 2.5 \text{ m})(0.75)\right]2 \text{ kN/m} = 5.625 \text{ kN}$$

Total Maximum Shear at C.

$$(V_C)_{\text{max}} = 3 \text{ kN} + 5.625 \text{ kN} = 8.625 \text{ kN} \qquad \textit{Ans.}$$

Notice that once the *positions* of the loads have been established using the influence line, Fig. 6–10c, this value of $(V_C)_{\text{max}}$ can *also* be determined using statics and the method of sections. Show that this is the case.

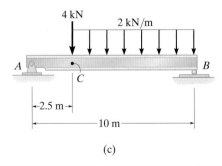

(c)

EXAMPLE 6.8

The frame structure shown in Fig. 6–11a is used to support a hoist for transferring loads for storage at points underneath it. It is anticipated that the load on the dolly is 3 kN and the beam CB has a mass of 24 kg/m. Assume the dolly has negligible size and can travel the entire length of the beam. Also, assume A is a pin and B is a roller. Determine the maximum vertical support reactions at A and B and the maximum moment in the beam at D.

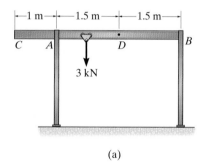

(a)

SOLUTION

Maximum Reaction at A. We first draw the influence line for A_y, Fig. 6–11b. Specifically, when a unit load is at A the reaction at A is 1 as shown. The ordinate at C, is 1.33. Here the maximum value for A_y occurs when the dolly is at C. Since the dead load (beam weight) must be placed over the entire length of the beam, we have,

$$(A_y)_{max} = 3000(1.33) + 24(9.81)\left[\tfrac{1}{2}(4)(1.33)\right]$$

$$= 4.63 \text{ kN} \qquad\qquad Ans.$$

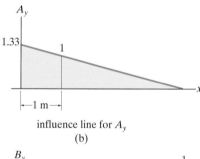

influence line for A_y
(b)

Maximum Reaction at B. The influence line (or beam) takes the shape shown in Fig. 6–11c. The values at C and B are determined by statics. Here the dolly must be at B. Thus,

$$(B_y)_{max} = 3000(1) + 24(9.81)\left[\tfrac{1}{2}(3)(1)\right] + 24(9.81)\left[\tfrac{1}{2}(1)(-0.333)\right]$$

$$= 3.31 \text{ kN} \qquad\qquad Ans.$$

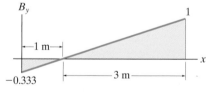

influence line for B_y
(c)

Maximum Moment at D. The influence line has the shape shown in Fig. 6–11d. The values at C and D are determined from statics. Here,

$$(M_D)_{max} = 3000(0.75) + 24(9.81)\left[\tfrac{1}{2}(1)(-0.5)\right] + 24(9.81)\left[\tfrac{1}{2}(3)(0.75)\right]$$

$$= 2.46 \text{ kN} \cdot \text{m} \qquad\qquad Ans.$$

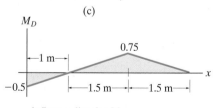

influence line for M_D
(d)

Fig. 6–11

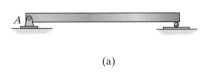

(a)

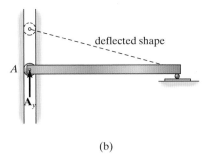

deflected shape

(b)

6.3   Qualitative Influence Lines

In 1886, Heinrich Müller-Breslau developed a technique for rapidly constructing the shape of an influence line. Referred to as the *Müller-Breslau principle, it states that the influence line for a function (reaction, shear, or moment) is to the same scale as the deflected shape of the beam when the beam is acted upon by the function.* In order to draw the deflected shape properly, the capacity of the beam to resist the applied function must be *removed* so the beam can deflect when the function is applied. For example, consider the beam in Fig. 6–12*a*. If the shape of the influence line for the *vertical reaction* at *A* is to be determined, the pin is first replaced by a *roller guide* as shown in Fig. 6–12*b*. A roller guide is necessary since the beam must still resist a horizontal force at *A* but *no vertical force*. When the positive (upward) force $\mathbf{A}_y$ is then applied at *A*, the beam deflects to the dashed position,* which represents the general shape of the influence line for A_y, Fig. 6–12*c*. (Numerical values for this specific case have been calculated in Example 6–1.) If the shape of the influence line for the *shear* at *C* is to be determined, Fig. 6–13*a*, the connection at *C* may be symbolized by a *roller guide* as shown in Fig. 6–13*b*. This device will resist a moment and axial force but no *shear*.[†] Applying a positive shear force $\mathbf{V}_C$ to the beam at *C* and allowing the beam to deflect to the dashed position, we find the influence-line shape as shown in Fig. 6–13*c*. Finally, if the shape of the influence line for the *moment* at *C*, Fig. 6–14*a*, is to be determined, an internal *hinge* or *pin* is placed at *C*, since this connection resists axial and shear forces but *cannot resist a moment*, Fig. 6–14*b*. Applying positive moments $\mathbf{M}_C$ to the beam, the beam then deflects to the dashed position, which is the shape of the influence line, Fig. 6–14*c*.

The proof of the Müller-Breslau principle can be established using the principle of virtual work. Recall that *work* is the product of either *a linear*

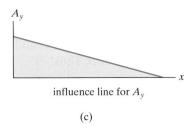

influence line for A_y

(c)

Fig. 6–12

Design of this bridge girder is based on influence lines that must be constructed for this train loading.

*Throughout the discussion all deflected positions are drawn to an exaggerated scale to illustrate the concept.

†Here the rollers *symbolize* supports that carry loads both in tension or compression. See Table 2–1, support (2).

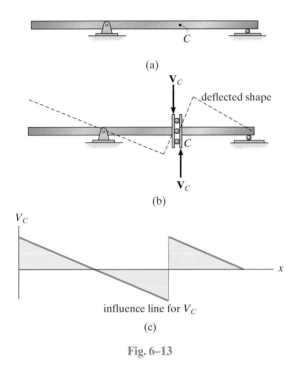

influence line for V_C

(c)

Fig. 6–13

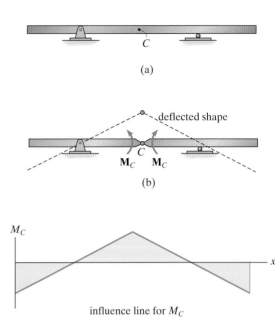

influence line for M_C

(c)

Fig. 6–14

displacement and force in the direction of the displacement or *a rotational displacement and moment in the direction of the displacement*. If a rigid body (beam) is in equilibrium, the sum of all the forces and moments on it must be equal to zero. Consequently, if the body is given an *imaginary* or *virtual displacement,* the work done by *all* these forces and couple moments must also be equal to zero. Consider, for example, the simply supported beam shown in Fig. 6–15a, which is subjected to a unit load placed at an arbitrary point along its length. If the beam is given a virtual (or imaginary) displacement δy at the support A, Fig. 6–15b, then only the support reaction $\mathbf{A}_y$ and the unit load do virtual work. Specifically, A_y does positive work $A_y \, \delta y$ and the unit load does negative work, $-1\delta y'$. (The support at B does not move and therefore the force at B does no work.) Since the beam is in equilibrium and therefore does not actually move, the virtual work sums to zero, i.e.,

$$A_y \, \delta y - 1 \, \delta y' = 0$$

If δy is set equal to 1, then

$$A_y = \delta y'$$

In other words, the value of A_y represents the ordinate of the influence line at the position of the unit load. Since this value is equivalent to the displacement $\delta y'$ at the position of the unit load, it shows that the *shape* of the influence line for the reaction at A has been established. This proves the Müller-Breslau principle for reactions.

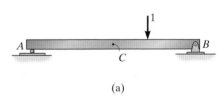

(a)

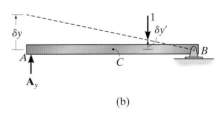

(b)

Fig. 6–15

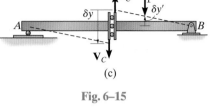

(c)

Fig. 6–15

In the same manner, if the beam is sectioned at C, and the beam undergoes a virtual displacement δy at this point, Fig. 6–15c, then only the internal shear at C and the unit load do work. Thus, the virtual work equation is

$$V_C\, \delta y - 1\, \delta y' = 0$$

Again, if $\delta y = 1$, then

$$V_C = \delta y'$$

and the *shape* of the influence line for the shear at C has been established.

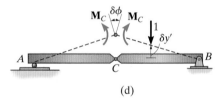

(d)

Lastly, assume a hinge or pin is introduced into the beam at point C, Fig. 6–15d. If a virtual rotation $\delta\phi$ is introduced at the pin, virtual work will be done only by the internal moment and the unit load. So

$$M_C\, \delta\phi - 1\, \delta y' = 0$$

Setting $\delta\phi = 1$, it is seen that

$$M_C = \delta y'$$

which indicates that the deflected beam has the same *shape* as the influence line for the internal moment at point C (see Fig. 6–14).

Obviously, the Müller-Breslau principle provides a quick method for establishing the *shape* of the influence line. Once this is known, the ordinates at the peaks can be determined by using the basic method discussed in Sec. 6–1. Also, by simply knowing the general shape of the influence line, it is possible to *locate* the live load on the beam and then determine the maximum value of the function by *using statics*. Example 6–12 illustrates this technique.

EXAMPLE 6.9

For each beam in Fig. 6–16a through 6–16c, sketch the influence line for the vertical reaction at A.

SOLUTION

The support is replaced by a roller guide at A since it will resist $\mathbf{A}_x$, but not $\mathbf{A}_y$. The force $\mathbf{A}_y$ is then applied.

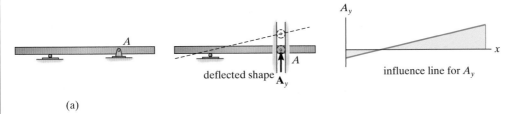

(a)

Fig. 6–16

Again, a roller guide is placed at A and the force $\mathbf{A}_y$ is applied.

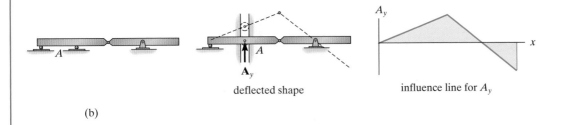

(b)

A *double-roller guide* must be used at A in this case, since this type of support will resist both a moment $\mathbf{M}_A$ at the fixed support and axial load $\mathbf{A}_x$, but will not resist $\mathbf{A}_y$.

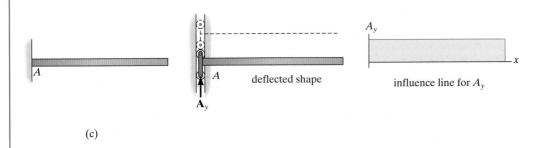

(c)

EXAMPLE | 6.10

For each beam in Figs. 6–17a through 6–17c, sketch the influence line for the shear at B.

SOLUTION

The roller guide is introduced at B and the positive shear $\mathbf{V}_B$ is applied. Notice that the right segment of the beam will *not deflect* since the roller at A actually constrains the beam from moving vertically, either up or down. [See support (2) in Table 2–1.]

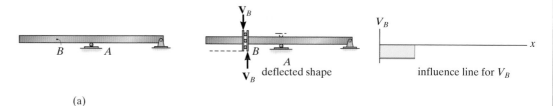

(a)

Fig. 6–17

Placing the roller guide at B and applying the positive shear at B yields the deflected shape and corresponding influence line.

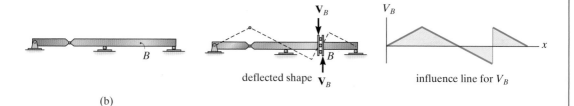

(b)

Again, the roller guide is placed at B, the positive shear is applied, and the deflected shape and corresponding influence line are shown. Note that the left segment of the beam does not deflect, due to the fixed support.

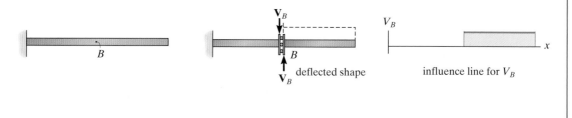

(c)

EXAMPLE 6.11

For each beam in Figs. 6–18a through 6–18c, sketch the influence line for the moment at B.

SOLUTION
A hinge is introduced at B and positive moments $\mathbf{M}_B$ are applied to the beam. The deflected shape and corresponding influence line are shown.

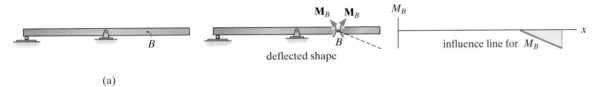

(a)

Fig. 6–18

Placing a hinge at B and applying positive moments $\mathbf{M}_B$ to the beam yields the deflected shape and influence line.

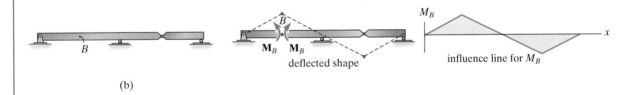

(b)

With the hinge and positive moment at B the deflected shape and influence line are shown. The left segment of the beam is constrained from moving due to the fixed wall at A.

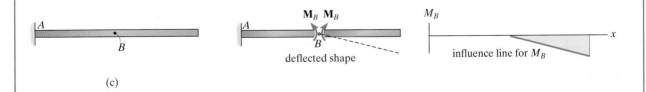

(c)

EXAMPLE 6.12

Determine the maximum positive moment that can be developed at point D in the beam shown in Fig. 6–19a due to a concentrated moving load of 16 kN, a uniform moving load of 3 kN/m, and a beam weight of 2 kN/m.

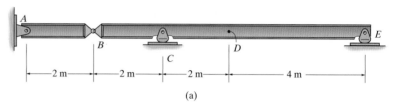

(a)

Fig. 6–19

SOLUTION

A hinge is placed at D and positive moments $\mathbf{M}_D$ are applied to the beam. The deflected shape and corresponding influence line are shown in Fig. 6–19b. Immediately one recognizes that the concentrated moving load of 16 kN creates a maximum *positive* moment at D when it is placed at D, i.e., the peak of the influence line. Also, the uniform moving load of 3 kN/m must extend from C to E in order to cover the region where the area of the influence line is positive. Finally, the uniform weight of 2 kN/m acts over the *entire length* of the beam. The loading is shown on the beam in Fig. 6–19c. Knowing the position of the loads, we can now determine the maximum moment at D using statics. In Fig. 6–19d the reactions on BE have been computed. Sectioning the beam at D and using segment DE, Fig. 6–19e, we have

$$\zeta + \Sigma M_D = 0; \quad -M_D - 20(2) + 19(4) = 0$$
$$M_D = 36 \text{ kN} \cdot \text{m} \qquad Ans.$$

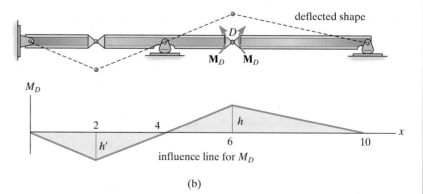

influence line for M_D

(b)

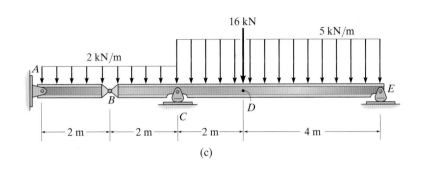

(c)

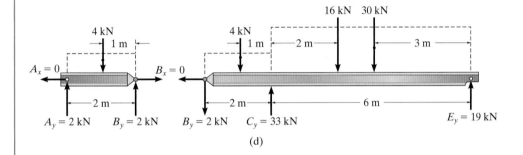

(d)

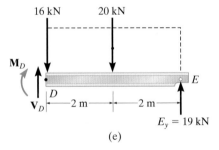

(e)

This problem can also be worked by using *numerical values* for the influence line as in Sec. 6–1. Actually, by inspection of Fig. 6–19b, only the peak value h at D must be determined. This requires placing a unit load on the beam at D in Fig. 6–19a and then solving for the internal moment in the beam at D. Show that the value obtained is $h = 1.333$. By proportional triangles, $h'/(4 - 2) = 1.333/(6 - 4)$ or $h' = 1.333$. Hence, with the loading on the beam as in Fig. 6–19c, using the areas and peak values of the influence line, Fig. 6–19b, we have

$$M_D = 5\left[\tfrac{1}{2}(10 - 4)(1.333)\right] + 16(1.333) - 2\left[\tfrac{1}{2}(4)(1.333)\right]$$

$$= 36 \text{ kN} \cdot \text{m} \qquad\qquad Ans.$$

FUNDAMENTAL PROBLEMS

F6–1. Use the Müller-Breslau principle to sketch the influence lines for the vertical reaction at A, the shear at C, and the moment at C.

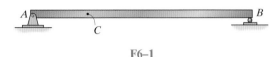

F6–1

F6–2. Use the Müller-Breslau principle to sketch the influence lines for the vertical reaction at A, the shear at D, and the moment at B.

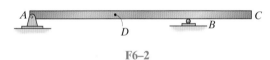

F6–2

F6–3. Use the Müller-Breslau principle to sketch the influence lines for the vertical reaction at A, the shear at D, and the moment at D.

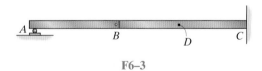

F6–3

F6–4. Use the Müller-Breslau principle to sketch the influence lines for the vertical reaction at A, the shear at B, and the moment at B.

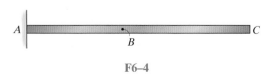

F6–4

F6–5. Use the Müller-Breslau principle to sketch the influence lines for the vertical reaction at A, the shear at C, and the moment at C.

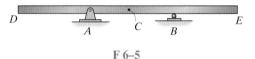

F 6–5

F6–6. Use the Müller-Breslau principle to sketch the influence lines for the vertical reaction at A, the shear just to the left of the roller support at E, and the moment at A.

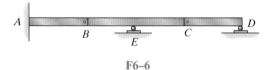

F6–6

F6–7. The beam supports a distributed live load of 1.5 kN/m and single concentrated load of 8 kN. The dead load is 2 kN/m. Determine (a) the maximum positive moment at C, (b) the maximum positive shear at C.

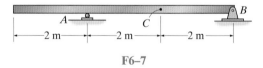

F6–7

F6–8. The beam supports a distributed live load of 2 kN/m and single concentrated load of 6 kN. The dead load is 4 kN/m. Determine (a) the maximum vertical positive reaction at C, (b) the maximum negative moment at A.

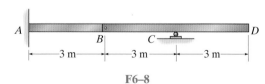

F6–8

PROBLEMS

6–1.   Draw the influence lines for (a) the moment at C, (b) the reaction at B, and (c) the shear at C. Assume A is pinned and B is a roller. Solve this problem using the basic method of Sec. 6–1.

6–2.   Solve Prob. 6–1 using the Müller-Breslau principle.

6–7.   Draw the influence line for (a) the moment at B, (b) the shear at C, and (c) the vertical reaction at B. Solve this problem using the basic method of Sec. 6–1. *Hint*: The support at A resists only a horizontal force and a bending moment.

*6–8.   Solve Prob. 6–7 using the Müller-Breslau principle.

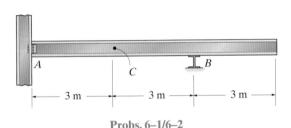

Probs. 6–1/6–2

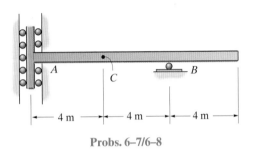

Probs. 6–7/6–8

6–3.   Draw the influence lines for (a) the vertical reaction at A, (b) the moment at A, and (c) the shear at B. Assume the support at A is fixed. Solve this problem using the basic method of Sec. 6–1.

*6–4.   Solve Prob. 6–3 using the Müller-Breslau principle.

6–9.   Draw the influence line for (a) the vertical reaction at A, (b) the shear at B, and (c) the moment at B. Assume A is fixed. Solve this problem using the basic method of Sec. 6–1.

6–10.   Solve Prob. 6–9 using the Müller-Breslau principle.

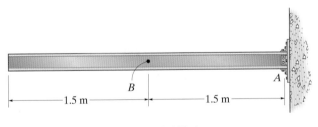

Probs. 6–3/6–4

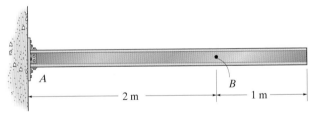

Probs. 6–9/6–10

6–5.   Draw the influence lines for (a) the vertical reaction at B, (b) the shear just to the right of the rocker at A, and (c) the moment at C. Solve this problem using the basic method of Sec. 6–1.

6–6.   Solve Prob. 6–5 using Müller-Breslau's principle.

6–11.   Draw the influence lines for (a) the vertical reaction at A, (b) the shear at C, and (c) the moment at C. Solve this problem using the basic method of Sec. 6–1.

*6–12.   Solve Prob. 6–11 using Müller-Breslau's principle.

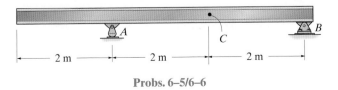

Probs. 6–5/6–6

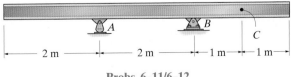

Probs. 6–11/6–12

6–13. Draw the influence lines for (a) the vertical reaction at A, (b) the vertical reaction at B, (c) the shear just to the right of the support at A, and (d) the moment at C. Assume the support at A is a pin and B is a roller. Solve this problem using the basic method of Sec. 6–1.

6–14. Solve Prob. 6–13 using the Müller-Breslau principle.

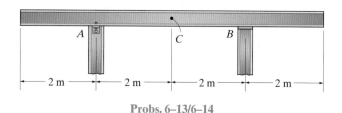

Probs. 6–13/6–14

6–15. The beam is subjected to a uniform dead load of 1.2 kN/m and a single live load of 40 kN. Determine (a) the maximum moment created by these loads at C, and (b) the maximum positive shear at C. Assume A is a pin. and B is a roller.

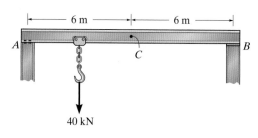

Prob. 6–15

*6–16. The beam supports a uniform dead load of 500 N/m and a single live concentrated force of 3000 N. Determine (a) the maximum positive moment at C, and (b) the maximum positive shear at C. Assume the support at A is a roller and B is a pin.

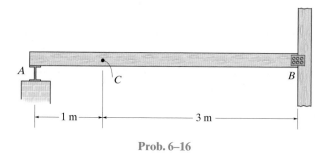

Prob. 6–16

6–17. A uniform live load of 4 kN/m and a single live concentrated force of 6 kN are to be placed on the beam. The beam has a weight of 2 kN/m. Determine (a) the maximum vertical reaction at support B, and (b) the maximum negative moment at point B. Assume the support at A is a pin and B is a roller.

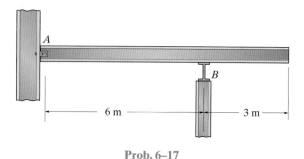

Prob. 6–17

6–18. The beam supports a uniform dead load of 8 kN/m, a live load of 30 kN/m, and a single live concentrated force of 32 kN. Determine (a) the maximum positive moment at C, and (b) the maximum positive vertical reaction at B. Assume A is a roller and B is a pin.

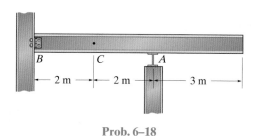

Prob. 6–18

6–19. The beam is used to support a dead load of 9 kN/m, a live load of 30 kN/m and a concentrated live load of 36 kN. Determine (a) the maximum positive (upward) reaction at A, (b) the maximum positive moment at C, and (c) the maximum positive shear just to the right of the support at A. Assume the support at A is a pin and B is a roller.

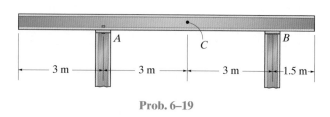

Prob. 6–19

*6–20. The compound beam is subjected to a uniform dead load of 1.5 kN/m and a single live load of 10 kN. Determine (a) the maximum negative moment created by these loads at *A*, and (b) the maximum positive shear at *B*. Assume *A* is a fixed support, *B* is a pin, and *C* is a roller.

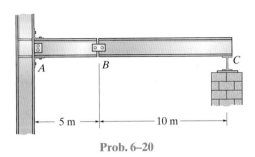

Prob. 6–20

6–21. Where should a single 2-kN live load be placed on the beam so it causes the largest moment at *D*? What is this moment? Assume the support at *A* is fixed, *B* is pinned, and *C* is a roller.

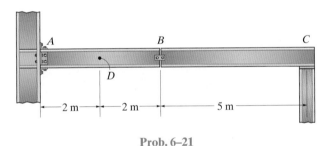

Prob. 6–21

6–22. Where should the beam *ABC* be loaded with a 6 kN/m uniform distributed live load so it causes (a) the largest moment at point *A* and (b) the largest shear at *D*? Calculate the values of the moment and shear. Assume the support at *A* is fixed, *B* is pinned and *C* is a roller.

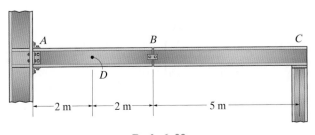

Prob. 6–22

6–23. The beam is used to support a dead load of 800 N/m, a live load of 4 kN/m, and a concentrated live load of 20 kN. Determine (a) the maximum positive (upward) reaction at *B*, (b) the maximum positive moment at *C*, and (c) the maximum negative shear at *C*. Assume *B* and *D* are pins.

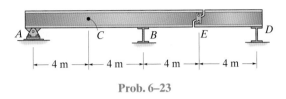

Prob. 6–23

*6–24. The beam is used to support a dead load of 8 kN/m, a live load of 40 kN/m, and a concentrated live load of 32 kN. Determine (a) the maximum positive vertical reaction at *A*, (b) the maximum positive shear just to the right of the support at *A*, and (c) the maximum negative moment at *C*. Assume *A* is a roller, *C* is fixed, and *B* is pinned.

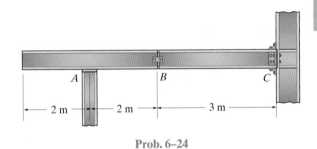

Prob. 6–24

6–25. The beam is used to support a dead load of 5 kN/m, a live load of 20 kN/m, and a concentrated live load of 32 kN. Determine (a) the maximum positive (upward) reaction at *A*, (b) the maximum positive moment at *E*, and (c) the maximum positive shear just to the right of the support at *C*. Assume *A* and *C* are rollers and *D* is a pin.

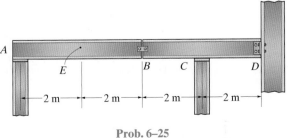

Prob. 6–25

6.4   Influence Lines for Floor Girders

Occasionally, floor systems are constructed as shown in Fig. 6–20a, where it can be seen that floor loads are transmitted from *slabs* to *floor beams*, then to *side girders*, and finally supporting *columns*. An idealized model of this system is shown in plane view, Fig. 6–20b. Here the slab is assumed to be a one-way slab and is segmented into simply supported spans resting on the floor beams. Furthermore, the girder is simply supported on the columns. Since the girders are main load-carrying members in this system, it is sometimes necessary to construct their shear and moment influence lines. This is especially true for industrial buildings subjected to heavy concentrated loads. In this regard, notice that a unit load on the floor slab is transferred to the girder only at points where it is in contact with the floor beams, i.e., points A, B, C, and D. These points are called *panel points*, and the region between these points is called a *panel*, such as BC in Fig. 6–20b.

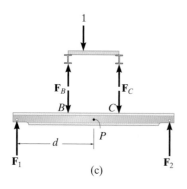

(a)

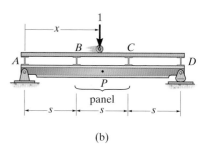

(b)

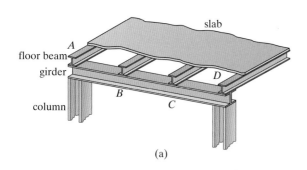

(c)

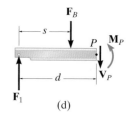

(d)

Fig. 6–20

The influence line for a specified point on the girder can be determined using the same statics procedure as in Sec. 6–1; i.e., place the unit load at various points x on the floor slab and always compute the function (shear or moment) at the specified point P in the girder, Fig. 6–20b. Plotting these values versus x yields the influence line for the function at P. In particular, the value for the internal moment in a girder panel will depend upon where point P is chosen for the influence line, since the magnitude of $\mathbf{M}_P$ depends upon the point's location from the end of the girder. For example, if the unit load acts on the floor slab as shown in Fig. 6–20c, one first finds the reactions $\mathbf{F}_B$ and $\mathbf{F}_C$ on the slab, then calculates the support reactions $\mathbf{F}_1$ and $\mathbf{F}_2$ on the girder. The internal moment at P is then determined by the method of sections, Fig. 6–20d. This gives $M_P = F_1 d - F_B(d - s)$. Using a similar analysis, the internal shear $\mathbf{V}_P$ can be determined. In this case, however, $\mathbf{V}_P$ will be *constant* throughout the panel $BC(V_P = F_1 - F_B)$ and so it does not depend upon the exact location d of P within the panel. For this reason, influence lines for shear in floor girders are specified for *panels* in the girder and not specific points along the girder. The shear is then referred to as *panel shear*. It should also be noted that since the girder is affected only by the loadings transmitted by the floor beams, the unit load is generally placed at each floor-beam location to establish the necessary data used to draw the influence line.

The following numerical examples should clarify the force analysis.

6

The design of the floor system of this warehouse building must account for critical locations of storage materials on the floor. Influence lines must be used for this purpose. (*Photo courtesy of Portland Cement Association.*)

EXAMPLE 6.13

Draw the influence line for the shear in panel CD of the floor girder in Fig. 6–21a.

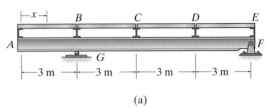

(a)

SOLUTION

Tabulate Values. The unit load is placed at each floor beam location and the shear in panel CD is calculated. A table of the results is shown in Fig. 6–21b. The details for the calculations when $x = 0$ and $x = 6$ m are given in Figs. 6–21c and 6–21d, respectively. Notice how in each case the reactions of the floor beams on the girder are calculated first, followed by a determination of the girder support reaction at F ($\mathbf{G}_y$ is not needed), and finally, a segment of the girder is considered and the internal panel shear V_{CD} is calculated. As an exercise, verify the values for V_{CD} when $x = 3$ m, 9 m, and 12 m.

Fig. 6–21

x	V_{CD}
0	0.333
3	0
6	−0.333
9	0.333
12	0

(b)

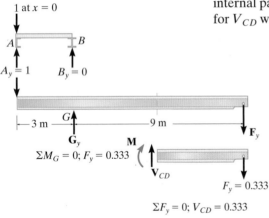

(c)

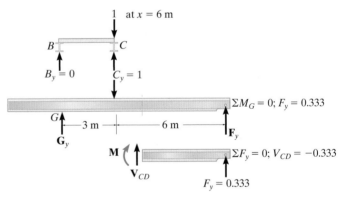

(d)

Influence Line. When the tabular values are plotted and the points connected with straight line segments, the resulting influence line for V_{CD} is as shown in Fig. 6–21e.

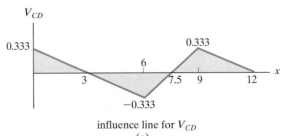

influence line for V_{CD}

(e)

EXAMPLE 6.14

Draw the influence line for the moment at point F for the floor girder in Fig. 6–22a.

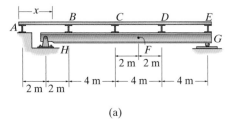

x	M_F
0	0
2	0.429
4	0.857
8	2.571
10	2.429
12	2.286
16	0

(a)   (b)

Fig. 6–22

SOLUTION

Tabulate Values. The unit load is placed at $x = 0$ and each panel point thereafter. The corresponding values for M_F are calculated and shown in the table, Fig. 6–22b. Details of the calculations for $x = 2$ m are shown in Fig. 6–22c. As in the previous example, it is first necessary to determine the reactions of the floor beams on the girder, followed by a determination of the girder support reaction $\mathbf{G}_y$ ($\mathbf{H}_y$ is not needed), and finally segment GF of the girder is considered and the internal moment $\mathbf{M}_F$ is calculated. As an exercise, determine the other values of M_F listed in Fig. 6–22b.

Influence Line. A plot of the tabular values yields the influence line for M_F, Fig. 6–22d.

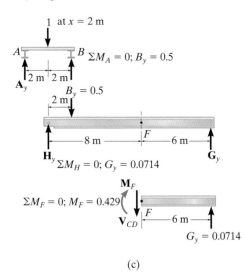

(c)

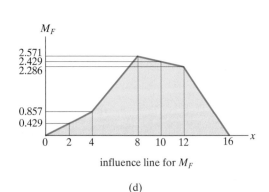

influence line for M_F

(d)

6.5   Influence Lines for Trusses

Trusses are often used as primary load-carrying elements for bridges. Hence, for design it is important to be able to construct the influence lines for each of its members. As shown in Fig. 6–23, the loading on the bridge deck is transmitted to stringers, which in turn transmit the loading to floor beams and then to the *joints* along the bottom cord of the truss. Since the truss members are affected only by the joint loading, we can therefore obtain the ordinate values of the influence line for a member by loading each joint along the deck with a unit load and then use the method of joints or the method of sections to calculate the force in the member. The data can be arranged in tabular form, listing "unit load at joint" versus "force in member." As a convention, if the member force is *tensile* it is considered a *positive* value; if it is *compressive* it is *negative*. The influence line for the member is constructed by plotting the data and drawing straight lines between the points.

The following examples illustrate the method of construction.

The members of this truss bridge were designed using influence lines in accordance with the AASHTO specifications.

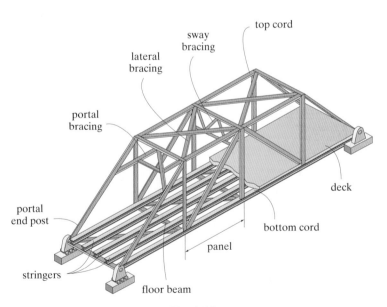

Fig. 6–23

EXAMPLE 6.15

Draw the influence line for the force in member GB of the bridge truss shown in Fig. 6–24a.

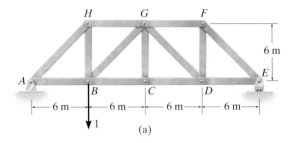

(a)

Fig. 6–24

SOLUTION

Tabulate Values.   Here each successive joint at the bottom cord is loaded with a unit load and the force in member GB is calculated using the method of sections, Fig. 6–24b. For example, placing the unit load at $x = 6$ m (joint B), the support reaction at E is calculated first, Fig. 6–24a, then passing a section through HG, GB, BC and isolating the right segment, the force in GB is determined, Fig. 6–24c. In the same manner, determine the other values listed in the table.

x	F_{GB}
0	0
6	0.354
12	−0.707
18	−0.354
24	0

(b)

Influence Line.   Plotting the tabular data and connecting the points yields the influence line for member GB, Fig. 6–24d. Since the influence line extends over the entire span of the truss, member GB is referred to as a *primary member*. This means GB is subjected to a force regardless of where the bridge deck (roadway) is loaded, except, of course, at $x = 8$ m. The point of zero force, $x = 8$ m, is determined by similar triangles between $x = 6$ m and $x = 12$ m, that is, $(0.354 + 0.707)/(12 - 6) = 0.354/x'$, $x' = 2$ m, so $x = 6 + 2 = 8$ m.

F_{HG}

$45°$

F_{GB}

F_{BC}

$\Sigma F_y = 0;\ \ 0.25 - F_{GB}\sin 45° = 0$
$F_{GB} = 0.354$

0.25

(c)

F_{GB}

0.354

8   12   18   24 x

6

−0.354

−0.707

influence line for F_{GB}

(d)

EXAMPLE 6.16

Draw the influence line for the force in member CG of the bridge truss shown in Fig. 6–25a.

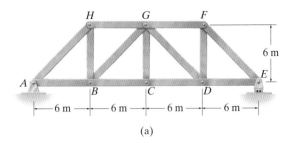

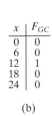

6 m

x	F_{GC}
0	0
6	0
12	1
18	0
24	0

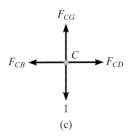

(a)   (b)   (c)

Fig. 6–25

SOLUTION

Tabulate Values.   A table of unit-load position at the joints of the bottom cord versus the force in member CG is shown in Fig. 6–25b. These values are easily obtained by isolating joint C, Fig. 6–25c. Here it is seen that CG is a zero-force member unless the unit load is applied at joint C, in which case $F_{CG} = 1$ (T).

Influence Line.   Plotting the tabular data and connecting the points yields the influence line for member CG as shown in Fig. 6–25d. In particular, notice that when the unit load is at $x = 9$ m, the force in member CG is $F_{CG} = 0.5$. This situation requires the unit load to be placed on the bridge deck *between* the joints. The transference of this load from the deck to the truss is shown in Fig. 6–25e. From this one can see that indeed $F_{CG} = 0.5$ by analyzing the equilibrium of joint C, Fig. 6–25f. Since the influence line for CG does *not* extend over the entire span of the truss, Fig. 6–25d, member CG is referred to as a *secondary member*.

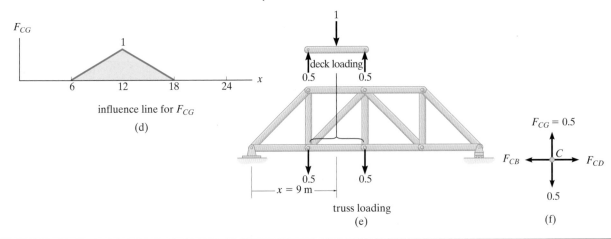

influence line for F_{CG}

(d)

deck loading

truss loading

(e)

(f)

EXAMPLE 6.17

In order to determine the maximum force in each member of the Warren truss, shown in the photo, we must first draw the influence lines for each of its members. If we consider a similar truss as shown in Fig. 6–26a, determine the largest force that can be developed in member BC due to a moving force of 100 kN and a moving distributed load of 12 kN/m. The loading is applied at the top cord.

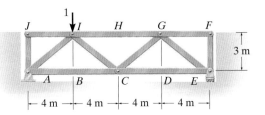

(a)

x	F_{BC}
0	0
4	1
8	0.667
12	0.333
16	0

(b)

Fig. 6–26

SOLUTION

Tabulate Values. A table of unit-load position x at the joints along the top cord versus the force in member BC is shown in Fig. 6–26b. The method of sections can be used for the calculations. For example, when the unit load is at joint I ($x = 4$ m), Fig. 6–26a, the reaction $\mathbf{E}_y$ is determined first ($E_y = 0.25$). Then the truss is sectioned through BC, IC, and HI, and the right segment is isolated, Fig. 6–26c. One obtains $\mathbf{F}_{BC}$ by summing moments about point I, to eliminate $\mathbf{F}_{HI}$ and $\mathbf{F}_{IC}$. In a similar manner determine the other values in Fig. 6–26b.

Influence Line. A plot of the tabular values yields the influence line, Fig. 6–26d. By inspection, BC is a primary member. Why?

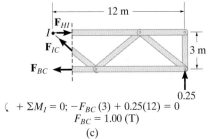

$$(\ + \Sigma M_I = 0; \ -F_{BC}(3) + 0.25(12) = 0$$
$$F_{BC} = 1.00 \text{ (T)}$$

(c)

Concentrated Live Force. The largest force in member BC occurs when the moving force of 100 kN is placed at $x = 20$ m. Thus,

$$F_{BC} = (1.00)(100) = 100 \text{ kN}$$

Distributed Live Load. The uniform live load must be placed over the entire deck of the truss to create the largest tensile force in BC.* Thus,

$$F_{BC} = \left[\tfrac{1}{2}(16)(1.00)\right]12 = 96 \text{ kN}$$

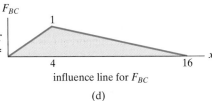

influence line for F_{BC}

(d)

Total Maximum Force.

$$(F_{BC})_{max} = 100 \text{ kN} + 96 \text{ kN} = 196 \text{ kN} \qquad Ans.$$

*The largest *tensile* force in member GB of Example 6–15 is created when the distributed load acts on the deck of the truss from $x = 0$ to $x = 8$ m, Fig. 6–24d.

PROBLEMS

6–26. A uniform live load of 1.8 kN/m and a single concentrated live force of 4 kN are placed on the floor beams. Determine (a) the maximum positive shear in panel BC of the girder and (b) the maximum moment in the girder at G.

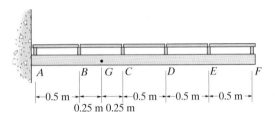

Prob. 6–26

6–27. A uniform live load of 2.8 kN/m and a single concentrated live force of 20 kN are placed on the floor beams. If the beams also support a uniform dead load of 700 N/m, determine (a) the maximum positive shear in panel BC of the girder and (b) the maximum positive moment in the girder at G.

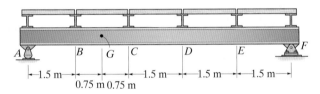

Prob. 6–27

*6–28. A uniform live load of 30 kN/m and a single concentrated live force of 30 kN are placed on the floor beams. If the beams also support a uniform dead load of 5.25 kN/m, determine (a) the maximum positive shear in panel CD of the girder and (b) the maximum negative moment in the girder at D. Assume the support at C is a roller and E is a pin.

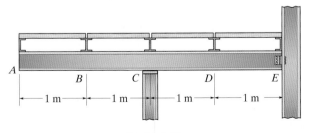

Prob. 6–28

6–29. Draw the influence line for (a) the shear in panel BC of the girder, and (b) the moment at D.

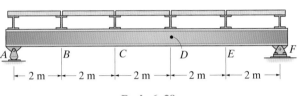

Prob. 6–29

6–30. A uniform live load of 5 kN/m and a single concentrated live force of 6 kN are to be placed on the floor beams. Determine (a) the maximum positive shear in panel AB, and (b) the maximum moment at D. Assume only vertical reaction occur at the supports.

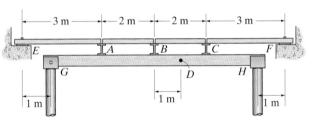

Prob. 6–30

6–31. A uniform live load of 15 kN/m and a single concentrated live force of 25 kN are to be placed on the top beams. Determine (a) the maximum positive shear in panel BC of the girder, and (b) the maximum positive moment at C. Assume the support at B is a roller and at D a pin.

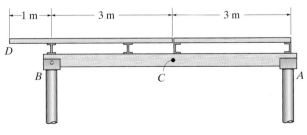

Prob. 6–31

*6–32. Draw the influence line for the moment at *F* in the girder. Determine the maximum positive live moment in the girder at *F* if a single concentrated live force of 8 kN moves across the top floor beams. Assume the supports for all members can only exert either upward or downward forces on the members.

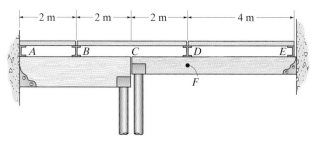

Prob. 6–32

6–33. A uniform live load of 40 kN/m and a single concentrated live force of 100 kN are placed on the floor beams. If the beams also support a uniform dead load of 7 kN/m, determine (a) the maximum negative shear in panel *DE* of the girder and (b) the maximum negative moment in the girder at *C*.

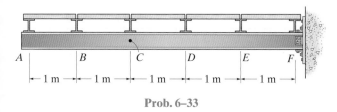

Prob. 6–33

6–34. A uniform live load of 3 kN/m and a single concentrated live force of 20 kN are placed on the floor beams. Determine (a) the maximum positive shear in panel *DE* of the girder, and (b) the maximum positive moment at *H*.

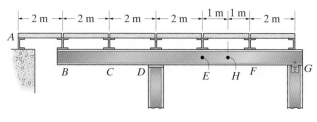

Prob. 6–34

6–35. Draw the influence line for the shear in panel *CD* of the girder. Determine the maximum negative live shear in panel *CD* due to a uniform live load of 6 kN/m acting on the top beams.

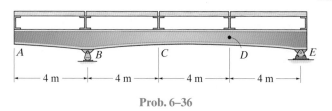

Prob. 6–35

*6–36. A uniform live load of 6.5 kN/m and a single concentrated live force of 15 kN are placed on the floor beams. If the beams also support a uniform dead load of 600 N/m, determine (a) the maximum positive shear in panel *CD* of the girder and (b) the maximum positive moment in the girder at *D*.

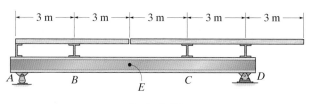

Prob. 6–36

6–37. A uniform live load of 1.75 kN/m and a single concentrated live force of 8 kN are placed on the floor beams. If the beams also support a uniform dead load of 250 N/m, determine (a) the maximum negative shear in panel *BC* of the girder and (b) the maximum positive moment at *B*.

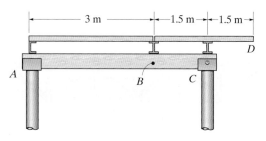

Prob. 6–37

6–38. Draw the influence line for the force in (a) member *KJ* and (b) member *CJ*.

6–39. Draw the influence line for the force in (a) member *JI*, (b) member *IE*, and (c) member *EF*.

6–45. Draw the influence line for the force in (a) member *EH* and (b) member *JE*.

6–46. Draw the influence line for the force in member *JI*.

6–47. Draw the influence line for the force in member *AL*.

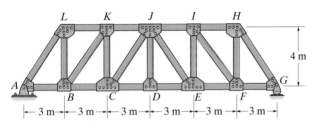

Probs. 6–38/6–39

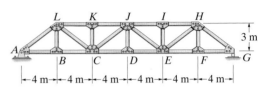

Probs. 6–45/6–46/6–47

*6–40. Draw the influence line for the force in member *KJ*.

6–41. Draw the influence line for the force in member *JE*.

*6–48. Draw the influence line for the force in member *BC* of the Warren truss. Indicate numerical values for the peaks. All members have the same length.

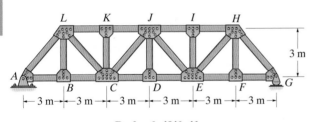

Probs. 6–40/6–41

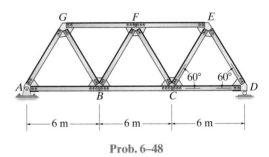

Prob. 6–48

6–42. Draw the influence line for the force in member *CD*.

6–43. Draw the influence line for the force in member *JK*.

*6–44. Draw the influence line for the force in member *DK*.

6–49. Draw the influence line for the force in member *BF* of the Warren truss. Indicate numerical values for the peaks. All members have the same length.

6–50. Draw the influence line for the force in member *FE* of the Warren truss. Indicate numerical values for the peaks. All members have the same length.

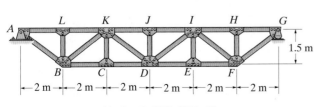

Probs. 6–42/6–43/6–44

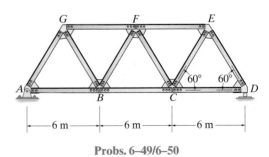

Probs. 6–49/6–50

6–51. Draw the influence line for the force in member *CL*.

*6–52. Draw the influence line for the force in member *DL*.

6–53. Draw the influence line for the force in member *CD*.

*6–56. Draw the influence line for the force in member *GD*, then determine the maximum force (tension or compression) that can be developed in this member due to a uniform live load of 3 kN/m that acts on the bridge deck along the bottom cord of the truss.

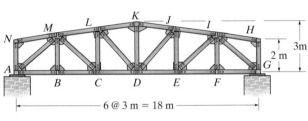

Probs. 6–51/6–52/6–53

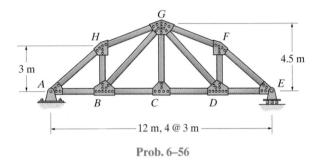

Prob. 6–56

6–57. Draw the influence line for the force in member *CD*, and then determine the maximum force (tension or compression) that can be developed in this member due to a uniform live load of 12 kN/m which acts along the bottom cord of the truss.

6–54. Draw the influence line for the force in member *CD*.

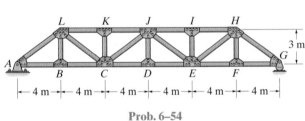

Prob. 6–54

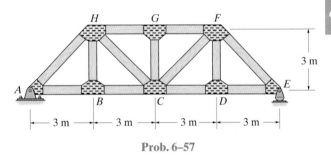

Prob. 6–57

6–58. Draw the influence line for the force in member *CF*, and then determine the maximum force (tension or compression) that can be developed in this member due to a uniform live load of 12 kN/m which is transmitted to the truss along the bottom cord.

6–55. Draw the influence line for the force in member *KJ*.

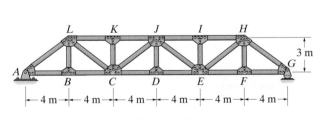

Prob. 6–55

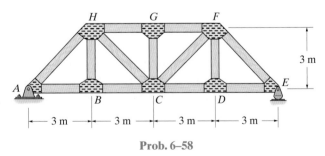

Prob. 6–58

As the train passes over this girder bridge the engine and its cars will exert vertical reactions on the girder. These along with the dead load of the bridge must be considered for design.

6.6 Maximum Influence at a Point due to a Series of Concentrated Loads

Once the influence line of a function has been established for a point in a structure, the maximum effect caused by a live concentrated force is determined by multiplying the peak ordinate of the influence line by the magnitude of the force. In some cases, however, *several* concentrated forces must be placed on the structure. An example would be the wheel loadings of a truck or train. In order to determine the maximum effect in this case, either a trial-and-error procedure can be used or a method that is based on the change in the function that takes place as the load is moved. Each of these methods will now be explained specifically as it applies to shear and moment.

Shear. Consider the simply supported beam with the associated influence line for the shear at point C in Fig. 6–27a. The maximum *positive shear* at point C is to be determined due to the series of concentrated (wheel) loads which move from right to left over the beam. The critical loading will occur when one of the loads is placed *just to the right* of point C, which is coincident with the positive peak of the influence line. By trial and error each of three possible cases can therefore be investigated, Fig. 6–27b. We have

Case 1: $(V_C)_1 = 4.5(0.75) + 18(0.625) + 18(0.5) = 23.63 \text{ kN}$
Case 2: $(V_C)_2 = 4.5(-0.125) + 18(0.75) + 18(0.625) = 24.19 \text{ kN}$
Case 3: $(V_C)_3 = 4.5(0) + 18(-0.125) + 18(0.75) = 11.25 \text{ kN}$

Case 2, with the 4.5-kN force located 1.5^+ m from the left support, yields the largest value for V_C and therefore represents the critical loading. Actually investigation of Case 3 is unnecessary, since by inspection such an arrangement of loads would yield a value of $(V_C)_3$ that would be less than $(V_C)_2$.

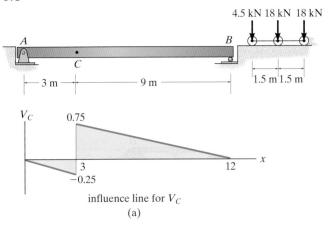

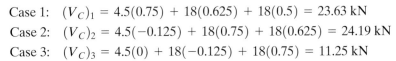

influence line for V_C

(a)

Fig. 6–27

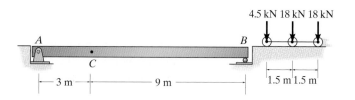

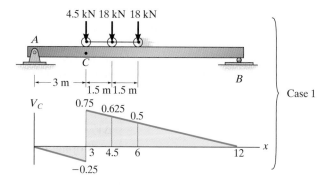

Case 1

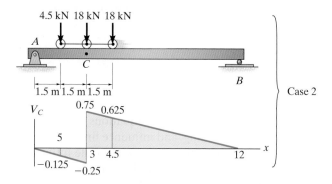

Case 2

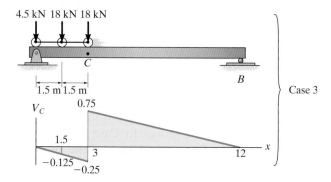

Case 3

(b)

Fig. 6–27

When many concentrated loads act on the span, as in the case of the E-72 load of Fig. 1–11, the trial-and-error computations used above can be tedious. Instead, the critical position of the loads can be determined in a more direct manner by finding the change in shear, ΔV, which occurs when the loads are moved from Case 1 to Case 2, then from Case 2 to Case 3, and so on. As long as each computed ΔV is *positive*, the new position will yield a larger shear in the beam at C than the previous position. Each movement is investigated until a negative change in shear is computed. When this occurs, the previous position of the loads will give the critical value. The change in shear ΔV for a load P that moves from position x_1 to x_2 over a beam can be determined by multiplying P by the change in the ordinate of the influence line, that is, $(y_2 - y_1)$. If the slope of the influence line is s, then $(y_2 - y_1) = s(x_2 - x_1)$, and therefore

$$\Delta V = Ps(x_2 - x_1)$$
Sloping Line

(6–1)

If the load moves past a point where there is a discontinuity or "jump" in the influence line, as point C in Fig. 6–27a, then the change in shear is simply

$$\Delta V = P(y_2 - y_1)$$
Jump

(6–2)

Use of the above equations will be illustrated with reference to the beam, loading, and influence line for V_C, shown in Fig. 6–28a. Notice that the magnitude of the slope of the influence line is $s = 0.75/(12 - 3) = 0.25/3 = 0.0833$, and the jump at C has a magnitude of $0.75 + 0.25 = 1$. Consider the loads of Case 1 moving 1.5 m to Case 2, Fig. 6–28b. When this occurs, the 4.5-kN load jumps *down* (-1) and *all* the loads move *up* the slope of the influence line. This causes a change of shear,

$$\Delta V_{1-2} = 4.5(-1) + [4.5 + 18 + 18](0.0833)(1.5) = +0.563 \text{ kN}$$

Since ΔV_{1-2} is positive, Case 2 will yield a larger value for V_C than Case 1. [Compare the answers for $(V_C)_1$ and $(V_C)_2$ previously computed, where indeed $(V_C)_2 = (V_C)_1 + 0.563$.] Investigating ΔV_{2-3}, which occurs when Case 2 moves to Case 3, Fig. 6–28b, we must account for the downward (negative) jump of the 18-kN load and the 1.5-m horizontal movement of all the loads *up* the slope of the influence line. We have

$$\Delta V_{2-3} = 18(-1) + (4.5 + 18 + 18)(0.0833)(1.5) = -12.94 \text{ kN}$$

Since ΔV_{2-3} is negative, Case 2 is the position of the critical loading, as determined previously.

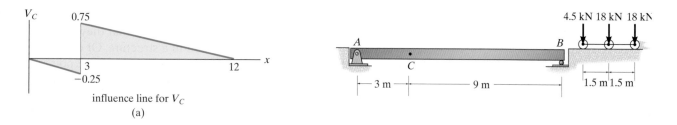

influence line for V_C

(a)

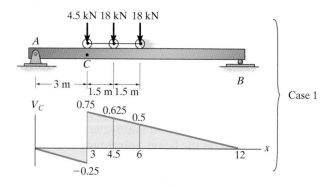

Case 1

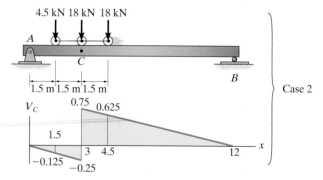

Case 2

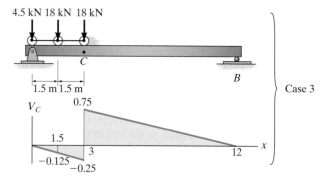

Case 3

(b)

Fig. 6–28

The girders of this bridge must resist the maximum moment caused by the weight of this jet plane as it passes over it.

Moment. We can also use the foregoing methods to determine the critical position of a series of concentrated forces so that they create the largest internal moment at a specific point in a structure. Of course, it is first necessary to draw the influence line for the moment at the point and determine the slopes s of its line segments. For a horizontal movement $(x_2 - x_1)$ of a concentrated force P, the change in moment, ΔM, is equivalent to the magnitude of the force times the change in the influence-line ordinate under the load, that is,

$$\Delta M = Ps(x_2 - x_1)$$
$$\text{Sloping Line}$$

(6–3)

As an example, consider the beam, loading, and influence line for the moment at point C in Fig. 6–29a. If each of the three concentrated forces is placed on the beam, coincident with the peak of the influence line, we will obtain the greatest influence from each force. The three cases of loading are shown in Fig. 6–29b. When the loads of Case 1 are moved 1.2 m to the left to Case 2, it is observed that the 9-kN load *decreases* ΔM_{1-2}, since the *slope* (2.25/3) is *downward,* Fig. 6–29a. Likewise, the 4-kN and 3-kN forces cause an *increase* of ΔM_{1-2}, since the *slope* $[2.25/(12-3)]$ is *upward.* We have

$$\Delta M_{1-2} = -9\left(\frac{2.25}{3}\right)(1.2) + (18 + 13.5)\left(\frac{2.25}{12-3}\right)(1.2) = 1.35 \text{ kN} \cdot \text{m}$$

Since ΔM_{1-2} is positive, we must further investigate moving the loads 1.8 m from Case 2 to Case 3.

$$\Delta M_{2-3} = -(9 + 18)\left(\frac{2.25}{3}\right)(1.8) + 13.5\left(\frac{2.25}{12-3}\right)(1.8) = -30.38 \text{ kN} \cdot \text{m}$$

Here the change is negative, so the greatest moment at C will occur when the beam is loaded as shown in Case 2, Fig. 6–29c. The maximum moment at C is therefore

$$(M_C)_{\text{max}} = 9(1.35) + 18(2.25) + 13.5(1.8) = 77.0 \text{ kN} \cdot \text{m}$$

The following examples further illustrate this method.

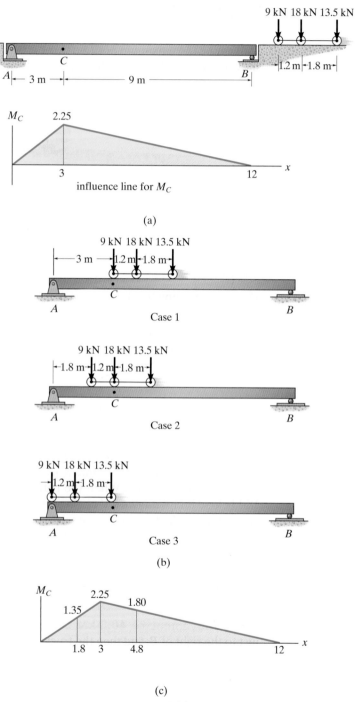

influence line for M_C

(a)

Case 1

Case 2

Case 3

(b)

(c)

Fig. 6–29

EXAMPLE 6.18

Determine the maximum positive shear created at point B in the beam shown in Fig. 6–30a due to the wheel loads of the moving truck.

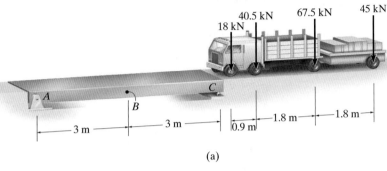

(a)

Fig. 6–30

SOLUTION
The influence line for the shear at B is shown in Fig. 6–30b.

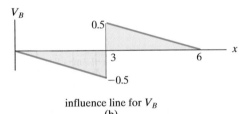

influence line for V_B
(b)

0.9-m Movement of 18-kN Load.   Imagine that the 18-kN load acts just to the right of point B so that we obtain its maximum positive influence. Since the beam segment BC is 3 m long, the 4.5-kN load is not as yet on the beam. When the truck moves 0.9 m to the left, the 18-kN load jumps *downward* on the influence line 1 unit and the 18-kN, 40.5-kN, and 67.5-kN loads create a positive increase in ΔV_B, since the slope is upward to the left. Although the 45-kN load also moves forward 0.9 m, it is still not on the beam. Thus,

$$\Delta V_B = 18(-1) + (18 + 40.5 + 67.5)\left(\frac{0.5}{3}\right)0.9 = +0.9 \text{ kN}$$

1.8-m Movement of 40.5-kN Load.   When the 40.5-kN load acts just to the right of B, and then the truck moves 1.8 m to the left, we have

$$\Delta V_B = 40.5(-1) + (18 + 40.5 + 67.5)\left(\frac{0.5}{3}\right)(1.8) + 45\left(\frac{0.5}{3}\right)(1.2)$$

$$= +6.3 \text{ kN}$$

Note in the calculation that the 45-kN load only moves 1.2 m on the beam.

1.8-m Movement of 67.5-kN Load.   If the 67.5-kN load is positioned just to the right of B and then the truck moves 1.8 m to the left, the 18-kN load moves only 0.3 m until it is off the beam, and likewise the 40.5-kN load moves only 1.2 m until it is off the beam. Hence,

$$\Delta V_B = 67.5(-1) + 18\left(\frac{0.5}{3}\right)(0.3) + 40.5\left(\frac{0.5}{3}\right)(1.2)$$

$$+ (67.5 + 45)\left(\frac{0.5}{3}\right)(1.8) \ = \ -24.8 \text{ kN}$$

Since ΔV_B is now negative, the correct position of the loads occurs when the 67.5-kN load is just to the right of point B, Fig. 6–30c. Consequently,

$$(V_B)_{max} = 18(-0.05) + 40.5(-0.2) + 67.5(0.5) + 45(0.2)$$

$$= 33.8 \text{ kN} \hspace{5cm} \textit{Ans.}$$

In practice one also has to consider motion of the truck from left to right and then choose the maximum value between these two situations.

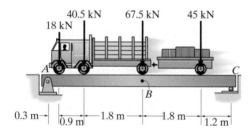

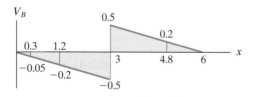

(c)

EXAMPLE 6.19

Determine the maximum positive moment created at point B in the beam shown in Fig. 6–31a due to the wheel loads of the crane.

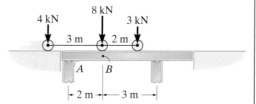

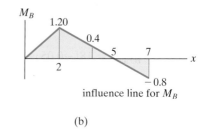

influence line for M_B

(b)

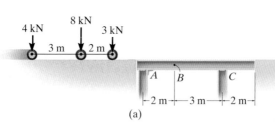

(a)

Fig. 6–31

SOLUTION

The influence line for the moment at B is shown in Fig. 6–31b.

2-m Movement of 3-kN Load.   If the 3-kN load is assumed to act at B and then moves 2 m to the right, Fig. 6–31b, the change in moment is

$$\Delta M_B = -3\left(\frac{1.20}{3}\right)(2) + 8\left(\frac{1.20}{3}\right)(2) = 7.20 \text{ kN} \cdot \text{m}$$

Why is the 4-kN load not included in the calculations?

3-m Movement of 8-kN Load.   If the 8-kN load is assumed to act at B and then moves 3 m to the right, the change in moment is

$$\Delta M_B = -3\left(\frac{1.20}{3}\right)(3) - 8\left(\frac{1.20}{3}\right)(3) + 4\left(\frac{1.20}{2}\right)(2)$$
$$= -8.40 \text{ kN} \cdot \text{m}$$

Notice here that the 4-kN load was initially 1 m off the beam, and therefore moves only 2 m on the beam.

Since there is a sign change in ΔM_B, the correct position of the loads for maximum positive moment at B occurs when the 8-kN force is at B, Fig. 6–31b. Therefore,

$$(M_B)_{\text{max}} = 8(1.20) + 3(0.4) = 10.8 \text{ kN} \cdot \text{m} \qquad Ans.$$

EXAMPLE 6.20

Determine the maximum compressive force developed in member *BG* of the side truss in Fig. 6–32a due to the right side wheel loads of the car and trailer. Assume the loads are applied directly to the truss and move only to the right.

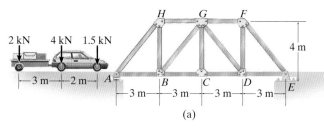

(a)

Fig. 6–32

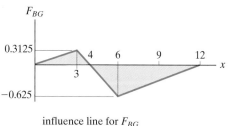

influence line for F_{BG}

(b)

SOLUTION

The influence line for the force in member *BG* is shown in Fig. 6–32b. Here a trial-and-error approach for the solution will be used. Since we want the greatest negative (compressive) force in *BG*, we begin as follows:

1.5-kN Load at Point C.   In this case

$$F_{BG} = 1.5 \text{ kN}(-0.625) + 4(0) + 2 \text{ kN}\left(\frac{0.3125}{3 \text{ m}}\right)(1 \text{ m})$$

$$= -0.729 \text{ kN}$$

4-kN Load at Point C.   By inspection this would seem a more reasonable case than the previous one.

$$F_{BG} = 4 \text{ kN}(-0.625) + 1.5 \text{ kN}\left(\frac{-0.625}{6 \text{ m}}\right)(4 \text{ m}) + 2 \text{ kN}(0.3125)$$

$$= -2.50 \text{ kN}$$

2-kN Load at Point C.   In this case all loads will create a compressive force in *BC*.

$$F_{BG} = 2 \text{ kN}(-0.625) + 4 \text{ kN}\left(\frac{-0.625}{6 \text{ m}}\right)(3 \text{ m}) + 1.5 \text{ kN}\left(\frac{-0.625}{6 \text{ m}}\right)(1 \text{ m})$$

$$= -2.66 \text{ kN} \qquad\qquad\qquad\qquad\qquad\qquad Ans.$$

Since this final case results in the largest answer, the critical loading occurs when the 2-kN load is at *C*.

6.7   Absolute Maximum Shear and Moment

In Sec. 6–6 we developed the methods for computing the maximum shear and moment at a *specified point* in a beam due to a series of concentrated moving loads. A more general problem involves the determination of both the *location of the point* in the beam *and the position of the loading* on the beam so that one can obtain the *absolute maximum* shear and moment caused by the loads. If the beam is cantilevered or simply supported, this problem can be readily solved.

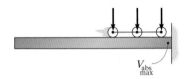

Shear.   For a *cantilevered beam* the absolute maximum shear will occur at a point located just next to the fixed support. The maximum shear is found by the method of sections, with the loads positioned anywhere on the span, Fig. 6–33.

Fig. 6–33

For *simply supported beams* the absolute maximum shear will occur just next to one of the supports. For example, if the loads are equivalent, they are positioned so that the first one in sequence is placed close to the support, as in Fig. 6–34.

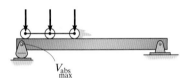

Moment.   The absolute maximum moment for a *cantilevered beam* occurs at the same point where absolute maximum shear occurs, although in this case the concentrated loads should be positioned at the *far end* of the beam, as in Fig. 6–35.

Fig. 6–34

For a *simply supported beam* the critical position of the loads and the associated absolute maximum moment cannot, in general, be determined by inspection. We can, however, determine the position analytically. For purposes of discussion, consider a beam subjected to the forces F_1, F_2, F_3 shown in Fig. 6–36a. Since the moment diagram for a series of concentrated forces consists of straight line segments having peaks at each force, the absolute maximum moment will occur under one of the forces. Assume this maximum moment occurs under F_2. The position of the loads F_1, F_2, F_3 on the beam will be specified by the distance x, measured from F_2 to the beam's centerline as shown. To determine a specific value of x, we first obtain the resultant force of the system, F_R, and its distance

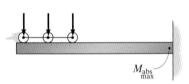

Fig. 6–35

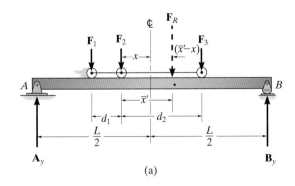

(a)

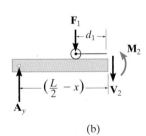

(b)

Fig. 6–36

$\bar{x}'$ measured from F_2. Once this is done, moments are summed about B, which yields the beam's left reaction, A_y, that is,

$$\Sigma M_B = 0; \qquad A_y = \frac{1}{L}(F_R)\left[\frac{L}{2} - (\bar{x}' - x)\right]$$

If the beam is sectioned just to the left of F_2, the resulting free-body diagram is shown in Fig. 6–36b. The moment M_2 under F_2 is therefore

$$\Sigma M = 0; \qquad M_2 = A_y\left(\frac{L}{2} - x\right) - F_1 d_1$$

$$= \frac{1}{L}(F_R)\left[\frac{L}{2} - (\bar{x}' - x)\right]\left(\frac{L}{2} - x\right) - F_1 d_1$$

$$= \frac{F_R L}{4} - \frac{F_R \bar{x}'}{2} - \frac{F_R x^2}{L} + \frac{F_R x \bar{x}'}{L} - F_1 d_1$$

For maximum M_2 we require

$$\frac{dM_2}{dx} = \frac{-2F_R x}{L} + \frac{F_R \bar{x}'}{L} = 0$$

or

$$x = \frac{\bar{x}'}{2}$$

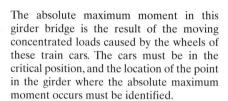

The absolute maximum moment in this girder bridge is the result of the moving concentrated loads caused by the wheels of these train cars. The cars must be in the critical position, and the location of the point in the girder where the absolute maximum moment occurs must be identified.

Hence, we may conclude that the *absolute maximum moment in a simply supported beam occurs under one of the concentrated forces, such that this force is positioned on the beam so that it and the resultant force of the system are equidistant from the beam's centerline.* Since there are a series of loads on the span (for example, F_1, F_2, F_3 in Fig. 6–36a), this principle will have to be applied to each load in the series and the corresponding maximum moment computed. By comparison, the largest moment is the absolute maximum. As a general rule, though, the absolute maximum moment often occurs under the largest force lying nearest the resultant force of the system.

Envelope of Maximum Influence-Line Values.

Rules or formulations for determining the absolute maximum shear or moment are difficult to establish for beams supported in ways other than the cantilever or simple support discussed here. An elementary way to proceed to solve this problem, however, requires constructing influence lines for the shear or moment at selected points along the entire length of the beam and then computing the maximum shear or moment in the beam for each point using the methods of Sec. 6–6. These values when plotted yield an "envelope of maximums," from which both the absolute maximum value of shear or moment and its location can be found. Obviously, a computer solution for this problem is desirable for complicated situations, since the work can be rather tedious if carried out by hand calculations.

EXAMPLE 6.21

Determine the absolute maximum moment in the simply supported bridge deck shown in Fig. 6–37a.

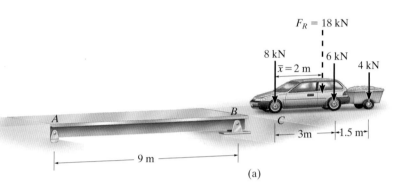

(a)

SOLUTION

The magnitude and position of the resultant force of the system are determined first, Fig. 6–37a. We have

$$+\downarrow F_R = \Sigma F; \qquad F_R = 8 + 6 + 4 = 18 \text{ kN}$$

$$\curvearrowright + M_{R_C} = \Sigma M_C; \qquad 18\bar{x} = 6(3) + 4(4.5)$$

$$\bar{x} = 2 \text{ m}$$

Let us first assume the absolute maximum moment occurs under the 6-kN load. The load and the resultant force are positioned equidistant from the beam's centerline, Fig. 6–37b. Calculating A_y first, Fig. 6–37b, we have

$$\downarrow + \Sigma M_B = 0; \qquad -A_y(9) + 18(5) = 0 \qquad A_y = 10 \text{ kN}$$

Now using the left section of the beam, Fig. 6–37c, yields

$$\downarrow + \Sigma M_S = 0; \qquad -10(5) + 8(3) + M_S = 0$$

$$M_S = 26.0 \text{ kN} \cdot \text{m}$$

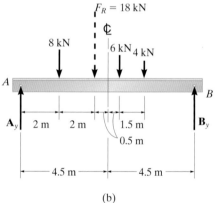

(b)

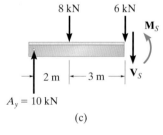

(c)

Fig. 6–37

There is a possibility that the absolute maximum moment may occur under the 8-kN load, since 8 kN > 6 kN and $\mathbf{F}_R$ is between both 8 kN and 6 kN. To investigate this case, the 8-kN load and $\mathbf{F}_R$ are positioned equidistant from the beam's centerline, Fig. 6–37d. Show that $A_y = 7$ kN as indicated in Fig. 6–37e and that

$$M_S = 24.5 \text{ kN} \cdot \text{m}$$

By comparison, the absolute maximum moment is

$$M_S = 26.0 \text{ kN} \cdot \text{m} \qquad\qquad Ans.$$

which occurs under the 6-kN load, when the loads are positioned on the beam as shown in Fig. 6–37b.

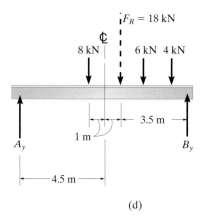

(d)

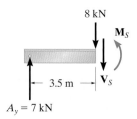

(e)

EXAMPLE **6.22**

The truck has a mass of 2 Mg and a center of gravity at G as shown in Fig. 6–38a. Determine the absolute maximum moment developed in the simply supported bridge deck due to the truck's weight. The bridge has a length of 10 m.

SOLUTION

As noted in Fig. 6–38a, the weight of the truck, $2(10^3)\,\text{kg}(9.81\,\text{m/s}^2) = 19.62\,\text{kN}$, and the wheel reactions have been calculated by statics. Since the largest reaction occurs at the front wheel, we will select this wheel along with the resultant force and position them *equidistant* from the centerline of the bridge, Fig. 6–38b. Using the resultant force rather than the wheel loads, the vertical reaction at B is then

$$\zeta + \Sigma M_A = 0; \qquad B_y(10) - 19.62(4.5) = 0$$
$$B_y = 8.829\,\text{kN}$$

The maximum moment occurs under the front wheel loading. Using the right section of the bridge deck, Fig. 6–38c, we have

$$\zeta + \Sigma M_s = 0; \qquad 8.829(4.5) - M_s = 0$$
$$M_s = 39.7\,\text{kN} \cdot \text{m} \qquad\qquad Ans.$$

19.62 kN

2 m   1 m

6.54 kN   13.08 kN

(a)

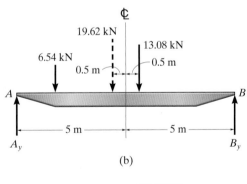

(b)

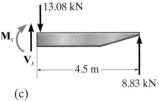

(c)

Fig. 6–38

PROBLEMS

6–59. Determine the maximum moment at point C on the single girder caused by the moving dolly that has a mass of 2 Mg and a mass center at G. Assume A is a roller.

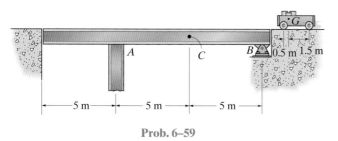

Prob. 6–59

*6–60. Determine the maximum moment in the suspended rail at point B if the rail supports the load of 10 kN on the trolley.

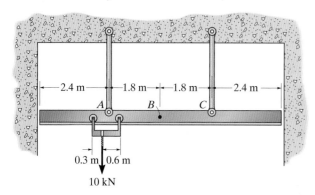

Prob. 6–60

6–61. Determine the maximum positive shear at point B if the rail supports the load of 10 kN on the trolley.

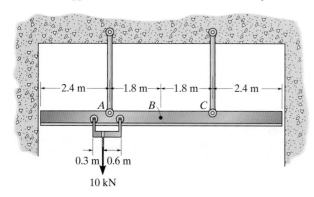

Prob. 6–61

6–62. Determine the maximum positive moment at the splice C on the side girder caused by the moving load which travels along the center of the bridge.

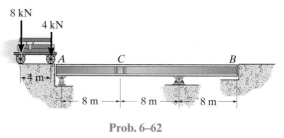

Prob. 6–62

6–63. Determine the maximum moment at C caused by the moving load.

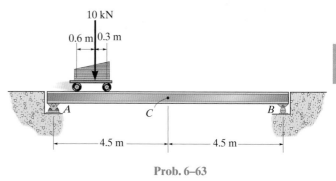

Prob. 6–63

*6–64. Draw the influence line for the force in member IH of the bridge truss. Determine the maximum force (tension or compression) that can be developed in this member due to a 324-kN truck having the wheel loads shown. Assume the truck can travel in *either direction* along the *center* of the deck, so that half its load is transferred to each of the two side trusses. Also assume the members are pin connected at the gusset plates.

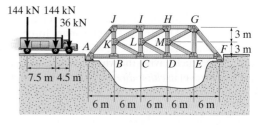

Prob. 6–64

6–65. Determine the maximum positive moment at point C on the single girder caused by the moving load.

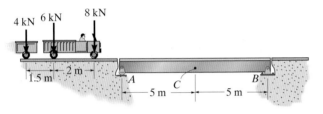

Prob. 6–65

6–66. The cart has a weight of 10 kN and a center of gravity at G. Determine the maximum positive moment created in the side girder at C as it crosses the bridge. Assume the car can travel in either direction along the center of the deck, so that *half* its load is transferred to each of the two side girders.

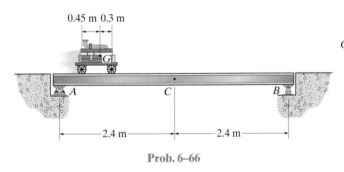

Prob. 6–66

6–67. Draw the influence line for the force in member BC of the bridge truss. Determine the maximum force (tension or compression) that can be developed in the member due to a 25-kN truck having the wheel loads shown. Assume the truck can travel in *either direction* along the *center* of the deck, so that *half* the load shown is transferred to each of the two side trusses. Also assume the members are pin connected at the gusset plates.

*6–68. Draw the influence line for the force in member IC of the bridge truss. Determine the maximum force (tension or compression) that can be developed in the member due to a 25-kN truck having the wheel loads shown. Assume the truck can travel in *either direction* along the *center* of the deck, so that half the load shown is transferred to each of the two side trusses. Also assume the members are pin connected at the gusset plates.

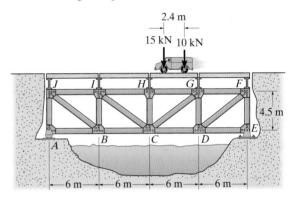

Probs. 6–67/6–68

6–69. The truck has a mass of 4 Mg and mass center at G_1, and the trailer has a mass of 1 Mg and mass center at G_2. Determine the absolute maximum live moment developed in the bridge.

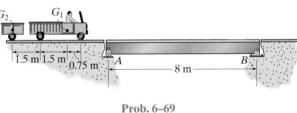

Prob. 6–69

6–70. Determine the absolute maximum live moment in the bridge in Problem 6–69 if the trailer is removed.

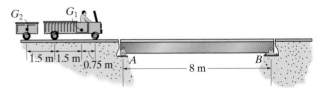

Prob. 6–70

6–71. Determine the absolute maximum live shear and absolute maximum moment in the jib beam *AB* due to the 10-kN loading. The end constraints require 0.1 m ≤ *x* ≤ 3.9 m.

6–73. Determine the absolute maximum moment in the girder bridge due to the truck loading shown. The load is applied directly to the girder.

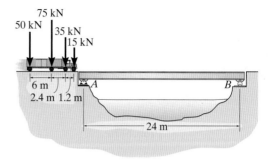

Prob. 6–73

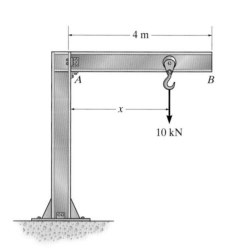

Prob. 6–71

6–74. Determine the absolute maximum shear in the beam due to the loading shown.

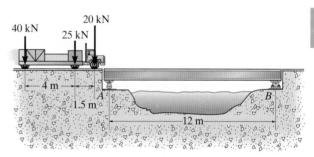

Prob. 6–74

6–72. Determine the maximum moment at *C* caused by the moving loads.

6–75. Determine the absolute maximum moment in the beam due to the loading shown.

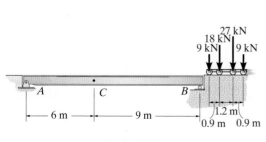

Prob. 6–72

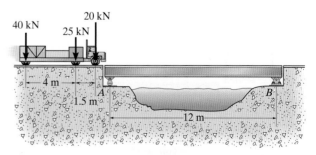

Prob. 6–75

*6–76. Determine the absolute maximum shear in the bridge girder due to the loading shown.

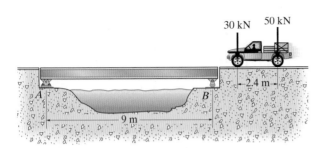

Prob. 6–76

6–77. Determine the absolute maximum moment in the bridge girder due to the loading shown.

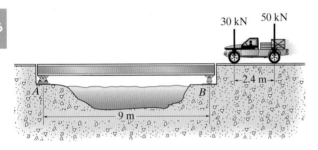

Prob. 6–77

6–78. Determine the absolute maximum moment in the girder due to the loading shown.

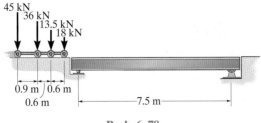

Prob. 6–78

6–79. Determine the absolute maximum shear in the beam due to the loading shown.

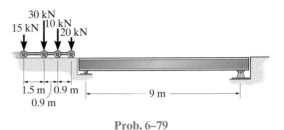

Prob. 6–79

*6–80. Determine the absolute maximum moment in the bridge due to the loading shown.

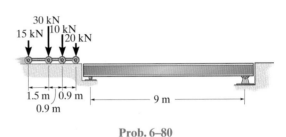

Prob. 6–80

6–81. The trolley rolls at C and D along the bottom and top flange of beam AB. Determine the absolute maximum moment developed in the beam if the load supported by the trolley is 10 kN. Assume the support at A is a pin and at B a roller.

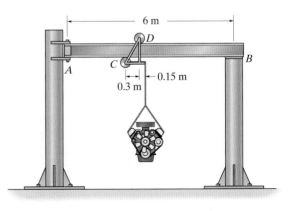

Prob. 6–81

PROJECT PROBLEMS

6–1P. The chain hoist on the wall crane can be placed anywhere along the boom (0.1 m < x < 3.4 m) and has a rated capacity of 28 kN. Use an impact factor of 0.3 and determine the absolute maximum bending moment in the boom and the maximum force developed in the tie rod BC. The boom is pinned to the wall column at its left end A. Neglect the size of the trolley at D.

6–2P. A simply supported pedestrian bridge is to be constructed in a city park and two designs have been proposed as shown in case a and case b. The truss members are to be made from timber. The deck consists of 1.5-m-long planks that have a mass of 20 kg/m². A local code states the live load on the deck is required to be 5 kPa with an impact factor of 0.2. Consider the deck to be simply supported on stringers. Floor beams then transmit the load to the bottom joints of the truss. (See Fig. 6–23.) In each case find the member subjected to the largest tension and largest compression load and suggest why you would choose one design over the other. Neglect the weights of the truss members.

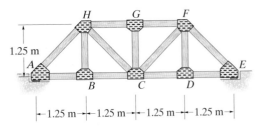

case a

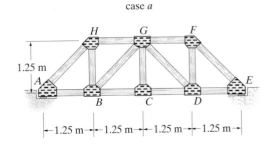

case b

Prob. 6–2P

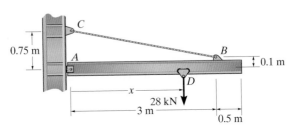

Prob. 6–1P

CHAPTER REVIEW

An influence line indicates the value of a reaction, shear, or moment at a specific point on a member, as a unit load moves over the member.

Once the influence line for a reaction, shear, or moment (function) is constructed, then it will be possible to locate the live load on the member to produce the maximum positive or negative value of the function.

A concentrated live force is applied at the positive (negative) peaks of the influence line. The value of the function is then equal to the product of the influence line ordinate and the magnitude of the force.

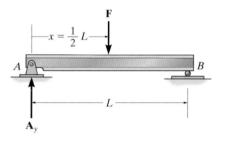
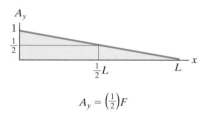

A uniform distributed load extends over a positive (negative) region of the influence line. The value of the function is then equal to the product of the area under the influence line for the region and the magnitude of the uniform load.

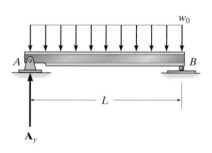
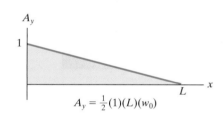

(Ref: Sections 6.1 & 6.2)

The general shape of the influence line can be established using the Müller-Breslau principle, which states that the influence line for a reaction, shear, or moment is to the same scale as the deflected shape of the member when it is acted upon by the reaction, shear, or moment.

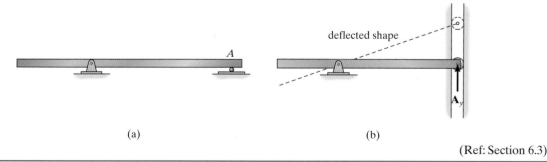

(a)   (b)

(Ref: Section 6.3)

Influence lines for floor girders and trusses can be established by placing the unit load at each panel point or joint, and calculating the value of the required reaction, shear, or moment.

(Ref: Sections 6.4 & 6.5)

When a series of concentrated loads pass over the member, then the various positions of the load on the member have to be considered to determine the largest shear or moment in the member. In general, place the loadings so that each contributes its maximum influence, as determined by multiplying each load by the ordinate of the influence line. This process of finding the actual position can be done using a trial-and-error procedure, or by finding the change in either the shear or moment when the loads are moved from one position to another. Each moment is investigated until a negative value of shear or moment occurs. Once this happens the previous position will define the critical loading.

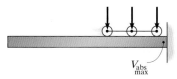

(Ref: Sections 6.6 & 6.7)

Absolute maximum *shear* in a cantilever or simply supported beam will occur at a support, when one of the loads is placed next to the support.

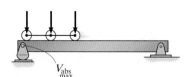

(Ref: Section 6.7)

Absolute maximum *moment* in a cantilevered beam occurs when the series of concentrated loads are placed at the farthest point away from the fixed support.

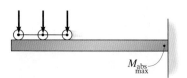

(Ref: Section 6.7)

To determine the absolute maximum moment in a simply supported beam, the resultant of the force system is first determined. Then it, along with one of the concentrated forces in the system is positioned so that these two forces are equidistant from the centerline of the beam. The maximum moment then occurs under the selected force. Each force in the system is selected in this manner, and by comparison the largest for all these cases is the absolute maximum moment.

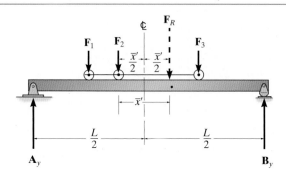

(Ref: Section 6.7)

The portal to this bridge must resist lateral loads due to wind and traffic. An approximate analysis can be made of the forces produced for a preliminary design of the members, before a more exact structural analysis is done.

Approximate Analysis of Statically Indeterminate Structures

7

In this chapter we will present some of the approximate methods used to analyze statically indeterminate trusses and frames. These methods were developed on the basis of structural behavior, and their accuracy in most cases compares favorably with more exact methods of analysis. Although not all types of structural forms will be discussed here, it is hoped that enough insight is gained from the study of these methods so that one can judge what would be the best approximations to make when performing an approximate force analysis of a statically indeterminate structure.

7.1   Use of Approximate Methods

When a *model* is used to represent any structure, the analysis of it must satisfy *both* the conditions of equilibrium and compatibility of displacement at the joints. As will be shown in later chapters of this text, the compatibility conditions for a *statically indeterminate* structure can be related to the loads provided we know the material's modulus of elasticity and the size and shape of the members. For an initial design, however, we will *not* know a member's size, and so a statically indeterminate analysis cannot be considered. For analysis a simpler model of the structure must be developed, one that is statically determinate. Once this model is specified, the analysis of it is called an *approximate analysis*. By performing an approximate analysis, a preliminary design of the members of a structure can be made, and when this is complete, the more exact indeterminate analysis can then be performed and the design refined. An approximate analysis also provides insight as to a structure's behavior under load and is beneficial when checking a more exact analysis or when time, money, or capability are not available for performing the more exact analysis.

Realize that, in a general sense, all methods of structural analysis are approximate, simply because the actual conditions of loading, geometry, material behavior, and joint resistance at the supports are never known in an *exact sense*. In this text, however, the statically indeterminate analysis of a structure will be referred to as an *exact analysis,* and the simpler statically determinate analysis will be referred to as the *approximate analysis*.

7.2   Trusses

A common type of truss often used for lateral bracing of a building or for the top and bottom cords of a bridge is shown in Fig. 7–1a. (Also see Fig. 3–4.) When used for this purpose, this truss is not considered a primary element for the support of the structure, and as a result it is often analyzed by approximate methods. In the case shown, it will be noticed that if a diagonal is removed from each of the three panels, it will render the truss statically determinate. Hence, the truss is statically indeterminate to the third degree (using Eq. 3–1, $b + r > 2j$, or $16 + 3 > 8(2)$) and therefore we must make three assumptions regarding the bar forces in order to reduce the truss to one that is statically determinate. These assumptions can be made with regard to the cross-diagonals, realizing that when one diagonal in a panel is in tension the corresponding cross-diagonal will be in compression. This is evident from Fig. 7–1b, where the "panel shear" **V** is carried by the *vertical component* of tensile force in member *a* and the *vertical component* of compressive force in member *b*. Two methods of analysis are generally acceptable.

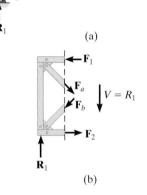

(a)

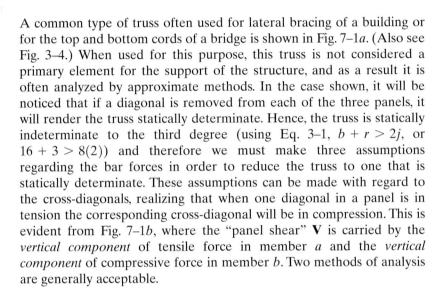

(b)

Fig. 7–1

Method 1: If the diagonals are intentionally designed to be *long and slender,* it is reasonable to assume that they *cannot* support a compressive force; otherwise, they may easily buckle. Hence the panel shear is resisted entirely by the *tension diagonal,* whereas the *compressive diagonal is assumed to be a zero-force member.*

Method 2: If the diagonal members are intended to be constructed from large rolled sections such as angles or channels, they may be equally capable of supporting a tensile and compressive force. Here we will assume that the tension and compression diagonals each carry *half* the panel shear.

Both of these methods of approximate analysis are illustrated numerically in the following examples.

An approximate method can be used to determine the forces in the cross bracing in each panel of this bascule railroad bridge. Here the cross members are thin and so we can assume they carry no compressive force.

EXAMPLE 7.1

Determine (approximately) the forces in the members of the truss shown in Fig. 7–2a. The diagonals are to be designed to support both tensile and compressive forces, and therefore each is assumed to carry half the panel shear. The support reactions have been computed.

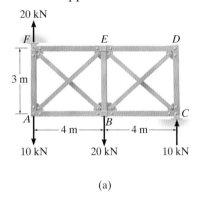

(a)

Fig. 7–2

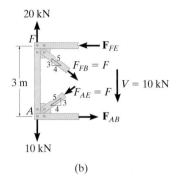

(b)

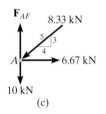

(c)

SOLUTION

By inspection the truss is statically indeterminate to the second degree. The two assumptions require the tensile and compressive diagonals to carry equal forces, that is, $F_{FB} = F_{AE} = F$. For a vertical section through the left panel, Fig. 7–2b, we have

$$+\uparrow \Sigma F_y = 0; \quad 20 - 10 - 2\left(\tfrac{3}{5}\right)F = 0 \qquad F = 8.33 \text{ kN} \qquad \textit{Ans.}$$

so that

$$F_{FB} = 8.33 \text{ kN (T)} \qquad \textit{Ans.}$$
$$F_{AE} = 8.33 \text{ kN (C)} \qquad \textit{Ans.}$$

$$\zeta + \Sigma M_A = 0; \quad -8.33\left(\tfrac{4}{5}\right)(3) + F_{FE}(3) = 0 \quad F_{FE} = 6.67 \text{ kN (C)} \quad \textit{Ans.}$$

$$\zeta + \Sigma M_F = 0; \quad -8.33\left(\tfrac{4}{5}\right)(3) + F_{AB}(3) = 0 \quad F_{AB} = 6.67 \text{ kN (T)} \quad \textit{Ans.}$$

From joint A, Fig. 7–2c,

$$+\uparrow \Sigma F_y = 0; \quad F_{AF} - 8.33\left(\tfrac{3}{5}\right) - 10 = 0 \qquad F_{AF} = 15 \text{ kN (T)} \quad \textit{Ans.}$$

A vertical section through the right panel is shown in Fig. 7–2d. Show that

$$F_{DB} = 8.33 \text{ kN (T)}, \quad F_{ED} = 6.67 \text{ kN (C)} \qquad \textit{Ans.}$$
$$F_{EC} = 8.33 \text{ kN (C)}, \quad F_{BC} = 6.67 \text{ kN (T)} \qquad \textit{Ans.}$$

Furthermore, using the free-body diagrams of joints D and E, Figs. 7–2e and 7–2f, show that

$$F_{DC} = 5 \text{ kN (C)} \qquad \textit{Ans.}$$
$$F_{EB} = 10 \text{ kN (T)} \qquad \textit{Ans.}$$

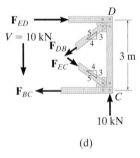

(d)

(e)

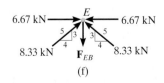

(f)

7

EXAMPLE 7.2

Cross bracing is used to provide lateral support for this bridge deck due to the wind and unbalanced traffic loads. Determine (approximately) the forces in the members of this truss. Assume the diagonals are slender and therefore will not support a compressive force. The loads and support reactions are shown in Fig. 7–3a.

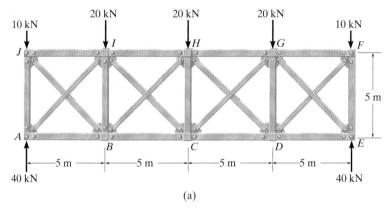

(a)

Fig. 7–3

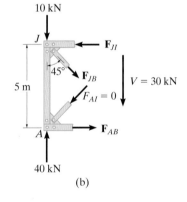

(b)

(c)

SOLUTION

By inspection the truss is statically indeterminate to the fourth degree. Thus the four assumptions to be used require that each compression diagonal sustain zero force. Hence, from a vertical section through the left panel, Fig. 7–3b, we have

$$F_{AI} = 0 \qquad Ans.$$

$$+\uparrow \Sigma F_y = 0; \qquad 40 - 10 - F_{JB} \cos 45° = 0$$

$$F_{JB} = 42.4 \text{ kN (T)} \qquad Ans.$$

$$\zeta + \Sigma M_A = 0; \qquad -42.4 \sin 45° (5) + F_{JI} (5) = 0$$

$$F_{JI} = 30 \text{ kN (C)} \qquad Ans.$$

$$\zeta + \Sigma M_J = 0; \qquad -F_{AB} (5) = 0$$

$$F_{AB} = 0 \qquad Ans.$$

From joint A, Fig. 7–3c,

$$F_{JA} = 40 \text{ kN (C)} \qquad Ans.$$

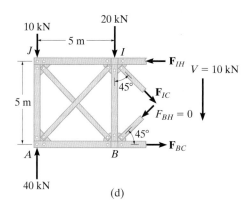

(d)

A vertical section of the truss through members IH, IC, BH, and BC is shown in Fig.7–3d. The panel shear is $V = \Sigma F_y = 40 - 10 - 20 = 10$ kN. We require

$$F_{BH} = 0 \qquad\qquad Ans.$$

$+\uparrow \Sigma F_y = 0; \qquad 40 - 10 - 20 - F_{IC} \cos 45° = 0$

$$F_{IC} = 14.1 \text{ kN (T)} \qquad Ans.$$

$\downarrow+\Sigma M_B = 0; \quad -40(5) + 10(5) - 14.1 \sin 45°(5) + F_{IH}(5) = 0$

$$F_{IH} = 40 \text{ kN (C)} \qquad Ans.$$

$\downarrow+\Sigma M_I = 0; \qquad -40(5) + 10(5) + F_{BC}(5) = 0$

$$F_{BC} = 30 \text{ kN (C)} \qquad Ans.$$

From joint B, Fig. 7–3e,

$+\uparrow \Sigma F_y = 0; \qquad\qquad 42.4 \sin 45° - F_{BI} = 0$

$$F_{BI} = 30 \text{ kN (C)} \qquad Ans.$$

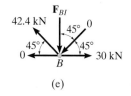

(e)

The forces in the other members can be determined by symmetry, except F_{CH}; however, from joint C, Fig. 7–3f, we have

$+\uparrow \Sigma F_y = 0; \qquad\qquad 2(14.1 \sin 45°) - F_{CH} = 0$

$$F_{CH} = 20 \text{ kN (C)} \qquad Ans.$$

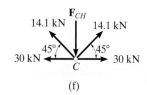

(f)

7

PROBLEMS

7–1. Determine (approximately) the force in each member of the truss. Assume the diagonals can support either a tensile or a compressive force.

7–2. Solve Prob. 7–1 assuming that the diagonals cannot support a compressive force.

7–5. Determine (approximately) the force in each member of the truss. Assume the diagonals can support either a tensile or a compressive force.

7–6. Solve Prob. 7–5 assuming that the diagonals cannot support a compressive force.

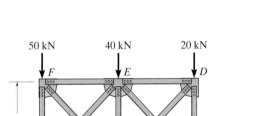

Probs. 7–1/7–2

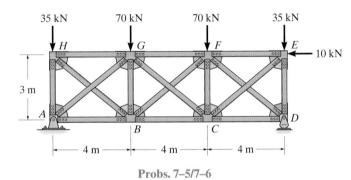

Probs. 7–5/7–6

7–3. Determine (approximately) the force in each member of the truss. Assume the diagonals can support either a tensile or a compressive force.

*7–4. Solve Prob. 7–3 assuming that the diagonals cannot support a compressive force.

7–7. Determine (approximately) the force in each member of the truss. Assume the diagonals can support either a tensile or compressive force.

*7–8. Solve Prob. 7–7 assuming that the diagonals cannot support a compressive force.

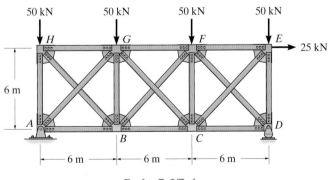

Probs. 7–3/7–4

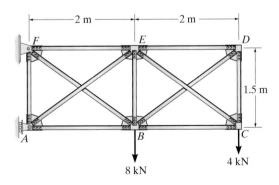

Probs. 7–7/7–8

7–9. Determine (approximately) the force in each member of the truss. Assume the diagonals can support both tensile and compressive forces.

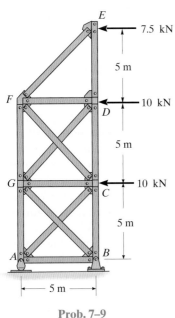

Prob. 7–9

7–10. Determine (approximately) the force in each member of the truss. Assume the diagonals DG and AC cannot support a compressive force.

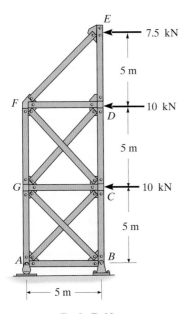

Prob. 7–10

7–11. Determine (approximately) the force in each member of the truss. Assume the diagonals can support either a tensile or compressive force.

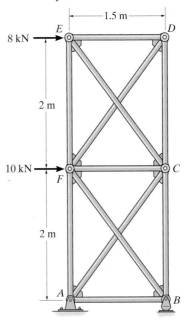

Prob. 7–11

***7–12.** Determine (approximately) the force in each member of the truss. Assume the diagonals cannot support a compressive force.

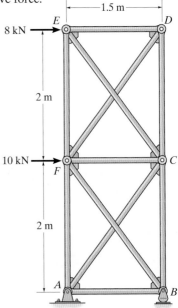

Prob. 7–12

7

7.3 Vertical Loads on Building Frames

Building frames often consist of girders that are *rigidly connected* to columns so that the entire structure is better able to resist the effects of lateral forces due to wind and earthquake. An example of such a rigid framework, often called a building bent, is shown in Fig. 7–4.

In practice, a structural engineer can use several techniques for performing an approximate analysis of a building bent. Each is based upon knowing how the structure *will deform under load*. One technique would be to consider only the members within a localized region of the structure. This is possible provided the deflections of the members within the region cause little disturbance to the members outside the region. Most often, however, the deflection curve of the entire structure is considered. From this, the approximate location of points of inflection, that is, the points where the member changes its curvature, can be specified. These points can be considered as *pins* since there is zero moment within the member at the points of inflection. We will use this idea in this section to analyze the forces on building frames due to vertical loads, and in Secs. 7–5 and 7–6 an approximate analysis for frames subjected to lateral loads will be presented. Since the frame can be subjected to both of these loadings simultaneously, then, provided the material remains elastic, the resultant loading is determined by superposition.

Assumptions for Approximate Analysis.

Consider a typical girder located within a building bent and subjected to a uniform vertical load, as shown in Fig. 7–5a. The column supports at A and B will each exert three reactions on the girder, and therefore the girder will be statically indeterminate to the third degree (6 reactions – 3 equations of equilibrium). To make the girder statically determinate, an approximate analysis will therefore require three assumptions. If the columns are extremely stiff, no rotation at A and B will occur, and the deflection

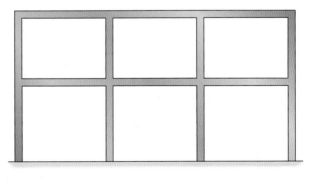

typical building frame

Fig. 7–4

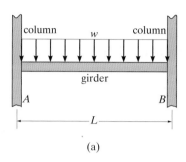

(a)

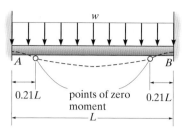

fixed supported
(b)

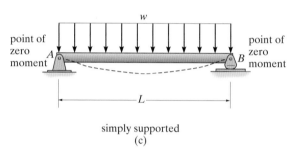

simply supported
(c)

approximate case
(d)

Fig. 7–5

curve for the girder will look like that shown in Fig. 7–5*b*. Using one of the methods presented in Chapters 9 through 11, an exact analysis reveals that for this case inflection points, or points of zero moment, occur at $0.21L$ from each support. If, however, the column connections at A and B are very flexible, then like a simply supported beam, zero moment will occur at the supports, Fig. 7–5*c*. In reality, however, the columns will provide some flexibility at the supports, and therefore we will assume that zero moment occurs at the *average point* between the two extremes, i.e., at $(0.21L + 0)/2 \approx 0.1L$ from each support, Fig. 7–5*d*. Furthermore, an exact analysis of frames supporting vertical loads indicates that the axial forces in the girder are negligible.

In summary then, each girder of length L may be modeled by a simply supported span of length $0.8L$ resting on two cantilevered ends, each having a length of $0.1L$, Fig. 7–5*e*. The following three assumptions are incorporated in this model:

1. There is zero moment in the girder, $0.1L$ from the left support.
2. There is zero moment in the girder, $0.1L$ from the right support.
3. The girder does not support an axial force.

By using statics, the internal loadings in the girders can now be obtained and a preliminary design of their cross sections can be made. The following example illustrates this numerically.

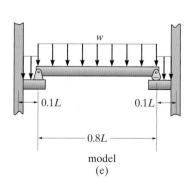

model
(e)

EXAMPLE 7.3

Determine (approximately) the moment at the joints E and C caused by members EF and CD of the building bent in Fig. 7–6a.

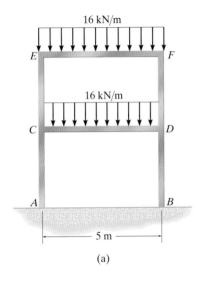

(a)

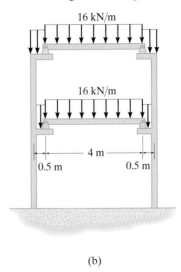

(b)

Fig. 7–6

SOLUTION

For an approximate analysis the frame is modeled as shown in Fig. 7–6b. Note that the cantilevered spans supporting the center portion of the girder have a length of $0.1L = 0.1(5) = 0.5$ m. Equilibrium requires the end reactions for the center portion of the girder to be 32 kN, Fig. 7–6c. The cantilevered spans are then subjected to a reaction moment of

$$M = 8(0.25) + 32(0.5) = 18 \text{ kN} \cdot \text{m} \qquad Ans.$$

This approximate moment, with opposite direction, acts on the joints at E and C, Fig. 7–6a. Using the results, the approximate moment diagram for one of the girders is shown in Fig. 7–6d.

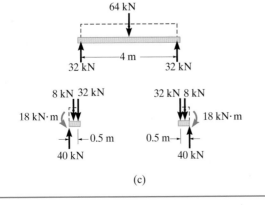

(c)

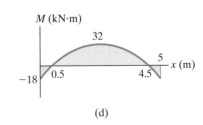

(d)

7.4 Portal Frames and Trusses

Frames. Portal frames are frequently used over the entrance of a bridge[*] and as a main stiffening element in building design in order to transfer horizontal forces applied at the top of the frame to the foundation. On bridges, these frames resist the forces caused by wind, earthquake, and unbalanced traffic loading on the bridge deck. Portals can be pin supported, fixed supported, or supported by partial fixity. The approximate analysis of each case will now be discussed for a simple three-member portal.

Pin Supported. A typical pin-supported portal frame is shown in Fig. 7–7a. Since four unknowns exist at the supports but only three equilibrium equations are available for solution, this structure is statically indeterminate to the first degree. Consequently, only one assumption must be made to reduce the frame to one that is statically determinate.

The elastic deflection of the portal is shown in Fig. 7–7b. This diagram indicates that a point of inflection, that is, where the moment changes from positive bending to negative bending, is located *approximately* at the girder's midpoint. Since the moment in the girder is zero at this point, we can *assume* a hinge exists there and then proceed to determine the reactions at the supports using statics. If this is done, it is found that the horizontal reactions (shear) at the base of each column are *equal* and the other reactions are those indicated in Fig. 7–7c. Furthermore, the moment diagrams for this frame are indicated in Fig. 7–7d.

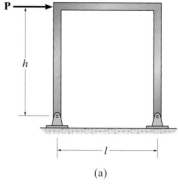

(a)

Fig. 7–7

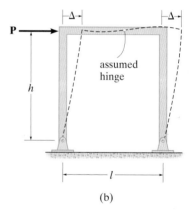

(b)

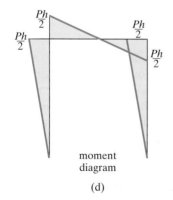

moment diagram

(d)

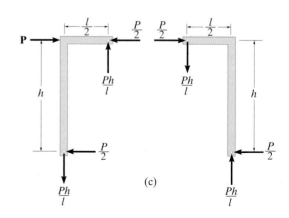

(c)

*See Fig. 3–4.

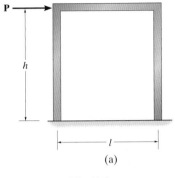

P

h

l

(a)

Fig. 7–8

Fixed Supported.   Portals with two fixed supports, Fig. 7–8a, are statically indeterminate to the third degree since there are a total of six unknowns at the supports. If the vertical members have equal lengths and cross-sectional areas, the frame will deflect as shown in Fig. 7–8b. For this case we will *assume* points of inflection occur at the midpoints of all three members, and therefore hinges are placed at these points. The reactions and moment diagrams for each member can therefore be determined by dismembering the frame at the hinges and applying the equations of equilibrium to each of the four parts. The results are shown in Fig. 7–8c. Note that, as in the case of the pin-connected portal, the horizontal reactions (shear) at the base of each column are *equal*. The moment diagram for this frame is indicated in Fig. 7–8d.

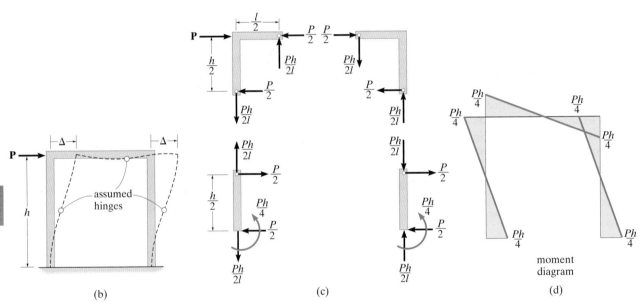

(b)

(c)

(d)

Partial Fixity.   Since it is both difficult and costly to construct a perfectly fixed support or foundation for a portal frame, it is conservative and somewhat realistic to assume a slight rotation occurs at the supports, Fig. 7–9a. As a result, the points of inflection on the columns lie somewhere between the case of having a pin-supported portal, Fig. 7–7a, where the "inflection points" are at the supports (base of columns), and a fixed-supported portal, Fig. 7–8a, where the inflection points are at the center of the columns. Many engineers arbitrarily define the location at $h/3$, Fig. 7–9b, and therefore place hinges at these points, and also at the center of the girder.

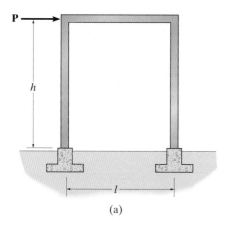

(a)

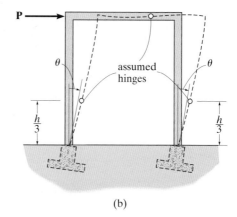

(b)

Fig. 7–9

Trusses.   When a portal is used to span large distances, a truss may be used in place of the horizontal girder. Such a structure is used on large bridges and as transverse bents for large auditoriums and mill buildings. A typical example is shown in Fig. 7–10a. In all cases, the suspended truss is assumed to be pin connected at its points of attachment to the columns. Furthermore, the truss keeps the columns straight within the region of attachment when the portal is subjected to the sidesway Δ, Fig. 7–10b. Consequently, we can analyze trussed portals using the same assumptions as those used for simple portal frames. For pin-supported columns, assume the horizontal reactions (shear) are equal, as in Fig. 7–7c. For fixed-supported columns, assume the horizontal reactions are equal and an inflection point (or hinge) occurs on each column, measured midway between the base of the column and the *lowest point* of truss member connection to the column. See Fig. 7–8c and Fig. 7–10b.

The following example illustrates how to determine the forces in the members of a trussed portal using the approximate method of analysis described above.

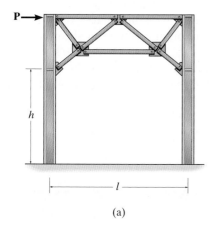

(a)

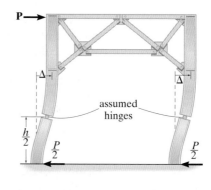

(b)

Fig. 7–10

EXAMPLE 7.4

Determine by approximate methods the forces acting in the members of the Warren portal shown in Fig. 7–11a.

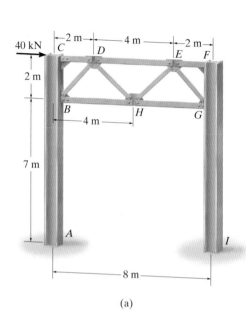

(a)

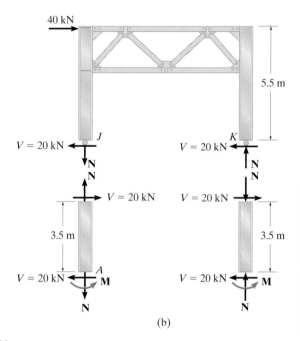

(b)

Fig. 7–11

SOLUTION

The truss portion B, C, F, G acts as a rigid unit. Since the supports are fixed, a point of inflection is assumed to exist $7 \text{ m}/2 = 3.5 \text{ m}$ above A and I, and equal horizontal reactions or shear act at the base of the columns, i.e., $\Sigma F_x = 0$; $V = 40 \text{ kN}/2 = 20 \text{ kN}$. With these assumptions, we can separate the structure at the hinges J and K, Fig. 7–11b, and determine the reactions on the columns as follows:

Lower Half of Column

$$\zeta + \Sigma M_A = 0; \qquad M - 3.5(20) = 0 \qquad M = 70 \text{ kN} \cdot \text{m}$$

Upper Portion of Column

$$\zeta + \Sigma M_J = 0; \qquad -40(5.5) + N(8) = 0 \qquad N = 27.5 \text{ kN}$$

Using the method of sections, Fig. 7–11c, we can now proceed to obtain the forces in members CD, BD, and BH.

$$+\uparrow \Sigma F_y = 0; \quad -27.5 + F_{BD} \sin 45° = 0 \qquad F_{BD} = 38.9 \text{ kN (T)} \quad Ans.$$

$$\zeta + \Sigma M_B = 0; \quad -20(3.5) - 40(2) + F_{CD}(2) = 0 \quad F_{CD} = 75 \text{ kN (C)} \quad Ans.$$

$$\zeta + \Sigma M_D = 0; \quad F_{BH}(2) - 20(5.5) + 27.5(2) = 0 \quad F_{BH} = 27.5 \text{ kN (T)} \quad Ans.$$

In a similar manner, show that one obtains the results on the free-body diagram of column FGI in Fig. 7–11d. Using these results, we can now find the force in each of the other truss members of the portal using the method of joints.

Joint D, Fig. 7–11e

$$+\uparrow \Sigma F_y = 0; \quad F_{DH} \sin 45° - 38.9 \sin 45° = 0 \quad F_{DH} = 38.9 \text{ kN (C)} \quad Ans.$$

$$\xrightarrow{+} \Sigma F_x = 0; \quad 75 - 2(38.9 \cos 45°) - F_{DE} = 0 \quad F_{DE} = 20 \text{ kN (C)} \quad Ans.$$

Joint H, Fig. 7–11f

$$+\uparrow \Sigma F_y = 0; \quad F_{HE} \sin 45° - 38.9 \sin 45° = 0 \quad F_{HE} = 38.9 \text{ kN (T)} \quad Ans.$$

These results are summarized in Fig. 7–11g.

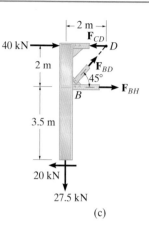

(c)

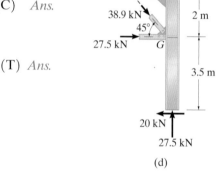

(d)

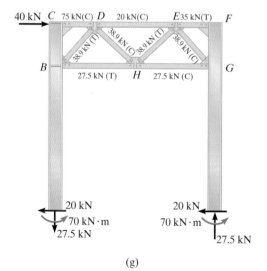

(g)

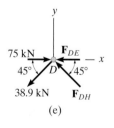

(e)

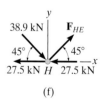

(f)

PROBLEMS

7–13. Determine (approximately) the internal moments at joints A and B of the frame.

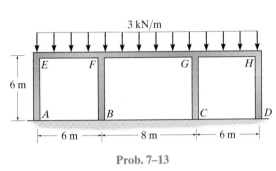

Prob. 7–13

7–15. Determine (approximately) the internal moment at A caused by the vertical loading.

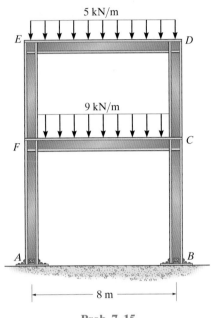

Prob. 7–15

7–14. Determine (approximately) the internal moments at joints F and D of the frame.

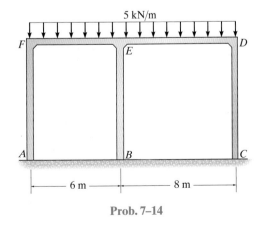

Prob. 7–14

*7–16. Determine (approximately) the internal moments at A and B caused by the vertical loading.

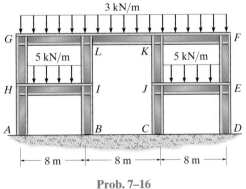

Prob. 7–16

7–17. Determine (approximately) the internal moments at joints *I* and *L*. Also, what is the internal moment at joint *H* caused by member *HG*?

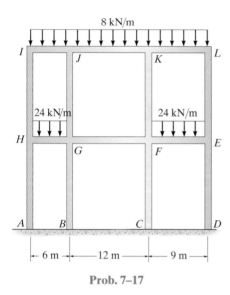

Prob. 7–17

7–19. Determine (approximately) the support reactions at *A* and *B* of the portal frame. Assume the supports are (a) pinned, and (b) fixed.

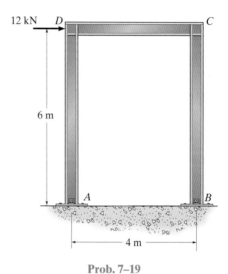

Prob. 7–19

7–18. Determine (approximately) the support actions at *A*, *B*, and *C* of the frame.

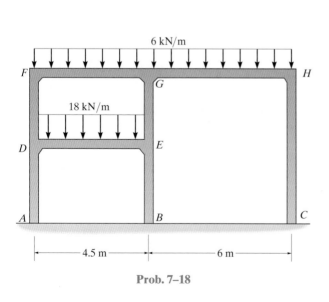

Prob. 7–18

***7–20.** Determine (approximately) the internal moment and shear at the ends of each member of the portal frame. Assume the supports at *A* and *D* are partially fixed, such that an inflection point is located at *h*/3 from the bottom of each column.

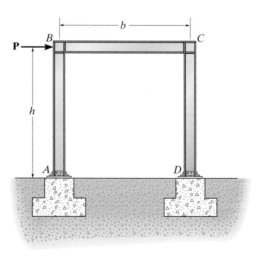

Prob. 7–20

7

7–21. Draw (approximately) the moment diagram for member *ACE* of the portal constructed with a *rigid* member *EG* and knee braces *CF* and *DH*. Assume that all points of connection are pins. Also determine the force in the knee brace *CF*.

7–22. Solve Prob. 7–21 if the supports at *A* and *B* are fixed instead of pinned.

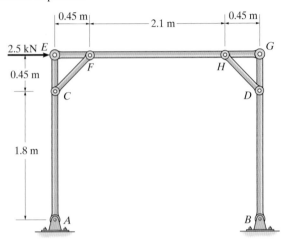

7–25. Draw (approximately) the moment diagram for column *AGF* of the portal. Assume all truss members and the columns to be pin connected at their ends. Also determine the force in all the truss members.

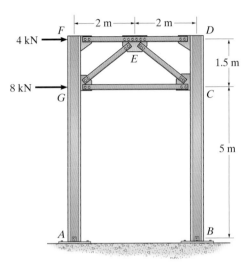

Prob. 7–25

Probs. 7–21/7–22

7–23. Determine (approximately) the force in each truss member of the portal frame. Also find the reactions at the fixed column supports *A* and *B*. Assume all members of the truss to be pin connected at their ends.

*__*7–24.__ Solve Prob. 7–23 if the supports at *A* and *B* are pinned instead of fixed.

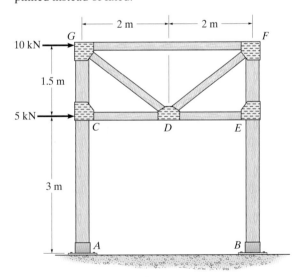

Probs. 7–23/7–24

7–26. Draw (approximately) the moment diagram for column *AGF* of the portal. Assume all the members of the truss to be pin connected at their ends. The columns are fixed at *A* and *B*. Also determine the force in all the truss members.

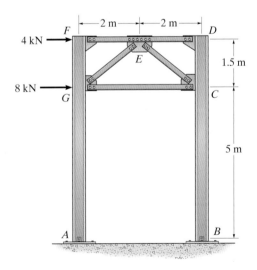

Prob. 7–26

7–27.   Determine (approximately) the force in each truss member of the portal frame. Also find the reactions at the fixed column supports A and B. Assume all members of the truss to be pin connected at their ends.

*7–28.   Solve Prob. 7–27 if the supports at A and B are pinned instead of fixed.

7–31.   Draw (approximately) the moment diagram for column ACD of the portal. Assume all truss members and the columns to be pin connected at their ends. Also determine the force in members FG, FH, and EH.

*7–32.   Solve Prob. 7–31 if the supports at A and B are fixed instead of pinned.

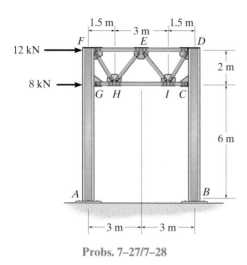

Probs. 7–27/7–28

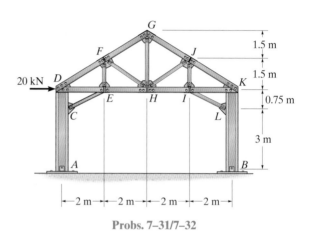

Probs. 7–31/7–32

7–29.   Determine (approximately) the force in members GF, GK, and JK of the portal frame. Also find the reactions at the fixed column supports A and B. Assume all members of the truss to be connected at their ends.

7–30.   Solve Prob. 7–29 if the supports at A and B are pin connected instead of fixed.

7–33.   Draw (approximately) the moment diagram for column AJI of the portal. Assume all truss members and the columns to be pin connected at their ends. Also determine the force in members HG, HL, and KL.

7–34.   Solve Prob. 7–33 if the supports at A and B are fixed instead of pinned.

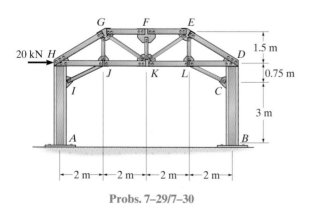

Probs. 7–29/7–30

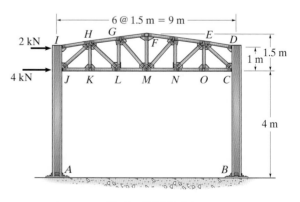

Probs. 7–33/7–34

7.5   Lateral Loads on Building Frames: Portal Method

In Sec. 7–4 we discussed the action of lateral loads on portal frames and found that for a frame fixed supported at its base, points of inflection occur at approximately the center of each girder and column and the columns carry equal shear loads, Fig. 7–8. A building bent deflects in the same way as a portal frame, Fig. 7–12a, and therefore it would be appropriate to assume inflection points occur at the center of the columns and girders. If we consider each bent of the frame to be composed of a series of portals, Fig. 7–12b, then as a further assumption, the *interior columns* would represent the effect of *two portal columns* and would therefore carry twice the shear V as the two exterior columns.

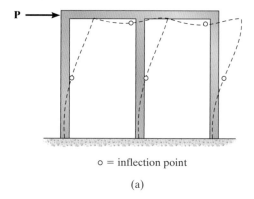

o = inflection point

(a)

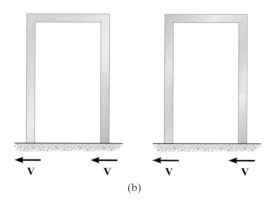

(b)

Fig. 7–12

In summary, the portal method for analyzing fixed-supported building frames requires the following assumptions:

1. A hinge is placed at the center of each girder, since this is assumed to be a point of zero moment.

2. A hinge is placed at the center of each column, since this is assumed to be a point of zero moment.

3. At a given floor level the shear at the interior column hinges is twice that at the exterior column hinges, since the frame is considered to be a superposition of portals.

These assumptions provide an adequate reduction of the frame to one that is statically determinate yet stable under loading.

By comparison with the more exact statically indeterminate analysis, *the portal method is most suitable for buildings having low elevation* and *uniform framing.* The reason for this has to do with the structure's action under load. In this regard, *consider the frame as acting like a cantilevered beam* that is fixed to the ground. Recall from mechanics of materials that *shear resistance* becomes more important in the design of *short* beams, whereas *bending* is more important if the beam is *long.* (See Sec. 7–6.) The portal method is based on the assumption related to shear as stated in item 3 above.

The following examples illustrate how to apply the portal method to analyze a building bent.

The portal method of analysis can be used to (approximately) perform a lateral-load analysis of this single-story frame.

EXAMPLE 7.5

Determine (approximately) the reactions at the base of the columns of the frame shown in Fig. 7–13a. Use the portal method of analysis.

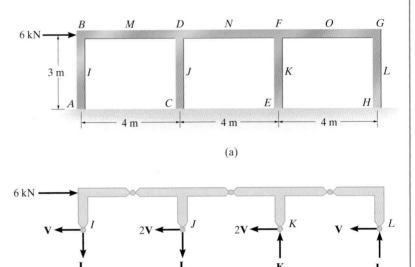

(a)

(b)

Fig. 7–13

SOLUTION

Applying the first two assumptions of the portal method, we place hinges at the centers of the girders and columns of the frame, Fig. 7–13a. A section through the column hinges at I, J, K, L yields the free-body diagram shown in Fig. 7–13b. Here the third assumption regarding the column shears applies. We require

$$\xrightarrow{+} \Sigma F_x = 0; \qquad 6 - 6V = 0 \qquad V = 1 \text{ kN}$$

Using this result, we can now proceed to dismember the frame at the hinges and determine their reactions. *As a general rule, always start this analysis at the corner or joint where the horizontal load is applied.* Hence, the free-body diagram of segment IBM is shown in Fig. 7–13c. The three reaction components at the hinges I_y, M_x, and M_y are determined by applying $\Sigma M_M = 0$, $\Sigma F_x = 0$, $\Sigma F_y = 0$, respectively. The adjacent segment MJN is analyzed next, Fig. 7–13d, followed by segment NKO, Fig. 7–13e, and finally segment OGL, Fig. 7–13f. Using these results, the free-body diagrams of the columns with their support reactions are shown in Fig. 7–13g.

If the horizontal segments of the girders in Figs. 7–13c, d, e and f are considered, show that the moment diagram for the girder looks like that shown in Fig. 7–13h.

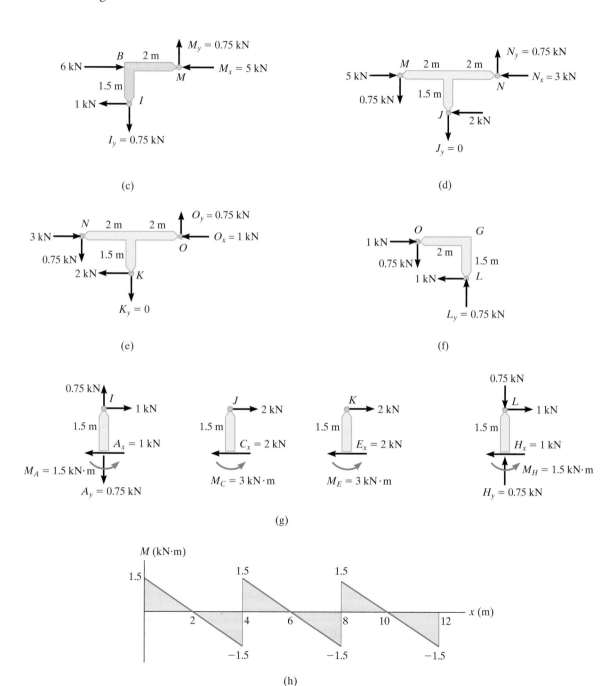

(c)

(d)

(e)

(f)

(g)

(h)

EXAMPLE 7.6

Determine (approximately) the reactions at the base of the columns of the frame shown in Fig. 7–14a. Use the portal method of analysis.

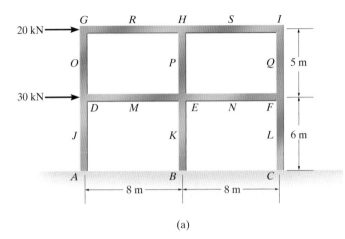

(a)

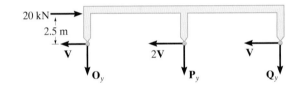

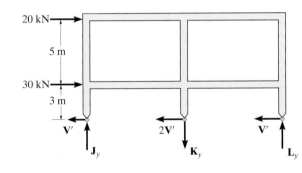

(b)

Fig. 7–14

SOLUTION

First hinges are placed at the *centers* of the girders and columns of the frame, Fig. 7–14a. A section through the hinges at O, P, Q and J, K, L yields the free-body diagrams shown in Fig. 7–14b. The column shears are calculated as follows:

$$\xrightarrow{+} \Sigma F_x = 0; \qquad 20 - 4V = 0 \qquad V = 5 \text{ kN}$$

$$\xrightarrow{+} \Sigma F_x = 0; \qquad 20 + 30 - 4V' = 0 \qquad V' = 12.5 \text{ kN}$$

Using these results, we can now proceed to analyze each part of the frame. The analysis starts with the *corner* segment *OGR*, Fig. 7–14c. The three unknowns O_y, R_x, and R_y have been calculated using the equations of equilibrium. With these results segment *OJM* is analyzed next, Fig. 7–14d; then segment *JA*, Fig. 7–14e; *RPS*, Fig. 7–14f; *PMKN*, Fig. 7–14g; and *KB*, Fig. 7–14h. Complete this example and analyze segments *SIQ*, then *QNL*, and finally *LC*, and show that $C_x = 12.5$ kN, $C_y = 15.625$ kN, and $M_C = 37.5$ kN·m. Also, use the results and show that the moment diagram for *DMENF* is given in Fig. 7–14i.

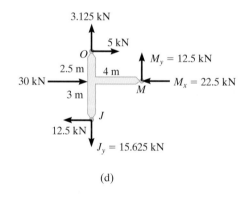

(c)

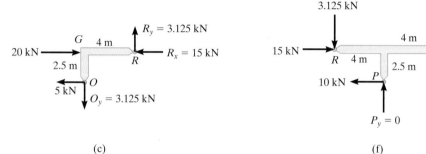

(f)

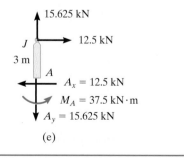

(d)

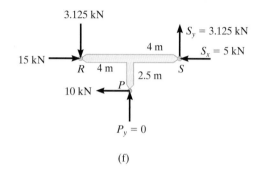

(g)

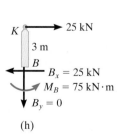

(e)

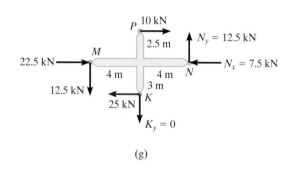

(h)

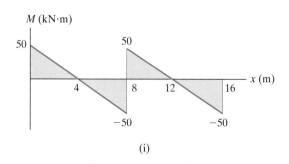

(i)

7.6   Lateral Loads on Building Frames: Cantilever Method

The cantilever method is based on the same action as a long cantilevered beam subjected to a transverse load. It may be recalled from mechanics of materials that such a loading causes a bending stress in the beam that varies linearly from the beam's neutral axis, Fig. 7–15*a*. In a similar manner, the lateral loads on a frame tend to tip the frame over, or cause a rotation of the frame about a "neutral axis" lying in a horizontal plane that passes through the columns between each floor. To counteract this tipping, the axial forces (or stress) in the columns will be tensile on one side of the neutral axis and compressive on the other side, Fig. 7–15*b*. Like the cantilevered beam, it therefore seems reasonable to assume this axial stress has a linear variation from the centroid of the column areas or neutral axis. *The cantilever method is therefore appropriate if the frame is tall and slender, or has columns with different cross-sectional areas.*

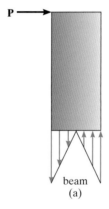

beam
(a)

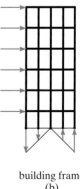

building frame
(b)

Fig. 7–15

In summary, using the cantilever method, the following assumptions apply to a fixed-supported frame.

1. A hinge is placed at the center of each girder, since this is assumed to be a point of zero moment.

2. A hinge is placed at the center of each column, since this is assumed to be a point of zero moment.

3. The axial *stress* in a column is proportional to its distance from the centroid of the cross-sectional areas of the columns at a given floor level. Since stress equals force per area, then in the special case of the *columns having equal cross-sectional areas,* the *force* in a column is also proportional to its distance from the centroid of the column areas.

These three assumptions reduce the frame to one that is both stable and statically determinate.

The following examples illustrate how to apply the cantilever method to analyze a building bent.

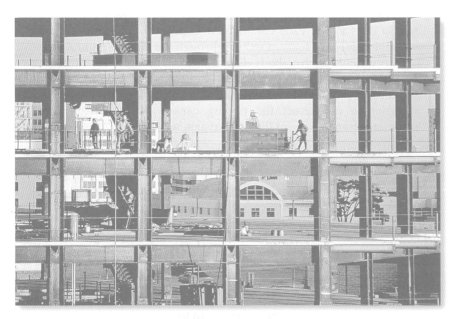

The building framework has rigid connections. A lateral-load analysis can be performed (approximately) by using the cantilever method of analysis.

EXAMPLE 7.7

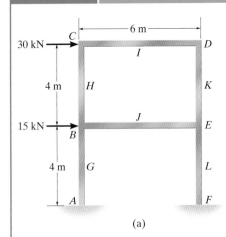

(a)

Fig. 7–16

Determine (approximately) the reactions at the base of the columns of the frame shown in Fig. 7–16a. The columns are assumed to have equal cross-sectional areas. Use the cantilever method of analysis.

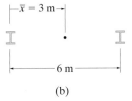

(b)

SOLUTION

First hinges are placed at the midpoints of the columns and girders. The locations of these points are indicated by the letters G through L in Fig. 7–16a. The centroid of the columns' cross-sectional areas can be determined by inspection, Fig. 7–16b, or analytically as follows:

$$\bar{x} = \frac{\Sigma \tilde{x} A}{\Sigma A} = \frac{0(A) + 6(A)}{A + A} = 3 \text{ m}$$

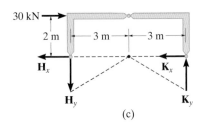

(c)

The axial *stress* in each column is thus proportional to its distance from this point. Here the columns have the same cross-sectional area, and so the force in each column is also proportional to its distance from the centroid. Hence, a section through the hinges H and K at the top story yields the free-body diagram shown in Fig. 7–16c. Note how the column to the left of the centroid must be subjected to tension and the one on the right is subjected to compression. This is necessary in order to counteract the tipping caused by the 30-kN force. Summing moments about the neutral axis, we have

$$\zeta + \Sigma M = 0; \qquad -30(2) + 3H_y + 3K_y = 0$$

The unknowns can be related by proportional triangles, Fig. 7–16c, that is,

$$\frac{H_y}{3} = \frac{K_y}{3} \qquad \text{or} \qquad H_y = K_y$$

Thus,

$$H_y = K_y = 10 \text{ kN}$$

In a similar manner, using a section of the frame through the hinges at G and L, Fig. 7–16d, we have

$$\zeta + \Sigma M = 0; \qquad -30(6) - 15(2) + 3G_y + 3L_y = 0$$

Since $G_y/3 = L_y/3$ or $G_y = L_y$, then

$$G_y = L_y = 35 \text{ kN}$$

Each part of the frame can now be analyzed using the above results. As in Examples 7–5 and 7–6, we begin at the upper corner where the applied loading occurs, i.e., segment HCI, Fig. 7–16a. Applying the three equations of equilibrium, $\Sigma M_1 = 0$, $\Sigma F_x = 0$, $\Sigma F_y = 0$, yields the results for H_x, I_x, and I_y, respectively, shown on the free-body diagram in Fig. 7–16e. Using these results, segment IDK is analyzed next, Fig. 7–16f; followed by HJG, Fig. 7–16g; then KJL, Fig. 7–16h; and finally the bottom portions of the columns, Fig. 7–16i and Fig. 7–16j. The moment diagrams for each girder are shown in Fig. 7–16k.

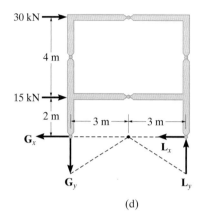

(d)

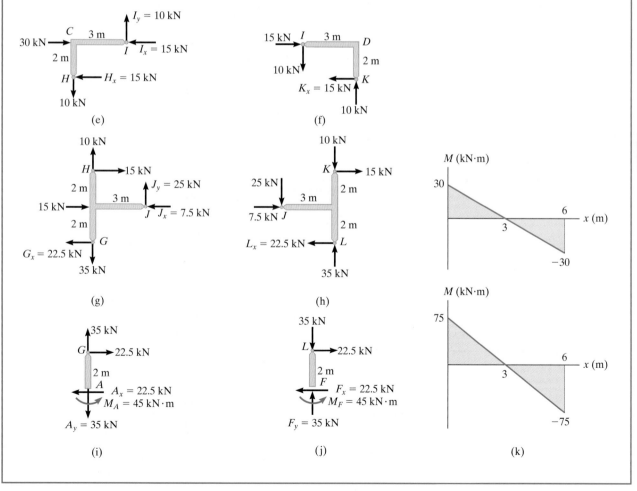

(e)

(f)

(g)

(h)

(i)

(j)

(k)

EXAMPLE 7.8

Show how to determine (approximately) the reactions at the base of the columns of the frame shown in Fig. 7–17a. The columns have the cross-sectional areas shown in Fig. 7–17b. Use the cantilever method of analysis.

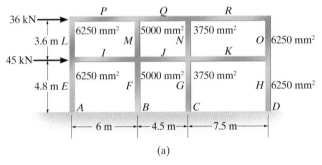

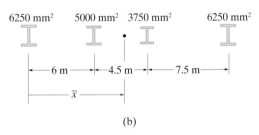

Fig. 7–17

SOLUTION

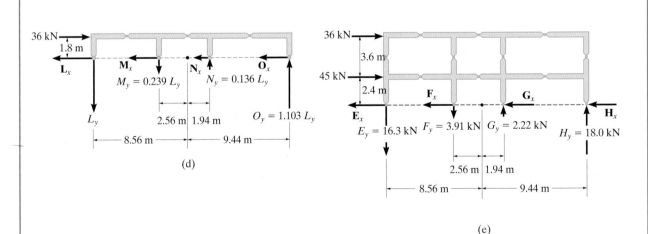

First, hinges are assumed to exist at the centers of the girders and columns of the frame, Fig. 7–17d and Fig. 7–17e. The centroid of the columns' cross-sectional areas is determined from Fig. 7–17b as follows:

$$\bar{x} = \frac{\Sigma \tilde{x} A}{\Sigma A} = \frac{0(6250) + 6(5000) + 10.5(3750) + 18(6250)}{6250 + 5000 + 3750 + 6250} = 8.559 \text{ m}$$

First we will consider the section through hinges at L, M, N and O.

In this problem the columns have *different* cross-sectional areas, so we must consider the *axial stress* in each column to be proportional to its distance from the neutral axis, located at $\bar{x} = 8.56$ m.

We can relate the column stresses by proportional triangles, Fig. 7–17c. Expressing the relations in terms of the force in each column, since $\sigma = F/A$, we have

$$\sigma_M = \frac{2.56 \text{ m}}{8.56 \text{ m}} \sigma_L; \quad \frac{M_y}{5000 \text{ mm}^2} = \frac{2.56}{8.56} \left(\frac{L_y}{6250 \text{ mm}^2}\right) \quad M_y = 0.239 L_y$$

$$\sigma_N = \frac{1.94 \text{ m}}{8.56 \text{ m}} \sigma_L; \quad \frac{N_y}{3750 \text{ mm}^2} = \frac{1.94}{8.56} \left(\frac{L_y}{6250 \text{ mm}^2}\right) \quad N_y = 0.136 L_y$$

$$\sigma_O = \frac{9.44 \text{ m}}{8.56 \text{ m}} \sigma_L; \quad \frac{O_y}{6250 \text{ mm}^2} = \frac{9.44}{8.56} \left(\frac{L_y}{6250 \text{ mm}^2}\right) \quad O_y = 1.103 L_y$$

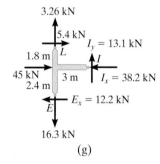

Now that each force is related to L_y, the free-body diagram is shown in Fig. 7–17d.

Note how the columns to the left of the centroid are subjected to tension and those on the right are subjected to compression. Why? Summing moments about the neutral axis, we have

$$\zeta + \Sigma M = 0; \quad -36 \text{ kN}(1.8\text{m}) + L_y(8.56 \text{ m}) + (0.239 L_y)(2.56 \text{ m})$$
$$+ (0.136 L_y)(1.94 \text{ m}) + (1.103 L_y)(9.44 \text{ m}) = 0$$

Solving,

$$L_y = 3.26 \text{ kN} \quad M_y = 0.78 \text{ kN} \quad N_y = 0.44 \text{ kN} \quad O_y = 3.60 \text{ kN}$$

Using this same method, show that one obtains the results in Fig. 7–17e for the columns at E, F, G, and H.

We can now proceed to analyze each part of the frame. As in the previous examples, we begin with the upper corner segment LP, Fig. 7–17f. Using the calculated results, segment LEI is analyzed next, Fig. 7–17g, followed by segment EA, Fig. 7–17h. One can continue to analyze the other segments in sequence, i.e., PQM, then $MJFI$, then FB, and so on.

PROBLEMS

7–35. Use the portal method of analysis and draw the moment diagram for girder *FED*.

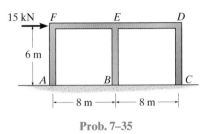

Prob. 7–35

*7–36. Use the portal method of analysis and draw the moment diagram for girder *JIHGF*.

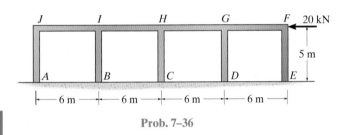

Prob. 7–36

7–37. Use the portal method and determine (approximately) the reactions at supports *A*, *B*, *C*, and *D*.

7–38. Use the cantilever method and determine (approximately) the reactions at supports *A*, *B*, *C*, and *D*. All columns have the same cross-sectional area.

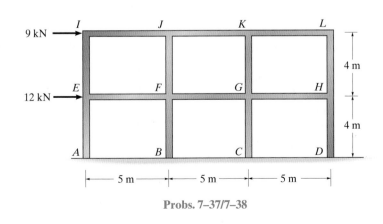

Probs. 7–37/7–38

7–39. Use the portal method of analysis and draw the moment diagram for column *AFE*.

*7–40. Solve Prob. 7–39 using the cantilever method of analysis. All the columns have the same cross-sectional area.

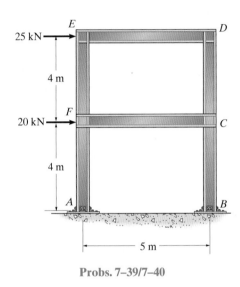

Probs. 7–39/7–40

7–41. Use the portal method and determine (approximately) the reactions at *A*.

7–42. Use the cantilever method and determine (approximately) the reactions at *A*. All of the columns have the same cross-sectional area.

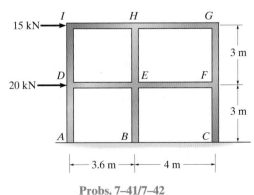

Probs. 7–41/7–42

7–43. Draw (approximately) the moment diagram for girder *PQRST* and column *BGLQ* of the building frame. Use the portal method.

*7–44. Draw (approximately) the moment diagram for girder *PQRST* and column *BGLQ* of the building frame. All columns have the same cross-sectional area. Use the cantilever method.

7–45. Draw the moment diagram for girder *IJKL* of the building frame. Use the portal method of analysis.

7–46. Solve Prob. 7–45 using the cantilever method of analysis. Each column has the cross-sectional area indicated.

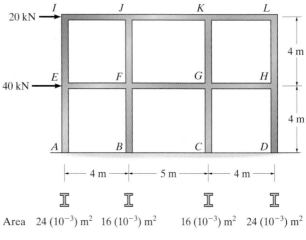

Area $24 \, (10^{-3}) \, \text{m}^2$ $16 \, (10^{-3}) \, \text{m}^2$ $16 \, (10^{-3}) \, \text{m}^2$ $24 \, (10^{-3}) \, \text{m}^2$

Probs. 7–45/7–46

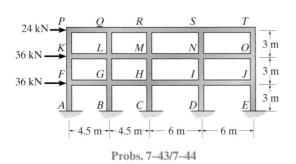

Probs. 7–43/7–44

PROJECT PROBLEMS

7–1P. The storage building bents shown in the photo are spaced 3 m apart and can be assumed pin connected at all points of support. Use the idealized model shown and determine the anticipated wind loading on the bent. Note that the wind loading is transmitted from the wall to the four purlins, then to the columns on the right side. Do an approximate analysis and determine the maximum axial load and maximum moment in column *AB*. Assume the columns and knee braces are pinned at their ends. The building is located on flat terrain in New Orleans, Louisiana, where *V* = 56 m/s.

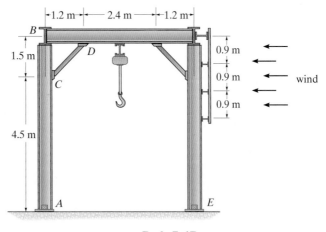

Prob. 7–1P

CHAPTER REVIEW

An approximate structural analysis is used to reduce a statically indeterminate structure to one that is statically determinate. By doing so a preliminary design of the members can be made, and once complete, the more exact indeterminate analysis can then be performed and the design refined.

Trusses having cross-diagonal bracing within their panels can be analyzed by assuming the tension diagonal supports the panel shear and the compressive diagonal is a zero-force member. This is reasonable if the members are long and slender. For larger cross sections, it is reasonable to assume each diagonal carries one-half the panel shear.

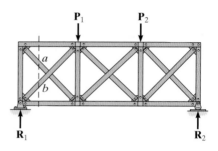

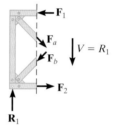

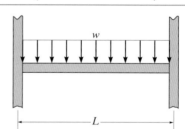

(Ref: Sections 7.1 & 7.2)

The approximate analysis of a vertical uniform load acting on a girder of length L of a fixed-connected building frame can be approximated by assuming that the girder does not support an axial load, and there are inflection points (hinges) located $0.1L$ from the supports.

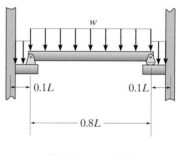

(Ref: Section 7.3)

Portal frames having fixed supports are approximately analyzed by assuming there are hinges at the midpoint of each column height, measured to the bottom of the truss bracing. Also, for these, and pin-supported frames, each column is assumed to support half the shear load on the frame.

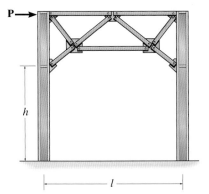

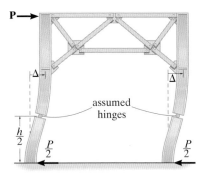

(Ref: Section 7.4)

For fixed-connected building frames subjected to lateral loads, we can assume there are hinges at the centers of the columns and girders. If the frame has a low elevation, shear resistance is important and so we can use the portal method, where the interior columns at any floor level carry twice the shear as that of the exterior columns. For tall slender frames, the cantilever method can be used, where the axial stress in a column is proportional to its distance from the centroid of the cross-sectional area of all the columns at a given floor level.

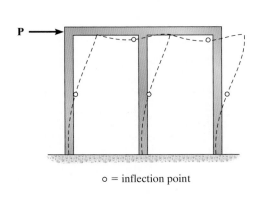

o = inflection point

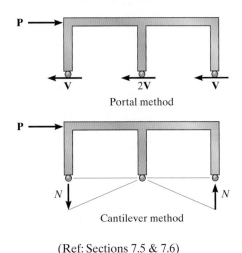

Portal method

Cantilever method

(Ref: Sections 7.5 & 7.6)

The deflection of this arch bridge must be carefully monitored while it is under construction.

Deflections

8

In this chapter we will show how to determine the elastic deflections of a beam using the method of double integration and two important geometrical methods, namely, the moment-area theorems and the conjugate-beam method. Double integration is used to obtain equations which define the slope and the elastic curve. The geometric methods provide a way to obtain the slope and deflection at specific points on the beam. Each of these methods has particular advantages or disadvantages, which will be discussed when each method is presented.

8.1 Deflection Diagrams and the Elastic Curve

Deflections of structures can occur from various sources, such as loads, temperature, fabrication errors, or settlement. In design, deflections must be limited in order to provide integrity and stability of roofs, and prevent cracking of attached brittle materials such as concrete, plaster or glass. Furthermore, a structure must not vibrate or deflect severely in order to "appear" safe for its occupants. More important, though, deflections at specified points in a structure must be determined if one is to analyze statically indeterminate structures.

The deflections to be considered throughout this text apply only to structures having *linear elastic material response*. Under this condition, a structure subjected to a load will return to its original undeformed position after the load is removed. The deflection of a structure is caused

TABLE 8-1

(1)

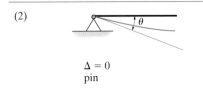

$\Delta = 0$
roller or rocker

(2)

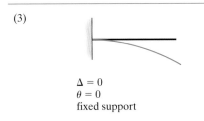

$\Delta = 0$
pin

(3)

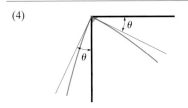

$\Delta = 0$
$\theta = 0$
fixed support

(4)

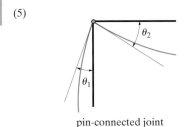

fixed-connected joint

(5)

pin-connected joint

by its internal loadings such as normal force, shear force, or bending moment. For *beams* and *frames,* however, the greatest deflections are most often caused by *internal bending,* whereas *internal axial forces* cause the deflections of a *truss.*

Before the slope or displacement of a point on a beam or frame is determined, it is often helpful to sketch the deflected shape of the structure when it is loaded in order to partially check the results. This *deflection diagram* represents the *elastic curve* or locus of points which defines the displaced position of the centroid of the cross section along the members. For most problems the elastic curve can be sketched without much difficulty. When doing so, however, it is necessary to know the restrictions as to slope or displacement that often occur at a support or a connection. With reference to Table 8–1, supports that *resist a force,* such as a pin, *restrict displacement;* and those that *resist moment,* such as a fixed wall, *restrict rotation.* Note also that deflection of frame members that are fixed connected (4) causes the joint to rotate the connected members by the same amount θ. On the other hand, if a pin connection is used at the joint, the members will each have a *different slope* or rotation at the pin, since the pin cannot support a moment (5).

The two-member frames support both the dead load of the roof and a live snow loading. The frame can be considered pinned at the wall, fixed at the ground, and having a fixed-connected joint.

If the elastic curve seems difficult to establish, it is suggested that the moment diagram for the beam or frame be drawn first. By our sign convention for moments established in Chapter 4, a *positive moment* tends to bend a beam or horizontal member *concave upward,* Fig. 8–1. Likewise, a *negative moment* tends to bend the beam or member *concave downward,* Fig. 8–2. Therefore, *if the shape of the moment diagram is known, it will be easy to construct the elastic curve and vice versa.* For example, consider the beam in Fig. 8–3 with its associated moment diagram. Due to the pin-and-roller support, the displacement at *A* and *D* must be zero. Within the region of negative moment, the elastic curve is concave downward; and within the region of positive moment, the elastic curve is concave upward. In particular, there must be an *inflection point* at the point where the curve changes from concave down to concave up, since this is a point of zero moment. Using these same principles, note how the elastic curve for the beam in Fig. 8–4 was drawn based on its moment diagram. In particular, realize that the positive moment reaction from the wall keeps the initial slope of the beam horizontal.

positive moment,
concave upward

Fig. 8–1

negative moment,
concave downward

Fig. 8–2

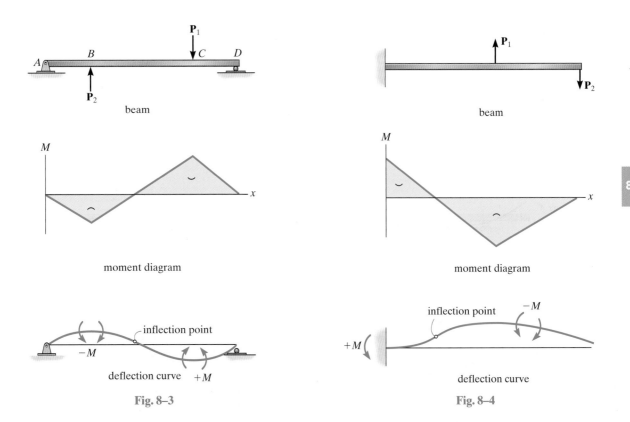

beam

moment diagram

deflection curve

Fig. 8–3

beam

moment diagram

deflection curve

Fig. 8–4

EXAMPLE 8.1

Draw the deflected shape of each of the beams shown in Fig. 8–5.

SOLUTION

In Fig. 8–5a the roller at A allows free rotation with no deflection while the fixed wall at B prevents both rotation and deflection. The deflected shape is shown by the bold line. In Fig. 8–5b, no rotation or deflection can occur at A and B. In Fig. 8–5c, the couple moment will rotate end A. This will cause deflections at both ends of the beam since no deflection is possible at B and C. Notice that segment CD remains undeformed (a straight line) since no internal load acts within it. In Fig. 8–5d, the pin (internal hinge) at B allows free rotation, and so the slope of the deflection curve will suddenly change at this point while the beam is constrained by its supports. In Fig. 8–5e, the compound beam deflects as shown. The slope abruptly changes on each side of the pin at B. Finally, in Fig. 8–5f, span BC will deflect concave upwards due to the load. Since the beam is continuous, the end spans will deflect concave downwards.

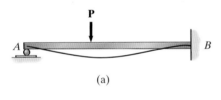

(a)

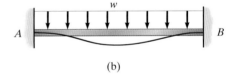

(b)

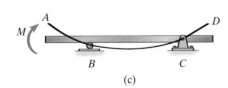

(c)

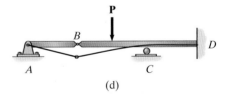

(d)

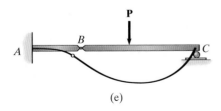

(e)

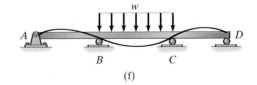

(f)

Fig. 8–5

EXAMPLE 8.2

Draw the deflected shapes of each of the frames shown in Fig. 8–6.

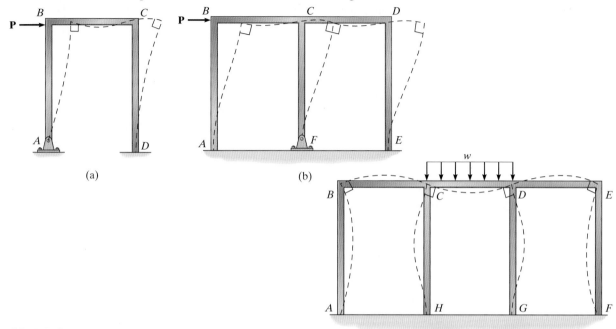

(a)

(b)

(c)

SOLUTION

In Fig. 8–6a, when the load **P** pushes joints *B* and *C* to the right, it will cause clockwise rotation of each column as shown. As a result, joints *B* and *C* must rotate clockwise. Since the 90° angle between the connected members must be maintained at these joints, the beam *BC* will deform so that its curvature is reversed from concave up on the left to concave down on the right. Note that this produces a point of inflection within the beam.

In Fig. 8–6b, **P** displaces joints *B*, *C*, and *D* to the right, causing each column to bend as shown. The fixed joints must maintain their 90° angles, and so *BC* and *CD* must have a reversed curvature with an inflection point near their midpoint.

In Fig. 8–6c, the vertical loading on this symmetric frame will bend beam *CD* concave upwards, causing clockwise rotation of joint *C* and counterclockwise rotation of joint *D*. Since the 90° angle at the joints must be maintained, the columns bend as shown. This causes spans *BC* and *DE* to be concave downwards, resulting in counterclockwise rotation at *B* and clockwise rotation at *E*. The columns therefore bend as shown. Finally, in Fig. 8–6d, the loads push joints *B* and *C* to the right, which bends the columns as shown. The fixed joint *B* maintains its 90° angle; however, no restriction on the relative rotation between the members at *C* is possible since the joint is a pin. Consequently, only beam *CD* does not have a reverse curvature.

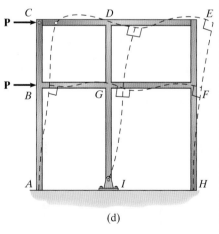

(d)

Fig. 8–6

FUNDAMENTAL PROBLEMS

F8–1. Draw the deflected shape of each beam. Indicate the inflection points.

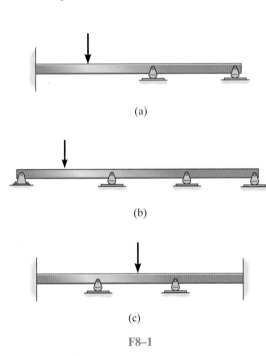

(a)

(b)

(c)

F8–1

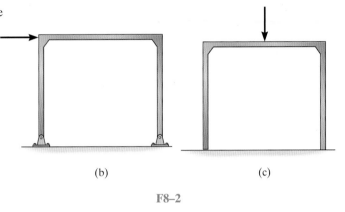

(b)   (c)

F8–2

F8–3. Draw the deflected shape of each frame. Indicate the inflection points.

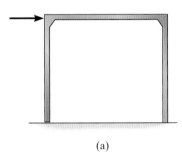

(a)

F8–2. Draw the deflected shape of each frame. Indicate the inflection points.

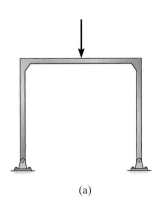

(a)

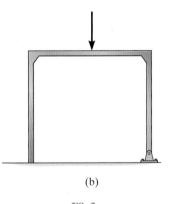

(b)

F8–3

8.2   Elastic-Beam Theory

In this section we will develop two important differential equations that relate the internal moment in a beam to the displacement and slope of its elastic curve. These equations form the basis for the deflection methods presented in this chapter, and for this reason the assumptions and limitations used in their development should be fully understood.

To derive these relationships, we will limit the analysis to the most common case of an initially straight beam that is elastically deformed by loads applied perpendicular to the beam's x axis and lying in the $x-v$ plane of symmetry for the beam's cross-sectional area, Fig. 8–7a. Due to the loading, the deformation of the beam is caused by both the internal shear force and bending moment. If the beam has a length that is much greater than its depth, the greatest deformation will be caused by bending, and therefore we will direct our attention to its effects. Deflections caused by shear will be discussed later in the chapter.

When the internal moment M deforms the element of the beam, each cross section remains plane and the angle between them becomes $d\theta$, Fig. 8–7b. The arc dx that represents a portion of the elastic curve intersects the neutral axis for each cross section. The *radius of curvature* for this arc is defined as the distance ρ, which is measured from the *center of curvature O'* to dx. Any arc on the element other than dx is subjected to a normal strain. For example, the strain in arc ds, located at a position y from the neutral axis, is $\epsilon = (ds' - ds)/ds$. However, $ds = dx = \rho\, d\theta$ and $ds' = (\rho - y)\, d\theta$, and so

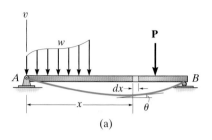

(a)

$$\epsilon = \frac{(\rho - y)\, d\theta - \rho\, d\theta}{\rho\, d\theta} \qquad \text{or} \qquad \frac{1}{\rho} = -\frac{\epsilon}{y}$$

If the material is homogeneous and behaves in a linear elastic manner, then Hooke's law applies, $\epsilon = \sigma/E$. Also, since the flexure formula applies, $\sigma = -My/I$. Combining these equations and substituting into the above equation, we have

$$\frac{1}{\rho} = \frac{M}{EI} \qquad\qquad (8\text{–}1)$$

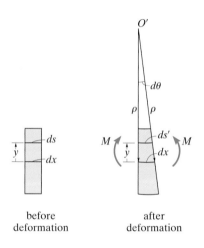

before
deformation

after
deformation

(b)

Fig. 8–7

Here

ρ = the radius of curvature at a specific point on the elastic curve ($1/\rho$ is referred to as the *curvature*)

M = the internal moment in the beam at the point where ρ is to be determined

E = the material's modulus of elasticity

I = the beam's moment of inertia computed about the neutral axis

The product EI in this equation is referred to as the *flexural rigidity*, and it is always a positive quantity. Since $dx = \rho\,d\theta$, then from Eq. 8–1,

$$d\theta = \frac{M}{EI}\,dx \qquad (8\text{–}2)$$

If we choose the v axis positive upward, Fig. 8–7a, and if we can express the curvature $(1/\rho)$ in terms of x and v, we can then determine the elastic curve for the beam. In most calculus books it is shown that this curvature relationship is

$$\frac{1}{\rho} = \frac{d^2v/dx^2}{[1 + (dv/dx)^2]^{3/2}}$$

Therefore,

$$\frac{M}{EI} = \frac{d^2v/dx^2}{[1 + (dv/dx)^2]^{3/2}} \qquad (8\text{–}3)$$

This equation represents a nonlinear second-order differential equation. Its solution, $v = f(x)$, gives the exact shape of the elastic curve—assuming, of course, that beam deflections occur only due to bending. In order to facilitate the solution of a greater number of problems, Eq. 8–3 will be modified by making an important simplification. Since the slope of the elastic curve for most structures is very small, we will use small deflection theory and assume $dv/dx \approx 0$. Consequently its square will be negligible compared to unity and therefore Eq. 8–3 reduces to

$$\boxed{\frac{d^2v}{dx^2} = \frac{M}{EI}} \qquad (8\text{–}4)$$

It should also be pointed out that by assuming $dv/dx \approx 0$, the original length of the beam's axis x and the *arc* of its elastic curve will be approximately the same. In other words, ds in Fig. 8–7b is approximately equal to dx, since

$$ds = \sqrt{dx^2 + dv^2} = \sqrt{1 + (dv/dx)^2}\,dx \approx dx$$

This result implies that points on the elastic curve will only be displaced vertically and not horizontally.

Tabulated Results.

In the next section we will show how to apply Eq. 8–4 to find the slope of a beam and the equation of its elastic curve. The results from such an analysis for some common beam loadings often encountered in structural analysis are given in the table on the inside front cover of this book. Also listed are the slope and displacement at critical points on the beam. Obviously, no single table can account for the many different cases of loading and geometry that are encountered in practice. When a table is not available or is incomplete, the displacement or slope of a specific point on a beam or frame can be determined by using the double integration method or one of the other methods discussed in this and the next chapter.

8.3 The Double Integration Method

Once M is expressed as a function of position x, then successive integrations of Eq. 8–4 will yield the beam's slope, $\theta \approx \tan\theta = dv/dx = \int (M/EI)\,dx$ (Eq. 8–2), and the equation of the elastic curve, $v = f(x) = \int \int (M/EI)\,dx$, respectively. For each integration it is necessary to introduce a "constant of integration" and then solve for the constants to obtain a unique solution for a particular problem. Recall from Sec. 4–2 that if the loading on a beam is discontinuous—that is, it consists of a series of several distributed and concentrated loads—then several functions must be written for the internal moment, each valid within the region between the discontinuities. For example, consider the beam shown in Fig. 8–8. The internal moment in regions AB, BC, and CD must be written in terms of the x_1, x_2, and x_3 coordinates. Once these functions are integrated through the application of Eq. 8–4 and the constants of integration determined, the functions will give the slope and deflection (elastic curve) for each region of the beam for which they are valid.

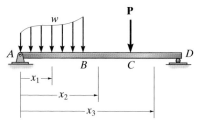

Fig. 8–8

(a)

Sign Convention.

When applying Eq. 8–4, it is important to use the proper sign for M as established by the sign convention that was used in the derivation of this equation, Fig. 8–9a. Furthermore, recall that positive deflection, v, is upward, and as a result, the positive slope angle θ will be measured counterclockwise from the x axis. The reason for this is shown in Fig. 8–9b. Here, positive increases dx and dv in x and v create an increase $d\theta$ that is counterclockwise. Also, since the slope angle θ will be very small, its value in radians can be determined directly from $\theta \approx \tan\theta = dv/dx$.

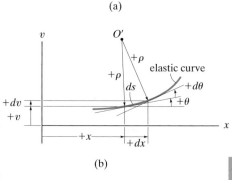

(b)

Fig. 8–9

Boundary and Continuity Conditions.

The constants of integration are determined by evaluating the functions for slope or displacement at a particular point on the beam where the value of the function is known. These values are called *boundary conditions*. For example, if the beam is supported by a roller or pin, then it is required that the displacement be zero at these points. Also, at a fixed support the slope and displacement are both zero.

If a single x coordinate cannot be used to express the equation for the beam's slope or the elastic curve, then continuity conditions must be used to evaluate some of the integration constants. Consider the beam in Fig. 8–10. Here the x_1 and x_2 coordinates are valid only within the regions AB and BC, respectively. Once the functions for the slope and deflection are obtained, they must give the same values for the slope and deflection at point B, $x_1 = x_2 = a$, so that the elastic curve is physically continuous. Expressed mathematically, this requires $\theta_1(a) = \theta_2(a)$ and $v_1(a) = v_2(a)$. These equations can be used to determine two constants of integration.

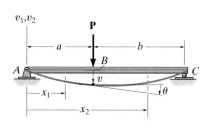

Fig. 8–10

Procedure for Analysis

The following procedure provides a method for determining the slope and deflection of a beam (or shaft) using the method of double integration. It should be realized that this method is suitable only for *elastic deflections* for which the beam's slope is very small. Furthermore, the method considers *only deflections due to bending*. Additional deflection due to shear generally represents only a few percent of the bending deflection, and so it is usually neglected in engineering practice.

Elastic Curve

- Draw an exaggerated view of the beam's elastic curve. Recall that points of zero slope and zero displacement occur at a fixed support, and zero displacement occurs at pin and roller supports.

- Establish the x and v coordinate axes. The x axis must be parallel to the undeflected beam and its origin at the left side of the beam, with a positive direction to the right.

- If several discontinuous loads are present, establish x coordinates that are valid for each region of the beam between the discontinuities.

- In all cases, the associated positive v axis should be directed upward.

Load or Moment Function

- For each region in which there is an x coordinate, express the internal moment M as a function of x.

- *Always* assume that M acts in the positive direction when applying the equation of moment equilibrium to determine $M = f(x)$.

Slope and Elastic Curve

- Provided EI is constant, apply the moment equation $EI\, d^2v/dx^2 = M(x)$, which requires two integrations. For each integration it is important to include a constant of integration. The constants are determined using the boundary conditions for the supports and the continuity conditions that apply to slope and displacement at points where two functions meet.

- Once the integration constants are determined and substituted back into the slope and deflection equations, the slope and displacement at *specific points* on the elastic curve can be determined. The numerical values obtained can be checked graphically by comparing them with the sketch of the elastic curve.

- *Positive* values for *slope* are *counterclockwise* and *positive displacement* is *upward*.

EXAMPLE 8.3

Each simply supported floor joist shown in the photo is subjected to a uniform design loading of 4 kN/m, Fig. 8–11a. Determine the maximum deflection of the joist. EI is constant.

Elastic Curve. Due to symmetry, the joist's maximum deflection will occur at its center. Only a single x coordinate is needed to determine the internal moment.

Moment Function. From the free-body diagram, Fig. 8–11b, we have

$$M = 20x - 4x\left(\frac{x}{2}\right) = 20x - 2x^2$$

Slope and Elastic Curve. Applying Eq. 8–4 and integrating twice gives

$$EI\frac{d^2v}{dx^2} = 20x - 2x^2$$

$$EI\frac{dv}{dx} = 10x^2 - 0.6667x^3 + C_1$$

$$EI\,v = 3.333x^3 - 0.1667x^4 + C_1x + C_2$$

Here $v = 0$ at $x = 0$ so that $C_2 = 0$, and $v = 0$ at $x = 10$, so that $C_1 = -166.7$. The equation of the elastic curve is therefore

$$EI\,v = 3.333x^3 - 0.1667x^4 - 166.7x$$

At $x = 5$ m, note that $dv/dx = 0$. The maximum deflection is therefore

$$v_{max} = -\frac{521}{EI} \qquad \qquad Ans.$$

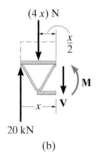

4 kN/m

10 m

20 kN   20 kN

(a)

$(4\,x)$ N

$\dfrac{x}{2}$

M

V

20 kN

(b)

Fig. 8–11

8

EXAMPLE 8.4

The cantilevered beam shown in Fig. 8–12a is subjected to a couple moment M_0 at its end. Determine the equation of the elastic curve. EI is constant.

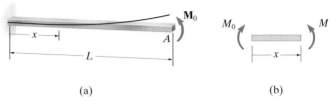

(a)   (b)

Fig. 8–12

SOLUTION

Elastic Curve.   The load tends to deflect the beam as shown in Fig. 8–9a. By inspection, the internal moment can be represented throughout the beam using a single x coordinate.

Moment Function.   From the free-body diagram, with **M** acting in the *positive direction*, Fig. 8–12b, we have

$$M = M_0$$

Slope and Elastic Curve.   Applying Eq. 8–4 and integrating twice yields

$$EI\frac{d^2v}{dx^2} = M_0 \qquad (1)$$

$$EI\frac{dv}{dx} = M_0 x + C_1 \qquad (2)$$

$$EIv = \frac{M_0 x^2}{2} + C_1 x + C_2 \qquad (3)$$

Using the boundary conditions $dv/dx = 0$ at $x = 0$ and $v = 0$ at $x = 0$, then $C_1 = C_2 = 0$. Substituting these results into Eqs. (2) and (3) with $\theta = dv/dx$, we get

$$\theta = \frac{M_0 x}{EI}$$

$$v = \frac{M_0 x^2}{2EI} \qquad \qquad Ans.$$

Maximum slope and displacement occur at A $(x = L)$, for which

$$\theta_A = \frac{M_0 L}{EI} \qquad (4)$$

$$v_A = \frac{M_0 L^2}{2EI} \qquad (5)$$

The *positive* result for θ_A indicates *counterclockwise* rotation and the *positive* result for v_A indicates that v_A is *upward*. This agrees with the results sketched in Fig. 8–12a.

In order to obtain some idea as to the actual *magnitude* of the slope and displacement at the end A, consider the beam in Fig. 8–12a to have a length of 3.6 m, support a couple moment of 20 kN · m, and be made of steel having $E_{st} = 200$ GPa. If this beam were designed *without* a factor of safety by assuming the allowable normal stress is equal to the yield stress $\sigma_{allow} = 250$ N/mm², then a $W6 \times 9$ would be found to be adequate $(I = 6.8(10^6)$ mm⁴$)$. From Eqs. (4) and (5) we get

$$\theta_A = \frac{20 \text{ kN} \cdot \text{m}(3.6 \text{ m})}{[200(10^6) \text{ kN/m}^2][6.8(10^6)(10^{-12})\text{m}^4]} = 0.0529 \text{ rad}$$

$$v_A = \frac{20 \text{ kN} \cdot \text{m}(3.6 \text{ m})^2}{2[200(10^6) \text{ kN/m}^2][6.8(10^6)(10^{-12}) \text{ m}^4]} = 0.0953 \text{ m} = 95.3 \text{ mm}$$

Since $\theta_A^2 = 0.00280$ rad² $\ll 1$, this justifies the use of Eq. 8–4, rather than applying the more exact Eq. 8–3, for computing the deflection of beams. Also, since this numerical application is for a *cantilevered beam,* we have obtained *larger values* for maximum θ and v than would have been obtained if the beam were supported using pins, rollers, or other supports.

EXAMPLE 8.5

The beam in Fig. 8–13a is subjected to a load **P** at its end. Determine the displacement at C. EI is constant.

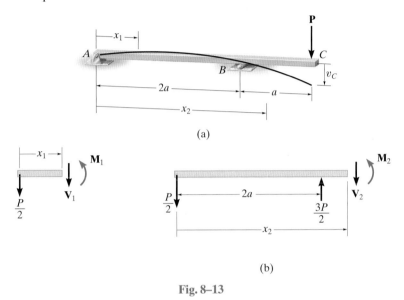

(a)

(b)

Fig. 8–13

SOLUTION

Elastic Curve. The beam deflects into the shape shown in Fig. 8–13a. Due to the loading, two x coordinates must be considered.

Moment Functions. Using the free-body diagrams shown in Fig. 8–13b, we have

$$M_1 = -\frac{P}{2}x_1 \qquad 0 \le x_1 \le 2a$$

$$M_2 = -\frac{P}{2}x_2 + \frac{3P}{2}(x_2 - 2a)$$

$$= Px_2 - 3Pa \qquad 2a \le x_2 \le 3a$$

Slope and Elastic Curve. Applying Eq. 8–4,

for x_1, $$EI\frac{d^2v_1}{dx_1^2} = -\frac{P}{2}x_1$$

$$EI\frac{dv_1}{dx_1} = -\frac{P}{4}x_1^2 + C_1 \qquad (1)$$

$$EIv_1 = -\frac{P}{12}x_1^3 + C_1x_1 + C_2 \qquad (2)$$

For x_2, $\quad EI\dfrac{d^2v_2}{dx_2^2} = Px_2 - 3Pa$

$$EI\dfrac{dv_2}{dx_2} = \dfrac{P}{2}x_2^2 - 3Pax_2 + C_3 \tag{3}$$

$$EIv_2 = \dfrac{P}{6}x_2^3 - \dfrac{3}{2}Pax_2^2 + C_3x_2 + C_4 \tag{4}$$

The *four* constants of integration are determined using *three* boundary conditions, namely, $v_1 = 0$ at $x_1 = 0$, $v_1 = 0$ at $x_1 = 2a$, and $v_2 = 0$ at $x_2 = 2a$, and *one* continuity equation. Here the continuity of slope at the roller requires $dv_1/dx_1 = dv_2/dx_2$ at $x_1 = x_2 = 2a$. (Note that continuity of displacement at B has been indirectly considered in the boundary conditions, since $v_1 = v_2 = 0$ at $x_1 = x_2 = 2a$.) Applying these four conditions yields

$v_1 = 0$ at $x_1 = 0$; $\quad 0 = 0 + 0 + C_2$

$v_1 = 0$ at $x_1 = 2a$; $\quad 0 = -\dfrac{P}{12}(2a)^3 + C_1(2a) + C_2$

$v_2 = 0$ at $x_2 = 2a$; $\quad 0 = \dfrac{P}{6}(2a)^3 - \dfrac{3}{2}Pa(2a)^2 + C_3(2a) + C_4$

$\dfrac{dv_1(2a)}{dx_1} = \dfrac{dv_2(2a)}{dx_2}$; $\quad -\dfrac{P}{4}(2a)^2 + C_1 = \dfrac{P}{2}(2a)^2 - 3Pa(2a) + C_3$

Solving, we obtain

$$C_1 = \dfrac{Pa^2}{3} \qquad C_2 = 0 \qquad C_3 = \dfrac{10}{3}Pa^2 \qquad C_4 = -2Pa^3$$

Substituting C_3 and C_4 into Eq. (4) gives

$$v_2 = \dfrac{P}{6EI}x_2^3 - \dfrac{3Pa}{2EI}x_2^2 + \dfrac{10Pa^2}{3EI}x_2 - \dfrac{2Pa^3}{EI}$$

The displacement at C is determined by setting $x_2 = 3a$. We get

$$v_C = -\dfrac{Pa^3}{EI} \qquad\qquad \textit{Ans.}$$

8

FUNDAMENTAL PROBLEMS

F8–4. Determine the equation of the elastic curve for the beam using the x coordinate that is valid for $0 < x < L$. EI is constant.

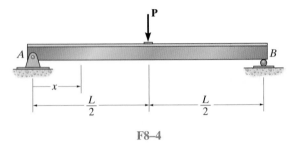

F8–4

F8–5. Determine the equation of the elastic curve for the beam using the x coordinate that is valid for $0 < x < L$. EI is constant.

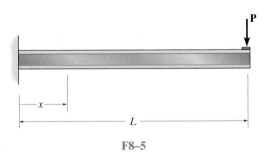

F8–5

F8–6. Determine the equation of the elastic curve for the beam using the x coordinate that is valid for $0 < x < L$. EI is constant.

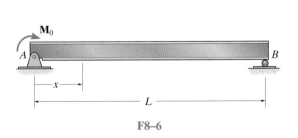

F8–6

F8–7. Determine the equation of the elastic curve for the beam using the x coordinate that is valid for $0 < x < L$. EI is constant.

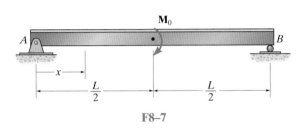

F8–7

F8–8. Determine the equation of the elastic curve for the beam using the x coordinate that is valid for $0 < x < L$. EI is constant.

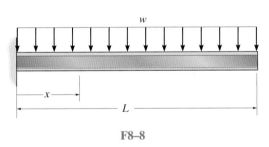

F8–8

F8–9. Determine the equation of the elastic curve for the beam using the x coordinate that is valid for $0 < x < L$. EI is constant.

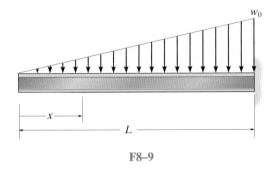

F8–9

PROBLEMS

8–1. Determine the equations of the elastic curve for the beam using the x_1 and x_2 coordinates. Specify the slope at A and the maximum deflection. EI is constant.

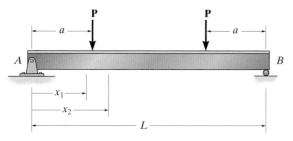

Prob. 8–1

8–2. The bar is supported by a roller constraint at B, which allows vertical displacement but resists axial load and moment. If the bar is subjected to the loading shown, determine the slope at A and the deflection at C. EI is constant.

8–3. Determine the deflection at B of the bar in Prob. 8–2.

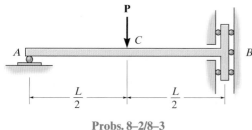

Prob. 8–2/8–3

***8–4.** Determine the equations of the elastic curve using the coordinates x_1 and x_2, specify the slope and deflection at B. EI is constant.

8–5. Determine the equations of the elastic curve using the coordinates x_1 and x_3, and specify the slope and deflection at point B. EI is constant.

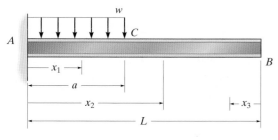

Probs. 8–4/8–5

8–6. Determine the maximum deflection between the supports A and B. EI is constant. Use the method of integration.

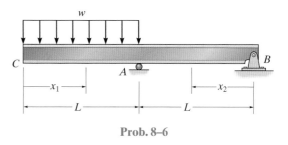

Prob. 8–6

8–7. Determine the elastic curve for the simply supported beam using the x coordinate $0 \le x \le L/2$. Also, determine the slope at A and the maximum deflection of the beam. EI is constant.

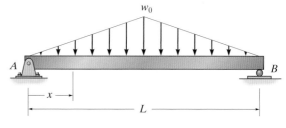

Prob. 8–7

***8–8.** Determine the equations of the elastic curve using the coordinates x_1 and x_2, and specify the slope at C and displacement at B. EI is constant.

8–9. Determine the equations of the elastic curve using the coordinates x_1 and x_3, and specify the slope at B and deflection at C. EI is constant.

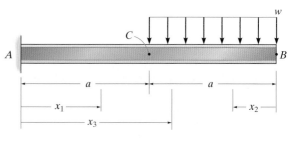

Probs. 8–8/8–9

8

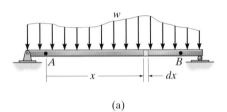

(a)

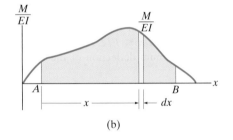

(b)

8.4   Moment-Area Theorems

The initial ideas for the two moment-area theorems were developed by Otto Mohr and later stated formally by Charles E. Greene in 1873. These theorems provide a semigraphical technique for determining the slope of the elastic curve and its deflection due to bending. They are particularly advantageous when used to solve problems involving beams, especially those subjected to a series of concentrated loadings or having segments with different moments of inertia.

To develop the theorems, reference is made to the beam in Fig. 8–14a. If we draw the moment diagram for the beam and then divide it by the flexural rigidity, EI, the "M/EI diagram" shown in Fig. 8–14b results. By Eq. 8–2,

$$d\theta = \left(\frac{M}{EI}\right) dx$$

Thus it can be seen that the change $d\theta$ in the slope of the tangents on either side of the element dx is equal to the lighter-shaded *area* under the M/EI diagram. Integrating from point A on the elastic curve to point B, Fig. 8–14c, we have

$$\theta_{B/A} = \int_A^B \frac{M}{EI} dx \qquad (8\text{--}5)$$

This equation forms the basis for the first moment-area theorem.

> **Theorem 1: The change in slope between any two points on the elastic curve equals the area of the M/EI diagram between these two points.**

The notation $\theta_{B/A}$ is referred to as the angle of the tangent at B measured with respect to the tangent at A. From the proof it should be evident that this angle is measured *counterclockwise* from tangent A to tangent B if the area of the M/EI diagram is *positive*, Fig. 8–14c. Conversely, if this area is *negative*, or below the x axis, the angle $\theta_{B/A}$ is measured *clockwise* from tangent A to tangent B. Furthermore, from the dimensions of Eq. 8–5, $\theta_{B/A}$ is measured in radians.

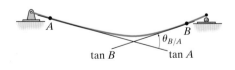

elastic curve
(c)

Fig. 8–14

The second moment-area theorem is based on the relative deviation of *tangents* to the elastic curve. Shown in Fig. 8–15c is a greatly exaggerated view of the *vertical deviation dt* of the tangents on each side of the differential element *dx*. This deviation is measured along a vertical line passing through point *A*. Since the slope of the elastic curve and its deflection are assumed to be very small, it is satisfactory to approximate the length of each tangent line by *x* and the arc *ds'* by *dt*. Using the circular-arc formula $s = \theta r$, where *r* is of length *x*, we can write $dt = x\, d\theta$. Using Eq. 8–2, $d\theta = (M/EI)\, dx$, the vertical deviation of the tangent at *A with respect to the tangent* at *B* can be found by integration, in which case

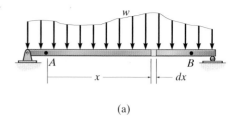

(a)

$$t_{A/B} = \int_A^B x\frac{M}{EI}\, dx \qquad (8\text{–}6)$$

Recall from statics that the centroid of an area is determined from $\bar{x}\int dA = \int x\, dA$. Since $\int M/EI\, dx$ represents an area of the *M/EI* diagram, we can also write

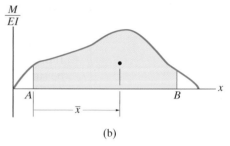

(b)

$$t_{A/B} = \bar{x}\int_A^B \frac{M}{EI}\, dx \qquad (8\text{–}7)$$

Here $\bar{x}$ is the distance from the vertical axis through *A* to the *centroid* of the area between *A* and *B*, Fig. 8–15b.

The second moment-area theorem can now be stated as follows:

> **Theorem 2: The vertical deviation of the tangent at a point (A) on the elastic curve with respect to the tangent extended from another point (B) equals the "moment" of the area under the M/EI diagram between the two points (A and B). This moment is computed about point A (the point on the elastic curve), where the deviation $t_{A/B}$ is to be determined.**

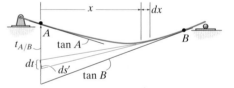

elastic curve

(c)

Provided the moment of a *positive M/EI* area from *A* to *B* is computed, as in Fig. 8–15b, it indicates that the tangent at point *A* is *above* the tangent to the curve extended from point *B*, Fig. 8–15c. Similarly, *negative M/EI* areas indicate that the tangent at *A* is *below* the tangent extended from *B*. Note that in general $t_{A/B}$ is not equal to $t_{B/A}$, which is shown in Fig. 8–15d. Specifically, the moment of the area under the *M/EI* diagram between *A* and *B* is computed about point *A* to determine $t_{A/B}$, Fig. 8–15b, and it is computed about point *B* to determine $t_{B/A}$.

It is important to realize that the moment-area theorems can only be used to determine the angles or deviations between two tangents on the beam's elastic curve. In general, they *do not* give a direct solution for the slope or displacement at a point on the beam. These unknowns must first be related to the angles or vertical deviations of tangents at points on the elastic curve. Usually the tangents at the supports are drawn in this regard since these points do not undergo displacement and/or have zero slope. Specific cases for establishing these geometric relationships are given in the example problems.

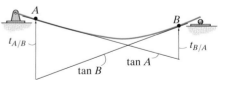

elastic curve

(d)

Fig. 8–15

Procedure for Analysis

The following procedure provides a method that may be used to determine the displacement and slope at a point on the elastic curve of a beam using the moment-area theorems.

M/EI Diagram

- Determine the support reactions and draw the beam's M/EI diagram.

- If the beam is loaded with concentrated forces, the M/EI diagram will consist of a series of straight line segments, and the areas and their moments required for the moment-area theorems will be relatively easy to compute.

- If the loading consists of a *series* of concentrated forces and distributed loads, it may be simpler to compute the required M/EI areas and their moments by drawing the M/EI diagram in parts, using the method of superposition as discussed in Sec. 4–5. In any case, the M/EI diagram will consist of parabolic or perhaps higher-order curves, and it is suggested that the table on the inside back cover be used to locate the area and centroid under each curve.

Elastic Curve

- Draw an exaggerated view of the beam's elastic curve. Recall that points of zero slope occur at fixed supports and zero displacement occurs at all fixed, pin, and roller supports.

- If it becomes difficult to draw the general shape of the elastic curve, use the moment (or M/EI) diagram. Realize that when the beam is subjected to a *positive moment* the beam bends *concave up*, whereas *negative moment* bends the beam *concave down*. Furthermore, an inflection point or change in curvature occurs where the moment in the beam (or M/EI) is zero.

- The displacement and slope to be determined should be indicated on the curve. Since the moment-area theorems apply only between two tangents, attention should be given as to which tangents should be constructed so that the angles or deviations between them will lead to the solution of the problem. In this regard, *the tangents at the points of unknown slope and displacement and at the supports should be considered,* since the beam usually has zero displacement and/or zero slope at the supports.

Moment-Area Theorems

- Apply Theorem 1 to determine the angle between two tangents, and Theorem 2 to determine vertical deviations between these tangents.

- Realize that Theorem 2 in general *will not* yield the displacement of a point on the elastic curve. When applied properly, it will only give the vertical distance or deviation of a tangent at point A on the elastic curve from the tangent at B.

- After applying either Theorem 1 or Theorem 2, the algebraic sign of the answer can be verified from the angle or deviation as indicated on the elastic curve.

EXAMPLE 8.6

Determine the slope at points B and C of the beam shown in Fig. 8–16a. Take $E = 200$ GPa and $I = 300(10^6)$ mm^4.

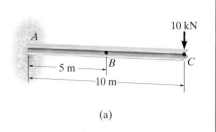

(a)

SOLUTION

M/EI Diagram.   This diagram is shown in Fig. 8–16b. It is easier to solve the problem in terms of EI and substitute the numerical data as a last step.

Elastic Curve.   The 10-kN load causes the beam to deflect as shown in Fig. 8–16c. (The beam is deflected concave down, since M/EI is negative.) Here the tangent at A (the support) is *always horizontal*. The tangents at B and C are also indicated. We are required to find θ_B and θ_C. By the construction, the angle between tan A and tan B, that is, $\theta_{B/A}$, is equivalent to θ_B.

$$\theta_B = \theta_{B/A}$$

Also,

$$\theta_C = \theta_{C/A}$$

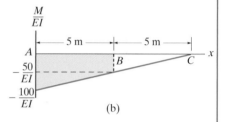

(b)

Moment-Area Theorem.   Applying Theorem 1, $\theta_{B/A}$ is equal to the area under the M/EI diagram between points A and B; that is,

$$\theta_B = \theta_{B/A} = -\left(\frac{50 \text{ kN} \cdot \text{m}}{EI}\right)(5 \text{ m}) - \frac{1}{2}\left(\frac{100 \text{ kN} \cdot \text{m}}{EI} - \frac{50 \text{ kN} \cdot \text{m}}{EI}\right)(5 \text{ m})$$

$$= -\frac{375 \text{ kN} \cdot \text{m}^2}{EI}$$

Substituting numerical data for E and I, we have

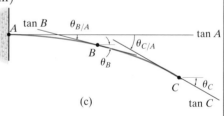

$$\theta_B = \frac{-375 \text{ kN} \cdot \text{m}^2}{[200(10^6) \text{ kN/m}^2][360(10^6)(10^{-12}) \text{ m}^4]}$$

$$= -0.00521 \text{ rad} \qquad\qquad Ans.$$

(c)

Fig. 8–16

The *negative sign* indicates that the angle is measured clockwise from A, Fig. 8–16c.

In a similar manner, the area under the M/EI diagram between points A and C equals $\theta_{C/A}$. We have

$$\theta_C = \theta_{C/A} = \frac{1}{2}\left(-\frac{100 \text{ kN} \cdot \text{m}}{EI}\right)(10 \text{ m}) = -\frac{500 \text{ kN} \cdot \text{m}^2}{EI}$$

Substituting numerical values for EI, we have

$$\theta_C = \frac{-500 \text{ kN} \cdot \text{m}^2}{[200(10^6) \text{ kN/m}^2][360(10^6)(10^{-12}) \text{ m}^4]}$$

$$= -0.00694 \text{ rad} \qquad\qquad Ans.$$

EXAMPLE 8.7

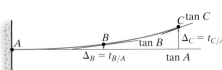

$I_{AB} = 8(10^6)$ mm^4 $I_{BC} = 4(10^6)$ mm^4
4 m   3 m

(a)

Determine the deflection at points B and C of the beam shown in Fig. 8–17a. Values for the moment of inertia of each segment are indicated in the figure. Take $E = 200$ GPa.

SOLUTION

M/EI Diagram.   By inspection, the moment diagram for the beam is a rectangle. Here we will construct the M/EI diagram relative to I_{BC}, realizing that $I_{AB} = 2I_{BC}$. Fig. 8–17b. Numerical data for EI_{BC} will be substituted as a last step.

Elastic Curve.   The couple moment at C causes the beam to deflect as shown in Fig. 8–17c. The tangents at A (the support), B, and C are indicated. We are required to find Δ_B and Δ_C. These displacements can be related *directly* to the deviations between the tangents, so that from the construction Δ_B is equal to the deviation of tan B relative to tan A; that is,

$$\Delta_B = t_{B/A}$$

Also,

$$\Delta_C = t_{C/A}$$

Moment-Area Theorem.   Applying Theorem 2, $t_{B/A}$ is equal to the moment of the area under the M/EI_{BC} diagram between A and B computed about point B, since this is the point where the tangential deviation is to be determined. Hence, from Fig. 8–17b,

$$\Delta_B = t_{B/A} = \left[\frac{250 \text{ N}\cdot\text{m}}{EI_{BC}}(4\text{ m})\right](2\text{ m}) = \frac{2000 \text{ N}\cdot\text{m}^3}{EI_{BC}}$$

Substituting the numerical data yields

$$\Delta_B = \frac{2000 \text{ N}\cdot\text{m}^3}{[200(10^9) \text{ N/m}^2][4(10^6) \text{ mm}^4(1 \text{ m}^4/(10^3)^4 \text{ mm}^4)]}$$

$$= 0.0025 \text{ m} = 2.5 \text{ mm.}\qquad\qquad Ans.$$

Likewise, for $t_{C/A}$ we must compute the moment of the entire M/EI_{BC} diagram from A to C about point C. We have

$$\Delta_C = t_{C/A} = \left[\frac{250 \text{ N}\cdot\text{m}}{EI_{BC}}(4\text{ m})\right](5\text{ m}) + \left[\frac{500 \text{ N}\cdot\text{m}}{EI_{BC}}(3\text{ m})\right](1.5\text{ m})$$

$$= \frac{7250 \text{ N}\cdot\text{m}^3}{EI_{BC}} = \frac{7250 \text{ N}\cdot\text{m}^3}{[200(10^9) \text{ N/m}^2][4(10^6)(10^{-12}) \text{ m}^4]}$$

$$= 0.00906 \text{ m} = 9.06 \text{ mm}\qquad\qquad Ans.$$

Since both answers are *positive*, they indicate that points B and C lie *above* the tangent at A.

$\dfrac{M}{EI_{BC}}$

$\dfrac{250}{EI_{BC}}$ $\dfrac{500}{EI_{BC}}$

A 2 m B C x
4 m   3 m

(b)

A B C tan C
tan B $\Delta_C = t_{C/A}$
$\Delta_B = t_{B/A}$ tan A

(c)

Fig. 8–17

EXAMPLE 8.8

Determine the slope at point C of the beam in Fig. 8–18a. $E = 200$ GPa, $I = 6(10^6)$ mm^4.

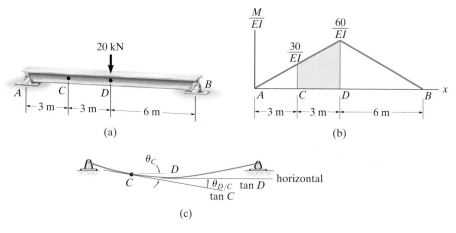

Fig. 8–18

SOLUTION

***M/EI* Diagram.**   Fig. 8–18b.

Elastic Curve.   Since the loading is applied symmetrically to the beam, the elastic curve is symmetric, as shown in Fig. 8–18c. We are required to find θ_C. This can easily be done, realizing that the tangent at D is *horizontal*, and therefore, by the construction, the angle $\theta_{D/C}$ between tan C and tan D is equal to θ_C; that is,

$$\theta_C = \theta_{D/C}$$

Moment-Area Theorem.   Using Theorem 1, $\theta_{D/C}$ is equal to the shaded area under the *M/EI* diagram between points C and D. We have

$$\theta_C = \theta_{D/C} = 3 \text{ m}\left(\frac{30 \text{ kN} \cdot \text{m}}{EI}\right) + \frac{1}{2}(3 \text{ m})\left(\frac{60 \text{ kN} \cdot \text{m}}{EI} - \frac{30 \text{ kN} \cdot \text{m}}{EI}\right)$$

$$= \frac{135 \text{ kN} \cdot \text{m}^2}{EI}$$

Thus,

$$\theta_C = \frac{135 \text{ kN} \cdot \text{m}^2}{[200(10^6) \text{ kN/m}^2][6(10^6)(10^{-12}) \text{ m}^4]} = 0.112 \text{ rad} \qquad Ans.$$

8

EXAMPLE 8.9

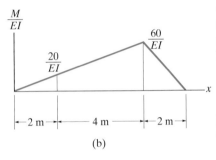

40 kN

(a)

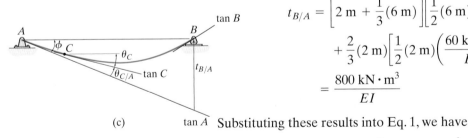

$\dfrac{M}{EI}$

$\dfrac{60}{EI}$

$\dfrac{20}{EI}$

(b)

(c)

tan B

tan C

tan A

Fig. 8–19

Determine the slope at point C of the beam in Fig. 8–19a. $E = 200$ GPa, $I = 360(10^6)$ mm^4.

SOLUTION

M/EI Diagram. Fig. 8–19b.

Elastic Curve. The elastic curve is shown in Fig. 8–19c. We are required to find θ_C. To do this, establish tangents at A, B (the supports), and C and note that $\theta_{C/A}$ is the angle between the tangents at A and C. Also, the angle ϕ in Fig. 8–19c can be found using $\phi = t_{B/A}/L_{AB}$. This equation is valid since $t_{B/A}$ is actually very small, so that $t_{B/A}$ can be approximated by the length of a circular arc defined by a radius of $L_{AB} = 8$ m and sweep of ϕ. (Recall that $s = \theta r$.) From the geometry of Fig. 8–19c, we have

$$\theta_C = \phi - \theta_{C/A} = \frac{t_{B/A}}{8} - \theta_{C/A} \qquad (1)$$

Moment-Area Theorems. Using Theorem 1, $\theta_{C/A}$ is equivalent to the area under the M/EI diagram between points A and C; that is,

$$\theta_{C/A} = \frac{1}{2}(2\text{ m})\left(\frac{20\text{ kN}\cdot\text{m}}{EI}\right) = \frac{20\text{ kN}\cdot\text{m}^2}{EI}$$

Applying Theorem 2, $t_{B/A}$ is equivalent to the moment of the area under the M/EI diagram between B and A about point B, since this is the point where the tangential deviation is to be determined. We have

$$t_{B/A} = \left[2\text{ m} + \frac{1}{3}(6\text{ m})\right]\left[\frac{1}{2}(6\text{ m})\left(\frac{60\text{ kN}\cdot\text{m}}{EI}\right)\right]$$
$$+ \frac{2}{3}(2\text{ m})\left[\frac{1}{2}(2\text{ m})\left(\frac{60\text{ kN}\cdot\text{m}}{EI}\right)\right]$$
$$= \frac{800\text{ kN}\cdot\text{m}^3}{EI}$$

Substituting these results into Eq. 1, we have

$$\theta_C = \frac{800\text{ kN}\cdot\text{m}^3}{(8\text{ m})\,EI} - \frac{20\text{ kN}\cdot\text{m}^2}{EI} = \frac{80\text{ kN}\cdot\text{m}^2}{EI}$$

so that

$$\theta_C = \frac{80\text{ kN}\cdot\text{m}^2}{[200(10^6)\text{ kN/m}^2][360(10^6)(10^{-12})\text{ m}^4]}$$
$$= 0.00111\text{ rad} \qquad Ans.$$

EXAMPLE 8.10

Determine the deflection at C of the beam shown in Fig. 8–20a. Take $E = 200$ GPa, $I = 4(10^6)$ mm^4.

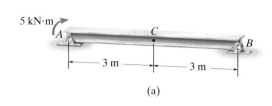

(a)

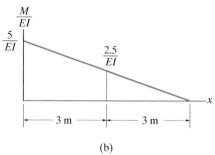

(b)

SOLUTION

M/EI Diagram. Fig. 8–20b.

Elastic Curve. Here we are required to find Δ_C, Fig. 8–20c. This is not necessarily the maximum deflection of the beam, since the loading and hence the elastic curve are *not symmetric*. Also indicated in Fig. 8–20c are the tangents at A, B (the supports), and C. If $t_{A/B}$ is determined, then Δ' can be found from proportional triangles, that is, $\Delta'/3 = t_{A/B}/6$ or $\Delta' = t_{A/B}/2$. From the construction in Fig. 8–20c, we have

$$\Delta_C = \frac{t_{A/B}}{2} - t_{C/B} \qquad (1)$$

Moment-Area Theorem. We will apply Theorem 2 to determine $t_{A/B}$ and $t_{C/B}$. Here $t_{A/B}$ is the moment of the M/EI diagram between A and B about point A,

$$t_{A/B} = \left[\frac{1}{3}(6 \text{ m})\right]\left[\frac{1}{2}(6 \text{ m})\left(\frac{5 \text{ kN} \cdot \text{m}}{EI}\right)\right] = \frac{30 \text{ kN} \cdot \text{m}^3}{EI}$$

and $t_{C/B}$ is the moment of the M/EI diagram between C and B about C.

$$t_{C/B} = \left[\frac{1}{3}(3 \text{ m})\right]\left[\frac{1}{2}(3 \text{ m})\left(\frac{2.5 \text{ kN} \cdot \text{m}}{EI}\right)\right] = \frac{3.75 \text{ kN} \cdot \text{m}^3}{EI}$$

Substituting these results into Eq. (1) yields

$$\Delta_C = \frac{1}{2}\left(\frac{30 \text{ kN} \cdot \text{m}^3}{EI}\right) - \frac{3.75 \text{ kN} \cdot \text{m}^3}{EI} = \frac{11.25 \text{ kN} \cdot \text{m}^3}{EI}$$

Working in units of kN and m, we have

$$\Delta_C = \frac{11.25 \text{ kN} \cdot \text{m}^3}{[200(10^6) \text{ kN/m}^2][4(10^6)(10^{-12}) \text{ m}^4]}$$

$$= 0.0141 \text{ m} = 14.1 \text{ mm} \qquad \textit{Ans.}$$

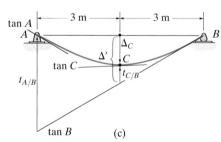

(c)

Fig. 8–20

8

EXAMPLE 8.11

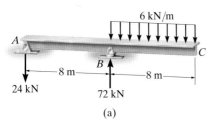

(a)

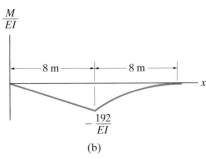

(b)

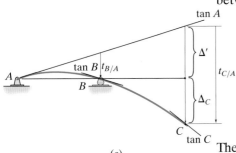

(c)

Fig. 8–21

Determine the deflection at point C of the beam shown in Fig. 8–21a. $E = 200$ GPa, $I = 250(10^6)$ mm^4.

SOLUTION

M/EI Diagram. As shown in Fig. 8–21b, this diagram consists of a triangular and a parabolic segment.

Elastic Curve. The loading causes the beam to deform as shown in Fig. 8–21c. We are required to find Δ_C. By constructing tangents at A, B (the supports), and C, it is seen that $\Delta_C = t_{C/A} - \Delta'$. However, Δ' can be related to $t_{B/A}$ by proportional triangles, that is, $\Delta'/16 = t_{B/A}/8$ or $\Delta' = 2t_{B/A}$. Hence

$$\Delta_C = t_{C/A} - 2t_{B/A} \qquad (1)$$

Moment-Area Theorem. We will apply Theorem 2 to determine $t_{C/A}$ and $t_{B/A}$. Using the table on the inside back cover for the parabolic segment and considering the moment of the M/EI diagram between A and C about point C, we have

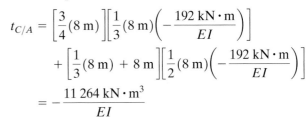

$$t_{C/A} = \left[\frac{3}{4}(8 \text{ m})\right]\left[\frac{1}{3}(8 \text{ m})\left(-\frac{192 \text{ kN} \cdot \text{m}}{EI}\right)\right]$$
$$+ \left[\frac{1}{3}(8 \text{ m}) + 8 \text{ m}\right]\left[\frac{1}{2}(8 \text{ m})\left(-\frac{192 \text{ kN} \cdot \text{m}}{EI}\right)\right]$$
$$= -\frac{11\,264 \text{ kN} \cdot \text{m}^3}{EI}$$

The moment of the M/EI diagram between A and B about point B gives

$$t_{B/A} = \left[\frac{1}{3}(8 \text{ m})\right]\left[\frac{1}{2}(8 \text{ m})\left(-\frac{192 \text{ kN} \cdot \text{m}}{EI}\right)\right] = -\frac{2048 \text{ kN} \cdot \text{m}^3}{EI}$$

Why are these terms negative? Substituting the results into Eq. (1) yields

$$\Delta_C = -\frac{11\,264 \text{ kN} \cdot \text{m}^3}{EI} - 2\left(-\frac{2048 \text{ kN} \cdot \text{m}^3}{EI}\right)$$
$$= -\frac{7168 \text{ kN} \cdot \text{m}^3}{EI}$$

Thus,

$$\Delta_C = \frac{-7168 \text{ kN} \cdot \text{m}^3}{[200(10^6) \text{ kN/m}^2][250(10^6)(10^{-12}) \text{ m}^4]}$$
$$= -0.143 \text{ m} \qquad \qquad Ans.$$

EXAMPLE 8.12

Determine the slope at the roller B of the double overhang beam shown in Fig. 8–22a. Take $E = 200$ GPa, $I = 18(10^6)$ mm^4.

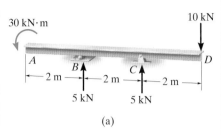

(a)

SOLUTION

M/EI Diagram.   The M/EI diagram can be simplified by drawing it in parts and considering the M/EI diagrams for the three loadings each acting on a cantilever beam fixed at D, Fig. 8–22b. (The 10-kN load is not considered since it produces no moment about D.)

Elastic Curve.   If tangents are drawn at B and C, Fig. 8–22c, the slope B can be determined by finding $t_{C/B}$, and for small angles,

$$\theta_B = \frac{t_{C/B}}{2\text{ m}} \qquad (1)$$

Moment Area Theorem.   To determine $t_{C/B}$ we apply the moment area theorem by finding the moment of the M/EI diagram between BC about point C. This only involves the shaded area under two of the diagrams in Fig. 8–22b. Thus,

$$t_{C/B} = (1\text{ m})\left[(2\text{ m})\left(\frac{-30\text{ kN}\cdot\text{m}}{EI}\right)\right] + \left(\frac{2\text{ m}}{3}\right)\left[\frac{1}{2}(2\text{ m})\left(\frac{10\text{ kN}\cdot\text{m}}{EI}\right)\right]$$

$$= \frac{53.33\text{ kN}\cdot\text{m}^3}{EI}$$

Substituting into Eq. (1),

$$\theta_B = \frac{53.33\text{ kN}\cdot\text{m}^3}{(2\text{ m})[200(10^6)\text{ kN/m}^3][18(10^6)(10^{-12})\text{ m}^4]}$$

$$= 0.00741\text{ rad} \qquad\qquad Ans.$$

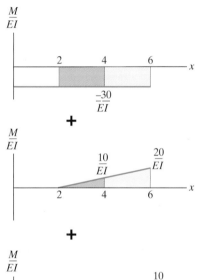

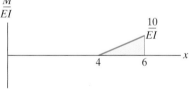

(b)

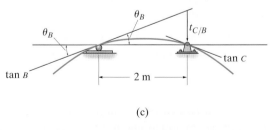

(c)

Fig. 8–22

8.5   Conjugate-Beam Method

The conjugate-beam method was developed by H. Müller-Breslau in 1865. Essentially, it requires the same amount of computation as the moment-area theorems to determine a beam's slope or deflection; however, this method relies only on the principles of statics, and hence its application will be more familiar.

The basis for the method comes from the *similarity* of Eq. 4–1 and Eq. 4–2 to Eq. 8–2 and Eq. 8–4. To show this similarity, we can write these equations as follows:

$$\frac{dV}{dx} = w \qquad \qquad \frac{d^2M}{dx^2} = w$$

$$\frac{d\theta}{dx} = \frac{M}{EI} \qquad \qquad \frac{d^2v}{dx^2} = \frac{M}{EI}$$

Or integrating,

$$V = \int w\,dx \qquad \qquad M = \int \left[\int w\,dx \right] dx$$

$$\updownarrow \qquad \updownarrow \qquad \qquad \updownarrow \qquad \updownarrow$$

$$\theta = \int \left(\frac{M}{EI} \right) dx \qquad v = \int \left[\int \left(\frac{M}{EI} \right) dx \right] dx$$

Here the *shear V* compares with the *slope θ*, the *moment M* compares with the *displacement v*, and the *external load w* compares with the *M/EI* diagram. To make use of this comparison we will now consider a beam having the same length as the real beam, but referred to here as the "conjugate beam," Fig. 8–23. The conjugate beam is "loaded" with the *M/EI* diagram derived from the load *w* on the real beam. From the above comparisons, we can state two theorems related to the conjugate beam, namely,

> **Theorem 1: The slope at a point in the real beam is numerically equal to the shear at the corresponding point in the conjugate beam.**
>
> **Theorem 2: The displacement of a point in the real beam is numerically equal to the moment at the corresponding point in the conjugate beam.**

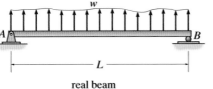

real beam

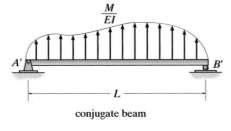

conjugate beam

Fig. 8–23

Conjugate-Beam Supports.

When drawing the conjugate beam it is important that the shear and moment developed at the supports of the conjugate beam account for the corresponding slope and displacement of the real beam at its supports, a consequence of Theorems 1 and 2. For example, as shown in Table 8–2, a pin or roller support at the end of the real beam provides *zero displacement,* but the beam has a nonzero slope. Consequently, from Theorems 1 and 2, the conjugate beam must be supported by a pin or roller, since this support has *zero moment* but has a shear or end reaction. When the real beam is fixed supported (3), both the slope and displacement at the support are zero. Here the conjugate beam has a free end, since at this end there is zero shear and zero moment. Corresponding real and conjugate-beam supports for other cases are listed in the table. Examples of real and conjugate beams are shown in Fig. 8–24. Note that, as a rule, neglecting axial force, statically determinate real beams have statically determinate conjugate beams; and statically indeterminate real beams, as in the last case in Fig. 8–24, become unstable conjugate beams. Although this occurs, the *M/EI* loading will provide the necessary "equilibrium" to hold the conjugate beam stable.

TABLE 8–2

Real Beam		Conjugate Beam	
1)	θ $\Delta = 0$ pin	V $M = 0$ pin	
2)	θ $\Delta = 0$ roller	V $M = 0$ roller	
3)	$\theta = 0$ $\Delta = 0$ fixed	$V = 0$ $M = 0$ free	
4)	θ Δ free	V M fixed	
5)	θ $\Delta = 0$ internal pin	V $M = 0$ hinge	
6)	θ $\Delta = 0$ internal roller	V $M = 0$ hinge	
7)	θ Δ hinge	V M internal roller	

8

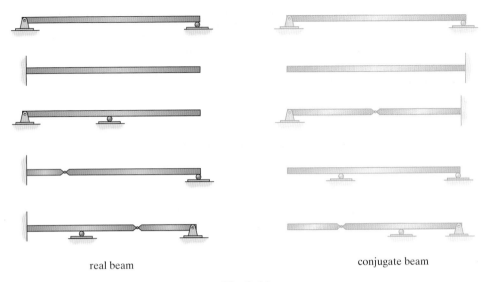

real beam   conjugate beam

Fig. 8–24

Procedure for Analysis

The following procedure provides a method that may be used to determine the displacement and slope at a point on the elastic curve of a beam using the conjugate-beam method.

Conjugate Beam

- Draw the conjugate beam for the real beam. This beam has the same length as the real beam and has corresponding supports as listed in Table 8–2.

- In general, if the real support allows a *slope*, the conjugate support must develop a *shear*; and if the real support allows a *displacement*, the conjugate support must develop a *moment*.

- The conjugate beam is loaded with the real beam's *M/EI* diagram. This loading is assumed to be *distributed* over the conjugate beam and is directed *upward* when *M/EI* is *positive* and *downward* when *M/EI* is *negative*. In other words, the loading always acts *away* from the beam.

Equilibrium

- Using the equations of equilibrium, determine the reactions at the conjugate beam's supports.

- Section the conjugate beam at the point where the slope θ and displacement Δ of the real beam are to be determined. At the section show the unknown shear V' and moment M' acting in their positive sense.

- Determine the shear and moment using the equations of equilibrium. V' and M' equal θ and Δ, respectively, for the real beam. In particular, if these values are *positive*, the *slope* is *counterclockwise* and the *displacement* is *upward*.

EXAMPLE 8.13

Determine the slope and deflection at point B of the steel beam shown in Fig. 8–25a. The reactions have been computed. $E = 200$ GPa, $I = 475(10^6)$ mm^4.

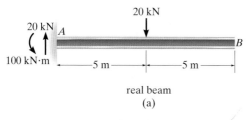

real beam
(a)

Fig. 8–25

SOLUTION

Conjugate Beam. The conjugate beam is shown in Fig. 8–25b. The supports at A' and B' correspond to supports A and B on the real beam, Table 8–2. It is important to understand why this is so. The M/EI diagram is *negative*, so the distributed load acts *downward*, i.e., away from the beam.

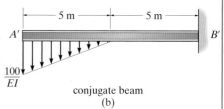

conjugate beam
(b)

Equilibrium. Since θ_B and Δ_B are to be determined, we must compute $V_{B'}$ and $M_{B'}$ in the conjugate beam, Fig. 8–25c.

$$+\uparrow \Sigma F_y = 0; \qquad -\frac{250 \text{ kN} \cdot \text{m}^2}{EI} - V_{B'} = 0$$

$$\theta_B = V_{B'} = -\frac{250 \text{ kN} \cdot \text{m}^2}{EI}$$

$$= \frac{-250 \text{ kN} \cdot \text{m}^2}{[200(10^6) \text{ kN/m}^2][475(10^6)(10^{-12}) \text{ m}^4]}$$

$$= -0.00263 \text{ rad} \qquad\qquad Ans.$$

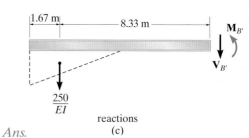

reactions
(c)

$$\zeta + \Sigma M_{B'} = 0; \qquad \frac{250 \text{ kN} \cdot \text{m}^2}{EI}(8.33 \text{ m}) + M_{B'} = 0$$

$$\Delta_B = M_{B'} = -\frac{2\,083 \text{ kN} \cdot \text{m}^3}{EI}$$

$$= \frac{-2\,083 \text{ kN} \cdot \text{m}^3}{[200(10^6) \text{ kN/m}^2][475(10^6)(10^{-12}) \text{ m}^4]}$$

$$= -0.0219 \text{ m} = -21.9 \text{ mm} \qquad\qquad Ans.$$

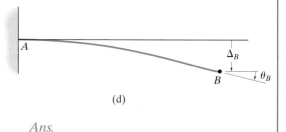

(d)

The negative signs indicate the slope of the beam is measured clockwise and the displacement is downward, Fig. 8–25d.

8

EXAMPLE 8.14

Determine the maximum deflection of the steel beam shown in Fig. 8–26a. The reactions have been computed. $E = 200$ GPa, $I = 60(10^6)$ mm^4.

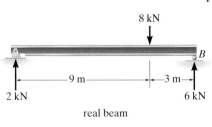

8 kN

B

9 m — 3 m

2 kN   6 kN

real beam

(a)

Fig. 8–26

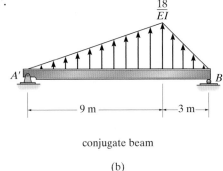

$\frac{18}{EI}$

A' B'

9 m — 3 m

conjugate beam

(b)

SOLUTION

Conjugate Beam. The conjugate beam loaded with the M/EI diagram is shown in Fig. 8–26b. Since the M/EI diagram is positive, the distributed load acts upward (away from the beam).

Equilibrium. The external reactions on the conjugate beam are determined first and are indicated on the free-body diagram in Fig. 8–26c. *Maximum deflection* of the real beam occurs at the point where the *slope* of the beam is *zero*. This corresponds to the same point in the conjugate beam where the *shear* is *zero*. Assuming this point acts within the region $0 \le x \le 9$ m from A', we can isolate the section shown in Fig. 8–26d. Note that the peak of the distributed loading was determined from proportional triangles, that is, $w/x = (18/EI)/9$. We require $V' = 0$ so that

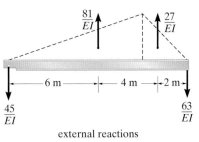

$\frac{81}{EI}$ $\frac{27}{EI}$

6 m — 4 m — 2 m

$\frac{45}{EI}$ $\frac{63}{EI}$

external reactions

(c)

$$+\uparrow \Sigma F_y = 0; \qquad -\frac{45}{EI} + \frac{1}{2}\left(\frac{2x}{EI}\right)x = 0$$

$$x = 6.71 \text{ m} \qquad (0 \le x \le 9 \text{ m}) \text{ OK}$$

Using this value for x, the maximum deflection in the real beam corresponds to the moment M'. Hence,

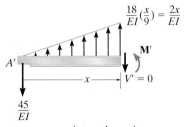

$\frac{18}{EI}\left(\frac{x}{9}\right) = \frac{2x}{EI}$

$\mathbf{M'}$

A'

x $V' = 0$

$\frac{45}{EI}$

internal reactions

(d)

$$\curvearrowleft + \Sigma M = 0; \qquad \frac{45}{EI}(6.71) - \left[\frac{1}{2}\left(\frac{2(6.71)}{EI}\right)6.71\right]\frac{1}{3}(6.71) + M' = 0$$

$$\Delta_{max} = M' = -\frac{201.2 \text{ kN} \cdot \text{m}^3}{EI}$$

$$= \frac{-201.2 \text{ kN} \cdot \text{m}^3}{[200(10^6) \text{ kN/m}^2][60(10^6) \text{ mm}^4(1 \text{ m}^4/(10^3)^4 \text{ mm}^4)]}$$

$$= -0.0168 \text{ m} = -16.8 \text{ mm} \qquad\qquad Ans.$$

The negative sign indicates the deflection is downward.

EXAMPLE 8.15

The girder in Fig. 8–27a is made from a continuous beam and reinforced at its center with cover plates where its moment of inertia is larger. The 4-m end segments have a moment of inertia of $I = 270(10^6)$ mm^4, and the center portion has a moment of inertia of $I' = 540(10^6)$ mm^4. Determine the deflection at the center C. Take $E = 200$ GPa. The reactions have been calculated.

SOLUTION

Conjugate Beam. The moment diagram for the beam is determined first, Fig. 8–27b. Since $I' = 2I$, for simplicity, we can express the load on the conjugate beam in terms of the constant EI, as shown in Fig. 8–27c.

Equilibrium. The reactions on the conjugate beam can be calculated by the symmetry of the loading or using the equations of equilibrium. The results are shown in Fig. 8–27d. Since the deflection at C is to be determined, we must compute the internal moment at C'. Using the method of sections, segment $A'C'$ is isolated and the resultants of the distributed loads and their locations are determined, Fig. 8–27e. Thus,

$$\zeta + \Sigma M_{C'} = 0; \quad \frac{620}{EI}(6) - \frac{400}{EI}(3.33) - \frac{200}{EI}(1) - \frac{20}{EI}(0.67) + M_{C'} = 0$$

$$M_{C'} = -\frac{2\,173 \text{ kN} \cdot \text{m}^3}{EI}$$

Substituting the numerical data for EI and converting units, we have

$$\Delta_C = M_{C'} = -\frac{2\,173 \text{ kN} \cdot \text{m}^3}{[200(10^6) \text{ kN/m}^2][270(10^6)(10^{-12}) \text{ m}^4]}$$

$$= -0.0402 \text{ m} = -40.2 \text{ mm} \qquad Ans.$$

The negative sign indicates that the deflection is downward.

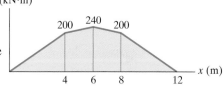

real beam

(a)

Fig. 8–27

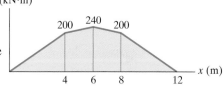

M (kN·m)

moment diagram

(b)

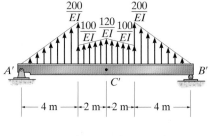

conjugate beam

(c)

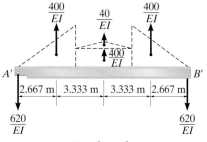

external reactions

(d)

internal reactions

(e)

8

EXAMPLE 8.16

Determine the displacement of the pin at B and the slope of each beam segment connected to the pin for the compound beam shown in Fig. 8–28a. $E = 200$ GPa, $I = 18(10^6)$ mm^4.

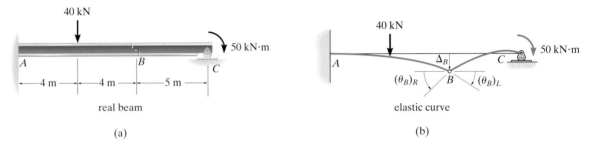

real beam

(a)

elastic curve

(b)

Fig. 8–28

SOLUTION

Conjugate Beam. The elastic curve for the beam is shown in Fig. 8–28b in order to identify the unknown displacement Δ_B and the slopes $(\theta_B)_L$ and $(\theta_B)_R$ to the left and right of the pin. Using Table 8–2, the conjugate beam is shown in Fig. 8–28c. For simplicity in calculation, the M/EI diagram has been drawn in *parts* using the principle of superposition as described in Sec. 4–5. In this regard, the real beam is thought of as cantilevered from the left support, A. The moment diagrams for the 40-kN load, the reactive force $C_y = 10$ kN, and the 50-kN · m loading are given. Notice that negative regions of this diagram develop a downward distributed load and positive regions have a distributed load that acts upward.

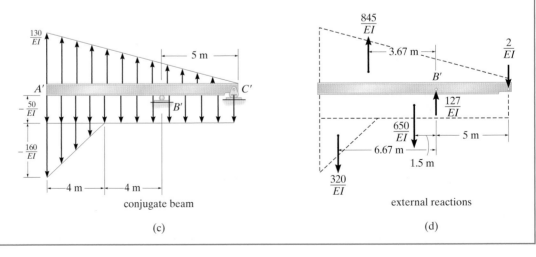

conjugate beam

(c)

external reactions

(d)

Equilibrium. The external reactions at B' and C' are calculated first and the results are indicated in Fig. 8–28d. In order to determine $(\theta_B)_R$, the conjugate beam is sectioned just to the *right* of B' and the shear force $(V_{B'})_R$ is computed, Fig. 8–28e. Thus,

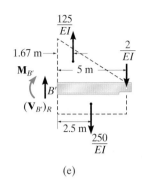

$$+\uparrow\Sigma F_y = 0; \qquad (V_{B'})_R + \frac{125}{EI} - \frac{250}{EI} - \frac{2}{EI} = 0$$

$$(\theta_B)_R = (V_{B'})_R = \frac{127\ \text{kN}\cdot\text{m}^2}{EI}$$

$$= \frac{127\ \text{kN}\cdot\text{m}^2}{[200(10^6)\ \text{kN/m}^2][18(10^6)(10^{-12})\ \text{m}^4]}$$

$$= 0.0353\ \text{rad} \qquad\qquad Ans.$$

(e)

The internal moment at B' yields the displacement of the pin. Thus,

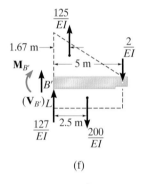

$$\zeta+\Sigma M_{B'} = 0; \qquad -M_{B'} + \frac{125}{EI}(1.67) - \frac{250}{EI}(2.5) - \frac{2}{EI}(5) = 0$$

$$\Delta_B = M_{B'} = -\frac{427\ \text{kN}\cdot\text{m}^3}{EI}$$

$$= \frac{-427\ \text{kN}\cdot\text{m}^3}{[200(10^6)\ \text{kN/m}^2][18(10^6)(10^{-12})\ \text{m}^4]}$$

$$= -0.119\ \text{m} = -119\ \text{mm} \qquad\qquad Ans.$$

(f)

The slope $(\theta_B)_L$ can be found from a section of beam just to the *left* of B', Fig. 8–28f. Thus,

$$+\uparrow\Sigma F_y = 0; \qquad (V_{B'})_L + \frac{127}{EI} + \frac{125}{EI} - \frac{250}{EI} - \frac{2}{EI} = 0$$

$$(\theta_B)_L = (V_{B'})_L = 0 \qquad\qquad Ans.$$

Obviously, $\Delta_B = M_{B'}$ for this segment is the *same* as previously calculated, since the moment arms are only slightly different in Figs. 8–28e and 8–28f.

8

FUNDAMENTAL PROBLEMS

F8–10. Use the moment-area theorems and determine the slope at A and deflection at A. EI is constant.

F8–11. Solve Prob. F8–10 using the conjugate beam method.

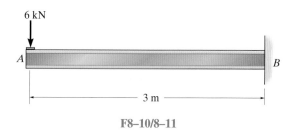

F8–10/8–11

F8–12. Use the moment-area theorems and determine the slope at B and deflection at B. EI is constant.

F8–13. Solve Prob. F8–12 using the conjugate beam method.

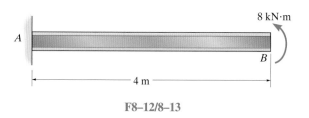

F8–12/8–13

F8–14. Use the moment-area theorems and determine the slope at A and displacement at C. EI is constant.

F8–15. Solve Prob. F8–14 using the conjugate beam method.

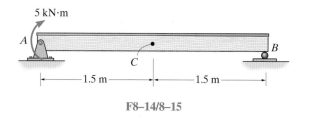

F8–14/8–15

F8–16. Use the moment-area theorems and determine the slope at A and displacement at C. EI is constant.

F8–17. Solve Prob. F8–16 using the conjugate beam method.

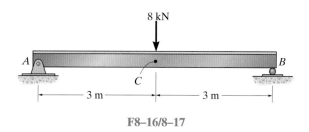

F8–16/8–17

F8–18. Use the moment-area theorems and determine the slope at A and displacement at C. EI is constant.

F8–19. Solve Prob. F8–18 using the conjugate beam method.

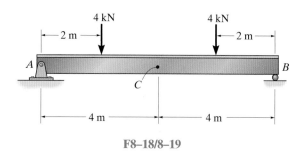

F8–18/8–19

F8–20. Use the moment-area theorems and determine the slope at B and displacement at B. EI is constant.

F8–21. Solve Prob. F8–20 using the conjugate beam method.

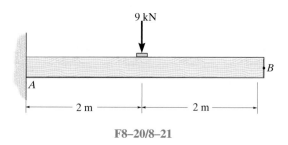

F8–20/8–21

PROBLEMS

8–10. Determine the slope at B and the maximum displacement of the beam. Use the moment-area theorems. Take $E = 200$ GPa, $I = 200(10^6)$ mm^4.

8–11. Solve Prob. 8–10 using the conjugate-beam method.

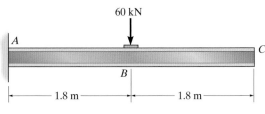

60 kN

1.8 m   1.8 m

Probs. 8–10/8–11

*8–12.** Determine the slope and displacement at C. EI is constant. Use the moment-area theorems.

8–13. Solve Prob. 8–12 using the conjugate-beam method.

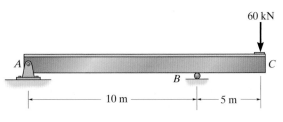

60 kN

10 m   5 m

Probs. 8–12/8–13

8–14. Determine the value of a so that the slope at A is equal to zero. EI is constant. Use the moment-area theorems.

8–15. Solve Prob. 8–14 using the conjugate-beam method.

*8–16.** Determine the value of a so that the displacement at C is equal to zero. EI is constant. Use the moment-area theorems.

8–17. Solve Prob. 8–16 using the conjugate-beam method.

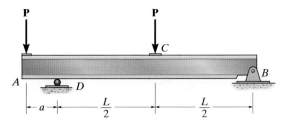

P   P

a $\dfrac{L}{2}$ $\dfrac{L}{2}$

Probs. 8–14/8–15/8–16/8–17

8–18. Determine the slope and the displacement at C. EI is constant. Use the moment-area theorems.

8–19. Solve Prob. 8–18 using the conjugate-beam method.

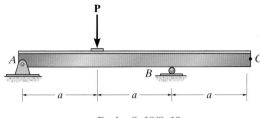

P

a a a

Probs. 8–18/8–19

*8–20.** Determine the slope and the displacement at the end C of the beam. $E = 200$ GPa, $I = 70(10^6)$ mm^4. Use the moment-area theorems.

8–21. Solve Prob. 8–20 using the conjugate-beam method.

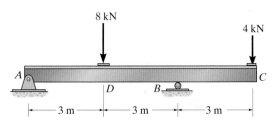

8 kN   4 kN

3 m   3 m   3 m

Probs. 8–20/8–21

8–22. At what distance a should the bearing supports at A and B be placed so that the displacement at the center of the shaft is equal to the deflection at its ends? The bearings exert only vertical reactions on the shaft. EI is constant. Use the moment-area theorems.

8–23. Solve Prob. 8–22 using the conjugate-beam method.

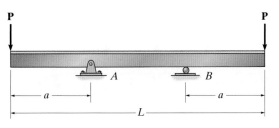

P   P

a a

L

Probs. 8–22/8–23

*8–24. Determine the displacement at *C* and the slope at *B*. *EI* is constant. Use the moment-area theorems.

8–25. Solve Prob. 8–24 using the conjugate-beam method.

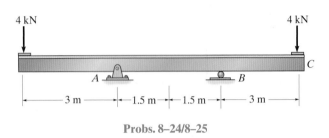

Probs. 8–24/8–25

8–26. Determine the displacement at *C* and the slope at *B*. *EI* is constant. Use the moment-area theorems.

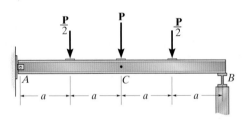

Prob. 8–26

8–27. Determine the displacement at *C* and the slope at *B*. *EI* is constant. Use the conjugate-beam method.

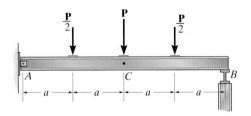

Prob. 8–27

*8–28. Determine the force **F** at the end of the beam *C* so that the displacement at *C* is zero. *EI* is constant. Use the moment-area theorems.

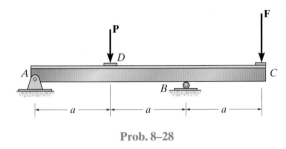

Prob. 8–28

8–29. Determine the force **F** at the end of the beam *C* so that the displacement at *C* is zero. *EI* is constant. Use the conjugate-beam method.

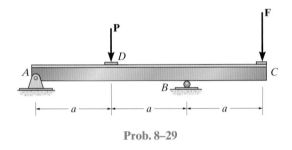

Prob. 8–29

8–30. Determine the slope at *B* and the displacement at *C*. *EI* is constant. Use the moment-area theorems.

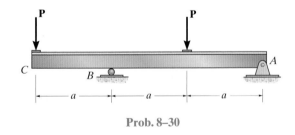

Prob. 8–30

8–31. Determine the slope at *B* and the displacement at *C*. *EI* is constant. Use the conjugate-beam method.

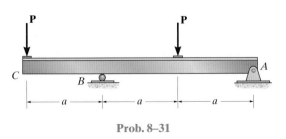

Prob. 8–31

*8–32.   Determine the maximum displacement and the slope at A. EI is constant. Use the moment-area theorems.

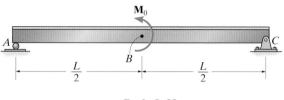

Prob. 8–32

8–33.   Determine the maximum displacement at B and the slope at A. EI is constant. Use the conjugate-beam method.

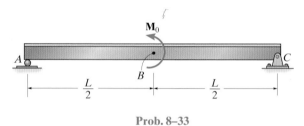

Prob. 8–33

8–34.   Determine the slope and displacement at C. EI is constant. Use the moment-area theorems.

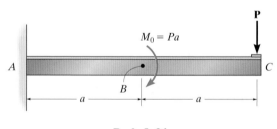

Prob. 8–34

8–35.   Determine the slope and displacement at C. EI is constant. Use the conjugate-beam method.

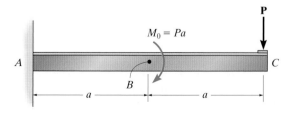

Prob. 8–35

*8–36.   Determine the displacement at C. Assume A is a fixed support, B is a pin, and D is a roller. EI is constant. Use the moment-area theorems.

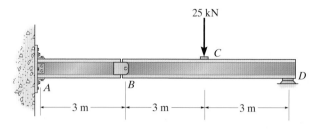

Prob. 8–36

8–37.   Determine the displacement at C. Assume A is a fixed support, B is a pin, and D is a roller. EI is constant. Use the conjugate-beam method.

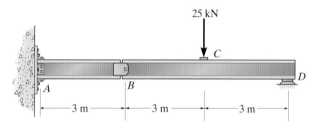

Prob. 8–37

8–38.   Determine the displacement at D and the slope at D. Assume A is a fixed support, B is a pin, and C is a roller. Use the moment-area theorems.

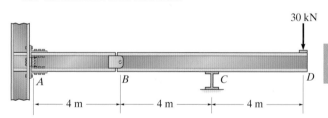

Prob. 8–38

8–39.   Determine the displacement at D and the slope at D. Assume A is a fixed support, B is a pin, and C is a roller. Use the conjugate-beam method.

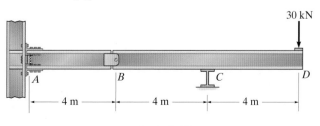

Prob. 8–39

8

CHAPTER REVIEW

The deflection of a member (or structure) can always be established provided the moment diagram is known, because positive moment will tend to bend the member concave upwards, and negative moment will tend to bend the member concave downwards. Likewise, the general shape of the moment diagram can be determined if the deflection curve is known.

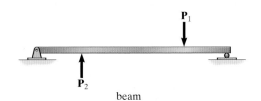

beam

deflection curve

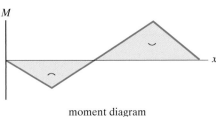

moment diagram

(Ref: Section 8.1)

Deflection of a beam due to bending can be determined by using double integration of the equation.

$$\frac{d^2v}{dx^2} = \frac{M}{EI}$$

Here the internal moment M must be expressed as a function of the x coordinates that extend across the beam. The constants of integration are obtained from the boundary conditions, such as zero deflection at a pin or roller support and zero deflection and slope at a fixed support. If several x coordinates are necessary, then the continuity of slope and deflection must be considered, where at $x_1 = x_2 = a$, $\theta_1(a) = \theta_2(a)$ and $v_1(a) = v_2(a)$.

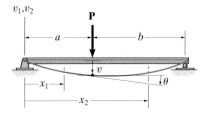

(Ref: Sections 8.2 & 8.3)

If the moment diagram has a simple shape, the moment-area theorems or the conjugate beam method can be used to determine the deflection and slope at a point on the beam.

The moment-area theorems consider the angles and vertical deviation between the tangents at two points A and B on the elastic curve. The change in slope is found from the area under the M/EI diagram between the two points, and the deviation is determined from the moment of the M/EI diagram area about the point where the deviation occurs.

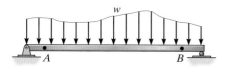

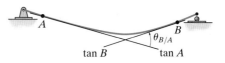

$$\theta_{B/A} = \text{Area of } M/EI \text{ diagram}$$

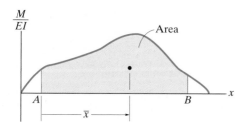

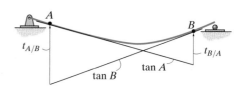

$$t_{A/B} = \bar{x}(\text{Area of } M/EI \text{ diagram})$$

(Ref: Section 8.4)

The conjugate beam method is very methodical and requires application of the principles of statics. Quite simply, one establishes the conjugate beam using Table 8–2, then considers the loading as the M/EI diagram. The slope (deflection) at a point on the real beam is then equal to the shear (moment) at the same point on the conjugate beam.

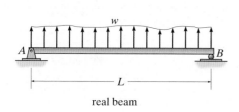

real beam

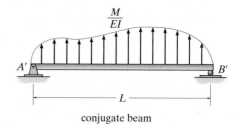

conjugate beam

(Ref: Section 8.5)

The displacement at the ends of this bridge deck, as it is being constructed, can be determined using energy methods.

Deflections Using Energy Methods

9

In this chapter, we will show how to apply energy methods to solve problems involving slope and deflection. The chapter begins with a discussion of work and strain energy, followed by a development of the principle of work and energy. The method of virtual work and Castigliano's theorem are then developed, and these methods are used to determine the displacements at points on trusses, beams, and frames.

9.1 External Work and Strain Energy

The semigraphical methods presented in the previous chapters are very effective for finding the displacements and slopes at points in *beams* subjected to rather simple loadings. For more complicated loadings or for structures such as trusses and frames, it is suggested that energy methods be used for the computations. Most energy methods are based on the *conservation of energy principle*, which states that the work done by all the external forces acting on a structure, U_e, is transformed into internal work or strain energy, U_i, which is developed when the structure deforms. If the material's elastic limit is not exceeded, the *elastic strain energy* will return the structure to its undeformed state when the loads are removed. The conservation of energy principle can be stated mathematically as

$$U_e = U_i \qquad (9\text{--}1)$$

Before developing any of the energy methods based on this principle, however, we will first determine the external work and strain energy caused by a force and a moment. The formulations to be presented will provide a basis for understanding the work and energy methods that follow.

External Work—Force. When a force **F** undergoes a displacement dx in the *same direction* as the force, the work done is $dU_e = F\,dx$. If the total displacement is x, the work becomes

$$U_e = \int_0^x F\,dx \qquad (9\text{–}2)$$

Consider now the effect caused by an axial force applied to the end of a bar as shown in Fig. 9–1a. As the magnitude of **F** is *gradually* increased from zero to some limiting value $F = P$, the final elongation of the bar becomes Δ. If the material has a linear elastic response, then $F = (P/\Delta)x$. Substituting into Eq. 9–2, and integrating from 0 to Δ, we get

$$U_e = \tfrac{1}{2}P\Delta \qquad (9\text{–}3)$$

which represents the shaded *triangular area* in Fig. 9–1a.

We may also conclude from this that as a force is gradually applied to the bar, and its magnitude builds linearly from zero to some value P, the work done is equal to the *average force magnitude* ($P/2$) times the displacement (Δ).

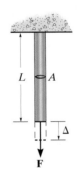

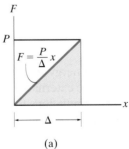

(a)

Fig. 9–1

Suppose now that **P** is already applied to the bar and that *another force* **F'** is now applied, so the bar deflects further by an amount Δ', Fig. 9–1*b*. The work done by **P** (not **F'**) when the bar undergoes the further deflection Δ' is then

$$U_e' = P\Delta' \qquad (9\text{–}4)$$

Here the work represents the shaded *rectangular area* in Fig. 9–1*b*. In this case **P** does not change its magnitude since Δ' is caused only by **F'**. Therefore, work is simply the force magnitude (P) times the displacement (Δ').

In summary, then, when a force **P** is applied to the bar, followed by application of the force **F'**, the total work done by both forces is represented by the triangular area ACE in Fig. 9–1*b*. The triangular area ABG represents the work of **P** that is caused by its displacement Δ, the triangular area BCD represents the work of **F'** since this force causes a displacement Δ', and lastly, the shaded rectangular area $BDEG$ represents the additional work done by **P** when displaced Δ' as caused by **F'**.

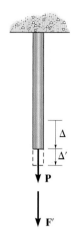

External Work—Moment.

The work of a moment is defined by the product of the magnitude of the moment **M** and the angle $d\theta$ through which it rotates, that is, $dU_e = M\,d\theta$, Fig. 9–2. If the total angle of rotation is θ radians, the work becomes

$$U_e = \int_0^{\theta} M\,d\theta \qquad (9\text{–}5)$$

As in the case of force, if the moment is applied *gradually* to a structure having linear elastic response from zero to M, the work is then

$$U_e = \tfrac{1}{2}M\theta \qquad (9\text{–}6)$$

However, if the moment is already applied to the structure and other loadings further distort the structure by an amount θ', then **M** rotates θ', and the work is

$$U_e' = M\theta' \qquad (9\text{–}7)$$

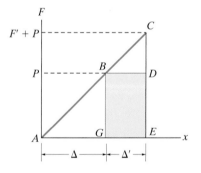

(b)

Fig. 9–1

9

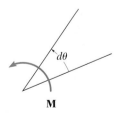

Fig. 9–2

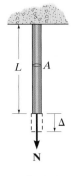

Fig. 9–3

Strain Energy—Axial Force.

When an axial force **N** is applied gradually to the bar in Fig. 9–3, it will strain the material such that the *external work* done by **N** will be converted into *strain energy*, which is stored in the bar (Eq. 9–1). Provided the material is *linearly elastic*, Hooke's law is valid, $\sigma = E\epsilon$, and if the bar has a constant cross-sectional area A and length L, the normal stress is $\sigma = N/A$ and the final strain is $\epsilon = \Delta/L$. Consequently, $N/A = E(\Delta/L)$, and the final deflection is

$$\Delta = \frac{NL}{AE} \tag{9–8}$$

Substituting into Eq. 9–3, with $P = N$, the strain energy in the bar is therefore

$$U_i = \frac{N^2 L}{2AE} \tag{9–9}$$

Strain Energy—Bending.

Consider the beam shown in Fig. 9–4a, which is distorted by the *gradually* applied loading **P** and w. These loads create an internal moment **M** in the beam at a section located a distance x from the left support. The resulting rotation of the differential element dx, Fig. 9–4b, can be found from Eq. 8–2, that is, $d\theta = (M/EI)\,dx$. Consequently, the strain energy, or work stored in the element, is determined from Eq. 9–6 since the internal moment is gradually developed. Hence,

$$dU_i = \frac{M^2\,dx}{2EI} \tag{9–10}$$

The strain energy for the beam is determined by integrating this result over the beam's entire length L. The result is

$$U_i = \int_0^L \frac{M^2\,dx}{2EI} \tag{9–11}$$

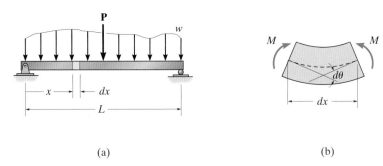

(a)   (b)

Fig. 9–4

9.2   Principle of Work and Energy

Now that the work and strain energy for a force and a moment have been formulated, we will illustrate how the conservation of energy or the principle of work and energy can be applied to determine the displacement at a point on a structure. To do this, consider finding the displacement Δ at the point where the force **P** is applied to the cantilever beam in Fig. 9–5. From Eq. 9–3, the external work is $U_e = \frac{1}{2}P\Delta$. To obtain the resulting strain energy, we must first determine the internal moment as a function of position x in the beam and then apply Eq. 9–11. In this case $M = -Px$, so that

$$U_i = \int_0^L \frac{M^2\,dx}{2EI} = \int_0^L \frac{(-Px)^2\,dx}{2EI} = \frac{1}{6}\frac{P^2L^3}{EI}$$

Equating the external work to internal strain energy and solving for the unknown displacement Δ, we have

$$U_e = U_i$$

$$\frac{1}{2}P\Delta = \frac{1}{6}\frac{P^2L^3}{EI}$$

$$\Delta = \frac{PL^3}{3EI}$$

Although the solution here is quite direct, application of this method is limited to only a few select problems. It will be noted that only *one load* may be applied to the structure, since if more than one load were applied, there would be an unknown displacement under each load, and yet it is possible to write only *one* "work" equation for the beam. Furthermore, *only the displacement under the force can be obtained,* since the external work depends upon both the force and its corresponding displacement. One way to circumvent these limitations is to use the method of virtual work or Castigliano's theorem, both of which are explained in the following sections.

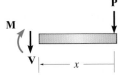

Fig. 9–5

9.3   Principle of Virtual Work

The principle of virtual work was developed by John Bernoulli in 1717 and is sometimes referred to as the unit-load method. It provides a general means of obtaining the displacement and slope at a specific point on a structure, be it a beam, frame, or truss.

Before developing the principle of virtual work, it is necessary to make some general statements regarding the principle of work and energy, which was discussed in the previous section. If we take a deformable structure of any shape or size and apply a series of *external loads* **P** to it, it will cause *internal* loads **u** at points throughout the structure. *It is necessary that the external and internal loads be related by the equations of equilibrium.* As a consequence of these loadings, external displacements Δ will occur at the **P** loads and internal displacements δ will occur at each point of internal load **u**. In general, *these displacements do not have to be elastic,* and they may not be related to the loads; however, *the external and internal displacements must be related by the compatibility of the displacements.* In other words, if the external displacements are known, the corresponding internal displacements are uniquely defined. In general, then, the principle of work and energy states:

$$\underset{\substack{\text{Work of}\\ \text{External Loads}}}{\Sigma P \Delta} = \underset{\substack{\text{Work of}\\ \text{Internal Loads}}}{\Sigma u \delta} \qquad (9\text{--}12)$$

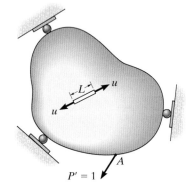

Apply virtual load $P' = 1$

(a)

Based on this concept, the principle of virtual work will now be developed. To do this, we will consider the structure (or body) to be of arbitrary shape as shown in Fig. 9–6*b*.* Suppose it is necessary to determine the displacement Δ of point *A* on the body caused by the "real loads" **P**₁, **P**₂, and **P**₃. It is to be understood that these loads cause no movement of the supports; in general, however, they can strain the material *beyond the elastic limit.* Since no external load acts on the body at *A* and in the direction of Δ, the displacement Δ can be determined by *first* placing on the body a *"virtual" load* such that this force **P'** acts in the *same direction* as Δ, Fig. 9–6*a*. For convenience, which will be apparent later, we will choose **P'** to have a "unit" magnitude, that is, $P' = 1$. The term "virtual" is used to describe the load, since *it is imaginary and does not actually exist as part of the real loading.* The unit load (**P'**) does, however, create an internal virtual load **u** in a representative element or fiber of the body, as shown in Fig. 9–6*a*. Here it is required that **P'** and **u** be related by the equations of equilibrium.[†]

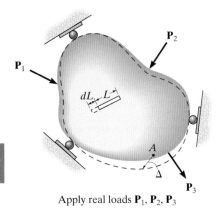

Apply real loads **P**₁, **P**₂, **P**₃

(b)

Fig. 9–6

---

*This arbitrary shape will later represent a specific truss, beam, or frame.
[†]Although these loads will cause virtual displacements, we will not be concerned with their magnitudes.

Once the virtual loadings are applied, *then* the body is subjected to the *real loads* $\mathbf{P_1}$, $\mathbf{P_2}$, and $\mathbf{P_3}$, Fig. 9–6b. Point A will be displaced an amount Δ, causing the element to deform an amount dL. As a result, the external virtual force $\mathbf{P'}$ and internal virtual load $\mathbf{u}$ "ride along" by Δ and dL, respectively, and therefore perform *external virtual work* of $1 \cdot \Delta$ on the body and internal virtual work of $u \cdot dL$ on the element. Realizing that the external virtual work is equal to the internal virtual work done on all the elements of the body, we can write the virtual-work equation as

$$\overset{\text{virtual loadings}}{1 \cdot \Delta = \Sigma u \cdot dL} \qquad (9\text{–}13)$$

$$\underset{\text{real displacements}}{}$$

where

$P' = 1 = $ external virtual unit load acting in the direction of Δ.

$u = $ internal virtual load acting on the element in the direction of dL.

$\Delta = $ external displacement caused by the real loads.

$dL = $ internal deformation of the element caused by the real loads.

By choosing $P' = 1$, it can be seen that the solution for Δ follows directly, since $\Delta = \Sigma u \, dL$.

In a similar manner, if the rotational displacement or slope of the tangent at a point on a structure is to be determined, a virtual *couple moment* $\mathbf{M'}$ having a "unit" magnitude is applied at the point. As a consequence, this couple moment causes a virtual load $\mathbf{u_\theta}$ in one of the elements of the body. Assuming that the real loads deform the element an amount dL, the rotation θ can be found from the virtual-work equation

$$\overset{\text{virtual loadings}}{1 \cdot \theta = \Sigma u_\theta \cdot dL} \qquad (9\text{–}14)$$

$$\underset{\text{real displacements}}{}$$

where

$M' = 1 = $ external virtual unit couple moment acting in the direction of θ.

$u_\theta = $ internal virtual load acting on an element in the direction of dL.

$\theta = $ external rotational displacement or slope in radians caused by the real loads.

$dL = $ internal deformation of the element caused by the real loads.

This method for applying the principle of virtual work is often referred to as the *method of virtual forces,* since a virtual force is applied resulting in the calculation of a *real displacement*. The equation of virtual work in this case represents a *compatibility requirement* for the structure. Although not important here, realize that we can also apply the principle of virtual work as a *method of virtual displacements*. In this case virtual displacements are imposed on the structure while the structure is subjected to *real loadings*. This method can be used to determine a force on or in a structure,* so that the equation of virtual work is then expressed as an *equilibrium requirement*.

9.4   Method of Virtual Work: Trusses

We can use the method of virtual work to determine the displacement of a truss joint when the truss is subjected to an external loading, temperature change, or fabrication errors. Each of these situations will now be discussed.

External Loading.   For the purpose of explanation let us consider the vertical displacement Δ of joint B of the truss in Fig. 9–7a. Here a typical element of the truss would be one of its *members* having a length L, Fig. 9–7b. If the applied loadings $\mathbf{P}_1$ and $\mathbf{P}_2$ cause a *linear elastic material response,* then this element deforms an amount $\Delta L = NL/AE$, where N is the normal or axial force in the member, caused by the loads. Applying Eq. 9–13, the virtual-work equation for the truss is therefore

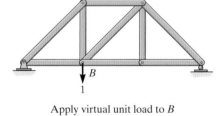

Apply virtual unit load to B

(a)

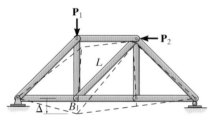

Apply real loads $\mathbf{P}_1$, $\mathbf{P}_2$

(b)

Fig. 9–7

$$1 \cdot \Delta = \Sigma \frac{nNL}{AE} \qquad (9\text{–}15)$$

where

1 = external virtual unit load acting on the truss joint in the stated direction of Δ.

n = internal virtual normal force in a truss member caused by the external virtual unit load.

Δ = external joint displacement caused by the real loads on the truss.

N = internal normal force in a truss member caused by the real loads.

L = length of a member.

A = cross-sectional area of a member.

E = modulus of elasticity of a member.

The formulation of this equation follows naturally from the development in Sec. 9–3. Here the external virtual unit load creates internal virtual forces $\mathbf{n}$ in each of the truss members. The real loads then cause the truss joint to be displaced Δ in the same direction as the virtual unit load, and each member is displaced NL/AE in the same direction as its respective $\mathbf{n}$ force. Consequently, the external virtual work $1 \cdot \Delta$ equals the internal virtual work or the internal (virtual) strain energy stored in *all* the truss members, that is, $\Sigma nNL/AE$.

*It was used in this manner in Sec. 6–3 with reference to the Müller-Breslau principle.

Temperature.   In some cases, truss members may change their length due to temperature. If α is the coefficient of thermal expansion for a member and ΔT is the change in its temperature, the change in length of a member is $\Delta L = \alpha \, \Delta T \, L$. Hence, we can determine the displacement of a selected truss joint due to this temperature change from Eq. 9–13, written as

$$\boxed{1 \cdot \Delta = \Sigma n \alpha \, \Delta T \, L} \qquad (9\text{–}16)$$

where

1 = external virtual unit load acting on the truss joint in the stated direction of Δ.

n = internal virtual normal force in a truss member caused by the external virtual unit load.

Δ = external joint displacement caused by the temperature change.

α = coefficient of thermal expansion of member.

ΔT = change in temperature of member.

L = length of member.

Fabrication Errors and Camber.   Occasionally, errors in fabricating the lengths of the members of a truss may occur. Also, in some cases truss members must be made slightly longer or shorter in order to give the truss a camber. Camber is often built into a bridge truss so that the bottom cord will curve upward by an amount equivalent to the downward deflection of the cord when subjected to the bridge's full dead weight. If a truss member is shorter or longer than intended, the displacement of a truss joint from its expected position can be determined from direct application of Eq. 9–13, written as

$$\boxed{1 \cdot \Delta = \Sigma n \, \Delta L} \qquad (9\text{–}17)$$

where

1 = external virtual unit load acting on the truss joint in the stated direction of Δ.

n = internal virtual normal force in a truss member caused by the external virtual unit load.

Δ = external joint displacement caused by the fabrication errors.

ΔL = difference in length of the member from its intended size as caused by a fabrication error.

A combination of the right sides of Eqs. 9–15 through 9–17 will be necessary if both external loads act on the truss and some of the members undergo a thermal change or have been fabricated with the wrong dimensions.

Procedure for Analysis

The following procedure may be used to determine a specific displacement of any joint on a truss using the method of virtual work.

Virtual Forces n

- Place the unit load on the truss at the joint where the desired displacement is to be determined. The load should be in the same direction as the specified displacement, e.g., horizontal or vertical.
- With the unit load so placed, and all the real loads *removed* from the truss, use the method of joints or the method of sections and calculate the internal **n** force in each truss member. Assume that tensile forces are positive and compressive forces are negative.

Real Forces N

- Use the method of sections or the method of joints to determine the **N** force in each member. These forces are caused only by the real loads acting on the truss. Again, assume tensile forces are positive and compressive forces are negative.

Virtual-Work Equation

- Apply the equation of virtual work, to determine the desired displacement. It is important to retain the algebraic sign for each of the corresponding **n** and **N** forces when substituting these terms into the equation.
- If the resultant sum $\Sigma nNL/AE$ is positive, the displacement Δ is in the same direction as the unit load. If a negative value results, Δ is opposite to the unit load.
- When applying $1 \cdot \Delta = \Sigma n\alpha \, \Delta T L$, realize that if any of the members undergoes an *increase in temperature*, ΔT will be *positive*, whereas a *decrease in temperature* results in a *negative* value for ΔT.
- For $1 \cdot \Delta = \Sigma n \, \Delta L$, when a fabrication error *increases the length* of a member, ΔL is *positive*, whereas a *decrease in length* is *negative*.
- When applying any formula, attention should be paid to the units of each numerical quantity. In particular, the virtual unit load can be assigned any arbitrary unit (N, kN, etc.), since the **n** forces will have these *same units*, and as a result the units for both the virtual unit load and the **n** forces will cancel from both sides of the equation.

EXAMPLE 9.1

Determine the vertical displacement of joint C of the steel truss shown in Fig. 9–8a. The cross-sectional area of each member is $A = 300$ mm^2 and $E = 200$ GPa.

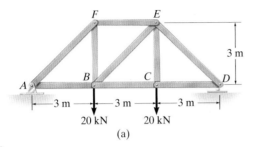

3 m

3 m

A

3 m — 3 m — 3 m

20 kN   20 kN

(a)

SOLUTION

Virtual Forces n. Only a vertical 1-kN load is placed at joint C, and the force in each member is calculated using the method of joints. The results are shown in Fig. 9–8b. Positive numbers indicate tensile forces and negative numbers indicate compressive forces.

Real Forces N. The real forces in the members are calculated using the method of joints. The results are shown in Fig. 9–8c.

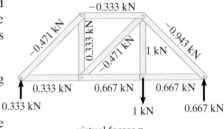

−0.333 kN

−0.471 kN   0.333 kN   −0.471 kN   1 kN   −0.943 kN

0.333 kN   0.667 kN   0.667 kN

0.333 kN   1 kN   0.667 kN

virtual forces **n**

(b)

Virtual-Work Equation. Arranging the data in tabular form, we have

Member	n (kN)	N (kN)	L (m)	nNL (kN² · m)
AB	0.333	20	3	20
BC	0.667	20	3	40
CD	0.667	20	3	40
DE	−0.943	−28.3	4.24	113
FE	−0.333	−20	3	20
EB	−0.471	0	4.24	0
BF	0.333	20	3	20
AF	−0.471	−28.3	4.24	56.6
CE	1	20	3	60
				Σ369.6

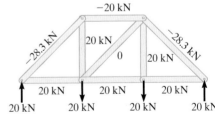

−20 kN

−28.3 kN   20 kN   0   20 kN   −28.3 kN

20 kN   20 kN   20 kN

20 kN   20 kN   20 kN   20 kN

real forces **N**

(c)

Fig. 9–8

Thus
$$1 \text{ k} \cdot \Delta_{C_v} = \sum \frac{nNL}{AE} = \frac{369.6 \text{ kN}^2 \cdot \text{m}}{AE}$$

Substituting the numerical values for A and E, we have

$$1 \text{ k} \cdot \Delta_{C_v} = \frac{369.6 \text{ kN}^2 \cdot \text{m}}{[300(10^{-6}) \text{ m}^2][200(10^6) \text{ kN/m}^2]}$$

$$\Delta_{C_v} = 0.00616 \text{ m} = 6.16 \text{ mm} \qquad Ans.$$

9

EXAMPLE 9.2

The cross-sectional area of each member of the truss shown in Fig. 9–9a is $A = 400 \text{ mm}^2$ and $E = 200 \text{ GPa}$. (a) Determine the vertical displacement of joint C if a 4-kN force is applied to the truss at C. (b) If no loads act on the truss, what would be the vertical displacement of joint C if member AB were 5 mm too short?

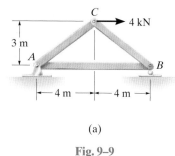

(a)

Fig. 9–9

SOLUTION

Part (a)

Virtual Forces n.   Since the *vertical displacement* of joint C is to be determined, a virtual force of 1 kN is applied at C in the vertical direction. The units of this force are the *same* as those of the real loading. The support reactions at A and B are calculated and the **n** force in each member is determined by the method of joints as shown on the free-body diagrams of joints A and B, Fig. 9–9b.

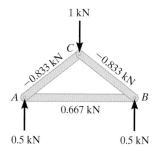

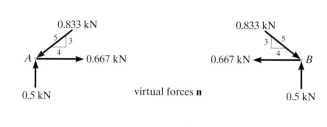

(b)

Real Forces N.   The joint analysis of A and B when the real load of 4 kN is applied to the truss is given in Fig. 9–9c.

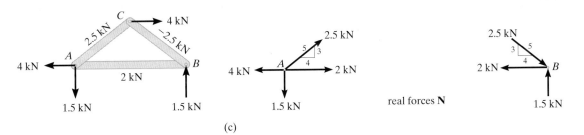

(c)

Virtual-Work Equation. Since AE is constant, each of the terms nNL can be arranged in tabular form and computed. Here positive numbers indicate tensile forces and negative numbers indicate compressive forces.

Member	n (kN)	N (kN)	L (m)	$n\,NL$ (kN$^2 \cdot$ m)
AB	0.667	2	8	10.67
AC	−0.833	2.5	5	−10.41
CB	−0.833	−2.5	5	10.41
				Σ10.67

Thus,

$$1 \text{ kN} \cdot \Delta_{C_v} = \sum \frac{nNL}{AE} = \frac{10.67 \text{ kN}^2 \cdot \text{m}}{AE}$$

Substituting the values $A = 400 \text{ mm}^2 = 400(10^{-6}) \text{ m}^2$, $E = 200 \text{ GPa} = 200(10^6) \text{ kN/m}^2$, we have

$$1 \text{ kN} \cdot \Delta_{C_v} = \frac{10.67 \text{ kN}^2 \cdot \text{m}}{400(10^{-6}) \text{ m}^2 (200(10^6) \text{ kN/m}^2)}$$

$$\Delta_{C_v} = 0.000133 \text{ m} = 0.133 \text{ mm} \qquad \textit{Ans.}$$

Part (b). Here we must apply Eq. 9–17. Since the vertical displacement of C is to be determined, we can use the results of Fig. 9–7b. Only member AB undergoes a change in length, namely, of $\Delta L = -0.005$ m. Thus,

$$1 \cdot \Delta = \Sigma n \, \Delta L$$

$$1 \text{ kN} \cdot \Delta_{C_v} = (0.667 \text{ kN})(-0.005 \text{ m})$$

$$\Delta_{C_v} = -0.00333 \text{ m} = -3.33 \text{ mm} \qquad \textit{Ans.}$$

The negative sign indicates joint C is displaced *upward,* opposite to the 1-kN vertical load. Note that if the 4-kN load and fabrication error are both accounted for, the resultant displacement is then $\Delta_{C_v} = 0.133 - 3.33 = -3.20$ mm (upward).

EXAMPLE 9.3

Determine the vertical displacement of joint C of the steel truss shown in Fig. 9–10a. Due to radiant heating from the wall, member AD is subjected to an *increase* in temperature of $\Delta T = +60°C$. Take $\alpha = 1.08(10^{-5})/°C$ and $E = 200$ GPa. The cross-sectional area of each member is indicated in the figure.

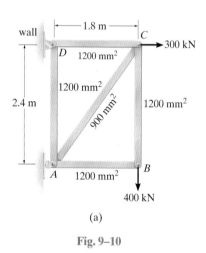

(a)

Fig. 9–10

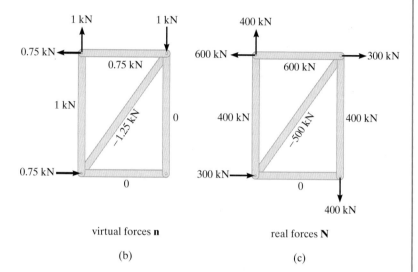

virtual forces **n**

(b)

real forces **N**

(c)

SOLUTION

Virtual Forces n. A *vertical* 1-kN load is applied to the truss at joint C, and the forces in the members are computed, Fig. 9–10b.

Real Forces N. Since the **n** forces in members AB and BC are *zero*, the **N** forces in these members do *not* have to be computed. Why? For completion, though, the entire real-force analysis is shown in Fig. 9–10c.

Virtual-Work Equation. Both loads and temperature affect the deformation; therefore, Eqs. 9–15 and 9–16 are combined. Working in units of kips and inches, we have

$$1 \cdot \Delta_{C_v} = \sum \frac{nNL}{AE} + \Sigma n\alpha\,\Delta T\,L$$

$$= \frac{(0.75)(600)(1.8)}{1200(10^{-6})[200(10^6)]} + \frac{(1)(400)(2.4)}{1200(10^{-6})[200(10^6)]}$$

$$+ \frac{(-1.25)(-500)(3)}{900(10^{-6})[200(10^6)]} + (1)[1.08(10^{-5})](60)(2.4)$$

$$\Delta_{C_v} = 0.00193 \text{ m} = 19.3 \text{ mm} \qquad\qquad Ans.$$

9.5   Castigliano's Theorem

In 1879 Alberto Castigliano, an Italian railroad engineer, published a book in which he outlined a method for determining the deflection or slope at a point in a structure, be it a truss, beam, or frame. This method, which is referred to as *Castigliano's second theorem,* or the *method of least work,* applies only to structures that have constant temperature, unyielding supports, and *linear elastic* material response. If the displacement of a point is to be determined, the theorem states that it is equal to the first partial derivative of the strain energy in the structure with respect to a force acting at the point and in the direction of displacement. In a similar manner, the slope at a point in a structure is equal to the first partial derivative of the strain energy in the structure with respect to a couple moment acting at the point and in the direction of rotation.

To derive Castigliano's second theorem, consider a body (structure) of any arbitrary shape which is subjected to a series of n forces $P_1, P_2, \ldots, P_n$. Since the external work done by these loads is equal to the internal strain energy stored in the body, we can write

$$U_i = U_e$$

The external work is a function of the external loads $(U_e = \Sigma \int P \, dx)$. Thus,

$$U_i = U_e = f(P_1, P_2, \ldots, P_n)$$

Now, if any one of the forces, say P_i, is increased by a differential amount dP_i, the internal work is also increased such that the new strain energy becomes

$$U_i + dU_i = U_i + \frac{\partial U_i}{\partial P_i} dP_i \qquad (9\text{--}18)$$

This value, however, should not depend on the sequence in which the n forces are applied to the body. For example, if we apply dP_i to the body *first,* then this will cause the body to be displaced a differential amount $d\Delta_i$ in the direction of dP_i. By Eq. 9–3 $\left(U_e = \frac{1}{2} P \Delta \right)$, the increment of strain energy would be $\frac{1}{2} dP_i \, d\Delta_i$. This quantity, however, is a second-order differential and may be neglected. Further application of the loads $P_1, P_2, \ldots, P_n$, which displace the body $\Delta_1, \Delta_2, \ldots, \Delta_n$, yields the strain energy.

$$U_i + dU_i = U_i + dP_i \Delta_i \qquad (9\text{--}19)$$

Here, as before, U_i is the internal strain energy in the body, caused by the loads $P_1, P_2, \ldots, P_n$, and $dU_i = dP_i \Delta_i$ is the *additional* strain energy caused by dP_i. (Eq. 9–4, $U_e = P\Delta'$.)

In summary, then, Eq. 9–18 represents the strain energy in the body determined by first applying the loads $P_1, P_2, \ldots, P_n$, *then* dP_i, and Eq. 9–19 represents the strain energy determined by first applying dP_i and *then* the loads P_1, P_2, ..., P_n. Since these two equations must be equal, we require

$$\Delta_i = \frac{\partial U_i}{\partial P_i} \qquad\qquad (9\text{--}20)$$

which proves the theorem; i.e., the displacement Δ_i in the direction of P_i is equal to the first partial derivative of the strain energy with respect to P_i.*

It should be noted that Eq. 9–20 is a statement regarding the *structure's compatibility*. Also, the above derivation requires that *only conservative forces* be considered for the analysis. These forces do work that is independent of the path and therefore create no energy loss. Since forces causing a linear elastic response are conservative, the theorem is restricted to *linear elastic behavior* of the material. This is unlike the method of virtual force discussed in the previous section, which applied to *both* elastic and inelastic behavior.

9.6   Castigliano's Theorem for Trusses

The strain energy for a member of a truss is given by Eq. 9–9, $U_i = N^2 L / 2AE$. Substituting this equation into Eq. 9–20 and omitting the subscript i, we have

$$\Delta = \frac{\partial}{\partial P} \sum \frac{N^2 L}{2AE}$$

It is generally easier to perform the differentiation prior to summation. In the general case L, A, and E are constant for a given member, and therefore we may write

$$\boxed{\Delta = \sum N \left(\frac{\partial N}{\partial P} \right) \frac{L}{AE}} \qquad\qquad (9\text{--}21)$$

where

Δ = external joint displacement of the truss.

P = external force applied to the truss joint in the direction of Δ.

N = internal force in a member caused by *both* the force P and the loads on the truss.

L = length of a member.

A = cross-sectional area of a member.

E = modulus of elasticity of a member.

---

*Castigliano's first theorem is similar to his second theorem; however, it relates the load P_i to the partial derivative of the strain energy with respect to the corresponding displacement, that is, $P_i = \partial U_i / \partial \Delta_i$. The proof is similar to that given above and, like the method of virtual displacement, Castigliano's first theorem applies to both elastic and inelastic material behavior. This theorem is another way of expressing the *equilibrium requirements* for a structure, and since it has very limited use in structural analysis, it will not be discussed in this book.

This equation is similar to that used for the method of virtual work, Eq. 9–15 ($1 \cdot \Delta = \Sigma nNL/AE$), except n is replaced by $\partial N/\partial P$. Notice that in order to determine this partial derivative it will be necessary to treat P as a *variable* (not a specific numerical quantity), and furthermore, each member force N must be expressed as a function of P. As a result, computing $\partial N/\partial P$ generally requires slightly more calculation than that required to compute each n force directly. These terms will of course be the same, since n or $\partial N/\partial P$ is simply the change of the internal member force with respect to the load P, or the change in member force per unit load.

Procedure for Analysis

The following procedure provides a method that may be used to determine the displacement of any joint of a truss using Castigliano's theorem.

External Force P

- Place a force P on the truss at the joint where the desired displacement is to be determined. This force is assumed to have a *variable magnitude* in order to obtain the change $\partial N/\partial P$. Be sure **P** is directed along the line of action of the displacement.

Internal Forces N

- Determine the force N in each member caused by both the real (numerical) loads and the variable force P. Assume tensile forces are positive and compressive forces are negative.
- Compute the respective partial derivative $\partial N/\partial P$ for each member.
- After N and $\partial N/\partial P$ have been determined, assign P its numerical value if it has replaced a real force on the truss. Otherwise, set P equal to zero.

Castigliano's Theorem

- Apply Castigliano's theorem to determine the desired displacement Δ. It is important to retain the algebraic signs for corresponding values of N and $\partial N/\partial P$ when substituting these terms into the equation.
- If the resultant sum $\Sigma N(\partial N/\partial P)L/AE$ is positive, Δ is in the same direction as P. If a negative value results, Δ is opposite to P.

9

EXAMPLE 9.4

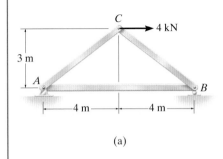

(a)

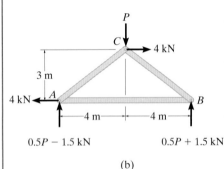

(b)

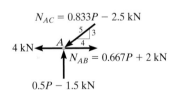

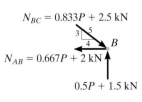

(c)

Fig. 9–11

Determine the vertical displacement of joint C of the truss shown in Fig. 9–11a. The cross-sectional area of each member is $A = 400 \text{ mm}^2$ and $E = 200 \text{ GPa}$.

SOLUTION

External Force P. A vertical force $\mathbf{P}$ is applied to the truss at joint C, since this is where the vertical displacement is to be determined, Fig. 9–11b.

Internal Forces N. The reactions at the truss supports at A and B are determined and the results are shown in Fig. 9–11b. Using the method of joints, the N forces in each member are determined, Fig. 9–11c.* For convenience, these results along with the partial derivatives $\partial N / \partial P$ are listed in tabular form as follows:

Member	N	$\dfrac{\partial N}{\partial P}$	$N\,(P = 0)$	L	$N\!\left(\dfrac{\partial N}{\partial P}\right)\!L$
AB	$0.667P + 2$	0.667	2	8	10.67
AC	$-(0.833P - 2.5)$	-0.833	2.5	5	-10.42
BC	$-(0.833P + 2.5)$	-0.833	-2.5	5	10.42

$$\Sigma = 10.67 \text{ kN} \cdot \text{m}$$

Since P does not actually exist as a real load on the truss, we require $P = 0$ in the table above.

Castigliano's Theorem. Applying Eq. 9–21, we have

$$\Delta_{C_v} = \Sigma N \left(\frac{\partial N}{\partial P} \right) \frac{L}{AE} = \frac{10.67 \text{ kN} \cdot \text{m}}{AE}$$

Substituting $A = 400 \text{ mm}^2 = 400(10^{-6}) \text{ m}^2$, $E = 200 \text{ GPa} = 200(10^9) \text{ Pa}$, and converting the units of N from kN to N, we have

$$\Delta_{C_v} = \frac{10.67(10^3) \text{ N} \cdot \text{m}}{400(10^{-6}) \text{ m}^2 (200(10^9) \text{ N/m}^2)} = 0.000133 \text{ m} = 0.133 \text{ mm}$$

Ans.

This solution should be compared with the virtual-work method of Example 9–2.

---

*It may be more convenient to analyze the truss with just the 4-kN load on it, then analyze the truss with the P load on it. The results can then be added together to give the N forces.

EXAMPLE 9.5

Determine the horizontal displacement of joint D of the truss shown in Fig. 9–12a. Take $E = 200$ GPa. The cross-sectional area of each member is indicated in the figure.

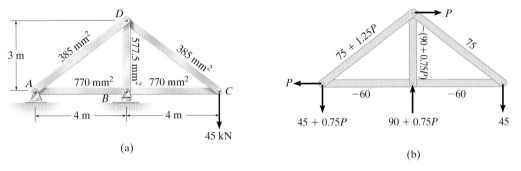

(a)   (b)

Fig. 9–12

SOLUTION

External Force P. Since the horizontal displacement of D is to be determined, a horizontal variable force P is applied to joint D, Fig. 9–12b.

Internal Forces N. Using the method of joints, the force N in each member is computed.* Again, when applying Eq. 9–21, we set $P = 0$ since this force does not actually exist on the truss. The results are shown in Fig. 9–12b. Arranging the data in tabular form, we have

Member	N	$\dfrac{\partial N}{\partial P}$	$N\,(P = 0)$	L	$N\left(\dfrac{\partial N}{\partial P}\right)L$
AB	-60	0	-60	4	0
BC	-60	0	-60	4	0
CD	75	0	75	5	0
DA	$75 + 1.25P$	1.25	75	5	468.75
BD	$-(90 + 0.75P)$	-0.75	-90	3	202.50

Castigliano's Theorem. Applying Eq. 9–21, we have

$$\Delta_{D_h} = \sum N\left(\frac{\partial N}{\partial P}\right)\frac{L}{AE} = 0 + 0 + 0 + \frac{468.75 \ \text{kN} \cdot \text{m}}{385(10^{-6}) \ \text{m}^2[200(10^6) \ \text{kN/m}^2]} + \frac{202.50 \ \text{kN} \cdot \text{m}}{577.5(10^{-6}) \ \text{m}^2[200(10^6) \ \text{kN/m}^2]}$$

$$= 0.00784 \ \text{m} = 7.84 \ \text{mm} \qquad\qquad Ans.$$

*As in the preceding example, it may be preferable to perform a separate analysis of the truss loaded with 45 kN and loaded with P and then superimpose the results.

EXAMPLE 9.6

Determine the vertical displacement of joint C of the truss shown in Fig. 9–13a. Assume that $A = 300$ mm^2 and $E = 200$ GPa.

SOLUTION

External Force P. The 20-kN force at C is replaced with a *variable force P* at joint C, Fig. 9–13b.

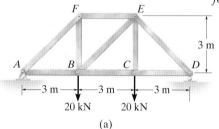

(a)

Internal Forces N. The method of joints is used to determine the force N in each member of the truss. The results are summarized in Fig. 9–13b. Here $P = 20$ kN when we apply Eq. 9–21. The required data can be arranged in tabulated form as follows:

Member	N	$\dfrac{\partial N}{\partial P}$	N $(P = 20$ kN)	L	$N\left(\dfrac{\partial N}{\partial P}\right)L$
AB	$0.333P + 13.33$	0.333	20	3	20
BC	$0.667P + 6.67$	0.667	20	3	40
CD	$0.667P + 6.67$	0.667	20	3	20
DE	$-(0.943P + 9.43)$	-0.943	-28.28	4.243	113.2
EF	$-(0.333P + 13.33)$	-0.333	-20	3	20
FA	$-(0.471P + 18.86)$	-0.471	-28.28	4.243	56.6
BF	$0.333P + 13.33$	0.333	20	3	20
BE	$-0.471P + 9.43$	-0.471	0	4.243	0
CE	P	1	20	3	60

$$\Sigma = 369.7 \text{ kN} \cdot \text{m}$$

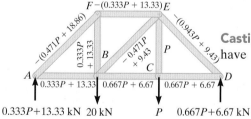

(b)

Fig. 9–13

Castigliano's Theorem. Substituting the data into Eq. 9–21, we have

$$\Delta_{C_v} = \sum N\left(\frac{\partial N}{\partial P}\right)\frac{L}{AE} = \frac{369.7 \text{ kN} \cdot \text{m}}{AE}$$

Converting the units of member length to millimeters and substituting the numerical value for AE, we have

$$\Delta_{C_v} = \frac{(369.7 \text{ kN} \cdot \text{m})(1000 \text{ mm/m})}{[300(10^{-6})\text{ m}^2][200(10^6)\text{ kN/m}^2]} = 6.16 \text{ mm} \quad \textit{Ans.}$$

The similarity between this solution and that of the virtual-work method, Example 9–1, should be noted.

F9–1. Determine the vertical displacement of joint B. AE is constant. Use the principle of virtual work.

F9–2. Solve Prob. F9–2 using Castigliano's theorem.

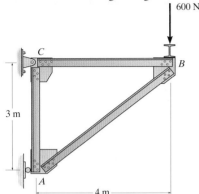

F9–1/9–2

F9–3. Determine the horizontal displacement of joint A. AE is constant. Use the principle of virtual work.

F9–4. Solve Prob. F9–3 using Castigliano's theorem.

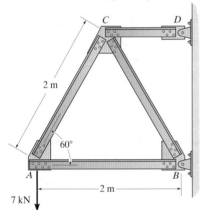

F9–3/9–4

F9–5. Determine the horizontal displacement of joint D. AE is constant. Use the principle of virtual work.

F9–6. Solve Prob. F9–5 using Castigliano's theorem.

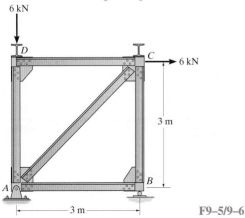

F9–5/9–6

F9–7. Determine the vertical displacement of joint D. AE is constant. Use the principle of virtual work.

F9–8. Solve Prob. F9–7 using Castigliano's theorem.

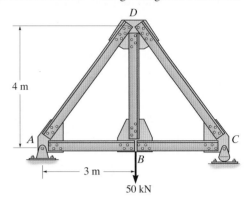

F9–7/9–8

F9–9. Determine the vertical displacement of joint B. AE is constant. Use the principle of virtual work.

F9–10. Solve Prob. F9–9 using Castigliano's theorem.

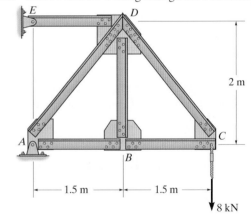

F9–9/9–10

9

F9–11. Determine the vertical displacement of joint C. AE is constant. Use the principle of virtual work.

F9–12. Solve Prob. F9–11 using Castigliano's theorem.

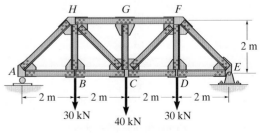

F9–11/9–12

361

PROBLEMS

9–1. Determine the vertical displacement of joint A. Each bar is made of steel and has a cross-sectional area of 600 mm². Take $E = 200$ GPa. Use the method of virtual work.

9–2. Solve Prob. 9–1 using Castigliano's theorem.

9–7. Determine the vertical displacement of joint D. Use the method of virtual work. AE is constant. Assume the members are pin connected at their ends.

*9–8. Solve Prob. 9–7 using Castigliano's theorem.

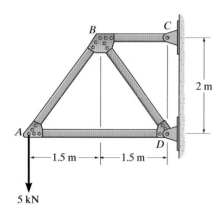

Probs. 9–1/9–2

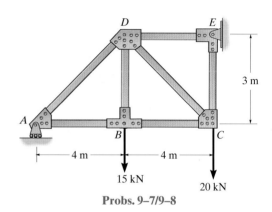

Probs. 9–7/9–8

9–3. Determine the vertical displacement of joint B. For each member $A = 400$ mm², $E = 200$ GPa. Use the method of virtual work.

*9–4. Solve Prob. 9–3 using Castigliano's theorem.

9–5. Determine the vertical displacement of joint E. For each member $A = 400$ mm², $E = 200$ GPa. Use the method of virtual work.

9–6. Solve Prob. 9–5 using Castigliano's theorem.

9–9. Determine the vertical displacement of joint F. Use the method of virtual work. AE is constant.

9–10. Solve Prob. 9–9 using Castigliano's theorem.

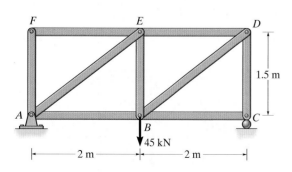

Probs. 9–3/9–4/9–5/9–6

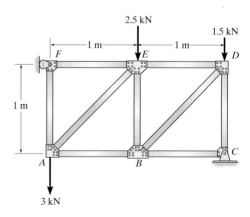

Probs. 9–9/9–10

9–11. Determine the vertical displacement of joint A. The cross-sectional area of each member is indicated in the figure. Assume the members are pin connected at their end points. $E = 200$ GPa. Use the method of virtual work.

*9–12. Solve Prob. 9–11 using Castigliano's theorem.

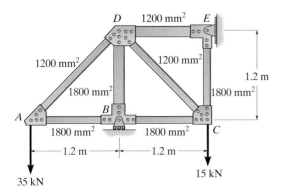

Probs. 9–11/9–12

9–13. Determine the horizontal displacement of joint D. Assume the members are pin connected at their end points. AE is constant. Use the method of virtual work.

9–14. Solve Prob. 9–13 using Castigliano's theorem.

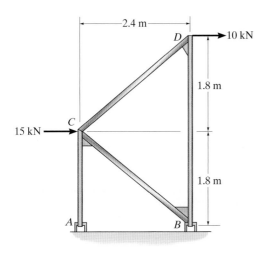

Probs. 9–13/9–14

9–15. Determine the vertical displacement of joint C of the truss. Each member has a cross-sectional area of $A = 300$ mm^2. $E = 200$ GPa. Use the method of virtual work.

*9–16. Solve Prob. 9–15 using Castigliano's theorem.

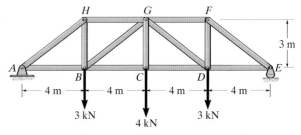

Probs. 9–15/9–16

9–17. Determine the vertical displacement of joint A. Assume the members are pin connected at their end points. Take $A = 1200$ mm^2 and $E = 200$ GPa for each member. Use the method of virtual work.

9–18. Solve Prob. 9–17 using Castigliano's theorem.

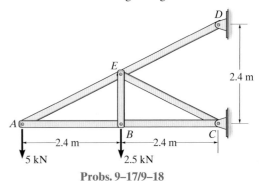

Probs. 9–17/9–18

9–19. Determine the vertical displacement of joint A if members AB and BC experience a temperature increase of $\Delta T = 110°C$. Take $A = 1200$ mm^2 and $E = 200$ GPa. Also, $\alpha = 12.0 (10^{-6})/°C$.

*9–20. Determine the vertical displacement of joint A if member AE is fabricated 12 mm too short.

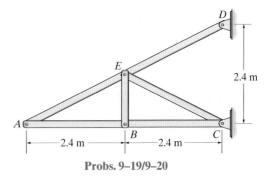

Probs. 9–19/9–20

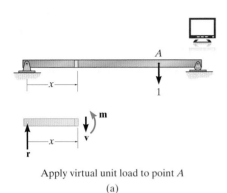

Apply virtual unit load to point A

(a)

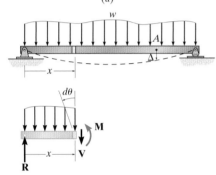

Apply real load w

(b)

Fig. 9–14

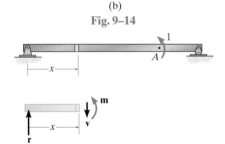

Apply virtual unit couple moment to point A

(a)

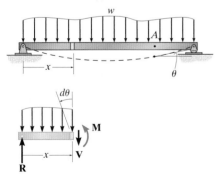

Apply real load w

Fig. 9–15

9.7 Method of Virtual Work: Beams and Frames

The method of virtual work can also be applied to deflection problems involving beams and frames. Since strains due to *bending* are the *primary cause* of beam or frame deflections, we will discuss their effects first. Deflections due to shear, axial and torsional loadings, and temperature will be considered in Sec. 9–8.

The principle of virtual work, or more exactly, the method of virtual force, may be formulated for beam and frame deflections by considering the beam shown in Fig. 9–14b. Here the displacement Δ of point A is to be determined. To compute Δ a virtual unit load acting in the direction of Δ is placed on the beam at A, and the *internal virtual moment* $\mathbf{m}$ is determined by the method of sections at an arbitrary location x from the left support, Fig. 9–14a. When the real loads act on the beam, Fig. 9–14b, point A is displaced Δ. Provided these loads cause *linear elastic material response,* then from Eq. 8–2, the element dx deforms or rotates $d\theta = (M/EI)\,dx$.* Here M is the internal moment at x caused by the real loads. Consequently, the *external virtual work* done by the unit load is $1 \cdot \Delta$, and the *internal virtual work* done by the moment $\mathbf{m}$ is $m\,d\theta = m(M/EI)\,dx$. Summing the effects on all the elements dx along the beam requires an integration, and therefore Eq. 9–13 becomes

$$1 \cdot \Delta = \int_0^L \frac{mM}{EI}\,dx \qquad (9\text{–}22)$$

where

1 = external virtual unit load acting on the beam or frame in the direction of Δ.

m = internal virtual moment in the beam or frame, expressed as a function of x and caused by the external virtual unit load.

Δ = external displacement of the point caused by the real loads acting on the beam or frame.

M = internal moment in the beam or frame, expressed as a function of x and caused by the real loads.

E = modulus of elasticity of the material.

I = moment of inertia of cross-sectional area, computed about the neutral axis.

In a similar manner, if the tangent rotation or slope angle θ at a point A on the beam's elastic curve is to be determined, Fig. 9–15, a unit couple moment is first applied at the point, and the corresponding internal moments m_θ have to be determined. Since the work of the unit couple is $1 \cdot \theta$, then

$$1 \cdot \theta = \int_0^L \frac{m_\theta M}{EI}\,dx \qquad (9\text{–}23)$$

*Recall that if the material is strained beyond its elastic limit, the principle of virtual work can still be applied, although in this case a nonlinear or plastic analysis must be used.

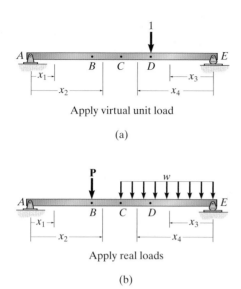

Apply virtual unit load

(a)

Apply real loads

(b)

Fig. 9–16

When applying Eqs. 9–22 and 9–23, it is important to realize that the definite integrals on the right side actually represent the amount of virtual strain energy that is *stored* in the beam. If concentrated forces or couple moments act on the beam or the distributed load is discontinuous, a single integration cannot be performed across the beam's entire length. Instead, separate x coordinates will have to be chosen within regions that have no discontinuity of loading. Also, it is not necessary that each x have the same origin; however, the x selected for determining the real moment M in a particular region must be the *same* x as that selected for determining the virtual moment m or m_θ within the same region. For example, consider the beam shown in Fig. 9–16. In order to determine the displacement of D, four regions of the beam must be considered, and therefore four integrals having the form $\int (mM/EI)\, dx$ must be evaluated. We can use x_1 to determine the strain energy in region AB, x_2 for region BC, x_3 for region DE, and x_4 for region DC. In any case, each x coordinate should be selected so that both M and m (or m_θ) can be easily formulated.

Integration Using Tables.   When the structure is subjected to a relatively simple loading, and yet the solution for a displacement requires several integrations, a *tabular method* may be used to perform these integrations. To do so the moment diagrams for each member are drawn first for both the real and virtual loadings. By matching these diagrams for m and M with those given in the table on the inside front cover, the integral $\int mM\, dx$ can be determined from the appropriate formula. Examples 9–8 and 9–10 illustrate the application of this method.

Procedure for Analysis

The following procedure may be used to determine the displacement and/or the slope at a point on the elastic curve of a beam or frame using the method of virtual work.

Virtual Moments m or m_θ

- Place a *unit load* on the beam or frame at the point and in the direction of the desired *displacement*.
- If the *slope* is to be determined, place a *unit couple moment* at the point.
- Establish appropriate x coordinates that are valid within regions of the beam or frame where there is no discontinuity of real or virtual load.
- With the virtual load in place, and all the real loads *removed* from the beam or frame, calculate the internal moment m or m_θ as a function of each x coordinate.
- Assume m or m_θ acts in the conventional positive direction as indicated in Fig. 4–1.

Real Moments

- Using the *same* x coordinates as those established for m or m_θ, determine the internal moments M caused only by the real loads.
- Since m or m_θ was assumed to act in the conventional positive direction, *it is important that positive M acts in this same direction*. This is necessary since positive or negative internal work depends upon the directional sense of load (defined by $\pm m$ or $\pm m_\theta$) and displacement (defined by $\pm M\,dx/EI$).

Virtual-Work Equation

- Apply the equation of virtual work to determine the desired displacement Δ or rotation θ. It is important to retain the algebraic sign of each integral calculated within its specified region.
- If the algebraic sum of all the integrals for the entire beam or frame is positive, Δ or θ is in the same direction as the virtual unit load or unit couple moment, respectively. If a negative value results, the direction of Δ or θ is opposite to that of the unit load or unit couple moment.

EXAMPLE 9.7

Determine the displacement of point B of the steel beam shown in Fig. 9–17a. Take $E = 200$ GPa, $I = 500(10^6)$ mm⁴.

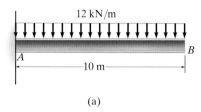

12 kN/m

A

B

10 m

(a)

SOLUTION

Virtual Moment m. The vertical displacement of point B is obtained by placing a virtual unit load of 1 kN at B, Fig. 9–17b. By inspection there are no discontinuities of loading on the beam for *both* the real and virtual loads. Thus, a *single x* coordinate can be used to determine the virtual strain energy. This coordinate will be selected with its origin at B, since then the reactions at A do not have to be determined in order to find the internal moments m and M. Using the method of sections, the internal moment m is formulated as shown in Fig. 9–17b.

Real Moment M. Using the *same x* coordinate, the internal moment M is formulated as shown in Fig. 9–17c.

Virtual-Work Equation. The vertical displacement of B is thus

$$1 \text{ kN} \cdot \Delta_B = \int_0^L \frac{mM}{EI} dx = \int_0^{10} \frac{(-1x)(-6x^2)\, dx}{EI}$$

$$1 \text{ kN} \cdot \Delta_B = \frac{15(10^3) \text{ kN}^2 \cdot \text{m}^3}{EI}$$

or

$$\Delta_B = \frac{15(10^3) \text{ kN} \cdot \text{m}^3}{200(10^6) \text{ kN/m}^2 (500(10^6) \text{ mm}^4)(10^{-12} \text{ m}^4/\text{mm}^4)}$$

$$= 0.150 \text{ m} = 150 \text{ mm} \qquad\qquad\qquad Ans.$$

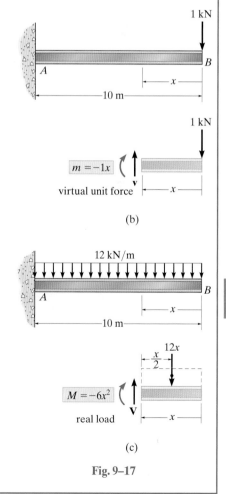

1 kN

A

B

10 m

x

1 kN

$m = -1x$

virtual unit force

x

(b)

12 kN/m

A

B

10 m

x

12x

$\dfrac{x}{2}$

$M = -6x^2$

real load

x

(c)

Fig. 9–17

9

EXAMPLE 9.8

Determine the slope θ at point B of the steel beam shown in Fig. 9–18a. Take $E = 200$ GPa, $I = 60(10^6)$ mm^4.

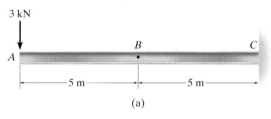

3 kN

(a)

Fig. 9–18

SOLUTION

Virtual Moment m_θ.   The slope at B is determined by placing a virtual unit couple moment of 1 kN · m at B, Fig. 9–18b. Here two x coordinates must be selected in order to determine the total virtual strain energy in the beam. Coordinate x_1 accounts for the strain energy within segment AB and coordinate x_2 accounts for that in segment BC. The internal moments m_θ within each of these segments are computed using the method of sections as shown in Fig. 9–18b.

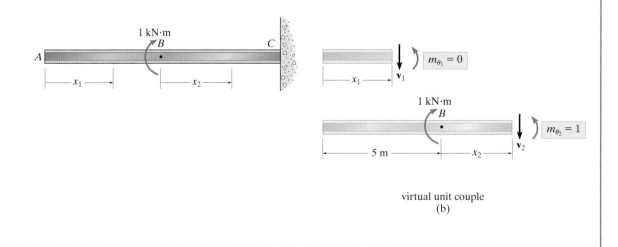

virtual unit couple
(b)

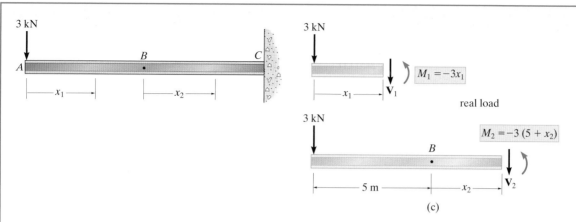

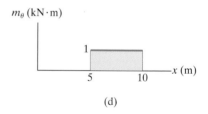

(c)

Real Moments M. Using the *same* coordinates x_1 and x_2, the internal moments M are computed as shown in Fig. 9–18c.

Virtual-Work Equation. The slope at B is thus given by

$$1 \cdot \theta_B = \int_0^L \frac{m_\theta M}{EI} dx$$

$$= \int_0^5 \frac{(0)(-3x_1)\, dx_1}{EI} + \int_0^5 \frac{(1)[-3(5 + x_2)]\, dx_2}{EI}$$

$$\theta_B = \frac{-112.5\ \text{kN} \cdot \text{m}^2}{EI} \tag{1}$$

We can also evaluate the integrals $\int m_\theta M\, dx$ graphically, using the table given on the inside front cover of the book. To do so it is first necessary to draw the moment diagrams for the beams in Figs. 9–18b and 9–18c. These are shown in Figs. 9–18d and 9–18e, respectively. Since there is no moment m for $0 \le x < 5$ m, we use only the shaded rectangular and trapezoidal areas to evaluate the integral. Finding these shapes in the appropriate row and column of the table, we have

$$\int_5^{10} m_\theta M\, dx = \tfrac{1}{2} m_\theta (M_1 + M_2)L = \tfrac{1}{2}(1)(-15 - 30)5$$

$$= -112.5\ \text{kN}^2 \cdot \text{m}^3$$

This is the same value as that determined in Eq. 1. Thus,

$$(1\ \text{kN} \cdot \text{m}) \cdot \theta_B = \frac{-112.5\ \text{kN}^2 \cdot \text{m}^3}{200(10^6)\ \text{kN/m}^2[60(10^6)\ \text{mm}^4](10^{-12}\ \text{m}^4/\text{mm}^4)}$$

$$\theta_B = -0.00938\ \text{rad} \qquad\qquad Ans.$$

The *negative sign* indicates θ_B is *opposite* to the direction of the virtual couple moment shown in Fig. 9–18b.

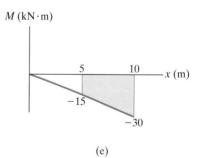

(e)

EXAMPLE 9.9

Determine the displacement at D of the steel beam in Fig. 9–19a. Take $E = 200$ GPa, $I = 300(10^6)$ mm^4.

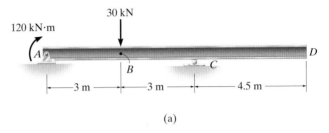

(a)

Fig. 9–19

SOLUTION

Virtual Moments m.   The beam is subjected to a virtual unit load at D as shown in Fig. 9–19b. By inspection, *three coordinates,* such as x_1, x_2, and x_3, must be used to cover all the regions of the beam. Notice that these coordinates cover regions where no discontinuities in either real or virtual load occur. The internal moments m have been computed in Fig. 9–19b using the method of sections.

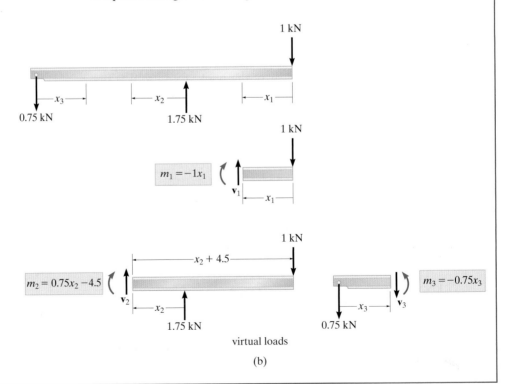

virtual loads

(b)

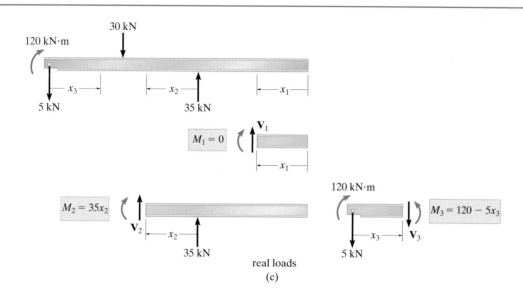

Real Moments M.   The reactions on the beam are computed first; then, using the *same x* coordinates as those used for *m*, the internal moments *M* are determined as shown in Fig. 9–19c.

Virtual-Work Equation.   Applying the equation of virtual work to the beam using the data in Figs. 9–19b and 9–19c, we have

$$1 \cdot \Delta_D = \int_0^L \frac{mM}{EI} dx$$

$$= \int_0^{4.5} \frac{(-1x_1)(0)\, dx_1}{EI} + \int_0^3 \frac{(0.75x_2 - 4.5)(35x_2)\, dx_2}{EI}$$

$$+ \int_0^3 \frac{(-0.75x_3)(120 - 5x_3)\, dx_3}{EI}$$

$$\Delta_D = \frac{0}{EI} - \frac{472.5}{EI} - \frac{371.25}{EI} = -\frac{843.75 \text{ kN} \cdot \text{m}^3}{EI}$$

or

$$\Delta_D = \frac{-843.75 \text{ kN} \cdot \text{m}^3}{[200(10^6) \text{ kN/m}^2][300(10^6)(10^{-12}) \text{ m}^4]}$$

$$= -0.0141 \text{ m} = -14.1 \text{ mm} \qquad \textit{Ans.}$$

The negative sign indicates the displacement is upward, opposite to the downward unit load, Fig. 9–19b. Also note that m_1 did not actually have to be calculated since $M_1 = 0$.

EXAMPLE 9.10

Determine the horizontal displacement of point C on the frame shown in Fig. 9–20a. Take $E = 200$ GPa and $I = 235(10^6)$ mm^4 for both members.

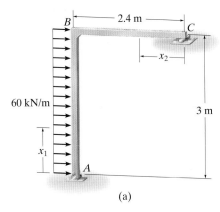

(a)

Fig. 9–20

SOLUTION

Virtual Moments m.   For convenience, the coordinates x_1 and x_2 in Fig. 9–20a will be used. A *horizontal* unit load is applied at C, Fig. 9–20b. Why? The support reactions and internal virtual moments are computed as shown.

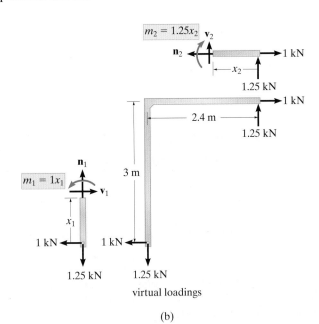

virtual loadings

(b)

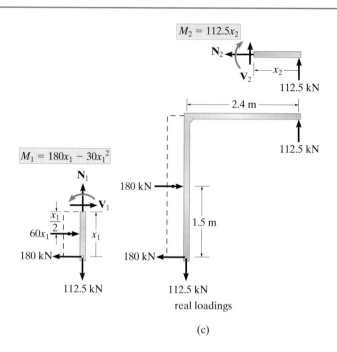

$M_2 = 112.5x_2$

$M_1 = 180x_1 - 30x_1^2$

real loadings

(c)

Real Moments M. In a similar manner the support reactions and real moments are computed as shown in Fig. 9–20c.

Virtual-Work Equation. Using the data in Figs. 9–20b and 9–20c, we have

$$1 \cdot \Delta_{C_h} = \int_0^L \frac{mM}{EI} dx = \int_0^3 \frac{(1x_1)(180x_1 - 30x_1^2) \, dx_1}{EI} + \int_0^{2.4} \frac{(1.25x_2)(112.5x_2) \, dx_2}{EI}$$

$$\Delta_{C_h} = \frac{1012.5}{EI} + \frac{648}{EI} = \frac{1660.5 \text{ kN} \cdot \text{m}^3}{EI} \tag{1}$$

If desired, the integrals $\int mM/dx$ can also be evaluated graphically using the table on the inside front cover. The moment diagrams for the frame in Figs. 9–20b and 9–20c are shown in Figs. 9–20d and 9–20e, respectively. Thus, using the formulas for similar shapes in the table yields

$$\int mM \, dx = \tfrac{5}{12}(3)(270)(3) + \tfrac{1}{3}(3)(270)(2.4)$$

$$= 1012.5 + 648 = 1660.5 \text{ kN} \cdot \text{m}^3$$

This is the same as that calculated in Eq. 1. Thus

$$\Delta_{C_h} = \frac{1660.5 \text{ kN} \cdot \text{m}^3}{[200(10^6) \text{ kN/m}^2][235(10^6)(10^{-12} \text{ m}^4)]}$$

$$= 0.0353 \text{ m} = 35.3 \text{ mm} \qquad \textit{Ans.}$$

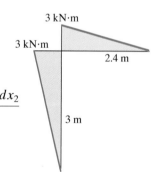

(d)

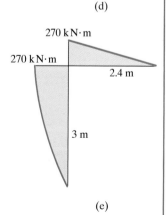

(e)

EXAMPLE 9.11

Determine the tangential rotation at point C of the frame shown in Fig. 9–21a. Take $E = 200$ GPa, $I = 15(10^6)$ mm^4.

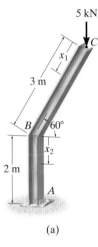

(a)

Fig. 9–21

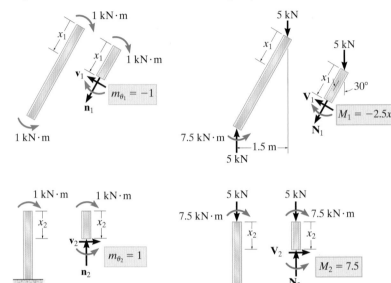

virtual loads

(b)

real loads

(c)

SOLUTION

Virtual Moments m_θ.   The coordinates x_1 and x_2 shown in Fig. 9–21a will be used. A unit couple moment is applied at C and the internal moments m_θ are calculated, Fig. 9–21b.

Real Moments M.   In a similar manner, the real moments M are calculated as shown in Fig. 9–21c.

Virtual-Work Equation.   Using the data in Figs. 9–21b and 9–21c, we have

$$1 \cdot \theta_C = \int_0^L \frac{m_\theta M}{EI} dx = \int_0^3 \frac{(-1)(-2.5x_1)\,dx_1}{EI} + \int_0^2 \frac{(1)(7.5)\,dx_2}{EI}$$

$$\theta_C = \frac{11.25}{EI} + \frac{15}{EI} = \frac{26.25 \text{ kN} \cdot \text{m}^2}{EI}$$

or

$$\theta_C = \frac{26.25 \text{ kN} \cdot \text{m}^2}{200(10^6) \text{ kN/m}^2[15(10^6) \text{ mm}^4](10^{-12} \text{ m}^4/\text{mm}^4)}$$
$$= 0.00875 \text{ rad} \qquad\qquad Ans.$$

9.8 Virtual Strain Energy Caused by Axial Load, Shear, Torsion, and Temperature

Although deflections of beams and frames are caused primarily by bending strain energy, in some structures the additional strain energy of axial load, shear, torsion, and perhaps temperature may become important. Each of these effects will now be considered.

Axial Load.   Frame members can be subjected to axial loads, and the virtual strain energy caused by these loadings has been established in Sec. 9–4. For members having a constant cross-sectional area, we have

$$U_n = \frac{nNL}{AE} \qquad (9\text{–}24)$$

where

n = internal virtual axial load caused by the external virtual unit load.

N = internal axial force in the member caused by the real loads.

E = modulus of elasticity for the material.

A = cross-sectional area of the member.

L = member's length.

Shear.   In order to determine the virtual strain energy in a beam due to shear, we will consider the beam element dx shown in Fig. 9–22. The shearing distortion dy of the element as caused by the *real loads* is $dy = \gamma\, dx$. If the shearing strain γ is caused by *linear elastic material response*, then Hooke's law applies, $\gamma = \tau/G$. Therefore, $dy = (\tau/G)\, dx$. We can express the shear stress as $\tau = K(V/A)$, where K is a *form factor* that depends upon the shape of the beam's cross-sectional area A. Hence, we can write $dy = K(V/GA)\, dx$. The internal virtual work done by a virtual shear force v, acting on the element *while* it is deformed dy, is therefore $dU_s = v\, dy = v(KV/GA)\, dx$. For the entire beam, the virtual strain energy is determined by integration.

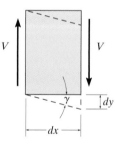

Fig. 9–22

$$U_s = \int_0^L K\!\left(\frac{vV}{GA}\right) dx \qquad (9\text{–}25)$$

where

v = internal virtual shear in the member, expressed as a function of x and caused by the external virtual unit load.

V = internal shear in the member, expressed as a function of x and caused by the real loads.

A = cross-sectional area of the member.

K = form factor for the cross-sectional area:

   $K = 1.2$ for rectangular cross sections.

   $K = 10/9$ for circular cross sections.

   $K \approx 1$ for wide-flange and I-beams, where A is the area of the web.

G = shear modulus of elasticity for the material.

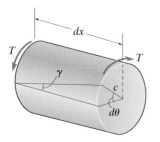

Fig. 9–23

Torsion. Often three-dimensional frameworks are subjected to torsional loadings. If the member has a *circular* cross-sectional area, no warping of its cross section will occur when it is loaded. As a result, the virtual strain energy in the member can easily be derived. To do so consider an element dx of the member that is subjected to an applied torque $\mathbf{T}$, Fig. 9–23. This torque causes a shear strain of $\gamma = (c\,d\theta)/dx$. Provided *linear elastic material response* occurs, then $\gamma = \tau/G$, where $\tau = Tc/J$. Thus, the angle of twist $d\theta = (\gamma\,dx)/c = (\tau/Gc)\,dx = (T/GJ)\,dx$. If a virtual unit load is applied to the structure that causes an internal virtual torque $\mathbf{t}$ in the member, then after applying the real loads, the virtual strain energy in the member of length dx will be $dU_t = t\,d\theta = tT\,dx/GJ$. Integrating over the length L of the member yields

$$U_t = \frac{tTL}{GJ} \tag{9–26}$$

where

t = internal virtual torque caused by the external virtual unit load.

T = internal torque in the member caused by the real loads.

G = shear modulus of elasticity for the material.

J = polar moment of inertia for the cross section, $J = \pi c^4/2$, where c is the radius of the cross-sectional area.

L = member's length.

The virtual strain energy due to torsion for members having noncircular cross-sectional areas is determined using a more rigorous analysis than that presented here.

Temperature. In Sec. 9–4 we considered the effect of a *uniform temperature change* ΔT on a truss member and indicated that the member will elongate or shorten by an amount $\Delta L = \alpha\,\Delta T L$. In some cases, however, a structural member can be subjected to a *temperature difference across its depth,* as in the case of the beam shown in Fig. 9–24a. If this occurs, it is possible to determine the displacement of points along the elastic curve of the beam by using the principle of virtual work. To do so we must first compute the amount of *rotation* of a differential element dx of the beam as caused by the thermal gradient that acts over the beam's cross section. For the sake of discussion we will choose the most common case of a beam having a neutral axis located at the mid-depth (c) of the beam. If we plot the temperature profile, Fig. 9–24b, it will be noted that the mean temperature is $T_m = (T_1 + T_2)/2$. If $T_1 > T_2$, the temperature difference at the top of the element causes strain elongation, while that at the bottom causes strain contraction. In both cases the difference in temperature is $\Delta T_m = T_1 - T_m = T_m - T_2$.

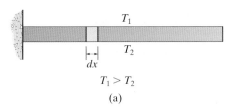

$T_1 > T_2$

(a)

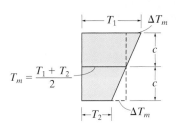

$T_m = \dfrac{T_1 + T_2}{2}$

temperature profile

(b)

Since the thermal change of length at the top and bottom is $\delta x = \alpha\, \Delta T_m\, dx$, Fig. 9–24c, then the rotation of the element is

$$d\theta = \frac{\alpha\, \Delta T_m\, dx}{c}$$

If we apply a virtual unit load at a point on the beam where a displacement is to be determined, or apply a virtual unit couple moment at a point where a rotational displacement of the tangent is to be determined, then this loading creates a virtual moment **m** in the beam at the point where the element dx is located. When the temperature gradient is imposed, the virtual strain energy in the beam is then

$$U_{\text{temp}} = \int_0^L \frac{m\alpha\, \Delta T_m\, dx}{c} \qquad (9\text{–}27)$$

where

m = internal virtual moment in the beam expressed as a function of x and caused by the external virtual unit load or unit couple moment.

α = coefficient of thermal expansion.

ΔT_m = temperature difference between the mean temperature and the temperature at the top or bottom of the beam.

c = mid-depth of the beam.

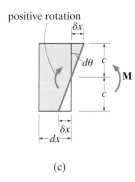

positive rotation

(c)

Fig. 9–24

Unless otherwise stated, *this text will consider only beam and frame deflections due to bending.* In general, though, beam and frame members may be subjected to several of the other loadings discussed in this section. However, as previously mentioned, the additional deflections caused by shear and axial force alter the deflection of beams by only a few percent and are therefore generally ignored for even "small" two- or three-member frame analysis of one-story height. If these and the other effects of torsion and temperature are to be considered for the analysis, then one simply adds their virtual strain energy as defined by Eqs. 9–24 through 9–27 to the equation of virtual work defined by Eq. 9–22 or Eq. 9–23. The following examples illustrate application of these equations.

9

EXAMPLE 9.12

Determine the horizontal displacement of point C on the frame shown in Fig. 9–25a. Take $E = 200$ GPa, $G = 80$ GPa, $I = 235(10^6)$ mm^4, and $A = 50(10^6)$ mm^4 for both members. The cross-sectional area is rectangular. Include the internal strain energy due to axial load and shear.

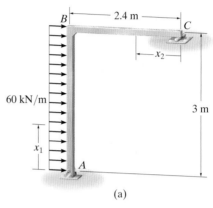

(a)

Fig. 9–25

SOLUTION

Here we must apply a horizontal unit load at C. The necessary free-body diagrams for the real and virtual loadings are shown in Figs. 9–25b and 9–25c.

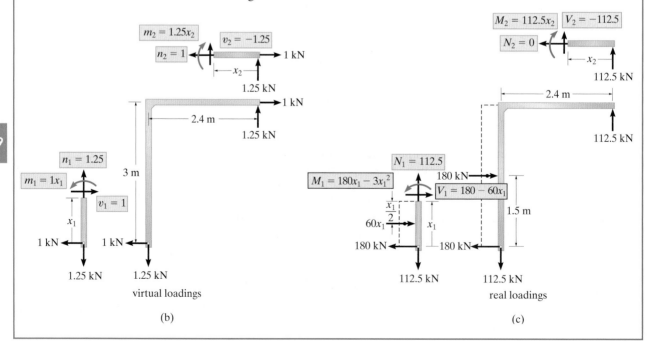

virtual loadings

(b)

real loadings

(c)

Bending.   The virtual strain energy due to bending has been determined in Example 9–10. There it was shown that

$$U_b = \int_0^L \frac{mM\ dx}{EI} = \frac{1660.5\ \text{kN}^2 \cdot \text{m}^3}{EI} = \frac{1660.5\ \text{kN}^2 \cdot \text{m}^3}{[200(10^6)\ \text{kN/m}^2][235(10^6)(10^{-12})\ \text{m}^4]}$$

$$= 0.0353\ \text{kN} \cdot \text{m} = 35.3\ \text{kN} \cdot \text{mm}$$

Axial load.   From the data in Figs. 9–25b and 9–25c, we have

$$U_a = \sum \frac{nNL}{AE}$$

$$= \frac{1.25\ \text{kN}(112.5\ \text{kN})(3\ \text{m})}{[50(10^3)(10^{-6})\ \text{m}^2][80(10^6)\ \text{kN/m}^2]} + \frac{1\ \text{kN}(0)(2.4\ \text{m})}{[50(10^3)(10^{-6})\ \text{m}^2][80(10^6)\ \text{kN/m}^2]}$$

$$= 0.000105\ \text{kN} \cdot \text{m} = 0.105\ \text{kN} \cdot \text{mm}$$

Shear.   Applying Eq. 9–25 with $K = 1.2$ for rectangular cross sections, and using the shear functions shown in Figs. 9–25b and 9–25c, we have

$$U_s = \int_0^L K\left(\frac{vV}{GA}\right) dx$$

$$= \int_0^3 \frac{1.2(1)(180 - 60x_1)\ dx_1}{GA} + \int_0^{2.4} \frac{1.2(-1.25)(-112.5)\ dx_2}{GA}$$

$$= \frac{729\ \text{kN}^2 \cdot \text{m}}{[80(10^6)\ \text{kN/m}^2][50(10^3)(10^{-6})\ \text{m}^2]} = 0.000182\ \text{kN} \cdot \text{m} = 0.182\ \text{kN} \cdot \text{mm}$$

Applying the equation of virtual work, we have

$$1\ \text{kN} \cdot \Delta_{C_h} = 35.3\ \text{kN} \cdot \text{mm} + 0.105\ \text{kN} \cdot \text{mm} + 0.182\ \text{kN} \cdot \text{mm}$$

$$\Delta_{C_h} = 35.59\ \text{mm} \qquad\qquad\qquad\qquad \textit{Ans.}$$

Including the effects of shear and axial load contributed only a 0.8% increase in the answer to that determined only from bending.

EXAMPLE | 9.13

The beam shown in Fig. 9–26a is used in a building subjected to two different thermal environments. If the temperature at the top surface of the beam is 25°C and at the bottom surface is 70°C, determine the vertical deflection of the beam at its midpoint due to the temperature gradient. Take $\alpha = 11.7(10^{-6})/°C$.

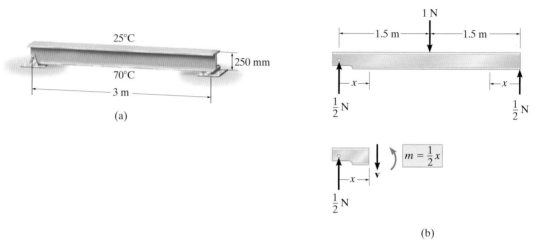

Fig. 9–26

SOLUTION

Since the deflection at the center of the beam is to be determined, a virtual unit load is placed there and the internal virtual moment in the beam is calculated, Fig. 9–26b.

The mean temperature at the center of the beam is $(70° + 25°)/2 = 47.5°C$, so that for application of Eq. 9–27, $\Delta T_m = 47.5°C - 25°C = 22.5°C$. Also, $c = 250\ mm/2 = 125\ mm$. Applying the principle of virtual work, we have

$$1\ \text{N} \cdot \Delta_{C_v} = \int_0^L \frac{m\alpha\, \Delta T_m\, dx}{c}$$

$$= 2\int_0^{1500\ mm} \frac{\left(\tfrac{1}{2}x\right)11.7(10^{-6})/°C(22.5°C)}{125\ mm}\,dx$$

$$\Delta_{C_v} = 2.37\ \text{mm} \qquad\qquad Ans.$$

The result indicates a very negligible deflection.

9.9   Castigliano's Theorem for Beams and Frames

The internal bending strain energy for a beam or frame is given by Eq. 9–11 ($U_i = \int M^2 \, dx/2EI$). Substituting this equation into Eq. 9–20 ($\Delta_i = \partial U_i/\partial P_i$) and omitting the subscript i, we have

$$\Delta = \frac{\partial}{\partial P} \int_0^L \frac{M^2 \, dx}{2EI}$$

Rather than squaring the expression for internal moment M, integrating, and then taking the partial derivative, it is generally easier to differentiate prior to integration. Provided E and I are constant, we have

$$\Delta = \int_0^L M\left(\frac{\partial M}{\partial P}\right)\frac{dx}{EI} \qquad (9\text{–}28)$$

where

Δ = external displacement of the point caused by the real loads acting on the beam or frame.

P = external force applied to the beam or frame in the direction of Δ.

M = internal moment in the beam or frame, expressed as a function of x and caused by both the force P and the real loads on the beam.

E = modulus of elasticity of beam material.

I = moment of inertia of cross-sectional area computed about the neutral axis.

If the slope θ at a point is to be determined, we must find the partial derivative of the internal moment M with respect to an *external couple moment M'* acting at the point, i.e.,

$$\theta = \int_0^L M\left(\frac{\partial M}{\partial M'}\right)\frac{dx}{EI} \qquad (9\text{–}29)$$

The above equations are similar to those used for the method of virtual work, Eqs. 9–22 and 9–23, except $\partial M/\partial P$ and $\partial M/\partial M'$ replace m and m_θ, respectively. As in the case for trusses, slightly more calculation is generally required to determine the partial derivatives and apply Castigliano's theorem rather than use the method of virtual work. Also, recall that this theorem applies only to material having a linear elastic response. If a more complete accountability of strain energy in the structure is desired, the strain energy due to shear, axial force, and torsion must be included. The derivations for shear and torsion follow the same development as Eqs. 9–25 and 9–26. The strain energies and their derivatives are, respectively,

9

$$U_s = K \int_0^L \frac{V^2 \, dx}{2AG} \qquad \frac{\partial U_s}{\partial P} = \int_0^L \frac{V}{AG} \left(\frac{\partial V}{\partial P} \right) dx$$

$$U_t = \int_0^L \frac{T^2 \, dx}{2JG} \qquad \frac{\partial U_t}{\partial P} = \int_0^L \frac{T}{JG} \left(\frac{\partial T}{\partial P} \right) dx$$

These effects, however, will not be included in the analysis of the problems in this text since beam and frame deflections are caused mainly by bending strain energy. Larger frames, or those with unusual geometry, can be analyzed by computer, where these effects can readily be incorporated into the analysis.

Procedure for Analysis

The following procedure provides a method that may be used to determine the deflection and/or slope at a point in a beam or frame using Castigliano's theorem.

External Force P or Couple Moment M'

- Place a force **P** on the beam or frame at the point and in the direction of the desired displacement.
- If the slope is to be determined, place a couple moment **M'** at the point.
- It is assumed that both P and M' have a *variable magnitude* in order to obtain the changes $\partial M / \partial P$ or $\partial M / \partial M'$.

Internal Moments M

- Establish appropriate x coordinates that are valid within regions of the beam or frame where there is no discontinuity of force, distributed load, or couple moment.
- Calculate the internal moment M as a function of P or M' and each x coordinate. Also, compute the partial derivative $\partial M / \partial P$ or $\partial M / \partial M'$ for each coordinate x.
- After M and $\partial M / \partial P$ or $\partial M / \partial M'$ have been determined, assign P or M' its numerical value if it has replaced a real force or couple moment. Otherwise, set P or M' equal to zero.

Castigliano's Theorem

- Apply Eq. 9–28 or 9–29 to determine the desired displacement Δ or slope θ. It is important to retain the algebraic signs for corresponding values of M and $\partial M / \partial P$ or $\partial M / \partial M'$.
- If the resultant sum of all the definite integrals is positive, Δ or θ is in the same direction as **P** or **M'**.

EXAMPLE 9.14

Determine the displacement of point B of the beam shown in Fig. 9–27a. Take $E = 200$ GPa, $I = 500(10^6)$ mm⁴.

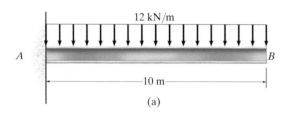

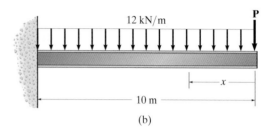

(a)

(b)

SOLUTION

External Force P.   A vertical force $\mathbf{P}$ is placed on the beam at B as shown in Fig. 9–27b.

Internal Moments M.   A single x coordinate is needed for the solution, since there are no discontinuities of loading between A and B. Using the method of sections, Fig. 9–27c, we have

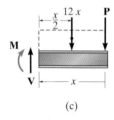

$$\zeta + \Sigma M = 0; \qquad -M - (12x)\left(\frac{x}{2}\right) - Px = 0$$

$$M = -6x^2 - Px \qquad \frac{\partial M}{\partial P} = -x$$

(c)

Fig. 9–27

Setting $P = 0$, its actual value, yields

$$M = -6x^2 \qquad \frac{\partial M}{\partial P} = -x$$

Castigliano's Theorem.   Applying Eq. 9–28, we have

$$\Delta_B = \int_0^L M\left(\frac{\partial M}{\partial P}\right)\frac{dx}{EI} = \int_0^{10} \frac{(-6x^2)(-x)\,dx}{EI} = \frac{15(10^3)\ \text{kN}\cdot\text{m}^3}{EI}$$

or

$$\Delta_B = \frac{15(10^3)\ \text{kN}\cdot\text{m}^3}{200(10^6)\ \text{kN/m}^2[500(10^6)\ \text{mm}^4](10^{-12}\ \text{m}^4/\text{mm}^4)}$$

$$= 0.150\ \text{m} = 150\ \text{mm} \qquad\qquad Ans.$$

The similarity between this solution and that of the virtual-work method, Example 9–7, should be noted.

EXAMPLE 9.15

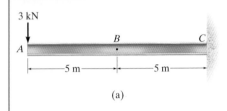

(a)

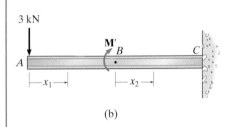

(b)

(c)

Fig. 9–28

Determine the slope at point B of the beam shown in Fig. 9–28a. Take $E = 200$ GPa, $I = 60(10^6)$ mm^4.

SOLUTION

External Couple Moment $M^{\bar{O}}$. Since the slope at point B is to be determined, an external couple $\mathbf{M'}$ is placed on the beam at this point, Fig. 9–28b.

Internal Moments M. Two coordinates, x_1 and x_2, must be used to determine the internal moments within the beam since there is a discontinuity, M', at B. As shown in Fig. 9–28b, x_1 ranges from A to B and x_2 ranges from B to C. Using the method of sections, Fig. 9–28c, the internal moments and the partial derivatives are computed as follows:

For x_1:

$$\zeta + \Sigma M = 0; \qquad \qquad M_1 + 3x_1 = 0$$
$$M_1 = -3x_1$$
$$\frac{\partial M_1}{\partial M'} = 0$$

For x_2:

$$\zeta + \Sigma M = 0; \qquad \qquad M_2 - M' + 3(5 + x_2) = 0$$
$$M_2 = M' - 3(5 + x_2)$$
$$\frac{\partial M_2}{\partial M'} = 1$$

Castigliano's Theorem. Setting $M' = 0$, its actual value, and applying Eq. 9–29, we have

$$\theta_B = \int_0^L M\left(\frac{\partial M}{\partial M'}\right)\frac{dx}{EI}$$

$$= \int_0^5 \frac{(-3x_1)(0)\, dx_1}{EI} + \int_0^5 \frac{-3(5 + x_2)(1)\, dx_2}{EI} = -\frac{112.5 \text{ kN} \cdot \text{m}^2}{EI}$$

or

$$\theta_B = \frac{-112.5 \text{ kN} \cdot \text{m}^2}{200(10^6) \text{ kN/m}^2[60(10^6) \text{ mm}^4](10^{-12} \text{ m}^4/\text{mm}^4)}$$
$$= -0.00938 \text{ rad} \qquad \qquad Ans.$$

The negative sign indicates that θ_B is opposite to the direction of the couple moment $\mathbf{M'}$. Note the similarity between this solution and that of Example 9–8.

EXAMPLE 9.16

Determine the vertical displacement of point C of the beam shown in Fig. 9–29a. Take $E = 200$ GPa, $I = 150(10^6)$ mm⁴.

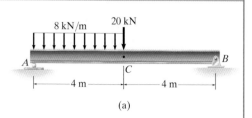

8 kN/m   20 kN

A C B

4 m   4 m

(a)

SOLUTION

External Force P.   A vertical force **P** is applied at point C, Fig. 9–29b. Later this force will be set equal to a fixed value of 20 kN.

Internal Moments M.   In this case two x coordinates are needed for the integration, Fig. 9–29b, since the load is discontinuous at C. Using the method of sections, Fig. 9–29c, we have

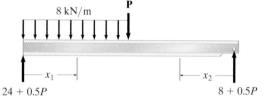

8 kN/m   **P**

x_1 x_2

$24 + 0.5P$ $8 + 0.5P$

(b)

For x_1:

$$\zeta + \Sigma M = 0; \quad -(24 + 0.5P)x_1 + 8x_1\left(\frac{x_1}{2}\right) + M_1 = 0$$

$$M_1 = (24 + 0.5P)x_1 - 4x_1^2$$

$$\frac{\partial M_1}{\partial P} = 0.5x_1$$

For x_2:

$$\zeta + \Sigma M = 0; \quad -M_2 + (8 + 0.5P)x_2 = 0$$

$$M_2 = (8 + 0.5P)x_2$$

$$\frac{\partial M_2}{\partial P} = 0.5x_2$$

$8x_1 \frac{x_1}{2}$

M_1 M_2

x_1 V_1 V_2 x_2

$24 + 0.5P$ $8 + 0.5P$

(c)

Fig. 9–29

Castigliano's Theorem.   Setting $P = 20$ kN, its actual value, and applying Eq. 9–28 yields

$$\Delta_{C_v} = \int_0^L M\left(\frac{\partial M}{\partial P}\right)\frac{dx}{EI}$$

$$= \int_0^4 \frac{(34x_1 - 4x_1^2)(0.5x_1)\,dx_1}{EI} + \int_0^4 \frac{(18x_2)(0.5x_2)\,dx_2}{EI}$$

$$= \frac{234.7 \text{ kN} \cdot \text{m}^3}{EI} + \frac{192 \text{ kN} \cdot \text{m}^3}{EI} = \frac{426.7 \text{ kN} \cdot \text{m}^3}{EI}$$

or

$$\Delta_{C_v} = \frac{426.7 \text{ kN} \cdot \text{m}^3}{200(10^6) \text{ kN/m}^2[150(10^6) \text{ mm}^4](10^{-12} \text{ m}^4/\text{mm}^4)}$$

$$= 0.0142 \text{ m} = 14.2 \text{ mm} \qquad\qquad Ans.$$

9

EXAMPLE 9.17

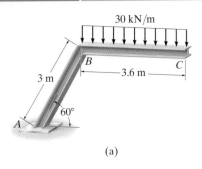

(a)

Determine the slope at point C of the two-member frame shown in Fig. 9–30a. The support at A is fixed. Take $E = 200$ GPa, $I = 235(10^6)$ mm^4.

SOLUTION

External Couple Moment M′. A variable moment **M′** is applied to the frame at point C, since the slope at this point is to be determined, Fig. 9–30b. Later this moment will be set equal to zero.

Internal Moments M. Due to the discontinuity of internal loading at B, two coordinates, x_1 and x_2, are chosen as shown in Fig. 9–30b. Using the method of sections, Fig. 9–30c, we have

For x_1:

$$\zeta + \Sigma M = 0; \qquad -M_1 - 30x_1\left(\frac{x_1}{2}\right) - M' = 0$$

$$M_1 = -\left(15x_1^2 + M'\right)$$

$$\frac{\partial M_1}{\partial M'} = -1$$

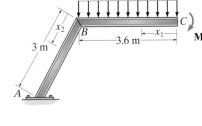

(b)

For x_2:

$$\zeta + \Sigma M = 0; \qquad -M_2 - 108(x_2 \cos 60° + 1.8) - M' = 0$$

$$M_2 = -108(x_2 \cos 60° + 1.8) - M'$$

$$\frac{\partial M_2}{\partial M'} = -1$$

Castigliano's Theorem. Setting $M' = 0$ and applying Eq. 9–29 yields

$$\theta_C = \int_0^L M\left(\frac{\partial M}{\partial M'}\right)\frac{dx}{EI}$$

$$= \int_0^{3.6} \frac{\left(-15x_1^2\right)(-1)\,dx_1}{EI} + \int_0^3 \frac{-108(x_2 \cos 60° + 1.8)(-1)\,dx_2}{EI}$$

$$= \frac{233.28 \text{ kN} \cdot \text{m}^2}{EI} + \frac{826.2 \text{ kN} \cdot \text{m}^2}{EI} = \frac{1059.48 \text{ kN} \cdot \text{m}^2}{EI}$$

$$\theta_C = \frac{1059.48 \text{ kN} \cdot \text{m}^2}{[200(10^6) \text{ kN/m}^2][235(10^6)(10^{-12}) \text{ m}^4]} = 0.0225 \text{ rad} \quad Ans.$$

(c)

Fig. 9–30

FUNDAMENTAL PROBLEMS

F9–13. Determine the slope and displacement at point *A*. *EI* is constant. Use the principle of virtual work.

F9–14. Solve Prob. F9–13 using Castigliano's theorem.

F9–19. Determine the slope at *A* and displacement at point *C*. *EI* is constant. Use the principle of virtual work.

F9–20. Solve Prob. F9–19 using Castigliano's theorem.

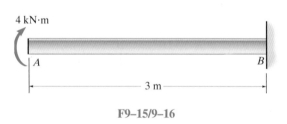

F9–13/9–14

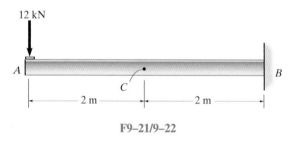

F9–19/9–20

F9–15. Determine the slope and displacement at point *A*. *EI* is constant. Use the principle of virtual work.

F9–16. Solve Prob. F9–15 using Castigliano's theorem.

F9–21. Determine the slope and displacement at point *C*. *EI* is constant. Use the principle of virtual work.

F9–22. Solve Prob. F9–21 using Castigliano's theorem.

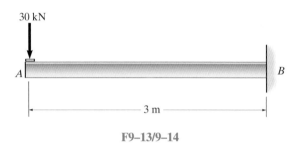

F9–15/9–16

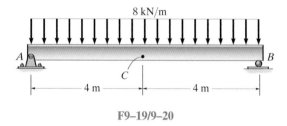

F9–21/9–22

F9–17. Determine the slope and displacement at point *B*. *EI* is constant. Use the principle of virtual work.

F9–18. Solve Prob. F9–17 using Castigliano's theorem.

F9–23. Determine the displacement at point *C*. *EI* is constant. Use the principle of virtual work.

F9–24. Solve Prob. F9–23 using Castigliano's theorem.

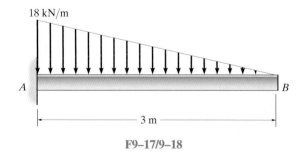

F9–17/9–18

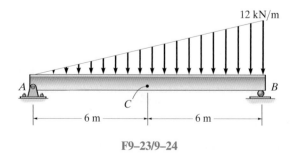

F9–23/9–24

PROBLEMS

9–21. Determine the displacement of point C and the slope at point B. EI is constant. Use the principle of virtual work.

9–22. Solve Prob. 9–21 using Castigliano's theorem.

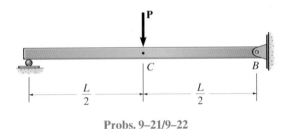

Probs. 9–21/9–22

9–23. Determine the displacement at point C. EI is constant. Use the method of virtual work.

*9–24. Solve Prob. 9–23 using Castigliano's theorem.

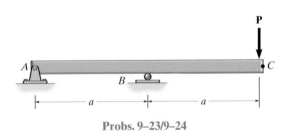

Probs. 9–23/9–24

9–25. Determine the slope at point C. EI is constant. Use the method of virtual work.

9–26. Solve Prob. 9–25 using Castigliano's theorem.

9–27. Determine the slope at point A. EI is constant. Use the method of virtual work.

*9–28. Solve Prob. 9–27 using Castigliano's theorem.

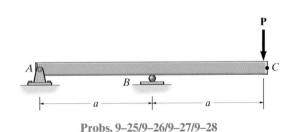

Probs. 9–25/9–26/9–27/9–28

9–29. Determine the slope and displacement at point C. Use the method of virtual work. $E = 200$ GPa, $I = 480(10^6)$ mm^4.

9–30. Solve Prob. 9–29 using Castigliano's theorem.

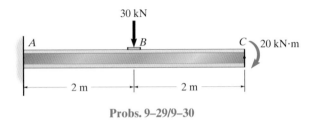

Probs. 9–29/9–30

9–31. Determine the displacement and slope at point C of the cantilever beam. The moment of inertia of each segment is indicated in the figure. Take $E = 200$ GPa. Use the principle of virtual work.

*9–32. Solve Prob. 9–31 using Castigliano's theorem.

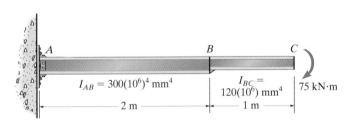

Probs. 9–31/9–32

9–33. Determine the slope and displacement at point B. EI is constant. Use the method of virtual work.

9–34. Solve Prob. 9–33 using Castigliano's theorem.

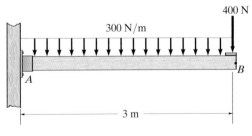

Probs. 9–33/9–34

9–35. Determine the slope and displacement at point B. Assume the support at A is a pin and C is a roller. Take $E = 200$ GPa, $I = 120(10^6)$ mm⁴. Use the method of virtual work.

***9–36.** Solve Prob. 9–35 using Castigliano's theorem.

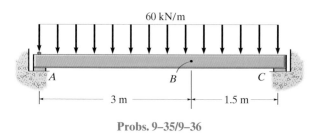

Probs. 9–35/9–36

9–37. Determine the slope and displacement at point B. Assume the support at A is a pin and C is a roller. Account for the additional strain energy due to shear. Take $E = 200$ GPa, $I = 120(10^6)$ mm⁴, $G = 80$ GPa and assume AB has a cross-sectional area of $A = 4.69(10^3)$ mm². Use the method of virtual work.

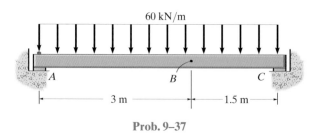

Prob. 9–37

9–38. Determine the displacement of point C. Use the method of virtual work. EI is constant.

9–39. Solve Prob. 9–38 using Castigliano's theorem.

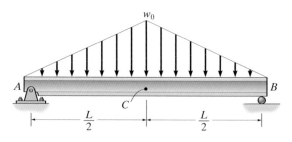

Probs. 9–38/9–39

***9–40.** Determine the slope and displacement at point A. Assume C is pinned. Use the principle of virtual work. EI is constant.

9–41. Solve Prob. 9–40 using Castigliano's theorem.

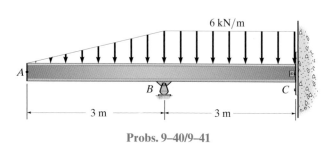

Probs. 9–40/9–41

9–42. Determine the displacement at point D. Use the principle of virtual work. EI is constant.

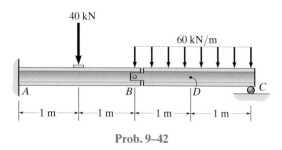

Prob. 9–42

9–43. Determine the displacement at point D. Use Castigliano's theorem. EI is constant.

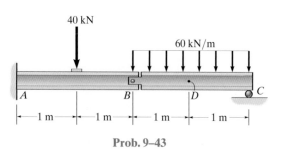

Prob. 9–43

*9–44. Use the method of virtual work and determine the vertical deflection at the rocker support D. EI is constant.

9–45. Solve Prob. 9–44 using Castigliano's theorem.

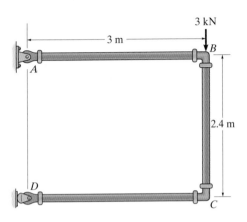

Probs. 9–44/9–45

9–46. The L-shaped frame is made from two segments, each of length L and flexural stiffness EI. If it is subjected to the uniform distributed load, determine the horizontal displacement of the end C. Use the method of virtual work.

9–47. The L-shaped frame is made from two segments, each of length L and flexural stiffness EI. If it is subjected to the uniform distributed load, determine the vertical displacement of point B. Use the method of virtual work.

*9–48. Solve Prob. 9–47 using Castigliano's theorem.

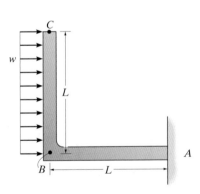

Probs. 9–46/9–47/9–48

9–49. Determine the horizontal displacement of point C. EI is constant. Use the method of virtual work.

9–50. Solve Prob. 9–49 using Castigliano's theorem.

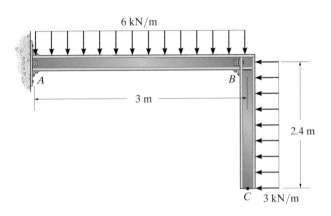

Probs. 9–49/9–50

9–51. Determine the vertical deflection at C. The cross-sectional area and moment of inertia of each segment is shown in the figure. Take $E = 200$ GPa. Assume A is a fixed support. Use the method of virtual work.

*9–52. Solve Prob. 9–51, including the effect of shear and axial strain energy.

9–53. Solve Prob. 9–51 using Castigliano's theorem.

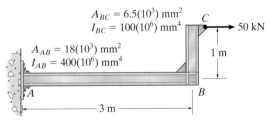

Probs. 9–51/9–52/9–53

9–54. Determine the slope at *A*. Take *E* = 200 GPa. The moment of inertia of each segment of the frame is indicated in the figure. Assume *D* is a pin support. Use the method of virtual work.

9–55. Solve Prob. 9–54 using Castigliano's theorem.

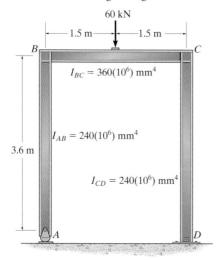

Probs. 9–54/9–55

***9–56.** Use the method of virtual work and determine the horizontal deflection at *C*. The cross-sectional area of each member is indicated in the figure. Assume the members are pin connected at their end points. *E* = 200 GPa.

9–57. Solve Prob. 9–56 using Castigliano's theorem.

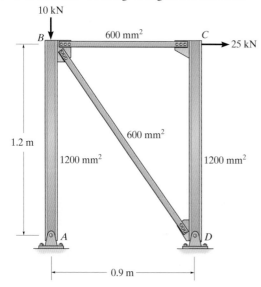

Probs. 9–56/9–57

9–58. Use the method of virtual work and determine the horizontal deflection at *C*. *EI* is constant. There is a pin at *A*. Assume *C* is a roller and *B* is a fixed joint.

9–59. Solve Prob. 9–58 using Castigliano's theorem.

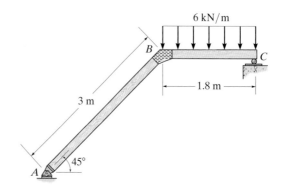

Probs. 9–58/9–59

***9–60.** The frame is subjected to the load of 20 kN. Determine the vertical displacement at *C*. Assume that the members are pin connected at *A*, *C*, and *E*, and fixed connected at the knee joints *B* and *D*. *EI* is constant. Use the method of virtual work.

9–61. Solve Prob. 9–60 using Castigliano's theorem.

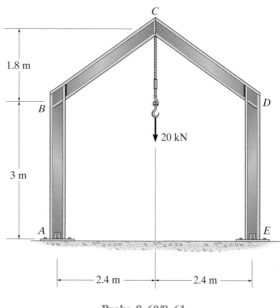

Probs. 9–60/9–61

9

CHAPTER REVIEW

All energy methods are based on the conservation of energy principle, which states that the work done by all external forces acting on the structure, U_e, is transformed into internal work or strain energy, U_i, developed in the members when the structure deforms.

$$\boxed{U_e = U_i}$$

(Ref: Section 9.1)

A force (moment) does work U when it undergoes a displacement (rotation) in the direction of the force (moment).

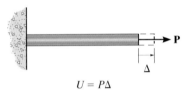

$$U = P\Delta$$

$$U = M\theta$$

(Ref: Section 9.1)

The principle of virtual work is based upon the work done by a "virtual" or imaginary unit force. If the deflection (rotation) at a point on the structure is to be obtained, a unit virtual force (couple moment) is applied to the structure at the point. This causes internal virtual loadings in the structure. The virtual work is then developed when the real loads are placed on the structure causing it to deform.

Truss displacements are found using

$$\boxed{1 \cdot \Delta = \sum \frac{nNL}{AE}}$$

If the displacement is caused by temperature, or fabrication errors, then

$$\boxed{1 \cdot \Delta = \sum n\alpha \Delta T L} \qquad \boxed{1 \cdot \Delta = \sum n\,\Delta L}$$

(Ref: Sections 9.3 & 9.4)

For beams and frames, the displacement (rotation) is defined from

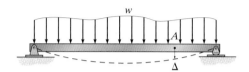

$$1 \cdot \Delta = \int_0^L \frac{mM}{EI} dx$$

$$1 \cdot \theta = \int_0^L \frac{m_\theta M}{EI} dx$$

(Ref: Section 9.7)

Castigliano's second theorem, called the method of least work, can be used to determine the deflections in structures that respond elastically. It states that the displacement (rotation) at a point on a structure is equal to the first partial derivative of the strain energy in the structure with respect to a force P (couple moment M') acting at the point and in the direction of the displacement (rotation). For a truss

$$\Delta = \sum N \left(\frac{\partial N}{\partial P} \right) \frac{L}{AE}$$

(Ref: Section 9.6)

For beams and frames

$$\Delta = \int_0^L M \left(\frac{\partial M}{\partial P} \right) \frac{dx}{EI}$$

$$\theta = \int_0^L M \left(\frac{\partial M}{\partial M'} \right) \frac{dx}{EI}$$

(Ref: Section 9.9)

9

The fixed-connected joints of this concrete framework make this a statically indeterminate structure.

Analysis of Statically Indeterminate Structures by the Force Method

10

In this chapter we will apply the *force* or *flexibility* method to analyze statically indeterminate trusses, beams, and frames. At the end of the chapter we will present a method for drawing the influence line for a statically indeterminate beam or frame.

10.1 Statically Indeterminate Structures

Recall from Sec. 2–4 that a structure of any type is classified as *statically indeterminate* when the number of unknown reactions or internal forces exceeds the number of equilibrium equations available for its analysis. In this section we will discuss the merits of using indeterminate structures and two fundamental ways in which they may be analyzed. Realize that most of the structures designed today are statically indeterminate. This indeterminacy may arise as a result of added supports or members, or by the general form of the structure. For example, reinforced concrete buildings are almost always statically indeterminate since the columns and beams are poured as continuous members through the joints and over supports.

Advantages and Disadvantages. Although the analysis of a statically indeterminate structure is more involved than that of a statically determinate one, there are usually several very important reasons for choosing this type of structure for design. Most important, for a given loading the maximum stress and deflection of an indeterminate structure are generally *smaller* than those of its statically determinate counterpart. For example, the statically indeterminate, fixed-supported beam in Fig. 10–1a will be subjected to a maximum moment of $M_{\max} = PL/8$, whereas the same beam, when simply supported, Fig. 10–1b, will be subjected to twice the moment, that is, $M_{\max} = PL/4$. As a result, the fixed-supported beam has one fourth the deflection and one half the stress at its center of the one that is simply supported.

Another important reason for selecting a statically indeterminate structure is because it has a tendency to redistribute its load to its redundant supports in cases where faulty design or overloading occurs. In these cases, the structure maintains its stability and collapse is prevented. This is particularly important when *sudden* lateral loads, such as wind or earthquake, are imposed on the structure. To illustrate, consider again the fixed-end beam in Fig. 10–1a. As **P** is increased, the beam's material at the walls and at the center of the beam begins to *yield* and forms localized "plastic hinges," which causes the beam to deflect as if it were hinged or pin connected at these points. Although the deflection becomes large, the walls will develop horizontal force and moment reactions that will hold the beam and thus prevent it from totally collapsing. In the case of the simply supported beam, Fig. 10–1b, an excessive load **P** will cause the "plastic hinge" to form only at the center of the beam, and due to the large vertical deflection, the supports will not develop the horizontal force and moment reactions that may be necessary to prevent total collapse.

Although statically indeterminate structures can support a loading with thinner members and with increased stability compared to their statically determinate counterparts, there are cases when these advantages may instead become disadvantages. The cost savings in material must be compared with the added cost necessary to fabricate the structure, since oftentimes it becomes more costly to construct the supports and joints of an indeterminate structure compared to one that is determinate. More important, though, because statically indeterminate structures have redundant support reactions, one has to be very careful to prevent differential displacement of the supports, since this effect will introduce internal stress in the structure. For example, if the wall at one end of the fixed-end beam in Fig. 10–1a were to settle, stress would be developed in the beam because of this "forced" deformation. On the other hand, if the beam were simply supported or statically determinate, Fig. 10–1b, then any settlement of its end would not cause the beam to deform, and therefore no stress would be developed in the beam. In general, then, any deformation, such as that caused by relative support displacement, or changes in member lengths caused by temperature or fabrication errors, will introduce additional stresses in the structure, which must be considered when designing indeterminate structures.

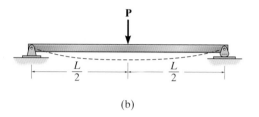

(a)   (b)

Fig. 10–1

Methods of Analysis.

When analyzing any indeterminate structure, it is necessary to satisfy *equilibrium*, *compatibility*, and *force-displacement* requirements for the structure. *Equilibrium* is satisfied when the reactive forces hold the structure at rest, and *compatibility* is satisfied when the various segments of the structure fit together without intentional breaks or overlaps. The *force-displacement* requirements depend upon the way the material responds; in this text we have assumed linear elastic response. In general there are two different ways to satisfy these requirements when analyzing a statically indeterminate structure: the *force* or *flexibility method*, and the *displacement* or *stiffness method*.

Force Method.

The force method was originally developed by James Clerk Maxwell in 1864 and later refined by Otto Mohr and Heinrich Müller-Breslau. This method was one of the first available for the analysis of statically indeterminate structures. Since compatibility forms the basis for this method, it has sometimes been referred to as the *compatibility method* or the *method of consistent displacements*. This method consists of writing equations that satisfy the *compatibility* and *force-displacement requirements* for the structure in order to determine the redundant *forces*. Once these forces have been determined, the remaining reactive forces on the structure are determined by satisfying the equilibrium requirements. The fundamental principles involved in applying this method are easy to understand and develop, and they will be discussed in this chapter.

Displacement Method.

The displacement method of analysis is based on first writing force-displacement relations for the members and then satisfying the *equilibrium requirements* for the structure. In this case the *unknowns* in the equations are *displacements*. Once the displacements are obtained, the forces are determined from the compatibility and force-displacement equations. We will study some of the classical techniques used to apply the displacement method in Chapters 11 and 12. Since almost all present day computer software for structural analysis is developed using this method we will present a matrix formulation of the displacement method in Chapters 14, 15, and 16.

Each of these two methods of analysis, which are outlined in Fig. 10–2, has particular advantages and disadvantages, depending upon the geometry of the structure and its degree of indeterminacy. A discussion of the usefulness of each method will be given after each has been presented.

10

	Unknowns	Equations Used for Solution	Coefficients of the Unknowns
Force Method	Forces	Compatibility and Force Displacement	Flexibility Coefficients
Displacement Method	Displacements	Equilibrium and Force Displacement	Stiffness Coefficients

Fig. 10–2

10.2 Force Method of Analysis: General Procedure

Perhaps the best way to illustrate the principles involved in the force method of analysis is to consider the beam shown in Fig. 10–3a. If its free-body diagram were drawn, there would be four unknown support reactions; and since three equilibrium equations are available for solution, the beam is indeterminate to the first degree. Consequently, one additional equation is necessary for solution. To obtain this equation, we will use the principle of superposition and consider the *compatibility of displacement* at one of the supports. This is done by choosing one of the support reactions as "redundant" and temporarily removing its effect on the beam so that the beam then becomes statically determinate and stable. This beam is referred to as the *primary structure*. Here we will remove the restraining action of the rocker at B. As a result, the load $\mathbf{P}$ will cause B to be displaced downward by an amount Δ_B as shown in Fig. 10–3b. By superposition, however, the unknown reaction at B, i.e., $\mathbf{B}_y$, causes the beam at B to be displaced Δ'_{BB} upward, Fig. 10–3c. Here the first letter in this double-subscript notation refers to the point (B) where the deflection is specified, and the second letter refers to the point (B) where the unknown reaction acts. Assuming positive displacements act upward, then from Figs. 10–3a through 10–3c we can write the necessary compatibility equation at the rocker as

$$(+\uparrow) \qquad\qquad 0 = -\Delta_B + \Delta'_{BB}$$

Let us now denote the displacement at B caused by a *unit load* acting in the direction of $\mathbf{B}_y$ as the *linear flexibility coefficient* f_{BB}, Fig. 10–3d. Using the same scheme for this double-subscript notation as above, f_{BB} is the deflection at B caused by a unit load at B. Since the material behaves in a linear-elastic manner, a force of $\mathbf{B}_y$ acting at B, instead of the unit load, will cause a proportionate increase in f_{BB}. Thus we can write

$$\Delta'_{BB} = B_y f_{BB}$$

When written in this format, it can be seen that the linear flexibility coefficient f_{BB} is a *measure of the deflection per unit force*, and so its units are m/N, mm/N, etc. The compatibility equation above can therefore be written in terms of the unknown B_y as

$$0 = -\Delta_B + B_y f_{BB}$$

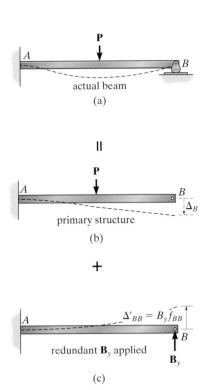

$\mathbf{P}$

A ⎯⎯⎯⎯⎯ B

actual beam

(a)

=

$\mathbf{P}$

A ⎯⎯⎯⎯⎯ B
Δ_B

primary structure

(b)

+

$\Delta'_{BB} = B_y f_{BB}$

A ⎯⎯⎯⎯⎯ B

redundant $\mathbf{B}_y$ applied
$\mathbf{B}_y$

(c)

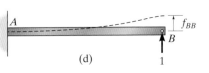

f_{BB}

A ⎯⎯⎯⎯⎯ B

(d)   1

Fig. 10–3

Using the methods of Chapter 8 or 9, or the deflection table on the inside front cover of the book, the appropriate load-displacement relations for the deflection Δ_B, Fig. 10–3b, and the flexibility coefficient f_{BB}, Fig. 10–3d, can be obtained and the solution for B_y determined, that is, $B_y = \Delta_B/f_{BB}$. Once this is accomplished, the three reactions at the wall A can then be found from the equations of equilibrium.

As stated previously, the choice of the redundant is *arbitrary*. For example, the moment at A, Fig. 10–4a, can be determined *directly* by removing the capacity of the beam to support a moment at A, that is, by replacing the fixed support by a pin. As shown in Fig. 10–4b, the rotation at A caused by the load $\mathbf{P}$ is θ_A, and the rotation at A caused by the redundant $\mathbf{M}_A$ at A is θ'_{AA}, Fig. 10–4c. If we denote an *angular flexibility coefficient* α_{AA} as the angular displacement at A caused by a unit couple moment applied to A, Fig. 10–4d, then

$$\theta'_{AA} = M_A \alpha_{AA}$$

Thus, the angular flexibility coefficient measures the angular displacement per unit couple moment, and therefore it has units of rad/N·m or rad/kN·m, etc. The compatibility equation for rotation at A therefore requires

$(\curvearrowright+)$ $\qquad\qquad 0 = \theta_A + M_A \alpha_{AA}$

In this case, $M_A = -\theta_A/\alpha_{AA}$, a negative value, which simply means that $\mathbf{M}_A$ acts in the opposite direction to the unit couple moment.

actual beam
(a)

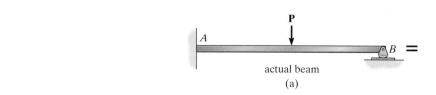

primary structure
(b)

$+$

$\theta'_{AA} = M_A \alpha_{AA}$
redundant $\mathbf{M}_A$ applied
(c)

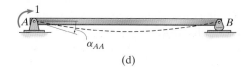

α_{AA}

(d)

Fig. 10–4

Fig. 10–5

A third example that illustrates application of the force method is given in Fig. 10–5a. Here the beam is indeterminate to the second degree and therefore two compatibility equations will be necessary for the solution. We will choose the vertical forces at the roller supports, B and C, as redundants. The resultant statically determinate beam deflects as shown in Fig. 10–5b when the redundants are removed. Each redundant force, which is *assumed* to act downward, deflects this beam as shown in Figs. 10–5c and 10–5d, respectively. Here the flexibility coefficients* f_{BB} and f_{CB} are found from a unit load acting at B, Fig. 10–5e; and f_{CC} and f_{BC} are found from a unit load acting at C, Fig. 10–5f. By superposition, the compatibility equations for the deflection at B and C, respectively, are

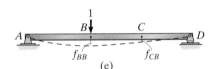

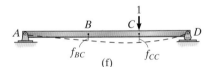

$$1+\downarrow 2 \qquad 0 = \Delta_B + B_y f_{BB} + C_y f_{BC}$$
$$1+\downarrow 2 \qquad 0 = \Delta_C + B_y f_{CB} + C_y f_{CC} \tag{10-1}$$

Once the load-displacement relations are established using the methods of Chapter 8 or 9, these equations may be solved simultaneously for the two unknown forces B_y and C_y.

Having illustrated the application of the force method of analysis by example, we will now discuss its application in general terms and then we will use it as a basis for solving problems involving trusses, beams, and frames. For all these cases, however, realize that since the method depends on superposition of displacements, it is necessary that *the material remain linear elastic when loaded*. Also, recognize that *any* external reaction or internal loading at a point in the structure can be directly determined by first releasing the capacity of the structure to support the loading and then writing a compatibility equation at the point. See Example 10–4.

---

*f_{BB} is the deflection at B caused by a unit load at B; f_{CB} the deflection at C caused by a unit load at B.

Procedure for Analysis

The following procedure provides a general method for determining the reactions or internal loadings of statically indeterminate structures using the force or flexibility method of analysis.

Principle of Superposition

Determine the number of degrees n to which the structure is indeterminate. Then specify the n unknown redundant forces or moments that must be removed from the structure in order to make it statically determinate and stable. Using the principle of superposition, draw the statically indeterminate structure and show it to be equal to a series of corresponding statically *determinate* structures. The primary structure supports the same external loads as the statically indeterminate structure, and each of the other structures added to the primary structure shows the structure loaded with a separate redundant force or moment. Also, sketch the elastic curve on each structure and indicate symbolically the displacement or rotation at the point of each redundant force or moment.

Compatibility Equations

Write a compatibility equation for the displacement or rotation at each point where there is a redundant force or moment. These equations should be expressed in terms of the unknown redundants and their corresponding flexibility coefficients obtained from unit loads or unit couple moments that are collinear with the redundant forces or moments.

Determine all the deflections and flexibility coefficients using the table on the inside front cover or the methods of Chapter 8 or 9.* Substitute these load-displacement relations into the compatibility equations and solve for the unknown redundants. In particular, if a numerical value for a redundant is negative, it indicates the redundant acts opposite to its corresponding unit force or unit couple moment.

Equilibrium Equations

Draw a free-body diagram of the structure. Since the redundant forces and/or moments have been calculated, the remaining unknown reactions can be determined from the equations of equilibrium.

It should be realized that once all the support reactions have been obtained, the shear and moment diagrams can then be drawn, and the deflection at any point on the structure can be determined using the same methods outlined previously for statically determinate structures.

10

*It is suggested that if the *M/EI* diagram for a beam consists of simple segments, the moment-area theorems or the conjugate-beam method be used. Beams with complicated *M/EI* diagrams, that is, those with many curved segments (parabolic, cubic, etc.), can be readily analyzed using the method of virtual work or by Castigliano's second theorem.

10.3   Maxwell's Theorem of Reciprocal Displacements; Betti's Law

When Maxwell developed the force method of analysis, he also published a theorem that relates the flexibility coefficients of any two points on an elastic structure—be it a truss, a beam, or a frame. This theorem is referred to as the theorem of reciprocal displacements and may be stated as follows: *The displacement of a point B on a structure due to a unit load acting at point A is equal to the displacement of point A when the unit load is acting at point B, that is,* $f_{BA} = f_{AB}$.

Proof of this theorem is easily demonstrated using the principle of virtual work. For example, consider the beam in Fig. 10–6. When a real unit load acts at A, assume that the internal moments in the beam are represented by m_A. To determine the flexibility coefficient at B, that is, f_{BA}, a virtual unit load is placed at B, Fig. 10–7, and the internal moments m_B are computed. Then applying Eq. 9–18 yields

$$f_{BA} = \int \frac{m_B m_A}{EI} dx$$

Likewise, if the flexibility coefficient f_{AB} is to be determined when a real unit load acts at B, Fig. 10–7, then m_B represents the internal moments in the beam due to a real unit load. Furthermore, m_A represents the internal moments due to a virtual unit load at A, Fig. 10–6. Hence,

$$f_{AB} = \int \frac{m_A m_B}{EI} dx$$

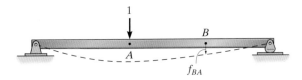

Fig. 10–6

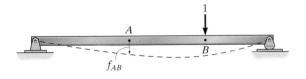

Fig. 10–7

Both integrals obviously give the same result, which proves the theorem. The theorem also applies for reciprocal rotations, and may be stated as follows: *The rotation at point B on a structure due to a unit couple moment acting at point A is equal to the rotation at point A when the unit couple moment is acting at point B.* Furthermore, using a unit force and unit couple moment, applied at separate points on the structure, we may also state: *The rotation in radians at point B on a structure due to a unit load acting at point A is equal to the displacement at point A when a unit couple moment is acting at point B.*

As a consequence of this theorem, some work can be saved when applying the force method to problems that are statically indeterminate to the second degree or higher. For example, only one of the two flexibility coefficients f_{BC} or f_{CB} has to be calculated in Eqs. 10–1, since $f_{BC} = f_{CB}$. Furthermore, the theorem of reciprocal displacements has applications in structural model analysis and for constructing influence lines using the Müller-Breslau principle (see Sec. 10–10).

When the theorem of reciprocal displacements is formalized in a more general sense, it is referred to as *Betti's law.* Briefly stated: The virtual work δU_{AB} done by a system of forces $\Sigma \mathbf{P}_B$ that undergo a displacement caused by a system of forces $\Sigma \mathbf{P}_A$ is equal to the virtual work δU_{BA} caused by the forces $\Sigma \mathbf{P}_A$ when the structure deforms due to the system of forces $\Sigma \mathbf{P}_B$ In other words, $\delta U_{AB} = \delta U_{BA}$. The proof of this statement is similar to that given above for the reciprocal-displacement theorem.

10.4 Force Method of Analysis: Beams

The force method applied to beams was outlined in Sec. 10–2. Using the "procedure for analysis" also given in Sec. 10–2, we will now present several examples that illustrate the application of this technique.

These bridge girders are statically indeterminate since they are continuous over their piers.

EXAMPLE 10.1

Determine the reaction at the roller support B of the beam shown in Fig. 10–8a. EI is constant.

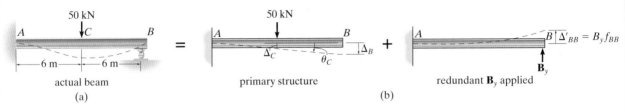

actual beam
(a)

primary structure

redundant $\mathbf{B}_y$ applied
(b)

Fig. 10–8

SOLUTION

Principle of Superposition. By inspection, the beam is statically indeterminate to the first degree. The redundant will be taken as $\mathbf{B}_y$ so that this force can be determined *directly*. Figure 10–8b shows application of the principle of superposition. Notice that removal of the redundant requires that the roller support or the constraining action of the beam in the direction of $\mathbf{B}_y$ be removed. Here we have assumed that $\mathbf{B}_y$ acts upward on the beam.

Compatibility Equation. Taking positive displacement as upward, Fig. 10–8b, we have

$$(+\uparrow) \qquad\qquad 0 = -\Delta_B + B_y f_{BB} \qquad\qquad (1)$$

The terms Δ_B and f_{BB} are easily obtained using the table on the inside front cover. In particular, note that $\Delta_B = \Delta_C + \theta_C(6\text{ m})$. Thus,

$$
\begin{aligned}
\Delta_B &= \frac{P(L/2)^3}{3EI} + \frac{P(L/2)^2}{2EI}\left(\frac{L}{2}\right) \\
&= \frac{(50\text{ kN})(6\text{ m})^3}{3EI} + \frac{(50\text{ kN})(6\text{ m})^2}{2EI}(6\text{ m}) = \frac{9000\text{ kN}\cdot\text{m}^3}{EI}\downarrow
\end{aligned}
$$

$$
f_{BB} = \frac{PL^3}{3EI} = \frac{1(12\text{ m})^3}{3EI} = \frac{576\text{ m}^3}{EI}\uparrow
$$

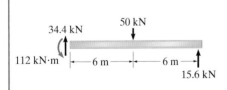

(c)

Substituting these results into Eq. (1) yields

$$(+\uparrow) \qquad 0 = -\frac{9000}{EI} + B_y\left(\frac{576}{EI}\right) \qquad B_y = 15.6\text{ kN} \qquad \textit{Ans.}$$

If this reaction is placed on the free-body diagram of the beam, the reactions at A can be obtained from the three equations of equilibrium, Fig. 10–8c.

Having determined all the reactions, the moment diagram can be constructed as shown in Fig. 10–8d.

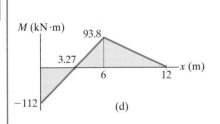

(d)

EXAMPLE | 10.2

Draw the shear and moment diagrams for the beam shown in Fig. 10–9 a. The support at B settles 40 mm. Take $E = 200$ GPa, $I = 500(10^6)$ mm^4.

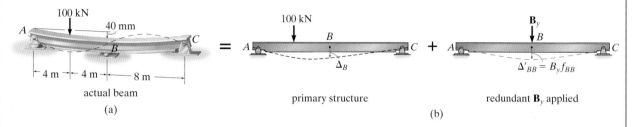

actual beam

(a)

primary structure

redundant $\mathbf{B}_y$ applied

(b)

Fig. 10–9

SOLUTION

Principle of Superposition.   By inspection, the beam is indeterminate to the first degree. The center support B will be chosen as the redundant, so that the roller at B is removed, Fig. 10–9b. Here $\mathbf{B}_y$ is assumed to act downward on the beam.

Compatibility Equation.   With reference to point B in Fig. 10–9b, using units of meters, we require

$$(+\downarrow) \qquad\qquad 0.04 \text{ m} = \Delta_B + B_y f_{BB} \qquad\qquad (1)$$

We will use the table on the inside front cover. Note that for Δ_B the equation for the deflection curve requires $0 < x < a$. Since $x = 8$ m, then $a = 12$ m. Thus,

$$\Delta_B = \frac{Pbx}{6LEI}(L^2 - b^2 - x^2) = \frac{100(4)(8)}{6(16)EI}[(16)^2 - (4)^2 - (8)^2]$$

$$= \frac{5866.7 \text{ kN} \cdot \text{m}^3}{EI}$$

$$f_{BB} = \frac{PL^3}{48EI} = \frac{1(16)^3}{48\,EI} = \frac{85.33 \text{ kN} \cdot \text{m}^3}{EI}$$

Substituting these values into Eq. (1), we get

$$0.04 \text{ m } [200(10^6) \text{ kN/m}^2][500(10^6)(10^{-12}) \text{ m}^4]$$

$$= 5866.7 \text{ kN} \cdot \text{m}^3 + B_y(85.33 \text{ kN} \cdot \text{m}^3)$$

$$B_y = -21.88 \text{ kN}$$

The negative sign indicates that $\mathbf{B}_y$ acts *upward* on the beam.

10

EXAMPLE | 10.2 (Continued)

Equilibrium Equations. From the free-body diagram shown in Fig. 10–9c we have

$$\zeta + \Sigma M_A = 0; \qquad -100(4) + 21.88(8) + C_y(16) = 0$$
$$C_y = 14.06 \text{ kN}$$

$$+\uparrow \Sigma F_y = 0; \qquad A_y - 100 + 21.88 + 14.06 = 0$$
$$A_y = 64.06 \text{ kN}$$

Using these results, verify the shear and moment diagrams shown in Fig. 10–9d.

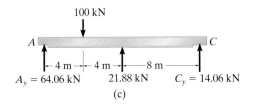

(c)

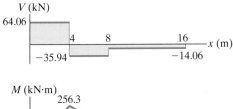

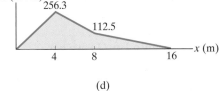

(d)

EXAMPLE 10.3

Draw the shear and moment diagrams for the beam shown in Figure 10–10a. EI is constant. Neglect the effects of axial load.

SOLUTION

Principle of Superposition. Since axial load is neglected, the beam is indeterminate to the second degree. The two end moments at A and B will be considered as the redundants. The beam's capacity to resist these moments is removed by placing a pin at A and a rocker at B. The principle of superposition applied to the beam is shown in Fig. 10–10b.

Compatibility Equations. Reference to points A and B, Fig. 10–10b, requires

$(\curvearrowright+)$ $$0 = \theta_A + M_A \alpha_{AA} + M_B \alpha_{AB} \qquad (1)$$

$(\curvearrowleft+)$ $$0 = \theta_B + M_A \alpha_{BA} + M_B \alpha_{BB} \qquad (2)$$

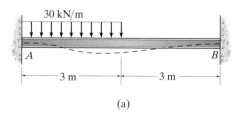

(a)

actual beam

||

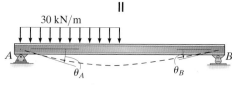

primary structure

+

$\theta'_{AA} = M_A \alpha_{AA}$ $\theta'_{BA} = M_A \alpha_{BA}$

redundant moment $\mathbf{M}_A$ applied

+

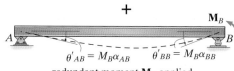

$\theta'_{AB} = M_B \alpha_{AB}$ $\theta'_{BB} = M_B \alpha_{BB}$

redundant moment $\mathbf{M}_B$ applied

(b)

Fig. 10–10

10

EXAMPLE   10.3 (Continued)

The required slopes and angular flexibility coefficients can be determined using the table on the inside front cover. We have

$$\theta_A = \frac{3wL^3}{128EI} = \frac{3(30)(6)^3}{128EI} = \frac{151.9}{EI}$$

$$\theta_B = \frac{7wL^3}{384EI} = \frac{7(30)(6)^3}{384EI} = \frac{118.1}{EI}$$

$$\alpha_{AA} = \frac{ML}{3EI} = \frac{1(6)}{3EI} = \frac{2}{EI}$$

$$\alpha_{BB} = \frac{ML}{3EI} = \frac{1(6)}{3EI} = \frac{2}{EI}$$

$$\alpha_{AB} = \frac{ML}{6EI} = \frac{1(6)}{6EI} = \frac{1}{EI}$$

Note that $\alpha_{BA} = \alpha_{AB}$, a consequence of Maxwell's theorem of reciprocal displacements.

Substituting the data into Eqs. (1) and (2) yields

$$0 = \frac{151.9}{EI} + M_A\left(\frac{2}{EI}\right) + M_B\left(\frac{1}{EI}\right)$$

$$0 = \frac{118.1}{EI} + M_A\left(\frac{1}{EI}\right) + M_B\left(\frac{2}{EI}\right)$$

Canceling EI and solving these equations simultaneously, we have

$$M_A = -61.9 \text{ kN} \cdot \text{m} \qquad M_B = -28.1 \text{ kN} \cdot \text{m}$$

Using these results, the end shears are calculated, Fig. 10–10c, and the shear and moment diagrams plotted.

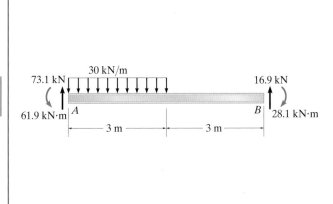

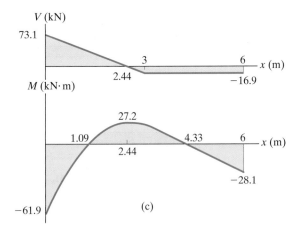

(c)

EXAMPLE 10.4

Determine the reactions at the supports for the beam shown in Fig. 10–11a. EI is constant.

SOLUTION

Principle of Superposition.   By inspection, the beam is indeterminate to the first degree. Here, for the sake of illustration, we will choose the internal moment at support B as the redundant. Consequently, the beam is cut open and end pins or an internal hinge are placed at B in order to release *only* the capacity of the beam to resist moment at this point, Fig. 10–11b. The internal moment at B is applied to the beam in Fig. 10–11c.

Compatibility Equations.   From Fig. 10–11a we require the relative rotation of one end of one beam with respect to the end of the other beam to be zero, that is,

$(\curvearrowright+)$
$$\theta_B + M_B \alpha_{BB} = 0$$

where

$$\theta_B = \theta'_B + \theta''_B$$

and

$$\alpha_{BB} = \alpha'_{BB} + \alpha''_{BB}$$

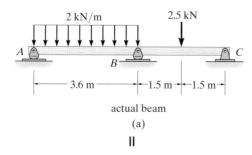

2 kN/m   2.5 kN

A   B   C

3.6 m —— 1.5 m — 1.5 m

actual beam
(a)

‖

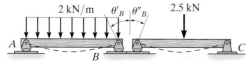

2 kN/m θ'_B θ''_B 2.5 kN

A   B   C

primary structure
(b)

+

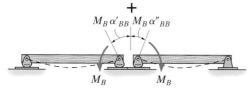

$M_B \alpha'_{BB}$ $M_B \alpha''_{BB}$

M_B M_B

redundant $\mathbf{M}_B$ applied
(c)   **Fig. 10–11**

10

EXAMPLE 10.4 (Continued)

The slopes and angular flexibility coefficients can be determined from the table on the inside front cover, that is,

$$\theta'_B = \frac{wL^3}{24EI} = \frac{2(3.6)^3}{24EI} = \frac{3.888 \text{ kN} \cdot \text{m}^2}{EI}$$

$$\theta''_B = \frac{PL^2}{16EI} = \frac{2.5(3)^2}{16EI} = \frac{1.406 \text{ kN} \cdot \text{m}^2}{EI}$$

$$\alpha'_{BB} = \frac{ML}{3EI} = \frac{1(3.6)}{3EI} = \frac{1.2 \text{ m}}{EI}$$

$$\alpha''_{BB} = \frac{ML}{3EI} = \frac{1(3)}{3EI} = \frac{1 \text{ m}}{EI}$$

Thus

$$\frac{3.888 \text{ kN} \cdot \text{m}^2}{EI} + \frac{1.406 \text{ kN} \cdot \text{m}^2}{EI} + M_B\left(\frac{1.2 \text{ m}}{EI} + \frac{1 \text{ m}}{EI}\right) = 0$$

$$M_B = -2.406 \text{ kN} \cdot \text{m}$$

The negative sign indicates M_B acts in the opposite direction to that shown in Fig. 10–11c. Using this result, the reactions at the supports are calculated as shown in Fig. 10–11d. Furthermore, the shear and moment diagrams are shown in Fig. 10–11e.

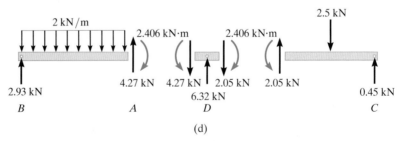

(d)

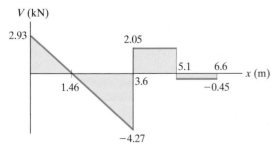
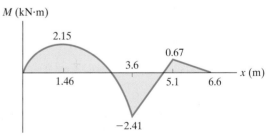

(e)

10.5 Force Method of Analysis: Frames

The force method is very useful for solving problems involving statically indeterminate frames that have a single story and unusual geometry, such as gabled frames. Problems involving multistory frames, or those with a high degree of indeterminacy, are best solved using the slope-deflection, moment-distribution, or the stiffness method discussed in later chapters.

The following examples illustrate the application of the force method using the procedure for analysis outlined in Sec. 10–2.

EXAMPLE | **10.5**

The frame, or bent, shown in the photo is used to support the bridge deck. Assuming EI is constant, a drawing of it along with the dimensions and loading is shown in Fig. 10–12a. Determine the support reactions.

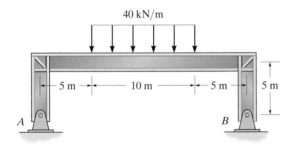

(a)

Fig. 10–12

SOLUTION

Principle of Superposition.   By inspection the frame is statically indeterminate to the first degree. We will choose the horizontal reaction at A to be the redundant. Consequently, the pin at A is replaced by a rocker, since a rocker will not constrain A in the horizontal direction. The principle of superposition applied to the idealized model of the frame is shown in Fig. 10–12b. Notice how the frame deflects in each case.

10

EXAMPLE 10.5 (Continued)

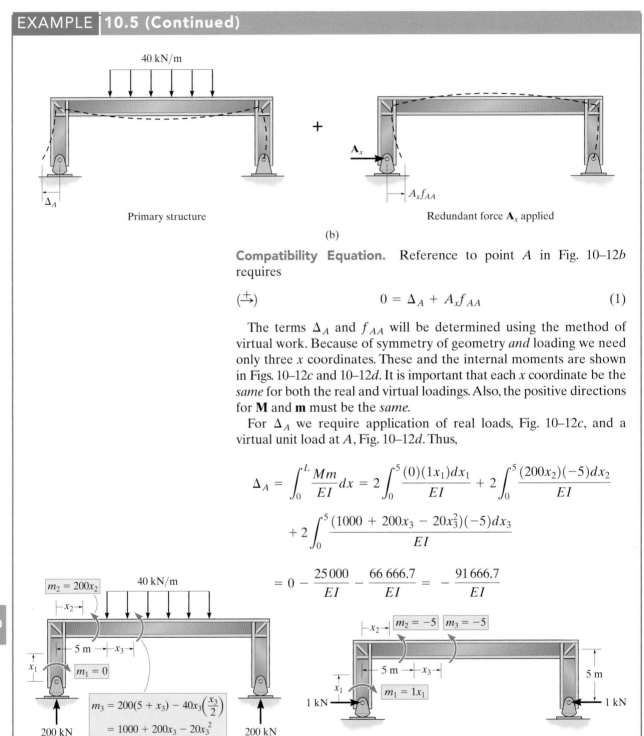

40 kN/m

Primary structure

+

$\mathbf{A}_x$

$A_x f_{AA}$

Δ_A

Redundant force $\mathbf{A}_x$ applied

(b)

Compatibility Equation. Reference to point A in Fig. 10–12b requires

$(\xrightarrow{+})$ $\qquad\qquad\qquad 0 = \Delta_A + A_x f_{AA}$ $\qquad\qquad$ (1)

The terms Δ_A and f_{AA} will be determined using the method of virtual work. Because of symmetry of geometry *and* loading we need only three x coordinates. These and the internal moments are shown in Figs. 10–12c and 10–12d. It is important that each x coordinate be the *same* for both the real and virtual loadings. Also, the positive directions for $\mathbf{M}$ and $\mathbf{m}$ must be the *same*.

For Δ_A we require application of real loads, Fig. 10–12c, and a virtual unit load at A, Fig. 10–12d. Thus,

$$\Delta_A = \int_0^L \frac{Mm}{EI}\,dx = 2\int_0^5 \frac{(0)(1x_1)dx_1}{EI} + 2\int_0^5 \frac{(200x_2)(-5)dx_2}{EI}$$

$$+ 2\int_0^5 \frac{(1000 + 200x_3 - 20x_3^2)(-5)dx_3}{EI}$$

$$= 0 - \frac{25\,000}{EI} - \frac{66\,666.7}{EI} = -\frac{91\,666.7}{EI}$$

$m_2 = 200x_2$

40 kN/m

$\mapsto x_2 \mapsto$

5 m $\mapsto x_3 \mapsto$

x_1

$m_1 = 0$

$m_3 = 200(5 + x_3) - 40x_3\left(\dfrac{x_3}{2}\right)$
$= 1000 + 200x_3 - 20x_3^2$

200 kN

200 kN

(c)

$\mapsto x_2 \mapsto$ $\boxed{m_2 = -5}$ $\boxed{m_3 = -5}$

5 m $\mapsto x_3 \mapsto$

x_1

$\boxed{m_1 = 1x_1}$

5 m

1 kN

1 kN

(d)

For f_{AA} we require application of a real unit load and a virtual unit load acting at A, Fig. 10–12d. Thus,

$$f_{AA} = \int_0^L \frac{mm}{EI}dx = 2\int_0^5 \frac{(1x_1)^2 dx_1}{EI} + 2\int_0^5 (5)^2 dx_2 + 2\int_0^5 (5)^2 dx_3$$

$$= \frac{583.33}{EI}$$

Substituting the results into Eq. (1) and solving yields

$$0 = \frac{-91\,666.7}{EI} + A_x\left(\frac{583.33}{EI}\right)$$

$$A_x = 157 \text{ kN} \hspace{4cm} Ans.$$

Equilibrium Equations.   Using this result, the reactions on the idealized model of the frame are shown in Fig. 10–12e.

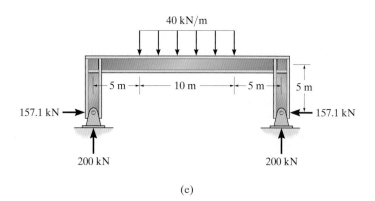

(e)

EXAMPLE 10.6

Determine the moment at the fixed support A for the frame shown in Fig. 10–13a. EI is constant.

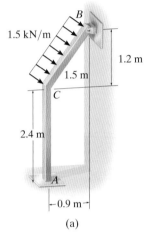

(a)   Fig. 10–13

SOLUTION

Principle of Superposition. The frame is indeterminate to the first degree. A direct solution for $\mathbf{M}_A$ can be obtained by choosing this as the redundant. Thus the capacity of the frame to support a moment at A is removed and therefore a pin is used at A for support. The principle of superposition applied to the frame is shown in Fig. 10–13b.

Compatibility Equation. Reference to point A in Fig. 10–13b requires

$(\curvearrowright+)$ $$0 = \theta_A + M_A\alpha_{AA} \qquad (1)$$

As in the preceding example, θ_A and α_{AA} will be computed using the method of virtual work. The frame's x coordinates and internal moments are shown in Figs. 10–13c and 10–13d.

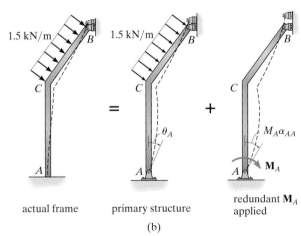

actual frame   primary structure   redundant $\mathbf{M}_A$ applied

(b)

For θ_A we require application of the real loads, Fig. 10–13c, and a virtual unit couple moment at A, Fig. 10–13d. Thus,

$$\theta_A = \sum \int_0^L \frac{M m_\theta \, dx}{EI}$$

$$= \int_0^{2.4} \frac{(0.13x_1)(1 - 0.278x_1) \, dx_1}{EI}$$

$$+ \int_0^{1.5} \frac{(1.34x_2 - 0.75x_2^2)(0.222x_2) \, dx_2}{EI}$$

$$= \frac{0.209}{EI} + \frac{0.124}{EI} = \frac{0.333}{EI}$$

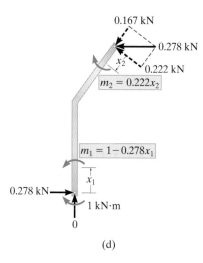

1.0 kN

2.25 kN   1.67 kN

1.34 kN

$M_2 = 1.34x_2 - 0.75x_2^2$

$M_1 = 0.13x_1$

x_1

0.13 kN

1.35 kN

(c)

For α_{AA} we require application of a real unit couple moment and a virtual unit couple moment acting at A, Fig. 10–13d. Thus,

$$\alpha_{AA} = \sum \int_0^L \frac{m_\theta m_\theta}{EI} dx$$

$$= \int_0^{2.4} \frac{(1 - 0.278x_1)^2 \, dx_1}{EI} + \int_0^{1.5} \frac{(0.222x_2)^2 \, dx_2}{EI}$$

$$= \frac{1.155}{EI} + \frac{0.555}{EI} = \frac{1.21}{EI}$$

0.167 kN

0.278 kN

0.222 kN

$m_2 = 0.222x_2$

$m_1 = 1 - 0.278x_1$

0.278 kN

x_1

1 kN·m

0

(d)

Substituting these results into Eq. (1) and solving yields

$$0 = \frac{0.333}{EI} + M_A \left(\frac{1.21}{EI} \right) \qquad M_A = -0.275 \text{ kN} \cdot \text{m} \quad Ans.$$

The negative sign indicates $\mathbf{M}_A$ acts in the opposite direction to that shown in Fig. 10–13b.

10

FUNDAMENTAL PROBLEMS

F10–1.   Determine the reactions at the fixed support at A and the roller at B. EI is constant.

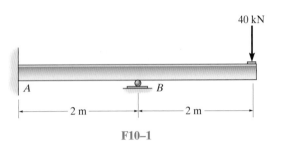

40 kN

2 m   2 m

F10–1

F10–4.   Determine the reactions at the pin at A and the rollers at B and C.

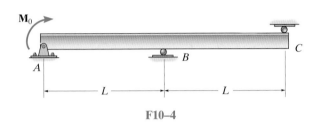

M_0

L L

F10–4

F10–2.   Determine the reactions at the fixed supports at A and the roller at B. EI is constant.

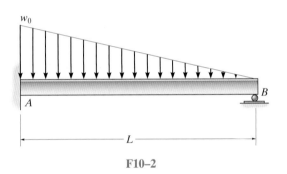

w_0

L

F10–2

F10–5.   Determine the reactions at the pin A and the rollers at B and C on the beam. EI is constant.

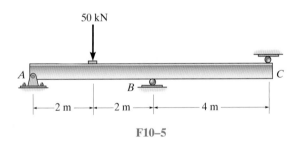

50 kN

2 m   2 m   4 m

F10–5

F10–3.   Determine the reactions at the fixed support at A and the roller at B. Support B settles 5 mm. Take $E = 200$ GPa and $I = 300(10^6)$ mm^4.

F10–6.   Determine the reactions at the pin at A and the rollers at B and C on the beam. Support B settles 5 mm. Take $E = 200$ GPa, $I = 300(10^6)$ mm^4.

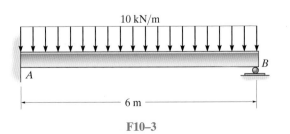

10 kN/m

6 m

F10–3

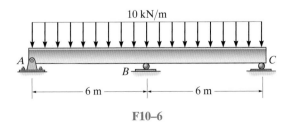

10 kN/m

6 m   6 m

F10–6

PROBLEMS

10–1. Determine the reactions at the supports A and B. EI is constant.

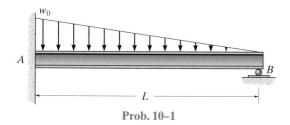

Prob. 10–1

10–2. Determine the reactions at the supports A, B, and C, then draw the shear and moment diagrams. EI is constant.

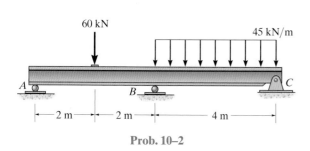

Prob. 10–2

10–3. Determine the reactions at the supports A and B. EI is constant.

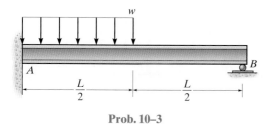

Prob. 10–3

*10–4. Determine the reactions at the supports A, B, and C; then draw the shear and moment diagram. EI is constant.

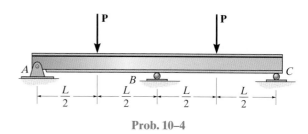

Prob. 10–4

10–5. Determine the reactions at the supports, then draw the shear and moment diagram. EI is constant.

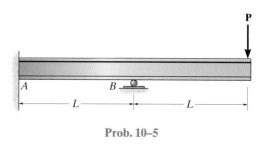

Prob. 10–5

10–6. Determine the reactions at the supports, then draw the moment diagram. Assume B and C are rollers and A is pinned. The support at B settles downward 75 mm. Take $E = 200$ GPa, $I = 200(10^6)$ mm^4.

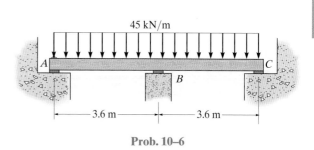

Prob. 10–6

10

10–7. Determine the deflection at the end B of the clamped A-36 steel strip. The spring has a stiffness of $k = 2$ N/mm. The strip is 5 mm wide and 10 mm high. Also, draw the shear and moment diagrams for the strip.

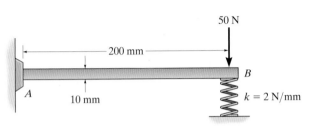

50 N

200 mm

B

A

10 mm

$k = 2$ N/mm

Prob. 10–7

*10–8. Determine the reactions at the supports. The moment of inertia for each segment is shown in the figure. Assume the support at B is a roller. Take $E = 200$ GPa.

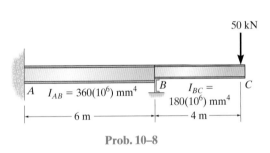

50 kN

A $I_{AB} = 360(10^6)$ mm^4

B $I_{BC} = 180(10^6)$ mm^4 C

6 m

4 m

Prob. 10–8

10–9. The simply supported beam is subjected to the loading shown. Determine the deflection at its center C. EI is constant.

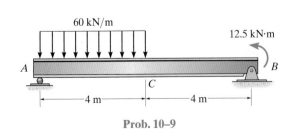

60 kN/m

12.5 kN·m

A

C

B

4 m

4 m

Prob. 10–9

10–10. Determine the reactions at the supports, then draw the moment diagram. Assume the support at B is a roller. EI is constant.

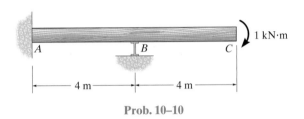

1 kN·m

A B C

4 m   4 m

Prob. 10–10

10–11. Determine the reactions at the supports, then draw the moment diagram. Assume A is a pin and B and C are rollers. EI is constant.

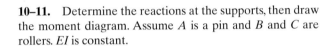

9 kN/m

A B C

5 m   5 m

Prob. 10–11

*10–12. Determine the reactions at the supports, then draw the moment diagram. Assume the support at A is a pin and B and C are rollers. EI is constant.

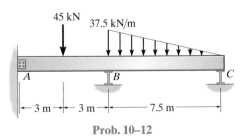

45 kN   37.5 kN/m

A B C

3 m   3 m   7.5 m

Prob. 10–12

10

10–13. Determine the reactions at the supports. Assume *A* and *C* are pins and the joint at *B* is fixed connected. *EI* is constant.

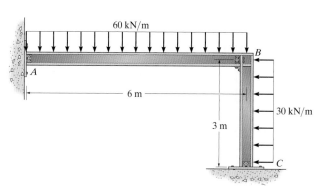

60 kN/m

B

A

6 m

3 m

30 kN/m

C

Prob. 10–13

10–14. Determine the reactions at the supports. *EI* is constant.

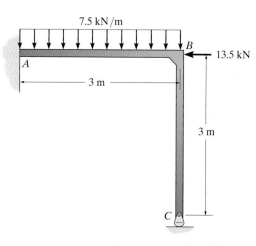

7.5 kN/m

B

13.5 kN

A

3 m

3 m

C

Prob. 10–14

10–15. Determine the reactions at the supports, then draw the moment diagram for each member. *EI* is constant.

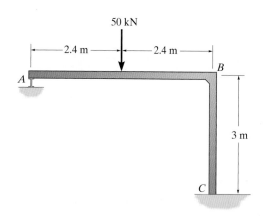

50 kN

2.4 m   2.4 m

A   *B*

3 m

C

Prob. 10–15

*****10–16.** Determine the reactions at the supports. Assume *A* is fixed connected. *E* is constant.

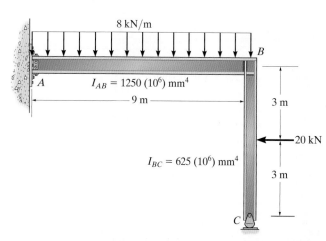

8 kN/m

B

A $I_{AB} = 1250 \,(10^6) \text{ mm}^4$

9 m

3 m

20 kN

$I_{BC} = 625 \,(10^6) \text{ mm}^4$

3 m

C

Prob. 10–16

10

10–17. Determine the reactions at the supports. EI is constant.

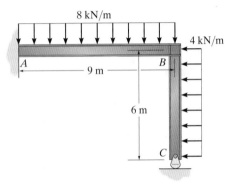

8 kN/m

4 kN/m

A

9 m

B

6 m

C

Prob. 10–17

10–18. Determine the reactions at the supports A and D. The moment of inertia of each segment of the frame is listed in the figure. Take $E = 200$ GPa.

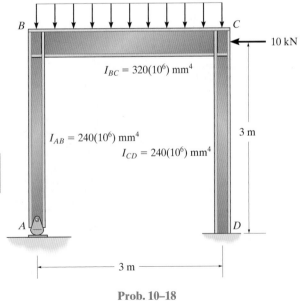

50 kN/m

B C

$I_{BC} = 320(10^6)$ mm^4

10 kN

3 m

$I_{AB} = 240(10^6)$ mm^4

$I_{CD} = 240(10^6)$ mm^4

A D

3 m

Prob. 10–18

10–19. The steel frame supports the loading shown. Determine the horizontal and vertical components of reaction at the supports A and D. Draw the moment diagram for the frame members. E is constant.

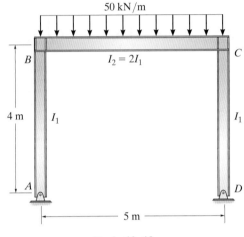

50 kN/m

B $I_2 = 2I_1$ C

4 m I_1 I_1

A D

5 m

Prob. 10–19

*10–20. Determine the reactions at the supports. Assume A and B are pins and the joints at C and D are fixed connections. EI is constant.

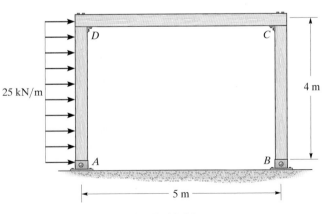

D C

4 m

25 kN/m

A B

5 m

Prob. 10–20

10–21.   Determine the reactions at the supports. Assume A and D are pins. EI is constant.

10–23.   Determine the reactions at the supports. Assume A and B are pins. EI is constant.

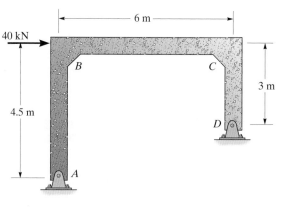

Prob. 10–21

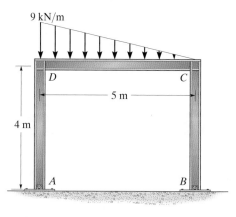

Prob. 10–23

10–22.   Determine the reactions at the supports. Assume A and B are pins. EI is constant.

*10–24.   Two boards each having the same EI and length L are crossed perpendicular to each other as shown. Determine the vertical reactions at the supports. Assume the boards just touch each other before the load **P** is applied.

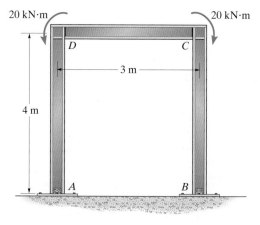

Prob. 10–22

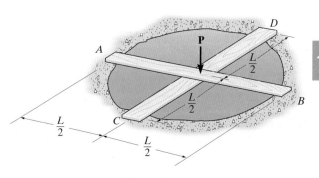

Prob. 10–24

10

10.6   Force Method of Analysis: Trusses

The degree of indeterminacy of a truss can usually be determined by inspection; however, if this becomes difficult, use Eq. 3–1, $b + r > 2j$. Here the unknowns are represented by the number of bar forces (b) plus the support reactions (r), and the number of available equilibrium equations is $2j$ since two equations can be written for each of the (j) joints.

The force method is quite suitable for analyzing trusses that are statically indeterminate to the first or second degree. The following examples illustrate application of this method using the procedure for analysis outlined in Sec. 10–2.

EXAMPLE | 10.7

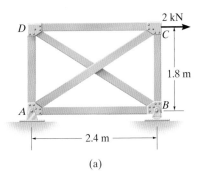

(a)

Fig. 10–14

Determine the force in member AC of the truss shown in Fig. 10–14a. AE is the same for all the members.

SOLUTION

Principle of Superposition. By inspection the truss is indeterminate to the first degree.* Since the force in member AC is to be determined, member AC will be chosen as the redundant. This requires "cutting" this member so that it cannot sustain a force, thereby making the truss statically determinate and stable. The principle of superposition applied to the truss is shown in Fig. 10–14b.

Compatibility Equation. With reference to member AC in Fig. 10–14b, we require the relative displacement Δ_{AC}, which occurs at the ends of the cut member AC due to the 2-kN load, plus the relative displacement $F_{AC}f_{AC\,AC}$ caused by the redundant force acting alone, to be equal to zero, that is,

$$0 = \Delta_{AC} + F_{AC}f_{AC\,AC} \qquad (1)$$

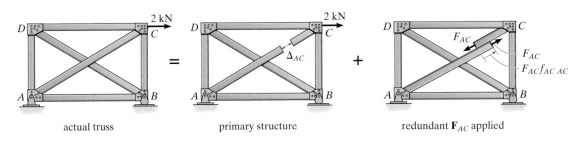

actual truss   primary structure   redundant $\mathbf{F}_{AC}$ applied

(b)

*Applying Eq. 3–1, $b + r > 2j$ or $6 + 3 > 2(4)$, $9 > 8$, $9 - 8 = $ 1st degree.

Here the flexibility coefficient $f_{AC\,AC}$ represents the relative displacement of the cut ends of member AC caused by a "real" unit load acting at the cut ends of member AC. This term, $f_{AC\,AC}$, and Δ_{AC} will be computed using the method of virtual work. The force analysis, using the method of joints, is summarized in Fig. 10–14c and 10–14d.

For Δ_{AC} we require application of the real load of 2 kN, Fig. 10–14c, and a virtual unit force acting at the cut ends of member AC, Fig. 10–14d. Thus,

$$\Delta_{AC} = \sum \frac{nNL}{AE}$$

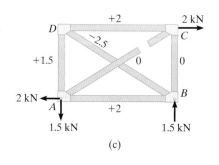

(c)

$$= 2\left[\frac{(-0.8)(2)(2.4)}{AE}\right] + \frac{(-0.6)(0)(1.8)}{AE} + \frac{(-0.6)(1.5)(1.8)}{AE}$$

$$+ \frac{(1)(-2.5)(3)}{AE} + \frac{(1)(0)(3)}{AE}$$

$$= -\frac{16.8}{AE}$$

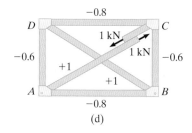

(d)

For $f_{AC\,AC}$ we require application of real unit forces and virtual unit forces acting on the cut ends of member AC, Fig. 10–14d. Thus,

$$f_{AC\,AC} = \sum \frac{n^2 L}{AE}$$

$$= 2\left[\frac{(-0.8)^2(2.4)}{AE}\right] + 2\left[\frac{(-0.6)^2(1.8)}{AE}\right] + 2\left[\frac{(1)^2(3)}{AE}\right]$$

$$= \frac{10.37}{AE}$$

Substituting the data into Eq. (1) and solving yields

$$0 = -\frac{16.8}{AE} + \frac{10.37}{AE} F_{AC}$$

$$F_{AC} = 1.62 \text{ kN (T)} \qquad\qquad Ans.$$

Since the numerical result is positive, AC is subjected to tension as assumed, Fig. 10–14b. Using this result, the forces in the other members can be found by equilibrium, using the method of joints.

10

EXAMPLE 10.8

Determine the force in each member of the truss shown in Fig. 10–15a if the turnbuckle on member AC is used to shorten the member by 12.5 mm. Each bar has a cross-sectional area of 125 mm², and E = 200 GPa.

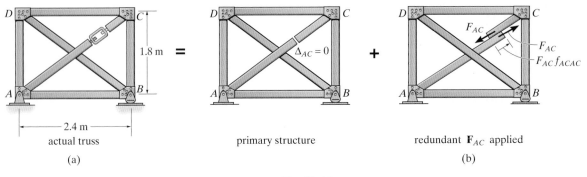

actual truss

(a)

primary structure

redundant **F**$_{AC}$ applied

(b)

Fig. 10–15

SOLUTION

Principle of Superposition. This truss has the same geometry as that in Example 10–7. Since AC has been shortened, we will choose it as the redundant, Fig. 10–15b.

Compatibility Equation. Since no external loads act on the primary structure (truss), there will be no relative displacement between the ends of the sectioned member caused by load; that is, $\Delta_{AC} = 0$. The flexibility coefficient $f_{AC\,AC}$ has been determined in Example 10–7, so

$$f_{AC\,AC} = \frac{10.37}{AE}$$

Assuming the amount by which the bar is shortened is positive, the compatibility equation for the bar is therefore

$$0.0125 \text{ m} = 0 + \frac{10.37}{AE} F_{AC}$$

Realizing that $f_{AC\,AC}$ is a measure of displacement per unit force, we have

$$0.0125 \text{ m} = 0 + \frac{10.37 \text{ m}}{(125 \text{ mm}^2)[200 \text{ kN}/\text{mm}^2]} F_{AC}$$

Thus,

$$F_{AC} = 30.14 \text{ kN} = 30.1 \text{ kN (T)} \qquad \qquad Ans.$$

Since no external forces act on the truss, the external reactions are zero. Therefore, using F_{AC} and analyzing the truss by the method of joints yields the results shown in Fig. 10–15c.

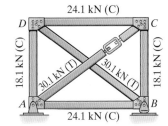

24.1 kN (C)

18.1 kN (C) 30.1 kN (T) 30.1 kN (T) 18.1 kN (C)

24.1 kN (C)

(c)

10.7 Composite Structures

Composite structures are composed of some members subjected only to axial force, while other members are subjected to bending. If the structure is statically indeterminate, the force method can conveniently be used for its analysis. The following example illustrates the procedure.

EXAMPLE 10.9

The simply supported queen-post trussed beam shown in the photo is to be designed to support a uniform load of 2 kN/m. The dimensions of the structure are shown in Fig. 10–16a. Determine the force developed in member CE. Neglect the thickness of the beam and assume the truss members are pin connected to the beam. Also, neglect the effect of axial compression and shear in the beam. The cross-sectional area of each strut is 400 mm^2, and for the beam $I = 20110^6 2$ mm^4. Take $E = 200$ GPa.

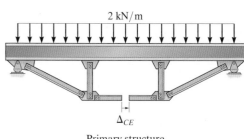

Actual structure

(a)

Fig. 10–16

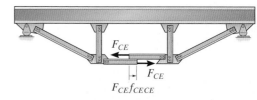

‖

2 kN/m

Δ_{CE}

Primary structure

+

F_{CE}

F_{CE}

$F_{CE}f_{CECE}$

Redundant $\mathbf{F}_{CE}$ applied

(b)

SOLUTION

Principle of Superposition. If the force in one of the truss members is known, then the force in all the other members, as well as in the beam, can be determined by statics. Hence, the structure is indeterminate to the first degree. For solution the force in member CE is chosen as the redundant. This member is therefore sectioned to eliminate its capacity to sustain a force. The principle of superposition applied to the structure is shown in Fig. 10–16b.

Compatibility Equation. With reference to the relative displacement of the cut ends of member CE, Fig. 10–16b, we require

$$0 = \Delta_{CE} + F_{CE}f_{CECE} \qquad (1)$$

EXAMPLE 10.9 (Continued)

The method of virtual work will be used to find Δ_{CE} and f_{CECE}. The necessary force analysis is shown in Figs. 10–16c and 10–16d.

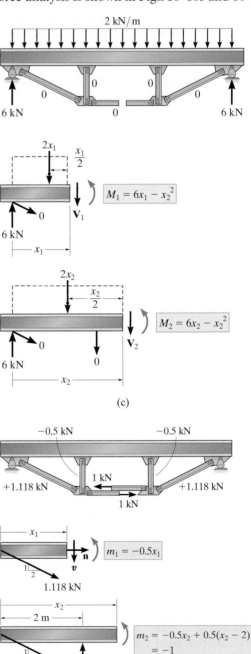

(c)

(d)

For Δ_{CE} we require application of the real loads, Fig. 10–16c, and a virtual unit load applied to the cut ends of member CE, Fig. 10–16d. Here we will use symmetry of *both* loading and geometry, and only consider the bending strain energy in the beam and, of course, the axial strain energy in the truss members. Thus,

$$\Delta_{CE} = \int_0^L \frac{Mm}{EI}dx + \sum \frac{nNL}{AE} = 2\int_0^2 \frac{(6x_1 - x_1^2)(-0.5x_1)dx_1}{EI}$$

$$+ 2\int_2^3 \frac{(6x_2 - x_2^2)(-1)dx_2}{EI} + 2\left(\frac{(1.118)(0)(\sqrt{5})}{AE}\right)$$

$$+ 2\left(\frac{(-0.5)(0)(1)}{AE}\right) + \left(\frac{1(0)2}{AE}\right)$$

$$= -\frac{12}{EI} - \frac{17.33}{EI} + 0 + 0 + 0$$

$$= \frac{-29.33(10^3)}{200(10^9)(20)(10^{-6})} = -7.333(10^{-3})\,\text{m}$$

For $f_{CE\,CE}$ we require application of a real unit load and a virtual unit load at the cut ends of member CE, Fig. 10–16d. Thus,

$$f_{CE\,CE} = \int_0^L \frac{m^2 dx}{EI} + \sum \frac{n^2 L}{AE} = 2\int_0^2 \frac{(-0.5x_1)^2 dx_1}{EI} + 2\int_2^3 \frac{(-1)^2 dx_2}{EI}$$

$$+ 2\left(\frac{(1.118)^2(\sqrt{5})}{AE}\right) + 2\left(\frac{(-0.5)^2(1)}{AE}\right) + \left(\frac{(1)^2(2)}{AE}\right)$$

$$= \frac{1.3333}{EI} + \frac{2}{EI} + \frac{5.590}{AE} + \frac{0.5}{AE} + \frac{2}{AE}$$

$$= \frac{3.333(10^3)}{200(10^9)(20)(10^{-6})} + \frac{8.090(10^3)}{400(10^{-6})(200(10^9))}$$

$$= 0.9345(10^{-3})\,\text{m/kN}$$

Substituting the data into Eq. (1) yields

$$0 = -7.333(10^{-3})\,\text{m} + F_{CE}(0.9345(10^{-3})\,\text{m/kN})$$

$$F_{CE} = 7.85\,\text{kN} \qquad\qquad\qquad \textit{Ans.}$$

10

10.8   Additional Remarks on the Force Method of Analysis

Now that the basic ideas regarding the force method have been developed, we will proceed to generalize its application and discuss its usefulness.

When computing the flexibility coefficients, f_{ij} (or α_{ij}), for the structure, it will be noticed that they depend only on the material and geometrical properties of the members and *not* on the loading of the primary structure. Hence these values, once determined, can be used to compute the reactions for any loading.

For a structure having n redundant reactions, $\mathbf{R}_n$, we can write n compatibility equations, namely:

$$\Delta_1 + f_{11}R_1 + f_{12}R_2 + \cdots + f_{1n}R_n = 0$$
$$\Delta_2 + f_{21}R_1 + f_{22}R_2 + \cdots + f_{2n}R_n = 0$$
$$\vdots$$
$$\Delta_n + f_{n1}R_1 + f_{n2}R_2 + \cdots + f_{nn}R_n = 0$$

Here the displacements, $\Delta_1, \ldots, \Delta_n$, are caused by *both* the *real loads* on the primary structure and by *support settlement* or *dimensional changes* due to temperature differences or fabrication errors in the members. To simplify computation for structures having a large degree of indeterminacy, the above equations can be recast into a matrix form,

$$\begin{bmatrix} f_{11} & f_{12} & \cdots & f_{1n} \\ f_{21} & f_{22} & \cdots & f_{2n} \\ & & \vdots & \\ f_{n1} & f_{n2} & \cdots & f_{nn} \end{bmatrix} \begin{bmatrix} R_1 \\ R_2 \\ \vdots \\ R_n \end{bmatrix} = - \begin{bmatrix} \Delta_1 \\ \Delta_2 \\ \vdots \\ \Delta_n \end{bmatrix} \qquad (10\text{--}2)$$

or simply

$$\mathbf{fR} = -\Delta$$

In particular, note that $f_{ij} = f_{ji}(f_{12} = f_{21}$, etc.), a consequence of Maxwell's theorem of reciprocal displacements (or Betti's law). Hence the *flexibility matrix* will be *symmetric,* and this feature is beneficial when solving large sets of linear equations, as in the case of a highly indeterminate structure.

Throughout this chapter we have determined the flexibility coefficients using the method of virtual work as it applies to the *entire structure*. It is possible, however, to obtain these coefficients for *each member* of the structure, and then, using transformation equations, to obtain their values for the entire structure. This approach is covered in books devoted to matrix analysis of structures, and will not be covered in this text.*

*See, for example, H. C. Martin, *Introduction to Matrix Methods of Structural Analysis,* McGraw-Hill, New York.

Although the details for applying the force method of analysis using computer methods will also be omitted here, we can make some general observations and comments that apply when using this method to solve problems that are highly indeterminate and thus involve large sets of equations. In this regard, numerical accuracy for the solution is improved if the flexibility coefficients located near the main diagonal of the $\mathbf{f}$ matrix are larger than those located off the diagonal. To achieve this, some thought should be given to selection of the primary structure. To facilitate computations of $\mathbf{f}_{ij}$, it is also desirable to choose the primary structure so that it is somewhat symmetric. This will tend to yield some flexibility coefficients that are similar or may be zero. Lastly, the deflected shape of the primary structure should be *similar* to that of the actual structure. If this occurs, then the redundants will induce only *small* corrections to the primary structure, which results in a more accurate solution of Eq. 10–2.

10.9   Symmetric Structures

A structural analysis of any highly indeterminate structure, or for that matter, even a statically determinate structure, can be simplified provided the designer or analyst can recognize those structures that are symmetric and support either symmetric or antisymmetric loadings. In a general sense, a structure can be classified as being *symmetric* provided half of it develops the same internal loadings and deflections as its mirror image reflected about its central axis. Normally symmetry requires the material composition, geometry, supports, and loading to be the same on each side of the structure. However, this does not always have to be the case. Notice that for horizontal stability a pin is required to support the beam and truss in Figs. 10–17a and 10–17b. Here the horizontal reaction at the pin is zero, and so both of these structures will deflect and produce the same internal loading as their reflected counterpart. As a result, they can be classified as being symmetric. Realize that this would not be the case for the frame, Fig. 10–17c, if the fixed support at A was replaced by a pin, since then the deflected shape and internal loadings would not be the same on its left and right sides.

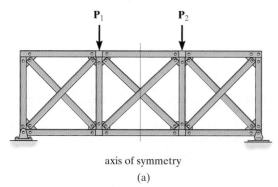

axis of symmetry

(a)

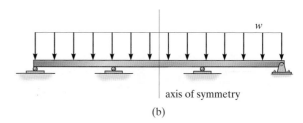

axis of symmetry

(b)

Fig. 10–17

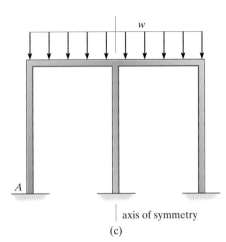

axis of symmetry

(c)

Fig. 10–17

Sometimes a symmetric structure supports an antisymmetric loading, that is, the loading on its reflected side has the opposite direction, such as shown by the two examples in Fig. 10–18. Provided the structure is symmetric and its loading is either symmetric or antisymmetric, then a structural analysis will only have to be performed on half the members of the structure since the same (symmetric) or opposite (antisymmetric) results will be produced on the other half. If a structure is symmetric and its applied loading is unsymmetrical, then it is possible to transform this loading into symmetric and antisymmetric components. To do this, *the loading is first divided in half, then it is reflected to the other side of the structure and both symmetric and antisymmetric components are produced*. For example, the loading on the beam in Fig. 10–19a is divided by two and reflected about the beam's axis of symmetry. From this, the symmetric and antisymmetric components of the load are produced as shown in Fig. 10–19b. When added together these components produce the original loading. A separate structural analysis can now be performed using the symmetric and antisymmetric loading components and the results superimposed to obtain the actual behavior of the structure.

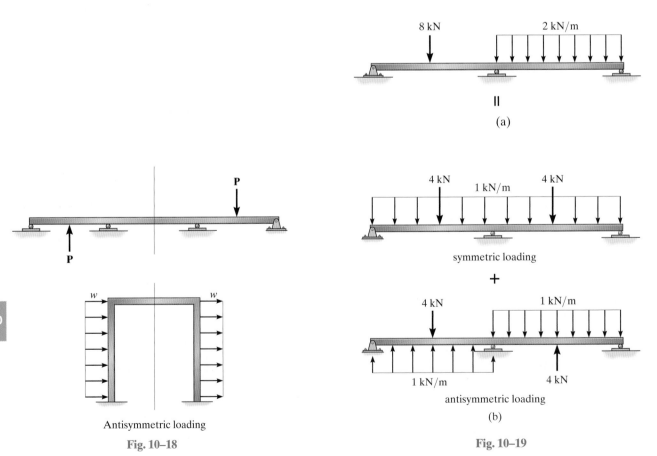

Antisymmetric loading

Fig. 10–18

symmetric loading

antisymmetric loading

(b)

Fig. 10–19

PROBLEMS

10–25. Determine the force in each member of the truss. *AE* is constant.

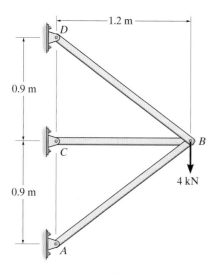

Prob. 10–25

10–26. Determine the force in each member of the truss. The cross-sectional area of each member is indicated in the figure. $E = 200$ GPa. Assume the members are pin connected at their ends.

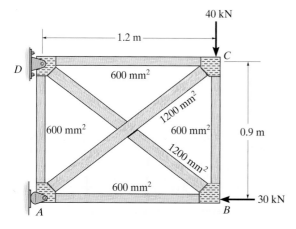

Prob. 10–26

10–27. Determine the force in member *AC* of the truss. *AE* is constant.

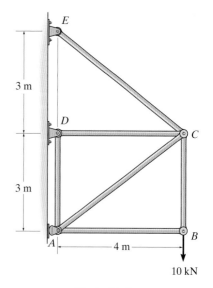

Prob. 10–27

*10–28. Determine the force in member *AD* of the truss. The cross-sectional area of each member is shown in the figure. Assume the members are pin connected at their ends. Take $E = 200$ GPa.

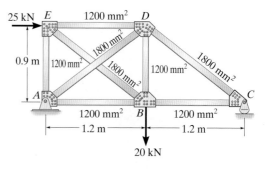

Prob. 10–28

10

10–29. Determine the force in each member of the truss. Assume the members are pin connected at their ends. AE is constant.

10–31. Determine the force in member CD of the truss. AE is constant.

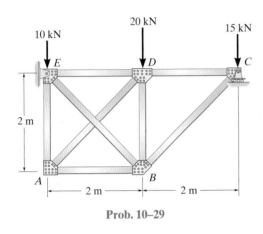

Prob. 10–29

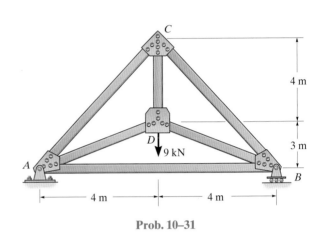

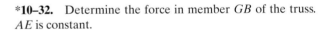

Prob. 10–31

10–30. Determine the force in each member of the pin-connected truss. AE is constant.

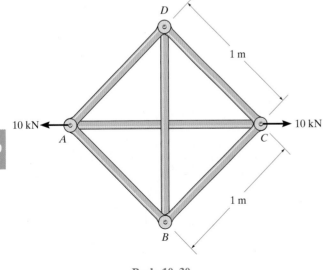

Prob. 10–30

*10–32. Determine the force in member GB of the truss. AE is constant.

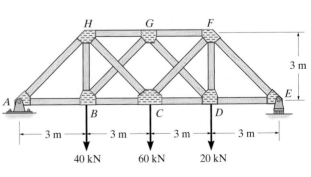

Prob. 10–32

10–33. The cantilevered beam AB is additionally supported using two tie rods. Determine the force in each of these rods. Neglect axial compression and shear in the beam. For the beam, $I_b = 200(10^6)$ mm^4, and for each tie rod, $A = 100$ mm^2. Take $E = 200$ GPa.

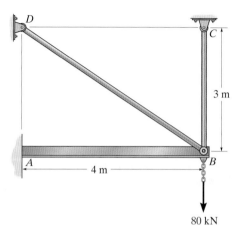

Prob. 10–33

10–34. Determine the force in member AB, BC and BD which is used in conjunction with the beam to carry the 150-kN load. The beam has a moment of inertia of $I = 240(10^6)$ mm^4 the members AB and BC each have a cross-sectional area of 1250 mm^2, and BD has a cross-sectional area of 2500 mm^2. Take $E = 200$ GPa. Neglect the thickness of the beam and its axial compression, and assume all members are pin connected. Also assume the support at A is a pin and E is a roller.

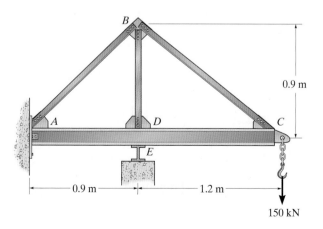

Prob. 10–34

10–35. The trussed beam supports the uniform distributed loading. If all the truss members have a cross-sectional area of 750 mm^2, determine the force in member BC. Neglect both the depth and axial compression in the beam. Take $E = 200$ GPa for all members. Also, for the beam $I_{AD} = 300(10^6)$ mm^4. Assume A is a pin and D is a rocker.

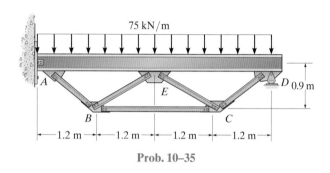

Prob. 10–35

*10–36. The trussed beam supports a concentrated force of 400 kN at its center. Determine the force in each of the three struts and draw the bending-moment diagram for the beam. The struts each have a cross-sectional area of 1250 mm^2. Assume they are pin connected at their end points. Neglect both the depth of the beam and the effect of axial compression in the beam. Take $E = 200$ GPa for both the beam and struts. Also, for the beam $I = 150(10^6)$ mm^4.

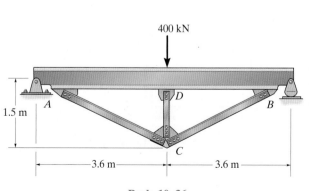

Prob. 10–36

10–37. Determine the reactions at support C. EI is constant for both beams.

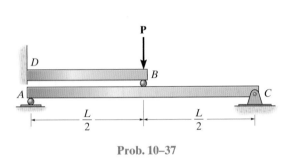

Prob. 10–37

10–38. The beam AB has a moment of inertia $I = 190(10^6)$ mm^4 and rests on the smooth supports at its ends. A 19-mm-diameter rod CD is welded to the center of the beam and to the fixed support at D. If the temperature of the rod is decreased by 65°C, determine the force developed in the rod. The beam and rod are both made of steel for which $E = 200$ GPa and $\alpha = 11.7(10^{-6})/°C$.

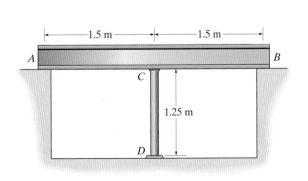

Prob. 10–38

10–39. The contilevered beam is supported at one end by a 12.5-mm-diameter suspender rod AC and fixed at the other end B. Determine the force in the rod due to a uniform loading of 60 kN/m. $E = 200$ GPa for both the beam and rod.

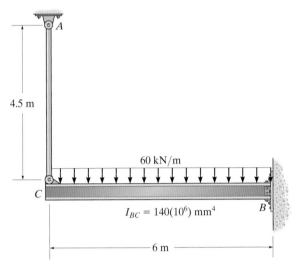

Prob. 10–39

*10–40. The structural assembly supports the loading shown. Draw the moment diagrams for each of the beams. Take $I = 100(10^6)$ mm^4 for the beams and $A = 200$ mm^2 for the tie rod. All members are made of steel for which $E = 200$ GPa.

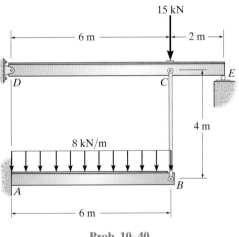

Prob. 10–40

10.10   Influence Lines for Statically Indeterminate Beams

In Sec. 6–3 we discussed the use of the Müller-Breslau principle for drawing the influence line for the reaction, shear, and moment at a point in a statically determinate beam. In this section we will extend this method and apply it to statically indeterminate beams.

Recall that, for a beam, *the Müller-Breslau principle states that the influence line for a function (reaction, shear, or moment) is to the same scale as the deflected shape of the beam when the beam is acted upon by the function*. To draw the deflected shape properly, the capacity of the beam to resist the applied function must be *removed* so the beam can deflect when the function is applied. For *statically determinate beams,* the deflected shapes (or the influence lines) will be a series of *straight line segments.* For *statically indeterminate beams, curves* will result. Construction of each of the three types of influence lines (reaction, shear, and moment) will now be discussed for a statically indeterminate beam. In each case we will illustrate the validity of the Müller-Breslau principle using Maxwell's theorem of reciprocal displacements.

Reaction at A.   To determine the influence line for the reaction at A in Fig. 10–20a, a unit load is placed on the beam at successive points, and at each point the reaction at A must be determined. A plot of these results yields the influence line. For example, when the load is at point D, Fig. 10–20a, the reaction at A, which represents the ordinate of the influence line at D, can be determined by the force method. To do this, the principle of superposition is applied, as shown in Figs. 10–20a through 10–20c. The compatibility equation for point A is thus $0 = f_{AD} + A_y f_{AA}$ or $A_y = -f_{AD}/f_{AA}$; however, by Maxwell's theorem of reciprocal displacements $f_{AD} = -f_{DA}$, Fig. 10–20d, so that we can also compute A_y (or the ordinate of the influence line at D) using the equation

$$A_y = \left(\frac{1}{f_{AA}}\right) f_{DA}$$

By comparison, the Müller-Breslau principle requires removal of the support at A and application of a vertical unit load. The resulting deflection curve, Fig. 10–20d, is to some scale the shape of the influence line for A_y. From the equation above, however, it is seen that the scale factor is $1/f_{AA}$.

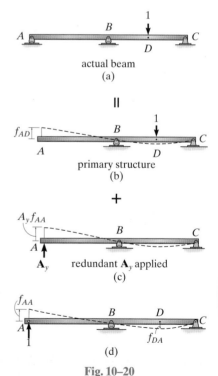

actual beam
(a)

$\parallel$

primary structure
(b)

$+$

redundant $\mathbf{A}_y$ applied
(c)

(d)

Fig. 10–20

Shear at E. If the influence line for the shear at point E of the beam in Fig. 10–21a is to be determined, then by the Müller-Breslau principle the beam is imagined cut open at this point and a *sliding device* is inserted at E, Fig. 10–21b. This device will transmit a moment and normal force but no shear. When the beam deflects due to positive unit shear loads acting at E, the slope on each side of the guide remains the same, and the deflection curve represents to some scale the influence line for the shear at E, Fig. 10–21c. Had the basic method for establishing the influence line for the shear at E been applied, it would then be necessary to apply a unit load at each point D and compute the internal shear at E, Fig. 10–21a. This value, V_E, would represent the ordinate of the influence line at D. Using the force method and Maxwell's theorem of reciprocal displacements, as in the previous case, it can be shown that

$$V_E = \left(\frac{1}{f_{EE}}\right) f_{DE}$$

This again establishes the validity of the Müller-Breslau principle, namely, a positive unit shear load applied to the beam at E, Fig. 10–21c, will cause the beam to deflect into the *shape* of the influence line for the shear at E. Here the scale factor is $(1/f_{EE})$.

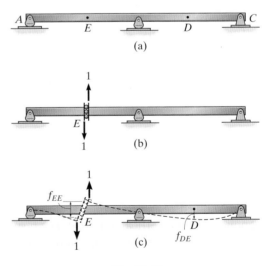

Fig. 10–21

Moment at E. The influence line for the moment at E in Fig. 10–22a can be determined by placing a *pin* or *hinge* at E, since this connection transmits normal and shear forces but cannot resist a moment, Fig. 10–22b. Applying a positive unit couple moment, the beam then deflects to the dashed position in Fig. 10–22c, which yields to some scale the influence line, again a consequence of the Müller-Breslau principle. Using the force method and Maxwell's reciprocal theorem, we can show that

$$M_E = \left(\frac{1}{\alpha_{EE}}\right)f_{DE}$$

The scale factor here is $(1/\alpha_{EE})$.

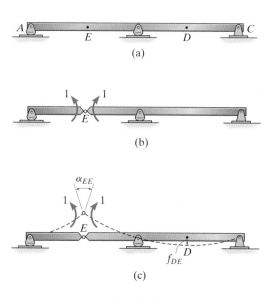

(a)

(b)

(c)

Fig. 10–22

Procedure for Analysis

The following procedure provides a method for establishing the influence line for the reaction, shear, or moment at a point on a beam using the Müller-Breslau technique.

Qualitative Influence Line

At the point on the beam for which the influence line is to be determined, place a connection that will remove the capacity of the beam to support the function of the influence line. If the function is a vertical *reaction,* use a vertical *roller guide;* if the function is *shear,* use a *sliding device;* or if the function is *moment,* use a *pin* or *hinge.* Place a unit load at the connection acting on the beam in the "positive direction" of the function. Draw the deflection curve for the beam. This curve represents to some scale the shape of the influence line for the beam.

Quantitative Influence Line

If numerical values of the influence line are to be determined, compute the *displacement* of successive points along the beam when the beam is subjected to the unit load placed at the connection mentioned above. Divide each value of displacement by the displacement determined at the point where the unit load acts. By applying this scalar factor, the resulting values are the ordinates of the influence line.

Influence lines for the continuous girder of this trestle were constructed in order to properly design the girder.

10.11   Qualitative Influence Lines for Frames

The Müller-Breslau principle provides a quick method and is of great value for establishing the general shape of the influence line for building frames. Once the influence-line *shape* is known, one can immediately specify the *location* of the live loads so as to create the greatest influence of the function (reaction, shear, or moment) in the frame. For example, the shape of the influence line for the *positive* moment at the center *I* of girder *FG* of the frame in Fig. 10–23*a* is shown by the dashed lines. Thus, uniform loads would be placed only on girders *AB*, *CD*, and *FG* in order to create the largest positive moment at *I*. With the frame loaded in this manner, Fig. 10–23*b*, an indeterminate analysis of the frame could then be performed to determine the critical moment at *I*.

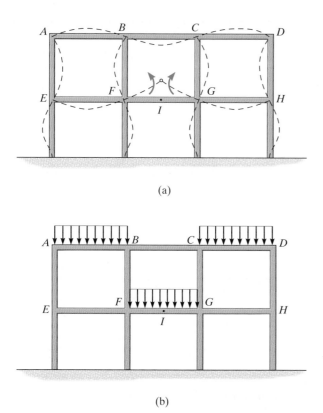

(a)

(b)

Fig. 10–23

EXAMPLE 10.10

Draw the influence line for the vertical reaction at A for the beam in Fig. 10–24a. EI is constant. Plot numerical values every 2 m.

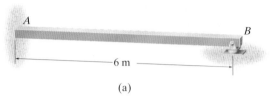

(a)

SOLUTION

The capacity of the beam to resist the reaction $\mathbf{A}_y$ is removed. This is done using a vertical roller device shown in Fig. 10–24b. Applying a vertical unit load at A yields the shape of the influence line shown in Fig. 10–24c.

In order to determine ordinates of the influence line we will use the conjugate-beam method. The reactions at A and B on the "real beam," when subjected to the unit load at A, are shown in Fig. 10–24b. The corresponding conjugate beam is shown in Fig. 10–24d. Notice that the support at A' remains the *same* as that for A in Fig. 10–24b. This is because a vertical roller device on the conjugate beam supports a moment but no shear, corresponding to a displacement but no slope at A on the real beam, Fig. 10–24c. The reactions at the supports of the conjugate beam have been computed and are shown in Fig. 10–24d. The displacements of points on the real beam, Fig. 10–24b, will now be computed.

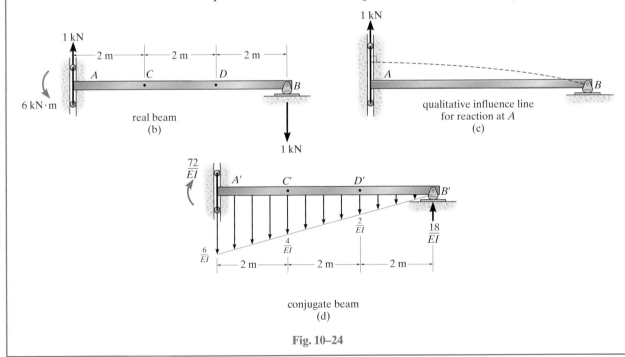

real beam
(b)

qualitative influence line
for reaction at A
(c)

conjugate beam
(d)

Fig. 10–24

For B', since no moment exists on the conjugate beam at B', Fig. 10–24d, then

$$\Delta_B = M_{B'} = 0$$

For D', Fig. 10–24e:

$$\Sigma M_{D'} = 0; \qquad \Delta_D = M_{D'} = \frac{18}{EI}(2) - \frac{1}{2}\left(\frac{2}{EI}\right)(2)\left(\frac{2}{3}\right) = \frac{34.67}{EI}$$

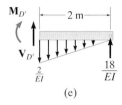

(e)

For C', Fig. 10–24f:

$$\Sigma M_{C'} = 0; \qquad \Delta_C = M_{C'} = \frac{18}{EI}(4) - \frac{1}{2}\left(\frac{4}{EI}\right)(4)\left(\frac{4}{3}\right) = \frac{61.33}{EI}$$

For A', Fig. 10–24d:

$$\Delta_A = M_{A'} = \frac{72}{EI}$$

Since a vertical 1-kN load acting at A on the beam in Fig.10–24a will cause a vertical reaction at A of 1 kN, the displacement at A, $\Delta_A = 72/EI$, should correspond to a numerical value of 1 for the influence-line ordinate at A. Thus, dividing the other computed displacements by this factor, we obtain

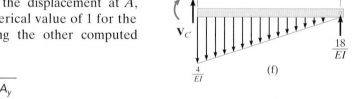

(f)

x	A_y
A	1
C	0.852
D	0.481
B	0

A plot of these values yields the influence line shown in Fig. 10–24g.

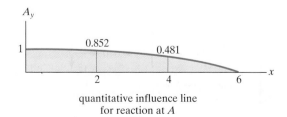

quantitative influence line
for reaction at A
(g)

EXAMPLE 10.11

Draw the influence line for the shear at D for the beam in Fig. 10–25a. EI is constant. Plot numerical values every 3 m.

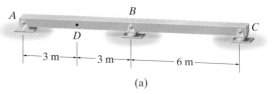

(a)

Fig. 10–25

SOLUTION

The capacity of the beam to resist shear at D is removed. This is done using the roller device shown in Fig. 10–25b. Applying a positive unit shear at D yields the shape of the influence line shown in Fig. 10–25c.

The support reactions at A, B, and C on the "real beam" when subjected to the unit shear at D are shown in Fig. 10–25b. The corresponding conjugate beam is shown in Fig. 10–25d. Here an external couple moment $\mathbf{M}_{D'}$ must be applied at D' in order to cause a different *internal moment* just to the left and just to the right of D'. These internal moments correspond to the displacements just to the left and just to the right of D on the real beam, Fig. 10–25c. The reactions at the supports A', B', C' and the external moment $\mathbf{M}_{D'}$ on the conjugate beam have been computed and are shown in Fig. 10–25e. As an exercise verify the calculations.

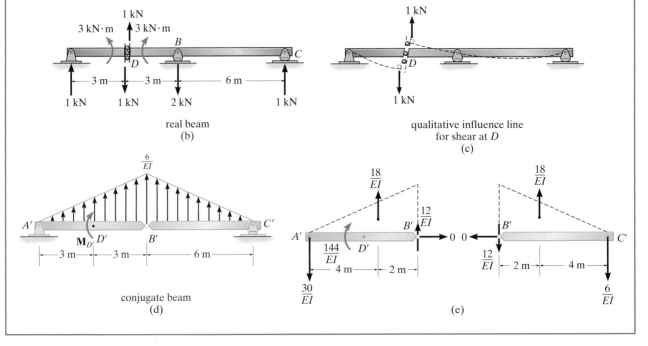

Since there is a *discontinuity* of moment at D', the internal moment just to the left and right of D' will be computed. Just to the left of D', Fig. 10–25f, we have

$$\Sigma M_{D'_L} = 0; \qquad \Delta_{D_L} = M_{D'_L} = \frac{4.5}{EI}(1) - \frac{30}{EI}(3) = -\frac{85.5}{EI}$$

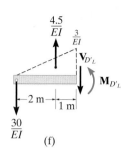

(f)

Just to the right of D', Fig. 10–25g, we have

$$\Sigma M_{D'_R} = 0; \qquad \Delta_{D_R} = M_{D'_R} = \frac{4.5}{EI}(1) - \frac{30}{EI}(3) + \frac{144}{EI} = \frac{58.5}{EI}$$

From Fig. 10–25e,

$$\Delta_A = M_{A'} = 0 \qquad \Delta_B = M_{B'} = 0 \qquad \Delta_C = M_{C'} = 0$$

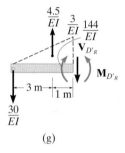

(g)

For point E, Fig. 10–25b, using the method of sections at the corresponding point E' on the conjugate beam, Fig. 10–25h, we have

$$\Sigma M_{E'} = 0; \qquad \Delta_E = M_{E'} = \frac{4.5}{EI}(1) - \frac{6}{EI}(3) = -\frac{13.5}{EI}$$

The ordinates of the influence line are obtained by dividing each of the above values by the scale factor $M_{D'} = 144/EI$. We have

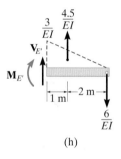

(h)

x	V_D
A	0
D_L	−0.594
D_R	0.406
B	0
E	−0.0938
C	0

A plot of these values yields the influence line shown in Fig. 10–25i.

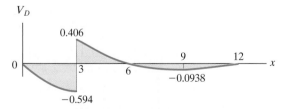

quantitative influence line
for shear at D
(i)

EXAMPLE 10.12

Draw the influence line for the moment at D for the beam in Fig. 10–26a. EI is constant. Plot numerical values every 3 m.

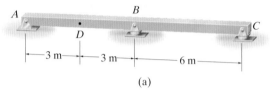

(a)

Fig. 10–26

SOLUTION

A hinge is inserted at D in order to remove the capacity of the beam to resist moment at this point, Fig. 10–26b. Applying positive unit couple moments at D yields the influence line shown in Fig. 10–26c.

The reactions at A, B, and C on the "real beam" when subjected to the unit couple moments at D are shown in Fig. 10–26b. The corresponding conjugate beam and its reactions are shown in Fig. 10–26d. It is suggested that the reactions be verified in both cases. From Fig. 10–26d, note that

$$\Delta_A = M_{A'} = 0 \qquad \Delta_B = M_{B'} = 0 \qquad \Delta_C = M_{C'} = 0$$

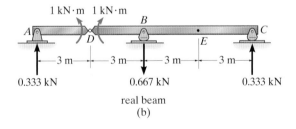

real beam
(b)

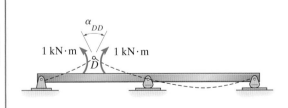

qualitative influence line for moment at D
(c)

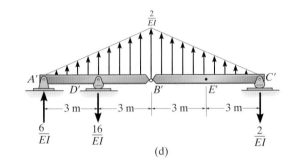

(d)

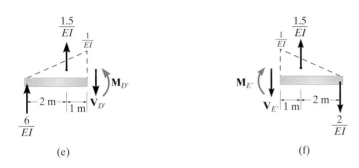

(e)   (f)

For point D', Fig. 10–26e:

$$\Sigma M_{D'} = 0; \qquad \Delta_D = M_{D'} = \frac{1.5}{EI}(1) + \frac{6}{EI}(3) = \frac{19.5}{EI}$$

For point E', Fig. 10–26f:

$$\Sigma M_{E'} = 0; \qquad \Delta_E = M_{E'} = \frac{1.5}{EI}(1) - \frac{2}{EI}(3) = -\frac{4.5}{EI}$$

The angular displacement α_{DD} at D of the "real beam" in Fig. 10–26c is defined by the reaction at D' on the conjugate beam. This factor, $D'_y = 16/EI$, is divided into the above values to give the ordinates of the influence line, that is,

x	M_D
A	0
D	1.219
B	0
E	−0.281
C	0

A plot of these values yields the influence line shown in Fig. 10–26g.

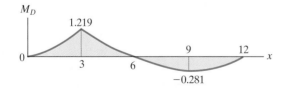

quantitative influence line
for moment at D
(g)

PROBLEMS

10–41. Draw the influence line for the reaction at C. Plot numerical values at the peaks. Assume A is a pin and B and C are rollers. EI is constant.

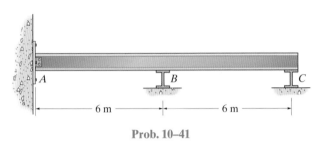

Prob. 10–41

10–42. Draw the influence line for the moment at A. Plot numerical values at the peaks. Assume A is fixed and the support at B is a roller. EI is constant.

10–43. Draw the influence line for the vertical reaction at B. Plot numerical values at the peaks. Assume A is fixed and the support at B is a roller. EI is constant.

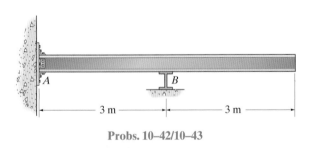

Probs. 10–42/10–43

*10–44. Draw the influence line for the shear at C. Plot numerical values every 1.5 m. Assume A is fixed and the support at B is a roller. EI is constant.

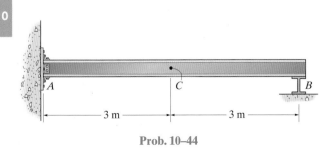

Prob. 10–44

10–45. Draw the influence line for the reaction at C. Plot the numerical values every 5 m. EI is constant.

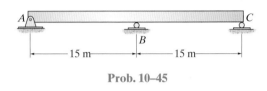

Prob. 10–45

10–46. Sketch the influence line for (a) the moment at E, (b) the reaction at C, and (c) the shear at E. In each case, indicate on a sketch of the beam where a uniform distributed live load should be placed so as to cause a maximum positive value of these functions. Assume the beam is fixed at D.

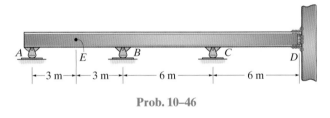

Prob. 10–46

10–47. Sketch the influence line for (a) the vertical reaction at C, (b) the moment at B, and (c) the shear at E. In each case, indicate on a sketch of the beam where a uniform distributed live load should be placed so as to cause a maximum positive value of these functions. Assume the beam is fixed at F.

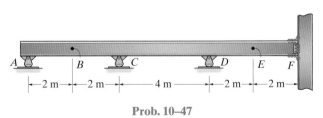

Prob. 10–47

*10–48. Use the Müller-Breslau principle to sketch the general shape of the influence line for (a) the moment at A and (b) the shear at B.

10–50. Use the Müller-Breslau principle to sketch the general shape of the influence line for (a) the moment at A and (b) the shear at B.

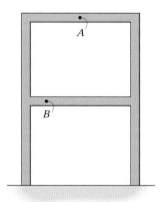

Prob. 10–48

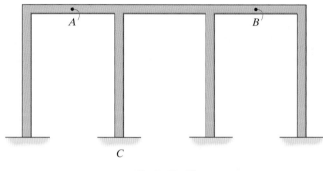

Prob. 10–50

10–51. Use the Müller-Breslau principle to sketch the general shape of the influence line for (a) the moment at A and (b) the shear at B.

10–49. Use the Müller-Breslau principle to sketch the general shape of the influence line for (a) the moment at A and (b) the shear at B.

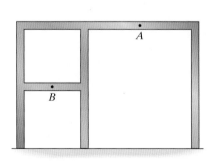

Prob. 10–49

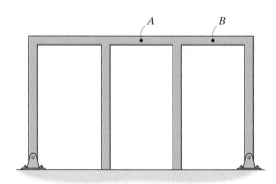

Prob. 10–51

10

CHAPTER REVIEW

The analysis of a statically indeterminate structure requires satisfying equilibrium, compatibility, and the force-displacement relationships for the structure. A force method of analysis consists of writing equations that satisfy compatibility and the force-displacement requirements, which then gives a direct solution for the redundant reactions. Once obtained, the remaining reactions are found from the equilibrium equations.

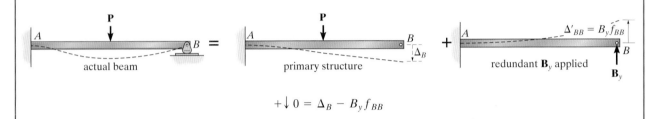

actual beam   =   primary structure   +   redundant $\mathbf{B}_y$ applied

$$+\downarrow 0 = \Delta_B - B_y f_{BB}$$

(Ref: Section 10.2)

Simplification of the force method is possible, using Maxwell's theorem of reciprocal displacements, which states that the displacement of a point B on a structure due to a unit load acting at point A, f_{BA}, is equal to the displacement of point A when the load acts at B, f_{AB}.

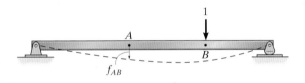

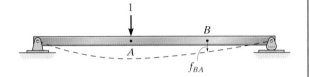

$$f_{BA} = f_{AB}$$

(Ref: Section 10.3)

The analysis of a statically indeterminate structure can be simplified if the structure has symmetry of material, geometry, and loading about its central axis. In particular, structures having an asymmetric loading can be replaced with a superposition of a symmetric and antisymmetric load.

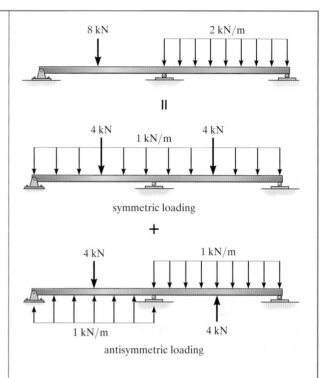

symmetric loading

antisymmetric loading

(Ref: Section 10.9)

Influence lines for statically indeterminate structures will consist of *curved lines*. They can be sketched using the Müller-Breslau principle, which states that the influence line shape for either the reaction, shear, or moment is to the same scale as the deflected shape of the structure when it is acted upon by the reaction, shear, or moment, respectively. By using Maxwell's theorem of reciprocal deflections, it is possible to obtain specific values of the ordinates of an influence line.

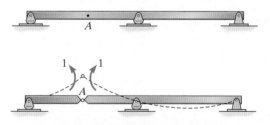

influence line shape for moment at *A*

10

(Ref: Section 10.10)

The members of this building frame are all fixed connected, so the framework is statically indeterminate.

Displacement Method of Analysis: Slope-Deflection Equations

11

In this chapter we will briefly outline the basic ideas for analyzing structures using the displacement method of analysis. Once these concepts have been presented, we will develop the general equations of slope deflection and then use them to analyze statically indeterminate beams and frames.

11.1 Displacement Method of Analysis: General Procedures

All structures must satisfy equilibrium, load-displacement, and compatibility of displacements requirements in order to ensure their safety. It was stated in Sec. 10–1 that there are two different ways to satisfy these requirements when analyzing a statically indeterminate structure. The force method of analysis, discussed in the previous chapter, is based on identifying the unknown redundant forces and then satisfying the structure's compatibility equations. This is done by expressing the displacements in terms of the loads by using the load-displacement relations. The solution of the resultant equations yields the redundant reactions, and then the equilibrium equations are used to determine the remaining reactions on the structure.

The *displacement method* works the opposite way. It first requires satisfying equilibrium equations for the structure. To do this the unknown displacements are written in terms of the loads by using the load-displacement relations, then these equations are solved for the displacements. Once the displacements are obtained, the unknown loads are determined from the compatibility equations using the load-displacement relations. Every displacement method follows this general procedure. In this chapter, the procedure will be generalized to produce the slope-deflection equations. In Chapter 12, the moment-distribution method will be developed. This method sidesteps the calculation of the displacements and instead makes it possible to apply a series of converging corrections that allow direct calculation of the end moments. Finally, in Chapters 14, 15, and 16, we will illustrate how to apply this method using matrix analysis, making it suitable for use on a computer.

In the discussion that follows we will show how to identify the unknown displacements in a structure and we will develop some of the important load-displacement relations for beam and frame members. The results will be used in the next section and in later chapters as the basis for applying the displacement method of analysis.

Degrees of Freedom.

When a structure is loaded, specified points on it, called *nodes*, will undergo unknown displacements. These displacements are referred to as the *degrees of freedom* for the structure, and in the displacement method of analysis it is important to specify these degrees of freedom since they become the unknowns when the method is applied. The number of these unknowns is referred to as the degree in which the structure is kinematically indeterminate.

To determine the kinematic indeterminacy we can imagine the structure to consist of a series of members connected to nodes, which are usually located at *joints, supports,* at the *ends* of a member, or where the members have a sudden *change in cross section.* In three dimensions, each node on a frame or beam can have at most three linear displacements and three rotational displacements; and in two dimensions, each node can have at most two linear displacements and one rotational displacement. Furthermore, nodal displacements may be restricted by the supports, or due to assumptions based on the behavior of the structure. For example, if the structure is a beam and only deformation due to bending is considered, then there can be no linear displacement along the axis of the beam since this displacement is caused by axial-force deformation.

To clarify these concepts we will consider some examples, beginning with the beam in Fig. 11–1a. Here any load **P** applied to the beam will cause node A only to rotate (neglecting axial deformation), while node B is completely restricted from moving. Hence the beam has only one unknown degree of freedom, θ_A, and is therefore kinematically indeterminate to the first degree. The beam in Fig. 11–1b has nodes at A, B, and C, and so has four degrees of freedom, designated by the rotational displacements θ_A, θ_B, θ_C, and the vertical displacement Δ_C. It is kinematically indeterminate to the fourth degree. Consider now the frame in Fig. 11–1c. Again, if we neglect axial deformation of the members, an arbitrary loading **P** applied to the frame can cause nodes B and C to rotate, and these nodes can be displaced horizontally by an *equal* amount. The frame therefore has three degrees of freedom, θ_B, θ_C, Δ_B, and thus it is kinematically indeterminate to the third degree.

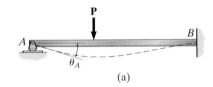

(a)

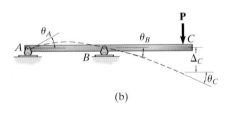

(b)

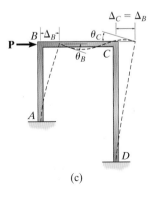

(c)

Fig. 11–1

In summary, specifying the kinematic indeterminacy or the number of unconstrained degrees of freedom for the structure is a necessary first step when applying a displacement method of analysis. It identifies the number of unknowns in the problem, based on the assumptions made regarding the deformation behavior of the structure. Furthermore, once these nodal displacements are known, the deformation of the structural members can be completely specified, and the loadings within the members obtained.

11.2   Slope-Deflection Equations

As indicated previously, the method of consistent displacements studied in Chapter 10 is called a force method of analysis, because it requires writing equations that relate the unknown forces or moments in a structure. Unfortunately, its use is limited to structures which are *not* highly indeterminate. This is because much work is required to set up the compatibility equations, and furthermore each equation written involves *all the unknowns,* making it difficult to solve the resulting set of equations unless a computer is available. By comparison, the slope-deflection method is not as involved. As we shall see, it requires less work both to write the necessary equations for the solution of a problem and to solve these equations for the unknown displacements and associated internal loads. Also, the method can be easily programmed on a computer and used to analyze a wide range of indeterminate structures.

The slope-deflection method was originally developed by Heinrich Manderla and Otto Mohr for the purpose of studying secondary stresses in trusses. Later, in 1915, G. A. Maney developed a refined version of this technique and applied it to the analysis of indeterminate beams and framed structures.

General Case.   The slope-deflection method is so named since it relates the unknown slopes and deflections to the applied load on a structure. In order to develop the general form of the slope-deflection equations, we will consider the typical span AB of a continuous beam as shown in Fig. 11–2, which is subjected to the arbitrary loading and has a constant EI. We wish to relate the beam's internal end moments M_{AB} and M_{BA} in terms of its three degrees of freedom, namely, its angular displacements θ_A and θ_B, and linear displacement Δ which could be caused by a relative settlement between the supports. Since we will be developing a formula, *moments* and *angular displacements* will be considered *positive* when they act *clockwise on the span,* as shown in Fig. 11–2. Furthermore, the *linear displacement* Δ is considered *positive* as shown, since this displacement causes the cord of the span and the span's cord angle ψ to rotate *clockwise.*

The slope-deflection equations can be obtained by using the principle of superposition by considering *separately* the moments developed at each support due to each of the displacements, θ_A, θ_B, and Δ, and then the loads.

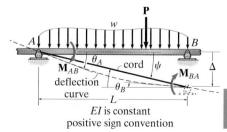

EI is constant
positive sign convention

Fig. 11–2

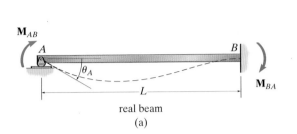

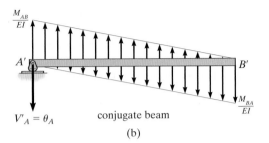

real beam
(a)

conjugate beam
(b)

Fig. 11–3

Angular Displacement at A, θ_A.

Consider node A of the member shown in Fig. 11–3a to rotate θ_A while its far-end node B is *held fixed*. To determine the moment M_{AB} needed to cause this displacement, we will use the conjugate-beam method. For this case the conjugate beam is shown in Fig. 11–3b. Notice that the end shear at A' acts downward on the beam, since θ_A is clockwise. The deflection of the "real beam" in Fig. 11–3a is to be zero at A and B, and therefore the corresponding sum of the *moments* at each end A' and B' of the conjugate beam must also be zero. This yields

$$\zeta + \Sigma M_{A'} = 0; \qquad \left[\frac{1}{2}\left(\frac{M_{AB}}{EI}\right)L\right]\frac{L}{3} - \left[\frac{1}{2}\left(\frac{M_{BA}}{EI}\right)L\right]\frac{2L}{3} = 0$$

$$\zeta + \Sigma M_{B'} = 0; \qquad \left[\frac{1}{2}\left(\frac{M_{BA}}{EI}\right)L\right]\frac{L}{3} - \left[\frac{1}{2}\left(\frac{M_{AB}}{EI}\right)L\right]\frac{2L}{3} + \theta_A L = 0$$

from which we obtain the following load-displacement relationships.

$$\boxed{M_{AB} = \frac{4EI}{L}\theta_A} \qquad\qquad (11\text{–}1)$$

$$\boxed{M_{BA} = \frac{2EI}{L}\theta_A} \qquad\qquad (11\text{–}2)$$

Angular Displacement at B, θ_B.

In a similar manner, if end B of the beam rotates to its final position θ_B, while end A is *held fixed*, Fig. 11–4, we can relate the applied moment M_{BA} to the angular displacement θ_B and the reaction moment M_{AB} at the wall. The results are

$$\boxed{M_{BA} = \frac{4EI}{L}\theta_B} \qquad\qquad (11\text{–}3)$$

$$\boxed{M_{AB} = \frac{2EI}{L}\theta_B} \qquad\qquad (11\text{–}4)$$

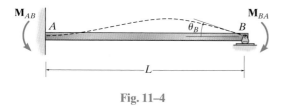

Fig. 11–4

Relative Linear Displacement, Δ.

If the far node B of the member is displaced relative to A, so that the cord of the member rotates clockwise (positive displacement) and yet both ends do not rotate, then equal but opposite moment and shear reactions are developed in the member, Fig. 11–5a. As before, the moment $\mathbf{M}$ can be related to the displacement Δ using the conjugate-beam method. In this case, the conjugate beam, Fig. 11–5b, is free at both ends, since the real beam (member) is fixed supported. However, due to the *displacement* of the real beam at B, the *moment* at the end B' of the conjugate beam must have a magnitude of Δ as indicated.* Summing moments about B', we have

$$\downarrow + \Sigma M_{B'} = 0; \qquad \left[\frac{1}{2}\frac{M}{EI}(L)\left(\frac{2}{3}L\right)\right] - \left[\frac{1}{2}\frac{M}{EI}(L)\left(\frac{1}{3}L\right)\right] - \Delta = 0$$

$$\boxed{M_{AB} = M_{BA} = M = \frac{-6EI}{L^2}\Delta} \qquad (11\text{--}5)$$

By our sign convention, this induced moment is negative since for equilibrium it acts counterclockwise on the member.

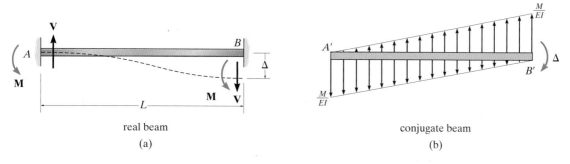

real beam
(a)

conjugate beam
(b)

Fig. 11–5

11

*The moment diagrams shown on the conjugate beam were determined by the method of superposition for a simply supported beam, as explained in Sec. 4–5.

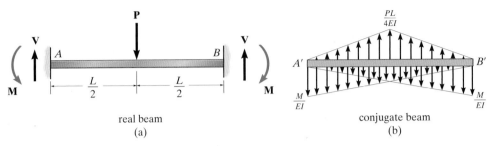

real beam
(a)

conjugate beam
(b)

Fig. 11–6

Fixed-End Moments.

In the previous cases we have considered relationships between the displacements and the necessary moments M_{AB} and M_{BA} acting at nodes A and B, respectively. In general, however, the linear or angular displacements of the nodes are caused by loadings acting on the *span* of the member, not by moments acting at its nodes. In order to develop the slope-deflection equations, we must transform these *span loadings* into equivalent moments acting at the nodes and then use the load-displacement relationships just derived. This is done simply by finding the reaction moment that each load develops at the nodes. For example, consider the fixed-supported member shown in Fig. 11–6a, which is subjected to a concentrated load **P** at its center. The conjugate beam for this case is shown in Fig. 11–6b. Since we require the slope at each end to be zero,

$$+\uparrow \Sigma F_y = 0; \qquad \left[\frac{1}{2}\left(\frac{PL}{4EI}\right)L\right] - 2\left[\frac{1}{2}\left(\frac{M}{EI}\right)L\right] = 0$$

$$M = \frac{PL}{8}$$

This moment is called a *fixed-end moment* (FEM). Note that according to our sign convention, it is negative at node A (counterclockwise) and positive at node B (clockwise). For convenience in solving problems, fixed-end moments have been calculated for other loadings and are tabulated on the inside back cover of the book. Assuming these FEMs have been determined for a specific problem (Fig. 11–7), we have

$$M_{AB} = (\text{FEM})_{AB} \qquad M_{BA} = (\text{FEM})_{BA} \qquad (11\text{–}6)$$

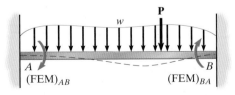

Fig. 11–7

Slope-Deflection Equation. If the end moments due to each displacement (Eqs. 11–1 through 11–5) and the loading (Eq. 11–6) are added together, the resultant moments at the ends can be written as

$$M_{AB} = 2E\left(\frac{I}{L}\right)\left[2\theta_A + \theta_B - 3\left(\frac{\Delta}{L}\right)\right] + (\text{FEM})_{AB}$$

$$M_{BA} = 2E\left(\frac{I}{L}\right)\left[2\theta_B + \theta_A - 3\left(\frac{\Delta}{L}\right)\right] + (\text{FEM})_{BA}$$

$$(11\text{–}7)$$

This pedestrian bridge has a reinforced concrete deck. Since it extends over all its supports, it is indeterminate to the second degree. The slope deflection equations provide a convenient method for finding the internal moments in each span.

Since these two equations are similar, the result can be expressed as a single equation. Referring to one end of the span as the near end (*N*) and the other end as the far end (*F*), and letting the *member stiffness* be represented as $k = I/L$, and the *span's cord rotation* as ψ (psi) $= \Delta/L$, we can write

$$\boxed{\begin{array}{c} M_N = 2Ek(2\theta_N + \theta_F - 3\psi) + (\text{FEM})_N \\ \text{For Internal Span or End Span with Far End Fixed} \end{array}}$$

$$(11\text{–}8)$$

where

$M_N =$ internal moment in the near end of the span; this moment is *positive clockwise* when acting on the span.

$E, k =$ modulus of elasticity of material and span stiffness $k = I/L$.

$\theta_N, \theta_F =$ near- and far-end slopes or angular displacements of the span at the supports; the angles are measured in *radians* and are *positive clockwise*.

$\psi =$ span rotation of its cord due to a linear displacement, that is, $\psi = \Delta/L$; this angle is measured in *radians* and is *positive clockwise*.

$(\text{FEM})_N =$ fixed-end moment at the near-end support; the moment is *positive clockwise* when acting on the span; refer to the table on the inside back cover for various loading conditions.

From the derivation Eq. 11–8 is both a compatibility and load-displacement relationship found by considering only the effects of bending and neglecting axial and shear deformations. It is referred to as the general *slope-deflection equation*. When used for the solution of problems, this equation is applied *twice* for each member span (*AB*); that is, application is from *A* to *B* and from *B* to *A* for span *AB* in Fig. 11–2.

11

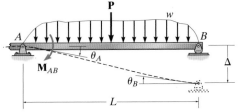

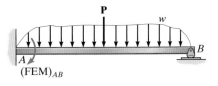

(a)

(b)

Fig. 11–8

Pin-Supported End Span. Occasionally an end span of a beam or frame is supported by a pin or roller at its *far end*, Fig. 11–8a. When this occurs, the moment at the roller or pin must be zero; and provided the angular displacement θ_B at this support does not have to be determined, we can modify the general slope-deflection equation so that it has to be applied *only once* to the span rather than twice. To do this we will apply Eq. 11–8 or Eqs. 11–7 to each end of the beam in Fig. 11–8. This results in the following two equations:

$$M_N = 2Ek(2\theta_N + \theta_F - 3\psi) + (FEM)_N$$
$$0 = 2Ek(2\theta_F + \theta_N - 3\psi) + 0 \qquad (11\text{–}9)$$

Here the $(FEM)_F$ is equal to zero since the far end is pinned, Fig. 11–8b. Furthermore, the $(FEM)_N$ can be obtained, for example, using the table in the right-hand column on the inside back cover of this book. Multiplying the first equation by 2 and subtracting the second equation from it *eliminates* the unknown θ_F and yields

$$\boxed{M_N = 3Ek(\theta_N - \psi) + (FEM)_N}$$
Only for End Span with Far End Pinned or Roller Supported   (11–10)

Since the moment at the far end is zero, only *one* application of this equation is necessary for the end span. This simplifies the analysis since the general equation, Eq. 11–8, would require *two* applications for this span and therefore involve the (extra) unknown angular displacement θ_B (or θ_F) at the end support.

To summarize application of the slope-deflection equations, consider the continuous beam shown in Fig. 11–9 which has four degrees of freedom. Here Eq. 11–8 can be applied twice to each of the three spans, i.e., from A to B, B to A, B to C, C to B, C to D, and D to C. These equations would involve the four unknown rotations, θ_A, θ_B, θ_C, θ_D. Since the end moments at A and D are zero, however, it is not necessary to determine θ_A and θ_D. A shorter solution occurs if we apply Eq. 11–10 from B to A and C to D and then apply Eq. 11–8 from B to C and C to B. These four equations will involve only the unknown rotations θ_B and θ_C.

11

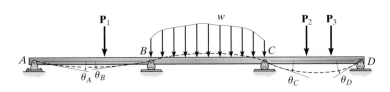

Fig. 11–9

11.3   Analysis of Beams

Procedure for Analysis

Degrees of Freedom

Label all the supports and joints (nodes) in order to identify the spans of the beam or frame between the nodes. By drawing the deflected shape of the structure, it will be possible to identify the number of degrees of freedom. Here each node can possibly have an angular displacement and a linear displacement. *Compatibility* at the nodes is maintained provided the members that are fixed connected to a node undergo the same displacements as the node. If these displacements are unknown, and in general they will be, then for convenience *assume* they act in the *positive direction* so as to cause *clockwise* rotation of a member or joint, Fig. 11–2.

Slope-Deflection Equations

The slope-deflection equations relate the unknown moments applied to the nodes to the displacements of the nodes for any span of the structure. If a load exists on the span, compute the FEMs using the table given on the inside back cover. Also, if a node has a linear displacement, Δ, compute $\psi = \Delta/L$ for the adjacent spans. Apply Eq. 11–8 to each end of the span, thereby generating *two* slope-deflection equations for each span. However, if a span at the *end* of a continuous beam or frame is pin supported, apply Eq. 11–10 only to the restrained end, thereby generating *one* slope-deflection equation for the span.

Equilibrium Equations

Write an equilibrium equation for each unknown degree of freedom for the structure. Each of these equations should be expressed in terms of unknown internal moments as specified by the slope-deflection equations. For beams and frames write the moment equation of equilibrium at each support, and for frames also write joint moment equations of equilibrium. If the frame sideways or deflects horizontally, column shears should be related to the moments at the ends of the column. This is discussed in Sec. 11.5.

   Substitute the slope-deflection equations into the equilibrium equations and solve for the unknown joint displacements. These results are then substituted into the slope-deflection equations to determine the internal moments at the ends of each member. If any of the results are *negative*, they indicate *counterclockwise* rotation; whereas *positive* moments and displacements are applied *clockwise*.

11

EXAMPLE 11.1

Draw the shear and moment diagrams for the beam shown in Fig. 11–10a. EI is constant.

6 kN/m

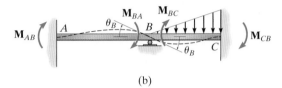

(a)

(b)

Fig. 11–10

SOLUTION

Slope-Deflection Equations.   Two spans must be considered in this problem. Since there is *no* span having the far end pinned or roller supported, Eq. 11–8 applies to the solution. Using the formulas for the FEMs tabulated for the triangular loading given on the inside back cover, we have

$$(\text{FEM})_{BC} = -\frac{wL^2}{30} = -\frac{6(6)^2}{30} = -7.2 \text{ kN} \cdot \text{m}$$

$$(\text{FEM})_{CB} = \frac{wL^2}{20} = \frac{6(6)^2}{20} = 10.8 \text{ kN} \cdot \text{m}$$

Note that $(\text{FEM})_{BC}$ is negative since it acts counterclockwise *on the beam* at B. Also, $(\text{FEM})_{AB} = (\text{FEM})_{BA} = 0$ since there is no load on span AB.

In order to identify the unknowns, the elastic curve for the beam is shown in Fig. 11–10b. As indicated, there are four unknown internal moments. Only the slope at B, θ_B, is unknown. Since A and C are fixed supports, $\theta_A = \theta_C = 0$. Also, since the supports do not settle, nor are they displaced up or down, $\psi_{AB} = \psi_{BC} = 0$. For span AB, considering A to be the near end and B to be the far end, we have

$$M_N = 2E\left(\frac{I}{L}\right)(2\theta_N + \theta_F - 3\psi) + (\text{FEM})_N$$

$$M_{AB} = 2E\left(\frac{I}{8}\right)[2(0) + \theta_B - 3(0)] + 0 = \frac{EI}{4}\theta_B \qquad (1)$$

Now, considering B to be the near end and A to be the far end, we have

$$M_{BA} = 2E\left(\frac{I}{8}\right)[2\theta_B + 0 - 3(0)] + 0 = \frac{EI}{2}\theta_B \qquad (2)$$

In a similar manner, for span BC we have

$$M_{BC} = 2E\left(\frac{I}{6}\right)[2\theta_B + 0 - 3(0)] - 7.2 = \frac{2EI}{3}\theta_B - 7.2 \qquad (3)$$

$$M_{CB} = 2E\left(\frac{I}{6}\right)[2(0) + \theta_B - 3(0)] + 10.8 = \frac{EI}{3}\theta_B + 10.8 \qquad (4)$$

11

Equilibrium Equations. The above four equations contain five unknowns. The necessary fifth equation comes from the condition of moment equilibrium at support B. The free-body diagram of a segment of the beam at B is shown in Fig. 11–10c. Here $\mathbf{M}_{BA}$ and $\mathbf{M}_{BC}$ are assumed to act in the positive direction to be consistent with the slope-deflection equations.* The beam shears contribute negligible moment about B since the segment is of differential length. Thus,

$$\zeta + \Sigma M_B = 0; \qquad M_{BA} + M_{BC} = 0 \qquad\qquad (5)$$

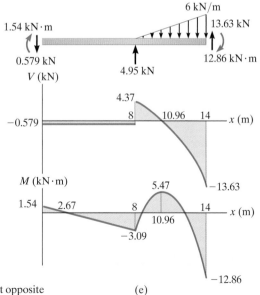

(c)

To solve, substitute Eqs. (2) and (3) into Eq. (5), which yields

$$\theta_B = \frac{6.17}{EI}$$

Resubstituting this value into Eqs. (1)–(4) yields

$$M_{AB} = 1.54 \ \text{kN} \cdot \text{m}$$

$$M_{BA} = 3.09 \ \text{kN} \cdot \text{m}$$

$$M_{BC} = -3.09 \ \text{kN} \cdot \text{m}$$

$$M_{CB} = 12.86 \ \text{kN} \cdot \text{m}$$

The negative value for M_{BC} indicates that this moment acts counter-clockwise on the beam, not clockwise as shown in Fig. 11–10b.

Using these results, the shears at the end spans are determined from the equilibrium equations, Fig. 11–10d. The free-body diagram of the entire beam and the shear and moment diagrams are shown in Fig. 11–10e.

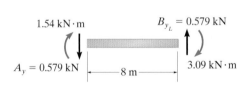

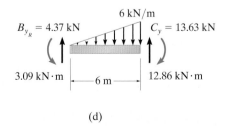

(d)

*Clockwise on the beam segment, but—by the principle of action, equal but opposite reaction—counterclockwise on the support.

(e)

11

EXAMPLE 11.2

Draw the shear and moment diagrams for the beam shown in Fig. 11–11a. EI is constant.

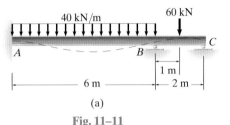

(a)

Fig. 11–11

SOLUTION

Slope-Deflection Equations. Two spans must be considered in this problem. Equation 11–8 applies to span AB. We can use Eq. 11–10 for span BC since the *end C is on a roller.* Using the formulas for the FEMs tabulated on the inside back cover, we have

$$(\text{FEM})_{AB} = -\frac{wL^2}{12} = -\frac{1}{12}(40)(6)^2 = -120 \text{ kN} \cdot \text{m}$$

$$(\text{FEM})_{BA} = \frac{wL^2}{12} = \frac{1}{12}(40)(6)^2 = 120 \text{ kN} \cdot \text{m}$$

$$(\text{FEM})_{BC} = -\frac{3PL}{16} = -\frac{3(60)(2)}{16} = -22.5 \text{ kN} \cdot \text{m}$$

Note that $(\text{FEM})_{AB}$ and $(\text{FEM})_{BC}$ are negative since they act counterclockwise on the beam at A and B, respectively. Also, since the supports do not settle, $\psi_{AB} = \psi_{BC} = 0$. Applying Eq. 11–8 for span AB and realizing that $\theta_A = 0$, we have

$$M_N = 2E\left(\frac{I}{L}\right)(2\theta_N + \theta_F - 3\psi) + (\text{FEM})_N$$

$$M_{AB} = 2E\left(\frac{I}{6}\right)[2(0) + \theta_B - 3(0)] - 120$$

$$M_{AB} = 0.3333EI\theta_B - 120 \qquad (1)$$

$$M_{BA} = 2E\left(\frac{I}{6}\right)[2\theta_B + 0 - 3(0)] + 120$$

$$M_{BA} = 0.667EI\theta_B + 120 \qquad (2)$$

Applying Eq. 11–10 with B as the near end and C as the far end, we have

$$M_N = 3E\left(\frac{I}{L}\right)(\theta_N - \psi) + (\text{FEM})_N$$

$$M_{BC} = 3E\left(\frac{I}{2}\right)(\theta_B - 0) - 22.5$$

$$M_{BC} = 1.5EI\theta_B - 22.5 \qquad (3)$$

Remember that Eq. 11–10 is *not* applied from C (near end) to B (far end).

Equilibrium Equations. The above three equations contain four unknowns. The necessary fourth equation comes from the conditions of equilibrium at the support B. The free-body diagram is shown in Fig. 11–11b. We have

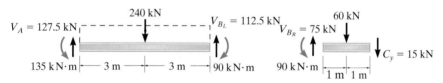

(b)

$$\downarrow + \Sigma M_B = 0; \qquad M_{BA} + M_{BC} = 0 \qquad\qquad (4)$$

To solve, substitute Eqs. (2) and (3) into Eq. (4), which yields

$$\theta_B = -\frac{45}{EI}$$

Since θ_B is negative (counterclockwise) the elastic curve for the beam has been correctly drawn in Fig. 11–11a. Substituting θ_B into Eqs. (1)–(3), we get

$$M_{AB} = -135 \text{ kN} \cdot \text{m}$$

$$M_{BA} = 90 \text{ kN} \cdot \text{m}$$

$$M_{BC} = -90 \text{ kN} \cdot \text{m}$$

Using these data for the moments, the shear reactions at the ends of the beam spans have been determined in Fig. 11–11c. The shear and moment diagrams are plotted in Fig. 11–11d.

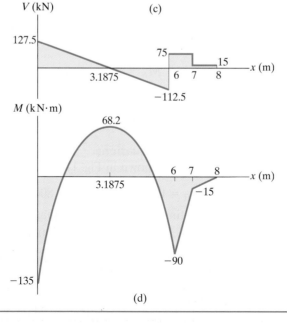

(d)

EXAMPLE 11.3

Determine the moment at A and B for the beam shown in Fig. 11–12a. The support at B is displaced (settles) 80 mm. Take $E = 200$ GPa, $I = 5(10^6)$ mm^4.

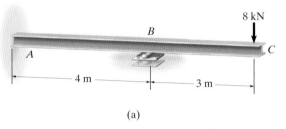

(a)

Fig. 11–12

SOLUTION

Slope-Deflection Equations. Only one span (AB) must be considered in this problem since the moment $\mathbf{M}_{BC}$ due to the overhang can be calculated from statics. Since there is no loading on span AB, the FEMs are zero. As shown in Fig. 11–12b, the downward displacement (settlement) of B causes the cord for span AB to rotate clockwise. Thus,

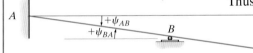

(b)

$$\psi_{AB} = \psi_{BA} = \frac{0.08 \text{ m}}{4} = 0.02 \text{ rad}$$

The stiffness for AB is

$$k = \frac{I}{L} = \frac{5(10^6) \text{ mm}^4(10^{-12}) \text{ m}^4/\text{mm}^4}{4 \text{ m}} = 1.25(10^{-6}) \text{ m}^3$$

Applying the slope-deflection equation, Eq. 11–8, to span AB, with $\theta_A = 0$, we have

$$M_N = 2E\left(\frac{I}{L}\right)(2\theta_N + \theta_F - 3\psi) + (\text{FEM})_N$$

$$M_{AB} = 2(200(10^9) \text{ N/m}^2)[1.25(10^{-6}) \text{ m}^3][2(0) + \theta_B - 3(0.02)] + 0 \quad (1)$$

$$M_{BA} = 2(200(10^9) \text{ N/m}^2)[1.25(10^{-6}) \text{ m}^3][2\theta_B + 0 - 3(0.02)] + 0 \quad (2)$$

Equilibrium Equations. The free-body diagram of the beam at support B is shown in Fig. 11–12c. Moment equilibrium requires

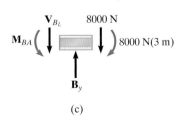

(c)

$$\zeta + \Sigma M_B = 0; \qquad M_{BA} - 8000 \text{ N}(3 \text{ m}) = 0$$

Substituting Eq. (2) into this equation yields

$$1(10^6)\theta_B - 30(10^3) = 24(10^3)$$

$$\theta_B = 0.054 \text{ rad}$$

Thus, from Eqs. (1) and (2),

$$M_{AB} = -3.00 \text{ kN} \cdot \text{m}$$

$$M_{BA} = 24.0 \text{ kN} \cdot \text{m}$$

EXAMPLE | 11.4

Determine the internal moments at the supports of the beam shown in Fig. 11–13a. The roller support at C is pushed downward 30 mm by the force **P**. Take $E = 200$ GPa, $I = 600(10^6)$ mm^4.

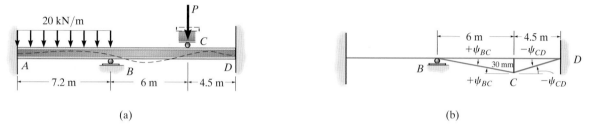

(a)   (b)

Fig. 11–13

SOLUTION

Slope-Deflection Equations.   Three spans must be considered in this problem. Equation 11–8 applies since the end supports A and D are fixed. Also, only span AB has FEMs.

$$(\text{FEM})_{AB} = -\frac{wL^2}{12} = -\frac{1}{12}(20)(7.2)^2 = -86.4 \text{ kN} \cdot \text{m}$$

$$(\text{FEM})_{BA} = \frac{wL^2}{12} = \frac{1}{12}(20)(7.2)^2 = 86.4 \text{ kN} \cdot \text{m}$$

As shown in Fig. 11–13b, the displacement (or settlement) of the support C causes ψ_{BC} to be positive, since the cord for span BC rotates clockwise, and ψ_{CD} to be negative, since the cord for span CD rotates counterclockwise. Hence,

$$\psi_{BC} = \frac{0.03 \text{ m}}{6 \text{ m}} = 0.005 \text{ rad} \qquad \psi_{CD} = -\frac{0.03 \text{ m}}{4.5 \text{ m}} = -0.00667 \text{ rad}$$

Also, expressing the units for the stiffness in meters, we have

$$k_{AB} = \frac{600(10^6)(10^{-12})}{7.2} = 83.33(10^{-6}) \text{ m}^3 \qquad k_{BC} = \frac{600(10^6)(10^{-12})}{6} = 100(10^{-6}) \text{ m}^3$$

$$k_{CD} = \frac{600(10^6)(10^{-12})}{4.5} = 133.33(10^{-6}) \text{ m}^3$$

Noting that $\theta_A = \theta_D = 0$ since A and D are fixed supports, and applying the slope-deflection Eq. 11–8 twice to each span, we have

11

EXAMPLE 11.4 (Continued)

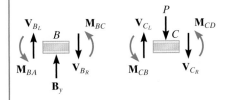

(a)

For span AB:

$$M_{AB} = 2[200(10^6)][83.33(10^{-6})][2(0) + \theta_B - 3(0)] - 86.4$$
$$M_{AB} = 33\,333.3\theta_B - 86.4 \tag{1}$$
$$M_{BA} = 2[200(10^6)][83.33(10^6)][2\theta_B + 0 - 3(0)] + 86.4$$
$$M_{BA} = 66\,666.7\theta_B + 86.4 \tag{2}$$

For span BC:

$$M_{BC} = 2[200(10^6)][100(10^{-6})][2\theta_B + \theta_C - 3(0.005)] + 0$$
$$M_{BC} = 80\,000\theta_B + 40\,000\theta_C - 600 \tag{3}$$
$$M_{CB} = 2[200(10^6)][100(10^{-6})][2\theta_C + \theta_B - 3(0.005)] + 0$$
$$M_{CB} = 80\,000\theta_C + 40\,000\theta_B - 600 \tag{4}$$

For span CD:

$$M_{CD} = 2[200(10^6)][133.33(10^{-6})][2\theta_C + 0 - 3(-0.00667)] + 0$$
$$M_{CD} = 106\,666.7\theta_C + 0 + 1066.7 \tag{5}$$
$$M_{DC} = 2[200(10^6)][133.33(10^{-6})][2(0) + \theta_C - 3(-0.00667)] + 0$$
$$M_{DC} = 53\,333.3\theta_C + 1066.7 \tag{6}$$

Equilibrium Equations.   These six equations contain eight unknowns. Writing the moment equilibrium equations for the supports at B and C, Fig. 10–13c, we have

$$\zeta + \Sigma M_B = 0; \qquad\qquad M_{BA} + M_{BC} = 0 \tag{7}$$
$$\zeta + \Sigma M_C = 0; \qquad\qquad M_{CB} + M_{CD} = 0 \tag{8}$$

In order to solve, substitute Eqs. (2) and (3) into Eq. (7), and Eqs. (4) and (5) into Eq. (8). This yields

$$\theta_C + 3.667\theta_B = 0.01284$$
$$-\theta_C - 0.214\theta_B = 0.00250$$

Thus,

$$\theta_B = 0.00444 \text{ rad} \qquad \theta_C = -0.00345 \text{ rad}$$

The negative value for θ_C indicates counterclockwise rotation of the tangent at C, Fig. 11–13a. Substituting these values into Eqs. (1)–(6) yields

$M_{AB} = 61.6 \text{ kN} \cdot \text{m}$	*Ans.*
$M_{BA} = 383 \text{ kN} \cdot \text{m}$	*Ans.*
$M_{BC} = -383 \text{ kN} \cdot \text{m}$	*Ans.*
$M_{CB} = -698 \text{ kN} \cdot \text{m}$	*Ans.*
$M_{CD} = 698 \text{ kN} \cdot \text{m}$	*Ans.*
$M_{DC} = 883 \text{ kN} \cdot \text{m}$	*Ans.*

Apply these end moments to spans BC and CD and show that $V_{C_L} = 180.2$ kN, $V_{C_R} = -351.3$ kN and the force on the roller is $P = 531.5$ kN.

(c)

11

PROBLEMS

11–1. Determine the moments at A, B, and C and then draw the moment diagram. EI is constant. Assume the support at B is a roller and A and C are fixed.

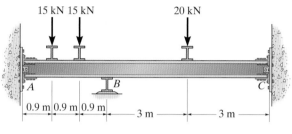

15 kN 15 kN   20 kN

A   B   C

0.9 m|0.9 m|0.9 m   3 m   3 m

Prob. 11–1

11–2. Determine the moments at A, B, and C, then draw the moment diagram for the beam. The moment of inertia of each span is indicated in the figure. Assume the support at B is a roller and A and C are fixed. $E = 200$ GPa.

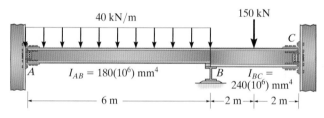

40 kN/m   150 kN

C

A $I_{AB} = 180(10^6)$ mm^4 B $I_{BC} = 240(10^6)$ mm^4

6 m   2 m   2 m

Prob. 11–2

11–3. Determine the moments at the supports A and C, then draw the moment diagram. Assume joint B is a roller. EI is constant.

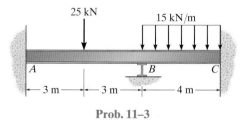

25 kN   15 kN/m

A   B   C

3 m   3 m   4 m

Prob. 11–3

*11–4. Determine the moments at the supports, then draw the moment diagram. Assume B is a roller and A and C are fixed. EI is constant.

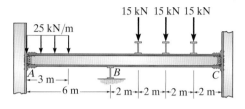

25 kN/m   15 kN 15 kN 15 kN

A   B   C

3 m

6 m   2 m 2 m 2 m 2 m

Prob. 11–4

11–5. Determine the moment at A, B, C and D, then draw the moment diagram for the beam. Assume the supports at A and D are fixed and B and C are rollers. EI is constant.

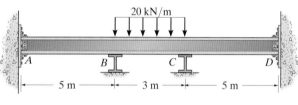

20 kN/m

A   B   C   D

5 m   3 m   5 m

Prob. 11–5

11–6. Determine the moments at A, B, C and D, then draw the moment diagram for the beam. Assume the supports at A and D are fixed and B and C are rollers. EI is constant.

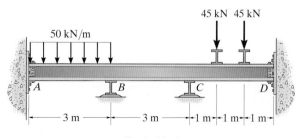

45 kN 45 kN

50 kN/m

A   B   C   D

3 m   3 m   1 m 1 m 1 m

Prob. 11–6

11–7. Determine the moment at B, then draw the moment diagram for the beam. Assume the supports at A and C are pins and B is a roller. EI is constant.

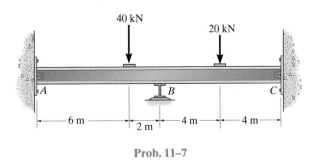

Prob. 11–7

11–10. Determine the moments at A and B, then draw the moment diagram for the beam. EI is constant.

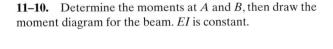
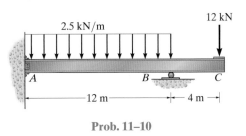

Prob. 11–10

*11–8. Determine the moments at A, B, and C, then draw the moment diagram. EI is constant. Assume the support at B is a roller and A and C are fixed.

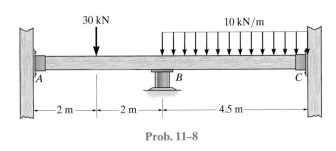

Prob. 11–8

11–11. Determine the moments at A, B, and C, then draw the moment diagram for the beam. Assume the support at A is fixed, B and C are rollers, and D is a pin. EI is constant.

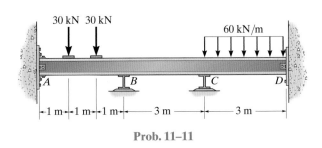

Prob. 11–11

11–9. Determine the moments at each support, then draw the moment diagram. Assume A is fixed. EI is constant.

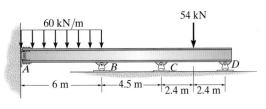

Prob. 11–9

*11–12. Determine the moments acting at A and B. Assume A is fixed supported, B is a roller, and C is a pin. EI is constant.

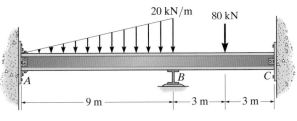

Prob. 11–12

11

11.4   Analysis of Frames: No Sidesway

A frame will not sidesway, or be displaced to the left or right, provided it is properly restrained. Examples are shown in Fig. 11–14. Also, no sidesway will occur in an unrestrained frame provided it is symmetric with respect to both loading and geometry, as shown in Fig. 11–15. For both cases the term ψ in the slope-deflection equations is equal to zero, since bending does not cause the joints to have a linear displacement.

The following examples illustrate application of the slope-deflection equations using the procedure for analysis outlined in Sec. 11–3 for these types of frames.

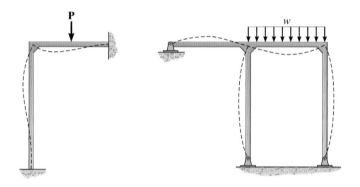

Fig. 11–14

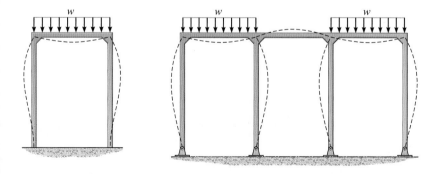

Fig. 11–15

EXAMPLE 11.5

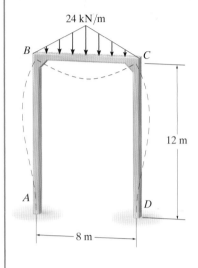

24 kN/m

12 m

8 m

(a)

Fig. 11–16

Determine the moments at each joint of the frame shown in Fig. 11–16a. EI is constant.

SOLUTION

Slope-Deflection Equations. Three spans must be considered in this problem: AB, BC, and CD. Since the spans are fixed supported at A and D, Eq. 11–8 applies for the solution.

From the table on the inside back cover, the FEMs for BC are

$$(FEM)_{BC} = -\frac{5wL^2}{96} = -\frac{5(24)(8)^2}{96} = -80 \text{ kN} \cdot \text{m}$$

$$(FEM)_{CB} = \frac{5wL^2}{96} = \frac{5(24)(8)^2}{96} = 80 \text{ kN} \cdot \text{m}$$

Note that $\theta_A = \theta_D = 0$ and $\psi_{AB} = \psi_{BC} = \psi_{CD} = 0$, since no sidesway will occur.

Applying Eq. 11–8, we have

$$M_N = 2Ek(2\theta_N + \theta_F - 3\psi) + (FEM)_N$$

$$M_{AB} = 2E\left(\frac{I}{12}\right)[2(0) + \theta_B - 3(0)] + 0$$

$$M_{AB} = 0.1667EI\theta_B \tag{1}$$

$$M_{BA} = 2E\left(\frac{I}{12}\right)[2\theta_B + 0 - 3(0)] + 0$$

$$M_{BA} = 0.333EI\theta_B \tag{2}$$

$$M_{BC} = 2E\left(\frac{I}{8}\right)[2\theta_B + \theta_C - 3(0)] - 80$$

$$M_{BC} = 0.5EI\theta_B + 0.25EI\theta_C - 80 \tag{3}$$

$$M_{CB} = 2E\left(\frac{I}{8}\right)[2\theta_C + \theta_B - 3(0)] + 80$$

$$M_{CB} = 0.5EI\theta_C + 0.25EI\theta_B + 80 \tag{4}$$

$$M_{CD} = 2E\left(\frac{I}{12}\right)[2\theta_C + 0 - 3(0)] + 0$$

$$M_{CD} = 0.333EI\theta_C \tag{5}$$

$$M_{DC} = 2E\left(\frac{I}{12}\right)[2(0) + \theta_C - 3(0)] + 0$$

$$M_{DC} = 0.1667EI\theta_C \tag{6}$$

11

Equilibrium Equations.   The preceding six equations contain eight unknowns. The remaining two equilibrium equations come from moment equilibrium at joints B and C, Fig. 11–16b. We have

$$M_{BA} + M_{BC} = 0 \tag{7}$$

$$M_{CB} + M_{CD} = 0 \tag{8}$$

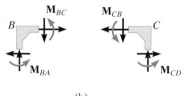

(b)

To solve these eight equations, substitute Eqs. (2) and (3) into Eq. (7) and substitute Eqs. (4) and (5) into Eq. (8). We get

$$0.833 EI\theta_B + 0.25 EI\theta_C = 80$$

$$0.833 EI\theta_C + 0.25 EI\theta_B = -80$$

Solving simultaneously yields

$$\theta_B = -\theta_C = \frac{137.1}{EI}$$

which conforms with the way the frame deflects as shown in Fig. 11–16a. Substituting into Eqs. (1)–(6), we get

$$M_{AB} = 22.9 \text{ kN} \cdot \text{m} \qquad \qquad Ans.$$
$$M_{BA} = 45.7 \text{ kN} \cdot \text{m} \qquad \qquad Ans.$$
$$M_{BC} = -45.7 \text{ kN} \cdot \text{m} \qquad \qquad Ans.$$
$$M_{CB} = 45.7 \text{ kN} \cdot \text{m} \qquad \qquad Ans.$$
$$M_{CD} = -45.7 \text{ kN} \cdot \text{m} \qquad \qquad Ans.$$
$$M_{DC} = -22.9 \text{ kN} \cdot \text{m} \qquad \qquad Ans.$$

Using these results, the reactions at the ends of each member can be determined from the equations of equilibrium, and the moment diagram for the frame can be drawn, Fig. 11–16c.

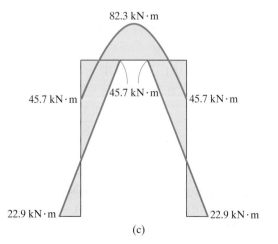

(c)

EXAMPLE 11.6

Determine the internal moments at each joint of the frame shown in Fig. 11–17a. The moment of inertia for each member is given in the figure. Take $E = 200$ GPa.

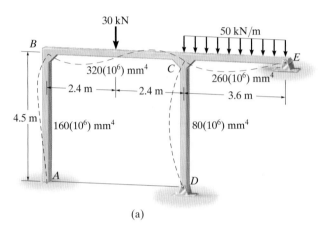

(a)

Fig. 11–17

SOLUTION

Slope-Deflection Equations. Four spans must be considered in this problem. Equation 11–8 applies to spans AB and BC, and Eq. 11–10 will be applied to CD and CE, because the ends at D and E are pinned. Computing the member stiffnesses, we have

$$k_{AB} = \frac{160(10^6)(10^{-12})}{4.5} = 35.56(10^{-6}) \text{ m}^3 \qquad k_{CD} = \frac{80(10^6)(10^{-12})}{4.5} = 17.78(10^{-6}) \text{ m}^3$$

$$k_{BC} = \frac{320(10^6)(10^{-12})}{4.8} = 66.67(10^{-6}) \text{ m}^3 \qquad k_{CE} = \frac{260(10^6)(10^{-12})}{3.6} = 72.23(10^{-6}) \text{ m}^3$$

The FEMs due to the loadings are

$$(\text{FEM})_{BC} = -\frac{PL}{8} = -\frac{30(4.8)}{8} = -18 \text{ kN} \cdot \text{m}$$

$$(\text{FEM})_{CB} = \frac{PL}{8} = \frac{30(4.8)}{8} = 18 \text{ kN} \cdot \text{m}$$

$$(\text{FEM})_{CE} = -\frac{wL^2}{8} = -\frac{50(3.6)^2}{8} = -81 \text{ kN} \cdot \text{m}$$

Applying Eqs. 11–8 and 11–10 to the frame and noting that $\theta_A = 0$, $\psi_{AB} = \psi_{BC} = \psi_{CD} = \psi_{CE} = 0$ since no sidesway occurs, we have

$$M_N = 2Ek(2\theta_N + \theta_F - 3\psi) + (\text{FEM})_N$$

$$M_{AB} = 2[200(10^6)](35.56)(10^{-6})[2(0) + \theta_B - 3(0)] + 0$$

$$M_{AB} = 14\,222.2\theta_B \tag{1}$$

$$M_{BA} = 2[200(10^6)](35.56)(10^{-6})[2\theta_B + 0 - 3(0)] + 0$$
$$M_{BA} = 28\,444.4\theta_B \qquad\qquad (2)$$

$$M_{BC} = 2[200(10^6)](66.67)(10^{-6})[2\theta_B + \theta_C - 3(0)] - 18$$
$$M_{BC} = 53\,333.3\theta_B + 26\,666.7\theta_C - 18 \qquad (3)$$

$$M_{CB} = 2[200(10^6)](66.67)(10^{-6})[2\theta_C + \theta_B - 3(0)] + 18$$
$$M_{CB} = 26\,666.7\theta_B + 53\,333.3\theta_C + 18 \qquad (4)$$

$$M_N = 3Ek(\theta_N - \psi) + (FEM)_N$$
$$M_{CD} = 3[200(10^6)](17.78)(10^{-6})[\theta_C - 0] + 0 \qquad (5)$$

$$M_{CD} = 10\,666.7\theta_C$$
$$M_{CE} = 3[200(10^6)](72.22)(10^{-6})[\theta_C - 0] - 81$$
$$M_{CE} = 43\,333.3\theta_C - 81 \qquad (6)$$

Equations of Equilibrium. These six equations contain eight unknowns. Two moment equilibrium equations can be written for joints B and C, Fig. 11–17b. We have

$$M_{BA} + M_{BC} = 0 \qquad (7)$$
$$M_{CB} + M_{CD} + M_{CE} = 0 \qquad (8)$$

In order to solve, substitute Eqs. (2) and (3) into Eq. (7), and Eqs. (4)–(6) into Eq. (8). This gives

$$81\,777.7\theta_B + 26\,666.7\theta_C = 18$$
$$26\,666.7\theta_B + 107\,333.3\theta_C = 63$$

Solving these equations simultaneously yields

$$\theta_B = 3.124(10^{-5})\ \text{rad} \qquad \theta_C = 5.792(10^{-4})\ \text{rad}$$

These values, being clockwise, tend to distort the frame as shown in Fig. 11–17a. Substituting these values into Eqs. (1)–(6) and solving, we get

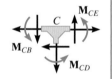

(b)

$$M_{AB} = 0.444\ \text{kN·m} \qquad\qquad Ans.$$
$$M_{BA} = 0.888\ \text{kN·m} \qquad\qquad Ans.$$
$$M_{BC} = -0.888\ \text{kN·m} \qquad\qquad Ans.$$
$$M_{CB} = 49.7\ \text{kN·m} \qquad\qquad Ans.$$
$$M_{CD} = 6.18\ \text{kN·m} \qquad\qquad Ans.$$
$$M_{CE} = -55.9\ \text{kN·m} \qquad\qquad Ans.$$

11

11.5   Analysis of Frames: Sidesway

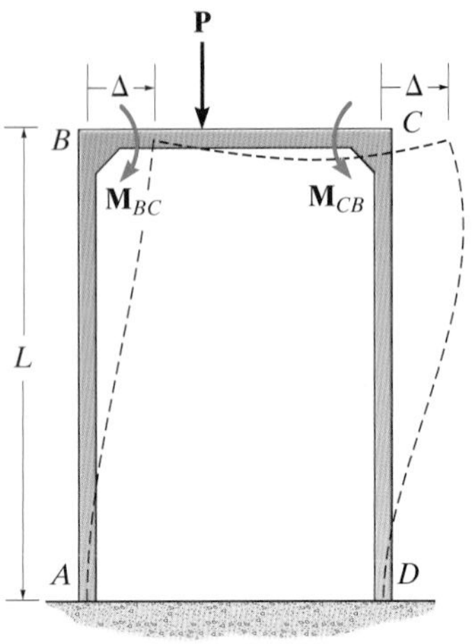

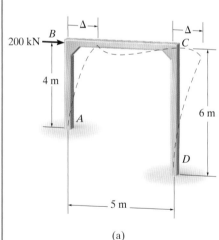

Fig. 11–18

A frame will sidesway, or be displaced to the side, when it or the loading acting on it is nonsymmetric. To illustrate this effect, consider the frame shown in Fig. 11–18. Here the loading **P** causes *unequal* moments $\mathbf{M}_{BC}$ and $\mathbf{M}_{CB}$ at the joints B and C, respectively. $\mathbf{M}_{BC}$ tends to displace joint B to the right, whereas $\mathbf{M}_{CB}$ tends to displace joint C to the left. Since $\mathbf{M}_{BC}$ is larger than $\mathbf{M}_{CB}$, the net result is a sidesway Δ of both joints B and C to the right, as shown in the figure.* When applying the slope-deflection equation to each column of this frame, we must therefore consider the column rotation ψ (since $\psi = \Delta/L$) as unknown in the equation. As a result an extra equilibrium equation must be included for the solution. In the previous sections it was shown that unknown *angular displacements* θ were related by joint *moment equilibrium equations*. In a similar manner, when unknown joint *linear displacements* Δ (or span rotations ψ) occur, we must write *force equilibrium equations* in order to obtain the complete solution. The unknowns in these equations, however, must only involve the internal *moments* acting at the ends of the columns, since the slope-deflection equations involve these moments. The technique for solving problems for frames with sidesway is best illustrated by examples.

EXAMPLE 11.7

Determine the moments at each joint of the frame shown in Fig. 11–19a. EI is constant.

SOLUTION

Slope-Deflection Equations.   Since the ends A and D are fixed, Eq. 11–8 applies for all three spans of the frame. Sidesway occurs here since both the applied loading and the geometry of the frame are nonsymmetric. Here the load is applied directly to joint B and therefore no FEMs act at the joints. As shown in Fig. 11–19a, both joints B and C are assumed to be displaced an *equal amount* Δ. Consequently, $\psi_{AB} = \Delta/4$ and $\psi_{DC} = \Delta/6$. Both terms are positive since the cords of members AB and CD "rotate" clockwise. Relating ψ_{AB} to ψ_{DC}, we have $\psi_{AB} = (6/4)\psi_{DC}$. Applying Eq. 11–8 to the frame, we have

Fig. 11–19

(a)

$$M_{AB} = 2E\left(\frac{I}{4}\right)\left[2(0) + \theta_B - 3\left(\frac{6}{4}\psi_{DC}\right)\right] + 0 = EI(0.5\theta_B - 2.25\psi_{DC}) \tag{1}$$

$$M_{BA} = 2E\left(\frac{I}{4}\right)\left[2\theta_B + 0 - 3\left(\frac{6}{4}\psi_{DC}\right)\right] + 0 = EI(1.0\theta_B - 2.25\psi_{DC}) \tag{2}$$

$$M_{BC} = 2E\left(\frac{I}{5}\right)[2\theta_B + \theta_C - 3(0)] + 0 = EI(0.8\theta_B + 0.4\theta_C) \tag{3}$$

*Recall that the deformation of all three members due to shear and axial force is neglected.

$$M_{CB} = 2E\left(\frac{I}{5}\right)[2\theta_C + \theta_B - 3(0)] + 0 = EI(0.8\theta_C + 0.4\theta_B) \qquad (4)$$

$$M_{CD} = 2E\left(\frac{I}{6}\right)[2\theta_C + 0 - 3\psi_{DC}] + 0 = EI(0.667\theta_C - 1.0\psi_{DC}) \qquad (5)$$

$$M_{DC} = 2E\left(\frac{I}{6}\right)[2(0) + \theta_C - 3\psi_{DC}] + 0 = EI(0.333\theta_C - 1.0\psi_{DC}) \qquad (6)$$

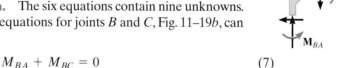

Equations of Equilibrium. The six equations contain nine unknowns. Two moment equilibrium equations for joints B and C, Fig. 11–19b, can be written, namely,

$$M_{BA} + M_{BC} = 0 \qquad (7)$$

$$M_{CB} + M_{CD} = 0 \qquad (8)$$

(b)

Since a horizontal displacement Δ occurs, we will consider summing forces on the *entire frame* in the x direction. This yields

$$\xrightarrow{+} \Sigma F_x = 0; \qquad 200 - V_A - V_D = 0$$

The horizontal reactions or column shears V_A and V_D can be related to the internal moments by considering the free-body diagram of each column separately, Fig. 11–19c. We have

$$\Sigma M_B = 0; \qquad V_A = -\frac{M_{AB} + M_{BA}}{4}$$

$$\Sigma M_C = 0; \qquad V_D = -\frac{M_{DC} + M_{CD}}{6}$$

Thus,

$$200 + \frac{M_{AB} + M_{BA}}{4} + \frac{M_{DC} + M_{CD}}{6} = 0 \qquad (9)$$

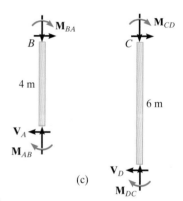

(c)

In order to solve, substitute Eqs. (2) and (3) into Eq. (7), Eqs. (4) and (5) into Eq. (8), and Eqs. (1), (2), (5), (6) into Eq. (9). This yields

$$1.8\theta_B + 0.4\theta_C - 2.25\psi_{DC} = 0$$
$$0.4\theta_B + 1.467\theta_C - \psi_{DC} = 0$$
$$1.5\theta_B + 0.667\theta_C - 5.833\psi_{DC} = -\frac{800}{EI}$$

Solving simultaneously, we have

$$EI\theta_B = 243.78 \qquad EI\theta_C = 75.66 \qquad EI\psi_{DC} = 208.48$$

Finally, using these results and solving Eqs. (1)–(6) yields

$$M_{AB} = -347 \text{ kN} \cdot \text{m} \qquad \textit{Ans.}$$
$$M_{BA} = -225 \text{ kN} \cdot \text{m} \qquad \textit{Ans.}$$
$$M_{BC} = 225 \text{ kN} \cdot \text{m} \qquad \textit{Ans.}$$
$$M_{CB} = 158 \text{ kN} \cdot \text{m} \qquad \textit{Ans.}$$
$$M_{CD} = -158 \text{ kN} \cdot \text{m} \qquad \textit{Ans.}$$
$$M_{DC} = -183 \text{ kN} \cdot \text{m} \qquad \textit{Ans.}$$

11

EXAMPLE 11.8

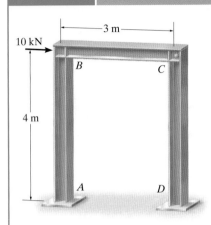

(a)

Fig. 11–20

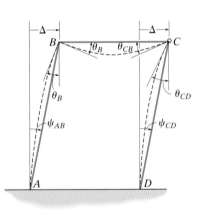

(b)

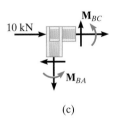

(c)

Determine the moments at each joint of the frame shown in Fig. 11–20a. The supports at A and D are fixed and joint C is assumed pin connected. EI is constant for each member.

SOLUTION

Slope-Deflection Equations.   We will apply Eq. 11–8 to member AB since it is fixed connected at both ends. Equation 11–10 can be applied from B to C and from D to C since the pin at C supports zero moment. As shown by the deflection diagram, Fig. 11–20b, there is an unknown linear displacement Δ of the frame and unknown angular displacement θ_B at joint B.* Due to Δ, the cord members AB and CD rotate clockwise, $\psi = \psi_{AB} = \psi_{DC} = \Delta/4$. Realizing that $\theta_A = \theta_D = 0$ and that there are no FEMs for the members, we have

$$M_N = 2E\left(\frac{I}{L}\right)(2\theta_N + \theta_F - 3\psi) + (\text{FEM})_N$$

$$M_{AB} = 2E\left(\frac{I}{4}\right)[2(0) + \theta_B - 3\psi] + 0 \tag{1}$$

$$M_{BA} = 2E\left(\frac{I}{4}\right)(2\theta_B + 0 - 3\psi) + 0 \tag{2}$$

$$M_N = 3E\left(\frac{I}{L}\right)(\theta_N - \psi) + (\text{FEM})_N$$

$$M_{BC} = 3E\left(\frac{I}{3}\right)(\theta_B - 0) + 0 \tag{3}$$

$$M_{DC} = 3E\left(\frac{I}{4}\right)(0 - \psi) + 0 \tag{4}$$

Equilibrium Equations.   Moment equilibrium of joint B, Fig. 11–20c, requires

$$M_{BA} + M_{BC} = 0 \tag{5}$$

If forces are summed for the *entire frame* in the horizontal direction, we have

$$\xrightarrow{+} \Sigma F_x = 0; \qquad 10 - V_A - V_D = 0 \tag{6}$$

As shown on the free-body diagram of each column, Fig. 11–20d, we have

$$\Sigma M_B = 0; \qquad V_A = -\frac{M_{AB} + M_{BA}}{4}$$

$$\Sigma M_C = 0; \qquad V_D = -\frac{M_{DC}}{4}$$

*The angular displacements θ_{CB} and θ_{CD} at joint C (pin) are not included in the analysis since Eq. 11–10 is to be used.

Thus, from Eq. (6),

$$10 + \frac{M_{AB} + M_{BA}}{4} + \frac{M_{DC}}{4} = 0 \qquad (7)$$

Substituting the slope-deflection equations into Eqs. (5) and (7) and simplifying yields

$$\theta_B = \frac{3}{4}\psi$$

$$10 + \frac{EI}{4}\left(\frac{3}{2}\theta_B - \frac{15}{4}\psi\right) = 0$$

Thus,

$$\theta_B = \frac{240}{21EI} \qquad \psi = \frac{320}{21EI}$$

Substituting these values into Eqs. (1)–(4), we have

$$M_{AB} = -17.1 \text{ kN} \cdot \text{m}, \quad M_{BA} = -11.4 \text{ kN} \cdot \text{m} \qquad \textit{Ans.}$$

$$M_{BC} = 11.4 \text{ kN} \cdot \text{m}, \quad M_{DC} = -11.4 \text{ kN} \cdot \text{m} \qquad \textit{Ans.}$$

Using these results, the end reactions on each member can be determined from the equations of equilibrium, Fig. 11–20e. The moment diagram for the frame is shown in Fig. 11–20f.

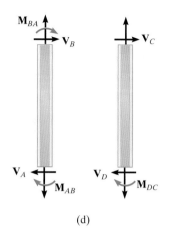

(d)

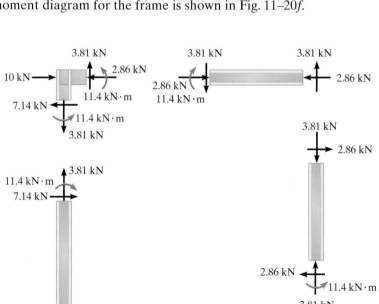

(e)

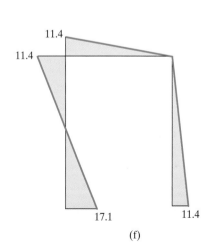

(f)

EXAMPLE 11.9

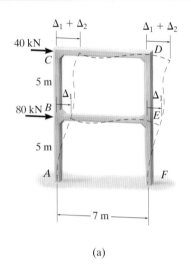

$\Delta_1 + \Delta_2$ $\Delta_1 + \Delta_2$

40 kN

C

5 m

Δ_1 Δ_1

80 kN B   E

5 m

A   F

7 m

(a)

Fig. 11–21

Explain how the moments in each joint of the two-story frame shown in Fig. 11–21a are determined. EI is constant.

SOLUTION

Slope-Deflection Equation. Since the supports at A and F are fixed, Eq. 11–8 applies for all six spans of the frame. No FEMs have to be calculated, since the applied loading acts at the joints. Here the loading displaces joints B and E an amount Δ_1, and C and D an amount $\Delta_1 + \Delta_2$. As a result, members AB and FE undergo rotations of $\psi_1 = \Delta_1/5$ and BC and ED undergo rotations of $\psi_2 = \Delta_2/5$.

Applying Eq. 11–8 to the frame yields

$$M_{AB} = 2E\left(\frac{I}{5}\right)[2(0) + \theta_B - 3\psi_1] + 0 \tag{1}$$

$$M_{BA} = 2E\left(\frac{I}{5}\right)[2\theta_B + 0 - 3\psi_1] + 0 \tag{2}$$

$$M_{BC} = 2E\left(\frac{I}{5}\right)[2\theta_B + \theta_C - 3\psi_2] + 0 \tag{3}$$

$$M_{CB} = 2E\left(\frac{I}{5}\right)[2\theta_C + \theta_B - 3\psi_2] + 0 \tag{4}$$

$$M_{CD} = 2E\left(\frac{I}{7}\right)[2\theta_C + \theta_D - 3(0)] + 0 \tag{5}$$

$$M_{DC} = 2E\left(\frac{I}{7}\right)[2\theta_D + \theta_C - 3(0)] + 0 \tag{6}$$

$$M_{BE} = 2E\left(\frac{I}{7}\right)[2\theta_B + \theta_E - 3(0)] + 0 \tag{7}$$

$$M_{EB} = 2E\left(\frac{I}{7}\right)[2\theta_E + \theta_B - 3(0)] + 0 \tag{8}$$

$$M_{ED} = 2E\left(\frac{I}{5}\right)[2\theta_E + \theta_D - 3\psi_2] + 0 \tag{9}$$

$$M_{DE} = 2E\left(\frac{I}{5}\right)[2\theta_D + \theta_E - 3\psi_2] + 0 \tag{10}$$

$$M_{FE} = 2E\left(\frac{I}{5}\right)[2(0) + \theta_E - 3\psi_1] + 0 \tag{11}$$

$$M_{EF} = 2E\left(\frac{I}{5}\right)[2\theta_E + 0 - 3\psi_1] + 0 \tag{12}$$

These 12 equations contain 18 unknowns.

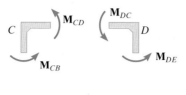

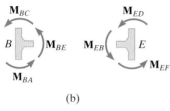

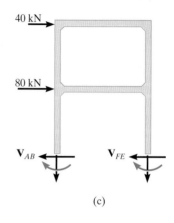

(b)

Equilibrium Equations. Moment equilibrium of joints B, C, D, and E, Fig. 11–21b, requires

$$M_{BA} + M_{BE} + M_{BC} = 0 \qquad (13)$$

$$M_{CB} + M_{CD} = 0 \qquad (14)$$

$$M_{DC} + M_{DE} = 0 \qquad (15)$$

$$M_{EF} + M_{EB} + M_{ED} = 0 \qquad (16)$$

(c)

As in the preceding examples, the shear at the base of all the columns for any story must balance the applied horizontal loads, Fig. 11–21c. This yields

$$\xrightarrow{+} \Sigma F_x = 0; \qquad 40 - V_{BC} - V_{ED} = 0$$

$$40 + \frac{M_{BC} + M_{CB}}{5} + \frac{M_{ED} + M_{DE}}{5} = 0 \qquad (17)$$

$$\xrightarrow{+} \Sigma F_x = 0; \qquad 40 + 80 - V_{AB} - V_{FE} = 0$$

$$120 + \frac{M_{AB} + M_{BA}}{5} + \frac{M_{EF} + M_{FE}}{5} = 0 \qquad (18)$$

Solution requires substituting Eqs. (1)–(12) into Eqs. (13)–(18), which yields six equations having six unknowns, $\psi_1, \psi_2, \theta_B, \theta_C, \theta_D$, and θ_E. These equations can then be solved simultaneously. The results are resubstituted into Eqs. (1)–(12), which yields the moments at the joints.

11

EXAMPLE | 11.10

Determine the moments at each joint of the frame shown in Fig. 11–22a. *EI* is constant for each member.

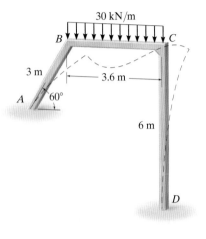

(a)

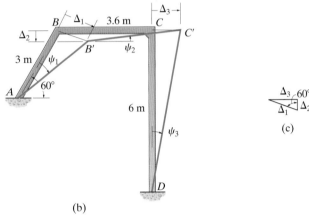

(b)

(c)

Fig. 11–22

SOLUTION

Slope-Deflection Equations.   Equation 11–8 applies to each of the three spans. The FEMs are

$$(\text{FEM})_{BC} = -\frac{wL^2}{12} = -\frac{30(3.6)^2}{12} = -32.4 \text{ kN} \cdot \text{m}$$

$$(\text{FEM})_{CB} = \frac{wL^2}{12} = \frac{30(3.6)^2}{12} = 32.4 \text{ kN} \cdot \text{m}$$

The sloping member *AB* causes the frame to sidesway to the right as shown in Fig. 11–22a. As a result, joints *B* and *C* are subjected to both rotational *and* linear displacements. The linear displacements are shown in Fig. 11–22b, where *B* moves Δ_1 to *B'* and *C* moves Δ_3 to *C'*. These displacements cause the members' cords to rotate ψ_1, ψ_3 (clockwise) and $-\psi_2$ (counterclockwise) as shown.* Hence,

$$\psi_1 = \frac{\Delta_1}{3} \qquad \psi_2 = -\frac{\Delta_2}{4} \qquad \psi_3 = \frac{\Delta_3}{6}$$

As shown in Fig. 11–22c, the three displacements can be related. For example, $\Delta_2 = 0.5\Delta_1$ and $\Delta_3 = 0.866\Delta_1$. Thus, from the above equations we have

$$\psi_2 = -0.417\psi_1 \qquad \psi_3 = 0.433\psi_1$$

Using these results, the slope-deflection equations for the frame are

---

*Recall that distortions due to axial forces are neglected and the arc displacements *BB'* and *CC'* can be considered as straight lines, since ψ_1 and ψ_3 are actually very small.

$$M_{AB} = 2E\left(\frac{I}{3}\right)[2(0) + \theta_B - 3\psi_1] + 0 \tag{1}$$

$$M_{BA} = 2E\left(\frac{I}{3}\right)[2\theta_B + 0 - 3\psi_1] + 0 \tag{2}$$

$$M_{BC} = 2E\left(\frac{I}{3.6}\right)[2\theta_B + \theta_C - 3(-0.417\psi_1)] - 32.4 \tag{3}$$

$$M_{CB} = 2E\left(\frac{I}{3.6}\right)[2\theta_C + \theta_B - 3(-0.417\psi_1)] + 32.4 \tag{4}$$

$$M_{CD} = 2E\left(\frac{I}{6}\right)[2\theta_C + 0 - 3(0.433\psi_1)] + 0 \tag{5}$$

$$M_{DC} = 2E\left(\frac{I}{6}\right)[2(0) + \theta_C - 3(0.433\psi_1)] + 0 \tag{6}$$

These six equations contain nine unknowns.

Equations of Equilibrium.   Moment equilibrium at joints B and C yields

$$M_{BA} + M_{BC} = 0 \tag{7}$$
$$M_{CD} + M_{CB} = 0 \tag{8}$$

The necessary third equilibrium equation can be obtained by summing moments about point O on the entire frame, Fig. 11–22d. This eliminates the unknown normal forces N_A and N_D, and therefore

$$\zeta + \Sigma M_O = 0;$$

$$M_{AB} + M_{DC} - \left(\frac{M_{AB} + M_{BA}}{3}\right)(10.2) - \left(\frac{M_{DC} + M_{CD}}{6}\right)(12.24) - 108(1.8) = 0$$

$$-2.4M_{AB} - 3.4M_{BA} - 2.04M_{CD} - 1.04M_{DC} - 194.4 = 0 \tag{9}$$

Substituting Eqs. (2) and (3) into Eq. (7), Eqs. (4) and (5) into Eq. (8), and Eqs. (1), (2), (5), and (6) into Eq. (9) yields

$$0.733\theta_B + 0.167\theta_C - 0.392\psi_1 = \frac{9.72}{EI}$$

$$0.167\theta_B + 0.533\theta_C + 0.0784\psi_1 = -\frac{9.72}{EI}$$

$$-1.840\theta_B - 0.512\theta_C + 3.880\psi_1 = \frac{58.32}{EI}$$

Solving these equations simultaneously yields

$$EI\theta_B = 35.51 \qquad EI\theta_C = -33.33 \qquad EI\psi_1 = 27.47$$

Substituting these values into Eqs. (1)–(6), we have

$M_{AB} = -31.3 \text{ kN} \cdot \text{m}$ $M_{BC} = 7.60 \text{ kN} \cdot \text{m}$ $M_{CD} = -34.2 \text{ kN} \cdot \text{m}$ *Ans.*

$M_{BA} = -7.60 \text{ kN} \cdot \text{m}$ $M_{CB} = 34.2 \text{ kN} \cdot \text{m}$ $M_{DC} = -23.0 \text{ kN} \cdot \text{m}$ *Ans.*

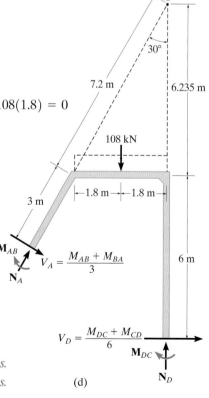

(d)

11

PROBLEMS

11–13. Determine the moments at A, B, and C, then draw the moment diagram for each member. Assume all joints are fixed connected. EI is constant.

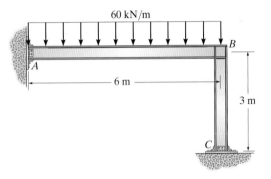

60 kN/m

B

A

6 m

3 m

C

Prob. 11–13

11–15. Determine the moment at B, then draw the moment diagram for each member of the frame. Assume the support at A is fixed and C is pinned. EI is constant.

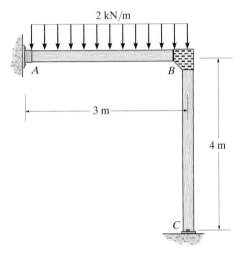

2 kN/m

A B

3 m

4 m

C

Prob. 11–15

11–14. Determine the moments at the supports, then draw the moment diagram. The members are fixed connected at the supports and at joint B. The moment of inertia of each member is given in the figure. Take $E = 200$ GPa.

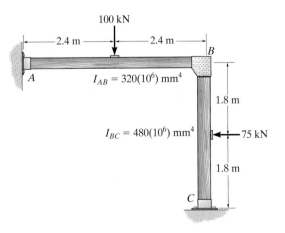

100 kN

2.4 m   2.4 m

B

A $I_{AB} = 320(10^6)$ mm^4

1.8 m

$I_{BC} = 480(10^6)$ mm^4 75 kN

1.8 m

C

Prob. 11–14

*11–16.** Determine the moments at B and D, then draw the moment diagram. Assume A and C are pinned and B and D are fixed connected. EI is constant.

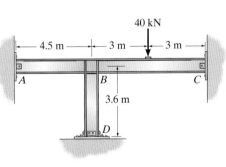

40 kN

4.5 m   3 m   3 m

A B C

3.6 m

D

Prob. 11–16

11–17. Determine the moment that each member exerts on the joint at B, then draw the moment diagram for each member of the frame. Assume the support at A is fixed and C is a pin. EI is constant.

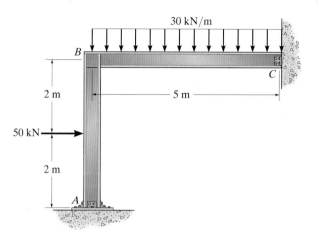

Prob. 11–17

11–18. Determine the moment that each member exerts on the joint at B, then draw the moment diagram for each member of the frame. Assume the supports at A, C, and D are pins. EI is constant.

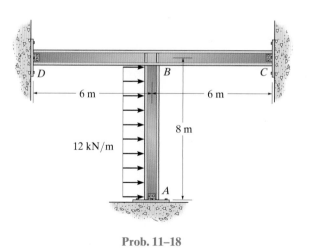

Prob. 11–18

11–19. Determine the moment at joints D and C, then draw the moment diagram for each member of the frame. Assume the supports at A and B are pins. EI is constant.

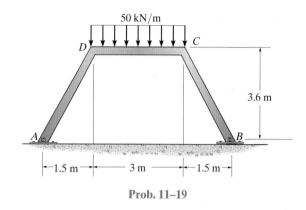

Prob. 11–19

*11–20. Determine the moment that each member exerts on the joints at B and D, then draw the moment diagram for each member of the frame. Assume the supports at A, C, and E are pins. EI is constant.

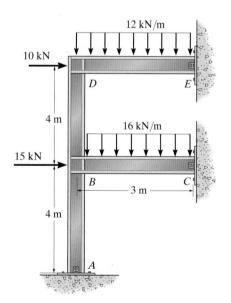

Prob. 11–20

11

11–21. Determine the moment at joints C and D, then draw the moment diagram for each member of the frame. Assume the supports at A and B are pins. EI is constant.

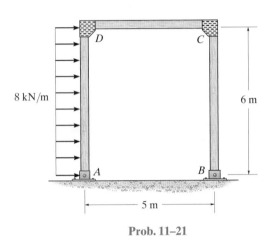

Prob. 11–21

11–23. Determine the moments acting at the supports A and D of the battered-column frame. Take $E = 200$ GPa, $I = 46.3(10^6)\,\text{mm}^4$.

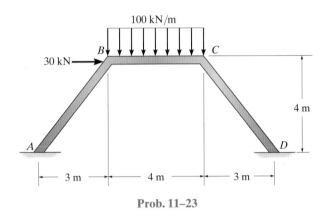

Prob. 11–23

11–22. Determine the moment at joints A, B, C, and D, then draw the moment diagram for each member of the frame. Assume the supports at A and B are fixed. EI is constant.

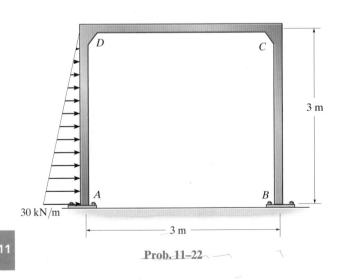

Prob. 11–22

*11–24. Wind loads are transmitted to the frame at joint E. If A, B, E, D, and F are all pin connected and C is fixed connected, determine the moments at joint C and draw the bending moment diagrams for the girder BCE. EI is constant.

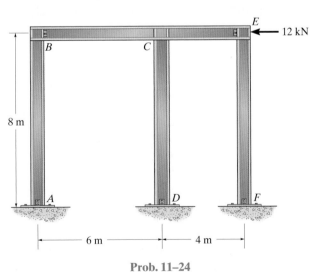

Prob. 11–24

11

PROJECT PROBLEM

11–1P. The roof is supported by joists that rest on two girders. Each joist can be considered simply supported, and the front girder can be considered attached to the three columns by a pin at A and rollers at B and C. Assume the roof will be made from 75 mm-thick cinder concrete, and each joist has a weight of 2.5 kN. According to code the roof will be subjected to a snow loading of 1.2 kN/m². The joists have a length of 8 m. Draw the shear and moment diagrams for the girder. Assume the supporting columns are rigid.

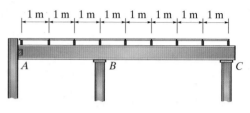

Project Prob. 11–1P

CHAPTER REVIEW

The unknown displacements of a structure are referred to as the degrees of freedom for the structure. They consist of either joint displacements or rotations. (Ref: Section 11.1)

The slope-deflection equations relate the unknown moments at each joint of a structural member to the unknown rotations that occur there. The following equation is applied twice to each member or span, considering each side as the "near" end and its counterpart as the far end.

$$M_N = 2Ek(2\theta_N + \theta_F - 3\psi) + (FEM)_N$$
For Internal Span or End Span with Far End Fixed

This equation is only applied once, where the "far" end is at the pin or roller support.

$$M_N = 3Ek(\theta_N - \psi) + (FEM)_N$$
Only for End Span with Far End Pinned or Roller Supported

(Ref: Section 11.2)

Once the slope-deflection equations are written, they are substituted into the equations of moment equilibrium at each joint and then solved for the unknown displacements. If the structure (frame) has sidesway, then an unknown horizontal displacement at each floor level will occur, and the unknown column shears must be related to the moments at the joints, using both the force and moment equilibrium equations. Once the unknown displacements are obtained, the unknown reactions are found from the load-displacement relations. (Ref: Sections 11.3 to 11.5)

11

The girders of this concrete building are all fixed connected, so the statically indeterminate analysis of the framework can be done using the moment distribution method.

Displacement Method of Analysis: Moment Distribution

12

The moment-distribution method is a displacement method of analysis that is easy to apply once certain elastic constants have been determined. In this chapter we will first state the important definitions and concepts for moment distribution and then apply the method to solve problems involving statically indeterminate beams and frames. Application to multistory frames is discussed in the last part of the chapter.

12.1 General Principles and Definitions

The method of analyzing beams and frames using moment distribution was developed by Hardy Cross, in 1930. At the time this method was first published it attracted immediate attention, and it has been recognized as one of the most notable advances in structural analysis during the twentieth century.

As will be explained in detail later, moment distribution is a method of successive approximations that may be carried out to any desired degree of accuracy. Essentially, the method begins by assuming each joint of a structure is fixed. Then, by unlocking and locking each joint in succession, the internal moments at the joints are "distributed" and balanced until the joints have rotated to their final or nearly final positions. It will be found that this process of calculation is both repetitive and easy to apply. Before explaining the techniques of moment distribution, however, certain definitions and concepts must be presented.

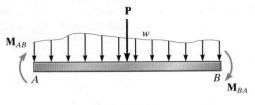

Fig. 12–1

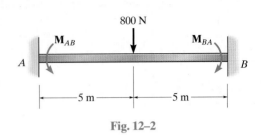

Fig. 12–2

Sign Convention. We will establish the same sign convention as that established for the slope-deflection equations: *Clockwise moments* that act *on the member* are considered *positive*, whereas *counterclockwise moments* are *negative*, Fig. 12–1.

Fixed-End Moments (FEMs). The moments at the "walls" or fixed joints of a loaded member are called *fixed-end moments*. These moments can be determined from the table given on the inside back cover, depending upon the type of loading on the member. For example, the beam loaded as shown in Fig. 12–2 has fixed-end moments of FEM $= PL/8 = 800(10)/8 = 1000$ N $\cdot$ m. Noting the action of these moments *on the beam* and applying our sign convention, it is seen that $M_{AB} = -1000$ N $\cdot$ m and $M_{BA} = +1000$ N $\cdot$ m.

Member Stiffness Factor. Consider the beam in Fig. 12–3, which is pinned at one end and fixed at the other. Application of the moment **M** causes the end A to rotate through an angle θ_A. In Chapter 11 we related M to θ_A using the conjugate-beam method. This resulted in Eq. 11–1, that is, $M = (4EI/L)\,\theta_A$. The term in parentheses

$$K = \frac{4EI}{L}$$
Far End Fixed

(12–1)

is referred to as the *stiffness factor* at A and can be defined as the amount of moment M required to rotate the end A of the beam $\theta_A = 1$ rad.

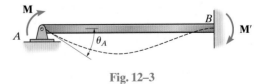

Fig. 12–3

Joint Stiffness Factor.

If several members are fixed connected to a joint and each of their far ends is fixed, then by the principle of superposition, the *total stiffness factor* at the joint is the sum of the member stiffness factors at the joint, that is, $K_T = \Sigma K$. For example, consider the frame joint A in Fig. 12–4a. The numerical value of each member stiffness factor is determined from Eq. 12–1 and listed in the figure. Using these values, the total stiffness factor of joint A is $K_T = \Sigma K = 4000 + 5000 + 1000 = 10\,000$. This value represents the amount of moment needed to rotate the joint through an angle of 1 rad.

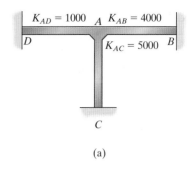

(a)

Distribution Factor (DF).

If a moment $\mathbf{M}$ is applied to a fixed connected joint, the connecting members will each supply a portion of the resisting moment necessary to satisfy moment equilibrium at the joint. That fraction of the total resisting moment supplied by the member is called the *distribution factor* (DF). To obtain its value, imagine the joint is fixed connected to n members. If an applied moment $\mathbf{M}$ causes the joint to rotate an amount θ, then each member i rotates by this same amount. If the stiffness factor of the ith member is K_i, then the moment contributed by the member is $M_i = K_i\theta$. Since equilibrium requires $M = M_1 + M_n = K_1\theta + K_n\theta = \theta\Sigma K_i$ then the distribution factor for the ith member is

$$DF_i = \frac{M_i}{M} = \frac{K_i\theta}{\theta\Sigma K_i}$$

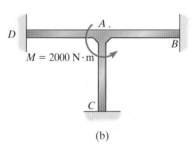

(b)

Canceling the common term θ, it is seen that the distribution factor for a member is equal to the stiffness factor of the member divided by the total stiffness factor for the joint; that is, in general,

$$\boxed{DF = \frac{K}{\Sigma K}} \qquad (12\text{–}2)$$

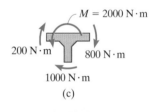

(c)

Fig. 12–4

For example, the distribution factors for members AB, AC, and AD at joint A in Fig. 12–4a are

$$DF_{AB} = 4000/10\,000 = 0.4$$
$$DF_{AC} = 5000/10\,000 = 0.5$$
$$DF_{AD} = 1000/10\,000 = 0.1$$

As a result, if $M = 2000\ \text{N}\cdot\text{m}$ acts at joint A, Fig. 12–4b, the equilibrium moments exerted by the members on the joint, Fig. 12–4c, are

$$M_{AB} = 0.4(2000) = 800\ \text{N}\cdot\text{m}$$
$$M_{AC} = 0.5(2000) = 1000\ \text{N}\cdot\text{m}$$
$$M_{AD} = 0.1(2000) = 200\ \text{N}\cdot\text{m}$$

The statically indeterminate loading in bridge girders that are continuous over their piers can be determined using the method of moment distribution.

Member Relative-Stiffness Factor. Quite often a continuous beam or a frame will be made from the same material so its modulus of elasticity E will be the *same* for all the members. If this is the case, the common factor $4E$ in Eq. 12–1 will *cancel* from the numerator and denominator of Eq. 12–2 when the distribution factor for a joint is determined. Hence, it is *easier* just to determine the member's *relative-stiffness factor*

$$K_R = \frac{I}{L}$$
Far End Fixed

(12–3)

and use this for the computations of the DF.

Carry-Over Factor. Consider again the beam in Fig. 12–3. It was shown in Chapter 11 that $M_{AB} = (4EI/L)\,\theta_A$ (Eq. 11–1) and $M_{BA} = (2EI/L)\,\theta_A$ (Eq. 11–2). Solving for θ_A and equating these equations we get $M_{BA} = M_{AB}/2$. In other words, the moment $\mathbf{M}$ at the pin induces a moment of $\mathbf{M}' = \frac{1}{2}\mathbf{M}$ at the wall. The carry-over factor represents the fraction of $\mathbf{M}$ that is "carried over" from the pin to the wall. Hence, in the case of a beam with *the far end fixed*, the carry-over factor is $+\frac{1}{2}$. The plus sign indicates both moments act in the same direction.

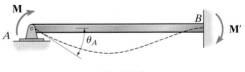

Fig. 12–3

12.2   Moment Distribution for Beams

Moment distribution is based on the principle of successively locking and unlocking the joints of a structure in order to allow the moments at the joints to be distributed and balanced. The best way to explain the method is by examples.

Consider the beam with a constant modulus of elasticity E and having the dimensions and loading shown in Fig. 12–5a. Before we begin, we must first determine the distribution factors at the two ends of each span. Using Eq. 12–1, $K = 4EI/L$, the stiffness factors on either side of B are

$$K_{BA} = \frac{4E(120)(10^6)}{3} = 4E(40)(10^6) \text{ mm}^4/\text{m}$$

$$K_{BC} = \frac{4E(240)(10^6)}{4} = 4E(60)(10^6) \text{ mm}^4/\text{m}$$

Thus, using Eq. 12–2, DF $= K/\Sigma K$, for the ends connected to joint B, we have

$$\text{DF}_{BA} = \frac{4E(40)}{4E(40) + 4E(60)} = 0.4$$

$$\text{DF}_{BC} = \frac{4E(60)}{4E(40) + 4E(60)} = 0.6$$

At the walls, joint A and joint C, the distribution factor depends on the member stiffness factor and the "stiffness factor" of the wall. Since in theory it would take an "infinite" size moment to rotate the wall one radian, the wall stiffness factor is infinite. Thus for joints A and C we have

$$\text{DF}_{AB} = \frac{4E(40)}{\infty + 4E(40)} = 0$$

$$\text{DF}_{CB} = \frac{4E(60)}{\infty + 4E(60)} = 0$$

Note that the above results could also have been obtained if the relative stiffness factor $K_R = I/L$ (Eq. 12–3) had been used for the calculations. Furthermore, as long as a *consistent* set of units is used for the stiffness factor, the DF will always be dimensionless, and at a joint, except where it is located at a fixed wall, the sum of the DFs will always equal 1.

Having computed the DFs, we will now determine the FEMs. Only span BC is loaded, and using the table on the inside back cover for a uniform load, we have

$$(\text{FEM})_{BC} = -\frac{wL^2}{12} = -\frac{6000(4)^2}{12} = -8000 \text{ N} \cdot \text{m}$$

$$(\text{FEM})_{CB} = \frac{wL^2}{12} = \frac{6000(4)^2}{12} = 8000 \text{ N} \cdot \text{m}$$

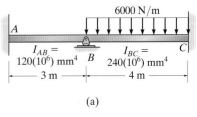

(a)

Fig. 12–5

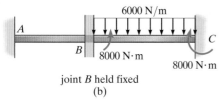

joint B held fixed
(b)

correction moment applied to joint B
(c)

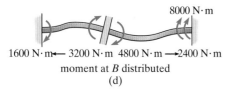

1600 N·m ← 3200 N·m   4800 N·m → 2400 N·m
moment at B distributed
(d)

Joint	A	B		C
Member	AB	BA	BC	CB
DF	0	0.4	0.6	0
FEM			−8000	8000
Dist,CO	1600 ← 3200		4800 → 2400	
ΣM	1600	3200	−3200	10 400

(e)

We begin by assuming joint B is fixed or locked. The fixed-end moment at B then holds span BC in this fixed or locked position as shown in Fig. 12–5b. This, of course, does not represent the actual equilibrium situation at B, since the moments on *each side* of this joint must be equal but opposite. To correct this, we will apply an equal, but opposite moment of 8000 N · m to the joint and allow the joint to rotate freely, Fig. 12–5c. As a result, portions of this moment are distributed in spans BC and BA in accordance with the DFs (or stiffness) of these spans at the joint. Specifically, the moment in BA is 0.4(8000) = 3200 N · m and the moment in BC is 0.6(8000) = 4800 N · m. Finally, due to the released rotation that takes place at B, these moments must be "carried over" since moments are developed at the far ends of the span. Using the carry-over factor of $+\frac{1}{2}$, the results are shown in Fig. 12–5d.

This example indicates the basic steps necessary when distributing moments at a joint: Determine the unbalanced moment acting at the initially "locked" joint, unlock the joint and apply an equal but opposite unbalanced moment to correct the equilibrium, distribute the moment among the connecting spans, and carry the moment in each span over to its other end. The steps are usually presented in tabular form as indicated in Fig. 12–5e. Here the notation Dist, CO indicates a line where moments are distributed, then carried over. In this particular case only *one cycle* of moment distribution is necessary, since the wall supports at A and C "absorb" the moments and no further joints have to be balanced or unlocked to satisfy joint equilibrium. Once distributed in this manner, the moments at each joint are summed, yielding the final results shown on the bottom line of the table in Fig. 12–5e. Notice that joint B is now in equilibrium. Since M_{BC} is negative, this moment is applied to span BC in a counterclockwise sense as shown on free-body diagrams of the beam spans in Fig. 12–5f. With the end moments known, the end shears have been computed from the equations of equilibrium applied to each of these spans.

Consider now the same beam, except the support at C is a rocker, Fig. 12–6a. In this case only *one member* is at joint C, so the distribution factor for member CB at joint C is

$$DF_{CB} = \frac{4E(60)}{4E(60)} = 1$$

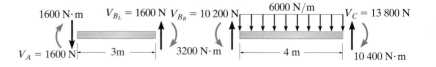

(f)

Fig. 12–5

The other distribution factors and the FEMs are the same as computed previously. They are listed on lines 1 and 2 of the table in Fig. 12–6b. Initially, we will assume joints B and C are locked. We begin by unlocking joint C and placing an equilibrating moment of $-8000\ \text{N} \cdot \text{m}$ at the joint. The entire moment is distributed in member CB since $(1)(-8000)\ \text{N} \cdot \text{m} = -8000\ \text{N} \cdot \text{m}$. The arrow on line 3 indicates that $\frac{1}{2}(-8000)\ \text{N} \cdot \text{m} = -4000\ \text{N} \cdot \text{m}$ is carried over to joint B since joint C has been allowed to rotate freely. Joint C is now *relocked*. Since the total moment at C is *balanced*, a line is placed under the -8000-$\text{N} \cdot \text{m}$ moment. We will now consider the unbalanced $-12\,000$-$\text{N} \cdot \text{m}$ moment at joint B. Here for equilibrium, a $+12\,000$-$\text{N} \cdot \text{m}$ moment is applied to B and this joint is unlocked such that portions of the moment are distributed into BA and BC, that is, $(0.4)(12\,000) = 4800\ \text{N} \cdot \text{m}$ and $(0.6)(12\,000) = 7200\ \text{N} \cdot \text{m}$ as shown on line 4. Also note that $+\frac{1}{2}$ of these moments must be carried over to the fixed wall A and roller C since joint B has rotated. Joint B is now *relocked*. Again joint C is unlocked and the unbalanced moment at the roller is distributed as was done previously. The results are on line 5. Successively locking and unlocking joints B and C will essentially diminish the size of the moment to be balanced until it becomes negligible compared with the original moments, line 14. Each of the steps on lines 3 through 14 should be thoroughly understood. Summing the moments, the final results are shown on line 15, where it is seen that the final moments now satisfy joint equilibrium.

Joint	A	B		C	
Member	AB	BA	BC	CB	
DF	0	0.4	0.6	1	1
FEM			-8000	8000	2
			-4000 ◄— -8000		3
	2400 ◄—	4800	7200 —► 3600		4
			-1800 ◄— -3600		5
	360 ◄—	720	1080 —► 540		6
			-270 ◄— -540		7
	54 ◄—	108	162 —► 81		8
			-40.5 ◄— -81		9
	8.1 ◄—	16.2	24.3 —► 12.2		10
			-6.1 ◄— -12.2		11
	1.2 ◄—	2.4	3.6 —► 1.8		12
			-0.9 ◄— -1.8		13
		0.4	0.5		14
ΣM	2823.3	5647.0	-5647.0	0	15

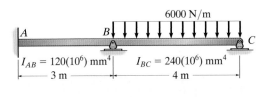

6000 N/m

$I_{AB} = 120(10^6)\ \text{mm}^4$ $I_{BC} = 240(10^6)\ \text{mm}^4$

3 m   4 m

(a)

(b)

Fig. 12–6

Rather than applying the moment distribution process successively to each joint, as illustrated here, it is also possible to apply it to all joints at the *same time*. This scheme is shown in the table in Fig. 12–6c. In this case, we start by fixing all the joints and then balancing and distributing the fixed-end moments at both joints B and C, line 3. Unlocking joints B and C simultaneously (joint A is always fixed), the moments are then carried over to the end of each span, line 4. Again the joints are relocked, and the moments are balanced and distributed, line 5. Unlocking the joints once again allows the moments to be carried over, as shown in line 6. Continuing, we obtain the final results, as before, listed on line 24. By comparison, this method gives a slower convergence to the answer than does the previous method; however, in many cases this method will be more efficient to apply, and for this reason we will use it in the examples that follow. Finally, using the results in either Fig. 12–6b or 12–6c, the free-body diagrams of each beam span are drawn as shown in Fig. 12–6d.

Although several steps were involved in obtaining the final results here, the work required is rather methodical since it requires application of a series of arithmetical steps, rather than solving a set of equations as in the slope deflection method. It should be noted, however, that the

Joint	A	B		C	
Member	AB	BA	BC	CB	
DF	0	0.4	0.6	1	1
FEM			−8000	8000	2
Dist.		3200	4800	−8000	3
CO	1600		−4000	2400	4
Dist.		1600	2400	−2400	5
CO	800		−1200	1200	6
Dist.		480	720	−1200	7
CO	240		−600	360	8
Dist.		240	360	−360	9
CO	120		−180	180	10
Dist.		72	108	−180	11
CO	36		−90	54	12
Dist.		36	54	−54	13
CO	18		−27	27	14
Dist.		10.8	16.2	−27	15
CO	5.4		−13.5	8.1	16
Dist.		5.4	8.1	−8.1	17
CO	2.7		−4.05	4.05	18
Dist.		1.62	2.43	−4.05	19
CO	0.81		−2.02	1.22	20
Dist.		0.80	1.22	−1.22	21
CO	0.40		−0.61	0.61	22
Dist.		0.24	0.37	−0.61	23
ΣM	2823	5647	−5647	0	24

(c)

Fig. 12–6

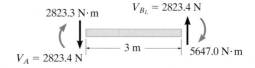

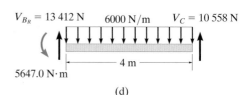

(d)

fundamental process of moment distribution follows the same procedure as any displacement method. There the process is to establish load-displacement relations at each joint and then satisfy joint equilibrium requirements by determining the correct angular displacement for the joint (compatibility). Here, however, the equilibrium and compatibility of rotation at the joint is satisfied *directly*, using a "moment balance" process that incorporates the load-deflection relations (stiffness factors). Further simplification for using moment distribution is possible, and this will be discussed in the next section.

Procedure for Analysis

The following procedure provides a general method for determining the end moments on beam spans using moment distribution.

Distribution Factors and Fixed-End Moments

The joints on the beam should be identified and the stiffness factors for each span at the joints should be calculated. Using these values the distribution factors can be determined from $DF = K/\Sigma K$. Remember that $DF = 0$ for a fixed end and $DF = 1$ for an *end* pin or roller support.

   The fixed-end moments for each loaded span are determined using the table given on the inside back cover. Positive FEMs act clockwise on the span and negative FEMs act counterclockwise. For convenience, these values can be recorded in tabular form, similar to that shown in Fig. 12–6c.

Moment Distribution Process

Assume that all joints at which the moments in the connecting spans must be determined are *initially locked*. Then:

1. Determine the moment that is needed to put each joint in equilibrium.

2. Release or "unlock" the joints and distribute the counterbalancing moments into the connecting span at each joint.

3. Carry these moments in each span over to its other end by multiplying each moment by the carry-over factor $+\frac{1}{2}$.

   By repeating this cycle of locking and unlocking the joints, it will be found that the moment corrections will diminish since the beam tends to achieve its final deflected shape. When a small enough value for the corrections is obtained, the process of cycling should be stopped with no "carry-over" of the last moments. Each column of FEMs, distributed moments, and carry-over moments should then be added. If this is done correctly, moment equilibrium at the joints will be achieved.

EXAMPLE 12.1

Determine the internal moments at each support of the beam shown in Fig. 12–7a. EI is constant.

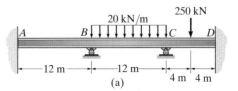

(a)

Fig. 12–7

SOLUTION

The distribution factors at each joint must be computed first.* The stiffness factors for the members are

$$K_{AB} = \frac{4EI}{12} \qquad K_{BC} = \frac{4EI}{12} \qquad K_{CD} = \frac{4EI}{8}$$

Therefore,

$$DF_{AB} = DF_{DC} = 0 \qquad DF_{BA} = DF_{BC} = \frac{4EI/12}{4EI/12 + 4EI/12} = 0.5$$

$$DF_{CB} = \frac{4EI/12}{4EI/12 + 4EI/8} = 0.4 \quad DF_{CD} = \frac{4EI/8}{4EI/12 + 4EI/8} = 0.6$$

The fixed-end moments are

$$(FEM)_{BC} = -\frac{wL^2}{12} = \frac{-20(12)^2}{12} = -240 \text{ kN} \cdot \text{m} \qquad (FEM)_{CB} = \frac{wL^2}{12} = \frac{20(12)^2}{12} = 240 \text{ kN} \cdot \text{m}$$

$$(FEM)_{CD} = -\frac{PL}{8} = \frac{-250(8)}{8} = -250 \text{ kN} \cdot \text{m} \qquad (FEM)_{DC} = \frac{PL}{8} = \frac{250(8)}{8} = 250 \text{ kN} \cdot \text{m}$$

Starting with the FEMs, line 4, Fig. 12–7b, the moments at joints B and C are distributed *simultaneously*, line 5. These moments are then carried over *simultaneously* to the respective ends of each span, line 6. The resulting moments are again simultaneously distributed and carried over, lines 7 and 8. The process is continued until the resulting moments are diminished an appropriate amount, line 13. The resulting moments are found by summation, line 14.

Placing the moments on each beam span and applying the equations of equilibrium yields the end shears shown in Fig. 12–7c and the bending-moment diagram for the entire beam, Fig. 12–7d.

*Here we have used the stiffness factor 4EI/L; however, the relative stiffness factor I/L could also have been used.

Joint	A	B		C		D	
Member	AB	BA	BC	CB	CD	DC	2
DF	0	0.5	0.5	0.4	0.6	0	3
FEM Dist.		120	−240 120	240 4	−250 6	250	4 5
CO Dist.	60	−1	2 −1	60 −24	−36	3	6 7
CO Dist.	−0.5	6	−12 6	−0.5 0.2	0.3	−18	8 9
CO Dist.	3	−0.05	0.1 −0.05	3 −1.2	−1.8	0.2	10 11
CO Dist.	−0.02	0.3	−0.6 0.3	−0.02 0.01	0.01	−0.9	12 13
ΣM	62.5	125.2	−125.2	281.5	−281.5	234.3	14

(b)

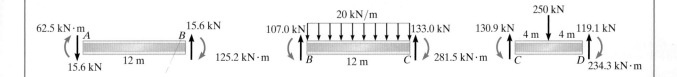

(c)

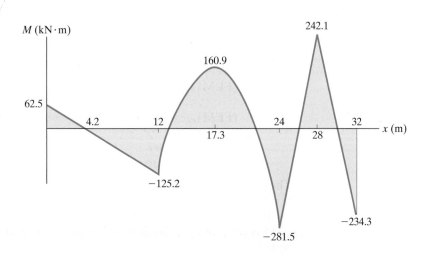

(d)

EXAMPLE 12.2

Determine the internal moment at each support of the beam shown in Fig. 12–8a. The moment of inertia of each span is indicated.

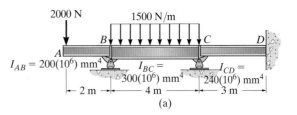

(a)

Fig. 12–8

SOLUTION

In this problem a moment does not get distributed in the overhanging span AB, and so the distribution factor $(DF)_{BA} = 0$. The stiffness of span BC is based on $4EI/L$ since the pin rocker is not at the far end of the beam. The stiffness factors, distribution factors, and fixed-end moments are computed as follows:

$$K_{BC} = \frac{4E(300)(10^6)}{4} = 300(10^6)E \qquad K_{CD} = \frac{4E(240)(10^6)}{3} = 320(10^6)E$$

$$DF_{BC} = 1 - (DF)_{BA} = 1 - 0 = 1$$

$$DF_{CB} = \frac{300E}{300E + 320E} = 0.484$$

$$DF_{CD} = \frac{320E}{300E + 320E} = 0.516$$

$$DF_{DC} = \frac{320E}{\infty + 320E} = 0$$

Due to the overhang,

$$(FEM)_{BA} = 2000 \text{ N}(2 \text{ m}) = 4000 \text{ N} \cdot \text{m}$$

$$(FEM)_{BC} = -\frac{wL^2}{12} = -\frac{1500(4)^2}{12} = -2000 \text{ N} \cdot \text{m}$$

$$(FEM)_{CB} = \frac{wL^2}{12} = \frac{1500(4)^2}{12} = 2000 \text{ N} \cdot \text{m}$$

These values are listed on the fourth line of the table, Fig. 12–8b. The overhanging span requires the internal moment to the left of B to be +4000 N · m. Balancing at joint B requires an internal moment of −4000 N · m to the right of B. As shown on the fifth line of the table −2000 N · m is added to BC in order to satisfy this condition. The distribution and carry-over operations proceed in the usual manner as indicated.

Since the internal moments are known, the moment diagram for the beam can be constructed (Fig. 12–8c).

Joint		B		C		D
Member		BC	CB	CD	DC
DF	0	1	0.484	0.516	0
FEM	4000	−2000	2000		
Dist.		−2000	−968	−1032	
CO		−484	−1000		−516
Dist.		484	484	516	
CO		242	242		258
Dist.		−242	−117.1	−124.9	
CO		−58.6	−121		−62.4
Dist.		58.6	58.6	62.4	
CO		29.3	29.3		31.2
Dist.		−29.3	−14.2	−15.1	
CO		−7.1	−14.6		−7.6
Dist.		7.1	7.1	7.6	
CO		3.5	3.5		3.8
Dist.		−3.5	−1.7	−1.8	
CO		−0.8	−1.8		−0.9
Dist.		0.8	0.9	0.9	
CO		0.4	0.4		0.4
Dist.		−0.4	−0.2	−0.2	
CO		−0.1	−0.2		−0.1
Dist.		0.1	0.1	0.1	
ΣM	4000	−4000	587.1	−587.1	−293.6

(b)

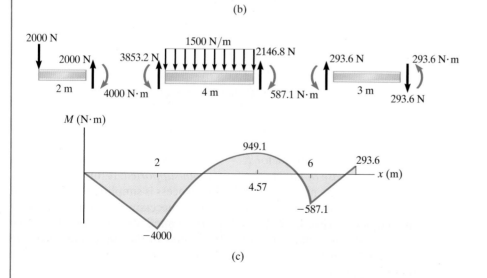

(c)

12

12.3 Stiffness-Factor Modifications

In the previous examples of moment distribution we have considered each beam span to be constrained by a fixed support (locked joint) at its far end when distributing and carrying over the moments. For this reason we have computed the stiffness factors, distribution factors, and the carry-over factors based on the case shown in Fig. 12–9. Here, of course, the stiffness factor is $K = 4EI/L$ (Eq. 12–1), and the carry-over factor is $+\frac{1}{2}$.

In some cases it is possible to modify the stiffness factor of a particular beam span and thereby simplify the process of moment distribution. Three cases where this frequently occurs in practice will now be considered.

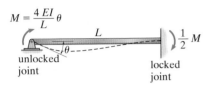

Fig. 12–9

Member Pin Supported at Far End. Many indeterminate beams have their far end span supported by an end pin (or roller) as in the case of joint B in Fig. 12–10a. Here the applied moment $\mathbf{M}$ rotates the end A by an amount θ. To determine θ, the shear in the conjugate beam at A' must be determined, Fig. 12–10b. We have

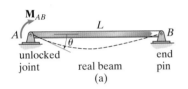

(a)

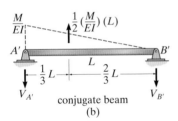

conjugate beam
(b)

Fig. 12–10

$$\curvearrowleft + \Sigma M_{B'} = 0; \qquad V'_A(L) - \frac{1}{2}\left(\frac{M}{EI}\right)L\left(\frac{2}{3}L\right) = 0$$

$$V'_A = \theta = \frac{ML}{3EI}$$

or

$$M = \frac{3EI}{L}\theta$$

Thus, the stiffness factor for this beam is

$$\boxed{K = \frac{3EI}{L}}$$
$$\text{Far End Pinned}$$
$$\text{or Roller Supported}$$

(12–4)

Also, note that *the carry-over factor is zero,* since the pin at B does not support a moment. By comparison, then, *if the far end was fixed supported, the stiffness factor $K = 4EI/L$ would have to be modified by $\frac{3}{4}$ to model the case of having the far end pin supported.* If this modification is considered, the moment distribution process is simplified since the end pin does *not* have to be unlocked–locked successively when distributing the moments. Also, since the end span is pinned, the fixed-end moments for the span are computed using the values in the right column of the table on the inside back cover. Example 12–4 illustrates how to apply these simplifications.

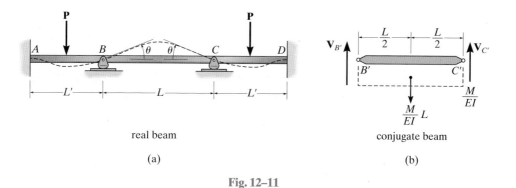

real beam

conjugate beam

(a)

(b)

Fig. 12–11

Symmetric Beam and Loading.

If a beam is symmetric with respect to both its loading and geometry, the bending-moment diagram for the beam will also be symmetric. As a result, a modification of the stiffness factor for the center span can be made, so that moments in the beam only have to be distributed through joints lying on either half of the beam. To develop the appropriate stiffness-factor modification, consider the beam shown in Fig. 12–11a. Due to the symmetry, the internal moments at B and C are equal. Assuming this value to be $\mathbf{M}$, the conjugate beam for span BC is shown in Fig. 12–11b. The slope θ at each end is therefore

$$\zeta + \Sigma M_{C'} = 0; \qquad -V_{B'}(L) + \frac{M}{EI}(L)\left(\frac{L}{2}\right) = 0$$

$$V_{B'} = \theta = \frac{ML}{2EI}$$

or

$$M = \frac{2EI}{L}\theta$$

The stiffness factor for the center span is therefore

$$\boxed{\begin{array}{c} K = \dfrac{2EI}{L} \\[2mm] \text{Symmetric Beam and Loading} \end{array}} \qquad (12\text{–}5)$$

Thus, moments for only half the beam can be distributed provided the stiffness factor for the center span is computed using Eq. 12–5. *By comparison, the center span's stiffness factor will be one half that usually determined using $K = 4EI/L$.*

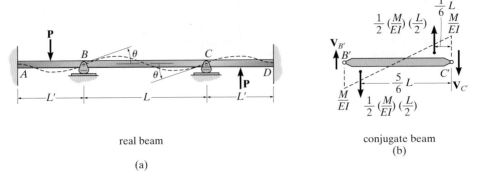

real beam

(a)

conjugate beam

(b)

Fig. 12–12

Symmetric Beam with Antisymmetric Loading.

If a symmetric beam is subjected to antisymmetric loading, the resulting moment diagram will be antisymmetric. As in the previous case, we can modify the stiffness factor of the center span so that only one half of the beam has to be considered for the moment-distribution analysis. Consider the beam in Fig. 12–12a. The conjugate beam for its center span BC is shown in Fig. 12–12b. Due to the antisymmetric loading, the internal moment at B is equal, but opposite to that at C. Assuming this value to be **M**, the slope θ at each end is determined as follows:

$$\zeta + \Sigma M_{C'} = 0; \quad -V_{B'}(L) + \frac{1}{2}\left(\frac{M}{EI}\right)\left(\frac{L}{2}\right)\left(\frac{5L}{6}\right) - \frac{1}{2}\left(\frac{M}{EI}\right)\left(\frac{L}{2}\right)\left(\frac{L}{6}\right) = 0$$

$$V_{B'} = \theta = \frac{ML}{6EI}$$

or

$$M = \frac{6EI}{L}\theta$$

The stiffness factor for the center span is, therefore,

$$K = \frac{6EI}{L}$$
Symmetric Beam with
Antisymmetric Loading   (12–6)

Thus, when the stiffness factor for the beam's center span is computed using Eq. 12–6, the moments in only half the beam have to be distributed. *Here the stiffness factor is one and a half times as large as that determined using $K = 4EI/L$.*

EXAMPLE 12.3

Determine the internal moments at the supports for the beam shown in Fig. 12–13a. *EI* is constant.

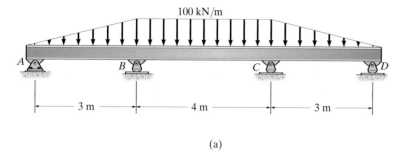

100 kN/m

3 m   4 m   3 m

(a)

Fig. 12–13

SOLUTION

By inspection, the beam and loading are symmetrical. Thus, we will apply $K = 2EI/L$ to compute the stiffness factor of the center span *BC* and therefore use only the left half of the beam for the analysis. The analysis can be shortened even further by using $K = 3EI/L$ for computing the stiffness factor of segment *AB* since the far end *A* is pinned. Furthermore, the distribution of moment at *A* can be skipped by using the FEM for a triangular loading on a span with one end fixed and the other pinned. Thus,

$$K_{AB} = \frac{3EI}{3} \quad \text{(using Eq. 12–4)}$$

$$K_{BC} = \frac{2EI}{4} \quad \text{(using Eq. 12–5)}$$

$$DF_{AB} = \frac{3EI/3}{3EI/3} = 1$$

$$DF_{BA} = \frac{3EI/3}{3EI/3 + 2EI/4} = 0.667$$

$$DF_{BC} = \frac{2EI/4}{3EI/3 + 2EI/4} = 0.333$$

$$(FEM)_{BA} = \frac{wL^2}{15} = \frac{100(3)^2}{15} = 60 \text{ kN} \cdot \text{m}$$

$$(FEM)_{BC} = -\frac{wL^2}{12} = -\frac{100(4)^2}{12} = -133.3 \text{ kN} \cdot \text{m}$$

Joint	A	B	
Member	AB	BA	BC
DF	1	0.667	0.333
FEM			−133.3
Dist.		60	24.4
		48.9	
ΣM	0	108.9	−108.9

(b)

These data are listed in the table in Fig. 12–13b. Computing the stiffness factors as shown above considerably reduces the analysis, since only joint *B* must be balanced and carry-overs to joints *A* and *C* are not necessary. Obviously, joint *C* is subjected to the same internal moment of 108.9 kN · m.

EXAMPLE 12.4

Determine the internal moments at the supports of the beam shown in Fig. 12–14a. The moment of inertia of the two spans is shown in the figure.

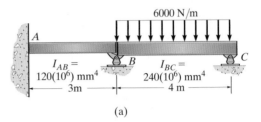

(a)

Fig. 12–14

SOLUTION

Since the beam is roller supported at its far end C, the stiffness of span BC will be computed on the basis of $K = 3EI/L$. We have

$$K_{AB} = \frac{4EI}{L} = \frac{4E(120)(10^6)}{3} = 160(10^6)E$$

$$K_{BC} = \frac{3EI}{L} = \frac{3E(240)(10^6)}{4} = 180(10^6)E$$

Thus,

$$DF_{AB} = \frac{160E}{\infty + 160E} = 0$$

$$DF_{BA} = \frac{160E}{160E + 180E} = 0.4706$$

$$DF_{BC} = \frac{180E}{160E + 180E} = 0.5294$$

$$DF_{CB} = \frac{180E}{180E} = 1$$

Further simplification of the distribution method for this problem is possible by realizing that a *single* fixed-end moment for the end span BC can be used. Using the right-hand column of the table on the inside back cover for a uniformly loaded span having one side fixed, the other pinned, we have

$$(\text{FEM})_{BC} = -\frac{wL^2}{8} = \frac{-6000(4)^2}{8} = -12\,000\ \text{N} \cdot \text{m}$$

The foregoing data are entered into the table in Fig. 12–14b and the moment distribution is carried out. By comparison with Fig. 12–6b, this method considerably simplifies the distribution.

Using the results, the beam's end shears and moment diagrams are shown in Fig. 12–14c.

Joint	A	B		C
Member	AB	BA	BC	CB
DF	0	0.4706	0.5294	1
FEM			−12 000	
Dist.		5647.2	6352.8	
CO	2823.6			
ΣM	2823.6	5647.2	−5647.2	0

(b)

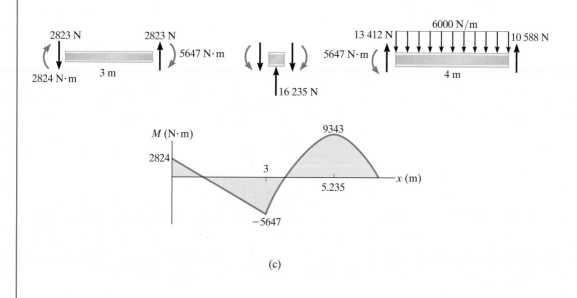

(c)

PROBLEMS

12–1. Determine the moments at B and C. EI is constant. Assume B and C are rollers and A and D are pinned.

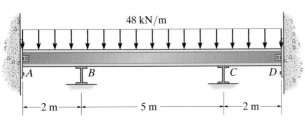

48 kN/m

A B C D

2 m — 5 m — 2 m

Prob. 12–1

12–2. Determine the moments at A, B, and C. Assume the support at B is a roller and A and C are fixed. EI is constant.

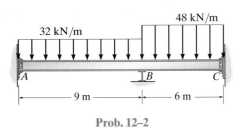

32 kN/m   48 kN/m

A B C

9 m — 6 m

Prob. 12–2

12–3. Determine the moments at A, B, and C, then draw the moment diagram. Assume the support at B is a roller and A and C are fixed. EI is constant.

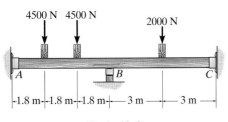

4500 N   4500 N   2000 N

A B C

1.8 m–1.8 m–1.8 m– 3 m — 3 m

Prob. 12–3

***12–4.** Determine the reactions at the supports and then draw the moment diagram. Assume A is fixed. EI is constant.

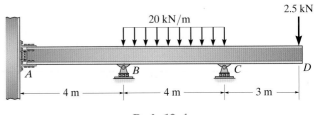

2.5 kN

20 kN/m

A B C D

4 m — 4 m — 3 m

Prob. 12–4

12–5. Determine the moments at B and C, then draw the moment diagram for the beam. Assume C is a fixed support. EI is constant.

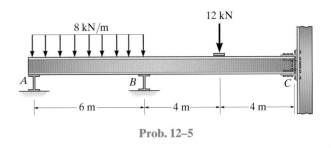

12 kN

8 kN/m

A B C

6 m — 4 m — 4 m

Prob. 12–5

12–6. Determine the moments at B and C, then draw the moment diagram for the beam. All connections are pins. Assume the horizontal reactions are zero. EI is constant.

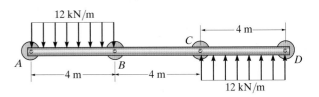

12 kN/m   4 m

A B C D

4 m — 4 m   12 kN/m

Prob. 12–6

12–7. Determine the reactions at the supports. Assume A is fixed and B and C are rollers that can either push or pull on the beam. EI is constant.

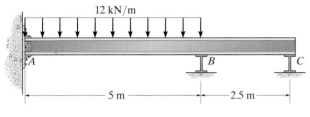

12 kN/m

5 m   2.5 m

Prob. 12–7

*12–8. Determine the moments at B and C, then draw the moment diagram for the beam. Assume the supports at B and C are rollers and A and D are pins. EI is constant.

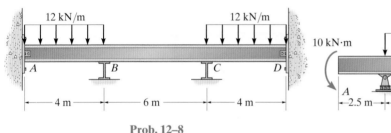

12 kN/m   12 kN/m

4 m   6 m   4 m

Prob. 12–8

12–9. Determine the moments at B and C, then draw the moment diagram for the beam. Assume the supports at B and C are rollers and A is a pin. EI is constant.

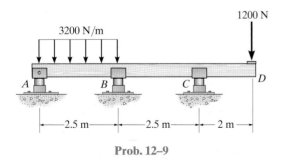

1200 N

3200 N/m

2.5 m   2.5 m   2 m

Prob. 12–9

12–10. Determine the moment at B, then draw the moment diagram for the beam. Assume the supports at A and C are rollers and B is a pin. EI is constant.

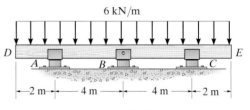

6 kN/m

2 m   4 m   4 m   2 m

Prob. 12–10

12–11. Determine the moments at B, C, and D, then draw the moment diagram for the beam. EI is constant.

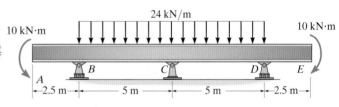

24 kN/m

10 kN·m   10 kN·m

2.5 m   5 m   5 m   2.5 m

Prob. 12–11

*12–12. Determine the moment at B, then draw the moment diagram for the beam. Assume the support at A is pinned, B is a roller and C is fixed. EI is constant.

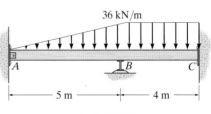

36 kN/m

5 m   4 m

Prob. 12–12

12.4   Moment Distribution for Frames: No Sidesway

Application of the moment-distribution method for frames having no sidesway follows the same procedure as that given for beams. To minimize the chance for errors, it is suggested that the analysis be arranged in a tabular form, as in the previous examples. Also, the distribution of moments can be shortened if the stiffness factor of a span can be modified as indicated in the previous section.

EXAMPLE | 12.5

Determine the internal moments at the joints of the frame shown in Fig. 12–15a. There is a pin at E and D and a fixed support at A. EI is constant.

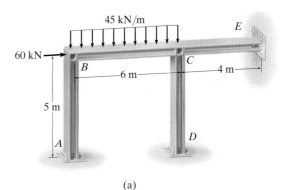

(a)

Joint	A	B		C			D	E
Member	AB	BA	BC	CB	CD	CE	DC	EC
DF	0	0.545	0.455	0.330	0.298	0.372	1	1
FEM			−135	135				
Dist.		73.6	61.4	−44.6	−40.2	−50.2		
CO	36.8		−22.3	30.7				
Dist.		12.2	10.1	−10.1	−9.1	−11.5		
CO	6.1		−5.1	5.1				
Dist.		2.8	2.3	−1.7	−1.5	−1.9		
CO	1.4		−0.8	1.2				
Dist.		0.4	0.4	−0.4	−0.4	−0.4		
CO	0.2		−0.2	0.2				
Dist.		0.1	0.1	−0.1	0.0	−0.1		
ΣM	44.5	89.1	−89.1	115	−51.2	−64.1		

(b)

Fig. 12–15

SOLUTION

By inspection, the pin at E will prevent the frame from sidesway. The stiffness factors of CD and CE can be computed using $K = 3EI/L$ since the far ends are pinned. Also, the 20-kN load does not contribute a FEM since it is applied at joint B. Thus,

$$K_{AB} = \frac{4EI}{5} \qquad K_{BC} = \frac{4EI}{6} \qquad K_{CD} = \frac{3EI}{5} \qquad K_{CE} = \frac{3EI}{4}$$

$$DF_{AB} = 0$$

$$DF_{BA} = \frac{4EI/5}{4EI/5 + 4EI/6} = 0.545$$

$$DF_{BC} = 1 - 0.545 = 0.455$$

$$DF_{CB} = \frac{4EI/6}{4EI/6 + 3EI/5 + 3EI/4} = 0.330$$

$$DF_{CD} = \frac{3EI/5}{4EI/6 + 3EI/5 + 3EI/4} = 0.298$$

$$DF_{CE} = 1 - 0.330 - 0.298 = 0.372$$

$$DF_{DC} = 1 \qquad DF_{EC} = 1$$

$$(FEM)_{BC} = \frac{-wL^2}{12} = \frac{-45(6)^2}{12} = -135 \text{ kN} \cdot \text{m}$$

$$(FEM)_{CB} = \frac{wL^2}{12} = \frac{45(6)^2}{12} = 135 \text{ kN} \cdot \text{m}$$

The data are shown in the table in Fig. 12–15b. Here the distribution of moments successively goes to joints B and C. The final moments are shown on the last line.

Using these data, the moment diagram for the frame is constructed in Fig. 12–15c.

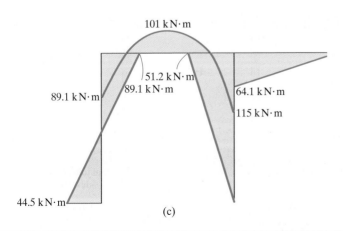

(c)

12.5 Moment Distribution for Frames: Sidesway

It has been shown in Sec. 11–5 that frames that are nonsymmetrical or subjected to nonsymmetrical loadings have a tendency to sidesway. An example of one such case is shown in Fig. 12–16a. Here the applied loading **P** will create unequal moments at joints B and C such that the frame will deflect an amount Δ to the right. To determine this deflection and the internal moments at the joints using moment distribution, we will use the principle of superposition. In this regard, the frame in Fig. 12–16b is first considered held from sidesway by applying an artificial joint support at C. Moment distribution is applied and then by statics the restraining force **R** is determined. The equal, but opposite, restraining force is then applied to the frame, Fig. 12–16c, and the moments in the frame are calculated. One method for doing this last step requires first *assuming* a numerical value for one of the internal moments, say $\mathbf{M}'_{BA}$. Using moment distribution and statics, the deflection Δ' and external force **R**$'$ corresponding to the assumed value of $\mathbf{M}'_{BA}$ can then be determined. Since linear elastic deformations occur, the force **R**$'$ develops moments in the frame that are *proportional* to those developed by **R**. For example, if $\mathbf{M}'_{BA}$ and **R**$'$ are known, the moment at B developed by **R** will be $M_{BA} = M'_{BA}(R/R')$. Addition of the joint moments for both cases, Fig. 12–16b and c, will yield the actual moments in the frame, Fig. 12–16a. Application of this technique is illustrated in Examples 12–6 through 12–8.

Multistory Frames. Quite often, multistory frameworks may have several *independent* joint displacements, and consequently the moment distribution analysis using the above techniques will involve more computation. Consider, for example, the two-story frame shown in Fig. 12–17a. This structure can have two independent joint displacements, since the sidesway Δ_1 of the first story is independent of any displacement

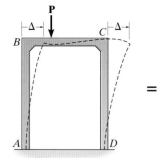

=

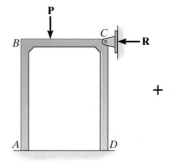

artificial joint applied
(no sidesway)

+

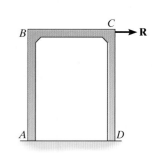

artificial joint removed
(sidesway)

(a)

(b)

(c)

Fig. 12–16

Δ_2 of the second story. Unfortunately, these displacements are not known initially, so the analysis must proceed on the basis of superposition, in the same manner as discussed previously. In this case, two restraining forces $\mathbf{R_1}$ and $\mathbf{R_2}$ are applied, Fig. 12–17b, and the fixed-end moments are determined and distributed. Using the equations of equilibrium, the numerical values of $\mathbf{R_1}$ and $\mathbf{R_2}$ are then determined. Next, the restraint at the floor of the first story is removed and the floor is given a displacement Δ'. This displacement causes fixed-end moments (FEMs) in the frame, which can be assigned specific numerical values. By distributing these moments and using the equations of equilibrium, the associated numerical values of $\mathbf{R_1'}$ and $\mathbf{R_2'}$ can be determined. In a similar manner, the floor of the second story is then given a displacement Δ'', Fig. 12–17d. Assuming numerical values for the fixed-end moments, the moment distribution and equilibrium analysis will yield specific values of $\mathbf{R_1''}$ and $\mathbf{R_2''}$. Since the last two steps associated with Figs. 12–17c and d depend on *assumed* values of the FEMs, correction factors C' and C'' must be applied to the distributed moments. With reference to the restraining forces in Figs. 12–17c and 12–17d, we require equal but opposite application of $\mathbf{R_1}$ and $\mathbf{R_2}$ to the frame, such that

$$R_2 = -C'R_2' + C''R_2''$$
$$R_1 = +C'R_1' - C''R_1''$$

Simultaneous solution of these equations yields the values of C' and C''. These correction factors are then multiplied by the internal joint moments found from the moment distribution in Figs. 12–17c and 12–17d. The resultant moments are then found by adding these corrected moments to those obtained for the frame in Fig. 12–17b.

Other types of frames having independent joint displacements can be analyzed using this same procedure; however, it must be admitted that the foregoing method does require quite a bit of numerical calculation. Although some techniques have been developed to shorten the calculations, it is best to solve these types of problems on a computer, preferably using a matrix analysis. The techniques for doing this will be discussed in Chapter 16.

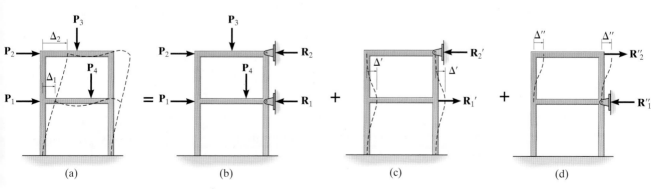

(a)   (b)   (c)   (d)

Fig. 12–17

EXAMPLE 12.6

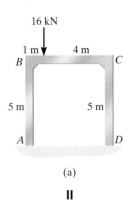

(a)

||

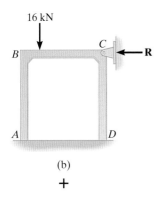

(b)

+

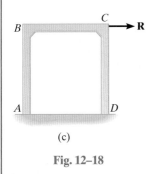

(c)

Fig. 12–18

Determine the moments at each joint of the frame shown in Fig. 12–18a. EI is constant.

SOLUTION

First we consider the frame held from sidesway as shown in Fig. 12–18b. We have

$$(\text{FEM})_{BC} = -\frac{16(4)^2(1)}{(5)^2} = -10.24 \text{ kN} \cdot \text{m}$$

$$(\text{FEM})_{CB} = \frac{16(1)^2(4)}{(5)^2} = 2.56 \text{ kN} \cdot \text{m}$$

The stiffness factor of each span is computed on the basis of $4EI/L$ or by using the relative-stiffness factor I/L. The DFs and the moment distribution are shown in the table, Fig. 12–18d. Using these results, the equations of equilibrium are applied to the free-body diagrams of the columns in order to determine $\mathbf{A}_x$ and $\mathbf{D}_x$, Fig. 12–18e. From the free-body diagram of the entire frame (not shown) the joint restraint $\mathbf{R}$ in Fig. 12–18b has a magnitude of

$$\Sigma F_x = 0; \qquad R = 1.73 \text{ kN} - 0.81 \text{ kN} = 0.92 \text{ kN}$$

An equal but opposite value of $R = 0.92$ kN must now be applied to the frame at C and the internal moments computed, Fig. 12–18c. To solve the problem of computing these moments, we will assume a force $\mathbf{R}'$ is applied at C, causing the frame to deflect $\mathbf{\Delta}'$ as shown in Fig. 12–18f. Here the joints at B and C are *temporarily restrained from rotating*, and as a result the fixed-end moments at the ends of the columns are determined from the formula for deflection found on the inside back cover, that is,

Joint	A	B		C		D
Member	AB	BA	BC	CB	CD	DC
DF	0	0.5	0.5	0.5	0.5	0
FEM			−10.24	2.56		
Dist.		5.12	5.12	−1.28	−1.28	
CO	2.56		−0.64	2.56		−0.64
Dist.		0.32	0.32	−1.28	−1.28	
CO	0.16		−0.64	0.16		−0.64
Dist.		0.32	0.32	−0.08	−0.08	
CO	0.16		−0.04	0.16		−0.04
Dist.		0.02	0.02	−0.08	−0.08	
ΣM	2.88	5.78	−5.78	2.72	−2.72	−1.32

(d)

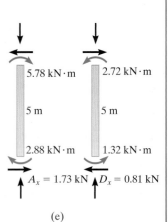

(e)

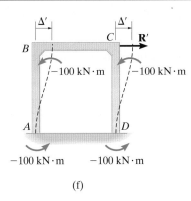

(f)

Joint	A	B		C		D
Member	AB	BA	BC	CB	CD	DC
DF	0	0.5	0.5	0.5	0.5	0
FEM	−100	−100			−100	−100
Dist.		50	50	50	50	
CO	25			25	25	25
Dist.		−12.5	−12.5	−12.5	−12.5	
CO	−6.25		−6.25	−6.25		−6.25
Dist.		3.125	3.125	3.125	3.125	
CO	1.56		1.56	1.56		1.56
Dist.		−0.78	−0.78	−0.78	−0.78	
CO	−0.39		−0.39	−0.39		−0.39
Dist.		0.195	0.195	0.195	0.195	
ΣM	−80.00	−60.00	60.00	60.00	−60.00	−80.00

(g)

$$M = \frac{6EI\Delta}{L^2}$$

Since *both* B and C happen to be displaced the same amount Δ', and AB and DC have the *same* E, I, and L, the FEM in AB will be the *same* as that in DC. As shown in Fig. 12–18f, we will arbitrarily *assume* this fixed-end moment to be

$$(\text{FEM})_{AB} = (\text{FEM})_{BA} = (\text{FEM})_{CD} = (\text{FEM})_{DC} = -100 \text{ kN} \cdot \text{m}$$

A *negative sign* is necessary since the moment must act *counterclockwise* on the column for deflection Δ' to the right. The value of **R′** associated with this −100 kN·m moment can now be determined. The moment distribution of the FEMs is shown in Fig. 12–18g. From equilibrium, the horizontal reactions at A and D are calculated, Fig. 12–18h. Thus, for the entire frame we require

$$\Sigma F_x = 0; \qquad R' = 28 + 28 = 56.0 \text{ kN}$$

Hence, $R' = 56.0$ kN creates the moments tabulated in Fig. 12–18g. Corresponding moments caused by $R = 0.92$ kN can be determined by proportion. Therefore, the resultant moment in the frame, Fig. 12–18a, is equal to the *sum* of those calculated for the frame in Fig. 12–18b plus the proportionate amount of those for the frame in Fig. 12–18c. We have

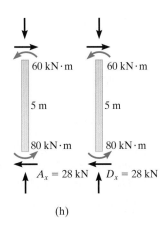

(h)

$$M_{AB} = 2.88 + \tfrac{0.92}{56.0}(-80) = 1.57 \text{ kN} \cdot \text{m} \qquad \textit{Ans.}$$

$$M_{BA} = 5.78 + \tfrac{0.92}{56.0}(-60) = 4.79 \text{ kN} \cdot \text{m} \qquad \textit{Ans.}$$

$$M_{BC} = -5.78 + \tfrac{0.92}{56.0}(60) = -4.79 \text{ kN} \cdot \text{m} \qquad \textit{Ans.}$$

$$M_{CB} = 2.72 + \tfrac{0.92}{56.0}(60) = 3.71 \text{ kN} \cdot \text{m} \qquad \textit{Ans.}$$

$$M_{CD} = -2.72 + \tfrac{0.92}{56.0}(-60) = -3.71 \text{ kN} \cdot \text{m} \qquad \textit{Ans.}$$

$$M_{DC} = -1.32 + \tfrac{0.92}{56.0}(-80) = -2.63 \text{ kN} \cdot \text{m} \qquad \textit{Ans.}$$

EXAMPLE | 12.7

Determine the moments at each joint of the frame shown in Fig. 12–19a. The moment of inertia of each member is indicated in the figure.

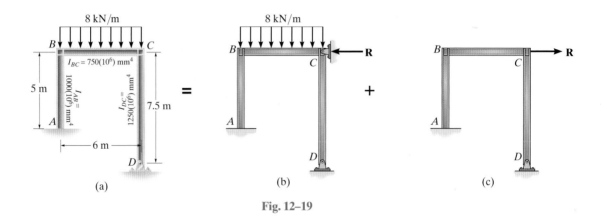

Fig. 12–19

SOLUTION

The frame is first held from sidesway as shown in Fig. 12–19b. The internal moments are computed at the joints as indicated in Fig. 12–19d. Here the stiffness factor of CD was computed using $3EI/L$ since there is a pin at D. Calculation of the horizontal reactions at A and D is shown in Fig. 12–19e. Thus, for the entire frame,

$$\Sigma F_x = 0; \qquad R = 5.784 - 2 = 3.78 \text{ kN}$$

Joint	A	B		C		D
Member	AB	BA	BC	CB	CD	DC
DF	0	0.615	0.385	0.5	0.5	1
FEM			−24	24		
Dist.		14.76	9.24	−12	−12	
CO	7.38		−6	4.62		
Dist.		3.69	2.31	−2.31	−2.31	
CO	1.84		−1.16	1.16		
Dist.		0.713	0.447	−0.58	−0.58	
CO	0.357		−0.29	0.224		
Dist.		0.18	0.11	−0.11	−0.11	
ΣM	9.58	19.34	−19.34	15.00	−15.00	0

(d)

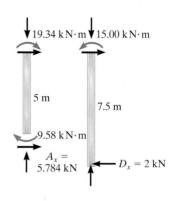

(e)

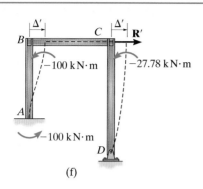

(f)

Joint	A	B		C		D
Member	AB	BA	BC	CB	CD	DC
DF	0	0.615	0.385	0.5	0.5	1
FEM	−100	−100			−27.78	
Dist.		61.5	38.5	13.89	13.89	
CO	30.75		6.94	19.25		
Dist.		−4.27	−2.67	−9.625	−9.625	
CO	−2.14		−4.81	−1.34		
Dist.		2.96	1.85	0.67	0.67	
CO	1.48		0.33	0.92		
Dist.		−0.20	−0.13	−0.46	−0.46	
ΣM	−69.91	−40.01	40.01	23.31	−23.31	0

(g)

The opposite force is now applied to the frame as shown in Fig. 12–19c. As in the previous example, we will consider a force **R′** acting as shown in Fig. 12–19f. As a result, joints B and C are displaced by the same amount Δ′. The fixed-end moments for BA are computed from

$$(\text{FEM})_{AB} = (\text{FEM})_{BA} = -\frac{6EI\Delta}{L^2} = -\frac{6E(1000)(10^6)\Delta'}{(5)^2}$$

However, from the table on the inside back cover, for CD we have

$$(\text{FEM})_{CD} = -\frac{3EI\Delta}{L^2} = -\frac{3E(1250)(10^6)\Delta'}{(7.5)^2}$$

Assuming the FEM for AB is −100 kN · m as shown in Fig. 12–19f, the *corresponding* FEM at C, causing the *same* Δ′, is found by comparison, i.e.,

$$\Delta' = -\frac{(-100)(5)^2}{6E(1000)(10^6)} = -\frac{(\text{FEM})_{CD}(7.5)^2}{3E(1250)(10^6)}$$

$$(\text{FEM})_{CD} = -27.78 \text{ kN} \cdot \text{m}$$

Moment distribution for these FEMs is tabulated in Fig. 12–19g. Computation of the horizontal reactions at A and D is shown in Fig. 12–19h. Thus, for the entire frame,

$$\Sigma F_x = 0; \qquad R' = 21.98 + 3.11 = 25.1 \text{ kN}$$

The resultant moments in the frame are therefore

$$M_{AB} = 9.58 + \left(\tfrac{3.78}{25.1}\right)(-69.91) = -0.948 \text{ kN} \cdot \text{m} \qquad Ans.$$

$$M_{BA} = 19.34 + \left(\tfrac{3.78}{25.1}\right)(-40.01) = 13.3 \text{ kN} \cdot \text{m} \qquad Ans.$$

$$M_{BC} = -19.34 + \left(\tfrac{3.78}{25.1}\right)(40.01) = -13.3 \text{ kN} \cdot \text{m} \qquad Ans.$$

$$M_{CB} = 15.00 + \left(\tfrac{3.78}{25.1}\right)(23.31) = 18.5 \text{ kN} \cdot \text{m} \qquad Ans.$$

$$M_{CD} = -15.00 + \left(\tfrac{3.78}{25.1}\right)(-23.31) = -18.5 \text{ kN} \cdot \text{m} \qquad Ans.$$

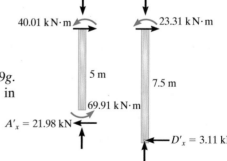

(h)

12

EXAMPLE 12.8

Determine the moments at each joint of the frame shown in Fig. 12–20a. EI is constant.

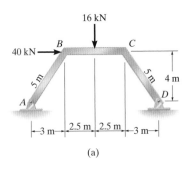

(a)

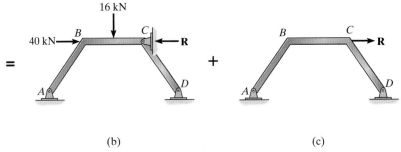

(b)   (c)

Fig. 12–20

SOLUTION

First sideway is prevented by the restraining force **R**, Fig. 12–20b. The FEMs for member BC are

$$(\text{FEM})_{BC} = -\frac{16(5)}{8} = -10 \text{ kN} \cdot \text{m} \qquad (\text{FEM})_{CB} = \frac{16(5)}{8} = 10 \text{ kN} \cdot \text{m}$$

Since spans AB and DC are pinned at their ends, the stiffness factor is computed using $3EI/L$. The moment distribution is shown in Fig. 12–20d.

Using these results, the *horizontal reactions* at A and D must be determined. This is done using an equilibrium analysis of *each member,* Fig. 12–20e. Summing moments about points B and C on each leg, we have

$$\zeta + \Sigma M_B = 0; \qquad -5.97 + A_x(4) - 8(3) = 0 \qquad A_x = 7.49 \text{ kN}$$
$$\zeta + \Sigma M_C = 0; \qquad 5.97 - D_x(4) + 8(3) = 0 \qquad D_x = 7.49 \text{ kN}$$

Thus, for the entire frame,

$$\Sigma F_x = 0; \qquad R = 7.49 - 7.49 + 40 = 40 \text{ kN}$$

Joint	A	B		C		D
Member	AB	BA	BC	CB	CD	DC
DF	1	0.429	0.571	0.571	0.429	1
FEM			−10	10		
Dist.		4.29	5.71	−5.71	−4.29	
CO			−2.86	2.86		
Dist.		1.23	1.63	−1.63	−1.23	
CO			−0.82	0.82		
Dist.		0.35	0.47	−0.47	−0.35	
CO			−0.24	0.24		
Dist.		0.10	0.13	−0.13	−0.10	
ΣM	0	5.97	−5.97	5.97	−5.97	0

(d)

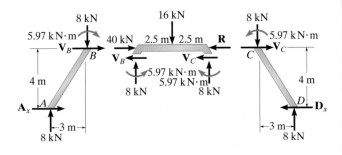

(e)

The opposite force **R** is now applied to the frame as shown in Fig. 12–20c. In order to determine the internal moments developed by **R** we will first consider the force **R′** acting as shown in Fig. 12–20f. Here the dashed lines do not represent the distortion of the frame members; instead, they are constructed as straight lines extended to the final positions B' and C' of points B and C, respectively. Due to the symmetry of the frame, the displacement $BB' = CC' = \Delta'$. Furthermore, these displacements cause BC to rotate. The vertical distance between B' and C' is $1.2\Delta'$, as shown on the displacement diagram, Fig. 12–20g. Since each span undergoes end-point displacements that cause the spans to rotate, fixed-end moments are induced in the spans. These are:
$(\text{FEM})_{BA} = (\text{FEM})_{CD} = -3EI\Delta'/(5)^2$, $(\text{FEM})_{BC} = (\text{FEM})_{CB} = 6EI(1.2\Delta')/(5)^2$.

Notice that for BA and CD the moments are *negative* since clockwise rotation of the span causes a *counterclockwise* FEM.

If we arbitrarily assign a value of $(\text{FEM})_{BA} = (\text{FEM})_{CD} = -100$ kN · m, then equating Δ' in the above formulas yields $(\text{FEM})_{BC} = (\text{FEM})_{CB} = 240$ kN · m. These moments are applied to the frame and distributed, Fig. 12–20h. Using these results, the equilibrium analysis is shown in Fig. 12–20i. For each leg, we have

$\zeta + \Sigma M_B = 0;\quad -A'_x(4) + 58.72(3) + 146.80 = 0 \qquad A'_x = 80.74$ kN
$\zeta + \Sigma M_C = 0;\quad -D'_x(4) + 58.72(3) + 146.80 = 0 \qquad D'_x = 80.74$ kN

Thus, for the entire frame,

$\Sigma F_x = 0;\qquad R' = 80.74 + 80.74 = 161.48$ kN

The resultant moments in the frame are therefore

$$M_{BA} = 5.97 + \left(\tfrac{40}{161.48}\right)(-146.80) = -30.4 \text{ kN} \cdot \text{m} \qquad Ans.$$

$$M_{BC} = -5.97 + \left(\tfrac{40}{161.48}\right)(146.80) = 30.4 \text{ kN} \cdot \text{m} \qquad Ans.$$

$$M_{CB} = 5.97 + \left(\tfrac{40}{161.48}\right)(146.80) = 42.3 \text{ kN} \cdot \text{m} \qquad Ans.$$

$$M_{CD} = -5.97 + \left(\tfrac{40}{161.48}\right)(-146.80) = -42.3 \text{ kN} \cdot \text{m} \qquad Ans.$$

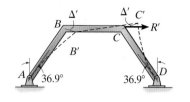

(f)

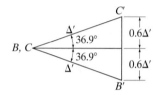

(g)

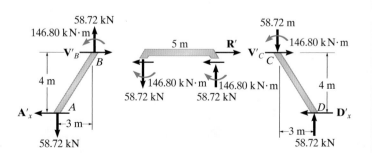

(i)

Joint	A	B		C		D
Member	AB	BA	BC	CB	CD	DC
DF	1	0.429	0.571	0.571	0.429	1
FEM		−100	240	240	−100	
Dist.		−60.06	−79.94	−79.94	−60.06	
CO			−39.97	−39.97		
Dist.		17.15	22.82	22.82	17.15	
CO			11.41	11.41		
Dist.		−4.89	−6.52	−6.52	−4.89	
CO			−3.26	−3.26		
Dist.		1.40	1.86	1.86	1.40	
CO			0.93	0.93		
Dist.		−0.40	−0.53	−0.53	−0.40	
ΣM	0	−146.80	146.80	146.80	−146.80	0

(h)

12–13. Determine the moment at B, then draw the moment diagram for each member of the frame. Assume the supports at A and C are pins. EI is constant.

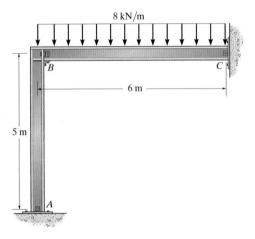

Prob. 12–13

12–14. Determine the moments at the ends of each member of the frame. Assume the joint at B is fixed, C is pinned, and A is fixed. The moment of inertia of each member is listed in the figure. $E = 200$ GPa.

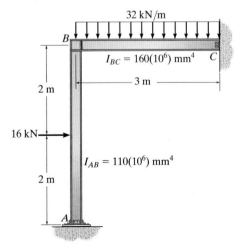

Prob. 12–14

12–15. Determine the reactions at A and D. Assume the supports at A and D are fixed and B and C are fixed connected. EI is constant.

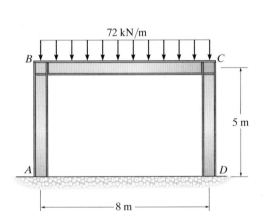

Prob. 12–15

*12–16. Determine the moments at D and C, then draw the moment diagram for each member of the frame. Assume the supports at A and B are pins and D and C are fixed joints. EI is constant.

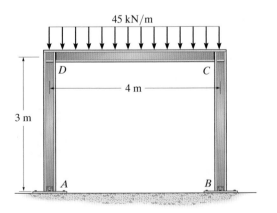

Prob. 12–16

12–17.   Determine the moments at the fixed support A and joint D and then draw the moment diagram for the frame. Assume B is pinned.

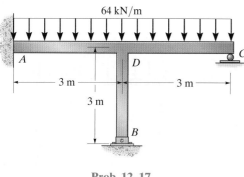

64 kN/m

A

D

C

3 m

3 m

3 m

B

Prob. 12–17

12–19.   The frame is made from pipe that is fixed connected. If it supports the loading shown, determine the moments developed at each of the joints. EI is constant.

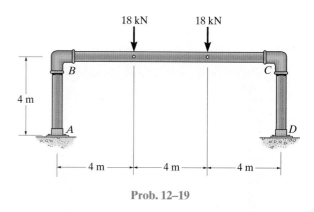

18 kN   18 kN

B

C

4 m

A

D

4 m   4 m   4 m

Prob. 12–19

12–18.   Determine the moments at each joint of the frame, then draw the moment diagram for member BCE. Assume B, C, and E are fixed connected and A and D are pins. $E = 200$ GPa.

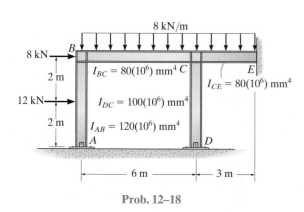

8 kN/m

B

8 kN

2 m

$I_{BC} = 80(10^6)$ mm^4 C

E

$I_{CE} = 80(10^6)$ mm^4

12 kN

$I_{DC} = 100(10^6)$ mm^4

2 m

$I_{AB} = 120(10^6)$ mm^4

A

D

6 m   3 m

Prob. 12–18

*****12–20.**   Determine the moments at B and C, then draw the moment diagram for each member of the frame. Assume the supports at A, E, and D are fixed. EI is constant.

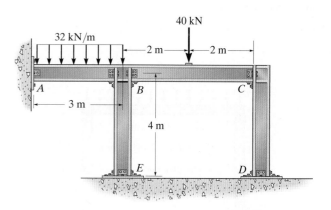

40 kN

32 kN/m

2 m   2 m

A

B

C

3 m

4 m

E

D

Prob. 12–20

12–21. Determine the moments at D and C, then draw the moment diagram for each member of the frame. Assume the supports at A and B are pins. EI is constant.

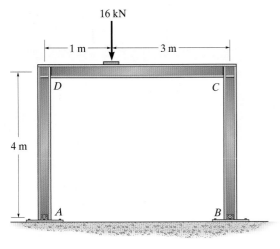

Prob. 12–21

12–23. Determine the moments acting at the ends of each member of the frame. EI is the constant.

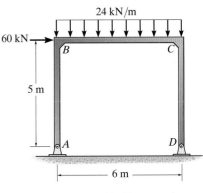

Prob. 12–23

12–22. Determine the moments acting at the ends of each member. Assume the supports at A and D are fixed. The moment of inertia of each member is indicated in the figure. $E = 200$ GPa.

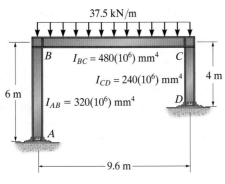

Prob. 12–22

*12–24.** Determine the moments acting at the ends of each member. Assume the joints are fixed connected and A and B are fixed supports. EI is constant.

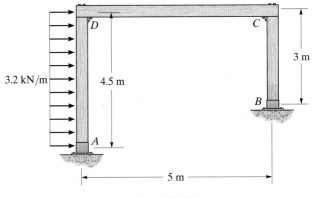

Prob. 12–24

12–25. Determine the moments at joints B and C, then draw the moment diagram for each member of the frame. The supports at A and D are pinned. EI is constant.

12–26. Determine the moments at C and D, then draw the moment diagram for each member of the frame. Assume the supports at A and B are pins. EI is constant.

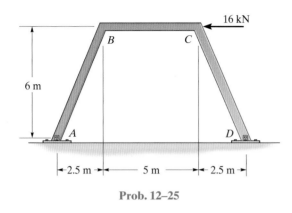

Prob. 12–25

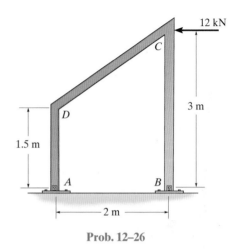

Prob. 12–26

CHAPTER REVIEW

Moment distribution is a method of successive approximations that can be carried out to any desired degree of accuracy. It initially requires locking all the joints of the structure. The equilibrium moment at each joint is then determined, the joints are unlocked and this moment is distributed onto each connecting member, and half its value is carried over to the other side of the span. This cycle of locking and unlocking the joints is repeated until the carry-over moments become acceptably small. The process then stops and the moment at each joint is the sum of the moments from each cycle of locking and unlocking.

(Ref: Section 12.1)

The process of moment distribution is conveniently done in tabular form. Before starting, the fixed-end moment for each span must be calculated using the table on the inside back cover of the book. The distribution factors are found by dividing a member's stiffness by the total stiffness of the joint. For members having a far end fixed, use $K = 4EI/L$; for a far-end pinned or roller supported member, $K = 3EI/L$; for a symmetric span and loading, $K = 2EI/L$; and for an antisymmetric loading, $K = 6EI/L$. Remember that the distribution factor for a fixed end is $DF = 0$, and for a pin or roller-supported end, $DF = 1$.

(Ref: Sections 12.2 & 12.3)

The use of variable-moment-of-inertia girders has reduced considerably the deadweight loading of each of these spans.

Beams and Frames Having Nonprismatic Members

13

In this chapter we will apply the slope-deflection and moment-distribution methods to analyze beams and frames composed of nonprismatic members. We will first discuss how the necessary carry-over factors, stiffness factors, and fixed-end moments are obtained. This is followed by a discussion related to using tabular values often published in design literature. Finally, the analysis of statically indeterminate structures using the slope-deflection and moment-distribution methods will be discussed.

13.1 Loading Properties of Nonprismatic Members

Often, to save material, girders used for long spans on bridges and buildings are designed to be nonprismatic, that is, to have a variable moment of inertia. The most common forms of structural members that are nonprismatic have haunches that are either stepped, tapered, or parabolic, Fig. 13–1. Provided we can express the member's moment of inertia as a function of the length coordinate x, then we can use the principle of virtual work or Castigliano's theorem as discussed in Chapter 9 to find its deflection. The equations are

$$\Delta = \int_0^l \frac{Mm}{EI}\,dx \quad \text{or} \quad \Delta = \int_0^l \frac{\partial M}{\partial P}\frac{M}{EI}\,dx$$

If the member's geometry and loading require evaluation of an integral that cannot be determined in closed form, then Simpson's rule or some other numerical technique will have to be used to carry out the integration.

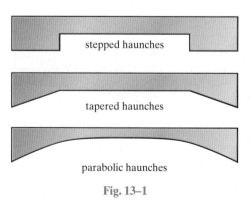

stepped haunches

tapered haunches

parabolic haunches

Fig. 13–1

13

If the slope deflection equations or moment distribution are used to determine the reactions on a nonprismatic member, then we must first calculate the following properties for the member.

Fixed-End Moments (FEM).
The end moment reactions on the member that is assumed fixed supported, Fig. 13–2a.

Stiffness Factor (K).
The magnitude of moment that must be applied to the end of the member such that the end rotates through an angle of $\theta = 1$ rad. Here the moment is applied at the pin support, while the other end is assumed fixed, Fig. 13–2b.

Carry-Over Factor (COF).
Represents the numerical fraction (C) of the moment that is "carried over" from the pin-supported end to the wall, Fig. 13.2c.

Once obtained, the computations for the stiffness and carry-over factors can be checked, in part, by noting an important relationship that exists between them. In this regard, consider the beam in Fig. 13–3 subjected to the loads and deflections shown. Application of the Maxwell-Betti reciprocal theorem requires the work done by the loads in Fig. 13–3a acting through the displacements in Fig. 13–3b be equal to the work of the loads in Fig. 13–3b acting through the displacements in Fig. 13–3a, that is,

$$U_{AB} = U_{BA}$$

$$K_A(0) + C_{AB}K_A(1) = C_{BA}K_B(1) + K_B(0)$$

or

$$C_{AB}K_A = C_{BA}K_B \qquad (13\text{--}1)$$

Hence, once determined, the stiffness and carry-over factors must satisfy Eq. 13–1.

The tapered concrete bent is used to support the girders of this highway bridge.

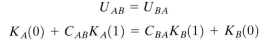

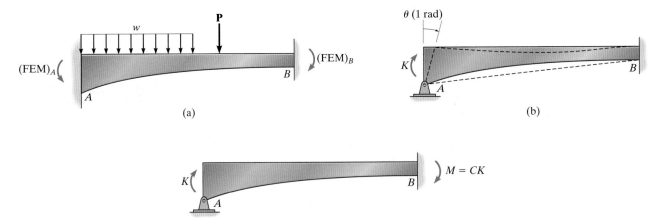

(a)

(b)

(c)

Fig. 13–2

Fig. 13–3

These properties can be obtained using, for example, the conjugate beam method or an energy method. However, considerable labor is often involved in the process. As a result, graphs and tables have been made available to determine this data for common shapes used in structural design. One such source is the *Handbook of Frame Constants,* published by the Portland Cement Association.* A portion of these tables, taken from this publication, is listed here as Tables 13–1 and 13–2. A more complete tabular form of the data is given in the PCA handbook along with the relevant derivations of formulas used.

The nomenclature is defined as follows:

a_A, a_B = ratio of the length of haunch at ends A and B to the length of span.

b = ratio of the distance from the concentrated load to end A to the length of span.

C_{AB}, C_{BA} = carry-over factors of member AB at ends A and B, respectively.

h_A, h_B = depth of member at ends A and B, respectively.

h_C = depth of member at minimum section.

I_C = moment of inertia of section at minimum depth.

k_{AB}, k_{BA} = stiffness factor at ends A and B, respectively.

L = length of member.

M_{AB}, M_{BA} = fixed-end moment at ends A and B, respectively; specified in tables for uniform load w or concentrated force P.

r_A, r_B = ratios for rectangular cross-sectional areas, where $r_A = (h_A - h_C)/h_C$, $r_B = (h_B - h_C)/h_C$.

As noted, the fixed-end moments and carry-over factors are found from the tables. The absolute stiffness factor can be determined using the tabulated stiffness factors and found from

$$K_A = \frac{k_{AB}EI_C}{L} \qquad K_B = \frac{k_{BA}EI_C}{L} \qquad (13\text{–}2)$$

Application of the use of the tables will be illustrated in Example 13–1.

Timber frames having a variable moment of inertia are often used in the construction of churches.

Handbook of Frame Constants. Portland Cement Association, Chicago, Illinois.

TABLE 13-1 Straight Haunches—Constant Width

Note: All carry-over factors are negative and all stiffness factors are positive.

Block 1: $a_A = 0.3$, $a_B = variable$; $r_A = 1.0$, $r_B = variable$

Right Haunch		Carry-over Factors		Stiffness Factors		Unif. Load FEM Coef.×wL^2		Conc. Load FEM—Coef.×PL										Haunch Load Left Coef×w_AL^2		Haunch Load Right Coef×w_BL^2	
a_B	r_B	C_{AB}	C_{BA}	k_{AB}	k_{BA}	M_{AB}	M_{BA}	M_{AB}(0.1)	M_{BA}	M_{AB}(0.3)	M_{BA}	M_{AB}(0.5)	M_{BA}	M_{AB}(0.7)	M_{BA}	M_{AB}(0.9)	M_{BA}	M_{AB}	M_{BA}	M_{AB}	M_{BA}
0.2	0.4	0.543	0.766	9.19	6.52	0.1194	0.0791	0.0935	0.0034	0.2185	0.0384	0.1955	0.1147	0.0889	0.1601	0.0096	0.0870	0.0133	0.0008	0.0006	0.0058
	0.6	0.576	0.758	9.53	7.24	0.1152	0.0851	0.0934	0.0038	0.2158	0.0422	0.1883	0.1250	0.0798	0.1729	0.0075	0.0898	0.0133	0.0009	0.0005	0.0060
	1.0	0.622	0.748	10.06	8.37	0.1089	0.0942	0.0931	0.0042	0.2118	0.0480	0.1771	0.1411	0.0668	0.1919	0.0047	0.0935	0.0132	0.0011	0.0004	0.0062
	1.5	0.660	0.740	10.52	9.38	0.1037	0.1018	0.0927	0.0047	0.2085	0.0530	0.1678	0.1550	0.0559	0.2078	0.0028	0.0961	0.0130	0.0012	0.0002	0.0064
	2.0	0.684	0.734	10.83	10.09	0.1002	0.1069	0.0924	0.0050	0.2062	0.0565	0.1614	0.1645	0.0487	0.2185	0.0019	0.0974	0.0129	0.0013	0.0001	0.0065
0.3	0.4	0.579	0.741	9.47	7.40	0.1175	0.0822	0.0934	0.0037	0.2164	0.0419	0.1909	0.1225	0.0856	0.1649	0.0100	0.0861	0.0133	0.0009	0.0022	0.0118
	0.6	0.629	0.726	9.98	8.64	0.1120	0.0902	0.0931	0.0042	0.2126	0.0477	0.1808	0.1379	0.0747	0.1807	0.0080	0.0888	0.0132	0.0010	0.0018	0.0124
	1.0	0.705	0.705	10.85	10.85	0.1034	0.1034	0.0924	0.0052	0.2063	0.0577	0.1640	0.1640	0.0577	0.2063	0.0052	0.0924	0.0131	0.0013	0.0013	0.0131
	1.5	0.771	0.689	11.70	13.10	0.0956	0.1157	0.0917	0.0062	0.2002	0.0675	0.1483	0.1892	0.0428	0.2294	0.0033	0.0953	0.0129	0.0015	0.0008	0.0137
	2.0	0.817	0.678	12.33	14.85	0.0901	0.1246	0.0913	0.0069	0.1957	0.0750	0.1368	0.2080	0.0326	0.2455	0.0022	0.0968	0.0128	0.0017	0.0006	0.0141

Block 2: $a_A = 0.2$, $a_B = variable$; $r_A = 1.5$, $r_B = variable$

a_B	r_B	C_{AB}	C_{BA}	k_{AB}	k_{BA}	M_{AB}	M_{BA}	M_{AB}(0.1)	M_{BA}	M_{AB}(0.3)	M_{BA}	M_{AB}(0.5)	M_{BA}	M_{AB}(0.7)	M_{BA}	M_{AB}(0.9)	M_{BA}	M_{AB}	M_{BA}	M_{AB}	M_{BA}
0.2	0.4	0.569	0.714	7.97	6.35	0.1166	0.0799	0.0966	0.0019	0.2186	0.0377	0.1847	0.1183	0.0821	0.1626	0.0088	0.0873	0.0064	0.0001	0.0006	0.0058
	0.6	0.603	0.707	8.26	7.04	0.1127	0.0858	0.0965	0.0021	0.2163	0.0413	0.1778	0.1288	0.0736	0.1752	0.0068	0.0901	0.0064	0.0001	0.0005	0.0060
	1.0	0.652	0.698	8.70	8.12	0.1069	0.0947	0.0963	0.0023	0.2127	0.0468	0.1675	0.1449	0.0616	0.1940	0.0043	0.0937	0.0064	0.0002	0.0004	0.0062
	1.5	0.691	0.691	9.08	9.08	0.1021	0.1021	0.0962	0.0025	0.2097	0.0515	0.1587	0.1587	0.0515	0.2097	0.0025	0.0962	0.0064	0.0002	0.0002	0.0064
	2.0	0.716	0.686	9.34	9.75	0.0990	0.1071	0.0960	0.0028	0.2077	0.0547	0.1528	0.1681	0.0449	0.2202	0.0017	0.0975	0.0064	0.0002	0.0001	0.0065
0.3	0.4	0.607	0.692	8.21	7.21	0.1148	0.0829	0.0965	0.0021	0.2168	0.0409	0.1801	0.1263	0.0789	0.1674	0.0091	0.0866	0.0064	0.0002	0.0020	0.0118
	0.6	0.659	0.678	8.65	8.40	0.1098	0.0907	0.0964	0.0024	0.2135	0.0464	0.1706	0.1418	0.0688	0.1831	0.0072	0.0892	0.0064	0.0002	0.0017	0.0123
	1.0	0.740	0.660	9.38	10.52	0.1018	0.1037	0.0961	0.0028	0.2078	0.0559	0.1550	0.1678	0.0530	0.2085	0.0047	0.0927	0.0064	0.0002	0.0012	0.0130
	1.5	0.809	0.645	10.09	12.66	0.0947	0.1156	0.0958	0.0033	0.2024	0.0651	0.1403	0.1928	0.0393	0.2311	0.0029	0.0950	0.0063	0.0003	0.0008	0.0137
	2.0	0.857	0.636	10.62	14.32	0.0897	0.1242	0.0955	0.0038	0.1985	0.0720	0.1296	0.2119	0.0299	0.2469	0.0020	0.0968	0.0063	0.0003	0.0005	0.0141

TABLE 13-2 Parabolic Haunches—Constant Width

Note: All carry-over factors are negative and all stiffness factors are positive.

Right Haunch a_B	r_B	Carry-over Factors C_{AB}	C_{BA}	Stiffness Factors k_{AB}	k_{BA}	Unif. Load FEM Coef.×wL^2 M_{AB}	M_{BA}	b=0.1 M_{AB}	M_{BA}	b=0.3 M_{AB}	M_{BA}	b=0.5 M_{AB}	M_{BA}	b=0.7 M_{AB}	M_{BA}	b=0.9 M_{AB}	M_{BA}	Left FEM Coef.×w_AL^2 M_{AB}	M_{BA}	Right FEM Coef.×w_BL^2 M_{AB}	M_{BA}
0.2	0.4	0.558	0.627	6.08	5.40	0.1022	0.0841	0.0938	0.0033	0.1891	0.0502	0.1572	0.1261	0.0715	0.1618	0.0073	0.0877	0.0032	0.0001	0.0002	0.0030
	0.6	0.582	0.624	6.21	5.80	0.0995	0.0887	0.0936	0.0036	0.1872	0.0535	0.1527	0.1339	0.0663	0.1708	0.0058	0.0902	0.0032	0.0001	0.0002	0.0031
	1.0	0.619	0.619	6.41	6.41	0.0956	0.0956	0.0935	0.0038	0.1844	0.0584	0.1459	0.1459	0.0584	0.1844	0.0038	0.0935	0.0032	0.0001	0.0001	0.0032
	1.5	0.649	0.614	6.59	6.97	0.0921	0.1015	0.0933	0.0041	0.1819	0.0628	0.1399	0.1563	0.0518	0.1962	0.0025	0.0958	0.0032	0.0001	0.0001	0.0032
	2.0	0.671	0.611	6.71	7.38	0.0899	0.1056	0.0932	0.0044	0.1801	0.0660	0.1358	0.1638	0.0472	0.2042	0.0017	0.0971	0.0032	0.0001	0.0000	0.0033
0.3	0.4	0.588	0.616	6.22	5.93	0.1002	0.0877	0.0937	0.0035	0.1873	0.0537	0.1532	0.1339	0.0678	0.1686	0.0073	0.0877	0.0032	0.0001	0.0007	0.0063
	0.6	0.625	0.609	6.41	6.58	0.0966	0.0942	0.0935	0.0039	0.1845	0.0587	0.1467	0.1455	0.0609	0.1808	0.0057	0.0902	0.0032	0.0001	0.0005	0.0065
	1.0	0.683	0.598	6.73	7.68	0.0911	0.1042	0.0932	0.0044	0.1801	0.0669	0.1365	0.1643	0.0502	0.2000	0.0037	0.0936	0.0031	0.0001	0.0004	0.0068
	1.5	0.735	0.589	7.02	8.76	0.0862	0.1133	0.0929	0.0050	0.1760	0.0746	0.1272	0.1819	0.0410	0.2170	0.0023	0.0959	0.0031	0.0001	0.0003	0.0070
	2.0	0.772	0.582	7.25	9.61	0.0827	0.1198	0.0927	0.0054	0.1730	0.0805	0.1203	0.1951	0.0345	0.2293	0.0016	0.0972	0.0031	0.0001	0.0002	0.0072
0.2	0.4	0.488	0.807	9.85	5.97	0.1214	0.0753	0.0929	0.0034	0.2131	0.0371	0.2021	0.1061	0.0979	0.1506	0.0105	0.0863	0.0171	0.0017	0.0003	0.0030
	0.6	0.515	0.803	10.10	6.45	0.1183	0.0795	0.0928	0.0036	0.2110	0.0404	0.1969	0.1136	0.0917	0.1600	0.0083	0.0892	0.0170	0.0018	0.0002	0.0030
	1.0	0.547	0.796	10.51	7.22	0.1138	0.0865	0.0926	0.0040	0.2079	0.0448	0.1890	0.1245	0.0809	0.1740	0.0056	0.0928	0.0168	0.0020	0.0001	0.0031
	1.5	0.571	0.786	10.90	7.90	0.1093	0.0922	0.0923	0.0043	0.2055	0.0485	0.1818	0.1344	0.0719	0.1862	0.0035	0.0951	0.0167	0.0021	0.0001	0.0032
	2.0	0.590	0.784	11.17	8.40	0.1063	0.0961	0.0922	0.0046	0.2041	0.0506	0.1764	0.1417	0.0661	0.1948	0.0025	0.0968	0.0166	0.0022	0.0001	0.0032
0.5	0.4	0.554	0.753	10.42	7.66	0.1170	0.0811	0.0926	0.0040	0.2087	0.0442	0.1924	0.1205	0.0898	0.1595	0.0107	0.0853	0.0169	0.0020	0.0042	0.0145
	0.6	0.606	0.730	10.96	9.12	0.1115	0.0889	0.0922	0.0046	0.2045	0.0506	0.1820	0.1360	0.0791	0.1738	0.0086	0.0878	0.0167	0.0022	0.0036	0.0152
	1.0	0.694	0.694	12.03	12.03	0.1025	0.1025	0.0915	0.0057	0.1970	0.0626	0.1639	0.1639	0.0626	0.1970	0.0057	0.0915	0.0164	0.0028	0.0028	0.0164
	1.5	0.781	0.664	13.12	15.47	0.0937	0.1163	0.0908	0.0070	0.1891	0.0759	0.1456	0.1939	0.0479	0.2187	0.0039	0.0940	0.0160	0.0034	0.0021	0.0174
	2.0	0.850	0.642	14.09	18.64	0.0870	0.1275	0.0901	0.0082	0.1825	0.0877	0.1307	0.2193	0.0376	0.2348	0.0027	0.0957	0.0157	0.0039	0.0016	0.0181

Top block: $a_A = 0.2$, a_B = variable, $r_A = 1.0$, r_B = variable.
Bottom block: $a_A = 0.5$, a_B = variable, $r_A = 1.0$, r_B = variable.

13.2 Moment Distribution for Structures Having Nonprismatic Members

Once the fixed-end moments and stiffness and carry-over factors for the nonprismatic members of a structure have been determined, application of the moment-distribution method follows the same procedure as outlined in Chapter 12. In this regard, recall that the distribution of moments may be shortened if a member stiffness factor is modified to account for conditions of end-span pin support and structure symmetry or antisymmetry. Similar modifications can also be made to nonprismatic members.

Beam Pin Supported at Far End. Consider the beam in Fig. 13–4a, which is pinned at its far end B. The absolute stiffness factor K'_A is the moment applied at A such that it rotates the beam at $A, \theta_A = 1$ rad. It can be determined as follows. First assume that B is temporarily fixed and a moment K_A is applied at A, Fig. 13–4b. The moment induced at B is $C_{AB}K_A$, where C_{AB} is the carry-over factor from A to B. Second, since B is not to be fixed, application of the opposite moment $C_{AB}K_A$ to the beam, Fig. 13–4c, will induce a moment $C_{BA}C_{AB}K_A$ at end A. By superposition, the result of these two applications of moment yields the beam loaded as shown in Fig. 13–4a. Hence it can be seen that the absolute stiffness factor of the beam at A is

$$K'_A = K_A(1 - C_{AB}C_{BA}) \qquad (13\text{–}3)$$

Here K_A is the absolute stiffness factor of the beam, assuming it to be fixed at the far end B. For example, in the case of a prismatic beam, $K_A = 4EI/L$ and $C_{AB} = C_{BA} = \frac{1}{2}$. Substituting into Eq. 13–3 yields $K'_A = 3EI/L$, the same as Eq. 12–4.

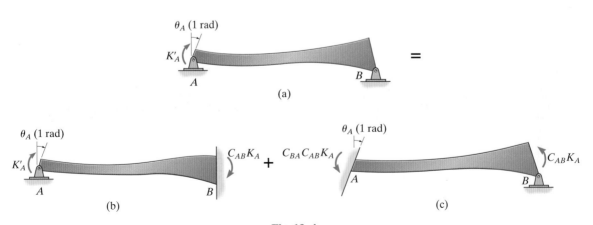

Fig. 13–4

Symmetric Beam and Loading. Here we must determine the moment K'_A needed to rotate end A, $\theta_A = +1$ rad, while $\theta_B = -1$ rad, Fig. 13–5a. In this case we first assume that end B is fixed and apply the moment K_A at A, Fig. 13–5b. Next we apply a negative moment K_B to end B assuming that end A is fixed. This results in a moment of $C_{BA}K_B$ at end A as shown in Fig. 13–5c. Superposition of these two applications of moment at A yields the results of Fig. 13–5a. We require

$$K'_A = K_A - C_{BA}K_B$$

Using Eq. 13–1 ($C_{BA}K_B = C_{AB}K_A$), we can also write

$$K'_A = K_A(1 - C_{AB}) \qquad (13\text{--}4)$$

In the case of a prismatic beam, $K_A = 4EI/L$ and $C_{AB} = \frac{1}{2}$, so that $K'_A = 2EI/L$, which is the same as Eq. 12–5.

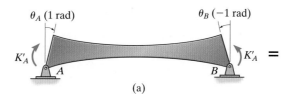

(a)

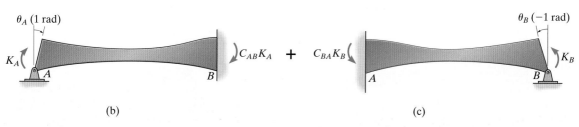

(b)   (c)

Fig. 13–5

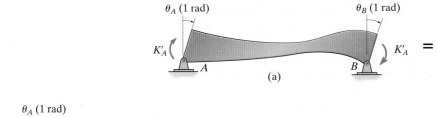

(a)

(b)   (c)

Fig. 13–6

Symmetric Beam with Antisymmetric Loading.   In the case of a symmetric beam with antisymmetric loading, we must determine K'_A such that equal rotations occur at the ends of the beam, Fig. 13–6a. To do this, we first fix end B and apply the moment K_A at A, Fig. 13–6b. Likewise, application of K_B at end B while end A is fixed is shown in Fig. 13–6c. Superposition of both cases yields the results of Fig. 13–6a. Hence,

$$K'_A = K_A + C_{BA}K_B$$

or, using Eq. 13–1 ($C_{BA}K_B = C_{AB}K_A$), we have for the absolute stiffness

$$K'_A = K_A(1 + C_{AB}) \qquad (13–5)$$

Substituting the data for a prismatic member, $K_A = 4EI/L$ and $C_{AB} = \frac{1}{2}$, yields $K'_A = 6EI/L$, which is the same as Eq. 12–6.

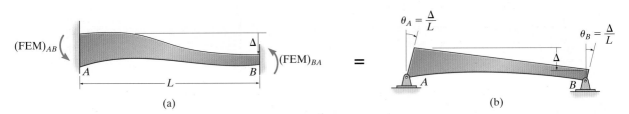

(a)   (b)

Fig. 13–7

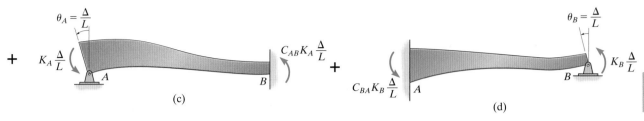

(c)   (d)

Fig. 13–7

Relative Joint Translation of Beam.

Fixed-end moments are developed in a nonprismatic member if it has a relative joint translation Δ between its ends A and B, Fig. 13–7a. In order to determine these moments, we proceed as follows. First consider the ends A and B to be pin connected and allow end B of the beam to be displaced a distance Δ such that the end rotations are $\theta_A = \theta_B = \Delta/L$, Fig. 13–7b. Second, assume that B is fixed and apply a moment of $M'_A = -K_A(\Delta/L)$ to end A such that it rotates the end $\theta_A = -\Delta/L$, Fig. 13–7c. Third, assume that A is fixed and apply a moment $M'_B = -K_B(\Delta/L)$ to end B such that it rotates the end $\theta_B = -\Delta/L$, Fig. 13–7d. Since the total sum of these three operations yields the condition shown in Fig. 13–7a, we have at A

$$(\text{FEM})_{AB} = -K_A\frac{\Delta}{L} - C_{BA}K_B\frac{\Delta}{L}$$

Applying Eq. 13–1 ($C_{BA}K_B = C_{AB}K_A$) yields

$$(\text{FEM})_{AB} = -K_A\frac{\Delta}{L}(1 + C_{AB}) \qquad (13\text{–}6)$$

A similar expression can be written for end B. Recall that for a prismatic member $K_A = 4EI/L$ and $C_{AB} = \frac{1}{2}$. Thus $(\text{FEM})_{AB} = -6EI\Delta/L^2$, which is the same as Eq. 11–5.

If end B is pinned rather than fixed, Fig. 13–8, the fixed-end moment at A can be determined in a manner similar to that described above. The result is

$$(\text{FEM})'_{AB} = -K_A\frac{\Delta}{L}(1 - C_{AB}C_{BA}) \qquad (13\text{–}7)$$

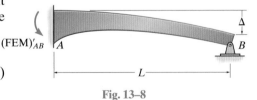

$(\text{FEM})'_{AB}$

Fig. 13–8

Here it is seen that for a prismatic member this equation gives $(\text{FEM})'_{AB} = -3EI\Delta/L^2$, which is the same as that listed on the inside back cover.

The following example illustrates application of the moment-distribution method to structures having nonprismatic members. Once the fixed-end moments and stiffness and carry-over factors have been determined, and the stiffness factor modified according to the equations given above, the procedure for analysis is the same as that discussed in Chapter 12.

EXAMPLE 13.1

Determine the internal moments at the supports of the beam shown in Fig. 13–9a. The beam has a thickness of 0.3 m and E is constant.

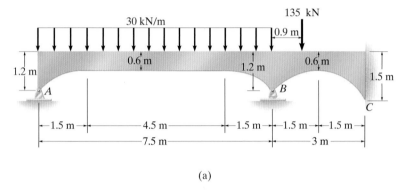

(a)

Fig. 13–9

SOLUTION

Since the haunches are parabolic, we will use Table 13–2 to obtain the moment-distribution properties of the beam.

Span AB

$$a_A = a_B = \frac{1.5}{7.5} = 0.2 \qquad r_A = r_B = \frac{1.2 - 0.6}{0.6} = 1.0$$

Entering Table 13–2 with these ratios, we find

$$C_{AB} = C_{BA} = 0.619$$

$$k_{AB} = k_{BA} = 6.41$$

Using Eqs. 13–2,

$$K_{AB} = K_{BA} = \frac{kEI_C}{L} = \frac{6.41E\left(\frac{1}{12}\right)(0.3)(0.6)^3}{7.5} = 4.615(10^{-3})E$$

Since the far end of span BA is pinned, we will modify the stiffness factor of BA using Eq. 13–3. We have

$$K'_{BA} = K_{BA}(1 - C_{AB}C_{BA}) = 4.615(10^{-3})E[1 - 0.619(0.619)] = 2.847(10^{-3})E$$

Uniform load, Table 13–2,

$$(FEM)_{AB} = -(0.0956)(30)(7.5)^2 = -161.33 \text{ kN} \cdot \text{m}$$

$$(FEM)_{BA} = 161.33 \text{ kN} \cdot \text{m}$$

Span BC

$$a_B = a_C = \frac{1.5}{3} = 0.5 \qquad r_B = \frac{1.2 - 0.6}{0.6} = 1.0$$

$$r_C = \frac{1.5 - 0.6}{0.6} = 1.5$$

From Table 13–2 we find

$$C_{BC} = 0.781 \qquad C_{CB} = 0.664$$

$$k_{BC} = 13.12 \qquad k_{CB} = 15.47$$

Thus, from Eqs. 13–2,

$$K_{BC} = \frac{kEI_C}{L} = \frac{13.12E\left(\frac{1}{12}\right)(0.3)(0.6)^3}{3} = 23.616(10^{-3})E$$

$$K_{CB} = \frac{kEI_C}{L} = \frac{15.47E\left(\frac{1}{12}\right)(0.3)(0.6)^3}{3} = 27.846(10^{-3})E$$

Concentrated load,

$$b = \frac{0.9}{3} = 0.3$$

$$(FEM)_{BC} = -0.1891(135)(3) = -76.59 \text{ kN} \cdot \text{m}$$

$$(FEM)_{CB} = 0.0759(135)(3) = 30.74 \text{ kN} \cdot \text{m}$$

Using the foregoing values for the stiffness factors, the distribution factors are computed and entered in the table, Fig. 13–9b. The moment distribution follows the same procedure outlined in Chapter 12. The results in kN · m are shown on the last line of the table.

Joint	A	B		C
Member	AB	BA	BC	CB
K	$4.615(10^{-3})E$	$2.847(10^{-3})E$	$23.616(10^{-3})E$	$27.846(10^{-3})E$
DF	1	0.107	0.893	0
COF	0.619	0.619	0.781	0.664
FEM Dist.	−161.33 161.33	161.33 −9.07	−76.59 −75.67	30.74
CO Dist.		99.86 −10.69	−89.18	−59.10
CO				−69.65
ΣM	0	241.43	−241.43	−98.01

(b)

Fig. 13–9

13

13.3   Slope-Deflection Equations for Nonprismatic Members

The slope-deflection equations for prismatic members were developed in Chapter 11. In this section we will generalize the form of these equations so that they apply as well to nonprismatic members. To do this, we will use the results of the previous section and proceed to formulate the equations in the same manner discussed in Chapter 11, that is, considering the effects caused by the loads, relative joint displacement, and each joint rotation separately, and then superimposing the results.

Loads.   Loads are specified by the fixed-end moments $(FEM)_{AB}$ and $(FEM)_{BA}$ acting at the ends A and B of the span. Positive moments act clockwise.

Relative Joint Translation.   When a relative displacement Δ between the joints occurs, the induced moments are determined from Eq. 13–6. At end A this moment is $-[K_A\Delta/L](1 + C_{AB})$ and at end B it is $-[K_B\Delta/L](1 + C_{BA})$.

Rotation at A.   If end A rotates θ_A, the required moment in the span at A is $K_A\theta_A$. Also, this induces a moment of $C_{AB}K_A\theta_A = C_{BA}K_B\theta_A$ at end B.

Rotation at B.   If end B rotates θ_B, a moment of $K_B\theta_B$ must act at end B, and the moment induced at end A is $C_{BA}K_B\theta_B = C_{AB}K_A\theta_B$.

The total end moments caused by these effects yield the generalized slope-deflection equations, which can therefore be written as

$$M_{AB} = K_A\left[\theta_A + C_{AB}\theta_B - \frac{\Delta}{L}(1 + C_{AB})\right] + (FEM)_{AB}$$

$$M_{BA} = K_B\left[\theta_B + C_{BA}\theta_A - \frac{\Delta}{L}(1 + C_{BA})\right] + (FEM)_{BA}$$

Since these two equations are similar, we can express them as a single equation. Referring to one end of the span as the near end (N) and the other end as the far end (F), and representing the member rotation as $\psi = \Delta/L$, we have

$$\boxed{M_N = K_N(\theta_N + C_N\theta_F - \psi(1 + C_N)) + (FEM)_N}\qquad (13\text{–}8)$$

Here

M_N = internal moment at the near end of the span; this moment is positive clockwise when acting on the span.

K_N = absolute stiffness of the near end determined from tables or by calculation.

θ_N, θ_F = near- and far-end slopes of the span at the supports; the angles are measured in *radians* and are *positive clockwise.*

ψ = span cord rotation due to a linear displacement, $\psi = \Delta/L$; this angle is measured in *radians* and is *positive clockwise.*

$(\text{FEM})_N$ = fixed-end moment at the near-end support; the moment is *positive* clockwise when acting on the span and is obtained from tables or by calculations.

Application of the equation follows the same procedure outlined in Chapter 11 and therefore will not be discussed here. In particular, note that Eq. 13–8 reduces to Eq. 11–8 when applied to members that are prismatic.

Light-weight metal buildings are often designed using frame members having variable moments of inertia.

A continuous, reinforced-concrete highway bridge.

PROBLEMS

13–1. Determine the moments at A, B, and C by the moment-distribution method. Assume the supports at A and C are fixed and a roller support at B is on a rigid base. The girder has a thickness of 1 m. Use Table 13–1. E is constant. The haunches are straight.

13–2. Solve Prob. 13–1 using the slope-deflection equations.

13–5. Use the moment-distribution method to determine the moment at each joint of the symmetric bridge frame. Supports at F and E are fixed and B and C are fixed connected. Use Table 13–2. Assume E is constant and the members are each 0.25 m thick.

13–6. Solve Prob. 13–5 using the slope-deflection equations.

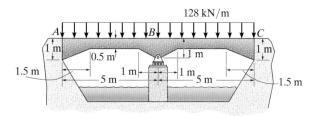

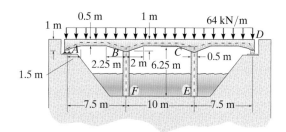

Probs. 13–1/13–2

Probs. 13–5/13–6

13–3. Apply the moment-distribution method to determine the moment at each joint of the parabolic haunched frame. Supports A and B are fixed. Use Table 13–2. The members are each 0.25 m thick. E is constant.

*13–4. Solve Prob. 13–3 using the slope-deflection equations.

13–7. Apply the moment-distribution method to determine the moment at each joint of the symmetric parabolic haunched frame. Supports A and D are fixed. Use Table 13–2. The members are each 0.2 m thick. E is constant.

*13–8. Solve Prob. 13–7 using the slope-deflection equations.

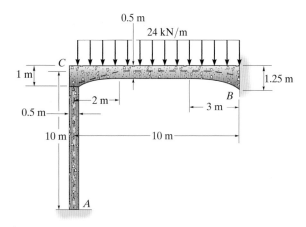

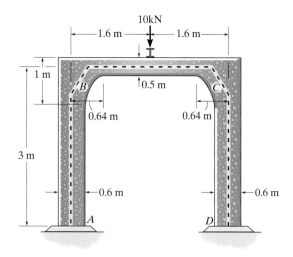

Probs. 13–3/13–4

Probs. 13–7/13–8

13–9. Use the moment-distribution method to determine the moment at each joint of the frame. The supports at *A* and *C* are pinned and the joints at *B* and *D* are fixed connected. Assume that *E* is constant and the members have a thickness of 0.25 m. The haunches are straight so use Table 13–1.

13–10. Solve Prob. 13–9 using the slope-deflection equations.

13–11. Use the moment-distribution method to determine the moment at each joint of the symmetric bridge frame. Supports *F* and *E* are fixed and *B* and *C* are fixed connected. The haunches are straight so use Table 13–2. Assume *E* is constant and the members are each 0.25 m thick.

*13–12. Solve Prob. 13–11 using the slope-deflection equations.

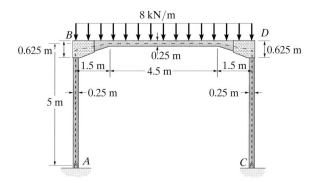

Probs. 13–9/13–10

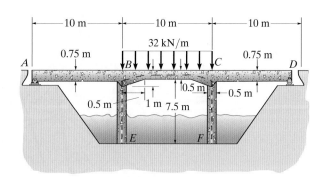

Probs. 13–11/13–12

CHAPTER REVIEW

Non-prismatic members having a variable moment of inertia are often used on long-span bridges and building frames to save material.

A structural analysis using non-prismatic members can be performed using either the slope-deflection equations or moment distribution. If this is done, it then becomes necessary to obtain the fixed-end moments, stiffness factors, and carry-over factors for the member. One way to obtain these values is to use the conjugate beam method, although the work is somewhat tedious. It is also possible to obtain these values from tabulated data, such as published by the Portland Cement Association.

If the moment distribution method is used, then the process can be simplified if the stiffness of some of the members is modified.

(Ref: Sections 13.1 to 13.3)

The space-truss analysis of electrical transmission towers can be performed using the stiffness method.

Truss Analysis Using the Stiffness Method

14

In this chapter we will explain the basic fundamentals of using the stiffness method for analyzing structures. It will be shown that this method, although tedious to do by hand, is quite suited for use on a computer. Examples of specific applications to planar trusses will be given. The method will then be expanded to include space-truss analysis. Beams and framed structures will be discussed in the next chapters.

14.1 Fundamentals of the Stiffness Method

There are essentially two ways in which structures can be analyzed using matrix methods. The stiffness method, to be used in this and the following chapters, is a displacement method of analysis. A force method, called the flexibility method, as outlined in Sec. 9–1, can also be used to analyze structures; however, this method will not be presented in this text. There are several reasons for this. Most important, the stiffness method can be used to analyze both statically determinate and indeterminate structures, whereas the flexibility method requires a different procedure for each of these two cases. Also, the stiffness method yields the displacements and forces directly, whereas with the flexibility method the displacements are not obtained directly. Furthermore, it is generally much easier to formulate the necessary matrices for the computer operations using the stiffness method; and once this is done, the computer calculations can be performed efficiently.

Application of the stiffness method requires subdividing the structure into a series of discrete *finite elements* and identifying their end points as *nodes*. For truss analysis, the finite elements are represented by each of the members that compose the truss, and the nodes represent the joints. The force-displacement properties of each element are determined and then related to one another using the force equilibrium equations written at the nodes. These relationships, for the entire structure, are then grouped together into what is called the *structure stiffness matrix* **K**. Once it is established, the unknown displacements of the nodes can then be determined for any given loading on the structure. When these displacements are known, the external and internal forces in the structure can be calculated using the force-displacement relations for each member.

Before developing a formal procedure for applying the stiffness method, it is first necessary to establish some preliminary definitions and concepts.

Member and Node Identification.   One of the first steps when applying the stiffness method is to identify the elements or members of the structure and their nodes. We will specify each member by a number enclosed within a square, and use a number enclosed within a circle to identify the nodes. Also, the "near" and "far" ends of the member must be identified. This will be done using an arrow written along the member, with the head of the arrow directed toward the far end. Examples of member, node, and "direction" identification for a truss are shown in Fig. 14–1*a*. These assignments have all been done *arbitrarily*.*

Global and Member Coordinates.   Since loads and displacements are vector quantities, it is necessary to establish a coordinate system in order to specify their correct sense of direction. Here we will use two different types of coordinate systems. A single *global* or *structure coordinate system, x, y*, will be used to specify the sense of each of the *external* force and displacement components at the nodes, Fig. 14–1*a*. A *local* or *member coordinate system* will be used for each member to specify the sense of direction of its displacements and *internal* loadings. This system will be identified using x', y' axes with the origin at the "near" node and the x' axis extending toward the "far" node. An example for truss member 4 is shown in Fig. 14–1*b*.

---

*For large trusses, matrix manipulations using **K** are actually more efficient using selective numbering of the members in a wave pattern, that is, starting from top to bottom, then bottom to top, etc.

Kinematic Indeterminacy. As discussed in Sec. 11–1, the unconstrained degrees of freedom for the truss represent the primary unknowns of any displacement method, and therefore these must be identified. As a general rule there are two degrees of freedom, or two possible displacements, for each joint (node). For application, each degree of freedom will be specified on the truss using a code number, shown at the joint or node, and referenced to its positive global coordinate direction using an associated arrow. For example, the truss in Fig. 14–1a has eight degrees of freedom, which have been identified by the "code numbers" 1 through 8 as shown. The truss is kinematically indeterminate to the fifth degree because of these eight possible displacements: 1 through 5 represent unknown or *unconstrained degrees of freedom*, and 6 through 8 represent *constrained degrees of freedom*. Due to the constraints, the displacements here are zero. For later application, *the lowest code numbers will always be used to identify the unknown displacements (unconstrained degrees of freedom) and the highest code numbers will be used to identify the known displacements (constrained degrees of freedom).* The reason for choosing this method of identification has to do with the convenience of later partitioning the structure stiffness matrix, so that the unknown displacements can be found in the most direct manner.

Once the truss is labeled and the code numbers are specified, the structure stiffness matrix **K** can then be determined. To do this we must first establish a *member stiffness matrix* **k′** for each member of the truss. This matrix is used to express the member's load-displacement relations in terms of the *local coordinates*. Since all the members of the truss are not in the same direction, we must develop a means for transforming these quantities from each member's local $x′$, $y′$ coordinate system to the structure's global x, y coordinate system. This can be done using *force and displacement transformation matrices*. Once established, the elements of the member stiffness matrix are transformed from local to global coordinates and then assembled to create the structure stiffness matrix. Using **K**, as stated previously, we can determine the node displacements first, followed by the support reactions and the member forces. We will now elaborate on the development of this method.

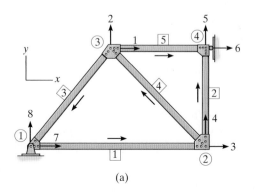

(a)   (b)

Fig. 14–1

14

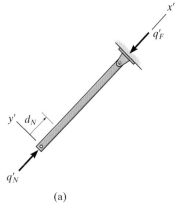

(a)

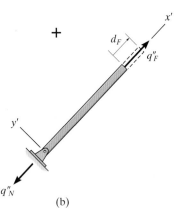

(b)

||

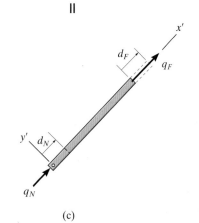

(c)

Fig. 14–2

14.2 Member Stiffness Matrix

In this section we will establish the stiffness matrix for a single truss member using local x', y' coordinates, oriented as shown in Fig. 14–2. The terms in this matrix will represent the load-displacement relations for the member.

A truss member can only be displaced along its axis (x' axis) since the loads are applied along this axis. Two independent displacements are therefore possible. When a positive displacement d_N is imposed on the near end of the member while the far end is held pinned, Fig. 14–2a, the forces developed at the ends of the members are

$$q_N' = \frac{AE}{L}d_N \qquad q_F' = -\frac{AE}{L}d_N$$

Note that q_F' is negative since for equilibrium it acts in the negative x' direction. Likewise, a positive displacement d_F at the far end, keeping the near end pinned, Fig. 14–2b, results in member forces of

$$q_N'' = -\frac{AE}{L}d_F \qquad q_F'' = \frac{AE}{L}d_F$$

By superposition, Fig. 14–2c, the resultant forces caused by both displacements are

$$q_N = \frac{AE}{L}d_N - \frac{AE}{L}d_F \qquad (14\text{–}1)$$

$$q_F = -\frac{AE}{L}d_N + \frac{AE}{L}d_F \qquad (14\text{–}2)$$

These load-displacement equations may be written in matrix form* as

$$\begin{bmatrix} q_N \\ q_F \end{bmatrix} = \frac{AE}{L}\begin{bmatrix} 1 & -1 \\ -1 & 1 \end{bmatrix}\begin{bmatrix} d_N \\ d_F \end{bmatrix}$$

or

$$\boxed{\mathbf{q} = \mathbf{k'd}} \qquad (14\text{–}3)$$

where

$$\mathbf{k'} = \frac{AE}{L}\begin{bmatrix} 1 & -1 \\ -1 & 1 \end{bmatrix} \qquad (14\text{–}4)$$

This matrix, $\mathbf{k'}$, is called the *member stiffness matrix*, and it is of the same form for each member of the truss. The four elements that comprise it are called *member stiffness influence coefficients*, k'_{ij}. Physically, k'_{ij} represents

*A review of matrix algebra is given in Appendix A.

the force at joint i when a *unit displacement* is imposed at joint j. For example, if $i = j = 1$, then k'_{11} is the force at the near joint when the far joint is held fixed, and the near joint undergoes a displacement of $d_N = 1$, i.e.,

$$q_N = k'_{11} = \frac{AE}{L}$$

Likewise, the force at the far joint is determined from $i = 2, j = 1$, so that

$$q_F = k'_{21} = -\frac{AE}{L}$$

These two terms represent the first column of the member stiffness matrix. In the same manner, the second column of this matrix represents the forces in the member only when the far end of the member undergoes a unit displacement.

14.3   Displacement and Force Transformation Matrices

Since a truss is composed of many members (elements), we will now develop a method for transforming the member forces **q** and displacements **d** defined in local coordinates to global coordinates. For the sake of convention, we will consider the global coordinates positive x to the right and positive y upward. The smallest angles between the *positive x, y* global axes and the *positive x'* local axis will be defined as θ_x and θ_y as shown in Fig. 14–3. The cosines of these angles will be used in the matrix analysis that follows. These will be identified as $\lambda_x = \cos \theta_x$, $\lambda_y = \cos \theta_y$. Numerical values for λ_x and λ_y can easily be generated by a computer once the x, y coordinates of the near end N and far end F of the member have been specified. For example, consider member NF of the truss shown in Fig. 14–4. Here the coordinates of N and F are (x_N, y_N) and (x_F, y_F), respectively.* Thus,

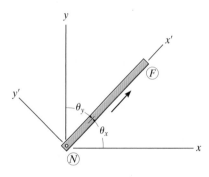

Fig. 14–3

$$\lambda_x = \cos \theta_x = \frac{x_F - x_N}{L} = \frac{x_F - x_N}{\sqrt{(x_F - x_N)^2 + (y_F - y_N)^2}} \quad (14\text{–}5)$$

$$\lambda_y = \cos \theta_y = \frac{y_F - y_N}{L} = \frac{y_F - y_N}{\sqrt{(x_F - x_N)^2 + (y_F - y_N)^2}} \quad (14\text{–}6)$$

The algebraic signs in these "generalized" equations will automatically account for members that are oriented in other quadrants of the x–y plane.

*The origin can be located at any convenient point. Usually, however, it is located where the x, y coordinates of all the nodes will be *positive,* as shown in Fig. 14–4.

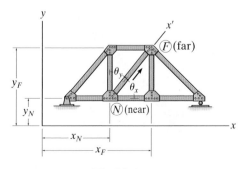

Fig. 14–4

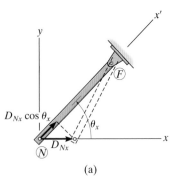

(a)

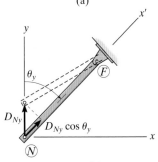

(b)

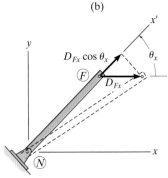

(c)

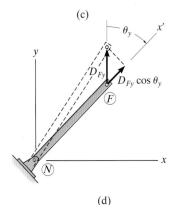

(d)

Fig. 14–5

Displacement Transformation Matrix. In global coordinates each end of the member can have two degrees of freedom or independent displacements; namely, joint N has D_{Nx} and D_{Ny}, Figs. 14–5a and 14–5b, and joint F has D_{Fx} and D_{Fy}, Figs. 14–5c and 14–5d. We will now consider each of these displacements separately, in order to determine its component displacement along the member. When the far end is held pinned and the near end is given a global displacement D_{Nx}, Fig. 14–5a, the corresponding displacement (deformation) along the member is $D_{Nx} \cos \theta_x$.* Likewise, a displacement D_{Ny} will cause the member to be displaced $D_{Ny} \cos \theta_y$ along the x' axis, Fig. 14–5b. The effect of *both* global displacements causes the member to be displaced

$$d_N = D_{Nx} \cos \theta_x + D_{Ny} \cos \theta_y$$

In a similar manner, positive displacements D_{Fx} and D_{Fy} successively applied at the far end F, while the near end is held pinned, Figs. 14–5c and 14–5d, will cause the member to be displaced

$$d_F = D_{Fx} \cos \theta_x + D_{Fy} \cos \theta_y$$

Letting $\lambda_x = \cos \theta_x$ and $\lambda_y = \cos \theta_y$ represent the *direction cosines* for the member, we have

$$d_N = D_{Nx}\lambda_x + D_{Ny}\lambda_y$$
$$d_F = D_{Fx}\lambda_x + D_{Fy}\lambda_y$$

which can be written in matrix form as

$$\begin{bmatrix} d_N \\ d_F \end{bmatrix} = \begin{bmatrix} \lambda_x & \lambda_y & 0 & 0 \\ 0 & 0 & \lambda_x & \lambda_y \end{bmatrix} \begin{bmatrix} D_{Nx} \\ D_{Ny} \\ D_{Fx} \\ D_{Fy} \end{bmatrix} \tag{14–7}$$

or

$$\boxed{\mathbf{d} = \mathbf{T}\mathbf{D}} \tag{14–8}$$

where

$$\mathbf{T} = \begin{bmatrix} \lambda_x & \lambda_y & 0 & 0 \\ 0 & 0 & \lambda_x & \lambda_y \end{bmatrix} \tag{14–9}$$

From the above derivation, $\mathbf{T}$ transforms the four global x, y displacements $\mathbf{D}$ into the two local x' displacements $\mathbf{d}$. Hence, $\mathbf{T}$ is referred to as the *displacement transformation matrix.*

---

*The change in θ_x or θ_y will be neglected, since it is very small.

Force Transformation Matrix. Consider now application of the force q_N to the near end of the member, the far end held pinned, Fig. 14–6a. Here the global force components of q_N at N are

$$Q_{Nx} = q_N \cos \theta_x \qquad Q_{Ny} = q_N \cos \theta_y$$

Likewise, if q_F is applied to the bar, Fig. 14–6b, the global force components at F are

$$Q_{Fx} = q_F \cos \theta_x \qquad Q_{Fy} = q_F \cos \theta_y$$

Using the direction cosines $\lambda_x = \cos \theta_x$, $\lambda_y = \cos \theta_y$, these equations become

$$Q_{Nx} = q_N \lambda_x \qquad Q_{Ny} = q_N \lambda_y$$
$$Q_{Fx} = q_F \lambda_x \qquad Q_{Fy} = q_F \lambda_y$$

which can be written in matrix form as

$$\begin{bmatrix} Q_{Nx} \\ Q_{Ny} \\ Q_{Fx} \\ Q_{Fy} \end{bmatrix} = \begin{bmatrix} \lambda_x & 0 \\ \lambda_y & 0 \\ 0 & \lambda_x \\ 0 & \lambda_y \end{bmatrix} \begin{bmatrix} q_N \\ q_F \end{bmatrix} \qquad (14\text{–}10)$$

or

$$\boxed{\mathbf{Q} = \mathbf{T}^T \mathbf{q}} \qquad (14\text{–}11)$$

where

$$\mathbf{T}^T = \begin{bmatrix} \lambda_x & 0 \\ \lambda_y & 0 \\ 0 & \lambda_x \\ 0 & \lambda_y \end{bmatrix} \qquad (14\text{–}12)$$

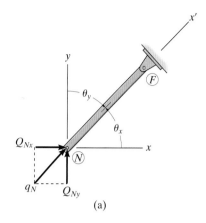

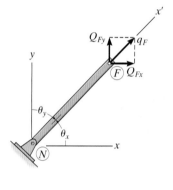

(a)

(b)

Fig. 14–6

In this case $\mathbf{T}^T$ transforms the two local (x') forces $\mathbf{q}$ acting at the ends of the member into the four global (x, y) force components $\mathbf{Q}$. By comparison, this *force transformation matrix* is the transpose of the displacement transformation matrix, Eq. 14–9.

14

14.4 Member Global Stiffness Matrix

We will now combine the results of the preceding sections and determine the stiffness matrix for a member which relates the member's global force components $\mathbf{Q}$ to its global displacements $\mathbf{D}$. If we substitute Eq. 14–8 ($\mathbf{d} = \mathbf{TD}$) into Eq. 14–3 ($\mathbf{q} = \mathbf{k'd}$), we can determine the member's forces $\mathbf{q}$ in terms of the global displacements $\mathbf{D}$ at its end points, namely,

$$\mathbf{q} = \mathbf{k'TD} \tag{14–13}$$

Substituting this equation into Eq. 14–11, $\mathbf{Q} = \mathbf{T}^T\mathbf{q}$, yields the final result,

$$\mathbf{Q} = \mathbf{T}^T\mathbf{k'TD}$$

or

$$\mathbf{Q} = \mathbf{kD} \tag{14–14}$$

where

$$\mathbf{k} = \mathbf{T}^T\mathbf{k'T} \tag{14–15}$$

The matrix $\mathbf{k}$ is the *member stiffness matrix* in global coordinates. Since $\mathbf{T}^T$, $\mathbf{T}$, and $\mathbf{k'}$ are known, we have

$$\mathbf{k} = \begin{bmatrix} \lambda_x & 0 \\ \lambda_y & 0 \\ 0 & \lambda_x \\ 0 & \lambda_y \end{bmatrix} \frac{AE}{L} \begin{bmatrix} 1 & -1 \\ -1 & 1 \end{bmatrix} \begin{bmatrix} \lambda_x & \lambda_y & 0 & 0 \\ 0 & 0 & \lambda_x & \lambda_y \end{bmatrix}$$

Performing the matrix operations yields

$$\mathbf{k} = \frac{AE}{L} \begin{matrix} & \begin{matrix} N_x & \quad N_y & \quad F_x & \quad F_y \end{matrix} \\ \begin{bmatrix} \lambda_x^2 & \lambda_x\lambda_y & -\lambda_x^2 & -\lambda_x\lambda_y \\ \lambda_x\lambda_y & \lambda_y^2 & -\lambda_x\lambda_y & -\lambda_y^2 \\ -\lambda_x^2 & -\lambda_x\lambda_y & \lambda_x^2 & \lambda_x\lambda_y \\ -\lambda_x\lambda_y & -\lambda_y^2 & \lambda_x\lambda_y & \lambda_y^2 \end{bmatrix} & \begin{matrix} N_x \\ N_y \\ F_x \\ F_y \end{matrix} \end{matrix} \tag{14–16}$$

The *location* of each element in this 4 × 4 symmetric matrix is referenced with each global degree of freedom associated with the near end N, followed by the far end F. This is indicated by the code number notation along the rows and columns, that is, N_x, N_y, F_x, F_y. Here **k** represents the force-displacement relations for the member when the components of force and displacement at the ends of the member are in the global or x, y directions. Each of the terms in the matrix is therefore a *stiffness influence coefficient* $\mathbf{k}_{ij}$, which denotes the x or y force component at i needed to cause an associated *unit x* or y displacement component at j. As a result, each identified column of the matrix represents the four force components developed at the ends of the member when the identified end undergoes a unit displacement related to its matrix column. For example, a unit displacement $D_{Nx} = 1$ will create the four force components on the member shown in the first column of the matrix.

14.5   Truss Stiffness Matrix

Once all the member stiffness matrices are formed in global coordinates, it becomes necessary to assemble them in the proper order so that the stiffness matrix **K** for the entire truss can be found. This process of combining the member matrices depends on careful identification of the elements in each member matrix. As discussed in the previous section, this is done by designating the rows and columns of the matrix by the *four* code numbers N_x, N_y, F_x, F_y used to identify the two global degrees of freedom that can occur at each end of the member (see Eq. 14–16). The structure stiffness matrix will then have an order that will be equal to the highest code number assigned to the truss, since this represents the total number of degrees of freedom for the structure. When the **k** matrices are assembled, each element in **k** will then be placed in its *same* row and column designation in the structure stiffness matrix **K**. In particular, when two or more members are *connected* to the *same* joint or node, then some of the elements of each member's **k** matrix will be assigned to the same position in the **K** matrix. When this occurs, the elements assigned to the common location must be added together algebraically. The reason for this becomes clear if one realizes that each element of the **k** matrix represents the resistance of the member to an applied force at its end. In this way, adding these resistances in the x or y direction when forming the **K** matrix determines the *total resistance* of each joint to a unit displacement in the x or y direction.

This method of assembling the member matrices to form the structure stiffness matrix will now be demonstrated by two numerical examples. Although this process is somewhat tedious when done by hand, it is rather easy to program on a computer.

EXAMPLE 14.1

Determine the structure stiffness matrix for the two-member truss shown in Fig. 14–7a. AE is constant.

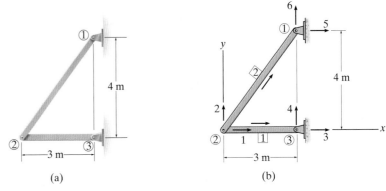

Fig. 14–7

SOLUTION

By inspection, ② will have two unknown displacement components, whereas joints ① and ③ are constrained from displacement. Consequently, the displacement components at joint ② are code numbered first, followed by those at joints ③ and ①, Fig. 14–7b. The origin of the global coordinate system can be located at any point. For convenience, we will choose joint ② as shown. The members are identified arbitrarily and arrows are written along the two members to identify the near and far ends of each member. The direction cosines and the stiffness matrix for each member can now be determined.

Member 1. Since ② is the near end and ③ is the far end, then by Eqs. 14–5 and 14–6, we have

$$\lambda_x = \frac{3-0}{3} = 1 \qquad \lambda_y = \frac{0-0}{3} = 0$$

Using Eq. 14–16, dividing each element by $L = 3$ m we have

$$
\mathbf{k}_1 = AE
\begin{array}{c}
\begin{array}{cccc} 1 & \;\;2 & \;\;\;3 & \;\;4 \end{array} \\
\left[
\begin{array}{cccc}
0.333 & 0 & -0.333 & 0 \\
0 & 0 & 0 & 0 \\
-0.333 & 0 & 0.333 & 0 \\
0 & 0 & 0 & 0
\end{array}
\right]
\begin{array}{c} 1 \\ 2 \\ 3 \\ 4 \end{array}
\end{array}
$$

The calculations can be checked in part by noting that $\mathbf{k}_1$ is *symmetric*. Note that the rows and columns in $\mathbf{k}_1$ are identified by the x, y degrees of freedom at the near end, followed by the far end, that is, 1, 2, 3, 4, respectively, for member 1, Fig. 14–7b. This is done in order to identify the elements for later assembly into the **K** matrix.

Member 2.   Since ② is the near end and ① is the far end, we have

$$\lambda_x = \frac{3 - 0}{5} = 0.6 \qquad \lambda_y = \frac{4 - 0}{5} = 0.8$$

Thus Eq. 14–16 with $L = 5$ m becomes

$$\mathbf{k}_2 = AE \begin{array}{cccc} 1 & 2 & 5 & 6 \\ \left[\begin{array}{cccc} 0.072 & 0.096 & -0.072 & -0.096 \\ 0.096 & 0.128 & -0.096 & -0.128 \\ -0.072 & -0.096 & 0.072 & 0.096 \\ -0.096 & -0.128 & 0.096 & 0.128 \end{array}\right] \begin{array}{c} 1 \\ 2 \\ 5 \\ 6 \end{array} \end{array}$$

Here the rows and columns are identified as 1, 2, 5, 6, since these numbers represent, respectively, the x, y degrees of freedom at the near and far ends of member 2.

Structure Stiffness Matrix.   This matrix has an order of 6×6 since there are six designated degrees of freedom for the truss, Fig. 14–7b. Corresponding elements of the above two matrices are added algebraically to form the structure stiffness matrix. Perhaps the assembly process is easier to see if the missing numerical columns and rows in $\mathbf{k}_1$ and $\mathbf{k}_2$ are expanded with zeros to form two 6×6 matrices. Then

$$\mathbf{K} = \mathbf{k}_1 + \mathbf{k}_2$$

$$\mathbf{K} = AE \begin{array}{cccccc} 1 & 2 & 3 & 4 & 5 & 6 \\ \left[\begin{array}{cccccc} 0.333 & 0 & -0.333 & 0 & 0 & 0 \\ 0 & 0 & 0 & 0 & 0 & 0 \\ -0.333 & 0 & 0.333 & 0 & 0 & 0 \\ 0 & 0 & 0 & 0 & 0 & 0 \\ 0 & 0 & 0 & 0 & 0 & 0 \\ 0 & 0 & 0 & 0 & 0 & 0 \end{array}\right] \begin{array}{c} 1 \\ 2 \\ 3 \\ 4 \\ 5 \\ 6 \end{array} \end{array} + AE \begin{array}{cccccc} 1 & 2 & 3 & 4 & 5 & 6 \\ \left[\begin{array}{cccccc} 0.072 & 0.096 & 0 & 0 & -0.072 & -0.096 \\ 0.096 & 0.128 & 0 & 0 & -0.096 & -0.128 \\ 0 & 0 & 0 & 0 & 0 & 0 \\ 0 & 0 & 0 & 0 & 0 & 0 \\ -0.072 & -0.096 & 0 & 0 & 0.072 & 0.096 \\ -0.096 & -0.128 & 0 & 0 & 0.096 & 0.128 \end{array}\right] \begin{array}{c} 1 \\ 2 \\ 3 \\ 4 \\ 5 \\ 6 \end{array} \end{array}$$

$$\mathbf{K} = AE \begin{bmatrix} 0.405 & 0.096 & -0.333 & 0 & -0.072 & -0.096 \\ 0.096 & 0.128 & 0 & 0 & -0.096 & -0.128 \\ -0.333 & 0 & 0.333 & 0 & 0 & 0 \\ 0 & 0 & 0 & 0 & 0 & 0 \\ -0.072 & -0.096 & 0 & 0 & 0.072 & 0.096 \\ -0.096 & -0.128 & 0 & 0 & 0.096 & 0.128 \end{bmatrix}$$

If a computer is used for this operation, generally one starts with $\mathbf{K}$ having all zero elements; then as the member global stiffness matrices are generated, they are placed directly into their respective element positions in the $\mathbf{K}$ matrix, rather than developing the member stiffness matrices, storing them, then assembling them.

EXAMPLE 14.2

Determine the structure stiffness matrix for the truss shown in Fig. 14–8a. AE is constant.

SOLUTION

Although the truss is statically indeterminate to the first degree, this will present no difficulty for obtaining the structure stiffness matrix. Each joint and member are arbitrarily identified numerically, and the near and far ends are indicated by the arrows along the members. As shown in Fig. 14–8b, the *unconstrained displacements* are *code numbered first*. There are eight degrees of freedom for the truss, and so **K** will be an 8×8 matrix. In order to keep all the joint coordinates positive, the origin of the global coordinates is chosen at ①. Equations 14–5, 14–6, and 14–16 will now be applied to each member.

Member 1. Here $L = 10$ m, so that

$$\lambda_x = \frac{10 - 0}{10} = 1 \qquad \lambda_y = \frac{0 - 0}{10} = 0$$

$$\mathbf{k}_1 = AE \begin{array}{c c c c}
1 & 2 & 6 & 5 \\
\begin{bmatrix}
0.1 & 0 & -0.1 & 0 \\
0 & 0 & 0 & 0 \\
-0.1 & 0 & 0.1 & 0 \\
0 & 0 & 0 & 0
\end{bmatrix} & \begin{array}{c} 1 \\ 2 \\ 6 \\ 5 \end{array}
\end{array}$$

Member 2. Here $L = 10\sqrt{2}$ m, so that

$$\lambda_x = \frac{10 - 0}{10\sqrt{2}} = 0.707 \qquad \lambda_y = \frac{10 - 0}{10\sqrt{2}} = 0.707$$

$$\mathbf{k}_2 = AE \begin{array}{c c c c}
1 & 2 & 7 & 8 \\
\begin{bmatrix}
0.035 & 0.035 & -0.035 & -0.035 \\
0.035 & 0.035 & -0.035 & -0.035 \\
-0.035 & -0.035 & 0.035 & 0.035 \\
-0.035 & -0.035 & 0.035 & 0.035
\end{bmatrix} & \begin{array}{c} 1 \\ 2 \\ 7 \\ 8 \end{array}
\end{array}$$

Member 3. Here $L = 10$ m, so that

$$\lambda_x = \frac{0 - 0}{10} = 0 \qquad \lambda_y = \frac{10 - 0}{10} = 1$$

$$\mathbf{k}_3 = AE \begin{array}{c c c c}
1 & 2 & 3 & 4 \\
\begin{bmatrix}
0 & 0 & 0 & 0 \\
0 & 0.1 & 0 & -0.1 \\
0 & 0 & 0 & 0 \\
0 & -0.1 & 0 & 0.1
\end{bmatrix} & \begin{array}{c} 1 \\ 2 \\ 3 \\ 4 \end{array}
\end{array}$$

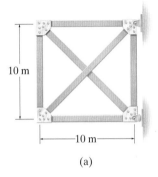

(a)

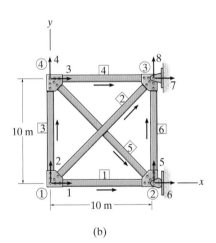

(b)

Fig. 14–8

Member 4.   Here $L = 10$ m, so that

$$\lambda_x = \frac{10 - 0}{10} = 1 \qquad \lambda_y = \frac{10 - 10}{10} = 0$$

$$\mathbf{k}_4 = AE \begin{array}{c} \\ \\ \end{array} \begin{matrix} 3 & 4 & 7 & 8 \\ \begin{bmatrix} 0.1 & 0 & -0.1 & 0 \\ 0 & 0 & 0 & 0 \\ -0.1 & 0 & 0.1 & 0 \\ 0 & 0 & 0 & 0 \end{bmatrix} & \begin{matrix} 3 \\ 4 \\ 7 \\ 8 \end{matrix} \end{matrix}$$

Member 5.   Here $L = 10\sqrt{2}$ m, so that

$$\lambda_x = \frac{10 - 0}{10\sqrt{2}} = 0.707 \qquad \lambda_y = \frac{0 - 10}{10\sqrt{2}} = -0.707$$

$$\mathbf{k}_5 = AE \begin{matrix} 3 & 4 & 6 & 5 \\ \begin{bmatrix} 0.035 & -0.035 & -0.035 & 0.035 \\ -0.035 & 0.035 & 0.035 & -0.035 \\ -0.035 & 0.035 & 0.035 & -0.035 \\ 0.035 & -0.035 & -0.035 & 0.035 \end{bmatrix} & \begin{matrix} 3 \\ 4 \\ 6 \\ 5 \end{matrix} \end{matrix}$$

Member 6.   Here $L = 10$ m, so that

$$\lambda_x = \frac{10 - 10}{10} = 0 \qquad \lambda_y = \frac{10 - 0}{10} = 1$$

$$\mathbf{k}_6 = AE \begin{matrix} 6 & 5 & 7 & 8 \\ \begin{bmatrix} 0 & 0 & 0 & 0 \\ 0 & 0.1 & 0 & -0.1 \\ 0 & 0 & 0 & 0 \\ 0 & -0.1 & 0 & 0.1 \end{bmatrix} & \begin{matrix} 6 \\ 5 \\ 7 \\ 8 \end{matrix} \end{matrix}$$

Structure Stiffness Matrix.   The foregoing six matrices can now be assembled into the 8×8 **K** matrix by algebraically adding their corresponding elements. For example, since $(k_{11})_1 = AE(0.1)$, $(k_{11})_2 = AE(0.035)$, $(k_{11})_3 = (k_{11})_4 = (k_{11})_5 = (k_{11})_6 = 0$, then, $K_{11} = AE(0.1 + 0.035) = AE(0.135)$, and so on. The final result is thus,

$$\mathbf{K} = AE \begin{matrix} 1 & 2 & 3 & 4 & 5 & 6 & 7 & 8 \\ \begin{bmatrix} 0.135 & 0.035 & 0 & 0 & 0 & -0.1 & -0.035 & -0.035 \\ 0.035 & 0.135 & 0 & -0.1 & 0 & 0 & -0.035 & -0.035 \\ 0 & 0 & 0.135 & -0.035 & 0.035 & -0.035 & -0.1 & 0 \\ 0 & -0.1 & -0.035 & 0.135 & -0.035 & 0.035 & 0 & 0 \\ 0 & 0 & 0.035 & -0.035 & 0.135 & -0.035 & 0 & -0.1 \\ -0.1 & 0 & -0.035 & 0.035 & -0.035 & 0.135 & 0 & 0 \\ -0.035 & -0.035 & -0.1 & 0 & 0 & 0 & 0.135 & 0.035 \\ -0.035 & -0.035 & 0 & 0 & -0.1 & 0 & 0.035 & 0.135 \end{bmatrix} & \begin{matrix} 1 \\ 2 \\ 3 \\ 4 \\ 5 \\ 6 \\ 7 \\ 8 \end{matrix} \end{matrix}$$

Ans.

14.6   Application of the Stiffness Method for Truss Analysis

Once the structure stiffness matrix is formed, the global force components $\mathbf{Q}$ acting on the truss can then be related to its global displacements $\mathbf{D}$ using

$$\mathbf{Q} = \mathbf{KD} \tag{14–17}$$

This equation is referred to as the *structure stiffness equation*. Since we have always assigned the lowest code numbers to identify the unconstrained degrees of freedom, this will allow us now to partition this equation in the following form*:

$$\left[\begin{array}{c} \mathbf{Q}_k \\ \hline \mathbf{Q}_u \end{array}\right] = \left[\begin{array}{c:c} \mathbf{K}_{11} & \mathbf{K}_{12} \\ \hdashline \mathbf{K}_{21} & \mathbf{K}_{22} \end{array}\right]\left[\begin{array}{c} \mathbf{D}_u \\ \hline \mathbf{D}_k \end{array}\right] \tag{14–18}$$

Here

$\mathbf{Q}_k, \mathbf{D}_k$ = *known* external loads and displacements; the loads here exist on the truss as part of the problem, and the displacements are generally specified as zero due to support constraints such as pins or rollers.

$\mathbf{Q}_u, \mathbf{D}_u$ = *unknown* loads and displacements; the loads here represent the unknown support reactions, and the displacements are at joints where motion is unconstrained in a particular direction.

$\mathbf{K}$ = *structure* stiffness matrix, which is partitioned to be compatible with the partitioning of $\mathbf{Q}$ and $\mathbf{D}$.

Expanding Eq. 14–18 yields

$$\mathbf{Q}_k = \mathbf{K}_{11}\mathbf{D}_u + \mathbf{K}_{12}\mathbf{D}_k \tag{14–19}$$

$$\mathbf{Q}_u = \mathbf{K}_{21}\mathbf{D}_u + \mathbf{K}_{22}\mathbf{D}_k \tag{14–20}$$

Most often $\mathbf{D}_k = \mathbf{0}$ since the supports are not displaced. Provided this is the case, Eq. 14–19 becomes

$$\mathbf{Q}_k = \mathbf{K}_{11}\mathbf{D}_u$$

Since the elements in the partitioned matrix $\mathbf{K}_{11}$ represent the *total resistance* at a truss joint to a unit displacement in either the x or y direction, then the above equation symbolizes the collection of all the *force equilibrium equations* applied to the joints where the external loads are zero or have a known value ($\mathbf{Q}_k$). Solving for $\mathbf{D}_u$, we have

$$\mathbf{D}_u = [\mathbf{K}_{11}]^{-1}\mathbf{Q}_k \tag{14–21}$$

From this equation we can obtain a direct solution for all the unknown joint displacements; then using Eq. 14–20 with $\mathbf{D}_k = \mathbf{0}$ yields

$$\mathbf{Q}_u = \mathbf{K}_{21}\mathbf{D}_u \tag{14–22}$$

from which the unknown support reactions can be determined. The member forces can be determined using Eq. 14–13, namely

$$\mathbf{q} = \mathbf{k'TD}$$

*This partitioning scheme will become obvious in the numerical examples that follow.

Expanding this equation yields

$$\begin{bmatrix} q_N \\ q_F \end{bmatrix} = \frac{AE}{L} \begin{bmatrix} 1 & -1 \\ -1 & 1 \end{bmatrix} \begin{bmatrix} \lambda_x & \lambda_y & 0 & 0 \\ 0 & 0 & \lambda_x & \lambda_y \end{bmatrix} \begin{bmatrix} D_{Nx} \\ D_{Ny} \\ D_{Fx} \\ D_{Fy} \end{bmatrix}$$

Since $q_N = -q_F$ for equilibrium, only one of the forces has to be found. Here we will determine q_F, the one that exerts tension in the member, Fig. 14–2c.

$$q_F = \frac{AE}{L} [-\lambda_x \quad -\lambda_y \quad \lambda_x \quad \lambda_y] \begin{bmatrix} D_{Nx} \\ D_{Ny} \\ D_{Fx} \\ D_{Fy} \end{bmatrix} \qquad (14\text{–}23)$$

In particular, if the computed result using this equation is negative, the member is then in compression.

Procedure for Analysis

The following method provides a means for determining the unknown displacements and support reactions for a truss using the stiffness method.

Notation

• Establish the x, y global coordinate system. The origin is usually located at the joint for which the coordinates for all the other joints are positive.

• Identify each joint and member numerically, and arbitrarily specify the near and far ends of each member symbolically by directing an arrow along the member with the head directed towards the far end.

• Specify the two code numbers at each joint, using the *lowest numbers* to identify *unconstrained degrees of freedom,* followed by the *highest numbers* to identify the *constrained degrees of freedom.*

• From the problem, establish D_k and Q_k.

Structure Stiffness Matrix

• For each member determine λ_x and λ_y and the member stiffness matrix using Eq. 14–16.

• Assemble these matrices to form the stiffness matrix for the entire truss as explained in Sec. 14–5. As a partial check of the calculations, the member and structure stiffness matrices should be *symmetric.*

Displacements and Loads

• Partition the structure stiffness matrix as indicated by Eq. 14–18.

• Determine the unknown joint displacements D_u using Eq. 14–21, the support reactions Q_u using Eq. 14–22, and each member force q_F using Eq. 14–23.

EXAMPLE 14.3

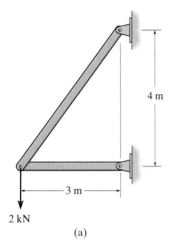

— 3 m —

2 kN

(a)

Fig. 14–9

4 m

Determine the force in each member of the two-member truss shown in Fig. 14–9a. AE is constant.

SOLUTION

Notation. The origin of x, y and the numbering of the joints and members are shown in Fig. 14–9b. Also, the near and far ends of each member are identified by arrows, and code numbers are used at each joint. By inspection it is seen that the known external displacements are $D_3 = D_4 = D_5 = D_6 = 0$. Also, the known external loads are $Q_1 = 0$, $Q_2 = -2$ kN. Hence,

$$\mathbf{D}_k = \begin{bmatrix} 0 \\ 0 \\ 0 \\ 0 \end{bmatrix} \begin{matrix} 3 \\ 4 \\ 5 \\ 6 \end{matrix} \qquad \mathbf{Q}_k = \begin{bmatrix} 0 \\ -2 \end{bmatrix} \begin{matrix} 1 \\ 2 \end{matrix}$$

Structure Stiffness Matrix. Using the same notation as used here, this matrix has been developed in Example 14–1.

Displacements and Loads. Writing Eq. 14–17, $\mathbf{Q} = \mathbf{KD}$, for the truss we have

$$\begin{bmatrix} 0 \\ -2 \\ Q_3 \\ Q_4 \\ Q_5 \\ Q_6 \end{bmatrix} = AE \begin{bmatrix} 0.405 & 0.096 & -0.333 & 0 & -0.072 & -0.096 \\ 0.096 & 0.128 & 0 & 0 & -0.096 & -0.128 \\ -0.333 & 0 & 0.333 & 0 & 0 & 0 \\ 0 & 0 & 0 & 0 & 0 & 0 \\ -0.072 & -0.096 & 0 & 0 & 0.072 & 0.096 \\ -0.096 & -0.128 & 0 & 0 & 0.096 & 0.128 \end{bmatrix} \begin{bmatrix} D_1 \\ D_2 \\ 0 \\ 0 \\ 0 \\ 0 \end{bmatrix} \quad (1)$$

From this equation we can now identify $\mathbf{K}_{11}$ and thereby determine $\mathbf{D}_u$. It is seen that the matrix multiplication, like Eq. 14–19, yields

$$\begin{bmatrix} 0 \\ -2 \end{bmatrix} = AE \begin{bmatrix} 0.405 & 0.096 \\ 0.096 & 0.128 \end{bmatrix} \begin{bmatrix} D_1 \\ D_2 \end{bmatrix} + \begin{bmatrix} 0 \\ 0 \end{bmatrix}$$

Here it is easy to solve by a direct expansion,

$$0 = AE(0.405D_1 + 0.096D_2)$$

$$-2 = AE(0.096D_1 + 0.128D_2)$$

Physically these equations represent $\Sigma F_x = 0$ and $\Sigma F_y = 0$ applied to joint ②. Solving, we get

$$D_1 = \frac{4.505}{AE} \qquad D_2 = \frac{-19.003}{AE}$$

By inspection of Fig. 14–9b, one would indeed expect a rightward and downward displacement to occur at joint ② as indicated by the positive and negative signs of these answers.

Using these results, the support reactions are now obtained from Eq. (1), written in the form of Eq. 14–20 (or Eq. 14–22) as

$$
\begin{bmatrix} Q_3 \\ Q_4 \\ Q_5 \\ Q_6 \end{bmatrix} = AE \begin{bmatrix} -0.333 & 0 \\ 0 & 0 \\ -0.072 & -0.096 \\ -0.096 & -0.128 \end{bmatrix} \frac{1}{AE} \begin{bmatrix} 4.505 \\ -19.003 \end{bmatrix} + \begin{bmatrix} 0 \\ 0 \\ 0 \\ 0 \end{bmatrix}
$$

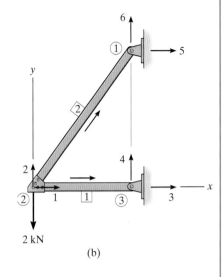

(b)

Expanding and solving for the reactions,

$$Q_3 = -0.333(4.505) = -1.5 \text{ kN}$$

$$Q_4 = 0$$

$$Q_5 = -0.072(4.505) - 0.096(-19.003) = 1.5 \text{ kN}$$

$$Q_6 = -0.096(4.505) - 0.128(-19.003) = 2.0 \text{ kN}$$

The force in each member is found from Eq. 14–23. Using the data for λ_x and λ_y in Example 14–1, we have

Member 1: $\lambda_x = 1, \lambda_y = 0, L = 3 \text{ m}$

$$
q_1 = \frac{AE}{3} \begin{bmatrix} 1 & 2 & 3 & 4 \\ -1 & 0 & 1 & 0 \end{bmatrix} \frac{1}{AE} \begin{bmatrix} 4.505 & 1 \\ -19.003 & 2 \\ 0 & 3 \\ 0 & 4 \end{bmatrix}
$$

$$= \frac{1}{3} [-4.505] = -1.5 \text{ kN} \qquad \qquad Ans.$$

Member 2: $\lambda_x = 0.6, \lambda_y = 0.8, L = 5 \text{ m}$

$$
q_2 = \frac{AE}{5} \begin{bmatrix} 1 & 2 & 5 & 6 \\ -0.6 & -0.8 & 0.6 & 0.8 \end{bmatrix} \frac{1}{AE} \begin{bmatrix} 4.505 & 1 \\ -19.003 & 2 \\ 0 & 5 \\ 0 & 6 \end{bmatrix}
$$

$$= \frac{1}{5} [-0.6(4.505) - 0.8(-19.003)] = 2.5 \text{ kN} \qquad Ans.$$

These answers can of course be verified by equilibrium, applied at joint ②.

EXAMPLE | 14.4

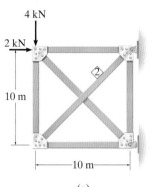

4 kN

2 kN

10 m

10 m

(a)

y

4 kN

2 kN

(b)

Fig. 14–10

Determine the support reactions and the force in member 2 of the truss shown in Fig. 14–10a. AE is constant.

SOLUTION

Notation. The joints and members are numbered and the origin of the x, y axes is established at ①, Fig. 14–10b. Also, arrows are used to reference the near and far ends of each member. Using the code numbers, where the lowest numbers denote the unconstrained degrees of freedom, Fig. 14–10b, we have

$$\mathbf{D}_k = \begin{bmatrix} 0 \\ 0 \\ 0 \\ 0 \end{bmatrix} \begin{matrix} 6 \\ 7 \\ 8 \end{matrix} \qquad \mathbf{Q}_k = \begin{bmatrix} 0 \\ 0 \\ 2 \\ -4 \\ 0 \end{bmatrix} \begin{matrix} 1 \\ 2 \\ 3 \\ 4 \\ 5 \end{matrix}$$

Structure Stiffness Matrix. This matrix has been determined in Example 14–2 using the same notation as in Fig. 14–10b.

Displacements and Loads. For this problem $\mathbf{Q} = \mathbf{KD}$ is

$$\begin{bmatrix} 0 \\ 0 \\ 2 \\ -4 \\ 0 \\ Q_6 \\ Q_7 \\ Q_8 \end{bmatrix} = AE \begin{bmatrix} 0.135 & 0.035 & 0 & 0 & 0 & -0.1 & -0.035 & -0.035 \\ 0.035 & 0.135 & 0 & -0.1 & 0 & 0 & -0.035 & -0.035 \\ 0 & 0 & 0.135 & -0.035 & 0.035 & -0.035 & -0.1 & 0 \\ 0 & -0.1 & -0.035 & 0.135 & -0.035 & 0.035 & 0 & 0 \\ 0 & 0 & 0.035 & -0.035 & 0.135 & -0.035 & 0 & -0.1 \\ -0.1 & 0 & -0.035 & 0.035 & -0.035 & 0.135 & 0 & 0 \\ -0.035 & -0.035 & -0.1 & 0 & 0 & 0 & 0.135 & 0.035 \\ -0.035 & -0.035 & 0 & 0 & -0.1 & 0 & 0.035 & 0.135 \end{bmatrix} \begin{bmatrix} D_1 \\ D_2 \\ D_3 \\ D_4 \\ D_5 \\ 0 \\ 0 \\ 0 \end{bmatrix} \quad (1)$$

Multiplying so as to formulate the unknown displacement equation 14–18, we get

$$\begin{bmatrix} 0 \\ 0 \\ 2 \\ -4 \\ 0 \end{bmatrix} = AE \begin{bmatrix} 0.135 & 0.035 & 0 & 0 & 0 \\ 0.035 & 0.135 & 0 & -0.1 & 0 \\ 0 & 0 & 0.135 & -0.035 & 0.035 \\ 0 & -0.1 & -0.035 & 0.135 & -0.035 \\ 0 & 0 & 0.035 & -0.035 & 0.135 \end{bmatrix} \begin{bmatrix} D_1 \\ D_2 \\ D_3 \\ D_4 \\ D_5 \end{bmatrix} + \begin{bmatrix} 0 \\ 0 \\ 0 \\ 0 \\ 0 \end{bmatrix}$$

Expanding and solving the equations for the displacements yields

$$\begin{bmatrix} D_1 \\ D_2 \\ D_3 \\ D_4 \\ D_5 \end{bmatrix} = \frac{1}{AE} \begin{bmatrix} 17.94 \\ -69.20 \\ -2.06 \\ -87.14 \\ -22.06 \end{bmatrix}$$

Developing Eq. 14–20 from Eq. (1) using the calculated results, we have

$$\begin{bmatrix} Q_6 \\ Q_7 \\ Q_8 \end{bmatrix} = AE \begin{bmatrix} -0.1 & 0 & -0.035 & 0.035 & -0.035 \\ -0.035 & -0.035 & -0.1 & 0 & 0 \\ -0.035 & -0.035 & 0 & 0 & -0.1 \end{bmatrix} \frac{1}{AE} \begin{bmatrix} 17.94 \\ -69.20 \\ -2.06 \\ -87.14 \\ -22.06 \end{bmatrix} + \begin{bmatrix} 0 \\ 0 \\ 0 \end{bmatrix}$$

Expanding and computing the support reactions yields

$$Q_6 = -4.0 \text{ kN} \qquad\qquad Ans.$$

$$Q_7 = 2.0 \text{ kN} \qquad\qquad Ans.$$

$$Q_8 = 4.0 \text{ kN} \qquad\qquad Ans.$$

The negative sign for Q_6 indicates that the rocker support reaction acts in the negative x direction. The force in member 2 is found from Eq. 14–23, where from Example 14–2, $\lambda_x = 0.707$, $\lambda_y = 0.707$, $L = 10\sqrt{2}$ m. Thus,

$$q_2 = \frac{AE}{10\sqrt{2}} [-0.707 \quad -0.707 \quad 0.707 \quad 0.707] \frac{1}{AE} \begin{bmatrix} 17.94 \\ -69.20 \\ 0 \\ 0 \end{bmatrix}$$

$$= 2.56 \text{ kN} \qquad\qquad Ans.$$

EXAMPLE 14.5

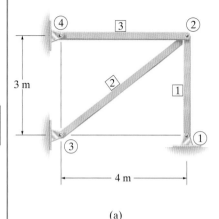

3 m

4 m

(a)

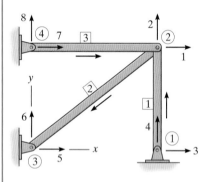

(b)

Fig. 14–11

Determine the force in member 2 of the assembly in Fig. 14–11a if the support at joint ① settles *downward* 25 mm. Take $AE = 8(10^3)$ kN.

SOLUTION

Notation.   For convenience the origin of the global coordinates in Fig. 14–11b is established at joint ③, and as usual the lowest code numbers are used to reference the unconstrained degrees of freedom. Thus,

$$
\mathbf{D}_k = \begin{bmatrix} 0 \\ -0.025 \\ 0 \\ 0 \\ 0 \\ 0 \end{bmatrix} \begin{matrix} 3 \\ 4 \\ 5 \\ 6 \\ 7 \\ 8 \end{matrix} \qquad \mathbf{Q}_k = \begin{bmatrix} 0 \\ 0 \end{bmatrix} \begin{matrix} 1 \\ 2 \end{matrix}
$$

Structure Stiffness Matrix.   Using Eq. 14–16, we have

Member 1: $\lambda_x = 0,\ \lambda_y = 1,\ L = 3$ m, so that

$$
\mathbf{k}_1 = AE \begin{matrix} & 3 & 4 & 1 & 2 \\ & \begin{bmatrix} 0 & 0 & 0 & 0 \\ 0 & 0.333 & 0 & -0.333 \\ 0 & 0 & 0 & 0 \\ 0 & -0.333 & 0 & 0.333 \end{bmatrix} & \begin{matrix} 3 \\ 4 \\ 1 \\ 2 \end{matrix} \end{matrix}
$$

Member 2: $\lambda_x = -0.8,\ \lambda_y = -0.6,\ L = 5$ m, so that

$$
\mathbf{k}_2 = AE \begin{matrix} & 1 & 2 & 5 & 6 \\ & \begin{bmatrix} 0.128 & 0.096 & -0.128 & -0.096 \\ 0.096 & 0.072 & -0.096 & -0.072 \\ -0.128 & -0.096 & 0.128 & 0.096 \\ -0.096 & -0.072 & 0.096 & 0.072 \end{bmatrix} & \begin{matrix} 1 \\ 2 \\ 5 \\ 6 \end{matrix} \end{matrix}
$$

Member 3: $\lambda_x = 1,\ \lambda_y = 0,\ L = 4$ m, so that

$$
\mathbf{k}_3 = AE \begin{matrix} & 7 & 8 & 1 & 2 \\ & \begin{bmatrix} 0.25 & 0 & -0.25 & 0 \\ 0 & 0 & 0 & 0 \\ -0.25 & 0 & 0.25 & 0 \\ 0 & 0 & 0 & 0 \end{bmatrix} & \begin{matrix} 7 \\ 8 \\ 1 \\ 2 \end{matrix} \end{matrix}
$$

By assembling these matrices, the structure stiffness matrix becomes

$$
\mathbf{K} = AE \begin{matrix} & 1 & 2 & 3 & 4 & 5 & 6 & 7 & 8 \\ & \begin{bmatrix} 0.378 & 0.096 & 0 & 0 & -0.128 & -0.096 & -0.25 & 0 \\ 0.096 & 0.405 & 0 & -0.333 & -0.096 & -0.072 & 0 & 0 \\ 0 & 0 & 0 & 0 & 0 & 0 & 0 & 0 \\ 0 & -0.333 & 0 & 0.333 & 0 & 0 & 0 & 0 \\ -0.128 & -0.096 & 0 & 0 & 0.128 & 0.096 & 0 & 0 \\ -0.096 & -0.072 & 0 & 0 & 0.096 & 0.072 & 0 & 0 \\ -0.25 & 0 & 0 & 0 & 0 & 0 & 0.25 & 0 \\ 0 & 0 & 0 & 0 & 0 & 0 & 0 & 0 \end{bmatrix} & \begin{matrix} 1 \\ 2 \\ 3 \\ 4 \\ 5 \\ 6 \\ 7 \\ 8 \end{matrix} \end{matrix}
$$

Displacements and Loads.   Here $\mathbf{Q} = \mathbf{KD}$ yields

$$
\begin{bmatrix} 0 \\ 0 \\ \hline Q_3 \\ Q_4 \\ Q_5 \\ Q_6 \\ Q_7 \\ Q_8 \end{bmatrix} = AE
\begin{bmatrix}
0.378 & 0.096 & 0 & 0 & -0.128 & -0.096 & -0.25 & 0 \\
0.096 & 0.405 & 0 & -0.333 & -0.096 & -0.072 & 0 & 0 \\
\hline
0 & 0 & 0 & 0 & 0 & 0 & 0 & 0 \\
0 & -0.333 & 0 & 0.333 & 0 & 0 & 0 & 0 \\
-0.128 & -0.096 & 0 & 0 & 0.128 & 0.096 & 0 & 0 \\
-0.096 & -0.072 & 0 & 0 & 0.096 & 0.072 & 0 & 0 \\
-0.25 & 0 & 0 & 0 & 0 & 0 & 0.25 & 0 \\
0 & 0 & 0 & 0 & 0 & 0 & 0 & 0
\end{bmatrix}
\begin{bmatrix} D_1 \\ D_2 \\ \hline 0 \\ -0.025 \\ 0 \\ 0 \\ 0 \\ 0 \end{bmatrix}
$$

Developing the solution for the displacements, Eq. 14–19, we have

$$
\begin{bmatrix} 0 \\ 0 \end{bmatrix} = AE \begin{bmatrix} 0.378 & 0.096 \\ 0.096 & 0.405 \end{bmatrix} \begin{bmatrix} D_1 \\ D_2 \end{bmatrix} + AE \begin{bmatrix} 0 & 0 & -0.128 & -0.096 & -0.25 & 0 \\ 0 & -0.333 & -0.096 & -0.072 & 0 & 0 \end{bmatrix} \begin{bmatrix} 0 \\ -0.025 \\ 0 \\ 0 \\ 0 \\ 0 \end{bmatrix}
$$

which yields

$$0 = AE[(0.378D_1 + 0.096D_2) + 0]$$
$$0 = AE[(0.096D_1 + 0.405D_2) + 0.00833]$$

Solving these equations simultaneously gives

$$D_1 = 0.00556 \text{ m}$$
$$D_2 = -0.021875 \text{ m}$$

Although the support reactions do not have to be calculated, if needed they can be found from the expansion defined by Eq. 14–20. Using Eq. 14–23 to determine the force in member 2 yields

Member 2: $\lambda_x = -0.8$, $\lambda_y = -0.6$, $L = 5$ m, $AE = 8(10^3)$ kN, so that

$$
q_2 = \frac{8(10^3)}{5} [0.8 \quad 0.6 \quad -0.8 \quad -0.6] \begin{bmatrix} 0.00556 \\ -0.021875 \\ 0 \\ 0 \end{bmatrix}
$$

$$
= \frac{8(10^3)}{5} (0.00444 - 0.0131) = -13.9 \text{ kN} \qquad \textit{Ans.}
$$

Using the same procedure, show that the force in member 1 is $q_1 = 8.34$ kN and in member 3, $q_3 = 11.1$ kN. The results are shown on the free-body diagram of joint ②, Fig. 14–11c, which can be checked to be in equilibrium.

(c)

14

14.7 Nodal Coordinates

On occasion a truss can be supported by a roller placed on an *incline*, and when this occurs the constraint of zero deflection at the support (node) *cannot* be directly defined using a single horizontal and vertical global coordinate system. For example, consider the truss in Fig. 14–12a. The condition of zero displacement at node ① is defined only along the y'' axis, and because the roller can displace along the x'' axis this node will have displacement *components* along *both* global coordinate axes, x, y. For this reason we cannot include the zero displacement condition at this node when writing the global stiffness equation for the truss using x, y axes without making some modifications to the matrix analysis procedure.

To solve this problem, so that it can easily be incorporated into a computer analysis, we will use a set of *nodal coordinates x'', y''* located at the inclined support. These axes are oriented such that the reactions and support displacements are along each of the coordinate axes, Fig. 14–12a. In order to determine the global stiffness equation for the truss, it then becomes necessary to develop force and displacement transformation matrices for each of the connecting members at this support so that the results can be summed within the same global x, y coordinate system. To show how this is done, consider truss member 1 in Fig. 14–12b, having a global coordinate system x, y at the near node Ⓝ, and a nodal coordinate system x'', y'' at the far node Ⓕ. When displacements **D** occur so that they have components along each of these axes as shown in Fig. 14–12c, the displacements **d** in the x' direction along the ends of the member become

$$d_N = D_{Nx} \cos \theta_x + D_{Ny} \cos \theta_y$$

$$d_F = D_{Fx''} \cos \theta_{x''} + D_{Fy''} \cos \theta_{y''}$$

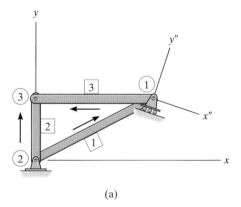

(a)

Fig. 14–12

These equations can be written in matrix form as

$$\begin{bmatrix} d_N \\ d_F \end{bmatrix} = \begin{bmatrix} \lambda_x & \lambda_y & 0 & 0 \\ 0 & 0 & \lambda_{x''} & \lambda_{y''} \end{bmatrix} \begin{bmatrix} D_{Nx} \\ D_{Ny} \\ D_{Fx''} \\ D_{Fy''} \end{bmatrix}$$

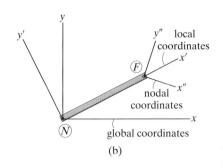

(b)

14

Likewise, forces $\mathbf{q}$ at the near and far ends of the member, Fig. 14–12d, have components $\mathbf{Q}$ along the global axes of

$$Q_{Nx} = q_N \cos \theta_x \qquad Q_{Ny} = q_N \cos \theta_y$$
$$Q_{Fx''} = q_F \cos \theta_{x''} \qquad Q_{Fy''} = q_F \cos \theta_{y''}$$

which can be expressed as

$$\begin{bmatrix} Q_{Nx} \\ Q_{Ny} \\ Q_{Fx''} \\ Q_{Fy''} \end{bmatrix} = \begin{bmatrix} \lambda_x & 0 \\ \lambda_y & 0 \\ 0 & \lambda_{x''} \\ 0 & \lambda_{y''} \end{bmatrix} \begin{bmatrix} q_N \\ q_F \end{bmatrix}$$

The displacement and force transformation matrices in the above equations are used to develop the member stiffness matrix for this situation. Applying Eq. 14–15, we have

$$\mathbf{k} = \mathbf{T}^T \mathbf{k}' \mathbf{T}$$

$$\mathbf{k} = \begin{bmatrix} \lambda_x & 0 \\ \lambda_y & 0 \\ 0 & \lambda_{x''} \\ 0 & \lambda_{y''} \end{bmatrix} \frac{AE}{L} \begin{bmatrix} 1 & -1 \\ -1 & 1 \end{bmatrix} \begin{bmatrix} \lambda_x & \lambda_y & 0 & 0 \\ 0 & 0 & \lambda_{x''} & \lambda_{y''} \end{bmatrix}$$

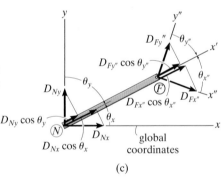

(c)

Performing the matrix operations yields,

$$\mathbf{k} = \frac{AE}{L} \begin{bmatrix} \lambda_x^2 & \lambda_x \lambda_y & -\lambda_x \lambda_{x''} & -\lambda_x \lambda_{y''} \\ \lambda_x \lambda_y & \lambda_y^2 & -\lambda_y \lambda_{x''} & -\lambda_y \lambda_{y''} \\ -\lambda_x \lambda_{x''} & -\lambda_y \lambda_{x''} & \lambda_{x''}^2 & \lambda_{x''} \lambda_{y''} \\ -\lambda_x \lambda_{y''} & -\lambda_y \lambda_{y''} & \lambda_{x''} \lambda_{y''} & \lambda_{y''}^2 \end{bmatrix} \qquad (14\text{–}24)$$

This stiffness matrix is then used for each member that is connected to an inclined roller support, and the process of assembling the matrices to form the structure stiffness matrix follows the standard procedure. The following example problem illustrates its application.

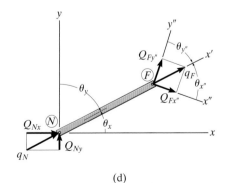

(d)

EXAMPLE | **14.6**

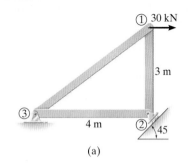

(a)

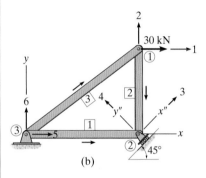

(b)

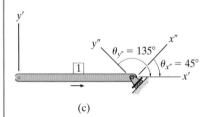

(c)

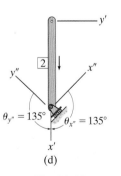

(d)

Fig. 14–13

Determine the support reactions for the truss shown in Fig. 14–13a.

SOLUTION

Notation. Since the roller support at ② is on an incline, we must use nodal coordinates at this node. The joints and members are numbered and the global x, y axes are established at node ③, Fig. 14–13b. Notice that the code numbers 3 and 4 are along the x'', y'' axes in order to use the condition that $D_4 = 0$.

Member Stiffness Matrices. The stiffness matrices for members 1 and 2 must be developed using Eq. 14–24 since these members have code numbers in the direction of global and nodal axes. The stiffness matrix for member 3 is determined in the usual manner.

Member 1. Fig. 14–13c, $\lambda_x = 1$, $\lambda_y = 0$, $\lambda_{x''} = 0.707$, $\lambda_{y''} = -0.707$

$$
\mathbf{k}_1 = AE
\begin{array}{c}
 \\
\begin{bmatrix}
0.25 & 0 & -0.17675 & 0.17675 \\
0 & 0 & 0 & 0 \\
-0.17675 & 0 & 0.125 & -0.125 \\
0.17675 & 0 & -0.125 & 0.125
\end{bmatrix}
\begin{array}{c} 5 \\ 6 \\ 3 \\ 4 \end{array}
\end{array}
$$

with columns labeled 5, 6, 3, 4

Member 2. Fig. 14–13d, $\lambda_x = 0$, $\lambda_y = -1$, $\lambda_{x''} = -0.707$, $\lambda_{y''} = -0.707$

$$
\mathbf{k}_2 = AE
\begin{bmatrix}
0 & 0 & 0 & 0 \\
0 & 0.3333 & -0.2357 & -0.2357 \\
0 & -0.2357 & 0.1667 & 0.1667 \\
0 & -0.2357 & 0.1667 & 0.1667
\end{bmatrix}
\begin{array}{c} 1 \\ 2 \\ 3 \\ 4 \end{array}
$$

with columns labeled 1, 2, 3, 4

Member 3. $\lambda_x = 0.8$, $\lambda_y = 0.6$

$$
\mathbf{k}_3 = AE
\begin{bmatrix}
0.128 & 0.096 & -0.128 & -0.096 \\
0.096 & 0.072 & -0.096 & -0.072 \\
-0.128 & -0.096 & 0.128 & 0.096 \\
-0.096 & -0.072 & 0.096 & 0.072
\end{bmatrix}
\begin{array}{c} 5 \\ 6 \\ 1 \\ 2 \end{array}
$$

with columns labeled 5, 6, 1, 2

Structure Stiffness Matrix.   Assembling these matrices to determine the structure stiffness matrix, we have

$$
\begin{bmatrix} 30 \\ 0 \\ 0 \\ \hline Q_4 \\ Q_5 \\ Q_6 \end{bmatrix} = AE \begin{bmatrix} 0.128 & 0.096 & 0 & 0 & -0.128 & -0.096 \\ 0.096 & 0.4053 & -0.2357 & -0.2357 & -0.096 & -0.072 \\ 0 & -0.2357 & 0.2917 & 0.0417 & -0.17675 & 0 \\ \hline 0 & -0.2357 & 0.0417 & 0.2917 & 0.17675 & 0 \\ -0.128 & -0.096 & -0.17675 & 0.17675 & 0.378 & 0.096 \\ -0.096 & -0.072 & 0 & 0 & 0.096 & 0.072 \end{bmatrix} \begin{bmatrix} D_1 \\ D_2 \\ D_3 \\ \hline 0 \\ 0 \\ 0 \end{bmatrix} \quad (1)
$$

Carrying out the matrix multiplication of the upper partitioned matrices, the three unknown displacements **D** are determined from solving the resulting simultaneous equations, i.e.,

$$D_1 = \frac{352.5}{AE}$$

$$D_2 = \frac{-157.5}{AE}$$

$$D_3 = \frac{-127.3}{AE}$$

The unknown reactions **Q** are obtained from the multiplication of the lower partitioned matrices in Eq. (1). Using the computed displacements, we have,

$Q_4 = 0(352.5) - 0.2357(-157.5) + 0.0417(-127.3)$

$= 31.8$ kN   *Ans.*

$Q_5 = -0.128(352.5) - 0.096(-157.5) - 0.17675(-127.3)$

$= -7.5$ kN   *Ans.*

$Q_6 = -0.096(352.5) - 0.072(-157.5) + 0(-127.3)$

$= -22.5$ kN   *Ans.*

14.8   Trusses Having Thermal Changes and Fabrication Errors

If some of the members of the truss are subjected to an increase or decrease in length due to thermal changes or fabrication errors, then it is necessary to use the method of superposition to obtain the solution. This requires three steps. First, the fixed-end forces necessary to *prevent* movement of the nodes as caused by temperature or fabrication are calculated. Second, the equal but opposite forces are placed on the truss at the nodes and the displacements of the nodes are calculated using the matrix analysis. Finally, the actual forces in the members and the reactions on the truss are determined by superposing these two results. This procedure, of course, is only necessary if the truss is statically indeterminate. If the truss is statically determinate, the displacements at the nodes can be found by this method; however, the temperature changes and fabrication errors will not affect the reactions and the member forces since the truss is free to adjust to these changes of length.

Thermal Effects.   If a truss member of length L is subjected to a temperature increase ΔT, the member will undergo an increase in length of $\Delta L = \alpha \Delta T L$, where α is the coefficient of thermal expansion. A compressive force q_0 applied to the member will cause a decrease in the member's length of $\Delta L' = q_0 L / AE$. If we equate these two displacements, then $q_0 = AE\alpha\Delta T$. This force will hold the nodes of the member fixed as shown in Fig. 14–14, and so we have

$$(q_N)_0 = AE\alpha\Delta T$$

$$(q_F)_0 = -AE\alpha\Delta T$$

Realize that if a temperature decrease occurs, then ΔT becomes negative and these forces reverse direction to hold the member in equilibrium.

We can transform these two forces into global coordinates using Eq. 14–10, which yields

$$\begin{bmatrix} (Q_{Nx})_0 \\ (Q_{Ny})_0 \\ (Q_{Fx})_0 \\ (Q_{Fy})_0 \end{bmatrix} = \begin{bmatrix} \lambda_x & 0 \\ \lambda_y & 0 \\ 0 & \lambda_x \\ 0 & \lambda_y \end{bmatrix} AE\alpha\Delta T \begin{bmatrix} 1 \\ -1 \end{bmatrix} = AE\alpha\Delta T \begin{bmatrix} \lambda_x \\ \lambda_y \\ -\lambda_x \\ -\lambda_y \end{bmatrix} \quad (14\text{–}25)$$

Fabrication Errors.   If a truss member is made too long by an amount ΔL before it is fitted into a truss, then the force q_0 needed to keep the member at its design length L is $q_0 = AE\Delta L / L$, and so for the member in Fig. 14–14, we have

$$(q_N)_0 = \frac{AE\Delta L}{L}$$

$$(q_F)_0 = -\frac{AE\Delta L}{L}$$

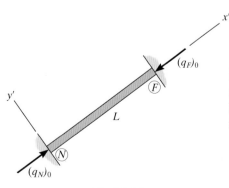

Fig. 14–14

If the member is originally too short, then ΔL becomes negative and these forces will reverse.

In global coordinates, these forces are

$$\begin{bmatrix} (Q_{Nx})_0 \\ (Q_{Ny})_0 \\ (Q_{Fx})_0 \\ (Q_{Fy})_0 \end{bmatrix} = \frac{AE\Delta L}{L} \begin{bmatrix} \lambda_x \\ \lambda_y \\ -\lambda_x \\ -\lambda_y \end{bmatrix} \qquad (14\text{--}26)$$

Matrix Analysis.   In the general case, with the truss subjected to applied forces, temperature changes, and fabrication errors, the initial force-displacement relationship for the truss then becomes

$$\mathbf{Q} = \mathbf{KD} + \mathbf{Q}_0 \qquad (14\text{--}27)$$

Here $\mathbf{Q}_0$ is a column matrix for the entire truss of the initial fixed-end forces caused by the temperature changes and fabrication errors of the members defined in Eqs. 14–25 and 14–26. We can partition this equation in the following form

$$\begin{bmatrix} \mathbf{Q}_k \\ \hline \mathbf{Q}_u \end{bmatrix} = \begin{bmatrix} \mathbf{K}_{11} & \mathbf{K}_{12} \\ \hline \mathbf{K}_{21} & \mathbf{K}_{22} \end{bmatrix} \begin{bmatrix} \mathbf{D}_u \\ \hline \mathbf{D}_k \end{bmatrix} + \begin{bmatrix} (\mathbf{Q}_k)_0 \\ \hline (\mathbf{Q}_u)_0 \end{bmatrix}$$

Carrying out the multiplication, we obtain

$$\mathbf{Q}_k = \mathbf{K}_{11}\mathbf{D}_u + \mathbf{K}_{12}\mathbf{D}_k + (\mathbf{Q}_k)_0 \qquad (14\text{--}28)$$

$$\mathbf{Q}_u = \mathbf{K}_{21}\mathbf{D}_u + \mathbf{K}_{22}\mathbf{D}_k + (\mathbf{Q}_u)_0 \qquad (14\text{--}29)$$

According to the superposition procedure described above, the unknown displacements $\mathbf{D}_u$ are determined from the first equation by subtracting $\mathbf{K}_{12}\mathbf{D}_k$ and $(\mathbf{Q}_k)_0$ from both sides and then solving for $\mathbf{D}_u$. This yields

$$\mathbf{D}_u = \mathbf{K}_{11}^{-1}(\mathbf{Q}_k - \mathbf{K}_{12}\mathbf{D}_k - (\mathbf{Q}_k)_0)$$

Once these nodal displacements are obtained, the member forces are then determined by superposition, i.e.,

$$\mathbf{q} = \mathbf{k}'\mathbf{TD} + \mathbf{q}_0$$

If this equation is expanded to determine the force at the far end of the member, we obtain

$$q_F = \frac{AE}{L}[-\lambda_x \ \ -\lambda_y \ \ \lambda_x \ \ \lambda_y] \begin{bmatrix} D_{Nx} \\ D_{Ny} \\ D_{Fx} \\ D_{Fy} \end{bmatrix} + (q_F)_0 \qquad (14\text{--}30)$$

This result is similar to Eq. 14–23, except here we have the additional term $(q_F)_0$ which represents the initial fixed-end member force due to temperature changes and/or fabrication error as defined previously. Realize that if the computed result from this equation is negative, the member will be in compression.

The following two examples illustrate application of this procedure.

14

EXAMPLE 14.7

Determine the force in members 1 and 2 of the pin-connected assembly of Fig. 14–15 if member 2 was made 0.01 m too short before it was fitted into place. Take $AE = 8(10^3)$ kN.

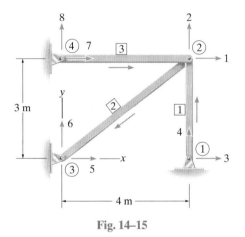

Fig. 14–15

SOLUTION

Since the member is short, then $\Delta L = -0.01$ m, and therefore applying Eq. 14–26 to member 2, with $\lambda_x = -0.8$, $\lambda_y = -0.6$, we have

$$
\begin{bmatrix} (Q_1)_0 \\ (Q_2)_0 \\ (Q_5)_0 \\ (Q_6)_0 \end{bmatrix} = \frac{AE(-0.01)}{5} \begin{bmatrix} -0.8 \\ -0.6 \\ 0.8 \\ 0.6 \end{bmatrix} = AE \begin{bmatrix} 0.0016 \\ 0.0012 \\ -0.0016 \\ -0.0012 \end{bmatrix} \begin{matrix} 1 \\ 2 \\ 5 \\ 6 \end{matrix}
$$

The structure stiffness matrix for this assembly has been established in Example 14–5. Applying Eq. 14–27, we have

$$
\begin{bmatrix} 0 \\ 0 \\ Q_3 \\ Q_4 \\ Q_5 \\ Q_6 \\ Q_7 \\ Q_8 \end{bmatrix} = AE \begin{bmatrix} 0.378 & 0.096 & 0 & 0 & -0.128 & -0.096 & -0.25 & 0 \\ 0.096 & 0.405 & 0 & -0.333 & -0.096 & -0.072 & 0 & 0 \\ 0 & 0 & 0 & 0 & 0 & 0 & 0 & 0 \\ 0 & -0.333 & 0 & 0.333 & 0 & 0 & 0 & 0 \\ -0.128 & -0.096 & 0 & 0 & 0.128 & 0.096 & 0 & 0 \\ -0.096 & -0.072 & 0 & 0 & 0.096 & 0.072 & 0 & 0 \\ -0.25 & 0 & 0 & 0 & 0 & 0 & 0.25 & 0 \\ 0 & 0 & 0 & 0 & 0 & 0 & 0 & 0 \end{bmatrix} \begin{bmatrix} D_1 \\ D_2 \\ 0 \\ 0 \\ 0 \\ 0 \\ 0 \\ 0 \end{bmatrix} + AE \begin{bmatrix} 0.0016 \\ 0.0012 \\ 0 \\ 0 \\ -0.0016 \\ -0.0012 \\ 0 \\ 0 \end{bmatrix} \quad (1)
$$

Partitioning the matrices as shown and carrying out the multiplication to obtain the equations for the unknown displacements yields

$$\begin{bmatrix} 0 \\ 0 \end{bmatrix} = AE \begin{bmatrix} 0.378 & 0.096 \\ 0.096 & 0.405 \end{bmatrix} \begin{bmatrix} D_1 \\ D_2 \end{bmatrix} + AE \begin{bmatrix} 0 & 0 & -0.128 & -0.096 & -0.25 & 0 \\ 0 & -0.333 & -0.096 & -0.072 & 0 & 0 \end{bmatrix} \begin{bmatrix} 0 \\ 0 \\ 0 \\ 0 \\ 0 \\ 0 \end{bmatrix} + AE \begin{bmatrix} 0.0016 \\ 0.0012 \end{bmatrix}$$

which gives

$$0 = AE[0.378D_1 + 0.096D_2] + AE[0] + AE[0.0016]$$

$$0 = AE[0.096D_1 + 0.405D_2] + AE[0] + AE[0.0012]$$

Solving these equations simultaneously,

$$D_1 = -0.003704 \text{ m}$$

$$D_2 = -0.002084 \text{ m}$$

Although not needed, the reactions **Q** can be found from the expansion of Eq. (1) following the format of Eq. 14–29.

In order to determine the force in members 1 and 2 we must apply Eq. 14–30, in which case we have

Member 1. $\lambda_x = 0$, $\lambda_y = 1$, $L = 3$ m, $AE = 8(10^3)$ kN, so that

$$q_1 = \frac{8(10^3)}{3} \begin{bmatrix} 0 & -1 & 0 & 1 \end{bmatrix} \begin{bmatrix} 0 \\ 0 \\ -0.003704 \\ -0.002084 \end{bmatrix} + [0]$$

$$q_1 = -5.56 \text{ kN} \qquad\qquad\qquad Ans.$$

Member 2. $\lambda_x = -0.8$, $\lambda_y = -0.6$, $L = 5$ m, $AE = 8(10^3)$ kN, so

$$q_2 = \frac{8(10^3)}{5} \begin{bmatrix} 0.8 & 0.6 & -0.8 & -0.6 \end{bmatrix} \begin{bmatrix} -0.003704 \\ -0.002084 \\ 0 \\ 0 \end{bmatrix} - \frac{8(10^3)\,(-0.01)}{5}$$

$$q_2 = 9.26 \text{ kN} \qquad\qquad\qquad Ans.$$

14

EXAMPLE **14.8**

Member 2 of the truss shown in Fig. 14–16 is subjected to an increase in temperature of 83.3°C. Determine the force developed in member 2. Take $\alpha = 11.7(10^{-6})/°C$, $E = 200$ GPa. Each member has a cross-sectional area of $A = 484$ mm^2.

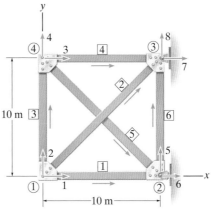

Fig. 14–16

SOLUTION

Since there is a temperature increase, $\Delta T = +83.3°C$. Applying Eq. 14–25 to member 2, where $\lambda_x = 0.7071$, $\lambda_y = 0.7071$, we have

$$\begin{bmatrix} (Q_1)_0 \\ (Q_2)_0 \\ (Q_7)_0 \\ (Q_8)_0 \end{bmatrix} = AE(11.7)\,(10^{-6})\,(83.3)\begin{bmatrix} 0.7071 \\ 0.7071 \\ -0.7071 \\ -0.7071 \end{bmatrix} = AE\begin{bmatrix} 0.000689325 \\ 0.000689325 \\ -0.000689325 \\ -0.000689325 \end{bmatrix}\begin{matrix} 1 \\ 2 \\ 7 \\ 8 \end{matrix}$$

The stiffness matrix for this truss has been developed in Example 14–2.

$$\begin{bmatrix} 0 \\ 0 \\ 0 \\ 0 \\ 0 \\ \hline Q_6 \\ Q_7 \\ Q_8 \end{bmatrix} = AE\begin{bmatrix} 0.135 & 0.035 & 0 & 0 & 0 & -0.1 & -0.035 & -0.035 \\ 0.035 & 0.135 & 0 & -0.1 & 0 & 0 & -0.035 & -0.035 \\ 0 & 0 & 0.135 & -0.035 & 0.035 & -0.035 & -0.1 & 0 \\ 0 & -0.1 & -0.035 & 0.135 & -0.035 & 0.035 & 0 & 0 \\ 0 & 0 & 0.035 & -0.035 & 0.135 & -0.035 & 0 & -0.1 \\ \hline -0.1 & 0 & -0.035 & 0.035 & -0.035 & 0.135 & 0 & 0 \\ -0.035 & -0.035 & -0.1 & 0 & 0 & 0 & 0.135 & 0.035 \\ -0.035 & -0.035 & 0 & 0 & -0.1 & 0 & 0.035 & 0.135 \end{bmatrix}\begin{bmatrix} D_1 \\ D_2 \\ D_3 \\ D_4 \\ D_5 \\ \hline 0 \\ 0 \\ 0 \end{bmatrix} + AE\begin{bmatrix} 0.000689325 \\ 0.000689325 \\ 0 \\ 0 \\ 0 \\ \hline 0 \\ -0.000689325 \\ -0.000689325 \end{bmatrix}\begin{matrix} 1 \\ 2 \\ 3 \\ 4 \\ 5 \\ 6 \\ 7 \\ 8 \end{matrix} \quad (1)$$

Expanding to determine the equations of the unknown displacements, and solving these equations simultaneously, yields

$$D_1 = -0.002027 \text{ m}$$
$$D_2 = -0.01187 \text{ m}$$
$$D_3 = -0.002027 \text{ m}$$
$$D_4 = -0.009848 \text{ m}$$
$$D_5 = -0.002027 \text{ m}$$

Using Eq. 14–30 to determine the force in member 2, we have

$$q_2 = \frac{484[200]}{10\sqrt{2}} \begin{bmatrix} -0.707 & -0.707 & 0.707 & 0.707 \end{bmatrix} \begin{bmatrix} -0.002027 \\ -0.01187 \\ 0 \\ 0 \end{bmatrix} - 484(200)[11.7(10^{-6})](83.3)$$

$$= -27.09 \text{ kN} \qquad\qquad\qquad Ans.$$

Note that the temperature increase of member 2 will not cause any reactions on the truss since externally the truss is statically determinate. To show this, consider the matrix expansion of Eq. (1) for determining the reactions. Using the results for the displacements, we have

$$Q_6 = AE[-0.1(-0.002027) + 0 - 0.035(-0.002027)$$

$$+ 0.035(-0.009848) - 0.035(-0.002027)] + AE[0] = 0$$

$$Q_7 = AE[-0.035(-0.002027) - 0.035(-0.01187)$$

$$- 0.1(-0.002027) + 0 + 0] + AE[-0.000689325] = 0$$

$$Q_8 = AE[-0.035(-0.002027) - 0.035(-0.01187) + 0$$

$$+ 0 - 0.1(-0.002027)] + AE[-0.000689325] = 0$$

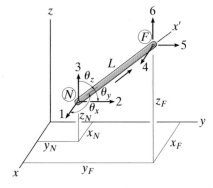

Fig. 14–17

14.9 Space-Truss Analysis

The analysis of both statically determinate and indeterminate space trusses can be performed by using the same procedure discussed previously. To account for the three-dimensional aspects of the problem, however, additional elements must be included in the transformation matrix **T**. In this regard, consider the truss member shown in Fig. 14–17. The stiffness matrix for the member defined in terms of the local coordinate x' is given by Eq. 14–4. Furthermore, by inspection of Fig. 14–17, the direction cosines between the global and local coordinates can be found using equations analogous to Eqs. 14–5 and 14–6, that is,

$$\lambda_x = \cos\theta_x = \frac{x_F - x_N}{L}$$

$$= \frac{x_F - x_N}{\sqrt{(x_F - x_N)^2 + (y_F - y_N)^2 + (z_F - z_N)^2}} \quad (14\text{–}31)$$

$$\lambda_y = \cos\theta_y = \frac{y_F - y_N}{L}$$

$$= \frac{y_F - y_N}{\sqrt{(x_F - x_N)^2 + (y_F - y_N)^2 + (z_F - z_N)^2}} \quad (14\text{–}32)$$

$$\lambda_z = \cos\theta_z = \frac{z_F - z_N}{L}$$

$$= \frac{z_F - z_N}{\sqrt{(x_F - x_N)^2 + (y_F - y_N)^2 + (z_F - z_N)^2}} \quad (14\text{–}33)$$

As a result of the third dimension, the transformation matrix, Eq. 14–9, becomes

$$\mathbf{T} = \begin{bmatrix} \lambda_x & \lambda_y & \lambda_z & 0 & 0 & 0 \\ 0 & 0 & 0 & \lambda_x & \lambda_y & \lambda_z \end{bmatrix}$$

Substituting this and Eq. 14–4 into Eq. 14–15, $\mathbf{k} = \mathbf{T}^T\mathbf{k}'\mathbf{T}$, yields

$$\mathbf{k} = \begin{bmatrix} \lambda_x & 0 \\ \lambda_y & 0 \\ \lambda_z & 0 \\ 0 & \lambda_x \\ 0 & \lambda_y \\ 0 & \lambda_z \end{bmatrix} \frac{AE}{L} \begin{bmatrix} 1 & -1 \\ -1 & 1 \end{bmatrix} \begin{bmatrix} \lambda_x & \lambda_y & \lambda_z & 0 & 0 & 0 \\ 0 & 0 & 0 & \lambda_x & \lambda_y & \lambda_z \end{bmatrix}$$

Carrying out the matrix multiplication yields the *symmetric* matrix

$$
\mathbf{k} = \frac{AE}{L}
\begin{array}{c}
\begin{array}{cccccc}
\quad N_x & \quad N_y & \quad N_z & \quad F_x & \quad F_y & \quad F_z
\end{array} \\
\begin{bmatrix}
\lambda_x^2 & \lambda_x\lambda_y & \lambda_x\lambda_z & -\lambda_x^2 & -\lambda_x\lambda_y & -\lambda_x\lambda_z \\
\lambda_y\lambda_x & \lambda_y^2 & \lambda_y\lambda_z & -\lambda_y\lambda_x & -\lambda_y^2 & -\lambda_y\lambda_z \\
\lambda_z\lambda_x & \lambda_z\lambda_y & \lambda_z^2 & -\lambda_z\lambda_x & -\lambda_z\lambda_y & -\lambda_z^2 \\
-\lambda_x^2 & -\lambda_x\lambda_y & -\lambda_x\lambda_z & \lambda_x^2 & \lambda_x\lambda_y & \lambda_x\lambda_z \\
-\lambda_y\lambda_x & -\lambda_y^2 & -\lambda_y\lambda_z & \lambda_y\lambda_x & \lambda_y^2 & \lambda_y\lambda_z \\
-\lambda_z\lambda_x & -\lambda_z\lambda_y & -\lambda_z^2 & \lambda_z\lambda_x & \lambda_z\lambda_y & \lambda_z^2
\end{bmatrix}
\begin{array}{c}
N_x \\ N_y \\ N_z \\ F_x \\ F_y \\ F_z
\end{array}
\end{array}
\qquad (14\text{–}34)
$$

This equation represents the *member stiffness matrix* expressed in *global coordinates*. The code numbers along the rows and columns reference the x, y, z directions at the near end, N_x, N_y, N_z, followed by those at the far end, F_x, F_y, F_z.

For computer programming, it is generally more efficient to use Eq. 14–34 than to carry out the matrix multiplication $\mathbf{T}^T\mathbf{k}'\mathbf{T}$ for each member. Computer storage space is saved if the "structure" stiffness matrix $\mathbf{K}$ is first initialized with all zero elements; then as the elements of each member stiffness matrix are generated, they are placed directly into their respective positions in $\mathbf{K}$. After the structure stiffness matrix has been developed, the same procedure outlined in Sec. 14–6 can be followed to determine the joint displacements, support reactions, and internal member forces.

The structural framework of this aircraft hangar is constructed entirely of trusses in order to reduce significantly the weight of the structure. *(Courtesy of Bethlehem Steel Corporation)*

CHAPTER REVIEW

The stiffness method is the preferred method for analyzing structures using a computer. It first requires identifying the number of structural elements and their nodes. The global coordinates for the entire structure are then established, and each member's local coordinate system is located so that its origin is at a selected near end, such that the positive x' axis extends towards the far end.

Formulation of the method first requires that each member stiffness matrix $\mathbf{k}'$ be constructed. It relates the loads at the ends of the member, $\mathbf{q}$, to their displacements, $\mathbf{d}$, where $\mathbf{q} = \mathbf{k}'\mathbf{d}$. Then, using the transformation matrix $\mathbf{T}$, the local displacements $\mathbf{d}$ are related to the global displacements $\mathbf{D}$, where $\mathbf{d} = \mathbf{TD}$. Also, the local forces $\mathbf{q}$ are transformed into the global forces $\mathbf{Q}$ using the transformation matrix $\mathbf{T}$, i.e., $\mathbf{Q} = \mathbf{T}^T\mathbf{q}$. When these matrices are combined, one obtains the member's stiffness matrix $\mathbf{K}$ in global coordinates, $\mathbf{k} = \mathbf{T}^T\mathbf{k}'\mathbf{T}$. Assembling all the member stiffness matrices yields the stiffness matrix $\mathbf{K}$ for the entire structure.

The displacements and loads on the structure are then obtained by partitioning $\mathbf{Q} = \mathbf{KD}$, such that the unknown displacements are obtained from $\mathbf{D}_u = [\mathbf{K}_{11}]^{-1}\mathbf{Q}_k$, provided the supports do not displace. Finally, the support reactions are obtained from $\mathbf{Q}_u = \mathbf{K}_{21}\mathbf{D}_u$ and each member force is found from $\mathbf{q} = \mathbf{k}'\mathbf{TD}$.

(Ref: Sections 14.1 to 14.6)

PROBLEMS

14–1. Determine the stiffness matrix **K** for the assembly. Take $A = 300 \text{ mm}^2$ and $E = 200$ GPa for each member.

14–2. Determine the horizontal and vertical displacements at joint ③ of the assembly in Prob. 14–1.

14–3. Determine the force in each member of the assembly in Prob. 14–1.

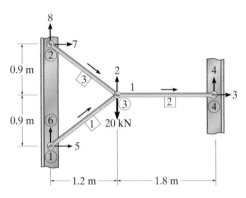

Probs. 14–1/14–2/14–3

14–7. Determine the stiffness matrix **K** for the truss. Take $A = 0.0015 \text{ m}^2$ and $E = 200$ GPa for each member.

*14–8. Determine the vertical displacement at joint ② and the force in member $\boxed{5}$. Take $A = 0.0015 \text{ m}^2$ and $E = 200$ GPa.

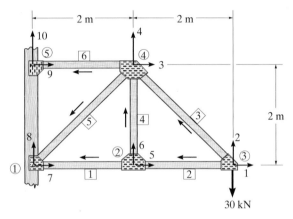

Probs. 14–7/14–8

*14–4. Determine the stiffness matrix **K** for the truss. Take $A = 1000 \text{ mm}^2, E = 200$ GPa.

14–5. Determine the horizontal displacement of joint ① and the force in member $\boxed{2}$. Take $A = 1000 \text{ mm}^2$, $E = 200$ GPa.

14–6. Determine the force in member $\boxed{2}$ if its temperature is increased by 55°C. Take $A = 1000 \text{ mm}^2$, $E = 200$ GPa, $\alpha = 11.7(10^{-6})/°C$.

14–9. Determine the stiffness matrix **K** for the truss. Take $A = 0.0015 \text{ m}^2$ and $E = 200$ GPa for each member.

14–10. Determine the force in member $\boxed{5}$. Take $A = 0.0015 \text{ m}^2$ and $E = 200$ GPa for each member.

14–11. Determine the vertical displacement of node ② if member $\boxed{6}$ was 10 mm too long before it was fitted into the truss. For the solution, remove the 20-kN load. Take $A = 0.0015 \text{ m}^2$ and $E = 200$ GPa for each member.

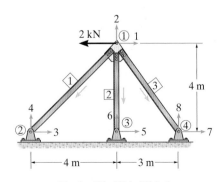

Probs. 14–4/14–5/14–6

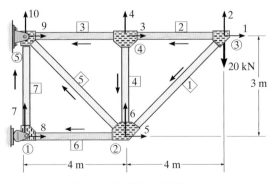

Probs. 14–9/14–10/14–11

*14–12. Determine the stiffness matrix **K** for the truss. Take $A = 0.0015 \text{ m}^2$, $E = 200 \text{ GPa}$.

14–13. Determine the horizontal displacement of joint ② and the force in member $\boxed{5}$. Take $A = 0.0015 \text{ m}^2$, $E = 200 \text{ GPa}$. Neglect the short link at ②.

14–14. Determine the force in member $\boxed{3}$ if this member was 1 mm too short before it was fitted onto the truss. Take $A = 0.0015 \text{ m}^2$. $E = 200 \text{ GPa}$. Neglect the short link at ②.

14–17. Use a computer program and determine the reactions on the truss and the force in each member. AE is constant.

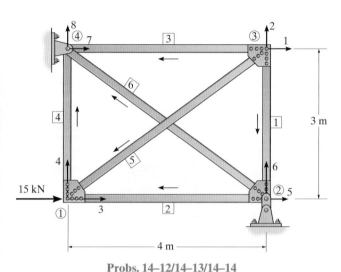

Probs. 14–12/14–13/14–14

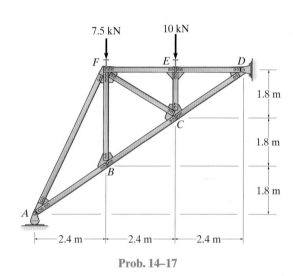

Prob. 14–17

14–15. Determine the stiffness matrix **K** for the truss. AE is constant.

*14–16. Determine the vertical displacement of joint ② and the support reactions. AE is constant.

14–18. Use a computer program and determine the reactions on the truss and the force in each member. AE is constant.

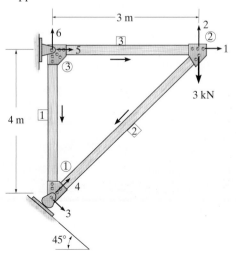

Probs. 14–15/14–16

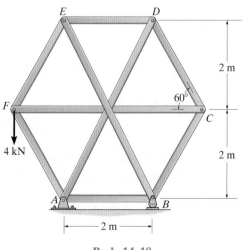

Prob. 14–18

The statically indeterminate loading in bridge girders that are continuous over their piers can be determined using the stiffness method.

Beam Analysis Using the Stiffness Method

15

The concepts presented in the previous chapter will be extended here and applied to the analysis of beams. It will be shown that once the member stiffness matrix and the transformation matrix have been developed, the procedure for application is exactly the same as that for trusses. Special consideration will be given to cases of differential settlement and temperature.

15.1 Preliminary Remarks

Before we show how the stiffness method applies to beams, we will first discuss some preliminary concepts and definitions related to these members.

Member and Node Identification. In order to apply the stiffness method to beams, we must first determine how to subdivide the beam into its component finite elements. In general, each element must be free from load and have a prismatic cross section. For this reason the nodes of each element are located at a support or at points where members are connected together, where an external force is applied, where the cross-sectional area suddenly changes, or where the vertical or rotational displacement at a point is to be determined. For example, consider the beam in Fig. 15–1a. Using the same scheme as that for trusses, four nodes are specified numerically within a circle, and the three elements are identified numerically within a square. Also, notice that the "near" and "far" ends of each element are identified by the arrows written alongside each element.

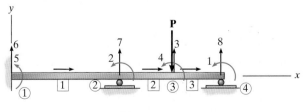

(a)

Fig. 15–1

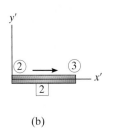

(b)

Fig. 15–1

15

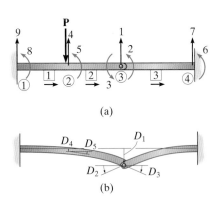

(a)

(b)

Fig. 15–2

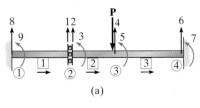

(a)

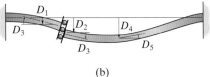

(b)

Fig. 15–3

Global and Member Coordinates. The global coordinate system will be identified using x, y, z axes that generally have their origin at a node and are positioned so that the nodes at other points on the beam all have positive coordinates, Fig. 15–1a. The local or member x', y', z' coordinates have their origin at the "near" end of each element, and the positive x' axis is directed towards the "far" end. Figure 15–1b shows these coordinates for element 2. In both cases we have used a right-handed coordinate system, so that if the fingers of the right hand are curled from the x (x') axis towards the y (y') axis, the thumb points in the positive direction of the z (z') axis, which is directed out of the page. Notice that for each beam element the x and x' axes will be collinear and the global and member coordinates will all be parallel. Therefore, unlike the case for trusses, here we will not need to develop transformation matrices between these coordinate systems.

Kinematic Indeterminacy. Once the elements and nodes have been identified, and the global coordinate system has been established, the degrees of freedom for the beam and its kinematic determinacy can be determined. If we consider the effects of both bending and shear, then *each node* on a beam can have two degrees of freedom, namely, a vertical displacement and a rotation. As in the case of trusses, these linear and rotational displacements will be identified by code numbers. The lowest code numbers will be used to identify the unknown displacements (unconstrained degrees of freedom), and the highest numbers are used to identify the known displacements (constrained degrees of freedom). Recall that the reason for choosing this method of identification has to do with the convenience of later partitioning the structure stiffness matrix, so that the unknown displacements can be found in the most direct manner.

To show an example of code-number labeling, consider again the continuous beam in Fig. 15–1a. Here the beam is kinematically indeterminate to the fourth degree. There are eight degrees of freedom, for which code numbers 1 through 4 represent the unknown displacements, and numbers 5 through 8 represent the known displacements, which in this case are all zero. As another example, the beam in Fig. 15–2a can be subdivided into three elements and four nodes. In particular, notice that the internal hinge at node 3 deflects the same for both elements 2 and 3; however, the rotation at the end of each element is different. For this reason three code numbers are used to show these deflections. Here there are nine degrees of freedom, five of which are unknown, as shown in Fig. 15–2b, and four known; again they are all zero. Finally, consider the slider mechanism used on the beam in Fig. 15–3a. Here the deflection of the beam is shown in Fig. 15–3b, and so there are five unknown deflection components labeled with the lowest code numbers. The beam is kinematically indeterminate to the fifth degree.

Development of the stiffness method for beams follows a similar procedure as that used for trusses. First we must establish the stiffness matrix for each element, and then these matrices are combined to form the beam or structure stiffness matrix. Using the structure matrix equation, we can then proceed to determine the unknown displacements at the nodes and from this determine the reactions on the beam and the internal shear and moment at the nodes.

15.2 Beam-Member Stiffness Matrix

In this section we will develop the stiffness matrix for a beam element or member having a constant cross-sectional area and referenced from the local x', y', z' coordinate system, Fig. 15–4. The origin of the coordinates is placed at the "near" end N, and the positive x' axis extends toward the "far" end F. There are two reactions at each end of the element, consisting of shear forces $q_{Ny'}$ and $q_{Fy'}$ and bending moments $q_{Nz'}$ and $q_{Fz'}$. These loadings all act in the positive coordinate directions. In particular, the moments $q_{Nz'}$ and $q_{Fz'}$ are positive *counterclockwise,* since by the right-hand rule the moment vectors are then directed along the positive z' axis, which is out of the page.

Linear and angular displacements associated with these loadings also follow this same positive sign convention. We will now impose each of these displacements separately and then determine the loadings acting on the member caused by each displacement.

positive sign convention

Fig. 15–4

y' Displacements.
When a positive displacement $d_{Ny'}$ is imposed while other possible displacements are prevented, the resulting shear forces and bending moments that are created are shown in Fig. 15–5a. In particular, the moment has been developed in Sec. 11–2 as Eq. 11–5. Likewise, when $d_{Fy'}$ is imposed, the required shear forces and bending moments are given in Fig. 15–5b.

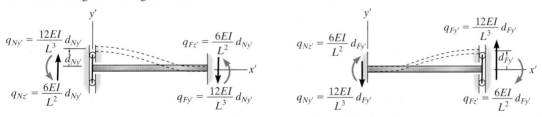

y' displacements

(a)   (b)

Fig. 15–5

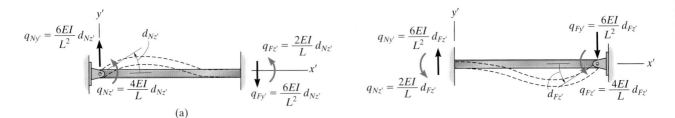

(a)

Fig. 15–6

z' Rotations.

If a positive rotation $d_{Nz'}$ is imposed while all other possible displacements are prevented, the required shear forces and moments necessary for the deformation are shown in Fig. 15–6a. In particular, the moment results have been developed in Sec. 11–2 as Eqs. 11–1 and 11–2. Likewise, when $d_{Fz'}$ is imposed, the resultant loadings are shown in Fig. 15–6b.

By superposition, if the above results in Figs. 15–5 and 15–6 are added, the resulting four load-displacement relations for the member can be expressed in matrix form as

$$
\begin{bmatrix} q_{Ny'} \\ q_{Nz'} \\ q_{Fy'} \\ q_{Fz'} \end{bmatrix} =
\begin{bmatrix}
\dfrac{12EI}{L^3} & \dfrac{6EI}{L^2} & -\dfrac{12EI}{L^3} & \dfrac{6EI}{L^2} \\[2mm]
\dfrac{6EI}{L^2} & \dfrac{4EI}{L} & -\dfrac{6EI}{L^2} & \dfrac{2EI}{L} \\[2mm]
-\dfrac{12EI}{L^3} & -\dfrac{6EI}{L^2} & \dfrac{12EI}{L^3} & -\dfrac{6EI}{L^2} \\[2mm]
\dfrac{6EI}{L^2} & \dfrac{2EI}{L} & -\dfrac{6EI}{L^2} & \dfrac{4EI}{L}
\end{bmatrix}
\begin{bmatrix} d_{Ny'} \\ d_{Nz'} \\ d_{Fy'} \\ d_{Fz'} \end{bmatrix}
\qquad (15\text{–}1)
$$

with column headings $N_{y'} \quad N_{z'} \quad F_{y'} \quad F_{z'}$

These equations can also be written in abbreviated form as

$$\mathbf{q} = \mathbf{kd} \qquad (15\text{–}2)$$

The symmetric matrix $\mathbf{k}$ in Eq. 15–1 is referred to as the *member stiffness matrix*. The 16 influence coefficients $\mathbf{k}_{ij}$ that comprise it account for the shear-force and bending-moment displacements of the member. Physically these coefficients represent the load on the member when the member undergoes a specified unit displacement. For example, if $d_{Ny'} = 1$, Fig. 15–5a, *while all other displacements are zero*, the member will be subjected only to the four loadings indicated in the first column of the $\mathbf{k}$ matrix. In a similar manner, the other columns of the $\mathbf{k}$ matrix are the member loadings for unit displacements identified by the degree-of-freedom code numbers listed above the columns. From the development, both equilibrium and compatibility of displacements have been satisfied. Also, it should be noted that this matrix is the *same* in both the local and global coordinates since the x', y', z' axes are parallel to x, y, z and, therefore, transformation matrices are not needed between the coordinates.

15.3   Beam-Structure Stiffness Matrix

Once all the member stiffness matrices have been found, we must assemble them into the structure stiffness matrix **K**. This process depends on first knowing the *location* of each element in the member stiffness matrix. Here the rows and columns of each **k** matrix (Eq. 15–1) are identified by the two code numbers at the near end of the member $(N_{y'}, N_{z'})$ followed by those at the far end $(F_{y'}, F_{z'})$. Therefore, when assembling the matrices, each element must be placed in the same location of the **K** matrix. In this way, **K** will have an order that will be equal to the highest code number assigned to the beam, since this represents the total number of degrees of freedom. Also, where several members are connected to a node, their member stiffness influence coefficients will have the same position in the **K** matrix and therefore must be algebraically added together to determine the nodal stiffness influence coefficient for the structure. This is necessary since each coefficient represents the nodal resistance of the structure in a particular direction $(y'$ or $z')$ when a unit displacement $(y'$ or $z')$ occurs either at the same or at another node. For example, $\mathbf{K}_{23}$ represents the load in the direction and at the location of code number "2" when a unit displacement occurs in the direction and at the location of code number "3."

15.4   Application of the Stiffness Method for Beam Analysis

After the structure stiffness matrix is determined, the loads at the nodes of the beam can be related to the displacements using the structure stiffness equation

$$\mathbf{Q} = \mathbf{KD}$$

Here **Q** and **D** are column matrices that represent both the known and unknown loads and displacements. Partitioning the stiffness matrix into the known and unknown elements of load and displacement, we have

$$\begin{bmatrix} \mathbf{Q}_k \\ \mathbf{Q}_u \end{bmatrix} = \begin{bmatrix} \mathbf{K}_{11} & \mathbf{K}_{12} \\ \mathbf{K}_{21} & \mathbf{K}_{22} \end{bmatrix} \begin{bmatrix} \mathbf{D}_u \\ \mathbf{D}_k \end{bmatrix}$$

which when expanded yields the two equations

$$\mathbf{Q}_k = \mathbf{K}_{11}\mathbf{D}_u + \mathbf{K}_{12}\mathbf{D}_k \qquad (15\text{--}3)$$

$$\mathbf{Q}_u = \mathbf{K}_{21}\mathbf{D}_u + \mathbf{K}_{22}\mathbf{D}_k \qquad (15\text{--}4)$$

The unknown displacements $\mathbf{D}_u$ are determined from the first of these equations. Using these values, the support reactions $\mathbf{Q}_u$ are computed for the second equation.

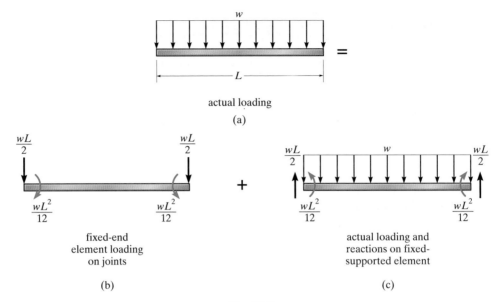

actual loading
(a)

fixed-end
element loading
on joints

(b)

actual loading and
reactions on fixed-
supported element

(c)

Fig. 15–7

Intermediate Loadings.

For application, it is important that the elements of the beam be free of loading along its length. This is necessary since the stiffness matrix for each element was developed for loadings applied only at its ends. (See Fig. 15–4.) Oftentimes, however, beams will support a distributed loading, and this condition will require modification in order to perform the matrix analysis.

To handle this case, we will use the principle of superposition in a manner similar to that used for trusses discussed in Sec. 14–8. To show its application, consider the beam element of length L in Fig. 15–7a, which is subjected to the uniform distributed load w. First we will apply fixed-end moments and reactions to the element, which will be used in the stiffness method, Fig. 15–7b. We will refer to these loadings as a column matrix $-\mathbf{q}_0$. Then the distributed loading and its reactions $\mathbf{q}_0$ are applied, Fig. 15–7c. The actual loading within the beam is determined by adding these two results. The fixed-end reactions for other cases of loading are given on the inside back cover. In addition to solving problems involving lateral loadings such as this, we can also use this method to solve problems involving temperature changes or fabrication errors.

Member Forces.

The shear and moment at the ends of each beam element can be determined using Eq. 15–2 and adding on any fixed-end reactions $\mathbf{q}_0$ if the element is subjected to an intermediate loading. We have

$$\mathbf{q} = \mathbf{kd} + \mathbf{q}_0 \qquad (15\text{–}5)$$

If the results are negative, it indicates that the loading acts in the opposite direction to that shown in Fig. 15–4.

Procedure for Analysis

The following method provides a means of determining the displacements, support reactions, and internal loadings for the members or finite elements of a statically determinate or statically indeterminate beam.

Notation

- Divide the beam into finite elements and arbitrarily identify each element and its nodes. Use a number written in a circle for a node and a number written in a square for a member. Usually an element extends between points of support, points of concentrated loads, and joints, or to points where internal loadings or displacements are to be determined. Also, E and I for the elements must be constants.

- Specify the near and far ends of each element symbolically by directing an arrow along the element, with the head directed toward the far end.

- At each nodal point specify numerically the y and z code numbers. In all cases use the *lowest code numbers* to identify all the unconstrained degrees of freedom, followed by the remaining or highest numbers to identify the degrees of freedom that are constrained.

- From the problem, establish the known displacements $\mathbf{D}_k$ and known external loads $\mathbf{Q}_k$. Include any *reversed* fixed-end loadings if an element supports an intermediate load.

Structure Stiffness Matrix

- Apply Eq. 15–1 to determine the stiffness matrix for each element expressed in global coordinates.

- After each member stiffness matrix is determined, and the rows and columns are identified with the appropriate code numbers, assemble the matrices to determine the structure stiffness matrix $\mathbf{K}$. As a partial check, the member *and* structure stiffness matrices should all be *symmetric*.

Displacements and Loads

- Partition the structure stiffness equation and carry out the matrix multiplication in order to determine the unknown displacements $\mathbf{D}_u$ and support reactions $\mathbf{Q}_u$.

- The internal shear and moment $\mathbf{q}$ at the ends of each beam element can be determined from Eq. 15–5, accounting for the additional fixed-end loadings.

15

EXAMPLE 15.1

Determine the reactions at the supports of the beam shown in Fig. 15–8a. EI is constant.

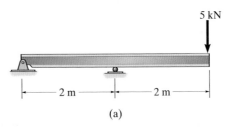

(a)

Fig. 15–8

SOLUTION

Notation.   The beam has two elements and three nodes, which are identified in Fig. 15–8b. The code numbers 1 through 6 are indicated such that the *lowest numbers 1–4 identify the unconstrained degrees of freedom.*

The known load and displacement matrices are

$$\mathbf{Q}_k = \begin{bmatrix} 0 \\ -5 \\ 0 \\ 0 \end{bmatrix} \begin{matrix} 1 \\ 2 \\ 3 \\ 4 \end{matrix} \qquad \mathbf{D}_k = \begin{bmatrix} 0 \\ 0 \end{bmatrix} \begin{matrix} 5 \\ 6 \end{matrix}$$

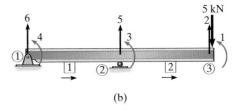

(b)

Member Stiffness Matrices.   Each of the two member stiffness matrices is determined from Eq. 15–1. Note carefully how the code numbers for each column and row are established.

$$\mathbf{k}_1 = EI \begin{array}{c} \begin{array}{cccc} 6 & 4 & 5 & 3 \end{array} \\ \begin{bmatrix} 1.5 & 1.5 & -1.5 & 1.5 \\ 1.5 & 2 & -1.5 & 1 \\ -1.5 & -1.5 & 1.5 & -1.5 \\ 1.5 & 1 & -1.5 & 2 \end{bmatrix} \begin{array}{c} 6 \\ 4 \\ 5 \\ 3 \end{array} \end{array} \qquad \mathbf{k}_2 = EI \begin{array}{c} \begin{array}{cccc} 5 & 3 & 2 & 1 \end{array} \\ \begin{bmatrix} 1.5 & 1.5 & -1.5 & 1.5 \\ 1.5 & 2 & -1.5 & 1 \\ -1.5 & -1.5 & 1.5 & -1.5 \\ 1.5 & 1 & -1.5 & 2 \end{bmatrix} \begin{array}{c} 5 \\ 3 \\ 2 \\ 1 \end{array} \end{array}$$

Displacements and Loads.   We can now assemble these elements into the structure stiffness matrix. For example, element $\mathbf{K}_{11} = 0 + 2 = 2$, $\mathbf{K}_{55} = 1.5 + 1.5 = 3$, etc. Thus,

$$\mathbf{Q} = \mathbf{KD}$$

$$
\begin{bmatrix}
0 \\
-5 \\
0 \\
0 \\
\hdashline
Q_5 \\
Q_6
\end{bmatrix}
= EI
\begin{array}{c}
\begin{array}{cccccc}
1 & \quad 2 & \quad 3 & \quad 4 & \quad 5 & \quad 6
\end{array} \\
\begin{bmatrix}
2 & -1.5 & 1 & 0 & 1.5 & 0 \\
-1.5 & 1.5 & -1.5 & 0 & -1.5 & 0 \\
1 & -1.5 & 4 & 1 & 0 & 1.5 \\
0 & 0 & 1 & 2 & -1.5 & 1.5 \\
\hdashline
1.5 & -1.5 & 0 & -1.5 & 3 & -1.5 \\
0 & 0 & 1.5 & 1.5 & -1.5 & 1.5
\end{bmatrix}
\end{array}
\begin{bmatrix}
D_1 \\
D_2 \\
D_3 \\
D_4 \\
0 \\
0
\end{bmatrix}
$$

The matrices are partitioned as shown. Carrying out the multiplication for the first four rows, we have

$$0 = 2D_1 - 1.5D_2 + D_3 + 0$$

$$-\frac{5}{EI} = -1.5D_1 + 1.5D_2 - 1.5D_3 + 0$$

$$0 = D_1 - 1.5D_2 + 4D_3 + D_4$$

$$0 = 0 + 0 + D_3 + 2D_4$$

Solving,

$$D_1 = -\frac{16.67}{EI}$$

$$D_2 = -\frac{26.67}{EI}$$

$$D_3 = -\frac{6.67}{EI}$$

$$D_4 = \frac{3.33}{EI}$$

Using these results, and multiplying the last two rows, gives

$$Q_5 = 1.5EI\left(-\frac{16.67}{EI}\right) - 1.5EI\left(-\frac{26.67}{EI}\right) + 0 - 1.5EI\left(\frac{3.33}{EI}\right)$$

$$= 10 \text{ kN} \qquad\qquad\qquad\qquad\qquad\qquad Ans.$$

$$Q_6 = 0 + 0 + 1.5EI\left(-\frac{6.67}{EI}\right) + 1.5EI\left(\frac{3.33}{EI}\right)$$

$$= -5 \text{ kN} \qquad\qquad\qquad\qquad\qquad\qquad Ans.$$

15

EXAMPLE 15.2

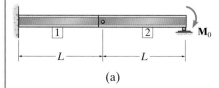

(a)

Fig. 15–9

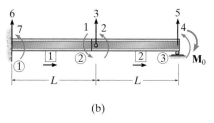

(b)

Determine the internal shear and moment in member 1 of the compound beam shown in Fig. 15–9a. EI is constant.

SOLUTION

Notation. When the beam deflects, the internal pin will allow a single deflection, however, the slope of each connected member will be different. Also, a slope at the roller will occur. These four unknown degrees of freedom are labeled with the code numbers 1, 2, 3, and 4, Fig. 15–9b.

$$\mathbf{Q}_k = \begin{bmatrix} 0 \\ 0 \\ 0 \\ -M_0 \end{bmatrix} \begin{matrix} 1 \\ 2 \\ 3 \\ 4 \end{matrix} \qquad \mathbf{D}_k = \begin{bmatrix} 0 \\ 0 \\ 0 \end{bmatrix} \begin{matrix} 5 \\ 6 \\ 7 \end{matrix}$$

Member Stiffness Matrices. Applying Eq. 15–1 to each member, in accordance with the code numbers shown in Fig. 15–9b, we have

$$\mathbf{k}_1 = EI \begin{bmatrix} \dfrac{12}{L^3} & \dfrac{6}{L^2} & -\dfrac{12}{L^3} & \dfrac{6}{L^2} \\ \dfrac{6}{L^2} & \dfrac{4}{L} & -\dfrac{6}{L^2} & \dfrac{2}{L} \\ -\dfrac{12}{L^3} & -\dfrac{6}{L^2} & \dfrac{12}{L^3} & -\dfrac{6}{L^2} \\ \dfrac{6}{L^2} & \dfrac{2}{L} & -\dfrac{6}{L^2} & \dfrac{4}{L} \end{bmatrix} \begin{matrix} 6 \\ 7 \\ 3 \\ 1 \end{matrix}$$

$$\mathbf{k}_2 = EI \begin{bmatrix} \dfrac{12}{L^3} & \dfrac{6}{L^2} & -\dfrac{12}{L^3} & \dfrac{6}{L^2} \\ \dfrac{6}{L^2} & \dfrac{4}{L} & -\dfrac{6}{L^2} & \dfrac{2}{L} \\ -\dfrac{12}{L^3} & -\dfrac{6}{L^2} & \dfrac{12}{L^3} & -\dfrac{6}{L^2} \\ \dfrac{6}{L^2} & \dfrac{2}{L} & -\dfrac{6}{L^2} & \dfrac{4}{L} \end{bmatrix} \begin{matrix} 3 \\ 2 \\ 5 \\ 4 \end{matrix}$$

(with column code numbers 6, 7, 3, 1 for $\mathbf{k}_1$ and 3, 2, 5, 4 for $\mathbf{k}_2$)

Displacements and Loads. The structure stiffness matrix is formed by assembling the elements of the member stiffness matrices. Applying the structure matrix equation, we have

$$\mathbf{Q} = \mathbf{KD}$$

$$
\begin{bmatrix} 0 \\ 0 \\ 0 \\ -M_0 \\ \hline Q_5 \\ Q_6 \\ Q_7 \end{bmatrix}
= EI
\begin{bmatrix}
\frac{4}{L} & 0 & -\frac{6}{L^2} & 0 & 0 & \frac{6}{L^2} & \frac{2}{L} \\
0 & \frac{4}{L} & \frac{6}{L^2} & \frac{2}{L} & -\frac{6}{L^2} & 0 & 0 \\
-\frac{6}{L^2} & \frac{6}{L^2} & \frac{24}{L^3} & \frac{6}{L^2} & -\frac{12}{L^3} & -\frac{12}{L^3} & -\frac{6}{L^2} \\
0 & \frac{2}{L} & \frac{6}{L^2} & \frac{4}{L} & -\frac{6}{L^2} & 0 & 0 \\
\hline
0 & -\frac{6}{L^2} & -\frac{12}{L^3} & -\frac{6}{L^2} & \frac{12}{L^3} & 0 & 0 \\
\frac{6}{L^2} & 0 & -\frac{12}{L^3} & 0 & 0 & \frac{12}{L^3} & \frac{6}{L^2} \\
\frac{2}{L} & 0 & -\frac{6}{L^2} & 0 & 0 & \frac{6}{L^2} & \frac{4}{L}
\end{bmatrix}
\begin{bmatrix} D_1 \\ D_2 \\ D_3 \\ D_4 \\ \hline 0 \\ 0 \\ 0 \end{bmatrix}
$$

Multiplying the first four rows to determine the displacement yields

$$0 = \frac{4}{L}D_1 - \frac{6}{L^2}D_3$$

$$0 = \frac{4}{L}D_2 + \frac{6}{L^2}D_3 + \frac{2}{L}D_4$$

$$0 = -\frac{6}{L^2}D_1 + \frac{6}{L^2}D_2 + \frac{24}{L^3}D_3 + \frac{6}{L^2}D_4$$

$$-M_0 = \frac{2}{L}D_2 + \frac{6}{L^2}D_3 + \frac{4}{L}D_4$$

So that

$$D_1 = \frac{M_0 L}{2EI}$$

$$D_2 = -\frac{M_0 L}{6EI}$$

$$D_3 = \frac{M_0 L^2}{3EI}$$

$$D_4 = -\frac{2M_0 L}{3EI}$$

Using these results, the reaction Q_5 is obtained from the multiplication of the fifth row.

$$Q_5 = -\frac{6EI}{L^2}\left(-\frac{M_0 L}{6EI}\right) - \frac{12EI}{L^3}\left(\frac{M_0 L^2}{3EI}\right) - \frac{6EI}{L^2}\left(-\frac{2M_0 L}{3EI}\right)$$

$$Q_5 = \frac{M_0}{L} \qquad\qquad\qquad Ans.$$

This result can be easily checked by statics applied to member $\boxed{2}$.

EXAMPLE 15.3

The beam in Fig. 15–10a is subjected to the two couple moments. If the center support ② settles 1.5 mm, determine the reactions at the supports. Assume the roller supports at ① and ③ can pull down or push up on the beam. Take $E = 200$ GPa and $I = 22(10^{-6})$ m^4.

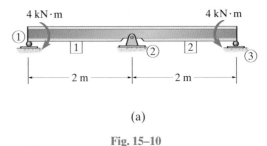

(a)

Fig. 15–10

SOLUTION

Notation.   The beam has two elements and three unknown degrees of freedom. These are labeled with the lowest code numbers, Fig. 15–10b. Here the known load and displacement matrices are

$$\mathbf{Q}_k = \begin{bmatrix} 4 \\ 0 \\ -4 \end{bmatrix} \begin{matrix} 1 \\ 2 \\ 3 \end{matrix} \qquad \mathbf{D}_k = \begin{bmatrix} 0 \\ -0.0015 \\ 0 \end{bmatrix} \begin{matrix} 4 \\ 5 \\ 6 \end{matrix}$$

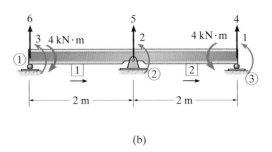

(b)

Member Stiffness Matrices.   The member stiffness matrices are determined using Eq. 15–1 in accordance with the code numbers and member directions shown in Fig. 15–10b. We have,

$$\mathbf{k}_1 = EI \begin{array}{c c c c} 6 & 3 & 5 & 2 \\ \begin{bmatrix} 1.5 & 1.5 & -1.5 & 1.5 \\ 1.5 & 2 & -1.5 & 1 \\ -1.5 & -1.5 & 1.5 & -1.5 \\ 1.5 & 1 & -1.5 & 2 \end{bmatrix} & \begin{array}{c} 6 \\ 3 \\ 5 \\ 2 \end{array} \end{array}$$

$$\mathbf{k}_2 = EI \begin{array}{c c c c} 5 & 2 & 4 & 1 \\ \begin{bmatrix} 1.5 & 1.5 & -1.5 & 1.5 \\ 1.5 & 2 & -1.5 & 1 \\ -1.5 & -1.5 & 1.5 & -1.5 \\ 1.5 & 1 & -1.5 & 2 \end{bmatrix} & \begin{array}{c} 5 \\ 2 \\ 4 \\ 1 \end{array} \end{array}$$

Displacements and Loads. Assembling the structure stiffness matrix and writing the stiffness equation for the structure, yields

$$\begin{bmatrix} 4 \\ 0 \\ -4 \\ \hline Q_4 \\ Q_5 \\ Q_6 \end{bmatrix} = EI \begin{array}{c c c c c c} 1 & 2 & 3 & 4 & 5 & 6 \\ \begin{bmatrix} 2 & 1 & 0 & -1.5 & 1.5 & 0 \\ 1 & 4 & 1 & -1.5 & 0 & 1.5 \\ 0 & 1 & 2 & 0 & -1.5 & 1.5 \\ \hline -1.5 & -1.5 & 0 & 1.5 & -1.5 & 0 \\ 1.5 & 0 & -1.5 & -1.5 & 3 & -1.5 \\ 0 & 1.5 & 1.5 & 0 & -1.5 & 1.5 \end{bmatrix} \end{array} \begin{bmatrix} D_1 \\ D_2 \\ D_3 \\ \hline 0 \\ -0.0015 \\ 0 \end{bmatrix}$$

Solving for the unknown displacements,

$$\frac{4}{EI} = 2D_1 + D_2 + 0D_3 - 1.5(0) + 1.5(-0.0015) + 0$$

$$0 = 1D_1 + 4D_2 + 1D_3 - 1.5(0) + 0 + 0$$

$$\frac{-4}{EI} = 0D_1 + 1D_2 + 2D_3 + 0 - 1.5(-0.0015) + 0$$

Substituting $EI = 200(10^6)(22)(10^{-6})$, and solving,

$$D_1 = 0.001580 \text{ rad}, \quad D_2 = 0, \quad D_3 = -0.001580 \text{ rad}$$

Using these results, the support reactions are therefore

$Q_4 = 200(10^6)22(10^{-6})[-1.5(0.001580) - 1.5(0) + 0 + 1.5(0) - 1.5(-0.0015) + 0] = -0.525 \text{ kN}$ *Ans.*

$Q_5 = 200(10^6)22(10^{-6})[1.5(0.001580) + 0 - 1.5(-0.001580) - 1.5(0) + 3(-0.0015) - 1.5(0)] = 1.05 \text{ kN}$ *Ans.*

$Q_6 = 200(10^6)22(10^{-6})[0 + 1.5(0) + 1.5(-0.001580) + 0 - 1.5(-0.0015) + 1.5(0)] = -0.525 \text{ kN}$ *Ans.*

EXAMPLE 15.4

Determine the moment developed at support A of the beam shown in Fig. 15–11a. Assume the roller supports can pull down or push up on the beam. Take $E = 200$ GPa, $I = 216(10^6)$ mm^4.

SOLUTION

Notation.   Here the beam has two unconstrained degrees of freedom, identified by the code numbers 1 and 2.

The matrix analysis requires that the external loading be applied at the nodes, and therefore the distributed and concentrated loads are replaced by their equivalent fixed-end moments, which are determined from the table on the inside back cover. (See Example 11–2.) Note that no external loads are placed at ① and no external vertical forces are placed at ② since the reactions at code numbers 3, 4 and 5 *are to be unknowns* in the load matrix. Using superposition, the results of the matrix analysis for the loading in Fig. 15–11b will later be modified by the loads shown in Fig. 15–11c. From Fig. 15–11b, the known displacement and load matrices are

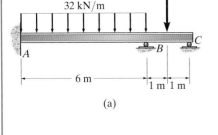

96 kN·m − 12 kN·m = 84 kN·m

12 kN·m

beam to be analyzed by stiffness method
(b)

$$\mathbf{D}_k = \begin{bmatrix} 0 \\ 0 \\ 0 \\ 0 \end{bmatrix} \begin{matrix} 3 \\ 4 \\ 5 \\ 6 \end{matrix} \qquad \mathbf{Q}_k = \begin{bmatrix} 12 \\ 84 \end{bmatrix} \begin{matrix} 1 \\ 2 \end{matrix}$$

Member Stiffness Matrices.   Each of the two member stiffness matrices is determined from Eq. 15–1.

Member 1:

$$\frac{12EI}{L^3} = \frac{12(200)(216)}{6^3} = 2400$$

$$\frac{6EI}{L^2} = \frac{6(200)(216)}{6^2} = 7200$$

$$\frac{4EI}{L} = \frac{4(200)(216)}{6} = 28\,800$$

$$\frac{2EI}{L} = \frac{2(200)(216)}{6} = 14\,400$$

$$\mathbf{k}_1 = \begin{array}{c} \begin{matrix} 4 \qquad\quad 3 \qquad\quad 5 \qquad\quad 2 \end{matrix} \\ \begin{bmatrix} 2400 & 7200 & -2400 & 7200 \\ 7200 & 28\,800 & -7200 & 14\,400 \\ -2400 & -7200 & 2400 & -7200 \\ 7200 & 14\,400 & -7200 & 28\,800 \end{bmatrix} \end{array} \begin{matrix} 4 \\ 3 \\ 5 \\ 2 \end{matrix}$$

Member 2:

$$\frac{12EI}{L^3} = \frac{12(200)(216)}{2^3} = 64\,800$$

$$\frac{6EI}{L^2} = \frac{6(200)(216)}{2^2} = 64\,800$$

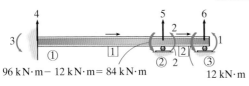

96 kN   32 kN/m   96 kN   24 kN / 48 kN / 24 kN

96 kN·m   12 kN·m

beam subjected to actual load and fixed-supported reactions
(c)

Fig. 15–11

$$\frac{4EI}{L} = \frac{4(200)(216)}{2} = 86\ 400$$

$$\frac{2EI}{L} = \frac{2(200)(216)}{2} = 43\ 200$$

$$\mathbf{k}_2 = \begin{bmatrix} \overset{5}{64\ 800} & \overset{2}{64\ 800} & \overset{6}{-64\ 800} & \overset{1}{64\ 800} \\ 64\ 800 & 86\ 400 & -64\ 800 & 43\ 200 \\ -64\ 800 & -64\ 800 & 64\ 800 & -64\ 800 \\ 64\ 800 & 43\ 200 & -64\ 800 & 86\ 400 \end{bmatrix} \begin{matrix} 5 \\ 2 \\ 6 \\ 1 \end{matrix}$$

Displacements and Loads.   We require

$$\mathbf{Q} = \mathbf{KD}$$

$$\begin{bmatrix} 12 \\ 84 \\ \hline Q_3 \\ Q_4 \\ Q_5 \\ Q_6 \end{bmatrix} = \begin{bmatrix} \overset{1}{86\ 400} & \overset{2}{43\ 200} & \overset{3}{0} & \overset{4}{0} & \overset{5}{64\ 800} & \overset{6}{-64\ 800} \\ 43\ 200 & 115\ 200 & 14\ 400 & 7200 & 57\ 600 & -64\ 800 \\ \hline 0 & 14\ 400 & 28\ 800 & 7200 & -7200 & 0 \\ 0 & 7200 & 7200 & 7200 & -2400 & 0 \\ 64\ 800 & 57\ 600 & -7200 & -2400 & 67\ 200 & -64\ 800 \\ -64\ 800 & -64\ 800 & 0 & 0 & -64\ 800 & 64\ 800 \end{bmatrix} \begin{bmatrix} D_1 \\ D_2 \\ 0 \\ 0 \\ 0 \\ 0 \end{bmatrix}$$

Solving in the usual manner,

$$12 = 86\ 400D_1 + 43\ 200D_2$$
$$48 = 43\ 200D_1 + 115\ 220D_2$$
$$D_1 = -0.2778(10^{-3})\ \text{m}$$
$$D_2 = 0.8333(10^{-3})\ \text{m}$$

Thus,

$$Q_3 = 0 + 14\ 400(0.8333)(10^{-3}) = 12\ \text{kN} \cdot \text{m}$$

The actual moment at A must include the fixed-supported *reaction* of $+96\ \text{kN} \cdot \text{m}$ shown in Fig. 15–11c, along with the calculated result for Q_3. Thus,

$$M_{AB} = 12\ \text{kN} \cdot \text{m} + 96\ \text{kN} \cdot \text{m} = 108\ \text{kN} \cdot \text{m}\!\uparrow \qquad Ans.$$

This result compares with that determined in Example 11–2.

Although not required here, we can determine the internal moment and shear at B by considering, for example, member 1, node 2, Fig. 15–11b. The result requires expanding

$$\mathbf{q}_1 = \mathbf{k}_1\mathbf{d} + (\mathbf{q}_0)_1$$

$$\begin{bmatrix} q_4 \\ q_3 \\ q_5 \\ q_2 \end{bmatrix} = \begin{bmatrix} \overset{4}{2400} & \overset{3}{7200} & \overset{5}{-2400} & \overset{2}{7200} \\ 7200 & 28\ 800 & -7200 & 14\ 400 \\ -2400 & -7200 & 2400 & -7200 \\ 7200 & 14\ 400 & -7200 & 28\ 800 \end{bmatrix} \begin{bmatrix} 0 \\ 0 \\ 0 \\ 0.8333 \end{bmatrix} (10^{-3}) + \begin{bmatrix} 96 \\ 96 \\ 96 \\ -96 \end{bmatrix}$$

EXAMPLE 15.5

Determine the deflection at ① and the reactions on the beam shown in Fig. 15–12a. *EI* is constant.

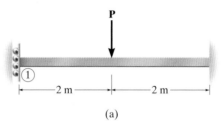

(a)

Fig. 15–12

SOLUTION

Notation. The beam is divided into two elements and the nodes and members are identified along with the directions from the near to far ends, Fig. 15–12b. The unknown deflections are shown in Fig. 15–12c. In particular, notice that a rotational displacement D_4 does not occur because of the roller constraint.

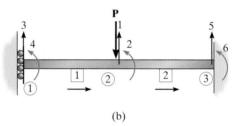

(b)

Member Stiffness Matrices. Since *EI* is constant and the members are of equal length, the member stiffness matrices are identical. Using the code numbers to identify each column and row in accordance with Eq. 15–1 and Fig. 15–12b, we have

$$\mathbf{k}_1 = EI \begin{array}{cccc} 3 & 4 & 1 & 2 \\ \left[\begin{array}{cccc} 1.5 & 1.5 & -1.5 & 1.5 \\ 1.5 & 2 & -1.5 & 1 \\ -1.5 & -1.5 & 1.5 & -1.5 \\ 1.5 & 1 & -1.5 & 2 \end{array}\right] & \begin{array}{c} 3 \\ 4 \\ 1 \\ 2 \end{array} \end{array}$$

$$\mathbf{k}_2 = EI \begin{array}{cccc} 1 & 2 & 5 & 6 \\ \left[\begin{array}{cccc} 1.5 & 1.5 & -1.5 & 1.5 \\ 1.5 & 2 & -1.5 & 1 \\ -1.5 & -1.5 & 1.5 & -1.5 \\ 1.5 & 1 & -1.5 & 2 \end{array}\right] & \begin{array}{c} 1 \\ 2 \\ 5 \\ 6 \end{array} \end{array}$$

Displacements and Loads. Assembling the member stiffness matrices into the structure stiffness matrix, and applying the structure stiffness matrix equation, we have

$$\mathbf{Q = KD}$$

$$
\begin{bmatrix} -P \\ 0 \\ 0 \\ \hline Q_4 \\ Q_5 \\ Q_6 \end{bmatrix} = EI
\begin{array}{c}
\begin{array}{cccccc} 1 & 2 & 3 & 4 & 5 & 6 \end{array} \\
\begin{bmatrix}
3 & 0 & -1.5 & -1.5 & -1.5 & 1.5 \\
0 & 4 & 1.5 & 1 & -1.5 & 1 \\
-1.5 & 1.5 & 1.5 & 1.5 & 0 & 0 \\
\hline
-1.5 & 1 & 1.5 & 2 & 0 & 0 \\
-1.5 & -1.5 & 0 & 0 & 1.5 & -1.5 \\
1.5 & 1 & 0 & 0 & -1.5 & 2
\end{bmatrix}
\end{array}
\begin{bmatrix} D_1 \\ D_2 \\ D_3 \\ 0 \\ 0 \\ 0 \end{bmatrix}
$$

Solving for the displacements yields

$$-\frac{P}{EI} = 3D_1 + 0D_2 - 1.5D_3$$

$$0 = 0D_1 + 4D_2 + 1.5D_3$$

$$0 = -1.5D_1 + 1.5D_2 + 1.5D_3$$

$$D_1 = -\frac{1.667P}{EI}$$

$$D_2 = \frac{P}{EI}$$

$$D_3 = -\frac{2.667P}{EI} \qquad\qquad Ans.$$

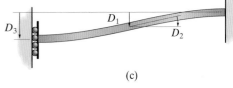

(c)

Note that the signs of the results match the directions of the deflections shown in Fig. 15–12c. Using these results, the reactions therefore are

$$Q_4 = -1.5EI\left(-\frac{1.667P}{EI}\right) + 1EI\left(\frac{P}{EI}\right) + 1.5EI\left(-\frac{2.667P}{EI}\right)$$

$$= -0.5P \qquad\qquad Ans.$$

$$Q_5 = -1.5EI\left(-\frac{1.667P}{EI}\right) - 1.5EI\left(\frac{P}{EI}\right) + 0\left(-\frac{2.667P}{EI}\right)$$

$$= P \qquad\qquad Ans.$$

$$Q_6 = 1.5EI\left(-\frac{1.667P}{EI}\right) + 1EI\left(\frac{P}{EI}\right) + 0\left(-\frac{2.667P}{EI}\right)$$

$$= -1.5P \qquad\qquad Ans.$$

PROBLEMS

15–1. Determine the moments at ① and ③. Assume ② is a roller and ① and ③ are fixed. *EI* is constant.

15–2. Determine the moments at ① and ③ if the support ② moves upward 5 mm. Assume ② is a roller and ① and ③ are fixed. $EI = 60(10^6)$ N·m².

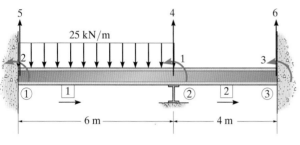

Probs. 15–1/15–2

15–5. Determine the support reactions. Assume ② and ③ are rollers and ① is a pin. *EI* is constant.

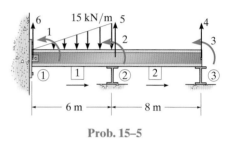

Prob. 15–5

15–3. Determine the reactions at the supports. Assume the rollers can either push or pull on the beam. *EI* is constant.

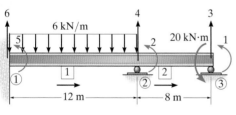

Prob. 15–3

15–6. Determine the reactions at the supports. Assume ① is fixed ② and ③ are rollers. *EI* is constant.

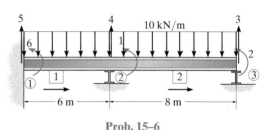

Prob. 15–6

*15–4. Determine the reactions at the supports. Assume ① is a pin and ② and ③ are rollers that can either push or pull on the beam. *EI* is constant.

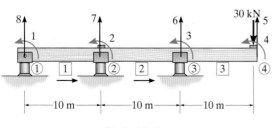

Prob. 15–4

15–7. Determine the reactions at the supports. Assume ① and ③ are fixed and ② is a roller. *EI* is constant.

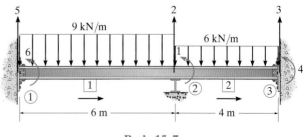

Prob. 15–7

*15–8. Determine the reactions at the supports. *EI* is constant.

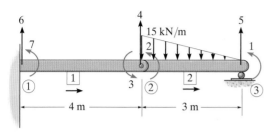

Prob. 15–8

15–9. Determine the moments at ② and ③. *EI* is constant. Assume ①, ②, and ③ are rollers and ④ is pinned.

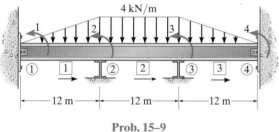

Prob. 15–9

15–10. Determine the reactions at the supports. Assume ② is pinned and ① and ③ are rollers. *EI* is constant.

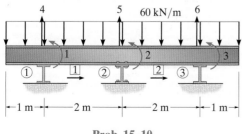

Prob. 15–10

15–11. Determine the reactions at the supports. There is a smooth slider at ①. *EI* is constant.

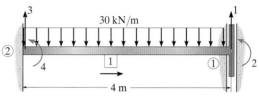

Prob. 15–11

*15–12. Use a computer program to determine the reactions on the beam. Assume *A* is fixed. *EI* is constant.

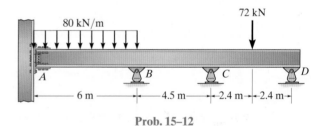

Prob. 15–12

15–13. Use a computer program to determine the reactions on the beam. Assume *A* and *D* are pins and *B* and *C* are rollers. *EI* is constant.

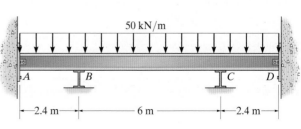

Prob. 15–13

15

The frame of this building is statically indeterminate. The force analysis can be done using the stiffness method.

Plane Frame Analysis Using the Stiffness Method

16

The concepts presented in the previous chapters on trusses and beams will be extended in this chapter and applied to the analysis of frames. It will be shown that the procedure for the solution is similar to that for beams, but will require the use of transformation matrices since frame members are oriented in different directions.

16.1 Frame-Member Stiffness Matrix

In this section we will develop the stiffness matrix for a prismatic frame member referenced from the local x', y', z' coordinate system, Fig. 16–1. Here the member is subjected to axial loads $q_{Nx'}$, $q_{Fx'}$, shear loads $q_{Ny'}$, $q_{Fy'}$, and bending moments $q_{Nz'}$, $q_{Fz'}$ at its near and far ends, respectively. These loadings all act in the positive coordinate directions along with their associated displacements. As in the case of beams, the moments $q_{Nz'}$ and $q_{Fz'}$ are positive counterclockwise, since by the right-hand rule the moment vectors are then directed along the positive z' axis, which is out of the page.

We have considered each of the load-displacement relationships caused by these loadings in the previous chapters. The axial load was discussed in reference to Fig. 14–2, the shear load in reference to Fig. 15–5, and the bending moment in reference to Fig. 15–6. By superposition, if these

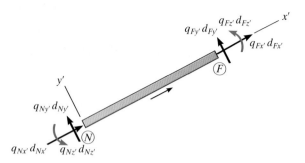

This pedestrian bridge takes the form of a "Vendreel truss." Strictly not a truss since the diagonals are absent, it forms a statically indeterminate box framework, which can be analyzed using the stiffness method.

positive sign convention

Fig. 16–1

results are added, the resulting six load-displacement relations for the member can be expressed in matrix form as

$$
\begin{bmatrix} q_{Nx'} \\ q_{Ny'} \\ q_{Nz'} \\ q_{Fx'} \\ q_{Fy'} \\ q_{Fz'} \end{bmatrix}
=
\begin{matrix}
 N_{x'} & N_{y'} & N_{z'} & F_{x'} & F_{y'} & F_{z'} \\
\end{matrix}
\begin{bmatrix}
\dfrac{AE}{L} & 0 & 0 & -\dfrac{AE}{L} & 0 & 0 \\[2mm]
0 & \dfrac{12EI}{L^3} & \dfrac{6EI}{L^2} & 0 & -\dfrac{12EI}{L^3} & \dfrac{6EI}{L^2} \\[2mm]
0 & \dfrac{6EI}{L^2} & \dfrac{4EI}{L} & 0 & -\dfrac{6EI}{L^2} & \dfrac{2EI}{L} \\[2mm]
-\dfrac{AE}{L} & 0 & 0 & \dfrac{AE}{L} & 0 & 0 \\[2mm]
0 & -\dfrac{12EI}{L^3} & -\dfrac{6EI}{L^2} & 0 & \dfrac{12EI}{L^3} & -\dfrac{6EI}{L^2} \\[2mm]
0 & \dfrac{6EI}{L^2} & \dfrac{2EI}{L} & 0 & -\dfrac{6EI}{L^2} & \dfrac{4EI}{L}
\end{bmatrix}
\begin{bmatrix} d_{Nx'} \\ d_{Ny'} \\ d_{Nz'} \\ d_{Fx'} \\ d_{Fy'} \\ d_{Fz'} \end{bmatrix}
$$

(16–1)

or in abbreviated form as

$$\mathbf{q} = \mathbf{k'd} \qquad (16\text{–}2)$$

The member stiffness matrix $\mathbf{k'}$ consists of thirty-six influence coefficients that physically represent the load on the member when the member undergoes a specified unit displacement. Specifically, each column in the matrix represents the member loadings for unit displacements identified by the degree-of-freedom coding listed above the columns. From the assembly, both equilibrium and compatibility of displacements have been satisfied.

16.2   Displacement and Force Transformation Matrices

As in the case for trusses, we must be able to transform the internal member loads $\mathbf{q}$ and deformations $\mathbf{d}$ from local x', y', z' coordinates to global x, y, z coordinates. For this reason transformation matrices are needed.

Displacement Transformation Matrix.   Consider the frame member shown in Fig. 16–2a. Here it is seen that a global coordinate displacement D_{Nx} creates local coordinate displacements

$$d_{Nx'} = D_{Nx} \cos \theta_x \qquad d_{Ny'} = -D_{Nx} \cos \theta_y$$

Likewise, a global coordinate displacement D_{Ny}, Fig. 16–2b, creates local coordinate displacements of

$$d_{Nx'} = D_{Ny} \cos \theta_y \qquad d_{Ny'} = D_{Ny} \cos \theta_x$$

Finally, since the z' and z axes are coincident, that is, directed out of the page, a rotation D_{Nz} about z causes a corresponding rotation $d_{Nz'}$ about z'. Thus,

$$d_{Nz'} = D_{Nz}$$

In a similar manner, if global displacements D_{Fx} in the x direction, D_{Fy} in the y direction, and a rotation D_{Fz} are imposed on the far end of the member, the resulting transformation equations are, respectively,

$$d_{Fx'} = D_{Fx} \cos \theta_x \qquad d_{Fy'} = -D_{Fx} \cos \theta_y$$
$$d_{Fx'} = D_{Fy} \cos \theta_y \qquad d_{Fy'} = D_{Fy} \cos \theta_x$$
$$d_{Fz'} = D_{Fz}$$

Letting $\lambda_x = \cos \theta_x$, $\lambda_y = \cos \theta_y$ represent the direction cosines of the member, we can write the superposition of displacements in matrix form as

$$
\begin{bmatrix} d_{Nx'} \\ d_{Ny'} \\ d_{Nz'} \\ d_{Fx'} \\ d_{Fy'} \\ d_{Fz'} \end{bmatrix}
=
\begin{bmatrix}
\lambda_x & \lambda_y & 0 & 0 & 0 & 0 \\
-\lambda_y & \lambda_x & 0 & 0 & 0 & 0 \\
0 & 0 & 1 & 0 & 0 & 0 \\
0 & 0 & 0 & \lambda_x & \lambda_y & 0 \\
0 & 0 & 0 & -\lambda_y & \lambda_x & 0 \\
0 & 0 & 0 & 0 & 0 & 1
\end{bmatrix}
\begin{bmatrix} D_{Nx} \\ D_{Ny} \\ D_{Nz} \\ D_{Fx} \\ D_{Fy} \\ D_{Fz} \end{bmatrix}
\qquad (16\text{–}3)
$$

or

$$\mathbf{d} = \mathbf{TD} \qquad (16\text{–}4)$$

By inspection, $\mathbf{T}$ transforms the six global x, y, z displacements $\mathbf{D}$ into the six local x', y', z' displacements $\mathbf{d}$. Hence $\mathbf{T}$ is referred to as the *displacement transformation matrix.*

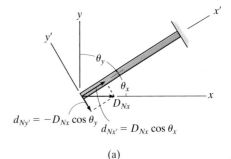

(a)

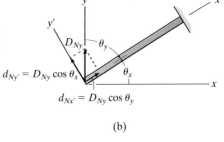

(b)

Fig. 16–2

16

Force Transformation Matrix.

If we now apply each component of load to the near end of the member, we can determine how to transform the load components from local to global coordinates. Applying $q_{Nx'}$, Fig. 16–3a, it can be seen that

$$Q_{Nx} = q_{Nx'} \cos \theta_x \qquad Q_{Ny} = q_{Nx'} \cos \theta_y$$

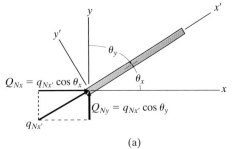

If $q_{Ny'}$ is applied, Fig. 16–3b, then its components are

$$Q_{Nx} = -q_{Ny'} \cos \theta_y \qquad Q_{Ny} = q_{Ny'} \cos \theta_x$$

Finally, since $q_{Nz'}$ is collinear with Q_{Nz}, we have

$$Q_{Nz} = q_{Nz'}$$

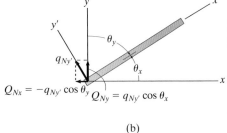

In a similar manner, end loads of $q_{Fx'}$, $q_{Fy'}$, $q_{Fz'}$ will yield the following respective components:

$$Q_{Fx} = q_{Fx'} \cos \theta_x \qquad Q_{Fy} = q_{Fx'} \cos \theta_y$$
$$Q_{Fx} = -q_{Fy'} \cos \theta_y \qquad Q_{Fy} = q_{Fy'} \cos \theta_x$$
$$Q_{Fz} = q_{Fz'}$$

(b)

Fig. 16–3

These equations, assembled in matrix form with $\lambda_x = \cos \theta_x$, $\lambda_y = \cos \theta_y$, yield

$$
\begin{bmatrix} Q_{Nx} \\ Q_{Ny} \\ Q_{Nz} \\ Q_{Fx} \\ Q_{Fy} \\ Q_{Fz} \end{bmatrix}
=
\begin{bmatrix}
\lambda_x & -\lambda_y & 0 & 0 & 0 & 0 \\
\lambda_y & \lambda_x & 0 & 0 & 0 & 0 \\
0 & 0 & 1 & 0 & 0 & 0 \\
0 & 0 & 0 & \lambda_x & -\lambda_y & 0 \\
0 & 0 & 0 & \lambda_y & \lambda_x & 0 \\
0 & 0 & 0 & 0 & 0 & 1
\end{bmatrix}
\begin{bmatrix} q_{Nx'} \\ q_{Ny'} \\ q_{Nz'} \\ q_{Fx'} \\ q_{Fy'} \\ q_{Fz'} \end{bmatrix}
\qquad (16\text{–}5)
$$

or

$$\mathbf{Q} = \mathbf{T}^T \mathbf{q} \qquad (16\text{–}6)$$

Here, as stated, $\mathbf{T}^T$ transforms the six member loads expressed in local coordinates into the six loadings expressed in global coordinates.

16.3 Frame-Member Global Stiffness Matrix

The results of the previous section will now be combined in order to determine the stiffness matrix for a member that relates the global loadings **Q** to the global displacements **D**. To do this, substitute Eq. 16–4 (**d** = **TD**) into Eq. 16–2 (**q** = **k′d**). We have

$$\mathbf{q} = \mathbf{k'TD} \qquad (16\text{–}7)$$

Here the member forces **q** are related to the global displacements **D**. Substituting this result into Eq. 16–6 (**Q** = **T**^{T}**q**) yields the final result,

$$\mathbf{Q} = \mathbf{T}^T\mathbf{k'TD} \qquad (16\text{–}8)$$

or

$$\mathbf{Q} = \mathbf{kD}$$

where

$$\mathbf{k} = \mathbf{T}^T\mathbf{k'T} \qquad (16\text{–}9)$$

Here **k** represents the global stiffness matrix for the member. We can obtain its value in generalized form using Eqs. 16–5, 16–1, and 16–3 and performing the matrix operations. This yields the final result,

$$
\mathbf{k} =
\begin{array}{cccccc}
 & N_x & N_y & N_z & F_x & F_y & F_z \\
\end{array}
$$

$$
\mathbf{k} =
\left[
\begin{array}{cccccc}
\left(\dfrac{AE}{L}\lambda_x^2 + \dfrac{12EI}{L^3}\lambda_y^2\right) & \left(\dfrac{AE}{L} - \dfrac{12EI}{L^3}\right)\lambda_x\lambda_y & -\dfrac{6EI}{L^2}\lambda_y & -\left(\dfrac{AE}{L}\lambda_x^2 + \dfrac{12EI}{L^3}\lambda_y^2\right) & -\left(\dfrac{AE}{L} - \dfrac{12EI}{L^3}\right)\lambda_x\lambda_y & -\dfrac{6EI}{L^2}\lambda_y \\[3mm]
\left(\dfrac{AE}{L} - \dfrac{12EI}{L^3}\right)\lambda_x\lambda_y & \left(\dfrac{AE}{L}\lambda_y^2 + \dfrac{12EI}{L^3}\lambda_x^2\right) & \dfrac{6EI}{L^2}\lambda_x & -\left(\dfrac{AE}{L} - \dfrac{12EI}{L^3}\right)\lambda_x\lambda_y & -\left(\dfrac{AE}{L}\lambda_y^2 + \dfrac{12EI}{L^3}\lambda_x^2\right) & \dfrac{6EI}{L^2}\lambda_x \\[3mm]
-\dfrac{6EI}{L^2}\lambda_y & \dfrac{6EI}{L^2}\lambda_x & \dfrac{4EI}{L} & \dfrac{6EI}{L^2}\lambda_y & -\dfrac{6EI}{L^2}\lambda_x & \dfrac{2EI}{L} \\[3mm]
-\left(\dfrac{AE}{L}\lambda_x^2 + \dfrac{12EI}{L^3}\lambda_y^2\right) & -\left(\dfrac{AE}{L} - \dfrac{12EI}{L^3}\right)\lambda_x\lambda_y & \dfrac{6EI}{L^2}\lambda_y & \left(\dfrac{AE}{L}\lambda_x^2 + \dfrac{12EI}{L^3}\lambda_y^2\right) & \left(\dfrac{AE}{L} - \dfrac{12EI}{L^3}\right)\lambda_x\lambda_y & \dfrac{6EI}{L^2}\lambda_y \\[3mm]
-\left(\dfrac{AE}{L} - \dfrac{12EI}{L^3}\right)\lambda_x\lambda_y & -\left(\dfrac{AE}{L}\lambda_y^2 + \dfrac{12EI}{L^3}\lambda_x^2\right) & -\dfrac{6EI}{L^2}\lambda_x & \left(\dfrac{AE}{L} - \dfrac{12EI}{L^3}\right)\lambda_x\lambda_y & \left(\dfrac{AE}{L}\lambda_y^2 + \dfrac{12EI}{L^3}\lambda_x^2\right) & -\dfrac{6EI}{L^2}\lambda_x \\[3mm]
-\dfrac{6EI}{L^2}\lambda_y & \dfrac{6EI}{L^2}\lambda_x & \dfrac{2EI}{L} & \dfrac{6EI}{L^2}\lambda_y & -\dfrac{6EI}{L^2}\lambda_x & \dfrac{4EI}{L}
\end{array}
\right]
\begin{array}{c}
N_x \\[3mm] N_y \\[3mm] N_z \\[3mm] F_x \\[3mm] F_y \\[3mm] F_z
\end{array}
$$

$$(16\text{–}10)$$

Note that this 6 × 6 matrix is *symmetric*. Furthermore, the location of each element is associated with the coding at the near end, N_x, N_y, N_z, followed by that of the far end, F_x, F_y, F_z, which is listed at the top of the columns and along the rows. Like the **k′** matrix, each column of the **k** matrix represents the coordinate loads on the member at the nodes that are necessary to resist a unit displacement in the direction defined by the coding of the column. For example, the first column of **k** represents the global coordinate loadings at the near and far ends caused by a *unit displacement* at the near end in the *x* direction, that is, D_{Nx}.

16.4   Application of the Stiffness Method for Frame Analysis

Once the member stiffness matrices are established, they may be assembled into the structure stiffness matrix in the usual manner. By writing the structure matrix equation, the displacements at the unconstrained nodes can be determined, followed by the reactions and internal loadings at the nodes. Lateral loads acting on a member, fabrication errors, temperature changes, inclined supports, and internal supports are handled in the same manner as was outlined for trusses and beams.

Procedure for Analysis

The following method provides a means of finding the displacements, support reactions, and internal loadings for members of statically determinate and indeterminate frames.

Notation

- Divide the structure into finite elements and arbitrarily identify each element and its nodes. Elements usually extend between points of support, points of concentrated loads, corners or joints, or to points where internal loadings or displacements are to be determined.

- Establish the x, y, z, global coordinate system, usually for convenience with the origin located at a nodal point on one of the elements and the axes located such that all the nodes have positive coordinates.

- At each nodal point of the frame, specify numerically the three x, y, z coding components. In all cases use the *lowest code numbers* to identify all the *unconstrained degrees of freedom*, followed by the remaining or *highest code numbers* to identify the *constrained degrees of freedom*.

- From the problem, establish the known displacements $\mathbf{D}_k$ and known external loads $\mathbf{Q}_k$. When establishing $\mathbf{Q}_k$ be sure to include any *reversed* fixed-end loadings if an element supports an intermediate load.

Structure Stiffness Matrix

- Apply Eq. 16–10 to determine the stiffness matrix for each element expressed in global coordinates. In particular, the direction cosines λ_x and λ_y are determined from the x, y coordinates of the ends of the element, Eqs. 14–5 and 14–6.

- After each member stiffness matrix is written, and the six rows and columns are identified with the near and far code numbers, merge the matrices to form the structure stiffness matrix $\mathbf{K}$. As a partial check, the element and structure stiffness matrices should all be *symmetric*.

Displacements and Loads

- Partition the stiffness matrix as indicated by Eq. 14–18. Expansion then leads to

$$\mathbf{Q}_k = \mathbf{K}_{11}\mathbf{D}_u + \mathbf{K}_{12}\mathbf{D}_k$$

$$\mathbf{Q}_u = \mathbf{K}_{21}\mathbf{D}_u + \mathbf{K}_{22}\mathbf{D}_k$$

The unknown displacements $\mathbf{D}_u$ are determined from the first of these equations. Using these values, the support reactions $\mathbf{Q}_u$ are computed from the second equation. Finally, the internal loadings **q** at the ends of the members can be computed from Eq. 16–7, namely

$$\mathbf{q} = \mathbf{k'TD}$$

If the results of any of the unknowns are calculated as negative quantities, it indicates they act in the negative coordinate directions.

EXAMPLE 16.1

Determine the loadings at the joints of the two-member frame shown in Fig. 16–4a. Take $I = 180(10^6)$ mm^4, $A = 6000$ mm^2, and $E = 200$ GPa for both members.

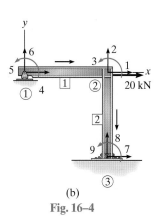

(a)

SOLUTION

Notation. By inspection, the frame has two elements and three nodes, which are identified as shown in Fig. 16–4b. The origin of the global coordinate system is located at ①. The code numbers at the nodes are specified with the *unconstrained degrees of freedom numbered first*. From the constraints at ① and ③, and the applied loading, we have

$$\mathbf{D}_k = \begin{bmatrix} 0 \\ 0 \\ 0 \\ 0 \end{bmatrix} \begin{matrix} 6 \\ 7 \\ 8 \\ 9 \end{matrix} \qquad \mathbf{Q}_k = \begin{bmatrix} 5 \\ 0 \\ 0 \\ 0 \\ 0 \end{bmatrix} \begin{matrix} 1 \\ 2 \\ 3 \\ 4 \\ 5 \end{matrix}$$

Structure Stiffness Matrix. The following terms are common to both element stiffness matrices:

$$\frac{AE}{L} = \frac{6(10^{-3})(200)(10^6)}{6} = 200(10^3)\ \text{kN/m}$$

$$\frac{12EI}{L^3} = \frac{12(200)(10^6)(180)(10^{-6})}{6^3} = 2(10^3)\ \text{kN/m}$$

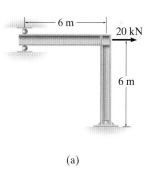

(b)

Fig. 16–4

EXAMPLE 16.1 (Continued)

$$\frac{6EI}{L^2} = \frac{6(200)(10^6)(180)(10^{-6})}{6^2} = 6(10^3) \text{ kN}$$

$$\frac{4EI}{L} = \frac{4(200)(10^6)(180)(10^{-6})}{6} = 24(10^3) \text{ kN} \cdot \text{m}$$

$$\frac{2EI}{L} = \frac{2(200)(10^6)(180)(10^{-6})}{6} = 12(10^3) \text{ kN} \cdot \text{m}$$

Member 1:

$$\lambda_x = \frac{6-0}{6} = 1 \qquad \lambda_y = \frac{0-0}{6} = 0$$

Substituting the data into Eq. 16–10, we have

	4	6	5	1	2	3	
$\mathbf{k_1} = (10^3)$	200	0	0	−200	0	0	4
	0	2	6	0	−2	6	6
	0	6	24	0	−6	12	5
	−200	0	0	200	0	0	1
	0	−2	−6	0	2	−6	2
	0	6	12	0	−6	24	3

The rows and columns of this 6 × 6 matrix are identified by the three x, y, z code numbers, first at the near end and followed by the far end, that is, 4, 6, 5, 1, 2, 3, respectively, Fig. 16–4b. This is done for later assembly of the elements.

Member 2:

$$\lambda_x = \frac{6-6}{6} = 0 \qquad \lambda_y = \frac{-6-0}{6} = -1$$

Substituting the data into Eq. 16–10 yields

	1	2	3	7	8	9	
$\mathbf{k_2} = (10^3)$	2	0	6	−2	0	6	1
	0	200	0	0	−200	0	2
	6	0	24	−6	0	12	3
	−2	0	−6	2	0	−6	7
	0	−200	0	0	200	0	8
	6	0	12	−6	0	24	9

As usual, column and row identification is referenced by the three code numbers in x, y, z sequence for the near and far ends, respectively, that is, 1, 2, 3, then 7, 8, 9, Fig. 16–4b.

The structure stiffness matrix is determined by assembling $\mathbf{k}_1$ and $\mathbf{k}_2$. The result, shown partitioned, as $\mathbf{Q} = \mathbf{KD}$, is

$$
\begin{bmatrix} 20 \\ 0 \\ 0 \\ 0 \\ 0 \\ \hline Q_6 \\ Q_7 \\ Q_8 \\ Q_9 \end{bmatrix} = (10^3)
\begin{array}{c}
\begin{matrix} 1 & \;2 & \;3 & \;4 & \;5 & \;6 & \;7 & \;8 & \;9 \end{matrix} \\
\left[\begin{array}{ccccc:cccc}
202 & 0 & 6 & -200 & 0 & 0 & -2 & 0 & 6 \\
0 & 202 & -6 & 0 & -6 & -2 & 0 & -200 & 0 \\
6 & -6 & 48 & 0 & 12 & 6 & -6 & 0 & 12 \\
-200 & 0 & 0 & 200 & 0 & 0 & 0 & 0 & 0 \\
0 & -6 & 12 & 0 & 24 & 6 & 0 & 0 & 0 \\
\hdashline
0 & -2 & 6 & 0 & 6 & 2 & 0 & 0 & 0 \\
-2 & 0 & -6 & 0 & 0 & 0 & 2 & 0 & -6 \\
0 & -200 & 0 & 0 & 0 & 0 & 0 & 200 & 0 \\
6 & 0 & 12 & 0 & 0 & 0 & -6 & 0 & 24
\end{array}\right]
\end{array}
\begin{bmatrix} D_1 \\ D_2 \\ D_3 \\ D_4 \\ D_5 \\ 0 \\ 0 \\ 0 \\ 0 \end{bmatrix} \quad (1)
$$

Displacements and Loads.   Expanding to determine the displacements yields

$$
\begin{bmatrix} 20 \\ 0 \\ 0 \\ 0 \\ 0 \end{bmatrix} = (10^3)
\begin{bmatrix}
202 & 0 & 6 & -200 & 0 \\
0 & 202 & -6 & 0 & -6 \\
6 & -6 & 48 & 0 & 12 \\
-200 & 0 & 0 & 200 & 0 \\
0 & -6 & 12 & 0 & 24
\end{bmatrix}
\begin{bmatrix} D_1 \\ D_2 \\ D_3 \\ D_4 \\ D_5 \end{bmatrix}
+
\begin{bmatrix} 0 \\ 0 \\ 0 \\ 0 \\ 0 \end{bmatrix}
$$

Solving, we obtain

$$
\begin{bmatrix} D_1 \\ D_2 \\ D_3 \\ D_4 \\ D_5 \end{bmatrix} =
\begin{bmatrix}
17.51(10^{-3})\ \text{m} \\
-37.47(10^{-6})\ \text{m} \\
-2.505(10^{-3})\ \text{rad} \\
17.51(10^{-3})\ \text{m} \\
1.243(10^{-3})\ \text{rad}
\end{bmatrix}
$$

Using these results, the support reactions are determined from Eq. (1) as follows:

$$
\begin{bmatrix} Q_6 \\ Q_7 \\ Q_8 \\ Q_9 \end{bmatrix} = (10^3)
\begin{array}{c}
\begin{matrix} 1 & \;\;2 & \;3 & \;4 & \;5 \end{matrix} \\
\begin{bmatrix}
0 & -2 & 6 & 0 & 6 \\
-2 & 0 & -6 & 0 & 0 \\
0 & -200 & 0 & 0 & 0 \\
6 & 0 & 12 & 0 & 0
\end{bmatrix}
\end{array}
\begin{bmatrix}
17.51(10^{-3}) \\
-37.47(10^{-6}) \\
-2.505(10^{-3}) \\
17.51(10^{-3}) \\
1.243(10^{-3})
\end{bmatrix}
+
\begin{bmatrix} 0 \\ 0 \\ 0 \\ 0 \end{bmatrix}
=
\begin{bmatrix}
-7.50\ \text{kN} \\
-20\ \text{kN} \\
7.50\ \text{kN} \\
75\ \text{kN} \cdot \text{m}
\end{bmatrix} \quad \textit{Ans.}
$$

16

EXAMPLE 16.1 (Continued)

The internal loadings at node ② can be determined by applying Eq. 16–7 to member 1. Here $\mathbf{k}_1'$ is defined by Eq. 16–1 and $\mathbf{T}_1$ by Eq. 16–3. Thus,

$$\mathbf{q}_1 = \mathbf{k}_1 \mathbf{T}_1 \mathbf{D} = (10^{-3})\begin{bmatrix} \overset{4}{200} & \overset{6}{0} & \overset{5}{0} & \overset{1}{-200} & \overset{2}{0} & \overset{3}{0} \\ 0 & 2 & 6 & 0 & -2 & 6 \\ 0 & 6 & 24 & 0 & -6 & 12 \\ -200 & 0 & 0 & 200 & 0 & 0 \\ 0 & -2 & -6 & 0 & 2 & -6 \\ 0 & 6 & 12 & 0 & -6 & 24 \end{bmatrix}\begin{bmatrix} 1 & 0 & 0 & 0 & 0 & 0 \\ 0 & 1 & 0 & 0 & 0 & 0 \\ 0 & 0 & 1 & 0 & 0 & 0 \\ 0 & 0 & 0 & 1 & 0 & 0 \\ 0 & 0 & 0 & 0 & 1 & 0 \\ 0 & 0 & 0 & 0 & 0 & 1 \end{bmatrix}\begin{bmatrix} 17.51(10^{-3}) \\ 0 \\ 1.243(10^{-3}) \\ 17.51(10^{-3}) \\ -37.47(10^{-6}) \\ -2.505(10^{-3}) \end{bmatrix}\begin{matrix} 4 \\ 6 \\ 5 \\ 1 \\ 2 \\ 3 \end{matrix}$$

Note the appropriate arrangement of the elements in the matrices as indicated by the code numbers alongside the columns and rows. Solving yields

$$\begin{bmatrix} q_4 \\ q_6 \\ q_5 \\ q_1 \\ q_2 \\ q_3 \end{bmatrix} = \begin{bmatrix} 0 \\ -7.50 \text{ kN} \\ 0 \\ 0 \\ 7.50 \text{ kN} \\ -45 \text{ kN} \cdot \text{m} \end{bmatrix} \qquad Ans.$$

The above results are shown in Fig. 16–4c. The directions of these vectors are in accordance with the positive directions defined in Fig. 16–1. Furthermore, the origin of the local x', y', z' axes is at the near end of the member. In a similar manner, the free-body diagram of member 2 is shown in Fig. 16–4d.

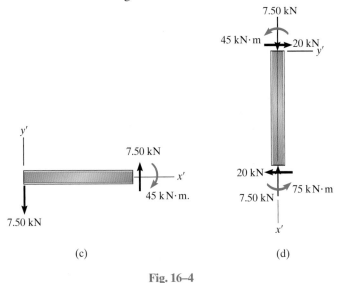

(c)

(d)

Fig. 16–4

EXAMPLE 16.2

Determine the loadings at the ends of each member of the frame shown in Fig. 16–5a. Take $I = 225(10^6)$ mm^4, $A = 7500$ mm^2, and $E = 200$ GPa for each member.

SOLUTION

Notation.   To perform a matrix analysis, the distributed loading acting on the horizontal member will be replaced by equivalent end moments and shears computed from statics and the table listed on the inside back cover. (Note that no external force of 150 kN or moment of 150 kN · m is placed at ③ since the reactions at code numbers 8 and 9 *are to be unknowns* in the load matrix.) Then using superposition, the results obtained for the frame in Fig. 16–5b will be modified for this member by the loads shown in Fig. 16–5c.

As shown in Fig. 16–5b, the nodes and members are numbered and the origin of the global coordinate system is placed at node ①. As usual, the code numbers are specified with numbers assigned first to the unconstrained degrees of freedom. Thus,

$$
\mathbf{D}_k = \begin{bmatrix} 0 \\ 0 \\ 0 \\ 0 \\ 0 \\ 0 \end{bmatrix} \begin{matrix} 4 \\ 5 \\ 6 \\ 7 \\ 8 \\ 9 \end{matrix}
\qquad
\mathbf{Q}_k = \begin{bmatrix} 0 \\ -150 \\ -150 \end{bmatrix} \begin{matrix} 1 \\ 2 \\ 3 \end{matrix}
$$

Structure Stiffness Matrix

Member 1:

$$
\frac{AE}{L} = \frac{7500(10^{-6})(200)(10^6)}{7.5} = 200(10^3) \text{ kN/m}
$$

$$
\frac{12EI}{L^3} = \frac{12(200)(10^6)(225)(10^{-6})}{(7.5)^3} = 1280 \text{ kN/m}
$$

$$
\frac{6EI}{L^2} = \frac{6(200)(225)}{(7.5)^2} = 4800 \text{ kN}
$$

$$
\frac{4EI}{L} = \frac{4(200)(225)}{7.5} = 24(10^3) \text{ kN} \cdot \text{m}
$$

$$
\frac{2EI}{L} = \frac{2(200)(225)}{7.5} = 12(10^3) \text{ kN} \cdot \text{m}
$$

$$
\lambda_x = \frac{6 - 0}{7.5} = 0.8 \qquad \lambda_y = \frac{4.5 - 0}{7.5} = 0.6
$$

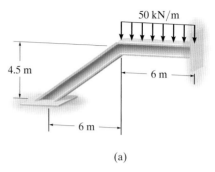

(a)

Fig. 16–5

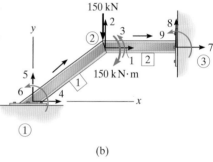

(b)

+

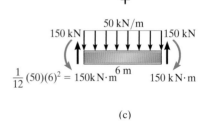

$\frac{1}{12}(50)(6)^2 = 150$ kN·m

(c)

16

EXAMPLE 16.2 (Continued)

Applying Eq. 16–10, we have

$$
\mathbf{k}_1 = (10^3)
\begin{array}{c}
\quad 4 \qquad\ 5 \qquad\ 6 \qquad\ 1 \qquad\ 2 \qquad\ 3 \\
\begin{bmatrix}
128.46 & 95.39 & -2.88 & -128.46 & -95.39 & -2.88 \\
95.39 & 72.82 & 3.84 & -95.39 & -72.82 & 3.84 \\
-2.88 & 3.84 & 24 & 2.88 & -3.84 & 12 \\
-128.46 & -95.39 & 2.88 & 128.46 & 95.39 & 2.88 \\
-95.39 & -72.82 & -3.84 & 95.39 & 72.82 & -3.84 \\
-2.88 & 3.84 & 12 & 2.88 & -3.84 & 24
\end{bmatrix}
\begin{array}{c} 4 \\ 5 \\ 6 \\ 1 \\ 2 \\ 3 \end{array}
\end{array}
$$

Member 2:

$$
\frac{AE}{L} = \frac{7500(10^{-6})(200)(10^6)}{6} = 250(10^3) \text{ kN/m}
$$

$$
\frac{12EI}{L^3} = \frac{12(200)(225)}{(6)^3} = 2500 \text{ kN/m}
$$

$$
\frac{6EI}{L^2} = \frac{6(200)(225)}{(6)^2} = 7500 \text{ kN/m}
$$

$$
\frac{4EI}{L} = \frac{4(200)(225)}{6} = 30(10^3) \text{ kN} \cdot \text{m}
$$

$$
\frac{2EI}{L} = \frac{2(200)(225)}{6} = 15(10^3) \text{ kN} \cdot \text{m}
$$

$$
\lambda_x = \frac{12 - 6}{6} = 1 \qquad \lambda_y = \frac{4.5 - 4.5}{6} = 0
$$

Thus, Eq. 16–10 becomes

$$
\mathbf{k}_2 = (10^3)
\begin{array}{c}
\quad 1 \qquad 2 \qquad 3 \qquad 7 \qquad 8 \qquad 9 \\
\begin{bmatrix}
250 & 0 & 0 & -250 & 0 & 0 \\
0 & 2.5 & 7.5 & 0 & -2.5 & 7.5 \\
0 & 7.5 & 30 & 0 & -7.5 & 15 \\
-250 & 0 & 0 & 250 & 0 & 0 \\
0 & -2.5 & -7.5 & 0 & 2.5 & -7.5 \\
0 & 7.5 & 15 & 0 & -7.5 & 30
\end{bmatrix}
\begin{array}{c} 1 \\ 2 \\ 3 \\ 7 \\ 8 \\ 9 \end{array}
\end{array}
$$

The structure stiffness matrix, included in $\mathbf{Q} = \mathbf{KD}$, becomes

$$
\begin{bmatrix}
0 \\
-150 \\
-150 \\
Q_4 \\
Q_5 \\
Q_6 \\
Q_7 \\
Q_8 \\
Q_9
\end{bmatrix}
= (10^3)
\begin{array}{ccccccccc}
1 & 2 & 3 & 4 & 5 & 6 & 7 & 8 & 9
\end{array}
\begin{bmatrix}
378.46 & 95.39 & 2.88 & -128.46 & -95.36 & 2.88 & -250 & 0 & 0 \\
95.39 & 75.32 & 3.66 & -95.36 & -72.82 & -3.84 & 0 & -2.5 & 7.5 \\
2.88 & 3.66 & 54 & -2.88 & 3.84 & 12 & 0 & -7.5 & 15 \\
-128.46 & -95.39 & -2.88 & 128.46 & 95.39 & -2.88 & 0 & 0 & 0 \\
-95.39 & -72.82 & 3.84 & 95.39 & 72.82 & 3.84 & 0 & 0 & 0 \\
2.88 & -3.84 & 12 & -2.88 & 3.84 & 24 & 0 & 0 & 0 \\
-250 & 0 & 0 & 0 & 0 & 0 & 250 & 0 & 0 \\
0 & -2.5 & -7.5 & 0 & 0 & 0 & 0 & 2.5 & -7.5 \\
0 & 7.5 & 15 & 0 & 0 & 0 & 0 & -7.5 & 30
\end{bmatrix}
\begin{bmatrix}
D_1 \\
D_2 \\
D_3 \\
0 \\
0 \\
0 \\
0 \\
0 \\
0
\end{bmatrix}
\quad (1)
$$

Displacements and Loads. Expanding to determine the displacements, and solving, we have

$$
\begin{bmatrix}
0 \\
-150 \\
-150
\end{bmatrix}
= (10^3)
\begin{bmatrix}
378.46 & 95.39 & 2.88 \\
95.39 & 75.32 & 3.66 \\
2.88 & 3.66 & 54
\end{bmatrix}
\begin{bmatrix}
D_1 \\
D_2 \\
D_3
\end{bmatrix}
+
\begin{bmatrix}
0 \\
0 \\
0
\end{bmatrix}
$$

$$
\begin{bmatrix}
D_1 \\
D_2 \\
D_3
\end{bmatrix}
=
\begin{bmatrix}
0.716 \text{ mm} \\
-2.76 \text{ mm} \\
-0.00261 \text{ rad}
\end{bmatrix}
$$

Using these results, the support reactions are determined from Eq. (1) as follows:

$$
\begin{bmatrix}
Q_4 \\
Q_5 \\
Q_6 \\
Q_7 \\
Q_8 \\
Q_9
\end{bmatrix}
=
\begin{bmatrix}
-128.46 & -95.39 & -2.88 \\
-95.39 & -72.82 & 3.84 \\
2.88 & -3.84 & 12 \\
-250 & 0 & 0 \\
0 & -2.5 & -7.5 \\
0 & 7.5 & 15
\end{bmatrix}
\begin{bmatrix}
0.716 \\
-2.76 \\
-0.00261(10^3)
\end{bmatrix}
+
\begin{bmatrix}
0 \\
0 \\
0 \\
0 \\
0 \\
0
\end{bmatrix}
=
\begin{bmatrix}
178.8 \text{ kN} \\
122.7 \text{ kN} \\
-18.7 \text{ kN} \cdot \text{m} \\
-179.0 \text{ kN} \\
26.5 \text{ kN} \\
-59.9 \text{ kN} \cdot \text{m}
\end{bmatrix}
$$

16

EXAMPLE 16.2 (Continued)

The internal loadings can be determined from Eq. 16–7 applied to members 1 and 2. In the case of member 1, $\mathbf{q} = \mathbf{k'_1 T_1 D}$, where $\mathbf{k'_1}$ is determined from Eq. 16–1, and $\mathbf{T_1}$ from Eq. 16–3. Thus,

$$
\begin{bmatrix} q_4 \\ q_5 \\ q_6 \\ q_1 \\ q_2 \\ q_3 \end{bmatrix} =
\begin{array}{cccccc} 4 & 5 & 6 & 1 & 2 & 3 \end{array}
\begin{bmatrix}
200 & 0 & 0 & -200 & 0 & 0 \\
0 & 1.28 & 4.8 & 0 & -1.28 & 4.8 \\
0 & 4.8 & 24 & 0 & -4.8 & 12 \\
-200 & 0 & 0 & 4.8 & 0 & 0 \\
0 & -1.28 & -4.8 & 0 & 1.28 & -4.8 \\
0 & 4.8 & 12 & 0 & -4.8 & 24
\end{bmatrix}
\begin{bmatrix}
0.8 & 0.6 & 0 & 0 & 0 & 0 \\
-0.6 & 0.8 & 0 & 0 & 0 & 0 \\
0 & 0 & 1 & 0 & 0 & 0 \\
0 & 0 & 0 & 0.8 & 0.6 & 0 \\
0 & 0 & 0 & -0.6 & 0.8 & 0 \\
0 & 0 & 0 & 0 & 0 & 1
\end{bmatrix}
\begin{bmatrix}
0 \\ 0 \\ 0 \\ 0.716 \\ -2.76 \\ -2.61
\end{bmatrix}
\begin{array}{c} 4 \\ 5 \\ 6 \\ 1 \\ 2 \\ 3 \end{array}
$$

Here the code numbers indicate the rows and columns for the near and far ends of the member, respectively, that is, 4, 5, 6, then 1, 2, 3, Fig. 16–5b. Thus,

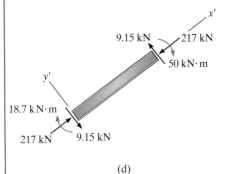

(d)

$$
\begin{bmatrix} q_4 \\ q_5 \\ q_6 \\ q_1 \\ q_2 \\ q_3 \end{bmatrix} =
\begin{bmatrix}
216.6 \text{ kN} \\
-9.15 \text{ kN} \\
-18.7 \text{ kN} \cdot \text{m} \\
-216.6 \text{ kN} \\
9.15 \text{ kN} \\
-50 \text{ kN} \cdot \text{m}
\end{bmatrix}
\qquad Ans.
$$

These results are shown in Fig. 16–5d.

A similar analysis is performed for member 2. The results are shown at the left in Fig. 16–5e. For this member we must superimpose the loadings of Fig. 16–5c, so that the final results for member 2 are shown to the right.

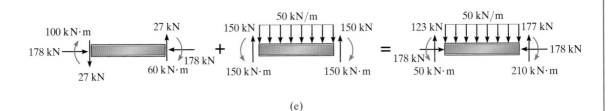

(e)

Fig. 16–5

PROBLEMS

16–1. Determine the structure stiffness matrix **K** for the frame. Assume ① and ③ are fixed. Take $E = 200$ GPa, $I = 300(10^6)$ mm^4, $A = 10(10^3)$ mm^2 for each member.

16–2. Determine the support reactions at the fixed supports ① and ③. Take $E = 200$ GPa, $I = 300(10^6)$ mm^4, $A = 10(10^3)$ mm^2 for each member.

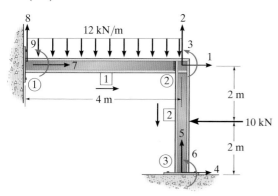

Probs. 16–1/16–2

16–3. Determine the structure stiffness matrix **K** for the frame. Assume ③ is pinned and ① is fixed. Take $E = 200$ MPa, $I = 300(10^6)$ mm^4, $A = 21(10^3)$ mm^2 for each member.

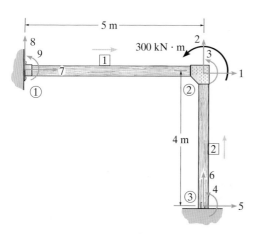

Prob. 16–3

*16–4. Determine the support reactions at ① and ③. Take $E = 200$ MPa, $I = 300(10^6)$ mm^4, $A = 21(10^3)$ mm^2 for each member.

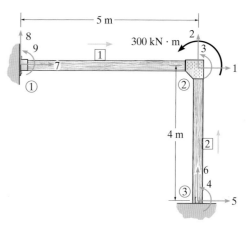

Prob. 16–4

16–5. Determine the structure stiffness matrix **K** for the frame. Take $E = 200$ GPa, $I = 350(10^6)$ mm^4, $A = 15(10^3)$ mm^2 for each member. Joints at ① and ③ are pins.

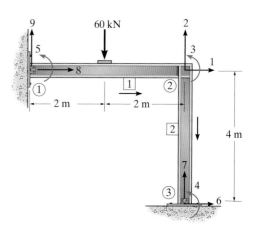

Prob. 16–5

16–6. Determine the support reactions at pins ① and ③. Take $E = 200$ GPa, $I = 350(10^6)$ mm^4, $A = 15(10^3)$ mm^2 for each member.

*16–8.** Determine the components of displacement at ①. Take $E = 200$ GPa, $I = 250(10^6)$ mm^4, $A = 12(10^3)$ mm^2 for each member.

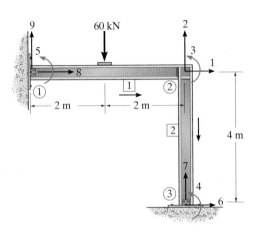

Prob. 16–6

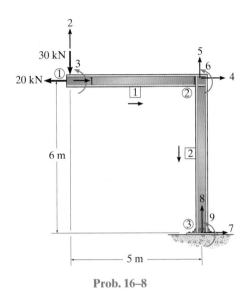

Prob. 16–8

16–7. Determine the structure stiffness matrix **K** for the frame. Take $E = 200$ GPa, $I = 250(10^6)$ mm^4, $A = 12(10^3)$ mm^2 for each member.

16–9. Determine the stiffness matrix **K** for the frame. Take $E = 200$ GPa, $I = 120(10^6)$ mm^4, $A = 6(10^3)$ mm^2 for each member.

16–10. Determine the support reactions at ① and ③. Take $E = 200$ GPa, $I = 120(10^6)$ mm^4, $A = 6(10^3)$ mm^2 for each member.

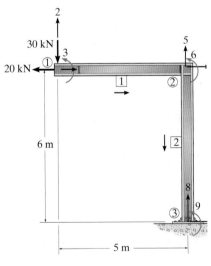

Prob. 16–7

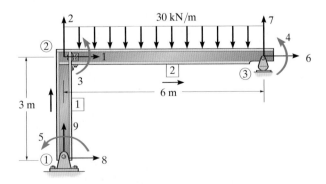

Probs. 16–9/16–10

16–11. Determine the structure stiffness matrix **K** for the frame. Take $E = 200$ GPa, $I = 280(10^6)$ mm^4, $A = 12(10^3)$ mm^2 for each member.

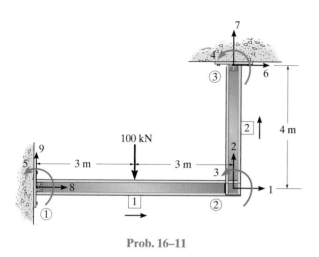

Prob. 16–11

16–13. Use a computer program to determine the reactions on the frame. AE and EI are constant.

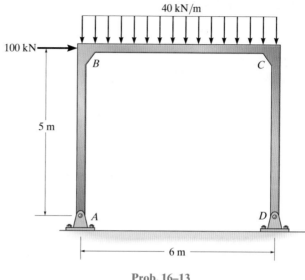

Prob. 16–13

*16–12. Determine the support reactions at the pins ① and ③. Take $E = 200$ GPa, $I = 280(10^6)$ mm^4, $A = 12(10^3)$ mm^2 for each member.

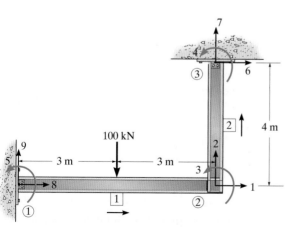

Prob. 16–12

16–14. Use a computer program to determine the reactions on the frame. Assume A, B, D, and F are pins. AE and EI are constant.

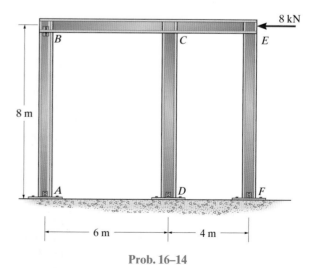

Prob. 16–14

APPENDIX

A

Matrix Algebra for Structural Analysis

A.1 Basic Definitions and Types of Matrices

With the accessibility of desk top computers, application of matrix algebra for the analysis of structures has become widespread. Matrix algebra provides an appropriate tool for this analysis, since it is relatively easy to formulate the solution in a concise form and then perform the actual matrix manipulations using a computer. For this reason it is important that the structural engineer be familiar with the fundamental operations of this type of mathematics.

Matrix. A *matrix* is a rectangular arrangement of numbers having m rows and n columns. The numbers, which are called *elements*, are assembled within brackets. For example, the **A** matrix is written as:

$$\mathbf{A} = \begin{bmatrix} a_{11} & a_{12} & \cdots & a_{1n} \\ a_{21} & a_{22} & \cdots & a_{2n} \\ & & \vdots & \\ a_{m1} & a_{m2} & \cdots & a_{mn} \end{bmatrix}$$

Such a matrix is said to have an *order* of $m \times n$ (m by n). Notice that the first subscript for an element denotes its row position and the second subscript denotes its column position. In general, then, a_{ij} is the element located in the ith row and jth column.

Row Matrix. If the matrix consists only of elements in a single row, it is called a *row matrix*. For example, a $1 \times n$ row matrix is written as

$$\mathbf{A} = [a_1 \quad a_2 \quad \cdots \quad a_n]$$

Here only a single subscript is used to denote an element, since the row subscript is always understood to be equal to 1, that is, $a_1 = a_{11}, a_2 = a_{12}$, and so on.

Column Matrix.   A matrix with elements stacked in a single column is called a *column matrix*. The $m \times 1$ column matrix is

$$\mathbf{A} = \begin{bmatrix} a_1 \\ a_2 \\ \vdots \\ a_m \end{bmatrix}$$

Here the subscript notation symbolizes $a_1 = a_{11}$, $a_2 = a_{21}$, and so on.

Square Matrix.   When the number of rows in a matrix equals the number of columns, the matrix is referred to as a *square matrix*. An $n \times n$ square matrix would be

$$\mathbf{A} = \begin{bmatrix} a_{11} & a_{12} & \cdots & a_{1n} \\ a_{21} & a_{22} & \cdots & a_{2n} \\ & & \vdots & \\ a_{n1} & a_{n2} & \cdots & a_{nn} \end{bmatrix}$$

Diagonal Matrix.   When all the elements of a square matrix are zero except along the main diagonal, running down from left to right, the matrix is called a *diagonal matrix*. For example,

$$\mathbf{A} = \begin{bmatrix} a_{11} & 0 & 0 \\ 0 & a_{22} & 0 \\ 0 & 0 & a_{33} \end{bmatrix}$$

Unit or Identity Matrix.   The *unit* or *identity matrix* is a diagonal matrix with all the diagonal elements equal to unity. For example,

$$\mathbf{I} = \begin{bmatrix} 1 & 0 & 0 \\ 0 & 1 & 0 \\ 0 & 0 & 1 \end{bmatrix}$$

Symmetric Matrix.   A *square matrix* is symmetric provided $a_{ij} = a_{ji}$. For example,

$$\mathbf{A} = \begin{bmatrix} 3 & 5 & 2 \\ 5 & -1 & 4 \\ 2 & 4 & 8 \end{bmatrix}$$

A.2   Matrix Operations

Equality of Matrices.   Matrices $\mathbf{A}$ and $\mathbf{B}$ are said to be equal if they are of the same order and each of their corresponding elements are equal, that is, $a_{ij} = b_{ij}$. For example, if

$$\mathbf{A} = \begin{bmatrix} 2 & 6 \\ 4 & -3 \end{bmatrix} \qquad \mathbf{B} = \begin{bmatrix} 2 & 6 \\ 4 & -3 \end{bmatrix}$$

then $\mathbf{A} = \mathbf{B}$.

Addition and Subtraction of Matrices.   Two matrices can be added together or subtracted from one another if they are of the same order. The result is obtained by adding or subtracting the corresponding elements. For example, if

$$\mathbf{A} = \begin{bmatrix} 6 & 7 \\ 2 & -1 \end{bmatrix} \qquad \mathbf{B} = \begin{bmatrix} -5 & 8 \\ 1 & 4 \end{bmatrix}$$

then

$$\mathbf{A} + \mathbf{B} = \begin{bmatrix} 1 & 15 \\ 3 & 3 \end{bmatrix} \qquad \mathbf{A} - \mathbf{B} = \begin{bmatrix} 11 & -1 \\ 1 & -5 \end{bmatrix}$$

Multiplication by a Scalar.   When a matrix is multiplied by a scalar, each element of the matrix is multiplied by the scalar. For example, if

$$\mathbf{A} = \begin{bmatrix} 4 & 1 \\ 6 & -2 \end{bmatrix} \qquad k = -6$$

then

$$k\mathbf{A} = \begin{bmatrix} -24 & -6 \\ -36 & 12 \end{bmatrix}$$

Matrix Multiplication.   Two matrices $\mathbf{A}$ and $\mathbf{B}$ can be multiplied together only if they are *conformable*. This condition is satisfied if the number of *columns* in $\mathbf{A}$ *equals* the number of *rows* in $\mathbf{B}$. For example, if

$$\mathbf{A} = \begin{bmatrix} a_{11} & a_{12} \\ a_{21} & a_{22} \end{bmatrix} \qquad \mathbf{B} = \begin{bmatrix} b_{11} & b_{12} & b_{13} \\ b_{21} & b_{22} & b_{23} \end{bmatrix} \qquad \text{(A--1)}$$

then $\mathbf{AB}$ can be determined since $\mathbf{A}$ has two columns and $\mathbf{B}$ has two rows. Notice, however, that $\mathbf{BA}$ is not possible. Why?

If matrix $\mathbf{A}$ having an order of $(m \times n)$ is multiplied by matrix $\mathbf{B}$ having an order of $(n \times q)$ it will yield a matrix $\mathbf{C}$ having an order of $(m \times q)$, that is,

$$
\begin{array}{ccc}
\mathbf{A} & \mathbf{B} & = & \mathbf{C} \\
(m \times n)(n \times q) & & (m \times q)
\end{array}
$$

The elements of matrix $\mathbf{C}$ are found using the elements a_{ij} of $\mathbf{A}$ and b_{ij} of $\mathbf{B}$ as follows:

$$
c_{ij} = \sum_{k=1}^{n} a_{ik} b_{kj} \tag{A–2}
$$

The methodology of this formula can be explained by a few simple examples. Consider

$$
\mathbf{A} = \begin{bmatrix} 2 & 4 & 3 \\ -1 & 6 & 1 \end{bmatrix} \qquad \mathbf{B} = \begin{bmatrix} 2 \\ 6 \\ 7 \end{bmatrix}
$$

By inspection, the product $\mathbf{C} = \mathbf{AB}$ is possible since the matrices are conformable, that is, $\mathbf{A}$ has three columns and $\mathbf{B}$ has three rows. By Eq. A–2, the multiplication will yield matrix $\mathbf{C}$ having two rows and one column. The results are obtained as follows:

c_{11}: Multiply the elements in the first row of $\mathbf{A}$ by corresponding elements in the column of $\mathbf{B}$ and add the results; that is,

$$
c_{11} = c_1 = 2(2) + 4(6) + 3(7) = 49
$$

c_{21}: Multiply the elements in the second row of $\mathbf{A}$ by corresponding elements in the column of $\mathbf{B}$ and add the results; that is,

$$
c_{21} = c_2 = -1(2) + 6(6) + 1(7) = 41
$$

Thus

$$
\mathbf{C} = \begin{bmatrix} 49 \\ 41 \end{bmatrix}
$$

As a second example, consider

$$\mathbf{A} = \begin{bmatrix} 5 & 3 \\ 4 & 1 \\ -2 & 8 \end{bmatrix} \qquad \mathbf{B} = \begin{bmatrix} 2 & 7 \\ -3 & 4 \end{bmatrix}$$

Here again the product $\mathbf{C} = \mathbf{AB}$ can be found since $\mathbf{A}$ has two columns and $\mathbf{B}$ has two rows. The resulting matrix $\mathbf{C}$ will have three rows and two columns. The elements are obtained as follows:

$c_{11} = 5(2) + 3(-3) = 1$ (first row of $\mathbf{A}$ times first column of $\mathbf{B}$)

$c_{12} = 5(7) + 3(4) = 47$ (first row of $\mathbf{A}$ times second column of $\mathbf{B}$)

$c_{21} = 4(2) + 1(-3) = 5$ (second row of $\mathbf{A}$ times first column of $\mathbf{B}$)

$c_{22} = 4(7) + 1(4) = 32$ (second row of $\mathbf{A}$ times second column of $\mathbf{B}$)

$c_{31} = -2(2) + 8(-3) = -28$ (third row of $\mathbf{A}$ times first column of $\mathbf{B}$)

$c_{32} = -2(7) + 8(4) = 18$ (third row of $\mathbf{A}$ times second column of $\mathbf{B}$)

The scheme for multiplication follows application of Eq. A–2. Thus,

$$\mathbf{C} = \begin{bmatrix} 1 & 47 \\ 5 & 32 \\ -28 & 18 \end{bmatrix}$$

Notice also that $\mathbf{BA}$ does not exist, since written in this manner the matrices are nonconformable.

The following rules apply to matrix multiplication.

1. In general the product of two matrices is not commutative:

$$\mathbf{AB} \neq \mathbf{BA} \qquad\qquad\qquad\qquad (\text{A–3})$$

2. The distributive law is valid:

$$\mathbf{A}(\mathbf{B} + \mathbf{C}) = \mathbf{AB} + \mathbf{AC} \qquad\qquad\qquad (\text{A–4})$$

3. The associative law is valid:

$$\mathbf{A}(\mathbf{BC}) = (\mathbf{AB})\mathbf{C} \qquad\qquad\qquad\qquad (\text{A–5})$$

Transposed Matrix. A matrix may be transposed by interchanging its rows and columns. For example, if

$$\mathbf{A} = \begin{bmatrix} a_{11} & a_{12} & a_{13} \\ a_{21} & a_{22} & a_{23} \\ a_{31} & a_{32} & a_{33} \end{bmatrix} \qquad \mathbf{B} = \begin{bmatrix} b_1 & b_2 & b_3 \end{bmatrix}$$

Then

$$\mathbf{A}^T = \begin{bmatrix} a_{11} & a_{21} & a_{31} \\ a_{12} & a_{22} & a_{32} \\ a_{13} & a_{23} & a_{33} \end{bmatrix} \qquad \mathbf{B}^T = \begin{bmatrix} b_1 \\ b_2 \\ b_3 \end{bmatrix}$$

Notice that $\mathbf{AB}$ is nonconformable and so the product does not exist. ($\mathbf{A}$ has three columns and $\mathbf{B}$ has one row.) Alternatively, multiplication $\mathbf{AB}^T$ is possible since here the matrices are conformable ($\mathbf{A}$ has three columns and $\mathbf{B}^T$ has three rows). The following properties for transposed matrices hold:

$$(\mathbf{A} + \mathbf{B})^T = \mathbf{A}^T + \mathbf{B}^T \tag{A–6}$$

$$(k\mathbf{A})^T = k\mathbf{A}^T \tag{A–7}$$

$$(\mathbf{AB})^T = \mathbf{B}^T\mathbf{A}^T \tag{A–8}$$

This last identity will be illustrated by example. If

$$\mathbf{A} = \begin{bmatrix} 6 & 2 \\ 1 & -3 \end{bmatrix} \qquad \mathbf{B} = \begin{bmatrix} 4 & 3 \\ 2 & 5 \end{bmatrix}$$

Then, by Eq. A–8,

$$\left(\begin{bmatrix} 6 & 2 \\ 1 & -3 \end{bmatrix} \begin{bmatrix} 4 & 3 \\ 2 & 5 \end{bmatrix} \right)^T = \begin{bmatrix} 4 & 2 \\ 3 & 5 \end{bmatrix} \begin{bmatrix} 6 & 1 \\ 2 & -3 \end{bmatrix}$$

$$\left(\begin{bmatrix} 28 & 28 \\ -2 & -12 \end{bmatrix} \right)^T = \begin{bmatrix} 28 & -2 \\ 28 & -12 \end{bmatrix}$$

$$\begin{bmatrix} 28 & -2 \\ 28 & -12 \end{bmatrix} = \begin{bmatrix} 28 & -2 \\ 28 & -12 \end{bmatrix}$$

Matrix Partitioning.   A matrix can be subdivided into submatrices by partitioning. For example,

$$\mathbf{A} = \begin{bmatrix} a_{11} & a_{12} & a_{13} & a_{14} \\ a_{21} & a_{22} & a_{23} & a_{24} \\ a_{31} & a_{32} & a_{33} & a_{34} \end{bmatrix} = \begin{bmatrix} \mathbf{A}_{11} & \mathbf{A}_{12} \\ \mathbf{A}_{21} & \mathbf{A}_{22} \end{bmatrix}$$

Here the submatrices are

$$\mathbf{A}_{11} = [a_{11}] \qquad \mathbf{A}_{12} = [a_{12} \quad a_{13} \quad a_{14}]$$

$$\mathbf{A}_{21} = \begin{bmatrix} a_{21} \\ a_{31} \end{bmatrix} \qquad \mathbf{A}_{22} = \begin{bmatrix} a_{22} & a_{23} & a_{24} \\ a_{32} & a_{33} & a_{34} \end{bmatrix}$$

The rules of matrix algebra apply to partitioned matrices provided the partitioning is conformable. For example, corresponding submatrices of **A** and **B** can be added or subtracted provided they have an equal number of rows and columns. Likewise, matrix multiplication is possible provided the respective number of columns and rows of both **A** and **B** and their submatrices are equal. For instance, if

$$
\mathbf{A} = \begin{bmatrix} 4 & 1 & -1 \\ -2 & 0 & -5 \\ \hline 6 & 3 & 8 \end{bmatrix} \qquad \mathbf{B} = \begin{bmatrix} 2 & -1 \\ 0 & 8 \\ \hline 7 & 4 \end{bmatrix}
$$

then the product **AB** exists, since the number of columns of **A** equals the number of rows of **B** (three). Likewise, the partitioned matrices are conformable for multiplication since **A** is partitioned into two columns and **B** is partitioned into two rows, that is,

$$
\mathbf{AB} = \begin{bmatrix} \mathbf{A}_{11} & \mathbf{A}_{12} \\ \mathbf{A}_{21} & \mathbf{A}_{22} \end{bmatrix} \begin{bmatrix} \mathbf{B}_{11} \\ \mathbf{B}_{21} \end{bmatrix} = \begin{bmatrix} \mathbf{A}_{11}\mathbf{B}_{11} + \mathbf{A}_{12}\mathbf{B}_{21} \\ \mathbf{A}_{21}\mathbf{B}_{11} + \mathbf{A}_{22}\mathbf{B}_{21} \end{bmatrix}
$$

Multiplication of the submatrices yields

$$
\mathbf{A}_{11}\mathbf{B}_{11} = \begin{bmatrix} 4 & 1 \\ -2 & 0 \end{bmatrix}\begin{bmatrix} 2 & -1 \\ 0 & 8 \end{bmatrix} = \begin{bmatrix} 8 & 4 \\ -4 & 2 \end{bmatrix}
$$

$$
\mathbf{A}_{12}\mathbf{B}_{21} = \begin{bmatrix} -1 \\ -5 \end{bmatrix}[7 \quad 4] = \begin{bmatrix} -7 & -4 \\ -35 & -20 \end{bmatrix}
$$

$$
\mathbf{A}_{21}\mathbf{B}_{11} = [6 \quad 3]\begin{bmatrix} 2 & -1 \\ 0 & 8 \end{bmatrix} = [12 \quad 18]
$$

$$
\mathbf{A}_{22}\mathbf{B}_{21} = [8][7 \quad 4] = [56 \quad 32]
$$

$$
\mathbf{AB} = \begin{bmatrix} \begin{bmatrix} 8 & 4 \\ -4 & 2 \end{bmatrix} + \begin{bmatrix} -7 & -4 \\ -35 & -20 \end{bmatrix} \\ [12 \quad 18] + [56 \quad 32] \end{bmatrix} = \begin{bmatrix} 1 & 0 \\ -39 & -18 \\ 68 & 50 \end{bmatrix}
$$

A.3   Determinants

In the next section we will discuss how to invert a matrix. Since this operation requires an evaluation of the determinant of the matrix, we will now discuss some of the basic properties of determinants.

A determinant is a square array of numbers enclosed within vertical bars. For example, an nth-order determinant, having n rows and n columns, is

$$
|A| = \begin{vmatrix} a_{11} & a_{12} & \cdots & a_{1n} \\ a_{21} & a_{22} & \cdots & a_{2n} \\ & & \vdots & \\ a_{n1} & a_{n2} & \cdots & a_{nn} \end{vmatrix} \tag{A–9}
$$

Evaluation of this determinant leads to a single numerical value which can be determined using *Laplace's expansion*. This method makes use of the determinant's minors and cofactors. Specifically, each element a_{ij} of a determinant of nth order has a *minor* M_{ij} which is a determinant of order $n-1$. This determinant (minor) remains when the ith row and jth column in which the a_{ij} element is contained is canceled out. If the minor is multiplied by $(-1)^{i+j}$ it is called the cofactor of a_{ij} and is denoted as

$$C_{ij} = (-1)^{i+j} M_{ij} \qquad \text{(A–10)}$$

For example, consider the third-order determinant

$$\begin{vmatrix} a_{11} & a_{12} & a_{13} \\ a_{21} & a_{22} & a_{23} \\ a_{31} & a_{32} & a_{33} \end{vmatrix}$$

The cofactors for the elements in the first row are

$$C_{11} = (-1)^{1+1}\begin{vmatrix} a_{22} & a_{23} \\ a_{32} & a_{33} \end{vmatrix} = \begin{vmatrix} a_{22} & a_{23} \\ a_{32} & a_{33} \end{vmatrix}$$

$$C_{12} = (-1)^{1+2}\begin{vmatrix} a_{21} & a_{23} \\ a_{31} & a_{33} \end{vmatrix} = -\begin{vmatrix} a_{21} & a_{23} \\ a_{31} & a_{33} \end{vmatrix}$$

$$C_{13} = (-1)^{1+3}\begin{vmatrix} a_{21} & a_{22} \\ a_{31} & a_{32} \end{vmatrix} = \begin{vmatrix} a_{21} & a_{22} \\ a_{31} & a_{32} \end{vmatrix}$$

Laplace's expansion for a determinant of order n, Eq. A–9, states that the numerical value represented by the determinant is equal to the sum of the products of the elements of any row or column and their respective cofactors, i.e.,

$$D = a_{i1}C_{i1} + a_{i2}C_{i2} + \cdots + a_{in}C_{in} \qquad (i = 1, 2, \ldots, \text{or } n)$$

or

$$D = a_{1j}C_{1j} + a_{2j}C_{2j} + \cdots + a_{nj}C_{nj} \qquad (j = 1, 2, \ldots, \text{or } n)$$

(A–11)

For application, it is seen that due to the cofactors the number D is defined in terms of n determinants (cofactors) of order $n-1$ each. These determinants can each be reevaluated using the same formula, whereby one must then evaluate $(n-1)$ determinants of order $(n-2)$, and so on. The process of evaluation continues until the remaining determinants to be evaluated reduce to the second order, whereby the cofactors of the elements are single elements of D. Consider, for example, the following second-order determinant

$$D = \begin{vmatrix} 3 & 5 \\ -1 & 2 \end{vmatrix}$$

We can evaluate D along the top row of elements, which yields

$$D = 3(-1)^{1+1}(2) + 5(-1)^{1+2}(-1) = 11$$

Or, for example, using the second column of elements, we have

$$D = 5(-1)^{1+2}(-1) + 2(-1)^{2+2}(3) = 11$$

Rather than using Eqs. A–11, it is perhaps easier to realize that the evaluation of a second-order determinant can be performed by multiplying the elements of the diagonal, from top left down to right, and subtract from this the product of the elements from top right down to left, i.e., follow the arrow,

$$D = \begin{vmatrix} 3 & 5 \\ -1 & 2 \end{vmatrix} = 3(2) - 5(-1) = 11$$

Consider next the third-order determinant

$$|D| = \begin{vmatrix} 1 & 3 & -1 \\ 4 & 2 & 6 \\ -1 & 0 & 2 \end{vmatrix}$$

Using Eq. A–11, we can evaluate $|D|$ using the elements either along the top row or the first column, that is

$$D = (1)(-1)^{1+1} \begin{vmatrix} 2 & 6 \\ 0 & 2 \end{vmatrix} + (3)(-1)^{1+2} \begin{vmatrix} 4 & 6 \\ -1 & 2 \end{vmatrix} + (-1)(-1)^{1+3} \begin{vmatrix} 4 & 2 \\ -1 & 0 \end{vmatrix}$$

$$= 1(4 - 0) - 3(8 + 6) - 1(0 + 2) = -40$$

$$D = 1(-1)^{1+1} \begin{vmatrix} 2 & 6 \\ 0 & 2 \end{vmatrix} + 4(-1)^{2+1} \begin{vmatrix} 3 & -1 \\ 0 & 2 \end{vmatrix} + (-1)(-1)^{3+1} \begin{vmatrix} 3 & -1 \\ 2 & 6 \end{vmatrix}$$

$$= 1(4 - 0) - 4(6 - 0) - 1(18 + 2) = -40$$

As an exercise try to evaluate $|D|$ using the elements along the second row.

A.4   Inverse of a Matrix

Consider the following set of three linear equations:

$$a_{11}x_1 + a_{12}x_2 + a_{13}x_3 = c_1$$

$$a_{21}x_1 + a_{22}x_2 + a_{23}x_3 = c_2$$

$$a_{31}x_1 + a_{32}x_2 + a_{33}x_3 = c_3$$

which can be written in matrix form as

$$\begin{bmatrix} a_{11} & a_{12} & a_{13} \\ a_{21} & a_{22} & a_{23} \\ a_{31} & a_{32} & a_{33} \end{bmatrix} \begin{bmatrix} x_1 \\ x_2 \\ x_3 \end{bmatrix} = \begin{bmatrix} c_1 \\ c_2 \\ c_3 \end{bmatrix} \tag{A–12}$$

$$\mathbf{Ax} = \mathbf{C} \tag{A–13}$$

One would think that a solution for x could be determined by dividing $\mathbf{C}$ by $\mathbf{A}$; however, division is not possible in matrix algebra. Instead, one multiplies by the inverse of the matrix. The *inverse* of the matrix $\mathbf{A}$ is another matrix of the same order and symbolically written as $\mathbf{A}^{-1}$. It has the following property,

$$\mathbf{A}\mathbf{A}^{-1} = \mathbf{A}^{-1}\mathbf{A} = \mathbf{I}$$

where $\mathbf{I}$ is an identity matrix. Multiplying both sides of Eq. A–13 by $\mathbf{A}^{-1}$, we obtain

$$\mathbf{A}^{-1}\mathbf{A}\mathbf{x} = \mathbf{A}^{-1}\mathbf{C}$$

Since $\mathbf{A}^{-1}\mathbf{A}\mathbf{x} = \mathbf{I}\mathbf{x} = \mathbf{x}$, we have

$$\mathbf{x} = \mathbf{A}^{-1}\mathbf{C} \tag{A–14}$$

Provided $\mathbf{A}^{-1}$ can be obtained, a solution for $\mathbf{x}$ is possible.

For hand calculation the method used to formulate $\mathbf{A}^{-1}$ can be developed using Cramer's rule. The development will not be given here; instead, only the results are given.* In this regard, the elements in the matrices of Eq. A–14 can be written as

$$\mathbf{x} = \mathbf{A}^{-1}\mathbf{C}$$

$$\begin{bmatrix} x_1 \\ x_2 \\ x_3 \end{bmatrix} = \frac{1}{|A|} \begin{bmatrix} C_{11} & C_{21} & C_{31} \\ C_{12} & C_{22} & C_{32} \\ C_{13} & C_{23} & C_{33} \end{bmatrix} \begin{bmatrix} c_1 \\ c_2 \\ c_3 \end{bmatrix} \tag{A–15}$$

Here $|A|$ is an evaluation of the determinant of the coefficient matrix $\mathbf{A}$, which is determined using the Laplace expansion discussed in Sec. A.3. The square matrix containing the cofactors C_{ij} is called the *adjoint matrix*. By comparison it can be seen that the inverse matrix $\mathbf{A}^{-1}$ is obtained from $\mathbf{A}$ by first replacing each element a_{ij} by its cofactor C_{ij}, then transposing the resulting matrix, yielding the adjoint matrix, and finally multiplying the adjoint matrix by $1/|A|$.

To illustrate how to obtain $\mathbf{A}^{-1}$ numerically, we will consider the solution of the following set of linear equations:

$$\begin{aligned} x_1 - \ x_2 + \ x_3 &= -1 \\ -x_1 + \ x_2 + \ x_3 &= -1 \\ x_1 + 2x_2 - 2x_3 &= \ \ 5 \end{aligned} \tag{A–16}$$

Here

$$\mathbf{A} = \begin{bmatrix} 1 & -1 & 1 \\ -1 & 1 & 1 \\ 1 & 2 & -2 \end{bmatrix}$$

*See Kreyszig, E., *Advanced Engineering Mathematics,* John Wiley & Sons, Inc., New York.

The cofactor matrix for **A** is

$$
\mathbf{C} = \begin{bmatrix}
\begin{vmatrix} 1 & 1 \\ 2 & -2 \end{vmatrix} & -\begin{vmatrix} -1 & 1 \\ 1 & -2 \end{vmatrix} & \begin{vmatrix} -1 & 1 \\ 1 & 2 \end{vmatrix} \\
-\begin{vmatrix} -1 & 1 \\ 2 & -2 \end{vmatrix} & \begin{vmatrix} 1 & 1 \\ 1 & -2 \end{vmatrix} & -\begin{vmatrix} 1 & -1 \\ 1 & 2 \end{vmatrix} \\
\begin{vmatrix} -1 & 1 \\ 1 & 1 \end{vmatrix} & -\begin{vmatrix} 1 & 1 \\ -1 & 1 \end{vmatrix} & \begin{vmatrix} 1 & -1 \\ -1 & 1 \end{vmatrix}
\end{bmatrix}
$$

Evaluating the determinants and taking the transpose, the adjoint matrix is

$$
\mathbf{C}^T = \begin{bmatrix}
-4 & 0 & -2 \\
-1 & -3 & -2 \\
-3 & -3 & 0
\end{bmatrix}
$$

Since

$$
A = \begin{vmatrix}
1 & -1 & 1 \\
-1 & 1 & 1 \\
1 & 2 & -2
\end{vmatrix} = -6
$$

The inverse of **A** is, therefore,

$$
\mathbf{A}^{-1} = -\frac{1}{6} \begin{bmatrix}
-4 & 0 & -2 \\
-1 & -3 & -2 \\
-3 & -3 & 0
\end{bmatrix}
$$

Solution of Eqs. A–16 yields

$$
\begin{bmatrix} x_1 \\ x_2 \\ x_3 \end{bmatrix} = -\frac{1}{6} \begin{bmatrix}
-4 & 0 & -2 \\
-1 & -3 & -2 \\
-3 & -3 & 0
\end{bmatrix} \begin{bmatrix} -1 \\ -1 \\ 5 \end{bmatrix}
$$

$$
x_1 = -\tfrac{1}{6}[(-4)(-1) + 0(-1) + (-2)(5)] = 1
$$

$$
x_2 = -\tfrac{1}{6}[(-1)(-1) + (-3)(-1) + (-2)(5)] = 1
$$

$$
x_3 = -\tfrac{1}{6}[(-3)(-1) + (-3)(-1) + (0)(5)] = -1
$$

Obviously, the numerical calculations are quite expanded for larger sets of equations. For this reason, computers are used in structural analysis to determine the inverse of matrices.

A.5   The Gauss Method for Solving Simultaneous Equations

When many simultaneous linear equations have to be solved, the Gauss elimination method may be used because of its numerical efficiency. Application of this method requires solving one of a set of n equations for an unknown, say x_1, in terms of all the other unknowns, $x_2, x_3, \ldots, x_n$. Substituting this so-called *pivotal equation* into the remaining equations leaves a set of $n - 1$ equations with $n - 1$ unknowns. Repeating the process by solving one of these equations for x_2 in terms of the $n - 2$ remaining unknowns $x_3, x_4, \ldots, x_n$ forms the second pivotal equation. This equation is then substituted into the other equations, leaving a set of $n - 3$ equations with $n - 3$ unknowns. The process is repeated until one is left with a pivotal equation having one unknown, which is then solved. The other unknowns are then determined by successive back substitution into the other pivotal equations. To improve the accuracy of solution, when developing each pivotal equation one should always select the equation of the set having the *largest* numerical coefficient for the unknown one is trying to eliminate. The process will now be illustrated by an example.

Solve the following set of equations using Gauss elimination:

$$-2x_1 + 8x_2 + 2x_3 = 2 \qquad\qquad \text{(A–17)}$$

$$2x_1 - x_2 + x_3 = 2 \qquad\qquad \text{(A–18)}$$

$$4x_1 - 5x_2 + 3x_3 = 4 \qquad\qquad \text{(A–19)}$$

We will begin by eliminating x_1. The largest coefficient of x_1 is in Eq. A–19; hence, we will take it to be the pivotal equation. Solving for x_1, we have

$$x_1 = 1 + 1.25x_2 - 0.75x_3 \qquad\qquad \text{(A–20)}$$

Substituting into Eqs. A–17 and A–18 and simplifying yields

$$2.75x_2 + 1.75x_3 = 2 \qquad\qquad \text{(A–21)}$$

$$1.5x_2 - 0.5x_3 = 0 \qquad\qquad \text{(A–22)}$$

Next we eliminate x_2. Choosing Eq. A–21 for the pivotal equation since the coefficient of x_2 is largest here, we have

$$x_2 = 0.727 - 0.636x_3 \qquad\qquad \text{(A–23)}$$

Substituting this equation into Eq. A–22 and simplifying yields the final pivotal equation, which can be solved for x_3. This yields $x_3 = 0.75$. Substituting this value into the pivotal Eq. A–23 gives $x_2 = 0.25$. Finally, from pivotal Eq. A–20 we get $x_1 = 0.75$.

PROBLEMS

A–1. If $\mathbf{A} = \begin{bmatrix} 3 & 6 \\ 2 & 7 \\ 4 & -2 \end{bmatrix}$ and $\mathbf{B} = \begin{bmatrix} -1 & 2 \\ 5 & 8 \\ -2 & 1 \end{bmatrix}$, determine $2\mathbf{A} - \mathbf{B}$ and $\mathbf{A} + 3\mathbf{B}$.

A–2. If $\mathbf{A} = \begin{bmatrix} 3 & 5 & -2 \\ 4 & 3 & 1 \\ 1 & -1 & 7 \end{bmatrix}$ and $\mathbf{B} = \begin{bmatrix} 6 & 4 & -3 \\ 3 & 2 & -2 \\ 5 & 1 & 6 \end{bmatrix}$, determine $3\mathbf{A} - 2\mathbf{B}$ and $\mathbf{A} - 2\mathbf{B}$.

A–3. If $\mathbf{A} = [2 \quad 5]$ and $\mathbf{B} = \begin{bmatrix} 4 & -1 \\ 2 & -2 \end{bmatrix}$, determine $\mathbf{AB}$.

*A–4. If $\mathbf{A} = \begin{bmatrix} 6 & 3 \\ 4 & 2 \end{bmatrix}$ and $\mathbf{B} = \begin{bmatrix} 6 & 2 \\ 5 & -1 \end{bmatrix}$, determine $\mathbf{AB}$.

A–5. If $\mathbf{A} = \begin{bmatrix} 2 \\ -5 \\ 6 \end{bmatrix}$ and $\mathbf{B} = [4 \quad 6 \quad -5]$, determine $\mathbf{AB}$.

A–6. If $\mathbf{A} = \begin{bmatrix} 2 \\ 5 \\ 6 \end{bmatrix}$ and $\mathbf{B} = \begin{bmatrix} -1 \\ 4 \\ 4 \end{bmatrix}$, show that

$(\mathbf{A} + \mathbf{B})^T = \mathbf{A}^T + \mathbf{B}^T$.

A–7. If $\mathbf{A} = \begin{bmatrix} 2 & 3 & 6 \\ 5 & 9 & 2 \\ -1 & 0 & 2 \end{bmatrix}$, determine $\mathbf{A} + \mathbf{A}^T$.

A–8. If $\mathbf{A} = \begin{bmatrix} 2 & 5 \\ 8 & -1 \end{bmatrix}$, determine $\mathbf{AA}^T$.

A–9. If $\mathbf{A} = \begin{bmatrix} 2 & 8 \\ -1 & 5 \end{bmatrix}$, determine $\mathbf{AA}^T$.

A–10. If $\mathbf{A} = \begin{bmatrix} 5 & 6 & 0 \\ -1 & 2 & 3 \end{bmatrix}$ and $\mathbf{B} = \begin{bmatrix} 2 \\ 0 \\ -1 \end{bmatrix}$, determine $\mathbf{AB}$.

A–11. If $\mathbf{A} = \begin{bmatrix} 2 & 5 & -1 \\ 3 & 2 & 5 \end{bmatrix}$ and $\mathbf{B} = \begin{bmatrix} 2 \\ 5 \\ -1 \end{bmatrix}$, determine $\mathbf{AB}$.

*A–12. If $\mathbf{A} = \begin{bmatrix} 6 & 5 & -1 \\ 0 & 3 & 2 \\ 2 & 1 & 4 \end{bmatrix}$ and $\mathbf{B} = \begin{bmatrix} 2 & -1 & -1 \\ 3 & 2 & 5 \\ 2 & 4 & 6 \end{bmatrix}$, determine $\mathbf{AB}$.

A–13. Show that the distributive law is valid, i.e., $\mathbf{A}(\mathbf{B} + \mathbf{C}) = \mathbf{AB} + \mathbf{AC}$ if $\mathbf{A} = \begin{bmatrix} 4 & 2 & -1 \\ 3 & 5 & 6 \end{bmatrix}$, $\mathbf{B} = \begin{bmatrix} 2 \\ -1 \\ 0 \end{bmatrix}$, $\mathbf{C} = \begin{bmatrix} 4 \\ 2 \\ 1 \end{bmatrix}$.

A–14. Show that the associative law is valid, i.e., $\mathbf{A}(\mathbf{BC}) = (\mathbf{AB})\mathbf{C}$, if $\mathbf{A} = \begin{bmatrix} 2 & 5 & 1 \\ -5 & 6 & 0 \end{bmatrix}$, $\mathbf{B} = \begin{bmatrix} 1 \\ -1 \\ 4 \end{bmatrix}$, $\mathbf{C} = [2 \quad -1 \quad 3]$.

A–15. Evaluate the determinants $\begin{vmatrix} 4 & 3 \\ -1 & 6 \end{vmatrix}$ and $\begin{vmatrix} 5 & 7 & 2 \\ 1 & 8 & 2 \\ -1 & 4 & 0 \end{vmatrix}$.

*A–16. If $\mathbf{A} = \begin{bmatrix} 2 & 5 \\ 4 & -1 \end{bmatrix}$, determine $\mathbf{A}^{-1}$.

A–17. If $\mathbf{A} = \begin{bmatrix} 3 & 5 & 7 \\ 4 & -1 & 2 \\ 0 & 3 & 1 \end{bmatrix}$, determine $\mathbf{A}^{-1}$.

A–18. Solve the equations $4x_1 + x_2 + x_3 = -1$, $-5x_1 + 4x_2 + 3x_3 = 4$, $x_1 - 2x_2 + x_3 = 2$ using the matrix equation $\mathbf{x} = \mathbf{A}^{-1}\mathbf{C}$.

A–19. Solve the equations in Prob. A–18 using the Gauss elimination method.

*A–20. Solve the equations $x_1 + 2x_2 - 2x_3 = 5$, $x_1 - x_2 + x_3 = -1$, $x_1 - x_2 - x_3 = 1$ using the matrix equation $\mathbf{x} = \mathbf{A}^{-1}\mathbf{C}$.

A–21. Solve the equations in Prob. A–20 using the Gauss elimination method.

General Procedure for Using Structural Analysis Software

Popular structural analysis software programs currently available, such as STAAD, RISA, SAP, etc. are all based on the stiffness method of matrix analysis, described in Chapters 13 through 15.* Although each program has a slightly different interface, they all require the operator to input data related to the structure.

A general procedure for using any of these programs is outlined below.

Preliminary Steps. Before using any program it is first necessary to numerically identify the members and joints, called *nodes*, of the structure and establish both global and local coordinate systems in order to specify the structure's geometry and loading. To do this, you may want to make a sketch of the structure and specify each member with a number enclosed within a square, and use a number enclosed within a circle to identify the nodes. In some programs, the "near" and "far" ends of the member must be identified. This is done using an arrow written along the member, with the head of the arrow directed toward the far end. Member, node, and "direction" identification for a plane truss, beam, and plane frame are shown in Figs. B–1, B–2, and B–3. In Fig. B–1, node ② is at the "near end" of member ④ and node ③ is at its "far end." These assignments can all be done arbitrarily. Notice, however, that the nodes on the truss are always at the joints, since this is where the loads are applied and the displacements and member forces are to be determined. For beams and frames, the nodes are at the supports, at a corner or joint, at an internal pin, or at a point where the linear or rotational displacement is to be determined, Figs. B–2 and B–3.

Since loads and displacements are vector quantities, it is necessary to establish a coordinate system in order to specify their correct sense of direction. Here we must use two types of coordinate systems.

Global Coordinates. A single *global or structure coordinate system*, using right-handed x, y, z axes, is used to specify the location of each node relative to the origin, and to identify the sense of each of the external load and displacement components at the nodes. It is convenient to locate the origin at a node so that all the other nodes have positive coordinates. See each figure.

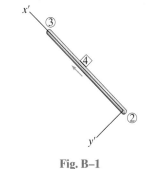

Fig. B–1

*A more complete coverage of this method including the effects of torsion in three-dimensional frames, is given in books related to matrix analysis.

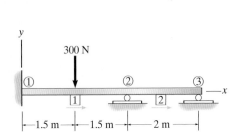

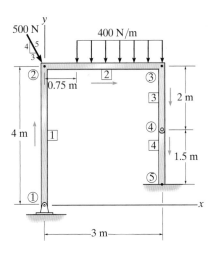

Fig. B–2

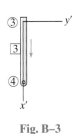

Fig. B–3

Local Coordinates.

A *local or member coordinate system* is used to specify the location and direction of external loadings acting on beam and frame members and for any structure, to provide a means of interperting the computed results of internal loadings acting at the nodes of each member. This system can be identified using right-handed x', y', z' axes with the origin at the "near" node and the x' axis extending along the member toward the "far" node. An example for truss member 4 and frame member 3 is shown in Figs. B–1 and B–3, respectively.

Program Operation.

When any program is excuted a menu should appear which allows various selections for inputing the data and getting the results. The following explains the items used for input data. For any problem, be sure to use a consistent set of units for numerical quantities.

General Structure Information.

This item should generally be selected first in order to assign a problem title and identify the type of structure to be analyzed–truss, beam, or frame.

Node Data.

Enter, in turn, each node number and its far and near end global coordinates.

Member Data.

Enter, in turn, each member number, the near and far node numbers, and the member properties, E (modulus of elasticity), A (cross-sectional area), and/or I (moment of inertia and/or the polar moment of inertia or other suitable torsional constant required for three-dimensional frames*). If these member properties are unknown

*Quite often a selected structural shape, e.g., a wide-flange or W shape, can be made when the program has a database of its geometric properties.

then provided the structure is statically determinate, these values can be set equal to one. If the structure is statically indeterminate then these must be no support settlement, and the members must have the same cross section and be made from the same material. The computed results will then give the correct reactions and internal forces, but not the correct displacements.

If an internal hinge or pin connects two members of a beam or frame, then the release of moment must be specified at that node. For example, member 3 of the frame in Fig. B–3 has a pin at the far node, 4. In a like manner, this pin can also be identified at the near node of member 4.

Support Data.
Enter, in turn, each node located at a support, and specify the called for global coordinate directions in which restraint occurs. For example, since node 5 of the frame in Fig. B–3 is a fixed support, a zero is entered for the x, y, and z (rotational) directions; however if this support settles downward 0.003 m then the value entered for y would be -0.003.

Load Data.
Loads are specified either at nodes, or on members. Enter the algebraic values of *nodal loadings* relative to the *global coordinates*. For example, for the truss in Fig. B–1 the loading at node 2 is in the y direction and has a value of -200. For beam and frame *members* the loadings and their location are referenced using the *local coordinates*. For example, the distributed loading on member 2 of the frame in Fig. B–3 is specified with an intensity of -400 N/m located 0.75 m from the near node 2 and -400 N/m located 3 m from this node.

Results.
Once all the data is entered, then the problem can be solved. One obtains the external reactions on the structure and the displacements and internal loadings at each node. As a partial check of the results a statics check is often given at each of the nodes. It is very important that you never fully trust the results obtained. Instead, it would be wise to perform an intuitive structural analysis to further check the output. After all, the structural engineer must take full responsibility for both the modeling and computing of final results.

B

Fundamental Problems
Partial Solutions and Answers

Chapter 2

F2–1. $\zeta + \Sigma M_A = 0;$ $60 - F_{BC}\left(\tfrac{3}{5}\right)(4) = 0$ $F_{BC} = 25.0$ kN

$\zeta + \Sigma M_B = 0;$ $60 - A_y(4) = 0$ $A_y = 15.0$ kN   *Ans*

$\xrightarrow{+} \Sigma F_x = 0;$ $A_x - 25.0\left(\tfrac{4}{5}\right) = 0$ $A_x = 20.0$ kN   *Ans*

$B_x = C_x = 25.0\left(\tfrac{4}{5}\right) = 20.0$ kN $B_y = C_y = 25.0\left(\tfrac{3}{5}\right) = 15.0$ kN   *Ans*

F2–2. $\zeta + \Sigma M_A = 0;$ $F_{BC} \sin 45°(4) - 10(4)(2) = 0$ $F_{BC} = \dfrac{20}{\sin 45°}$ kN

$\zeta + \Sigma M_B = 0;$ $10(4)(2) - A_y(4) = 0$ $A_y = 20.0$ kN   *Ans*

$\xrightarrow{+} \Sigma F_x = 0;$ $A_x - \left(\dfrac{20}{\sin 45°}\right)(\cos 45°) = 0$ $A_x = 20.0$ kN   *Ans*

$B_x = C_x = \left(\dfrac{20}{\sin 45°}\right)(\cos 45°) = 20.0$ kN   *Ans*

$B_y = C_y = \left(\dfrac{20}{\sin 45°}\right)(\sin 45°) = 20.0$ kN   *Ans*

F2–3. $\zeta + \Sigma M_A = 0;$ $F_{BC} \sin 60°(4) - 10(2)(1) = 0$ $F_{BC} = \dfrac{5}{\sin 60°}$ kN

$\zeta + \Sigma M_B = 0;$ $10(2)(3) - A_y(4) = 0$ $A_y = 15.0$ kN   *Ans*

$\xrightarrow{+} \Sigma F_x = 0;$ $\left(\dfrac{5}{\sin 60°}\right)(\cos 60°) - A_x = 0$ $A_x = 2.89$ kN   *Ans*

$B_x = C_x = \left(\dfrac{5}{\sin 60°}\right)(\cos 60°) = 2.89$ kN   *Ans*

$B_y = C_y = \left(\dfrac{5}{\sin 60°}\right)(\sin 60°) = 5.00$ kN   *Ans*

F2–4. Member AC

$\zeta + \Sigma M_C = 0;$ $10(3) - N_A(4) = 0$ $N_A = 7.50$ kN   *Ans*

$\zeta + \Sigma M_A = 0;$ $C_y(4) - 10(1) = 0$ $C_y = 2.50$ kN

Member BC

$\xrightarrow{+} \Sigma F_x = 0;$ $B_x = 0$ *Ans*

$+\uparrow \Sigma F_y = 0;$ $B_y - 2.50 - 8(2) = 0$ $B_y = 18.5$ kN   *Ans*

$\zeta + \Sigma M_B = 0;$ $2.50(2) + 8(2)(1) - M_B = 0$ $M_B = 21.0$ kN·m   *Ans*

F2–5. $\zeta + \Sigma M_A = 0$; $F_{BC}\left(\frac{3}{5}\right)(2) + F_{BC}\left(\frac{4}{5}\right)(1.5) - 1.5(1) = 0$ $F_{BC} = 0.625$ kN

$\xrightarrow{+} \Sigma F_x = 0$; $A_x - 0.625\left(\frac{4}{5}\right) = 0$ $A_x = 0.5$ kN   *Ans*

$+\uparrow \Sigma F_y = 0$; $A_y + 0.625\left(\frac{3}{5}\right) - 1.5 = 0$ $A_y = 1.125$ kN   *Ans*

$B_x = C_x = 0.625\left(\frac{4}{5}\right) = 0.5$ kN   *Ans*

$B_y = C_y = 0.625\left(\frac{3}{5}\right) = 0.375$ kN   *Ans*

F2–6. $\zeta + \Sigma M_C = 0$; $6(2) + 2(2) - N_A(4) = 0$ $N_A = 4.00$ kN   *Ans*

$\xrightarrow{+} \Sigma F_x = 0$; $C_x - 2 = 0$ $C_x = 2.00$ kN   *Ans*

$+\uparrow \Sigma F_y = 0$; $C_y + 4.00 - 6 = 0$ $C_y = 2.00$ kN   *Ans*

F2–7.

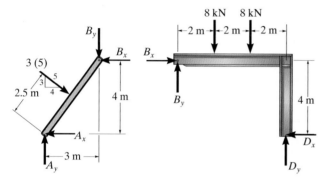

Member *AB*

$\zeta + \Sigma M_A = 0$; $B_x(4) - B_y(3) - 3(5)(2.5) = 0$

Member *BCD*

$\zeta + \Sigma M_D = 0$; $8(2) + 8(4) - B_x(4) - B_y(6) = 0$

$B_x = 10.25$ kN $B_y = 1.167$ kN $= 1.17$ kN   *Ans*

Member *AB*

$\xrightarrow{+} \Sigma F_x = 0$; $-A_x + 3(5)\left(\frac{4}{5}\right) - 10.25 = 0$ $A_x = 1.75$ kN   *Ans*

$+\uparrow \Sigma F_y = 0$; $A_y - (3)(5)\left(\frac{3}{5}\right) - 1.167 = 0$ $A_y = 10.167$ kN $= 10.2$ kN   *Ans*

Member *BCD*

$\xrightarrow{+} \Sigma F_x = 0$; $10.25 - D_x = 0$ $D_x = 10.25$ kN   *Ans*

$+\uparrow \Sigma F_y = 0$; $D_y + 1.167 - 8 - 8 = 0$ $D_y = 14.833$ kN $= 14.8$ kN   *Ans*

F2–8.

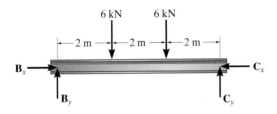

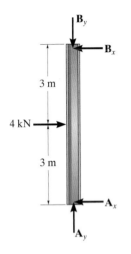

Member AB

$\downarrow + \Sigma M_A = 0;$ $\qquad$ $B_x(6) - 4(3) = 0$ $\quad B_x = 2.00 \text{ kN}$ $\qquad$ *Ans*

$\downarrow + \Sigma M_B = 0;$ $\qquad$ $4(3) - A_x(6) = 0$ $\quad A_x = 2.00 \text{ kN}$ $\qquad$ *Ans*

Member BC

$\xrightarrow{+} \Sigma F_x = 0;$ $\qquad$ $2.00 - C_x = 0$ $\quad C_x = 2.00 \text{ kN}$ $\qquad$ *Ans*

$\downarrow + \Sigma M_C = 0;$ $\qquad$ $6(2) + 6(4) - B_y(6) = 0$ $\quad B_y = 6.00 \text{ kN}$ $\qquad$ *Ans*

$\downarrow + \Sigma M_B = 0;$ $\qquad$ $C_y(6) - 6(2) - 6(4) = 0$ $\quad C_y = 6.00 \text{ kN}$ $\qquad$ *Ans*

Member AB

$+\uparrow \Sigma F_y = 0;$ $\qquad$ $A_y - 6.00 = 0$ $\quad A_y = 6.00 \text{ kN}$ $\qquad$ *Ans*

Member CD

$\xrightarrow{+} \Sigma F_x = 0;$ $\qquad$ $2.00 - D_x = 0$ $\quad D_x = 2.00 \text{ kN}$ $\qquad$ *Ans*

$+\uparrow \Sigma F_y = 0;$ $\qquad$ $D_y - 6.00 = 0$ $\quad D_y = 6.00 \text{ kN}$ $\qquad$ *Ans*

$\downarrow + \Sigma M_D = 0;$ $\qquad$ $M_D - 2.00(6) = 0$ $\quad M_D = 12.0 \text{ kN} \cdot \text{m}$ $\qquad$ *Ans*

F2–9.

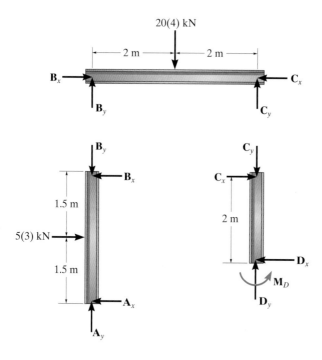

20(4) kN

Member AB

$\downarrow + \Sigma M_A = 0;$ $B_x(3) - 5(3)(1.5) = 0$ $B_x = 7.5$ kN   *Ans*

$\downarrow + \Sigma M_B = 0;$ $5(3)(1.5) - A_x(3) = 0$ $A_x = 7.5$ kN   *Ans*

Member BC

$\downarrow + \Sigma M_C = 0;$ $20(4)(2) - B_y(4) = 0$ $B_y = 40.0$ kN   *Ans*

$\downarrow + \Sigma M_B = 0;$ $C_y(4) - 20(4)(2) = 0$ $C_y = 40.0$ kN   *Ans*

$\xrightarrow{+} \Sigma F_x = 0;$ $7.5 - C_x = 0$ $C_x = 7.5$ kN   *Ans*

Member AB

$+\uparrow \Sigma F_y = 0;$ $A_y - 40.0 = 0$ $A_y = 40.0$ kN   *Ans*

Member CD

$\xrightarrow{+} \Sigma F_x = 0;$ $7.5 - D_x = 0$ $D_x = 7.5$ kN   *Ans*

$+\uparrow \Sigma F_y = 0;$ $D_y - 40.0 = 0$ $D_y = 40.0$ kN   *Ans*

$\downarrow + \Sigma M_D = 0;$ $M_D - 7.5(2) = 0$ $M_D = 15.0$ kN $\cdot$ m   *Ans*

F2–10.

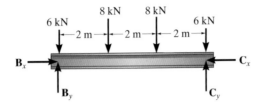

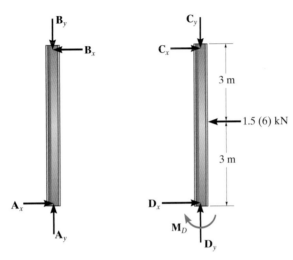

Member BC

$$\underset{\curvearrowleft}{} + \Sigma M_B = 0; \qquad C_y(6) - 8(2) - 8(4) - 6(6) = 0 \qquad C_y = 14.0 \text{ kN} \qquad\qquad Ans$$

$$\underset{\curvearrowleft}{} + \Sigma M_C = 0; \qquad 8(2) + 8(4) + 6(6) - B_y(6) = 0 \qquad B_y = 14.0 \text{ kN} \qquad\qquad Ans$$

Member AB

$$\underset{\curvearrowleft}{} + \Sigma M_A = 0; \qquad\qquad\qquad\qquad B_x = 0 \qquad\qquad\qquad\qquad\qquad Ans$$

$$\overset{+}{\rightarrow} \Sigma F_x = 0; \qquad\qquad\qquad\qquad A_x = 0 \qquad\qquad\qquad\qquad\qquad Ans$$

$$+ \uparrow \Sigma F_y = 0; \qquad\qquad\qquad A_y - 14.0 = 0 \qquad A_y = 14.0 \text{ kN} \qquad\qquad Ans$$

Member BC

$$\overset{+}{\rightarrow} \Sigma F_x = 0; \qquad\qquad\qquad\qquad C_x = 0 \qquad\qquad\qquad\qquad\qquad Ans$$

Member CD

$$\overset{+}{\rightarrow} \Sigma F_x = 0; \qquad\qquad D_x - 1.5(6) = 0 \qquad D_x = 9.00 \text{ kN} \qquad\qquad Ans$$

$$+ \uparrow \Sigma F_y = 0; \qquad\qquad D_y - 14.0 = 0 \qquad D_y = 14.0 \text{ kN} \qquad\qquad Ans$$

$$\underset{\curvearrowleft}{} + \Sigma M_D = 0; \qquad\quad 1.5(6)(3) - M_D = 0 \qquad M_D = 27.0 \text{ kN} \cdot \text{m} \qquad Ans$$

F3–1.   Joint C

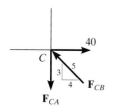

$$\xrightarrow{+}\ \Sigma F_x = 0; \qquad 40 - F_{CB}(\tfrac{4}{5}) = 0 \quad F_{CB} = 50.0 \text{ kN (C)} \qquad Ans$$

$$+\uparrow \Sigma F_y = 0; \qquad 50.0(\tfrac{3}{5}) - F_{CA} = 0 \quad F_{CA} = 30.0 \text{ kN (T)} \qquad Ans$$

Joint B

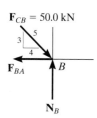

$$\xrightarrow{+}\ \Sigma F_x = 0; \qquad 50.0(\tfrac{4}{5}) - F_{BA} = 0 \quad F_{BA} = 40.0 \text{ kN (T)} \qquad Ans$$

$$+\uparrow \Sigma F_y = 0; \qquad N_B - 50.0(\tfrac{3}{5}) = 0 \quad N_B = 30.0 \text{ kN}$$

F3–2.   Joint B

$$+\uparrow \Sigma F_y = 0; \qquad F_{BC} \sin 45° - 6 = 0 \quad F_{BC} = 8.485 \text{ kN (T)} = 8.49 \text{ kN (T)} \qquad Ans$$

$$\xrightarrow{+}\ \Sigma F_x = 0; \qquad F_{BA} - 8.485 \cos 45° = 0 \quad F_{BA} = 6.00 \text{ kN (C)} \qquad Ans$$

Joint C

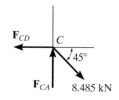

$$\xrightarrow{+}\ \Sigma F_x = 0; \qquad 8.485 \cos 45° - F_{CD} = 0 \quad F_{CD} = 6.00 \text{ kN (T)} \qquad Ans$$

$$+\uparrow \Sigma F_y = 0; \qquad F_{CA} - 8.485 \sin 45° = 0 \quad F_{CA} = 6.00 \text{ kN (C)} \qquad Ans$$

F3–3.   Joint C

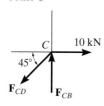

$$\xrightarrow{+}\ \Sigma F_x = 0; \qquad 10 - F_{CD} \cos 45° = 0 \quad F_{CD} = 14.14 \text{ kN (T)} = 14.1 \text{ kN (T)} \qquad Ans$$

$$+\uparrow \Sigma F_y = 0; \qquad F_{CB} - 14.14 \sin 45° = 0 \quad F_{CB} = 10.0 \text{ kN (C)} \qquad Ans$$

Joint D

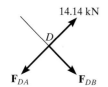

$$+\nearrow \Sigma F_{x'} = 0; \qquad 14.14 - F_{DA} = 0 \quad F_{DA} = 14.14 \text{ kN (T)} = 14.1 \text{ kN (T)} \qquad Ans$$

$$\nwarrow^{+} \Sigma F_{y'} = 0; \qquad F_{DB} = 0 \qquad Ans$$

Joint B

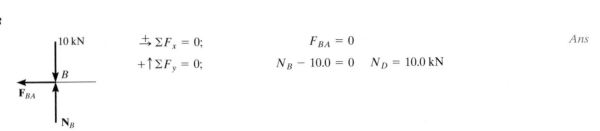

$$\xrightarrow{+} \Sigma F_x = 0; \qquad\qquad F_{BA} = 0 \qquad\qquad\qquad\qquad\qquad\qquad Ans$$

$$+\uparrow \Sigma F_y = 0; \qquad\qquad N_B - 10.0 = 0 \quad N_D = 10.0 \text{ kN}$$

F3–4.   Joint D

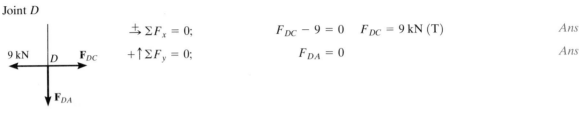

$$\xrightarrow{+} \Sigma F_x = 0; \qquad\qquad F_{DC} - 9 = 0 \quad F_{DC} = 9 \text{ kN (T)} \qquad\qquad Ans$$

$$+\uparrow \Sigma F_y = 0; \qquad\qquad\qquad F_{DA} = 0 \qquad\qquad\qquad\qquad\qquad Ans$$

Joint C

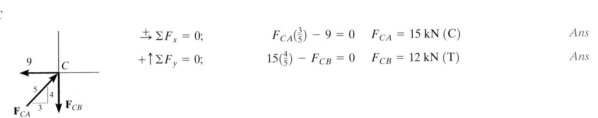

$$\xrightarrow{+} \Sigma F_x = 0; \qquad\qquad F_{CA}(\tfrac{3}{5}) - 9 = 0 \quad F_{CA} = 15 \text{ kN (C)} \qquad Ans$$

$$+\uparrow \Sigma F_y = 0; \qquad\qquad 15(\tfrac{4}{5}) - F_{CB} = 0 \quad F_{CB} = 12 \text{ kN (T)} \qquad Ans$$

Joint A

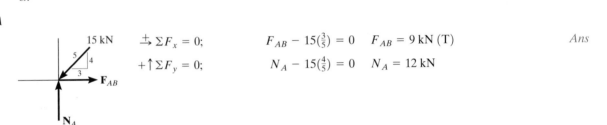

$$\xrightarrow{+} \Sigma F_x = 0; \qquad\qquad F_{AB} - 15(\tfrac{3}{5}) = 0 \quad F_{AB} = 9 \text{ kN (T)} \qquad Ans$$

$$+\uparrow \Sigma F_y = 0; \qquad\qquad N_A - 15(\tfrac{4}{5}) = 0 \quad N_A = 12 \text{ kN}$$

F3–5.   Joint D

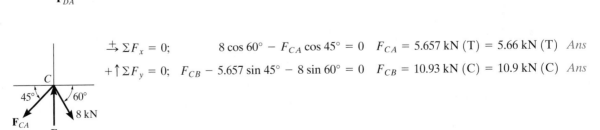

$$\xrightarrow{+} \Sigma F_x = 0; \qquad\qquad\qquad F_{DC} = 0 \qquad\qquad\qquad\qquad\qquad Ans$$

$$+\uparrow \Sigma F_y = 0; \qquad\qquad\qquad F_{DA} = 0 \qquad\qquad\qquad\qquad\qquad Ans$$

Joint C

$$\xrightarrow{+} \Sigma F_x = 0; \qquad 8 \cos 60° - F_{CA} \cos 45° = 0 \quad F_{CA} = 5.657 \text{ kN (T)} = 5.66 \text{ kN (T)} \quad Ans$$

$$+\uparrow \Sigma F_y = 0; \quad F_{CB} - 5.657 \sin 45° - 8 \sin 60° = 0 \quad F_{CB} = 10.93 \text{ kN (C)} = 10.9 \text{ kN (C)} \quad Ans$$

Joint B

$$\xrightarrow{+} \Sigma F_x = 0; \qquad\qquad F_{AB} = 0 \qquad\qquad Ans$$

$$+\uparrow \Sigma F_y = 0; \qquad\qquad N_B = 10.93 \text{ kN}$$

F3–6. Entire truss

$$\downarrow+\Sigma M_A = 0; \qquad E_y(8) - 600(2) - 800(4) - 600(6) = 0 \quad E_y = 1000 \text{ N}$$

Joint E

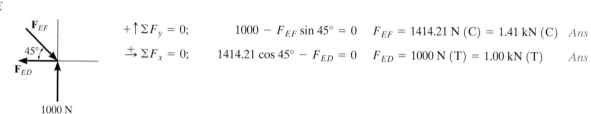

$$+\uparrow \Sigma F_y = 0; \qquad\quad 1000 - F_{EF} \sin 45° = 0 \quad F_{EF} = 1414.21 \text{ N (C)} = 1.41 \text{ kN (C)} \quad Ans$$

$$\xrightarrow{+} \Sigma F_x = 0; \qquad 1414.21 \cos 45° - F_{ED} = 0 \quad F_{ED} = 1000 \text{ N (T)} = 1.00 \text{ kN (T)} \quad Ans$$

Joint F

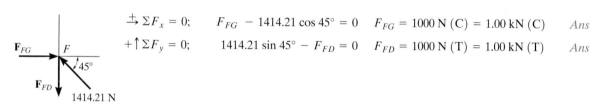

$$\xrightarrow{+} \Sigma F_x = 0; \qquad F_{FG} - 1414.21 \cos 45° = 0 \quad F_{FG} = 1000 \text{ N (C)} = 1.00 \text{ kN (C)} \quad Ans$$

$$+\uparrow \Sigma F_y = 0; \qquad 1414.21 \sin 45° - F_{FD} = 0 \quad F_{FD} = 1000 \text{ N (T)} = 1.00 \text{ kN (T)} \quad Ans$$

Joint D

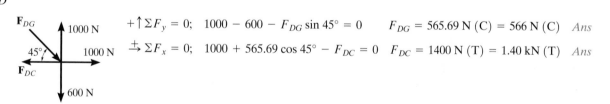

$$+\uparrow \Sigma F_y = 0; \quad 1000 - 600 - F_{DG} \sin 45° = 0 \qquad F_{DG} = 565.69 \text{ N (C)} = 566 \text{ N (C)} \quad Ans$$

$$\xrightarrow{+} \Sigma F_x = 0; \quad 1000 + 565.69 \cos 45° - F_{DC} = 0 \quad F_{DC} = 1400 \text{ N (T)} = 1.40 \text{ kN (T)} \quad Ans$$

Joint C

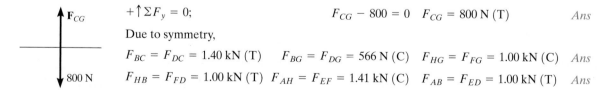

$$+\uparrow \Sigma F_y = 0; \qquad\qquad\qquad F_{CG} - 800 = 0 \quad F_{CG} = 800 \text{ N (T)} \qquad Ans$$

Due to symmetry,

$$F_{BC} = F_{DC} = 1.40 \text{ kN (T)} \qquad F_{BG} = F_{DG} = 566 \text{ N (C)} \qquad F_{HG} = F_{FG} = 1.00 \text{ kN (C)} \quad Ans$$

$$F_{HB} = F_{FD} = 1.00 \text{ kN (T)} \quad F_{AH} = F_{EF} = 1.41 \text{ kN (C)} \qquad F_{AB} = F_{ED} = 1.00 \text{ kN (T)} \quad Ans$$

F3–7.   For the entire truss

$$\zeta + \Sigma M_E = 0; \quad 10(1.5) + 10(3) + 10(4.5) - A_y(6) = 0 \quad A_y = 15.0 \text{ kN}$$

$$\overset{+}{\rightarrow} \Sigma F_x = 0; \qquad\qquad\qquad\qquad\qquad A_x = 0$$

For the left segment

$$+\uparrow \Sigma F_y = 0; \qquad 15.0 - 10 - F_{BG}\sin 45° = 0 \quad F_{BG} = 7.07 \text{ kN (C)} \qquad Ans$$

$$\zeta + \Sigma M_B = 0; \qquad\qquad F_{HG}(1.5) - 15(1.5) = 0 \quad F_{HG} = 15.0 \text{ kN (C)} \qquad Ans$$

$$\zeta + \Sigma M_G = 0; \qquad F_{BC}(1.5) + 10(1.5) - 15.0(3) = 0 \quad F_{BC} = 20.0 \text{ kN (T)} \qquad Ans$$

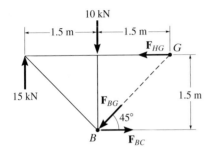

F3–8.   For the entire truss

$$\zeta + \Sigma M_E = 0; \qquad 3(4.8) + 3(3.6) + 3(2.4) + 3(1.2) - A_y(4.8) = 0 \quad A_y = 7.5 \text{ kN}$$

$$\overset{+}{\rightarrow} \Sigma F_x = 0; \qquad\qquad\qquad\qquad\qquad A_x = 0$$

For the left segment

$$\zeta + \Sigma M_C = 0; \qquad F_{HI}(0.9) + 3(1.2) + 3(2.4) - 7.5(2.4) = 0 \quad F_{HI} = 8 \text{ kN (C)}$$

$$\zeta + \Sigma M_I = 0; \qquad\qquad F_{BC}(0.9) + 3(1.2) - 7.5(1.2) = 0 \quad F_{BC} = 6 \text{ kN (T)} \qquad Ans$$

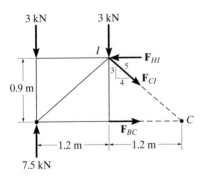

Joint *H*

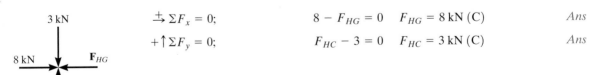

$$\overset{+}{\rightarrow} \Sigma F_x = 0; \qquad 8 - F_{HG} = 0 \quad F_{HG} = 8 \text{ kN (C)} \qquad Ans$$

$$+\uparrow \Sigma F_y = 0; \qquad F_{HC} - 3 = 0 \quad F_{HC} = 3 \text{ kN (C)} \qquad Ans$$

F3–9. For the entire truss

$$\zeta + \Sigma M_A = 0; \qquad\qquad N_C(4) - 8(2) - 6(2) = 0 \quad N_C = 7.00 \text{ kN}$$

Consider the right segment

$$+\uparrow \Sigma F_y = 0; \qquad\qquad 7.00 - F_{BD} \sin 45° = 0 \quad F_{BD} = 9.899 \text{ kN (T)} = 9.90 \text{ kN (T)} \qquad\qquad Ans$$

$$\zeta + \Sigma M_B = 0; \qquad\qquad 7.00(2) - 6(2) - F_{ED}(2) = 0 \quad F_{ED} = 1.00 \text{ kN (C)} \qquad\qquad Ans$$

$$\zeta + \Sigma M_D = 0; \qquad\qquad 0 - F_{BC}(2) = 0 \quad F_{BC} = 0 \qquad\qquad Ans$$

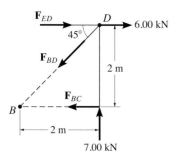

F3–10. For the entire truss

$$\zeta + \Sigma M_A = 0; \quad N_E(8) - 1800(8) - 1800(4) - 1800(6) - 1800(8) = 0 \quad N_E = 4500 \text{ N}$$

Consider the right segment

$$\zeta + \Sigma M_E = 0; \qquad\qquad 1800(2) - F_{CF}\left(\tfrac{3}{5}\right)(4) = 0 \quad F_{CF} = 1500 \text{ N (C)} \qquad\qquad Ans$$

$$\zeta + \Sigma M_C = 0; \quad 4500(4) - 1800(4) - 1800(2) - F_{GF}\left(\tfrac{3}{5}\right)(4) = 0 \quad F_{GF} = 3000 \text{ N (C)} \qquad\qquad Ans$$

$$\zeta + \Sigma M_F = 0; \qquad\qquad 4500(2) - 1800(2) - F_{CD}(1.5) = 0 \quad F_{CD} = 3600 \text{ N (T)} \qquad\qquad Ans$$

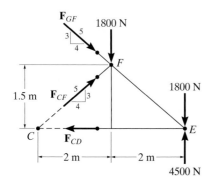

F3–11. For the entire truss

$\zeta+\Sigma M_A = 0;$ $\qquad N_D(6) - 2(6) - 4(3) = 0$ $\quad N_D = 4.00$ kN

Consider the right segment

$+\uparrow\Sigma F_y = 0;$ $\qquad 4.00 - 2 - F_{FC}\sin 45° = 0$ $\quad F_{FC} = 2.828$ kN (C) $= 2.83$ kN (C) $\qquad$ *Ans*

$\zeta+\Sigma M_F = 0;$ $\qquad 4.00(3) - 2(3) - F_{BC}(1.5) = 0$ $\quad F_{BC} = 4.00$ kN (T) $\qquad$ *Ans*

$\zeta+\Sigma M_C = 0;$ $\qquad 4.00(1.5) - 2(1.5) - F_{FE}(1.5) = 0$ $\quad F_{FE} = 2.00$ kN (C) $\qquad$ *Ans*

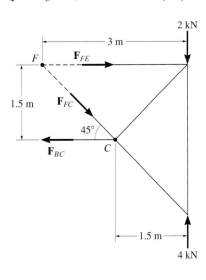

F3–12. For the entire truss

$\zeta+\Sigma M_A = 0;$ $\qquad N_E(4) - 2(1) - 2(2) - 2(3) = 0$ $\quad N_E = 3$ kN

Consider the right segment

$\zeta+\Sigma M_F = 0;$ $\qquad 3(1) - F_{CD}(0.75) = 0$ $\quad F_{CD} = 4$ kN (T) $\qquad$ *Ans*

$\zeta+\Sigma M_C = 0;$ $\qquad 3(2) - 2(1) - F_{GF}\left(\dfrac{1}{\sqrt{17}}\right)(4) = 0$ $\quad F_{GF} = 4.12$ kN (C) $\qquad$ *Ans*

$\zeta+\Sigma M_O = 0;$ $\qquad F_{CF}\left(\tfrac{3}{5}\right)(4) + 2(3) - 3(2) = 0$ $\quad F_{CF} = 0$ $\qquad$ *Ans*

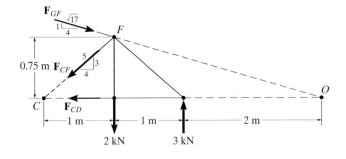

F4–1. $\downarrow + \Sigma M_A = 0;$ $\qquad B_y(2) + 20 - 10(4) = 0 \quad B_y = 10.0 \text{ kN}$

Segment CB

$\xrightarrow{+} \Sigma F_x = 0;$ $\qquad\qquad N_C = 0$ *Ans*

$+\uparrow \Sigma F_y = 0;$ $\qquad V_C + 10 - 10 = 0 \quad V_C = 0$ *Ans*

$\downarrow + \Sigma M_C = 0;$ $\qquad -M_C + 10(1) - 10(3) = 0 \quad M_C = -20 \text{ kN} \cdot \text{m}$ *Ans*

F4–2. $\downarrow + \Sigma M_A = 0;$ $B_y(3) - 4(1.5)(0.75) - 8(1.5)(2.25) = 0 \quad B_y = 10.5 \text{ kN}$

Segment CB

$\xrightarrow{+} \Sigma F_x = 0;$ $\qquad\qquad N_C = 0$ *Ans*

$+\uparrow \Sigma F_y = 0;$ $\qquad V_C + 10.5 - 8(1.5) = 0 \quad V_C = 1.50 \text{ kN}$ *Ans*

$\downarrow + \Sigma M_C = 0;$ $\qquad 10.5(1.5) - 8(1.5)(0.75) - M_C = 0 \quad M_C = 6.75 \text{ kN} \cdot \text{m}$ *Ans*

F4–3. $\downarrow + \Sigma M_B = 0;$ $\qquad \dfrac{1}{2}(6)(6)(3) - A_y(6) = 0 \quad A_y = 9.00 \text{ kN}$

$\xrightarrow{+} \Sigma F_x = 0;$ $\qquad\qquad A_x = 0$

Segment AC

$\xrightarrow{+} \Sigma F_x = 0;$ $\qquad\qquad N_C = 0$ *Ans*

$+\uparrow \Sigma F_y = 0;$ $\qquad 9.00 - \dfrac{1}{2}(3)(1.5) - V_C = 0 \quad V_C = 6.75 \text{ kN}$ *Ans*

$\downarrow + \Sigma M_C = 0;$ $\qquad M_C + \dfrac{1}{2}(3)(1.5)(0.5) - 9.00(1.5) = 0 \quad M_C = 12.4 \text{ kN} \cdot \text{m}$ *Ans*

F4–4. $\downarrow + \Sigma M_B = 0;$ $6(1)(0.5) - \dfrac{1}{2}(6)(1)(0.333) - A_y(1) = 0 \quad A_y = 2 \text{ kN}$

$\xrightarrow{+} \Sigma F_x = 0;$ $\qquad\qquad A_x = 0$

Segment AC

$\xrightarrow{+} \Sigma F_x = 0;$ $\qquad\qquad N_C = 0$ *Ans*

$+\uparrow \Sigma F_y = 0;$ $\qquad 2 - 6(0.5) - V_C = 0 \quad V_C = -1 \text{ kN}$ *Ans*

$\downarrow \Sigma M_C = 0;$ $\qquad M_C + 6(0.5)(0.25) - 2(0.5) = 0 \quad M_C = 0.25 \text{ kN} \cdot \text{m}$ *Ans*

F4–5. Reactions

$\downarrow + \Sigma M_A = 0;$ $\qquad F_B \sin 45°(3) - 5(6)(3) = 0 \quad F_B = 42.43 \text{ kN}$

$\xrightarrow{+} \Sigma F_x = 0;$ $\qquad 42.43 \cos 45° - A_x = 0 \quad A_x = 30.0 \text{ kN}$

$+\uparrow \Sigma F_y = 0;$ $\qquad 42.43 \sin 45° - 5(6) - A_y = 0 \quad A_y = 0$

Segment AC

$\xrightarrow{+} \Sigma F_x = 0;$ $\qquad N_C - 30.0 = 0 \quad N_C = 30.0 \text{ kN}$ *Ans*

$+\uparrow \Sigma F_y = 0;$ $\qquad -5(1.5) - V_C = 0 \quad V_C = -7.50 \text{ kN}$ *Ans*

$\downarrow + \Sigma M_C = 0;$ $\qquad M_C + 5(1.5)(0.75) = 0 \quad M_C = -5.625 \text{ kN} \cdot \text{m}$ *Ans*

F4–6.   Reactions

$\curvearrowleft + \Sigma M_A = 0;$ $B_y(5) - 4.5(3)(3.5) - 6(2) - 8(1) = 0$ $B_y = 13.45\ \text{kN}$

Segment CB

$\xrightarrow{+} \Sigma F_x = 0;$ $N_C = 0$ *Ans*

$+\uparrow \Sigma F_y = 0;$ $V_C + 13.45 - 4.5(2) = 0$ $V_C = -4.45\ \text{kN}$ *Ans*

$\curvearrowleft + \Sigma M_C = 0;$ $13.45(2) - 4.5(2)(1) - M_C = 0$ $M_C = 17.9\ \text{kN} \cdot \text{m}$ *Ans*

F4–7.   Left segment

$+\uparrow \Sigma F_y = 0;$ $-6 - \dfrac{1}{2}\left(\dfrac{18}{3}x\right)(x) - V = 0$ $V = \{-3x^2 - 6\}\ \text{kN}$ *Ans*

$\curvearrowleft + \Sigma M_O = 0;$ $M + \dfrac{1}{2}\left(\dfrac{18}{3}x\right)(x)\left(\dfrac{x}{3}\right) + 6x = 0$ $M = \{-x^3 - 6x\}\ \text{kN} \cdot \text{m}$ *Ans*

F4–8.   Reaction

$\curvearrowleft + \Sigma M_B = 0;$ $\dfrac{1}{2}(12)(6)(2) - A_y(6) = 0$ $A_y = 12.0\ \text{kN}$

Left segment

$+\uparrow \Sigma F_y = 0;$ $12.0 - \dfrac{1}{2}\left(\dfrac{12}{6}x\right)(x) - V = 0$ $V = \{12.0 - x^2\}\ \text{kN}$ *Ans*

$\curvearrowleft + \Sigma M_O = 0;$ $M + \dfrac{1}{2}\left(\dfrac{12}{6}x\right)(x)\left(\dfrac{x}{3}\right) - 12.0x = 0$ $M = \left\{12.0x - \dfrac{1}{3}x^3\right\}\ \text{kN} \cdot \text{m}$ *Ans*

F4–9.   Reactions

$\curvearrowleft + \Sigma M_A = 0;$ $B_y(8) - 8(4)(6) = 0$ $B_y = 24.0\ \text{kN}$

$\curvearrowleft + \Sigma M_B = 0;$ $8(4)(2) - A_y(8) = 0$ $A_y = 8.00\ \text{kN}$

$0 \le x < 4\ \text{m left segment}$

$+\uparrow \Sigma F_y = 0;$ $8.00 - V = 0$ $V = \{8\}\ \text{kN}$ *Ans*

$\curvearrowleft + \Sigma M_O = 0;$ $M - 8.00x = 0$ $M = \{8x\}\ \text{kN} \cdot \text{m}$ *Ans*

$4\ \text{m} < x < 8\ \text{m right segment}$

$+\uparrow \Sigma F_y = 0;$ $V + 24.0 - 8(8 - x) = 0$ $V = \{40 - 8x\}\ \text{kN}$ *Ans*

$\curvearrowleft + \Sigma M_O = 0;$ $24.0(8 - x) - 8(8 - x)\left(\dfrac{8 - x}{2}\right) - M = 0$ $M = \{-4x^2 + 40x - 64\}\ \text{kN} \cdot \text{m}$ *Ans*

F4–10. $0 \le x < 2\ \text{m}$

$+\uparrow \Sigma F_y = 0;$ $V = 0$ *Ans*

$\curvearrowleft + \Sigma M_O = 0;$ $M + 20 = 0$ $M = -20\ \text{kN} \cdot \text{m}$ *Ans*

$2\ \text{m} < x \le 4\ \text{m}$

$+\uparrow \Sigma F_y = 0;$ $-5(x - 2) - V = 0$ $V = \{10 - 5x\}\ \text{kN}$ *Ans*

$\curvearrowleft + \Sigma M_O = 0;$ $M + 5(x - 2)\left(\dfrac{x - 2}{2}\right) + 15 + 20 = 0$ $M = \left\{-\dfrac{5}{2}x^2 + 10x - 45\right\}\ \text{kN} \cdot \text{m}$ *Ans*

F4–11. Reactions

$+\uparrow \Sigma F_y = 0;$ $\qquad\qquad A_y - 5(2) - 15 = 0 \quad A_y = 25.0 \text{ kN}$

$\zeta + \Sigma M_A = 0;$ $\qquad M_A - 5(2)(1) - 15(4) = 0 \quad M_A = 70.0 \text{ kN} \cdot \text{m}$

$0 \le x < 2$ m left segment

$+\uparrow \Sigma F_y = 0;$ $\qquad\qquad 25.0 - 5x - V = 0 \quad V = \{25 - 5x\} \text{ kN}$ $\qquad$ *Ans*

$\zeta + \Sigma M_O = 0;$ $\qquad M + 5x\left(\dfrac{x}{2}\right) + 70.0 - 25.0x = 0 \quad M = \left\{-\dfrac{5}{2}x^2 + 25x - 70\right\} \text{ kN} \cdot \text{m}$ $\qquad$ *Ans*

2 m $< x \le 4$ m right segment

$+\uparrow \Sigma F_y = 0;$ $\qquad\qquad V - 15 = 0 \quad V = 15 \text{ kN}$ $\qquad$ *Ans*

$\zeta + \Sigma M_O = 0;$ $\qquad\qquad -M - 15(4 - x) = 0 \quad M = \{15x - 60\} \text{ kN} \cdot \text{m}$ $\qquad$ *Ans*

F4–12. Support reactions

$\zeta + \Sigma M_A = 0;$ $\qquad B_y(8) - 30(4)(2) - 90(4) = 0 \quad B_y = 75.0 \text{ kN}$

$\zeta + \Sigma M_B = 0;$ $\qquad 90(4) + 30(4)(6) - A_y(8) = 0 \quad A_y = 135.0 \text{ kN}$

$0 \le x < 4$ m left segment

$+\uparrow \Sigma F_y = 0;$ $\qquad\qquad 135.0 - 30x - V = 0 \quad V = \{135 - 30x\} \text{ kN}$ $\qquad$ *Ans*

$\zeta + \Sigma M_O = 0;$ $\qquad M + 30x\left(\dfrac{x}{2}\right) - 135.0x = 0 \quad M = \{-15x^2 + 135x\} \text{ kN} \cdot \text{m}$ $\qquad$ *Ans*

4 m $< x \le 8$ m right segment

$+\uparrow \Sigma F_y = 0;$ $\qquad\qquad V + 75.0 = 0 \quad V = \{-75 \text{ kN}\}$ $\qquad$ *Ans*

$\zeta + \Sigma M_O = 0;$ $\qquad\qquad 75.0(8 - x) - M = 0 \quad M = \{-75x + 600\} \text{ kN} \cdot \text{m}$ $\qquad$ *Ans*

F4–13.

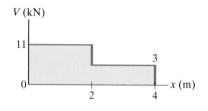
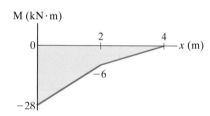

F4–14.

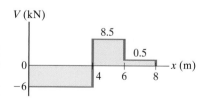
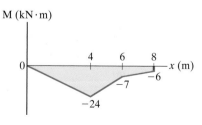

F4–15.

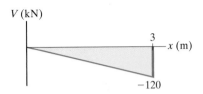

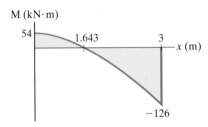

F4–16.

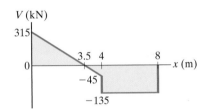

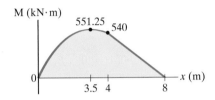

F4–17.

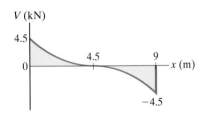

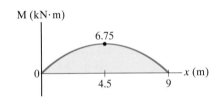

F4–18.

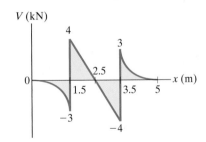

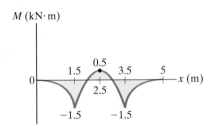

F4–19.

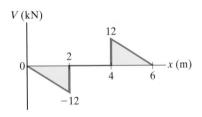

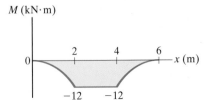

F4–20.

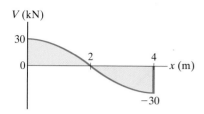

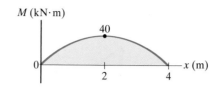

F6–1.

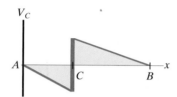

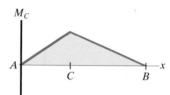

F6–2.

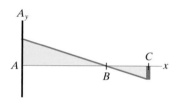

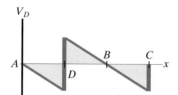

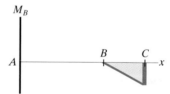

F6–3.

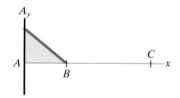

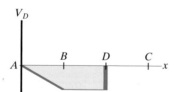

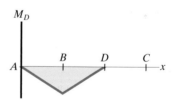

F6–4.

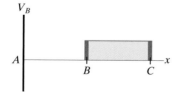

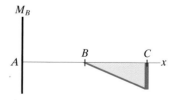

F6–5.

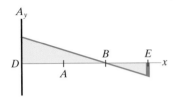
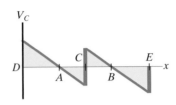
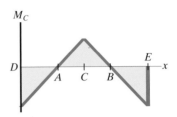

F6–6.

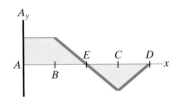
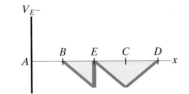
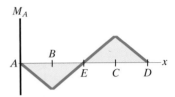

F6–7.

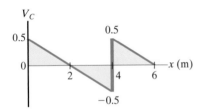
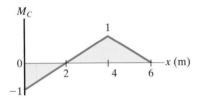

$$(M_C)_{\max(+)} = 8(1) + \left[\frac{1}{2}(6-2)(1)\right](1.5) + \left[\frac{1}{2}(2)(-1)\right](2) + \left[\frac{1}{2}(6-2)(1)\right](2)$$

$$= 13.0 \text{ kN} \cdot \text{m} \qquad\qquad Ans$$

$$(V_C)_{\max(+)} = 8(0.5) + \left[\frac{1}{2}(2)(0.5)\right](1.5) + \left[\frac{1}{2}(6-4)(0.5)\right](1.5)$$

$$+ \left[\frac{1}{2}(2)(0.5)\right](2) + \left[\frac{1}{2}(4-2)(-0.5)\right](2) + \left[\frac{1}{2}(6-4)(0.5)\right](2)$$

$$= 6.50 \text{ kN} \qquad\qquad Ans$$

F6–8.

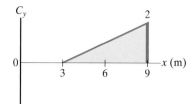 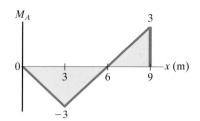

a) $(C_y)_{\max(+)} = 6(2) + \left[\frac{1}{2}(9-3)(2)\right](2) + \left[\frac{1}{2}(9-3)(2)\right](4) = 48 \text{ kN}$ *Ans*

b) $(M_A)_{\max(-)} = 6(-3) + \left[\frac{1}{2}(6-0)(-3)\right](2) + \left[\frac{1}{2}(6-0)(-3)\right](4) + \left[\frac{1}{2}(9-6)(3)\right](4)$

$\qquad = -54 \text{ kN} \cdot \text{m}$ *Ans*

F8–1.

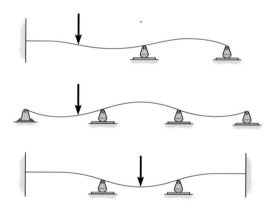

F8–2.

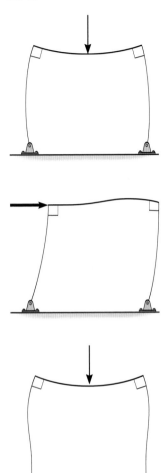

F8–3.

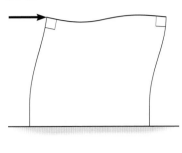

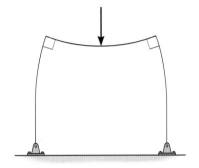

F8–4.   For $0 \le x_1 < \dfrac{L}{2}$

$$M_1 = \frac{P}{2}x_1$$

$$EI\frac{d^2 v_1}{dx_1^2} = \frac{P}{2}x_1$$

$$EI\frac{dv_1}{dx_1} = \frac{P}{4}x_1^2 + C_1 \qquad\qquad (1)$$

$$EI\,v_1 = \frac{P}{12}x_1^3 + C_1 x_1 + C_2 \qquad\qquad (2)$$

For $\dfrac{L}{2} < x_2 \le L$

$$M_2 = \frac{P}{2}(L - x_2) = \frac{PL}{2} - \frac{P}{2}x_2$$

$$EI\frac{d^2 v_2}{dx_2^2} = \frac{PL}{2} - \frac{P}{2}x_2$$

$$EI\frac{dv_2}{dx_2} = \frac{PL}{2}x_2 - \frac{P}{4}x_2^2 + C_3 \qquad\qquad (3)$$

$$EI\,v_2 = \frac{PL}{4}x_2^2 - \frac{P}{12}x_2^3 + C_3 x_2 + C_4 \qquad\qquad (4)$$

$v_1 = 0$ at $x_1 = 0$. From Eq (2), $C_2 = 0$

$\dfrac{dv_1}{dx_1} = 0$ at $x_1 = \dfrac{L}{2}$. From Eq (1), $C_1 = -\dfrac{PL^2}{16}$

$\dfrac{dv_2}{dx_2} = 0$ at $x_2 = \dfrac{L}{2}$. From Eq (3), $C_3 = -\dfrac{3PL^2}{16}$

$v_2 = 0$ at $x_2 = L$. From Eq (4), $C_4 = \dfrac{PL^3}{48}$

$$v_1 = \frac{Px_1}{48EI}(4x_1^2 - 3L^2) \qquad\qquad\qquad Ans$$

$$v_2 = \frac{P}{48EI}(-4x_2^3 + 12Lx_2^2 - 9L^2 x_2 + L^3) \qquad\qquad Ans$$

F8–5. $M = Px - PL$

$$EI\frac{d^2 v}{dx^2} = Px - PL$$

$$EI\frac{dv}{dx} = \frac{P}{2}x^2 - PLx + C_1 \qquad\qquad (1)$$

$$EI\,v = \frac{P}{6}x^3 - \frac{PL}{2}x^2 + C_1 x + C_2 \qquad\qquad (2)$$

$\dfrac{dv}{dx} = 0$ at $x = 0$. From Eq (1), $C_1 = 0$

$v = 0$ at $x = 0$. From Eq (2), $C_2 = 0$

$$v = \frac{Px}{6EI}(x^2 - 3Lx) \qquad\qquad\qquad Ans$$

F8–6. $M = M_0 - \dfrac{M_0}{L}x$

$EI\dfrac{d^2v}{dx^2} = M_0 - \dfrac{M_0}{L}x$

$EI\dfrac{dv}{dx} = M_0 x - \dfrac{M_0}{2L}x^2 + C_1$

$EI\,v = \dfrac{M_0}{2}x^2 - \dfrac{M_0}{6L}x^3 + C_1 x + C_2$ (1)

$v = 0$ at $x = 0$.   From Eq (1), $C_2 = 0$

$v = 0$ at $x = L$.   From Eq (1), $C_1 = -\dfrac{M_0 L}{3}$

$v = \dfrac{M_0}{6EIL}(-x^3 + 3Lx^2 - 2L^2 x)$ *Ans*

F8–7.   For $0 \le x_1 < \dfrac{L}{2}$

$M = -\dfrac{M_0}{L}x_1$

$EI\dfrac{d^2v_1}{dx_1^2} = -\dfrac{M_0}{L}x_1$

$EI\dfrac{dv_1}{dx_1} = -\dfrac{M_0}{2L}x_1^2 + C_1$ (1)

$EI\,v_1 = -\dfrac{M_0}{6L}x_1^3 + C_1 x_1 + C_2$ (2)

For $\dfrac{L}{2} < x_2 \le L$

$M = M_0 - \dfrac{M_0}{L}x_2$

$EI\dfrac{d^2v_2}{dy_2^2} = M_0 - \dfrac{M_0}{L}x_2$

$EI\dfrac{dv_2}{dx_2} = M_0 x_2 - \dfrac{M_0}{2L}x_2^2 + C_3$ (3)

$EI\,v_2 = \dfrac{M_0}{2}x_2^2 - \dfrac{M_0}{6L}x_2^3 + C_3 x_2 + C_4$ (4)

$v_1 = 0$ at $x_1 = 0$.   From Eq (2), $C_2 = 0$

$v_2 = 0$ at $x_2 = L$.   From Eq (4), $0 = C_3 L + C_4 + \dfrac{M_0 L^2}{3}$ (5)

$\dfrac{dv_1}{dx_1} = \dfrac{dv_2}{dx_2}$ at $x_1 = x_2 = \dfrac{L}{2}$.   From Eqs (1) and (3), $C_1 - C_3 = \dfrac{M_0 L}{2}$ (6)

$v_1 = v_2$ at $x_1 = x_2 = \dfrac{L}{2}$.   From Eqs (2) and (4), $C_1 L - C_3 L - 2C_4 = \dfrac{M_0 L^2}{4}$ (7)

Solving Eqs (5), (6) and (7)

$$C_4 = \frac{M_0 L^2}{8} \quad C_3 = -\frac{11 M_0 L}{24} \quad C_1 = \frac{M_0 L}{24}$$

$$v_1 = \frac{M_0}{24EIL}(-4x_1^3 + L^2 x_1)$$ *Ans*

$$v_2 = \frac{M_0}{24EIL}(-4x_2^3 + 12Lx_2^2 - 11L^2 x_2 + 3L^3)$$ *Ans*

F8–8. $M = -\dfrac{w}{2}x^2 + wLx - \dfrac{wL^2}{2}$

$$EI\frac{d^2 v}{dx^2} = -\frac{w}{2}x^2 + wLx - \frac{wL^2}{2}$$

$$EI\frac{dv}{dx} = -\frac{w}{6}x^3 + \frac{wL}{2}x^2 - \frac{wL^2}{2}x + C_1 \quad (1)$$

$$EI\,v = -\frac{w}{24}x^4 + \frac{wL}{6}x^3 - \frac{wL^2}{4}x^2 + C_1 x + C_2 \quad (2)$$

$\dfrac{dv}{dx} = 0$ at $x = 0$.   From Eq (1), $C_1 = 0$

$v = 0$ at $x = 0$.   From Eq (2), $C_2 = 0$

$$v = \frac{w}{24EI}(-x^4 + 4Lx^3 - 6L^2 x^2)$$ *Ans*

F8–9. $M = -\dfrac{w_0}{6L}x^3$

$$EI\frac{d^2 v}{dx^2} = -\frac{w_0}{6L}x^3$$

$$EI\frac{dv}{dx} = -\frac{w_0}{24L}x^4 + C_1 \quad (1)$$

$$EI\,v = -\frac{w_0}{120L}x^5 + C_1 x + C_2 \quad (2)$$

$\dfrac{dv}{dx} = 0$ at $x = L$.   From Eq (1), $C_1 = \dfrac{w_0 L^3}{24}$

$v = 0$ at $x = L$.   From Eq (2), $C_2 = -\dfrac{w_0 L^4}{30}$

$$v = \frac{w_0}{120EIL}(-x^5 + 5L^4 x - 4L^5)$$ *Ans*

F8–10.

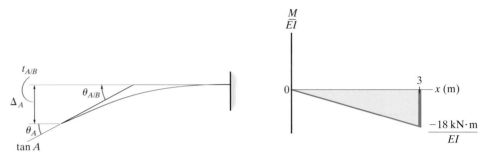

$$\theta_A = |\theta_{A/B}| = \left| \frac{1}{2} \left(\frac{-18 \text{ kN} \cdot \text{m}}{EI} \right) (3 \text{ m}) \right| = \frac{27 \text{ kN} \cdot \text{m}^2}{EI} \ \diagdown \qquad\qquad Ans$$

$$\Delta_A = |t_{A/B}| = \left\| \left[\frac{1}{2} \left(\frac{-18 \text{ kN} \cdot \text{m}}{EI} \right) (3 \text{ m}) \right] \left[\frac{2}{3} (3 \text{ m}) \right] \right\| = \frac{54 \text{ kN} \cdot \text{m}^2}{EI} \downarrow \qquad Ans$$

F8–11.

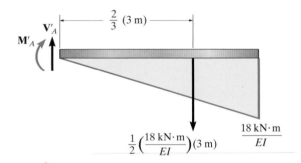

$$+\uparrow \Sigma F_y = 0; \quad V'_A - \frac{1}{2} \left(\frac{18 \text{ kN} \cdot \text{m}}{EI} \right) (3 \text{ m}) = 0 \qquad \theta_A = \frac{27 \text{ kN} \cdot \text{m}^2}{EI} \ \diagdown \qquad Ans$$

$$\zeta + \Sigma M_A = 0; \quad -M'_A - \left[\frac{1}{2} \left(\frac{18 \text{ kN} \cdot \text{m}}{EI} \right) (3 \text{ m}) \right] \left[\frac{2}{3} (3 \text{ m}) \right] = 0$$

$$M'_A = \Delta_A = -\frac{54 \text{ kN} \cdot \text{m}^3}{EI} = \frac{54 \text{ kN} \cdot \text{m}^3}{EI} \downarrow \qquad Ans$$

F8–12.

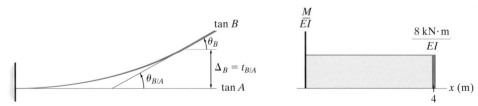

$$\theta_B = |\theta_{B/A}| = \left(\frac{8 \text{ kN} \cdot \text{m}}{EI} \right) (4 \text{ m}) = \frac{32 \text{ kN} \cdot \text{m}^2}{EI} \ \triangle \qquad\qquad Ans$$

$$\Delta_B = |t_{B/A}| = \left[\left(\frac{8 \text{ kN} \cdot \text{m}}{EI} \right) (4 \text{ m}) \right] \left[\frac{1}{2} (4 \text{ m}) \right] = \frac{64 \text{ kN} \cdot \text{m}^3}{EI} \uparrow \qquad Ans$$

F8–13.

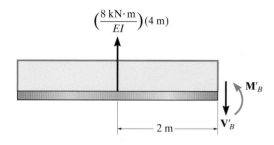

$$+\uparrow\Sigma F_y = 0; \qquad \left(\frac{8\ \text{kN}\cdot\text{m}}{EI}\right)(4\ \text{m}) - V'_B = 0 \qquad \theta_B = \frac{32\ \text{kN}\cdot\text{m}^2}{EI}\ \angle \qquad\qquad Ans$$

$$\zeta+\Sigma M_B = 0; \quad M'_B - \left[\left(\frac{8\ \text{kN}\cdot\text{m}}{EI}\right)(4\ \text{m})\right](2\ \text{m}) = 0 \quad M'_B = \Delta_B = \frac{64\ \text{kN}\cdot\text{m}^3}{EI}\uparrow \qquad Ans$$

F8–14.

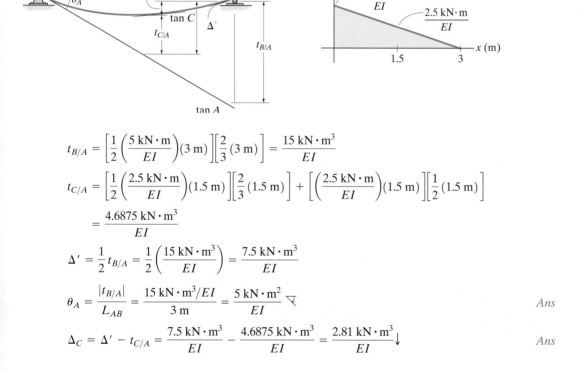

$$t_{B/A} = \left[\frac{1}{2}\left(\frac{5\ \text{kN}\cdot\text{m}}{EI}\right)(3\ \text{m})\right]\left[\frac{2}{3}(3\ \text{m})\right] = \frac{15\ \text{kN}\cdot\text{m}^3}{EI}$$

$$t_{C/A} = \left[\frac{1}{2}\left(\frac{2.5\ \text{kN}\cdot\text{m}}{EI}\right)(1.5\ \text{m})\right]\left[\frac{2}{3}(1.5\ \text{m})\right] + \left[\left(\frac{2.5\ \text{kN}\cdot\text{m}}{EI}\right)(1.5\ \text{m})\right]\left[\frac{1}{2}(1.5\ \text{m})\right]$$

$$= \frac{4.6875\ \text{kN}\cdot\text{m}^3}{EI}$$

$$\Delta' = \frac{1}{2}t_{B/A} = \frac{1}{2}\left(\frac{15\ \text{kN}\cdot\text{m}^3}{EI}\right) = \frac{7.5\ \text{kN}\cdot\text{m}^3}{EI}$$

$$\theta_A = \frac{|t_{B/A}|}{L_{AB}} = \frac{15\ \text{kN}\cdot\text{m}^3/EI}{3\ \text{m}} = \frac{5\ \text{kN}\cdot\text{m}^2}{EI}\ \triangledown \qquad\qquad Ans$$

$$\Delta_C = \Delta' - t_{C/A} = \frac{7.5\ \text{kN}\cdot\text{m}^3}{EI} - \frac{4.6875\ \text{kN}\cdot\text{m}^3}{EI} = \frac{2.81\ \text{kN}\cdot\text{m}^3}{EI}\downarrow \qquad\qquad Ans$$

F8–15.

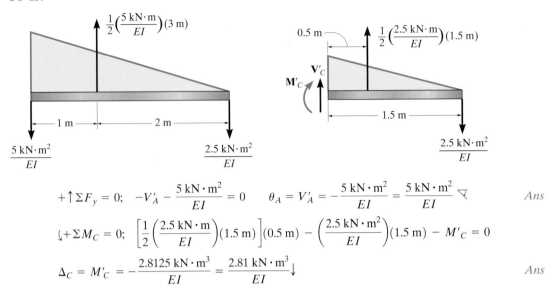

$$+\uparrow \Sigma F_y = 0; \quad -V'_A - \frac{5 \text{ kN} \cdot \text{m}^2}{EI} = 0 \qquad \theta_A = V'_A = -\frac{5 \text{ kN} \cdot \text{m}^2}{EI} = \frac{5 \text{ kN} \cdot \text{m}^2}{EI} \, \triangledown \qquad Ans$$

$$\left(+\Sigma M_C = 0; \quad \left[\frac{1}{2}\left(\frac{2.5 \text{ kN} \cdot \text{m}}{EI}\right)(1.5 \text{ m})\right](0.5 \text{ m}) - \left(\frac{2.5 \text{ kN} \cdot \text{m}^2}{EI}\right)(1.5 \text{ m}) - M'_C = 0$$

$$\Delta_C = M'_C = -\frac{2.8125 \text{ kN} \cdot \text{m}^3}{EI} = \frac{2.81 \text{ kN} \cdot \text{m}^3}{EI} \downarrow \qquad\qquad Ans$$

F8–16.

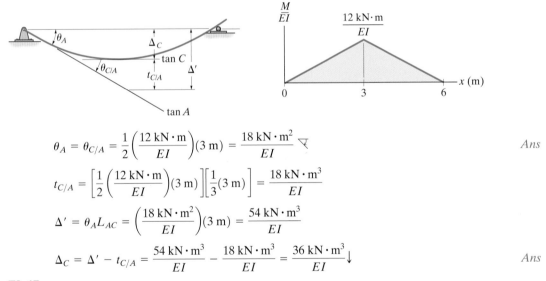

$$\theta_A = \theta_{C/A} = \frac{1}{2}\left(\frac{12 \text{ kN} \cdot \text{m}}{EI}\right)(3 \text{ m}) = \frac{18 \text{ kN} \cdot \text{m}^2}{EI} \, \triangledown \qquad\qquad Ans$$

$$t_{C/A} = \left[\frac{1}{2}\left(\frac{12 \text{ kN} \cdot \text{m}}{EI}\right)(3 \text{ m})\right]\left[\frac{1}{3}(3 \text{ m})\right] = \frac{18 \text{ kN} \cdot \text{m}^3}{EI}$$

$$\Delta' = \theta_A L_{AC} = \left(\frac{18 \text{ kN} \cdot \text{m}^2}{EI}\right)(3 \text{ m}) = \frac{54 \text{ kN} \cdot \text{m}^3}{EI}$$

$$\Delta_C = \Delta' - t_{C/A} = \frac{54 \text{ kN} \cdot \text{m}^3}{EI} - \frac{18 \text{ kN} \cdot \text{m}^3}{EI} = \frac{36 \text{ kN} \cdot \text{m}^3}{EI} \downarrow \qquad Ans$$

F8–17.

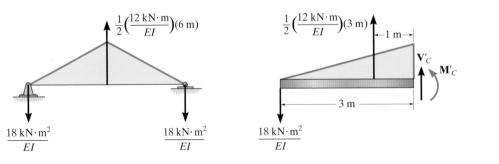

$$+\uparrow \Sigma F_y = 0; \quad -V'_A - \frac{18 \text{ kN} \cdot \text{m}^2}{EI} = 0 \qquad V'_A = \theta_A = -\frac{18 \text{ kN} \cdot \text{m}^2}{EI} = \frac{18 \text{ kN} \cdot \text{m}^2}{EI} \, \triangledown$$ *Ans*

$$\downdownarrows + \Sigma M_C = 0; \quad M'_C + \left(\frac{18 \text{ kN} \cdot \text{m}^2}{EI}\right)(3 \text{ m}) - \left[\frac{1}{2}\left(\frac{12 \text{ kN} \cdot \text{m}}{EI}\right)(3 \text{ m})\right](1 \text{ m}) = 0$$

$$M'_C = \Delta_C = -\frac{36 \text{ kN} \cdot \text{m}^3}{EI} = \frac{36 \text{ kN} \cdot \text{m}^3}{EI} \downarrow$$ *Ans*

F8–18.

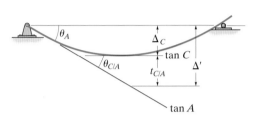

$$\theta_A = \theta_{C/A} = \frac{1}{2}\left(\frac{8 \text{ kN} \cdot \text{m}}{EI}\right)(2 \text{ m}) + \left(\frac{8 \text{ kN} \cdot \text{m}}{EI}\right)(2 \text{ m}) = \frac{24 \text{ kN} \cdot \text{m}^2}{EI} \, \triangledown$$ *Ans*

$$t_{C/A} = \left[\frac{1}{2}\left(\frac{8 \text{ kN} \cdot \text{m}}{EI}\right)(2 \text{ m})\right]\left[2 \text{ m} + \frac{1}{3}(2 \text{ m})\right] + \left[\left(\frac{8 \text{ kN} \cdot \text{m}}{EI}\right)(2 \text{ m})\right](1 \text{ m}) = \frac{37.33 \text{ kN} \cdot \text{m}^3}{EI}$$

$$\Delta' = \theta_A L_{AC} = \left(\frac{24 \text{ kN} \cdot \text{m}^2}{EI}\right)(4 \text{ m}) = \frac{96 \text{ kN} \cdot \text{m}^3}{EI}$$

$$\Delta_C = \Delta' - t_{C/A} = \frac{96 \text{ kN} \cdot \text{m}^2}{EI} - \frac{37.33 \text{ kN} \cdot \text{m}^3}{EI} = \frac{58.7 \text{ kN} \cdot \text{m}^3}{EI} \downarrow$$ *Ans*

F8–19.

$$\left(\frac{8 \text{ kN} \cdot \text{m}}{EI}\right)(4\text{m}) + \frac{1}{2}\left(\frac{8 \text{ kN} \cdot \text{m}}{EI}\right)(4\text{m}) = \frac{48 \text{ kN} \cdot \text{m}^2}{EI} \qquad \frac{1}{2}\left(\frac{8 \text{ kN} \cdot \text{m}}{EI}\right)(2 \text{ m}) \quad \left(\frac{8 \text{ kN} \cdot \text{m}}{EI}\right)(2 \text{ m})$$

2.667 m

V'_C

M'_C

1 m

4 m   4 m

4 m

$\dfrac{24 \text{ kN} \cdot \text{m}^2}{EI}$ $\dfrac{24 \text{ kN} \cdot \text{m}^2}{EI}$ $\dfrac{24 \text{ kN} \cdot \text{m}^2}{EI}$

$$+\uparrow \Sigma F_y = 0; \quad -V'_A - \frac{24 \text{ kN} \cdot \text{m}^2}{EI} = 0 \qquad \theta_A = V'_A = \frac{24 \text{ kN} \cdot \text{m}^2}{EI} \, \triangledown$$ *Ans*

$$\downdownarrows + \Sigma M_C = 0; \quad M'_C + \left(\frac{24 \text{ kN} \cdot \text{m}^2}{EI}\right)(4 \text{ m}) - \left[\frac{1}{2}\left(\frac{8 \text{ kN} \cdot \text{m}}{EI}\right)(2 \text{ m})\right](2.667 \text{ m}) - \left(\frac{8 \text{ kN} \cdot \text{m}}{EI}\right)(2 \text{ m})(1 \text{ m}) = 0$$

$$\Delta_C = M'_C = \frac{58.7 \text{ kN} \cdot \text{m}^3}{EI} \downarrow$$ *Ans*

F8–20.

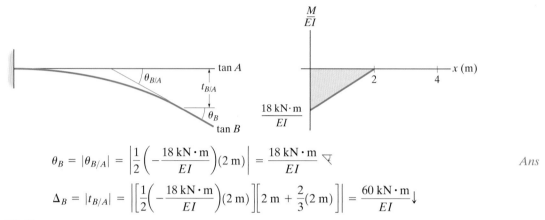

$$\theta_B = |\theta_{B/A}| = \left| \frac{1}{2} \left(-\frac{18 \text{ kN} \cdot \text{m}}{EI} \right) (2 \text{ m}) \right| = \frac{18 \text{ kN} \cdot \text{m}}{EI} \; \triangledown \qquad Ans$$

$$\Delta_B = |t_{B/A}| = \left| \left[\frac{1}{2} \left(-\frac{18 \text{ kN} \cdot \text{m}}{EI} \right) (2 \text{ m}) \right] \left[2 \text{ m} + \frac{2}{3} (2 \text{ m}) \right] \right| = \frac{60 \text{ kN} \cdot \text{m}}{EI} \downarrow$$

F8–21.

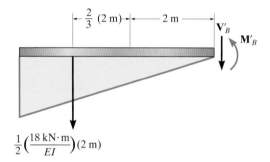

$$+\uparrow \Sigma F_y = 0; \quad -V'_B - \frac{1}{2} \left(\frac{18 \text{ kN} \cdot \text{m}}{EI} \right) (2 \text{ m}) = 0$$

$$\theta_B = -\frac{18 \text{ kN} \cdot \text{m}^2}{EI} = \frac{18 \text{ kN} \cdot \text{m}^2}{EI} \; \triangledown \qquad Ans$$

$$\zeta + \Sigma M_B = 0; \quad M'_B + \left[\frac{1}{2} \left(\frac{18 \text{ kN} \cdot \text{m}}{EI} \right) (2 \text{ m}) \right] \left[\frac{2}{3} (2 \text{ m}) + 2 \text{ m} \right] = 0$$

$$M'_B = \Delta_B = -\frac{60 \text{ kN} \cdot \text{m}^3}{EI} = \frac{60 \text{ kN} \cdot \text{m}^3}{EI} \downarrow \qquad Ans$$

F9–1.

Member	n (N)	N (N)	L (m)	nNL ($N^2 \cdot m$)
AB	−1.667	−1000	5	8333.33
AC	1	600	3	1800.00
BC	1.333	800	4	4266.67
				Σ 14 400

Thus,

$$1 \text{ N} \cdot \Delta_{B_v} = \sum \frac{nNL}{AE} = \frac{14\ 400\ N^2 \cdot m}{AE}$$

$$\Delta_{B_v} = \frac{14\ 400\ N \cdot m}{AE} \downarrow \qquad\qquad Ans$$

F9–2.

Member	N	$\dfrac{\delta N}{\delta P}$	N (P = 600 N)	L (m)	$N\left(\dfrac{\delta N}{\delta P}\right)L$ ($N \cdot m$)
AB	−1.667P	−1.667	−1000	5	8333.33
AC	P	1	600	3	1800.00
BC	1.333P	1.333	800	4	4266.67
					Σ 14 400

$$\Delta_{B_v} = \sum N\left(\frac{\delta N}{\delta P}\right)\frac{L}{AE} = \frac{14\ 400\ N \cdot m}{AE} \downarrow \qquad\qquad Ans$$

F9–3.

Member	n (kN)	N (kN)	L (m)	nNL ($kN^2 \cdot m$)
AB	1	−4.041	2	−8.0829
AC	0	8.0829	2	0
BC	0	−8.0829	2	0
CD	0	8.0829	1	0
				Σ −8.0829

Thus,

$$1 \text{ kN} \cdot \Delta_{A_h} = \sum \frac{nNL}{AE} = -\frac{8.0829\ kN^2 \cdot m}{AE}$$

$$\Delta_{A_h} = -\frac{8.08\ kN \cdot m}{AE} = \frac{8.08\ kN \cdot m}{AE} \rightarrow \qquad\qquad Ans$$

F9–4.

Member	N (kN)	$\dfrac{\delta N}{\delta P}$	N (P = 0) (kN)	L (m)	$N\left(\dfrac{\delta N}{\delta P}\right)L$ ($kN \cdot m$)
AB	P − 4.041	1	−4.041	2	−8.083
AC	8.083	0	8.083	2	0
BC	−8.083	0	−8.083	2	0
CD	8.083	0	8.083	1	0
					Σ −8.083

$$\Delta_{A_h} = \sum N\left(\frac{\delta N}{\delta P}\right)\frac{L}{AE} = -\frac{8.083\ kN \cdot m}{AE} = \frac{8.08\ kN \cdot m}{AE} \rightarrow \qquad\qquad Ans$$

F9–5.

Member	n (kN)	N (kN)	L	nNL (kN$^2 \cdot$m)
AB	0	0	3	0
AC	1.414	8.485	$3\sqrt{2}$	50.91
BC	-1	-6	3	18.00
AD	0	-6	3	0
CD	-1	0	3	0
				Σ 68.91

$$1 \text{ kN} \cdot \Delta_{D_h} = \Sigma \frac{nNL}{AE} = \frac{68.91 \text{ kN}^2 \cdot \text{m}}{AE}$$

$$\Delta_{D_h} = \frac{68.9 \text{ kN} \cdot \text{m}}{AE} \rightarrow$$ *Ans*

F9–6.

Member	N (kN)	$\dfrac{\delta N}{\delta P}$	N ($P = 0$) (kN)	L (m)	$N\left(\dfrac{\delta N}{\delta P}\right)L$ (kN$\cdot$m)
AB	0	0	0	3	0
AC	$\sqrt{2}(P + 6)$	$\sqrt{2}$	$6\sqrt{2}$	$3\sqrt{2}$	50.91
BC	$-(P + 6)$	-1	-6	3	18.00
AD	-6	0	-6	3	0
CD	$-P$	-1	0	3	0
					Σ 68.91

$$\Delta_{D_h} = \Sigma N\left(\frac{\delta N}{\delta P}\right)\frac{L}{AE}$$

$$= \frac{68.9 \text{ kN} \cdot \text{m}}{AE} \rightarrow$$ *Ans*

F9–7.

Member	n (kN)	N (kN)	L (m)	nNL (kN$^2 \cdot$m)
AB	0.375	18.75	3	21.09
BC	0.375	18.75	3	21.09
AD	-0.625	-31.25	5	97.66
CD	-0.625	-31.25	5	97.66
BD	0	50	4	0
				Σ 237.5

$$1 \text{ kN} \cdot \Delta_{D_v} = \Sigma \frac{nNL}{AE} = \frac{237.5 \text{ kN}^2 \cdot \text{m}}{AE}$$

$$\Delta_{D_v} = \frac{237.5 \text{ kN} \cdot \text{m}}{AE} \downarrow$$ *Ans*

F9–8.

Member	N (kN)	$\dfrac{\delta N}{\delta P}$	$N\ (P = 0)$ (kN)	L (m)	$N\left(\dfrac{\delta N}{\delta P}\right)L$ (kN·m)
AB	$\frac{3}{8}P + 18.75$	0.375	18.75	3	21.09
BC	$\frac{3}{8}P + 18.75$	0.375	18.75	3	21.09
AD	$-(\frac{5}{8}P + 31.25)$	-0.625	-31.25	5	97.66
CD	$-(\frac{5}{8}P + 31.25)$	-0.625	-31.25	5	97.66
BD	50	0	50	4	0

$$\Sigma\ 237.5$$

$$\Delta_{D_v} = \sum N\left(\frac{\delta N}{\delta P}\right)\frac{L}{AE} = \frac{237.5\ \text{kN·m}}{AE}\downarrow \qquad\qquad Ans$$

F9–9.

Member	n (kN)	N (kN)	L (m)	nNL (kN^2·m)
AB	0	-6	1.5	0
BC	0	-6	1.5	0
BD	1	0	2	0
CD	0	10	2.5	0
AD	-1.25	-10	2.5	31.25
DE	0.75	12	1.5	13.5

$$\Sigma\ 44.75$$

$$1\ \text{kN}\cdot\Delta_{B_v} = \sum\frac{nNL}{AE} = \frac{44.75\ \text{kN}^2\cdot\text{m}}{AE}$$

$$\Delta_{B_v} = \frac{44.75\ \text{kN·m}}{AE}\downarrow \qquad\qquad Ans$$

F9–10.

Member	N (kN)	$\dfrac{\delta N}{\delta P}$	$N\ (P = 0)$ (kN)	L (m)	$N\left(\dfrac{\delta N}{\delta P}\right)L$ (kN·m)
AB	-6	0	-6	1.5	0
BC	-6	0	-6	1.5	0
BD	P	1	0	2	0
CD	10	0	10	2.5	0
AD	$-(1.25P + 10)$	-1.25	-10	2.5	31.25
DE	$0.75P + 12$	0.75	12	1.5	13.5

$$\Sigma\ 44.75$$

$$\Delta_{B_v} = \sum N\left(\frac{\delta N}{\delta p}\right)\frac{L}{AE} = \frac{44.75\ \text{kN·m}}{AE}\downarrow \qquad\qquad Ans$$

F9–11.

Member	n (kN)	N (kN)	L (m)	nNL (kN$^2 \cdot$m)
AB	0.5	50	2	50.00
DE	0.5	50	2	50.00
BC	0.5	50	2	50.00
CD	0.5	50	2	50.00
AH	−0.7071	−70.71	$2\sqrt{2}$	141.42
EF	−0.7071	−70.71	$2\sqrt{2}$	141.42
BH	0	30	2	0
DF	0	30	2	0
CH	0.7071	28.28	$2\sqrt{2}$	56.57
CF	0.7071	28.28	$2\sqrt{2}$	56.57
CG	0	0	2	0
GH	−1	−70	2	140.00
FG	−1	−70	2	140.00

$$\Sigma\ 878.98$$

$$1\ \text{kN} \cdot \Delta_{C_v} = \sum \frac{nNL}{AE} = \frac{875.98\ \text{kN}^2 \cdot \text{m}}{AE}$$

$$\Delta_{C_v} = \frac{876\ \text{kN} \cdot \text{m}}{AE} \downarrow$$ *Ans*

F9–12.

Member	N (kN)	$\dfrac{\delta N}{\delta P}$	$N\ (P = 40\ \text{kN})$	L (m)	$N\left(\dfrac{\delta N}{\delta P}\right)L$ (kN·m)
AB	$0.5P + 30$	0.5	50	2	50.00
DE	$0.5P + 30$	0.5	50	2	50.00
BC	$0.5P + 30$	0.5	50	2	50.00
CD	$0.5P + 30$	0.5	50	2	50.00
AH	$-(0.7071P + 42.43)$	−0.7071	−70.71	$2\sqrt{2}$	141.42
EF	$-(0.7071P + 42.43)$	−0.7071	−70.71	$2\sqrt{2}$	141.42
BH	30	0	30	2	0
DF	30	0	30	2	0
CH	$0.7071P$	0.7071	28.28	$2\sqrt{2}$	56.57
CF	$0.7071P$	0.7071	28.28	$2\sqrt{2}$	56.57
CG	0	0	0	2	0
GH	$-(P + 30)$	−1	−70	2	140.00
FG	$-(P + 30)$	−1	−70	2	140.00

$$\Sigma\ 875.98$$

$$\Delta_{C_v} = \sum N\left(\frac{\delta N}{\delta P}\right)\frac{L}{AE} = \frac{875.98\ \text{kN} \cdot \text{m}}{AE}$$

$$\Delta_{C_v} = \frac{876\ \text{kN} \cdot \text{m}}{AE} \downarrow$$ *Ans*

F9–13. For the slope,

$$1 \text{ kN} \cdot \text{m} \cdot \theta_A = \int_0^2 \frac{m_\theta M}{EI} dx = \int_0^{3\text{m}} \frac{(-1)(-30x)}{EI} dx = \frac{135 \text{ kN}^2 \cdot \text{m}^3}{EI}$$

$$\theta_A = \frac{135 \text{ kN} \cdot \text{m}^2}{EI} \, \triangledown \qquad\qquad\qquad Ans$$

For the displacement,

$$1 \text{ kN} \cdot \Delta_{A_v} = \int_0^L \frac{mM}{EI} dx = \int_0^{3\text{m}} \frac{(-x)(-30x)}{EI} dx = \frac{270 \text{ kN}^2 \cdot \text{m}^3}{EI}$$

$$\Delta_{A_v} = \frac{270 \text{ kN} \cdot \text{m}^3}{EI} \downarrow \qquad\qquad\qquad Ans$$

F9–14. For the slope, $M = -30x - M'$. Then, $\dfrac{\partial M}{\partial M'} = -1$. Set $M' = 0$. Then, $M = (-30x) \text{ kN} \cdot \text{m}$.

$$\theta_A = \int_0^L M \left(\frac{\partial M}{\partial M'} \right) \frac{dx}{EI} = \int_0^{3\text{ m}} \frac{(-30x)(-1)dx}{EI} = \frac{135 \text{ kN} \cdot \text{m}^2}{EI} \, \triangledown \qquad Ans$$

For the displacement, $M = -Px$. Then $\dfrac{\partial M}{\partial P} = -x$.

Set $P = 30 \text{ kN}$. Then $M = (-30x) \text{ kN} \cdot \text{m}$.

$$\Delta_{A_v} = \int_0^L M \left(\frac{\partial M}{\partial P} \right) \frac{dx}{EI} = \int_0^{3\text{m}} \frac{(-30x)(-x)dx}{EI} = \frac{270 \text{ kN} \cdot \text{m}^3}{EI} \downarrow$$

F9–15. For the slope, $m_\theta = 1 \text{ kN} \cdot \text{m}$ and $M = 4 \text{ kN} \cdot \text{m}$.

$$1 \text{ kN} \cdot \text{m} \cdot \theta_A = \int_0^L \frac{m_\theta M}{EI} dx = \int_0^{3\text{m}} \frac{(1)(4)dx}{EI} = \frac{12 \text{ kN}^2 \cdot \text{m}^3}{EI}$$

$$\theta_A = \frac{12 \text{ kN} \cdot \text{m}^2}{EI} \, \triangle \qquad\qquad\qquad Ans$$

For the displacement, $m = x \text{ kN} \cdot \text{m}$ and $M = 4 \text{ kN} \cdot \text{m}$.

$$1 \text{ kN} \cdot \Delta_{A_v} = \int_0^L \frac{mM}{EI} dx = \int_0^{3\text{m}} \frac{x(4)dx}{EI} = \frac{18 \text{ kN}^2 \cdot \text{m}^3}{EI}$$

$$\Delta_{A_v} = \frac{18 \text{ kN} \cdot \text{m}^3}{EI} \uparrow \qquad\qquad\qquad Ans$$

F9–16. For the slope, $M = M'$. Then $\dfrac{\partial M}{\partial M} = 1$.

Set $M' = 4 \text{ kN} \cdot \text{m}$. Then $M = 4 \text{ kN} \cdot \text{m}$.

$$\theta_A = \int_0^L M \left(\frac{\partial M}{\partial M'} \right) \frac{dx}{EI} = \int_0^{3\text{m}} \frac{4(1)dx}{EI} = \frac{12 \text{ kN} \cdot \text{m}^2}{EI} \, \triangle \qquad Ans$$

For the displacement, $M = (Px + 4) \text{ kN} \cdot \text{m}$. Then $\dfrac{\partial M}{\partial P} = x$.

Set $P = 0$. Then $M = 4 \text{ kN} \cdot \text{m}$.

$$\Delta_{A_v} = \int_0^L M \left(\frac{\partial M}{\partial P} \right) \frac{dx}{EI} = \int_0^{3\text{m}} \frac{4(x)dx}{EI} = \frac{18 \text{ kN} \cdot \text{m}^3}{EI} \uparrow \qquad Ans$$

F9–17. For the slope, $m_\theta = -1$ kN $\cdot$ m and $M = (-x^3)$ kN $\cdot$ m.

$$1 \text{ kN} \cdot \text{m} \cdot \theta_B = \int_0^L \frac{m_\theta M}{EI} dx = \int_0^{3\text{m}} \frac{(-1)(-x^3)}{EI} dx = \frac{20.25 \text{ kN}^2 \cdot \text{m}^3}{EI}$$

$$\theta_B = \frac{20.25 \text{ kN} \cdot \text{m}^2}{EI} \; \triangledown$$ *Ans*

For the displacement, $m = (-x)$ kN $\cdot$ m and $M = (-x^3)$ kN $\cdot$ m.

$$1 \text{ kN} \cdot \Delta_{B_v} = \int_0^L \frac{mM}{EI} dx = \int_0^{3\text{m}} \frac{(-x)(-x^3)}{EI} dx = \frac{48.6 \text{ kN}^2 \cdot \text{m}^3}{EI}$$

$$\Delta_{B_v} = \frac{48.6 \text{ kN} \cdot \text{m}^3}{EI} \downarrow$$ *Ans*

F9–18. For the slope, $M = -(M' + x^3)$ kN $\cdot$ m. Then $\dfrac{\partial M}{\partial M'} = -1$.

Set $M' = 0$. Then $M = (-x^3)$ kN $\cdot$ m.

$$\theta_B = \int_0^L M\left(\frac{\partial M}{\partial M'}\right) \frac{dx}{EI} = \int_0^{3\text{m}} \frac{(-x^3)(-1) dx}{EI} = \frac{20.25 \text{ kN} \cdot \text{m}^2}{EI} \; \triangledown$$ *Ans*

For the displacement, $M = -(Px + x^3)$ kN $\cdot$ m.

Then $\dfrac{\partial M}{\partial P} = -x$. Set $P = 0$, then $M = (-x^3)$ kN $\cdot$ m.

$$\Delta_{B_v} = \int_0^L M\left(\frac{\delta M}{\partial P}\right) \frac{dx}{EI} = \int_0^{3\text{m}} \frac{(-x^3)(-x) dx}{EI} = \frac{48.6 \text{ kN} \cdot \text{m}^3}{EI} \downarrow$$ *Ans*

F9–19. For the slope, $m_\theta = (1 - 0.125x)$ kN $\cdot$ m and $M = (32x - 4x^2)$ kN $\cdot$ m.

$$1 \text{ kN} \cdot \text{m} \cdot \theta_A = \int_0^L \frac{m_\theta M}{EI} dx = \int_0^{8\text{m}} \frac{(1 - 0.125x)(32x - 4x^2)}{EI} dx = \frac{170.67 \text{ kN}^2 \cdot \text{m}^3}{EI}$$

$$\theta_A = \frac{171 \text{ kN} \cdot \text{m}^2}{EI} \; \triangledown$$ *Ans*

For the displacement, $m = (0.5x)$ kN $\cdot$ m and $M = (32x - 4x^2)$ kN $\cdot$ m.

$$1 \text{ kN} \cdot \Delta_{C_v} = \int \frac{mM}{EI} dx = 2\int_0^{4\text{m}} \frac{0.5x(32x - 4x^2)}{EI} dx = \frac{426.67 \text{ kN}^2 \cdot \text{m}^3}{EI}$$

$$\Delta_{C_v} = \frac{427 \text{ kN} \cdot \text{m}^3}{EI} \downarrow$$ *Ans*

F9–20. For the slope, $M = M' - 0.125M'x + 32x - 4x^2$. Then $\dfrac{\partial M}{\partial M'} = 1 - 0.125x$.

Set $M' = 0$, then $M = (32x - 4x^2)$ kN $\cdot$ m.

$$\theta_A = \int_0^L M\left(\frac{\partial M}{\partial M'}\right) \frac{dx}{EI} = \int_0^{8\text{m}} \frac{(32x - 4x^2)(1 - 0.125x)}{EI} dx$$

$$= \frac{170.67 \text{ kN} \cdot \text{m}^2}{EI} = \frac{171 \text{ kN} \cdot \text{m}^2}{EI} \; \triangledown$$ *Ans*

For the displacement, $M = 0.5Px + 32x - 4x^2$. Then $\dfrac{\partial M}{\partial P} = 0.5x$. Set $P = 0$, then $M = (32x - 4x^2)$ kN $\cdot$ m.

$$\Delta_{C_v} = \int M\left(\frac{\partial M}{\partial P}\right)\frac{dx}{EI} = 2\int_0^{4\text{m}} \frac{(32x - 4x^2)(0.5x)dx}{EI}$$

$$= \frac{426.67 \text{ kN} \cdot \text{m}^3}{EI} = \frac{427 \text{ kN} \cdot \text{m}^3}{EI} \downarrow \qquad\qquad Ans$$

F9–21.   For the slope, $(m_\theta)_1 = 0$, $(m_\theta)_2 = -1$ kN $\cdot$ m, $M_1 = (-12x_1)$ kN $\cdot$ m,

and $M_2 = -12(x_2 + 2)$ kN $\cdot$ m.

$$1 \text{ kN} \cdot \text{m} \cdot \theta_C = \int_0^L \frac{m_\theta M}{EI} dx = \int_0^{2\text{m}} \frac{0(-12x_1)}{EI} dx + \int_0^{2\text{m}} \frac{(-1)[-12(x_2 + 2)]}{EI} dx$$

$$1 \text{ kN} \cdot \text{m} \cdot \theta_C = \frac{72 \text{ kN}^2 \cdot \text{m}^3}{EI}$$

$$\theta_C = \frac{72 \text{ kN} \cdot \text{m}^2}{EI} \quad \rotatebox{180}{$\vartriangleright$} \qquad\qquad Ans$$

For the displacement, $m_1 = 0$, $m_2 = -x_2$, $M_1 = (-12x_1)$ kN $\cdot$ m,

and $M_2 = -12(x_2 + 2)$ kN $\cdot$ m.

$$1 \text{ kN} \cdot \Delta_C = \int_0^L \frac{mM}{EI} dx = \int_0^{2\text{m}} \frac{0(-12x_1)}{EI} dx + \int_0^{2\text{m}} \frac{(-x_2)[-12(x_2 + 2)]}{EI} dx$$

$$1 \text{ kN} \cdot \Delta_{C_v} = \frac{80 \text{ kN}^2 \cdot \text{m}^3}{EI}$$

$$\Delta_{C_v} = \frac{80 \text{ kN} \cdot \text{m}^3}{EI} \downarrow \qquad\qquad Ans$$

F9–22.   For the slope, $M_1 = (-12x_1)$ kN $\cdot$ m, and $M_2 = -12(x_2 + 2) - M'$.

Thus, $\dfrac{\partial M_1}{\partial M'} = 0$ and $\dfrac{\partial M_2}{\partial M'} = -1$. Set $M' = 0$, $M_2 = -12(x_2 + 2)$.

$$\theta_C = \int_0^L M\left(\frac{\partial M}{\partial M'}\right)\frac{dx}{EI} = \int_0^{2\text{m}} \frac{-12x_1(0)}{EI} dx + \int_0^2 \frac{[-12(x_2 + 2)](-1)}{EI} dx$$

$$= \frac{72 \text{ kN} \cdot \text{m}}{EI} \quad \rotatebox{180}{$\vartriangleright$} \qquad\qquad Ans$$

For the displacement, $M_1 = (-12x_1)$ kN $\cdot$ m and $M_2 = -12(x_2 + 2) - Px_2$.

Thus, $\dfrac{\partial M_1}{\partial P} = 0$ and $\dfrac{\partial M_2}{\partial P} = -x_2$. Set $P = 0$, $M_2 = -12(x_2 + 2)$ kN $\cdot$ m.

$$\Delta_C = \int_0^L M\left(\frac{\partial M}{\partial P}\right)\frac{dx}{EI} = \int_0^{2\text{m}} \frac{(-12x_1)(0)}{EI} dx + \int_0^{2\text{m}} \frac{[-12(x_2 + 2)](-x_2)}{EI} dx$$

$$= \frac{80 \text{ kN} \cdot \text{m}^3}{EI} \downarrow$$

F9–23. $M_1 = 0.5x_1$, $M_2 = 0.5x_2$, $M_1 = \left(24x_1 - \dfrac{1}{6}x_1^3\right)$ kN $\cdot$ m

and $M_2 = \left(48x_2 - 6x_2^2 + \dfrac{1}{6}x_2^3\right)$ kN $\cdot$ m.

$$1 \text{ kN} \cdot \Delta_{C_v} = \int_0^L \frac{mM}{EI}\,dx = \int_0^{6\text{m}} \frac{(0.5x_1)\left(24x_1 - \frac{1}{6}x_1^3\right)}{EI}\,dx_1 + \int_0^{6\text{m}} \frac{(0.5x_2)\left(48x_2 - 6x_2^2 + \frac{1}{6}x_2^3\right)}{EI}\,dx_2$$

$$= \frac{1620 \text{ kN}^2 \cdot \text{m}^3}{EI}$$

$$\Delta_{C_v} = \frac{1620 \text{ kN} \cdot \text{m}^3}{EI} \downarrow \qquad\qquad Ans$$

F9–24. $M_1 = 0.5Px_1 + 24x_1 - \dfrac{1}{6}x_1^3$, $M_2 = 0.5Px_2 + 48x_2 - 6x_2^2 + \dfrac{1}{6}x_2^3$.

Then $\dfrac{\partial M_1}{\partial P} = 0.5x_1$, $\dfrac{\partial M_2}{\partial P} = 0.5x_2$.

Set $P = 0$, $M_1 = \left(24x_1 - \dfrac{1}{6}x_1^3\right)$ kN $\cdot$ m and $M_2 = \left(48x_2 - 6x_2^2 + \dfrac{1}{6}x_2^3\right)$ kN $\cdot$ m

$$\Delta_{C_v} = \int_0^L M\left(\frac{\partial M}{\partial P}\right)\frac{dx}{EI} = \int_0^{6\text{m}} \frac{\left(24x_1 - \frac{1}{6}x_1^3\right)(0.5x_1)}{EI}\,dx_1 + \int_0^{6\text{m}} \frac{\left(48x_2 - 6x_2^2 + \frac{1}{6}x_2^3\right)(0.5x_2)}{EI}\,dx_2$$

$$= \frac{1620 \text{ kN} \cdot \text{m}^3}{EI} \downarrow \qquad\qquad Ans$$

F10–1.   Superposition

$$\Delta'_B = \frac{Px^2}{6EI}(3L - x) = \frac{40(2^2)}{6EI}[3(4) - 2] = \frac{266.67 \text{ kN} \cdot \text{m}^3}{EI} \downarrow$$

$$f_{BB} = \frac{(L/2)^3}{3EI} = \frac{L^3}{24EI} = \frac{4^3}{24EI} = \frac{2.667 \text{ m}^3}{EI} \uparrow$$

$$\Delta_B = \Delta'_B + B_y f_{BB}$$

$$(+\uparrow)\ 0 = -\frac{266.67 \text{ kN} \cdot \text{m}^3}{EI} + B_y\left(\frac{2.667 \text{ m}^3}{EI}\right)$$

$$B_y = 100 \text{ kN} \qquad\qquad Ans$$

Equilibrium

$\xrightarrow{+} \Sigma F_x = 0;$ $\qquad\qquad\qquad A_x = 0$ $\qquad\qquad\qquad Ans$

$+\uparrow \Sigma F_y = 0;$ $\qquad\qquad 100 - 40 - A_y = 0$ $\qquad A_y = 60 \text{ kN}$ $\qquad Ans$

$\downarrow+ \Sigma M_A = 0;$ $\qquad 100(2) - 40(4) - M_A = 0$ $\qquad M_A = 40 \text{ kN} \cdot \text{m}$ $\qquad Ans$

F10–2. Superposition

$$\Delta'_B = \int_0^L \frac{mM}{EI} dx = \int_0^L \frac{(-x)\left(-\frac{w_0}{6L}x^3\right)}{EI} dx = \frac{w_0 L^4}{30\,EI} \downarrow$$

$$f_{BB} = \int_0^L \frac{mm}{EI} dx = \int_0^L \frac{(-x)(-x)}{EI} dx = \frac{L^3}{3EI} \downarrow$$

$$\Delta_B = \Delta'_B + B_y f_{BB}$$

$$(+\downarrow)\ 0 = \frac{w_0 L^4}{30\,EI} + B_y\left(\frac{L^3}{3EI}\right) \qquad B_y = -\frac{w_0 L}{10} = \frac{w_0 L}{10} \uparrow \qquad\qquad Ans$$

Equilibrium

$$\xrightarrow{+}\ \Sigma F_x = 0; \qquad\qquad\qquad\qquad\qquad\qquad\qquad A_x = 0$$

$$+\uparrow \Sigma F_y = 0; \qquad\qquad A_y - \frac{1}{2}w_0 L + \frac{w_0 L}{10} = 0 \qquad A_y = \frac{2w_0 L}{5} \qquad\qquad Ans$$

$$\downarrow + \Sigma M_A = 0; \qquad M_A + \frac{w_0 L}{10}(L) - \left(\frac{1}{2}w_0 L\right)\left(\frac{L}{3}\right) = 0 \qquad M_A = \frac{w_0 L^2}{15} \qquad Ans$$

F10–3. Superposition

$$\Delta'_B = \frac{wL^4}{8EI} = \frac{10(6^4)}{8EI} = \frac{1620\ \text{kN}\cdot\text{m}^3}{EI} = \frac{1620(10^3)\ \text{N}\cdot\text{m}^3}{[200(10^4)\ \text{N/m}^2][300(10^{-6})\ \text{m}^4]} = 0.027\ \text{m}\downarrow$$

$$f_{BB} = \frac{L^3}{3EI} = \frac{6^3}{3EI} = \frac{72\ \text{m}^3}{EI} = \frac{72\ \text{m}^3}{[200(10^9)\ \text{N/m}^2][300(10^{-6})\ \text{m}^4]} = 1.2(10^{-6})\ \text{m/N}\uparrow$$

$$\Delta_B = \Delta'_B + B_y f_{BB}$$

$$(+\downarrow)\ 5(10^{-3})\ \text{m} = 0.027\,\text{m} + B_y[-1.2(10^{-6})\ \text{m/N}]$$

$$B_y = 18.33(10^3)\ \text{N} = 18.33\ \text{kN} = 18.3\ \text{kN} \qquad\qquad\qquad\qquad\qquad Ans$$

Equilibrium

$$\xrightarrow{+}\ \Sigma F_x = 0; \qquad\qquad\qquad\qquad A_x = 0 \qquad\qquad\qquad\qquad\qquad Ans$$

$$+\uparrow \Sigma F_y = 0; \qquad\qquad A_y + 18.33 - 60 = 0 \qquad A_y = 41.67\ \text{kN} = 41.7\ \text{kN} \qquad Ans$$

$$\downarrow + \Sigma M_A = 0; \qquad M_A + 18.33(6) - 60(3) = 0 \qquad\qquad M_A = 70.0\ \text{kN}\cdot\text{m} \qquad Ans$$

F10–4. Superposition

$$\Delta'_B = \frac{M_0 x}{6EIL_{AC}}(L_{AC}^2 - x^2) = \frac{M_0(L)}{6EI(2L)}[(2L)^2 - L^2] = \frac{M_0 L^2}{4EI} \downarrow$$

$$f_{BB} = \frac{L_{AC}^3}{48EI} = \frac{(2L)^3}{48EI} = \frac{L^3}{6EI} \uparrow$$

$$\Delta_B = \Delta'_B + B_y f_{BB}$$

$$(+\uparrow)\ 0 = -\frac{M_0 L^2}{4EI} + B_y\left(\frac{L^3}{6EI}\right) \qquad B_y = \frac{3M_0}{2L} \qquad\qquad\qquad\qquad Ans$$

Equilibrium

$$\xrightarrow{+}\ \Sigma F_x = 0; \qquad\qquad\qquad A_x = 0 \qquad\qquad\qquad\qquad\qquad\qquad\qquad Ans$$

$$\zeta + \Sigma M_A = 0; \qquad -C_y(2L) + \frac{3M_0}{2L}(L) - M_0 = 0 \qquad C_y = \frac{M_0}{4L} \qquad\qquad Ans$$

$$+\uparrow \Sigma F_y = 0; \qquad\qquad \frac{3M_0}{2L} - \frac{M_0}{4L} - A_y = 0 \qquad A_y = \frac{5M_0}{4L} \qquad\qquad Ans$$

F10–5. Superposition

$$\Delta'_B = \frac{Pbx}{6EIL_{AC}}(L_{AC}^2 - b^2 - x^2) = \frac{50(2)(4)}{6EI(8)}(8^2 - 2^2 - 4^2) = \frac{366.67\ kN \cdot m^3}{EI} \downarrow$$

$$f_{BB} = \frac{L_{AC}^3}{48EI} = \frac{8^3}{48EI} = \frac{10.667\ m^3}{EI} \uparrow$$

$$\Delta_B = \Delta'_B + B_y f_{BB}$$

$$(+\uparrow)\ 0 = -\frac{366.67\ kN \cdot m^3}{EI} + B_y\left(\frac{10.667\ m^3}{EI}\right)$$

$$B_y = 34.375\ kN = 34.4\ kN \qquad\qquad\qquad\qquad\qquad\qquad\qquad\qquad Ans$$

Equilibrium

$$\zeta + \Sigma M_A = 0; \qquad 34.375(4) - 50(2) - C_y(8) = 0 \qquad C_y = 4.6875\ kN = 4.69\ kN \qquad Ans$$

$$+\uparrow \Sigma F_y = 0; \qquad A_y + 34.375 - 50 - 4.6875 = 0 \qquad A_y = 20.3125\ kN = 20.3\ kN \qquad Ans$$

$$\xrightarrow{+}\ \Sigma F_x = 0; \qquad\qquad\qquad A_x = 0 \qquad\qquad\qquad\qquad\qquad\qquad\qquad Ans$$

F10–6. $\Delta'_B = \dfrac{5wL_{AC}^4}{384EI} = \dfrac{5(10)(12^4)}{384EI} = \dfrac{2700\ kN \cdot m^3}{EI} = \dfrac{2700(10^3)\ N \cdot m^3}{[200(10^9)\ N/m^2][300(10^{-6})\ m^4]} = 0.045\ m \downarrow$

$$f_{BB} = \frac{L_{AC}^3}{48EI} = \frac{12^3}{48EI} = \frac{36\ m^3}{EI} = \frac{36\ m^3}{[200(10^9)\ N/m^2][300(10^{-6})\ m^4]} = 0.6(10^{-6})\ m/N \uparrow$$

$$\Delta_B = \Delta'_B + B_y f_{BB}$$

$$(+\downarrow)\ 5(10^{-3})\ m = 0.045\ m + B_y[-0.6(10^{-6})\ m/N] \qquad B_y = 66.67(10^3)\ N = 66.7\ kN \qquad Ans$$

Equilibrium

$$\zeta + \Sigma M_A = 0; \qquad C_y(12) + 66.67(6) - 120(6) = 0 \qquad C_y = 26.67\ kN = 26.7\ kN \qquad Ans$$

$$+\uparrow \Sigma F_y = 0; \qquad A_y + 26.67 + 66.67 - 120 = 0 \qquad A_y = 26.67\ kN = 26.7\ kN \qquad Ans$$

$$\xrightarrow{+}\ \Sigma F_x = 0; \qquad\qquad\qquad A_x = 0 \qquad\qquad\qquad\qquad\qquad\qquad\qquad Ans$$

Answers to Selected Problems

Chapter 1

1–1. $F = 208$ kN
1–2. $F = 105$ kN
1–3. $w = 7.5$ kN/m
1–5. $F = 173$ kN
1–6. $w = 6.62$ kN/m
1–7. 6.20 kN/m
1–9. $w = 3.46$ kN/m
1–10. Total dead load $= 5.06$ kN/m^2
1–11. $F_s = 415.5$ kN
1–13. $L = 1.70$ kN/m^2
1–14. $L = 3.02$ kN/m^2
1–15. $p_{0-4.6} = 1348$ N/m^2
$p_{6.1} = 1409$ N/m^2
$p_{7.6} = 1457$ N/m^2
$p_{9.1} = 1502$ N/m^2
1–17. Windward:
$p_{0-4.6} = 1001$ N/m^2
$p_6 = 1044$ N/m^2
Leeward:
$p = -735$ N/m^2
1–18. $p = -892$ N/m^2
1–19. $F = 81.3$ kN
1–21. $p_f = 0.816$ kN/m^2
1–22. $p_f = 1.8$ kN/m^2

Chapter 2

2–1. on $BE = 14.2$ kN/m; on FED, $E_y = 35.6$ kN
2–2. on BE trapezodial load, peak 21.4 kN/m
on FED triangular loadings, peaks 10.7 kN/m,
with concentrated force of 26.7 kN at E.
2–3. on BF, 12.84 kN/m; on $ABCDE$ 3 forces of
57.78 kN
2–5. on BF = 9.63 kN/m,
on $ABCDE$ 3 forces of 28.89 kN
2–6. on BG = 3.308 kN/m,
on $ABCD$ 2 forces of 7.44 kN
2–7. on BG = 5.292 kN/m peak triangular
on $ABCD$ with 2 forces of 3.175 kN,
2.646 kN/m peak triangular
2–9. on BE trapezoidal loading, peak 59.31 kN/m,
on FED triangular loads, peaks 29.66 kN/m
and $E_y = 55.6$ kN

2–10. on BE 31.63 kN/m,
on FED force of 56.9 kN
2–11. **a.** Indeterminate to 2°
b. Unstable
c. Statically determinate
d. Statically determinate
e. Unstable
2–13. **a.** Statically determinate
b. Statically indeterminate to 1°
c. Statically indeterminate to 1°
2–14. **a.** Unstable
b. Stable and statically determinate
c. Stable and statically indeterminate to the
second degree
2–15. **a.** Unstable
b. Stable and statically indeterminate to the
first degree
c. Unstable
2–17. **a.** Unstable
b. Stable and statically indeterminate to the
sixth degree
c. Stable and statically determinate
d. Unstable
2–18. $B_y = 48.0$ kN
$A_y = 16.0$ kN
$A_x = 10.0$ kN
2–19. $F_B = 495$ kN
$A_x = 429$ kN
$A_y = 22.5$ kN
2–21. $N_A = 12$ kN
$B_x = 0$
$B_y = 30$ kN
$M_B = 84$ kN $\cdot$ m
2–22. $N_F = 20$ kN
$N_D = 30$ kN
$N_B = 75$ kN
$A_y = 35$ kN
$A_x = 0$
2–23. $N_A = 48.0$ kN
$C_x = 46.0$ kN
$N_B = 42.7$ kN
$C_y = 14.6$ kN

2–25. $C = 426$ N
$A_y = 213$ N
$A_y = 1791$ N

2–26. $B_y = 25.6$ kN
$A_y = 73.6$ kN
$B_x = 100$ kN

2–27. $C_y = 0$
$B_y = 7.50$ kN
$M_A = 45.0$ kN $\cdot$ m
$A_y = 7.50$ kN
$A_x = 0$

2–29. $A_y = 4.00$ kN
$M_B = 63.0$ kN $\cdot$ m
$B_y = 17.0$ kN
$B_x = 0$

2–30. $A_y = 2.00$ kN
$B_y = 12.0$ kN
$B_x = 0$
$M_B = 32.0$ kN $\cdot$ m

2–31. $w_1 = \dfrac{2P}{L}$

$w_2 = \dfrac{4P}{L}$

$w_1 = 1.33$ kN/m
$w_2 = 2.67$ kN/m

2–33. $A_x = 30.0$ kN
$A_y = 6.67$ kN
$C_x = 10.0$ kN
$C_y = 6.67$ kN

2–34. $N_A = 44.8$ kN
$B_x = 38.8$ kN
$B_y = 16.4$ kN

2–35. $B_y = 74.6$ kN
$A_x = 131$ kN
$A_y = 76.6$ kN

2–37. $A_y = 300$ N
$A_x = 300$ N
$C_x = 300$ N
$C_y = 300$ N

2–38. $T = 3.5$ kN
$A_y = 7$ kN
$A_x = 15.3$ kN
$D_x = 17.0$ kN
$D_y = 17.0$ kN

2–39. $F_{BE} = 6.12$ kN
$F_{CD} = 1.40$ kN

2–41. $A_x = 2.35$ kN
$A_y = 6.63$ kN

$C_x = 3.05$ kN
$C_y = 8.88$ kN

2–42. $C_x = 45.0$ kN
$D_x = 45.0$ kN
$C_y = 7.00$ kN
$A_y = 83.0$ kN
$A_x = 45.0$ kN
$D_y = 7.00$ kN

2–43. $B_x = 124$ kN
$B_y = 111$ kN
$C_x = 41$ kN
$C_y = 159$ kN

Chapter 3

3–1.   **a.** Unstable
  b. Statically indeterminate to 1°
  c. Statically determinate
  d. Statically determinate

3–2.   **a.** Statically determinate
  b. Statically determinate
  c. Unstable.

3–3.   **a.** Internally and externally stable (to 2°)
  b. Internally and externally stable (to 1°)
  c. Internally and externally stable (to 1°)

3–5. $F_{CD} = 3.9$ kN (C)
$F_{CB} = 3.6$ kN (T)
$F_{DB} = 0$
$F_{DE} = 3.9$ kN (C)
$F_{BE} = 1.48$ kN (T)
$F_{BA} = 3.61$ kN (T)

3–6. $F_{AH} = 17.9$ kN (C)
$F_{AB} = 16$ kN (T)
$F_{BC} = 16$ kN (T)
$F_{BH} = 0$
$F_{HC} = 8.94$ kN (C)
$F_{HG} = 8.94$ kN (C)
$F_{FG} = 0$
$F_{FE} = 6$ kN (C)
$F_{GE} = 8.94$ kN (C)
$F_{GC} = 0$
$F_{EC} = 8.94$ kN (T)
$F_{ED} = 14$ kN (C)
$F_{DC} = 0$

3–7. $F_{DC} = 9.24$ kN (T)
$F_{DE} = 4.62$ kN (C)
$F_{CE} = 9.24$ kN (C)
$F_{CB} = 9.24$ kN (T)
$F_{BE} = 9.24$ kN (C)

$F_{BA} = 9.24$ kN (T)
$F_{EA} = 4.62$ kN (C)

3–9. $F_{AF} = 15$ kN (T)
$F_{AB} = 12$ kN (C)
$F_{BF} = 40.5$ kN (C)
$F_{BC} = 12$ kN (C)
$F_{FC} = 22.5$ kN (T)
$F_{FE} = 6$ kN (C)
$F_{CE} = 13.5$ kN (C)
$F_{CD} = 6$ kN (T)
$F_{DE} = 7.5$ kN (C)

3–10. $F_{EF} = 58.6$ kN (C)
$F_{ED} = 44.4$ kN (T)
$F_{DF} = 0$
$F_{DC} = 44.4$ kN (T)
$F_{AH} = 10.7$ kN (C)
$F_{AB} = 6.88$ kN (T)
$F_{BH} = 0$
$F_{BC} = 6.88$ kN (T)
$F_{FC} = 20.2$ kN (C)
$F_{FG} = 38.4$ kN (C)
$F_{GC} = 15.0$ kN (C)
$F_{GH} = 38.4$ kN (C)
$F_{CH} = 29.3$ kN (T)

3–11. $F_{ED} = 8.33$ kN (T)
$F_{CD} = 6.67$ kN (C)
$F_{BC} = 6.67$ kN (C)
$F_{CE} = 5$ kN (T)
$F_{GF} = 20$ kN (T)
$F_{GA} = 15$ kN (T)
$F_{AF} = 18.0$ kN (C)
$F_{AB} = 10.0$ kN (C)
$F_{BE} = 4.17$ kN (C)
$F_{FB} = 7.50$ kN (T)
$F_{FE} = 12.5$ kN (T)

3–13. $F_{DE} = 16.3$ kN (C)
$F_{DC} = 8.40$ kN (T)
$F_{EA} = 8.85$ kN (C)
$F_{EC} = 6.20$ kN (C)
$F_{CF} = 8.77$ kN (T)
$F_{CB} = 2.20$ kN (T)
$F_{BA} = 3.11$ kN (T)
$F_{BF} = 6.20$ kN (C)
$F_{FA} = 6.20$ kN (T)

3–14. $F_{AK} = 42.9$ kN (C)
$F_{AB} = 41.1$ kN (T)
$F_{KB} = 4.00$ kN (C)
$F_{KJ} = 42.9$ kN (C)

$F_{BJ} = 7.94$ kN (T)
$F_{BC} = 34.3$ kN (T)
$F_{JI} = 35.7$ kN (C)
$F_{JC} = 6.00$ kN (C)
$F_{CI} = 9.11$ kN (T)
$F_{CD} = 27.4$ kN (T)
$F_{IH} = 35.7$ kN (C)
$F_{HD} = 6.00$ kN (C)
$F_{HE} = 7.94$ kN (T)
$F_{HG} = 42.9$ kN (C)
$F_{ED} = 34.3$ kN (T)
$F_{ID} = 9.11$ kN (T)
$F_{FG} = 42.9$ kN (C)
$F_{GE} = 4.00$ kN (C)
$F_{FE} = 41.1$ kN (T)

3–15. $F_{AH} = F_{FE} = 25$ kN (C)
$F_{AB} = F_{DE} = 20$ kN (T)
$F_{BC} = F_{CD} = 20$ kN (T)
$F_{BH} = F_{DF} = 10$ kN (T)
$F_{HG} = F_{GF} = 16.7$ kN (C)
$F_{HC} = F_{CF} = 8.33$ kN (C)
$F_{GC} = 20$ kN (T)

3–17. $F_{AG} = F_{DE} = 4.00$ kN (T)
$F_{AB} = F_{DC} = 3.46$ kN (C)
$F_{GF} = F_{EF} = 4.00$ kN (T)
$F_{GB} = F_{EC} = 0$
$F_{BF} = F_{CF} = 2.31$ kN (C)
$F_{BC} = 2.31$ kN (C)

3–18. $F_{DC} = 46.7$ kN (T)
$F_{FG} = 66.5$ kN (C)
$F_{FC} = 19.4$ kN (T)

3–19. $F_{CD} = 10.0$ kN (T)
$F_{JN} = 12.5$ kN (T)
$F_{JK} = 20.2$ kN (C)
Members *KN, NL, MB, BL, CL, IO, OH, GE, EH, HD* are zero force members.

3–21. $F_{CD} = 6.67$ kN (T)
$F_{GF} = 12.5$ kN (C)
$F_{GC} = 0$

3–22. $F_{BC} = 8.00$ kN (T)
$F_{HG} = 10.1$ kN (C)
$F_{BG} = 1.80$ kN (T)

3–23. $F_{GF} = 1.78$ kN (T)
$F_{CD} = 2.23$ kN (C)
$F_{CF} = 0$

3–25. $F_{IH} = 6.00$ kN (T)
$F_{ID} = 4.24$ kN (T)
$F_{CD} = 10.1$ kN (C)

3–26. $F_{JI} = 9.00$ kN (T)
$F_{IC} = 6.00$ kN (C)
$F_{CD} = 10.1$ kN (C)

3–27. $F_{KJ} = 115$ kN (C)
$F_{CJ} = 27.0$ kN (T)
$F_{CD} = 97.5$ kN (T)

3–29. $F_{AB} = 0$
$F_{AG} = 7.5$ kN (C)
$F_{GB} = 3.54$ kN (T)
$F_{GL} = 2.5$ kN (C)
$F_{GI} = 3.54$ kN (C)
$F_{LI} = 3.54$ kN (T)
$F_{LK} = 2.5$ kN (C)
$F_{IK} = 3.54$ kN (C)
$F_{IF} = 3.54$ kN (T)
$F_{BF} = 10.6$ kN (T)
$F_{BC} = 5$ kN (C)
$F_{FC} = 3.54$ kN (T)
$F_{FH} = 10.6$ kN (T)
$F_{KH} = 3.54$ kN (T)
$F_{KJ} = 7.5$ kN (C)
$F_{JH} = 10.6$ kN (T)
$F_{CD} = 0$
$F_{DE} = 2.5$ kN (C)
$F_{CE} = 3.54$ kN (C)
$F_{HE} = 3.54$ kN (T)
$F_{JE} = 7.5$ kN (C)

3–30. $F_{AD} = F_{BE} = 0$
$F_{AF} = F_{BC} = 4.00$ kN (C)
$F_{FD} = F_{CE} = 8.94$ kN (T)
$F_{FE} = F_{CD} = 11.3$ kN (C)
$F_{ED} = 16.0$ kN (C)

3–31. $F_{EF} = 1.15$ kN (T)
$F_{ED} = 3.46$ kN (C)
$F_{BA} = 1.15$ kN (T)
$F_{BC} = 1.15$ kN (T)
$F_{AD} = 4.24$ kN (T)
$F_{AF} = 1.58$ kN (T)
$F_{CF} = 1.41$ kN (C)
$F_{CD} = 4.73$ kN (C)
$F_{BE} = 1.15$ kN (T)

3–33. $F_{AB} = 6.46$ kN (T)
$F_{AC} = F_{AD} = 1.50$ kN (C)
$F_{BC} = F_{BD} = 3.70$ kN (C)
$F_{BE} = 4.80$ kN (T)

3–34. $F_{BC} = F_{BD} = 1.34$ kN (C)
$F_{AB} = 2.4$ kN (C)
$F_{AG} = F_{AE} = 1.01$ kN (T)

3–35. $F_{BG} = 1.80$ kN (T)
$F_{BE} = 1.80$ kN (T)
$F_{FE} = 0$
$F_{ED} = 0$

3–37. $F_{AB} = 4.00$ kN (T)
$F_{BE} = 5.66$ kN (T)
$F_{BD} = 2.00$ kN (C)
$F_{AC} = F_{AE} = F_{DE} = F_{DC} = F_{CE} = 0$

3–38. $F_{BC} = 1.15$ kN (C)
$F_{DF} = 4.16$ kN (C)
$F_{BE} = 4.16$ kN (T)

3–39. $F_{CF} = 0$
$F_{CD} = 2.31$ kN (T)
$F_{ED} = 3.46$ kN (T)

Chapter 4

4–1. $N_C = 0$
$V_C = 0.667$ kN
$M_C = 0.667$ kN · m
$N_D = 0$
$V_D = -5.33$ kN
$M_D = -9.33$ kN · m

4–2. $N_C = 0$
$V_C = 16.7$ kN
$M_C = 87.5$ kN · m
$N_D = 0$
$V_D = -33.3$ kN · m
$M_D = 137.5$ kN · m

4–3. $N_A = 0$
$V_A = 2025$ N
$M_A = -1.519$ kN · m
$N_B = 0$
$V_B = 3825$ N
$M_B = -8.539$ kN · m
$V_C = 0$
$N_C = -5.4$ kN
$M_C = -10.97$ kN · m

4–5. $w = 100$ N/m

4–6. $N_C = 0$
$V_C = -0.75$ kN
$M_C = -0.375$ kN · m
$N_D = 0$
$V_D = 1.25$ kN
$M_D = 1.875$ kN · m

4–7. $N_C = 0$
$V_C = 1.75$ kN
$M_C = 8.50$ kN · m

4–9. $N_C = 0$
$V_C = 1.25$ kN
$M_C = 3.50$ kN·m

4–10. $N_C = 0$
$V_C = -5.22$ kN
$M_C = 20.1$ kN·m

4–11. $N_D = 0$
$V_D = 5.58$ kN
$M_D = 19.76$ kN·m
$N_E = 0$
$V_E = 2.70$ kN
$M_E = -1.215$ kN·m

4–13. $0 \le x < 1$ m, $V = 4.50$ kN
$M = \{4.50x\}$ kN·m
1 m $< x < 3$ m, $V = 0.500$ kN
$M = \{0.5x + 4\}$ kN·m
3 m $< x \le 4$ m, $V = -5.50$ kN
$M = \{-5.50x + 22\}$ kN·m

4–14. $0 \le x < a$, $V = \dfrac{M_O}{L}$
$M = \dfrac{M_O}{L}x$
$a < x \le L$, $V = \dfrac{M_O}{L}$
$M = -\dfrac{M_O}{L}(L - x)$

4–15. $0 \le x < 2$ m, $V = 3.75$ kN
$M = \{3.75x\}$ kN·m
2 m $< x < 4$ m, $V = -3.25$ kN
$M = \{-3.25x + 14\}$ kN·m
4 m $< x < 8$ m,
$V = -3.25$ kN,
$M = \{-3.25x + 26\}$ kN·m

4–17. $0 \le x < 1$ m, $V = -4$ kN
$M = \{-4x\}$ kN·m
1 m $< x < 2$ m, $V = -12$ kN
$M = \{-12x + 8\}$ kN·m
2 m $< x \le 3$ m, $V = -20$ kN
$M = \{-20x + 24\}$ kN·m

4–18. $0 \le x < 3$ m, $V = \{150 - 20x\}$ kN
$M = \{-10x^2 + 150x - 540\}$ kN·m
3 m $< x \le 5$ m, $V = 40$ kN
$M = \{40x - 300\}$ kN·m

4–19. $0 \le x < 2$ m, $V = -1000$ N
$M = \{-1000x\}$ N·m
2 m $< x < 5$ m, $V = \{4200 - 1200x\}$ N
$M = \{-600x^2 + 4200x - 8000\}$ N·m
5 m $< x \le 7$ m, $V = 1000$ N
$M = \{1000x - 7000\}$ N·m

4–21. $V = \{-0.15x^2 - 4\}$ kN
$M = \{-0.05x^3 - 4x - 2\}$ kN·m

4–22. $V = \{0.444x^2 - 8x + 36\}$ kN
$M = \{0.148x^3 - 4x^2 + 36x - 108\}$ kN·m

4–23. $V_{max} = -1.93$ kN
$M_{max} = -6$ kN·m

4–25. $V_{max} = -4.89$ kN
$M_{max} = -20$ kN·m

4–26. $V_{max} = -50.5$ kN
$M_{max} = -100$ kN·m

4–27. $V_{max} = -15.2$ kN
$M_{max} = 19.3$ kN·m

4–29. $V_{max} = -1.25$ kN
$M_{max} = 0.521$ kN

4–30. $V_{max} = -2040$ N
$M_{max} = 3842$ N·m

4–31. $V_{max} = -\dfrac{3wL}{8}$
$M_{max} = \dfrac{9wL^2}{128}$

4–33. $V_{max} = -186$ kN
$M_{max} = 224$ kN·m

4–34. $V_{max} = \pm6$ kN
$M_{max} = 8$ kN·m

4–35. $V_{max} = -19$ kN
$M_{max} = -92$ kN·m

4–37. $V_{max} = 24.5$ kN
$M_{max} = 34.5$ kN·m

4–38. $V_{max} = 83$ kN
$M_{max} = -180$ kN·m

4–39. $V_{max} = -59.2$ kN
$M_{max} = -109.5$ kN·m

4–41. $V_{max} = 60$ kN
$M_{max} = -112.5$ kN·m

4–42. $V_{max} = 100$ kN
$M_{max} = -360$ kN·m

4–43. $V_{max} = -180$ kN
$M_{max} = 405$ kN·m

4–45. $V_{max} = 13.3$ kN
$M_{max} = 26.7$ kN·m

4–46. $V_{max} = -14.5$ kN
$M_{max} = 52.5$ kN·m

4–47. $V_{max} = 8.87$ kN
$M_{max} = 17.7$ kN·m

4–49. $V_{max} = 54$ kN
$M_{max} = -121.5$ kN·m

Chapter 5

5–1. $F_{BC} = 233.5$ kN $\quad F_{BA} = 415.0$ kN
$F_{CD} = 440.5$ kN
$l = 10.1$ m

5–2. $T_{max} = 6.41$ kN
$y_B = 2.43$ m

5–3. $T_{BC} = 1.60$ kN
$T_{CD} = 3.72$ kN
$T_{AB} = 2.99$ kN
$y_D = 2.10$ m

5–5. $P = 357$ N

5–6. $P_1 = 2.50$ kN
$P_2 = 6.25$ kN
$F_{max} = 12.5$ kN

5–7. $y = 0.0356x^2$
$T_O = 7.03$ kN
$T_B = 10.3$ kN

5–9. $T_{min} = 400$ kN
$T_{max} = 431$ kN

5–10. $w = 0.865$ kN/m

5–11. $T_{max} = 65.0$ kN
$T_{min} = 58.6$ kN

5–13. $T_{max} = 10.9$ kN

5–14. $T_{min} = 100$ kN
$T_{max} = 117$ kN
$T_{hanger} = 10$ kN

5–15. $V_{max} = \pm5$ kN
$M_{max} = 6.25$ kN · m

5–17. $T_{min} = 6.25$ MN
$T_{max} = 6.93$ MN

5–18. $y = (38.5x^2 + 577x)(10^{-3})$ m
$T_{max} = 5.20$ kN

5–19. $T_F = 7.0$ kN
$T_E = T_D = 8.75$ kN
$T_{hanger} = 1.31$ kN

5–21. $A_x = 0$
$C_y = 9.55$ kN
$A_y = 15.5$ kN
$T = 4.32$ kN

5–22. $F_B = 6.77$ kN
$F_A = 10.8$ kN
$F_C = 11.7$ kN

5–23. $M_D = 10.8$ kN · m

5–25. $B_x = 233$ kN, $B_y = 25$ kN
$A_x = 233$ kN, $A_y = 475$ kN
$C_x = 233$ kN, $C_y = 425$ kN

5–26. $h_1 = 13.125$ m
$h_2 = 22.5$ m
$h_3 = 28.125$ m

5–27. $B_y = 1.08$ kN, $B_x = 13.62$ kN
$A_x = 13.62$ kN, $A_y = 18.92$ kN
$C_x = 1.38$ kN, $C_y = 1.08$ kN

5–29. $A_x = 30$ kN
$D_y = 80.6$ kN
$A_y = 19.4$ kN
$T_{AD} = 40.8$ kN

Chapter 6

6–15. $(M_C)_{max} = 142$ kN · m
$(V_C)_{max} = 20$ kN

6–17. $(B_y)_{max(+)} = 49.5$ kN
$(M_B)_{max(-)} = -45$ kN · m

6–18. $(M_C)_{max(+)} = 90$ kN · m
$(B_y)_{max(+)} = 99$ kN

6–19. $(A_y)_{max(+)} = 316$ kN
$(M_C)_{max(+)} = 204$ kN · m
$(V_{A^+})_{max(+)} = 181$ kN

6–21. $(M_D)_{max} = -4$ kN · m

6–22. $(M_A)_{max} = -108$ kN · m
$(V_D)_{max} = 27$ kN

6–23. $(B_y)_{max(+)} = 87.6$ kN
$(M_C)_{max(+)} = 72.0$ kN · m
$(V_C)_{max(-)} = -23.6$ kN

6–25. $(A_y)_{max(+)} = 82$ kN
$(M_E)_{max(+)} = 82$ kN · m
$(V_{C^+})_{max(+)} = 132$ kN

6–26. $(V_{BC})_{max(+)} = 7.15$ kN
$(M_G)_{max(-)} = -9.81$ kN · m

6–27. $(V_{BC})_{max(+)} = 17.8$ kN
$(M_G)_{max(+)} = 46.7$ kN · m

6–30. $(V_{AB})_{max} = 10.92$ kN
$(M_D)_{max} = 49$ kN · m

6–31. $(V_{BC})_{max} = 35$ kN
$(M_C)_{max} = 105$ kN · m

6–33. $(V_{DE})_{max(-)} = -264.5$ kN
$(M_C)_{max(-)} = -294$ kN · m

6–34. $(V_{DE})_{max(+)} = 25.3$ kN
$(M_H)_{max(+)} = 32$ kN · m

6–35. $(V_{CD})_{max(-)} = -27$ kN

6–37. $(V_{BC})_{max(-)} = -8.21$ kN
$(M_B)_{max(+)} = 12.3$ kN · m

6–57. $(F_{CD})_{max(+)} = 54.0$ kN (T)

6–58. $(F_{CF})_{max(+)} = 33.9$ kN (T)

6–59. $(M_C)_{max} = 44.1$ kN · m

6–61. $(V_B)_{max} = 5.83$ kN

6–62. $(M_C)_{max(+)} = 20.0$ kN · m

6–63. $(M_C)_{max(+)} = 21$ kN · m

6–65. $(M_C)_{max(+)} = 34.0$ kN·m

6–66. $(M_C)_{max(+)} = 5.25$ kN·m

6–67. $(F_{BC})_{max} = 11.8$ kN (T)

6–69. $M_{max} = 67.8$ kN·m

6–70. $M_{max} = 64.5$ kN·m

6–71. $V_{max} = 10$ kN

$M_{max} = -39$ kN·m

6–73. $M_{max} = 832.6$ kN·m

6–74. $V_{max}^{abs} = 67.5$ kN

6–75. $M_{max} = 164$ kN·m

6–77. $M_{max} = 145.8$ kN·m

6–78. $M_{max} = 176$ kN·m

6–79. $V_{max}^{abs} = 62.5$ kN

6–81. $M_{max} = 15.8$ kN·m

Chapter 7

7–1. $F_{AE} = F_{BF} = 14.1$ kN (C)

$F_{EF} = 10.0$ kN (C)

$F_{AB} = 10.0$ kN (T)

$F_{CE} = F_{BD} = 14.1$ kN (C)

$F_{DE} = 10.0$ kN (C)

$F_{BC} = 10.0$ kN (T)

$F_{AF} = 60.0$ kN (C)

$F_{BE} = 20.0$ kN (C)

$F_{CD} = 30.0$ kN (C)

7–2. $F_{AE} = F_{CE} = 0$

$F_{BF} = 28.3$ kN (T)

$F_{EF} = 20.0$ kN (C)

$F_{AB} = 0$

$F_{BD} = 28.3$ kN (T)

$F_{DE} = 20.0$ kN (C)

$F_{BC} = 0$

$F_{AF} = 70.0$ kN (C)

$F_{BE} = 40.0$ kN (C)

$F_{CD} = 40.0$ kN (C)

7–3. $F_{HB} = 29.5$ kN (T)

$F_{AG} = 29.5$ kN (C)

$F_{AB} = 45.8$ kN (T)

$F_{AH} = 70.8$ kN (C)

$F_{HG} = 20.8$ kN (C)

$F_{GC} = 5.89$ kN (C)

$F_{BF} = 5.89$ kN (T)

$F_{GF} = 37.5$ kN (C)

$F_{GB} = 25.0$ kN (C)

$F_{BC} = 62.5$ kN (T)

$F_{EC} = 41.3$ kN (T)

$F_{DF} = 41.3$ kN (C)

$F_{CD} = 29.2$ kN (T)

$F_{ED} = 79.2$ kN (C)

$F_{FE} = 4.17$ kN (C)

$F_{FC} = 25.0$ kN (C)

7–5. $F_{AG} = 60.4$ kN (C)

$F_{BH} = 60.4$ kN (T)

$F_{AB} = 38.3$ kN (T)

$F_{GH} = 48.3$ kN (C)

$F_{BF} = 2.08$ kN (C)

$F_{CG} = 2.08$ kN (T)

$F_{FG} = 98.3$ kN (C)

$F_{BC} = 88.3$ kN (T)

$F_{DF} = 56.25$ kN (C)

$F_{CE} = 56.25$ kN (T)

$F_{EF} = 55.0$ kN (C)

$F_{CD} = 45.0$ kN (T)

$F_{AH} = 71.25$ kN (C)

$F_{BG} = 35.0$ kN (C)

$F_{DE} = 68.75$ kN (C)

7–6. $F_{AG} = F_{BF} = F_{DF} = 0$

$F_{CE} = 112.5$ kN (T)

$F_{DE} = 102.5$ kN (C)

$F_{BH} = 120.8$ kN (T)

$F_{GH} = 96.7$ kN (C)

$F_{AB} = 10.0$ kN (C)

$F_{CG} = 4.167$ kN (T)

$F_{FG} = 100.0$ kN (C)

$F_{BC} = 86.5$ kN (T)

$F_{EF} = 100.0$ kN (C)

$F_{CD} = 0$

$F_{AH} = 107.5$ kN (C)

$F_{BG} = 72.5$ kN (C)

$F_{CF} = 70.0$ kN (C)

$F_{DE} = 102.5$ kN (C)

7–7. $F_{EC} = 3.33$ kN (T)

$F_{BD} = 3.33$ kN (C)

$F_{ED} = 2.67$ kN (T)

$F_{BC} = 2.67$ kN (C)

$F_{CD} = 2.00$ kN (T)

$F_{FB} = 10.0$ kN (T)

$F_{AE} = 10.0$ kN (C)

$F_{FE} = 13.3$ kN (T)

$F_{AB} = 13.3$ kN (C)

$F_{BE} = 4.00$ kN (T)

$F_{AF} = 6.00$ kN (T)

7–9. $F_{DG} = 12.4$ kN (C)

$F_{CF} = 12.4$ kN (T)

$F_{FG} = 16.25$ kN (C)

$F_{CD} = 16.25$ kN (T)
$F_{AC} = 19.45$ kN (C)
$F_{BG} = 19.45$ kN (T)
$F_{BC} = 38.75$ kN (T)
$F_{AG} = 38.75$ kN (C)
$F_{EF} = 10.6$ kN (C)
$F_{DE} = 7.5$ kN (T)
$F_{DF} = 1.25$ kN (C)
$F_{CG} = 5.0$ kN (C)
$F_{AB} = 13.75$ k (T)

7–10. $F_{DG} = F_{AC} = 0$
$F_{CF} = 24.75$ kN (T)
$F_{CD} = 7.5$ kN (T)
$F_{FG} = 25.0$ kN (C)
$F_{BG} = 38.9$ kN (T)
$F_{BC} = 25.0$ kN (T)
$F_{AG} = 52.5$ kN (C)
$F_{EF} = 10.6$ kN (C)
$F_{DE} = 7.5$ kN (T)
$F_{DF} = 10.0$ kN (C)
$F_{CG} = 27.5$ kN (C)
$F_{AB} = 0$

7–11. $F_{DF} = 6.67$ kN (T)
$F_{CE} = 6.67$ kN (C)
$F_{CD} = 5.33$ kN (C)
$F_{AC} = 15.0$ kN (T)
$F_{BF} = 15.0$ kN (C)
$F_{EF} = 5.33$ kN (T)
$F_{BC} = 22.7$ kN (C)
$F_{AF} = 22.7$ kN (T)
$F_{DE} = 4.00$ kN (C)
$F_{CF} = 5.00$ kN (C)
$F_{AB} = 9.00$ kN (T)

7–13. $M_A = 4.86$ kN $\cdot$ m
$M_B = 3.78$ kN $\cdot$ m

7–14. $M_F = 8.1$ kN $\cdot$ m
$M_D = 14.4$ kN $\cdot$ m

7–15. $M_A = 40.3$ kN $\cdot$ m

7–17. $M_I = 12.96$ kN $\cdot$ m
$M_L = 29.16$ kN $\cdot$ m
$M_H = 38.88$ kN $\cdot$ m

7–18. $A_x = 0; B_x = 0; C_x = 0$
$A_y = 54$ kN; $B_y = 72$ kN; $C_y = 18$ kN
$M_A = 21.9$ kN $\cdot$ m; $M_B = 12.2$ kN $\cdot$ m; $M_C = 9.7$ kN $\cdot$ m

7–19. Pinned:
$A_x = 6.00$ kN

$A_y = 18.0$ kN
$B_x = 6.00$ kN
$B_y = 18.0$ kN
Fixed:
$A_x = 6.00$ kN
$A_y = 9.00$ kN
$M_A = 18.0$ kN $\cdot$ m
$B_x = 6.00$ kN
$B_y = 9.00$ kN
$M_B = 18.0$ kN $\cdot$ m

7–21. $F_{CF} = 8.84$ kN (T)

7–22. $F_{CE} = 5.30$ kN (T)

7–23. $A_x = 7.50$ kN
$A_y = 9.375$ kN
$M_A = 11.25$ kN $\cdot$ m
$B_x = 7.50$ kN
$B_y = 9.375$ kN
$M_B = 11.25$ kN $\cdot$ m
$F_{DG} = 15.625$ kN (C)
$F_{CD} = 10.0$ kN (T)
$F_{FG} = 5.0$ kN (C)
$F_{DF} = 15.625$ kN (T)
$F_{DE} = 15.0$ kN (C)

7–25. $F_{EG} = 27.5$ kN (T)
$F_{EF} = 24.0$ kN (C)
$F_{CG} = 4.00$ kN (C)
$F_{CE} = 27.5$ kN (C)
$F_{DE} = 20.0$ kN (T)

7–26. $F_{EG} = 15.0$ kN (T)
$F_{CG} = 4.00$ kN (C)
$F_{EF} = 14.0$ kN (C)
$F_{CE} = 15.0$ kN (C)
$F_{DE} = 10.0$ kN (T)

7–27. $A_x = 10.0$ kN
$A_y = 14.0$ kN
$M_A = 30.0$ kN $\cdot$ m
$B_x = 10.0$ kN
$B_y = 14.0$ kN
$M_B = 30.0$ kN $\cdot$ m
$F_{FH} = 17.5$ kN (C)
$F_{EF} = 16.5$ kN (C)
$F_{GH} = 17.0$ kN (T)
$F_{EH} = 17.5$ kN (T)
$F_{HI} = 4.00$ kN (C)
$F_{EI} = 17.5$ kN (C)

$F_{DE} = 4.50$ kN (T)
$F_{DI} = 17.5$ kN (T)
$F_{CI} = 25.0$ kN (C)

7–29. $A_x = 10.0$ kN
$A_y = 5.625$ kN
$M_A = 15.0$ kN·m
$B_x = 10.0$ kN
$B_y = 5.625$ kN
$M_B = 15.0$ kN·m
$F_{GK} = 9.375$ kN (C)
$F_{GF} = 0$
$F_{JK} = 2.5$ kN (C)

7–30. $A_x = 10.0$ kN
$A_y = 9.375$ kN
$B_x = 10.0$ kN
$B_y = 9.375$ kN
$F_{GK} = 15.625$ kN (C)
$F_{GF} = 0$
$F_{JK} = 2.50$ kN (T)

7–31. $F_{FG} = 0$
$F_{EH} = 2.50$ kN (T)
$F_{FH} = 15.625$ kN (C)

7–33. $F_{HG} = 4.02$ kN (C)
$F_{KL} = 5.29$ kN (T)
$F_{HL} = 5.43$ kN (C)

7–34. $F_{HG} = 2.52$ kN (C)
$F_{KL} = 1.86$ kN (T)
$F_{HL} = 2.99$ kN (C)

Chapter 8

8–1. $\theta_A = \dfrac{Pa(a - L)}{2EI}$

$v_1 = \dfrac{Px_1}{6EI}[x_1^2 + 3a(a - L)]$

$v_2 = \dfrac{Pa}{6EI}[3x_2(x_2 - L) + a^2]$

$v_{max} = \dfrac{Pa}{24EI}(4a^2 - 3L^2)$

8–2. $\theta_A = -\dfrac{3PL^2}{8EI}$

$v_C = \dfrac{-PL^3}{6EI}$

8–3. $v_B = -\dfrac{11PL^3}{48EI}$

8–5. $v_1 = \dfrac{w}{24EI}(-x_1^4 + 4ax_1^3 - 6a^2x_1^2)$

$v_3 = \dfrac{wa^3}{24EI}(4x_3 + a - 4L)$

$\theta_B = -\dfrac{wa^3}{6EI}$

$v_B = \dfrac{wa^3}{24EI}(a - 4L)$

8–6. $(v_2)_{max} = \dfrac{wL^4}{18\sqrt{3}EI}$

8–7. $\theta_A = \dfrac{5w_0L^3}{192EI}$

$v = \dfrac{w_0x}{960EIL}(40L^2x^2 - 16x^4 - 25L^4)$

$v_{max} = \dfrac{w_0L^4}{120EI}$

8–9. $\theta_B = -\dfrac{7wa^3}{6EI}$

$v_1 = \dfrac{wax_1}{12EI}(2x_1^2 - 9ax_1)$

$v_C = -\dfrac{7wa^4}{12EI}$

$v_3 = \dfrac{w}{24EI}(-x_3^4 + 8ax_3^3 - 24a^2x_3^2 + 4a^3x_3 - a^4)$

8–10. $\theta_B = 0.00243$ rad $\triangledown$
$\Delta_{max} = 7.29$ mm $\downarrow$

8–11. $\theta_B = 0.00243$ rad $\triangledown$
$\Delta_{max} = 7.29$ mm $\downarrow$

8–13. $\theta_C = \dfrac{1750 \text{ kN·m}^2}{EI}$ $\triangledown$

$\Delta_C = \dfrac{7500 \text{ kN·m}^3}{EI}$ $\downarrow$

8–14. $a = 0.153\,L$
8–15. $a = 0.153\,L$

8–17. $a = \dfrac{L}{3}$

8–18. $\theta_C = \dfrac{Pa^2}{4EI}$ $\triangle$

$\Delta_C = \dfrac{Pa^3}{4EI}$ $\uparrow$

8–19. $\theta_C = \dfrac{Pa^2}{4EI}$ $\triangle$

$\Delta_C = \dfrac{Pa^3}{4EI}$ $\uparrow$

8–21. $\theta_C = 0.00171$ rad $\triangledown$
$\Delta_C = 3.86$ mm $\downarrow$

8–22. $a = 0.152\,L$
8–23. $a = 0.152\,L$

8–25. $\theta_B = \dfrac{18 \text{ kN·m}^2}{EI}$ $\triangledown$

$\Delta_C = \dfrac{90 \text{ kN·m}^3}{EI}$ $\downarrow$

8–26. $\theta_B = \dfrac{7Pa^2}{4EI}$ ◁

$\Delta_C = \dfrac{9Pa^3}{4EI}\downarrow$

8–27. $\theta_B = \dfrac{7Pa^2}{4EI}$ ▽

$\Delta_C = \dfrac{9Pa^3}{4EI}\downarrow$

8–29. $F = \dfrac{P}{4}$

8–30. $\theta_B = \dfrac{5Pa^2}{12EI}$ ▽

$\Delta_C = \dfrac{3Pa^3}{4EI}\downarrow$

8–31. $\theta_B = \dfrac{5Pa^2}{12EI}$ ▽

$\Delta_C = \dfrac{3Pa^3}{4EI}\downarrow$

8–33. $\theta_A = \dfrac{M_0L}{24EI}$ ◁

$\Delta_{max} = \dfrac{0.00802M_0L^2}{EI}\downarrow$

8–34. $\theta_C = \dfrac{3Pa^2}{EI}$ ◁

$\Delta_C = \dfrac{25\,Pa^3}{6EI}\downarrow$

8–35. $\theta_C = \dfrac{3Pa^2}{EI}$ ◁

$\Delta_C = \dfrac{25\,Pa^3}{6EI}\downarrow$

8–37. $\theta_D = \dfrac{75\ \text{kN}\cdot\text{m}^2}{EI}$ ▽

$\Delta_C = \dfrac{169\ \text{kN}\cdot\text{m}^3}{EI}\downarrow$

8–38. $\theta_D = \dfrac{560\ \text{kN}\cdot\text{m}^2}{EI}$ ◁

$\Delta_D = \dfrac{1920\ \text{kN}\cdot\text{m}^3}{EI}\downarrow$

8–39. $\theta_D = \dfrac{560\ \text{kN}\cdot\text{m}^2}{EI}$ ◁

$\Delta_D = \dfrac{1920\ \text{kN}\cdot\text{m}^3}{EI}\downarrow$

Chapter 9

9–1. $\Delta_{A_v} = 0.536\ \text{mm}\downarrow$

9–2. $\Delta_{A_v} = 0.536\ \text{mm}\downarrow$

9–3. $\Delta_{B_v} = 3.38\ \text{mm}\downarrow$

9–5. $\Delta_{E_v} = 2.95\ \text{mm}\downarrow$

9–6. $\Delta_{E_v} = 2.95\ \text{mm}\downarrow$

9–7. $\Delta_{D_v} = \dfrac{199\ \text{kN}\cdot\text{m}}{AE}\downarrow$

9–9. $\Delta_{F_v} = \dfrac{79.04\ \text{kN}^2\cdot\text{m}}{AE}\downarrow$

9–10. $\Delta_{F_v} = \dfrac{79.04\ \text{kN}^2\cdot\text{m}}{AE}\downarrow$

9–11. $\Delta_{A_v} = 1.76\ \text{mm}\downarrow$

9–13. $\Delta_{D_h} = \dfrac{255\ \text{kN}\cdot\text{m}}{AE}\rightarrow$

9–14. $\Delta_{D_h} = \dfrac{255\ \text{kN}\cdot\text{m}}{AE}\rightarrow$

9–15. $\Delta_{C_v} = 4.91\ \text{mm}\downarrow$

9–17. $\Delta_{A_v} = 1.03\ \text{mm}\downarrow$

9–18. $\Delta_{A_v} = 1.03\ \text{mm}\downarrow$

9–19. $\Delta_{A_v} = 12.7\ \text{mm}\uparrow$

9–21. $\Delta_C = \dfrac{PL^3}{48EI}\downarrow$

$\theta_B = \dfrac{PL^2}{16EI}$ ◁

9–22. $\Delta_C = \dfrac{PL^3}{48EI}\downarrow$

$\theta_B = \dfrac{PL^2}{16EI}$ ◁

9–23. $\Delta_C = \dfrac{2Pa^3}{3EI}\downarrow$

9–25. $\theta_C = \dfrac{5Pa^2}{6EI}$ ◁

9–26. $\theta_C = \dfrac{5Pa^2}{6EI}$ ◁

9–27. $\theta_A = \dfrac{Pa^2}{6EI}$ ◁

9–29. $\theta_C = 0.00146\ \text{rad}$ ◁
$\Delta_C = 3.75\ \text{mm}\downarrow$

9–30. $\theta_C = 0.00146\ \text{rad}$ ◁
$\Delta_C = 3.75\ \text{mm}\downarrow$

9–31. $\theta_C = 0.00563\ \text{rad}$ ◁
$\Delta_C = 6.56\ \text{mm}\downarrow$

9–33. $\theta_B = \dfrac{3150\ \text{N}\cdot\text{m}^2}{EI}$ ◁

$\Delta_B = \dfrac{6637.5\ \text{N}\cdot\text{m}^3}{EI}\downarrow$

9–34. $\theta_B = \dfrac{3150\ \text{N}\cdot\text{m}^2}{EI}$ ◁

$\Delta_B = \dfrac{6637.5\ \text{N}\cdot\text{m}^3}{EI}\downarrow$

9–35. $\theta_B = 0.00457\ \text{rad}$ ◁
$\Delta_B = 11.6\ \text{mm}$

9–37. $\theta_B = 0.00457$ rad $\triangle$
$\Delta_B = 12.0$ mm $\downarrow$

9–38. $\Delta_C = \dfrac{w_0 L^4}{120EI} \downarrow$

9–39. $\Delta_C = \dfrac{w_0 L^4}{120EI} \downarrow$

9–41 $\theta_A = \dfrac{9 \text{ kN} \cdot \text{m}^2}{EI} \; \triangledown$
$\Delta_A = \dfrac{22.95 \text{ kN} \cdot \text{m}^3}{EI} \downarrow$

9–42. $\Delta_D = \dfrac{109 \text{ kN} \cdot \text{m}^3}{EI} \downarrow$

9–43. $\Delta_D = \dfrac{109 \text{ kN} \cdot \text{m}^3}{EI} \downarrow$

9–45. $(\Delta_D)_v = \dfrac{59.4 \text{ kN} \cdot \text{m}^3}{EI} \downarrow$

9–46. $\Delta_{C_h} = \dfrac{5wL^4}{8EI}$

9–47. $\Delta_{B_y} = \dfrac{wL^4}{4EI}$

9–49. $\Delta_{C_h} = \dfrac{139.5 \text{ kN} \cdot \text{m}^3}{EI} \leftarrow$

9–50. $\Delta_{C_h} = \dfrac{139.5 \text{ kN} \cdot \text{m}^3}{EI} \leftarrow$

9–51. $(\Delta_C)_v = 2.81$ mm $\downarrow$

9–53. $(\Delta_C)_v = 2.81$ mm $\downarrow$

9–54. $\theta_A = 0.469(10^{-3})$ rad

9–55. $\theta_A = 0.469(10^{-3})$ rad

9–57. $(\Delta_C)_h = 1.211$ mm $\rightarrow$

9–58. $(\Delta_C)_h = \dfrac{9.62 \text{ kN} \cdot \text{m}^3}{EI} \rightarrow$

9–59. $(\Delta_C)_h = \dfrac{9.62 \text{ kN} \cdot \text{m}^3}{EI} \rightarrow$

9–61. $(\Delta_C)_v = \dfrac{45 \text{ kN} \cdot \text{m}^3}{EI} \downarrow$

Chapter 10

10–1. $A_x = 0$
$A_y = \dfrac{2w_0 L}{5}$
$B_y = \dfrac{w_0 L}{10}$
$M_A = \dfrac{w_0 L^2}{15}$

10–2. $C_x = 0$
$C_y = 73.125$ kN
$B_y = 153.75$ kN
$A_y = 13.125$ kN

10–3. $A_x = 0$
$A_y = \dfrac{57wL}{128}$
$B_y = \dfrac{7wL}{128}$
$M_A = \dfrac{9wL^2}{128}$

10–5. $A_y = \dfrac{3P}{2}$
$A_x = 0$
$M_A = \dfrac{PL}{2}$
$B_y = \dfrac{5P}{2}$

10–6. $B_y = 183.3$ kN
$A_x = 0$
$A_y = 70.3$ kN
$C_y = 70.3$ kN

10–7. $\Delta_B = 1.50$ mm

10–9. $\Delta_C = \dfrac{1650 \text{ kN} \cdot \text{m}^3}{EI}$

10–10. $B_y = 0.375$ kN
$A_x = 0$
$A_y = 0.375$ kN
$M_A = 0.5$ kN $\cdot$ m

10–11. $B_y = 36$ kN
$A_y = 4.5$ kN
$A_x = 0$
$C_y = 4.5$ kN

10–13. $C_x = 18.75$ kN
$A_x = 108.75$ kN
$C_y = 212$ kN
$A_y = 148$ kN

10–14. $C_y = 8.44$ kN
$A_x = 13.5$ kN
$A_y = 14.1$ kN
$M_A = 8.44$ kN $\cdot$ m

10–15. $A_y = 21.7$ kN
$C_x = 0$
$C_y = 28.3$ kN
$M_C = 15.7$ kN $\cdot$ m

10–17. $C_y = 39.0$ kN
$A_y = 33.0$ kN
$A_x = 24.0$ kN
$M_A = 45.0$ kN $\cdot$ m

10–18. $A_y = -75.25$ kN
$D_x = 10$ kN
$D_y = 74.75$ kN
$M_D = 29.25$ kN $\cdot$ m

10–19. $D_x = -12.6$ kN
$D_y = 125$ kN
$A_y = 125$ kN
$A_x = 12.6$ kN

10–21. $D_x = 27.0$ kN
$A_x = 13.0$ kN
$D_y = 23.2$ kN
$A_y = 23.2$ kN

10–22. $B_x = 2.65$ kN
$A_x = 2.65$ kN
$B_y = 0$
$A_y = 0$

10–23. $B_x = 1.53$ kN
$A_x = 1.53$ kN
$B_y = 7.50$ kN
$A_y = 15.0$ kN

10–25. $F_{AB} = 3.33$ kN (C)
$F_{BD} = 3.33$ kN (T)
$F_{BC} = 0$

10–26. $F_{CB} = 15.3$ kN (C)
$F_{AC} = 41.2$ kN (C)
$F_{DC} = 32.9$ kN (T)
$F_{DB} = 25.2$ kN (T)
$F_{AB} = 50.4$ kN (C)
$F_{DA} = 24.7$ kN (T)

10–27. $F_{AC} = 7.91$ kN (C)

10–29. $F_{AD} = 8.54$ kN (C)
$F_{AE} = 6.04$ kN (T)
$F_{AB} = 6.04$ kN (T)
$F_{CB} = 14.1$ kN (T)
$F_{CD} = 10.0$ kN (C)
$F_{BE} = 5.61$ kN (T)
$F_{BD} = 14.0$ kN (C)
$F_{DE} = 3.96$ kN (C)

10–30. $F_{AC} = 7.07$ kN (T)
$F_{DC} = F_{CB} = 2.07$ kN (T)
$F_{AD} = F_{AB} = 2.07$ kN (T)
$F_{DB} = 2.93$ kN (C)

10–31. $F_{CD} = 4.63$ kN (T)

10–33. $F_{DB} = 19.2$ kN
$F_{CB} = 53.4$ kN

10–34. $F_{BD} = 112.1$ kN (C)
$F_{AB} = 90.6$ kN (T)
$F_{BC} = 80.1$ kN (T)

10–35. $F_{BC} = 121.7$ kN (T)

10–37. $C_x = 0$
$$C_y = \frac{P}{3}$$

10–38. $F_{CD} = 25.8$ kN

10–39. $F_{AC} = 126$ kN

10–45. At 20 m, $C_y = 0.241$ kN

Chapter 11

11–1. $M_A = -6.93$ kN $\cdot$ m
$M_B = -13.1$ kN $\cdot$ m
$M_C = -15.9$ kN $\cdot$ m

11–2. $M_{AB} = -127.5$ kN $\cdot$ m
$M_{BA} = 105$ kN $\cdot$ m
$M_{BC} = -105$ kN $\cdot$ m
$M_{CB} = 60$ kN $\cdot$ m

11–3. $M_{AB} = -18.5$ kN $\cdot$ m
$M_{CB} = 20.4$ kN $\cdot$ m
$M_{BA} = 19.25$ kN $\cdot$ m
$M_{BC} = -19.25$ kN $\cdot$ m

11–5. $M_{AB} = 4.09$ kN $\cdot$ m
$M_{BA} = 8.18$ kN $\cdot$ m
$M_{BC} = -8.18$ kN $\cdot$ m
$M_{CB} = 8.18$ kN $\cdot$ m
$M_{CD} = -8.18$ kN $\cdot$ m
$M_{DC} = -4.09$ kN $\cdot$ m

11–6. $M_{AB} = -49.5$ kN $\cdot$ m
$M_{BA} = 13.5$ kN $\cdot$ m
$M_{BC} = -13.5$ kN $\cdot$ m
$M_{CB} = 9$ kN $\cdot$ m
$M_{CD} = -9$ kN $\cdot$ m
$M_{DC} = 40.5$ kN $\cdot$ m

11–7. $M_{BA} = 41.25$ kN $\cdot$ m
$M_{BC} = -41.25$ kN $\cdot$ m

11–9. $M_{AB} = -225$ kN $\cdot$ m
$M_{BA} = 89.1$ kN $\cdot$ m
$M_{BC} = -89.1$ kN $\cdot$ m
$M_{CB} = 3.52$ kN $\cdot$ m
$M_{CD} = -3.52$ kN $\cdot$ m

11–10. $M_{AB} = -21 \text{ kN} \cdot \text{m}$
$M_{BA} = 48 \text{ kN} \cdot \text{m}$

11–11. $M_{AB} = -30.6 \text{ kN} \cdot \text{m}$
$M_{BA} = -1.15 \text{ kN} \cdot \text{m}$
$M_{BC} = 1.15 \text{ kN} \cdot \text{m}$
$M_{CB} = 34.0 \text{ kN} \cdot \text{m}$
$M_{CD} = -34.0 \text{ kN} \cdot \text{m}$

11–13. $M_{AB} = -210 \text{ kN} \cdot \text{m}$
$M_{BA} = 120 \text{ kN} \cdot \text{m}$
$M_{BC} = -120 \text{ kN} \cdot \text{m}$
$M_{CB} = -60 \text{ kN} \cdot \text{m}$

11–14. $M_{AB} = -64.4 \text{ kN} \cdot \text{m}$
$M_{CB} = 25.0 \text{ kN} \cdot \text{m}$

11–15. $M_{AB} = -1.98 \text{ kN} \cdot \text{m}$
$M_{BA} = 0.540 \text{ kN} \cdot \text{m}$
$M_{BC} = -0.540 \text{ kN} \cdot \text{m}$

11–17. $M_{AB} = -3.52 \text{ kN} \cdot \text{m}$
$M_{BA} = 68.0 \text{ kN} \cdot \text{m}$
$M_{BC} = -68.0 \text{ kN} \cdot \text{m}$

11–18. $M_{BA} = 69.8 \text{ kN} \cdot \text{m}$
$M_{BC} = -34.9 \text{ kN} \cdot \text{m}$
$M_{BD} = -34.9 \text{ kN} \cdot \text{m}$

11–19. $M_{DC} = -20.1 \text{ kN} \cdot \text{m}$
$M_{CD} = 20.1 \text{ kN} \cdot \text{m}$
$M_{DA} = 20.1 \text{ kN} \cdot \text{m}$
$M_{CB} = -20.1 \text{ kN} \cdot \text{m}$

11–21. $M_{DC} = 64.0 \text{ kN} \cdot \text{m}$
$M_{CD} = 80.0 \text{ kN} \cdot \text{m}$
$M_{DA} = -64.0 \text{ kN} \cdot \text{m}$
$M_{CB} = -80.0 \text{ kN} \cdot \text{m}$

11–22. $M_{AD} = -25.9 \text{ kN} \cdot \text{m}$
$M_{DA} = -3.32 \text{ kN} \cdot \text{m}$
$M_{DC} = 3.32 \text{ kN} \cdot \text{m}$
$M_{CD} = 6.32 \text{ kN} \cdot \text{m}$
$M_{BC} = -9.43 \text{ kN} \cdot \text{m}$
$M_{CB} = -6.32 \text{ kN} \cdot \text{m}$

11–23. $M_{AB} = 25.4 \text{ kN} \cdot \text{m}$
$M_{DC} = -56.7 \text{ kN} \cdot \text{m}$

Chapter 12

12–1. $M_B = M_C = -84.0 \text{ kN} \cdot \text{m}$

12–2. $M_A = -230 \text{ kN} \cdot \text{m}$
$M_B = -187 \text{ kN} \cdot \text{m}$
$M_C = -122 \text{ kN} \cdot \text{m}$

12–7. $A_x = 0$
$A_y = 33 \text{ kN}$
$B_y = 33 \text{ kN}$
$M_A = 30 \text{ kN} \cdot \text{m}$
$C_y = 6 \text{ kN}$

12–14. $M_{AB} = -2.30 \text{ kN} \cdot \text{m}$
$M_{BA} = 19.4 \text{ kN} \cdot \text{m}$
$M_{BC} = -19.4 \text{ kN} \cdot \text{m}$
$M_{CB} = 0$

12–15. $A_x = 87.8 \text{ kN}$
$A_y = 288 \text{ kN}$
$M_A = 146 \text{ kN} \cdot \text{m}$
$D_x = 87.8 \text{ kN}$
$D_y = 288 \text{ kN}$
$M_D = 146 \text{ kN} \cdot \text{m}$

12–19. $M_A = 20.6 \text{ kN} \cdot \text{m}$
$M_B = -41.1 \text{ kN} \cdot \text{m}$
$M_C = -41.1 \text{ kN} \cdot \text{m}$
$M_D = 20.6 \text{ kN} \cdot \text{m}$

12–22. $M_{AB} = 128 \text{ kN} \cdot \text{m}$
$M_{BA} = 218 \text{ kN} \cdot \text{m}$
$M_{BC} = -218 \text{ kN} \cdot \text{m}$
$M_{CB} = 175 \text{ kN} \cdot \text{m}$
$M_{CD} = -175 \text{ kN} \cdot \text{m}$
$M_{DC} = -55.7 \text{ kN} \cdot \text{m}$

12–23. $M_{BA} = -104 \text{ kN} \cdot \text{m}$
$M_{BC} = 104 \text{ kN} \cdot \text{m}$
$M_{CB} = 196 \text{ kN} \cdot \text{m}$
$M_{CD} = -196 \text{ kN} \cdot \text{m}$
$M_{AB} = M_{DC} = 0$

12–25. $M_{BA} = 24.0 \text{ kN} \cdot \text{m}$
$M_{BC} = -24.0 \text{ kN} \cdot \text{m}$
$M_{CB} = -24.0 \text{ kN} \cdot \text{m}$
$M_{CD} = 24.0 \text{ kN} \cdot \text{m}$

12–26. $M_{DA} = 14.2 \text{ kN} \cdot \text{m}$
$M_{DC} = -14.2 \text{ kN} \cdot \text{m}$
$M_{CD} = -7.54 \text{ kN} \cdot \text{m}$
$M_{CB} = 7.54 \text{ kN} \cdot \text{m}$

Chapter 13

13–1. $M_{AB} = -348 \text{ kN} \cdot \text{m}$
$M_{BA} = 301 \text{ kN} \cdot \text{m}$
$M_{BC} = -301 \text{ kN} \cdot \text{m}$
$M_{CB} = 348 \text{ kN} \cdot \text{m}$

13–2. $M_{AB} = -348$ kN $\cdot$ m
$M_{BA} = 301$ kN $\cdot$ m
$M_{BC} = -301$ kN $\cdot$ m
$M_{CB} = 348$ kN $\cdot$ m

13–3. $M_{AC} = 37.6$ kN $\cdot$ m
$M_{CA} = 75.1$ kN $\cdot$ m
$M_{CB} = -75.1$ kN $\cdot$ m
$M_{BC} = 369$ kN $\cdot$ m

13–5. $M_{AB} = 0$
$M_{BA} = 604$ kN $\cdot$ m
$M_{BC} = -609$ kN $\cdot$ m
$M_{BF} = 5.49$ kN $\cdot$ m
$M_{FB} = 2.76$ kN $\cdot$ m
$M_{CB} = 609$ kN $\cdot$ m
$M_{CD} = -604$ kN $\cdot$ m
$M_{CE} = -5.49$ kN $\cdot$ m
$M_{EC} = -2.76$ kN $\cdot$ m
$M_{DC} = 0$

13–6. $M_{AB} = 0$
$M_{BA} = 604$ kN $\cdot$ m
$M_{BC} = -610$ kN $\cdot$ m
$M_{BF} = 5.53$ kN $\cdot$ m
$M_{FB} = 2.77$ kN $\cdot$ m
$M_{CB} = 610$ kN $\cdot$ m
$M_{CD} = -604$ kN $\cdot$ m
$M_{CE} = -5.53$ kN $\cdot$ m
$M_{EC} = -2.77$ kN $\cdot$ m
$M_{DC} = 0$

13–7. $M_{AB} = 1.75$ kN $\cdot$ m
$M_{BA} = 3.51$ kN $\cdot$ m
$M_{BC} = -3.51$ kN $\cdot$ m
$M_{CB} = 3.51$ kN $\cdot$ m
$M_{CD} = -3.51$ kN $\cdot$ m
$M_{DC} = -1.75$ kN $\cdot$ m

13–9. $M_{BA} = 28.3$ kN $\cdot$ m
$M_{BD} = -28.3$ kN $\cdot$ m
$M_{DB} = 28.3$ kN $\cdot$ m
$M_{DC} = -28.3$ kN $\cdot$ m
$M_{AB} = M_{CD} = 0$

13–10. $M_{BA} = 28.3$ kN $\cdot$ m
$M_{BD} = -28.3$ kN $\cdot$ m
$M_{DB} = 28.3$ kN $\cdot$ m

$M_{DC} = -28.3$ kN $\cdot$ m
$M_{AB} = M_{CD} = 0$

13–11. $M_{CD} = M_{BA} = 180$ kN $\cdot$ m
$M_{CF} = M_{BE} = 94.6$ kN $\cdot$ m
$M_{CB} = M_{BC} = -274$ kN $\cdot$ m
$M_{FC} = M_{EB} = 47.3$ kN $\cdot$ m

Chapter 14

14–1.

$$\mathbf{K} = (10^3) \begin{bmatrix} 84.53 & 0 & -33.33 & 0 & -25.6 & -19.2 & -25.6 & 19.2 \\ 0 & 28.8 & 0 & 0 & -19.2 & -14.4 & 19.2 & -14.4 \\ -33.33 & 0 & 33.33 & 0 & 0 & 0 & 0 & 0 \\ 0 & 0 & 0 & 0 & 0 & 0 & 0 & 0 \\ -25.6 & -19.2 & 0 & 0 & 25.6 & 19.2 & 0 & 0 \\ -19.2 & -14.4 & 0 & 0 & 19.2 & 14.4 & 0 & 0 \\ -25.6 & 19.2 & 0 & 0 & 0 & 0 & 25.6 & -19.2 \\ 19.2 & -14.4 & 0 & 0 & 0 & 0 & -19.2 & 14.4 \end{bmatrix}$$

14–2. $D_1 = 0$
$D_2 = -0.694$ mm

14–3. $q_1 = 16.7$ kN (C)
$q_2 = 0$
$q_3 = 16.7$ kN (T)

14–5. $D_1 = -0.0624$ mm
$q_2 = 50.9$ kN (C)

14–6. $q_2 = 59.7$ kN (C)

14–7. $\mathbf{K} =$

$$\begin{bmatrix} 203.033 & -53.033 & -53.033 & 53.033 & -150 & 0 & 0 & 0 & 0 & 0 \\ -53.033 & 53.033 & 53.033 & -53.033 & 0 & 0 & 0 & 0 & 0 & 0 \\ -53.033 & 53.033 & 256.066 & 0 & 0 & 0 & -53.033 & -53.033 & -150 & 0 \\ 53.033 & -53.033 & 0 & 256.066 & 0 & -150 & -53.033 & -53.033 & 0 & 0 \\ -150 & 0 & 0 & 0 & 300 & 0 & -150 & 0 & 0 & 0 \\ 0 & 0 & 0 & -150 & 0 & 150 & 0 & 0 & 0 & 0 \\ 0 & 0 & -53.033 & -53.033 & -150 & 0 & 203.033 & 53.033 & 0 & 0 \\ 0 & 0 & -53.033 & -53.033 & 0 & 0 & 53.033 & 53.033 & 0 & 0 \\ 0 & 0 & -150 & 0 & 0 & 0 & 0 & 0 & 150 & 0 \\ 0 & 0 & 0 & 0 & 0 & 0 & 0 & 0 & 0 & 0 \end{bmatrix} (10^6)$$

14–9. $\mathbf{K} =$

$$\begin{bmatrix} 113.4 & 28.8 & -75 & 0 & -38.4 & -28.8 & 0 & 0 & 0 & 0 \\ 28.8 & 21.6 & 0 & 0 & -28.8 & -21.6 & 0 & 0 & 0 & 0 \\ -75 & 0 & 150 & 0 & 0 & 0 & 0 & 0 & -75 & 0 \\ 0 & 0 & 0 & 100 & 0 & -100 & 0 & 0 & 0 & 0 \\ -38.4 & -28.8 & 0 & 0 & 151.8 & 0 & 0 & -75 & -38.4 & 28.8 \\ -28.8 & -21.6 & 0 & -100 & 0 & 143.2 & 0 & 0 & 28.8 & -21.6 \\ 0 & 0 & 0 & 0 & 0 & 0 & 100 & 0 & 0 & -100 \\ 0 & 0 & 0 & 0 & -75 & 0 & 0 & 75 & 0 & 0 \\ 0 & 0 & -75 & 0 & -38.4 & 28.8 & 0 & 0 & 113.4 & -28.8 \\ 0 & 0 & 0 & 0 & 28.8 & -21.6 & -100 & 0 & -28.8 & 121.6 \end{bmatrix} (10^6)$$

$M_{DC} = -28.3$ kN $\cdot$ m
$M_{AB} = M_{CD} = 0$

14–10. $q_5 = 33.3$ kN

14–11. $D_6 = 0.0133$ m

14–13. $D_5 = 0.2197(10^{-3})$ m

$q_5 = 8.20$ kN (C)

14–14. $q_3 = 17.7$ kN (T)

14–15.

$$\mathbf{K} = AE \begin{bmatrix} 0.40533 & 0.096 & 0.01697 & -0.11879 & -0.33333 & 0 \\ 0.096 & 0.128 & 0.02263 & -0.15839 & 0 & 0 \\ 0.01697 & 0.02263 & 0.129 & -0.153 & 0 & 0.17678 \\ -0.11879 & -0.15839 & -0.153 & 0.321 & 0 & -0.17678 \\ -0.33333 & 0 & 0 & 0 & 0.33333 & 0 \\ 0 & 0 & 0.17678 & -0.17678 & 0 & 0.25 \end{bmatrix}$$

Chapter 15

15–1. $M_1 = 90$ kN $\cdot$ m

$M_3 = 22.5$ kN $\cdot$ m

15–2. $M_1 = 27.5$ kN $\cdot$ m

$M_3 = 116$ kN $\cdot$ m

15–3. $R_3 = 7.85$ kN

$R_4 = 40.2$ kN

$R_5 = 86.6$ kN $\cdot$ m

$R_6 = 39.6$ kN

15–5. $R_4 = 1.93$ kN

$R_5 = 34.5$ kN

$R_6 = 12.4$ kN

15–6. $R_3 = 32.25$ kN

$R_4 = 85.75$ kN

$R_5 = 22.0$ kN

$R_6 = 14.0$ kN $\cdot$ m

15–7. $R_2 = 41.4$ kN

$R_3 = 7.725$ kN

$R_4 = 2.30$ kN $\cdot$ m

$R_5 = 28.9$ kN

$R_6 = 30.8$ kN $\cdot$ m

15–9. $M_2 = M_3 = 44.2$ kN $\cdot$ m

15–10. $Q_4 = 127.5$ kN

$Q_5 = 105$ kN

$Q_6 = 127.5$ kN

15–11. $R_2 = 80$ kN $\cdot$ m

$R_3 = 120$ kN

$R_4 = 160$ kN $\cdot$ m

Chapter 16

16–1.

$$\mathbf{K} = \begin{bmatrix} 511.29 & 0 & 22.5 & -11.25 & 0 & 22.5 & -500 & 0 & 0 \\ 0 & 511.25 & -22.5 & 0 & -500 & 0 & 0 & -11.25 & -22.5 \\ 22.5 & -22.5 & 120 & -22.5 & 0 & 30 & 0 & 22.5 & 30 \\ -11.25 & 0 & -22.5 & 11.25 & 0 & -22.5 & 0 & 0 & 0 \\ 0 & -500 & 0 & 0 & 500 & 0 & 0 & 0 & 0 \\ 22.5 & 0 & 30 & -22.5 & 0 & 60 & 0 & 0 & 0 \\ -500 & 0 & 0 & 0 & 0 & 0 & 500 & 0 & 0 \\ 0 & -11.25 & 22.5 & 0 & 0 & 0 & 0 & 11.25 & 22.5 \\ 0 & -22.5 & 30 & 0 & 0 & 0 & 0 & 22.5 & 60 \end{bmatrix} (10^6)$$

16–2. $R_4 = 3.21$ kN $\rightarrow$

$R_5 = 21.6$ kN $\uparrow$

$R_6 = 2.72$ kN $\cdot$ m $\downarrow$

$R_7 = 6.79$ kN $\rightarrow$

$R_8 = 26.4$ kN $\uparrow$

$R_9 = 19.6$ kN $\cdot$ m $\nwarrow$

16–3.

$$\mathbf{K} = \begin{bmatrix} 851250 & 0 & 22500 & 22500 & -11250 & 0 & -440000 & 0 & 0 \\ 0 & 1055760 & -14400 & 0 & 0 & -1050000 & 0 & -5760 & -14400 \\ 22500 & -14400 & 108000 & 30000 & -22500 & 0 & 0 & 14400 & 24000 \\ 22500 & 0 & 30000 & 60000 & -22500 & 0 & 0 & 0 & 0 \\ -11250 & 0 & -22500 & -22500 & 11250 & 0 & 0 & 0 & 0 \\ 0 & -1050000 & 0 & 0 & 0 & 1050000 & 0 & 0 & 0 \\ -840000 & 0 & 0 & 0 & 0 & 0 & 140000 & 0 & 0 \\ 0 & -5760 & 14400 & 0 & 0 & 0 & 0 & 5760 & 14400 \\ 0 & -14400 & 24000 & 0 & 0 & 0 & 0 & 14400 & 48000 \end{bmatrix}$$

16–5.

$$\mathbf{K} = \begin{bmatrix} 763.125 & 0 & 26.25 & 26.25 & 0 & -13.125 & 0 & -750 & 0 \\ 0 & 763.125 & -26.25 & 0 & -26.25 & 0 & -750 & 0 & -13.125 \\ 26.25 & -26.25 & 140 & 35 & 35 & -26.25 & 0 & 0 & 26.25 \\ 26.25 & 0 & 35 & 70 & 0 & -26.25 & 0 & 0 & 0 \\ 0 & -26.25 & 35 & 0 & 70 & 0 & 0 & 0 & 26.25 \\ -13.125 & 0 & -26.25 & -26.25 & 0 & 13.175 & 0 & 0 & 0 \\ 0 & -750 & 0 & 0 & 0 & 0 & 750 & 0 & 0 \\ -750 & 0 & 0 & 0 & 0 & 0 & 0 & 750 & 0 \\ 0 & -13.125 & 26.25 & 0 & 26.25 & 0 & 0 & 0 & 13.125 \end{bmatrix} (10^6)$$

16–6. $R_6 = 5.54$ kN

$R_7 = 35.5$ kN

$R_8 = 5.54$ kN

$R_9 = 24.5$ kN

16–7.

$$\mathbf{K} = (10^3) \begin{bmatrix} 480 & 0 & 0 & -480 & 0 & 0 & 0 & 0 & 0 \\ 0 & 4.8 & 12 & 0 & -4.8 & 12 & 0 & 0 & 0 \\ 0 & 12 & 40 & 0 & -12 & 20 & 0 & 0 & 0 \\ -480 & 0 & 0 & 482.78 & 0 & 8.333 & -2.778 & 0 & 8.333 \\ 0 & -4.8 & -12 & 0 & 404.8 & -12 & 0 & -400 & 0 \\ 0 & 12 & 20 & 8.333 & -12 & 73.33 & -8.333 & 0 & 16.67 \\ 0 & 0 & 0 & -2.778 & 0 & -8.333 & 2.778 & 0 & -8.333 \\ 0 & 0 & 0 & 0 & -400 & 0 & 0 & 400 & 0 \\ 0 & 0 & 0 & 8.333 & 0 & 16.67 & -8.333 & 0 & 33.33 \end{bmatrix}$$

16–9. $\mathbf{K} = \begin{bmatrix} 210667 & 0 & 16000 & 0 & 16000 & -200000 & 0 & -10667 & 0 \\ 0 & 401333 & 4000 & 4000 & 0 & 0 & -1333 & 0 & -400000 \\ 16000 & 4000 & 48000 & 8000 & 16000 & 0 & -4000 & -16000 & 0 \\ 0 & 4000 & 8000 & 16000 & 0 & 0 & -4000 & 0 & 0 \\ 16000 & 0 & 16000 & 0 & 32000 & 0 & 0 & -16000 & 0 \\ -200000 & 0 & 0 & 0 & 0 & 200000 & 0 & 0 & 0 \\ 0 & -1333 & -4000 & -4000 & 0 & 0 & 1333 & 0 & 0 \\ -10667 & 0 & -16000 & 0 & -16000 & 0 & 0 & 10667 & 0 \\ 0 & -400000 & 0 & 0 & 0 & 0 & 0 & 0 & 400000 \end{bmatrix}$

16–10. $R_7 = 90 \text{ kN}$
$R_8 = 0$
$R_9 = 90 \text{ kN}$

16–11. $\mathbf{K} = \begin{bmatrix} 410500 & 0 & -21000 & -21000 & 0 & -10500 & 0 & -400000 & 0 \\ 0 & 603111 & -9333 & 0 & -9333 & 0 & -600000 & 0 & -3111 \\ -21000 & -9333 & 93333 & 28000 & 18667 & 21000 & 0 & 0 & 9333 \\ -21000 & 0 & 28000 & 56000 & 0 & 21000 & 0 & 0 & 0 \\ 0 & -9333 & 18667 & 0 & 37333 & 0 & 0 & 0 & 9333 \\ -10500 & 0 & 21000 & 21000 & 0 & 10500 & 0 & 0 & 0 \\ 0 & -600000 & 0 & 0 & 0 & 0 & 600000 & 0 & 0 \\ -400000 & 0 & 0 & 0 & 0 & 0 & 0 & 400000 & 0 \\ 0 & -3111 & 9333 & 0 & 9333 & 0 & 0 & 0 & 3111 \end{bmatrix}$

Appendix

A–1. $2\mathbf{A} - \mathbf{B} = \begin{bmatrix} 7 & 10 \\ -1 & 6 \\ 10 & -5 \end{bmatrix}$

$\mathbf{A} + 3\mathbf{B} = \begin{bmatrix} 0 & 12 \\ 17 & 31 \\ -2 & 1 \end{bmatrix}$

A–2. $3\mathbf{A} - 2\mathbf{B} = \begin{bmatrix} -3 & 7 & 0 \\ 6 & 5 & 7 \\ -7 & -5 & 9 \end{bmatrix}$

$\mathbf{A} - 2\mathbf{B} = \begin{bmatrix} -9 & -3 & 4 \\ -2 & -1 & 5 \\ -9 & -3 & -5 \end{bmatrix}$

A–3. $\mathbf{AB} = \begin{bmatrix} 18 & -12 \end{bmatrix}$

A–5. $\mathbf{AB} = \begin{bmatrix} 8 & 12 & -10 \\ -20 & -30 & 25 \\ 24 & 36 & -30 \end{bmatrix}$

A–6. $(\mathbf{A} + \mathbf{B})^T = \begin{bmatrix} 1 & 9 & 10 \end{bmatrix} = \mathbf{A}^T + \mathbf{B}^T$

A–7. $\mathbf{A} + \mathbf{A}^T = \begin{bmatrix} 4 & 8 & 5 \\ 8 & 18 & 2 \\ 5 & 2 & 4 \end{bmatrix}$

A–9. $\mathbf{A}^T = \begin{bmatrix} 68 & 38 \\ 38 & 26 \end{bmatrix}$

A–10. $\mathbf{AB} = \begin{bmatrix} 10 \\ -5 \end{bmatrix}$

A–11. $\mathbf{AB} = \begin{bmatrix} 30 \\ 11 \end{bmatrix}$

A–15. $|A| = 27$
$|B| = -30$

A–17. $\mathbf{A}^{-1} = \dfrac{1}{43} \begin{bmatrix} -7 & 16 & 17 \\ -4 & 3 & 22 \\ 12 & -9 & -23 \end{bmatrix}$

A–18. $x_1 = -\dfrac{4}{9}$
$x_2 = -\dfrac{5}{9}$
$x_3 = \dfrac{4}{3}$

A–19. $x_1 = -\dfrac{4}{9}$
$x_2 = -\dfrac{5}{9}$
$x_3 = \dfrac{4}{3}$

A–21. $x_1 = 1$
$x_2 = 1$
$x_3 = -1$

Index

Geometric Properties of Areas

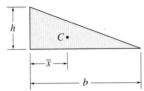

Triangle

$$A = \frac{1}{2} bh$$

$$\bar{x} = \frac{1}{3} b$$

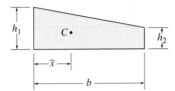

Trapezoid

$$A = \frac{1}{2} b (h_1 + h_2)$$

$$\bar{x} = \frac{b (2h_2 + h_1)}{3 (h_1 + h_2)}$$

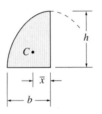

Semi Parabola

$$A = \frac{2}{3} bh$$

$$\bar{x} = \frac{3}{8} b$$

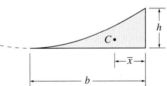

Parabolic spandrel

$$A = \frac{1}{3} bh$$

$$\bar{x} = \frac{1}{4} b$$

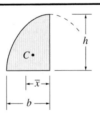

Semi-segment of nth degree curve

$$A = bh \left(\frac{n}{n + 1} \right)$$

$$\bar{x} = \frac{b (n + 1)}{2 (n + 2)}$$

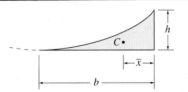

Spandrel of nth degree curve

$$A = bh \left(\frac{1}{n + 1} \right)$$

$$\bar{x} = \frac{b}{(n + 2)}$$

Fixed End Moments

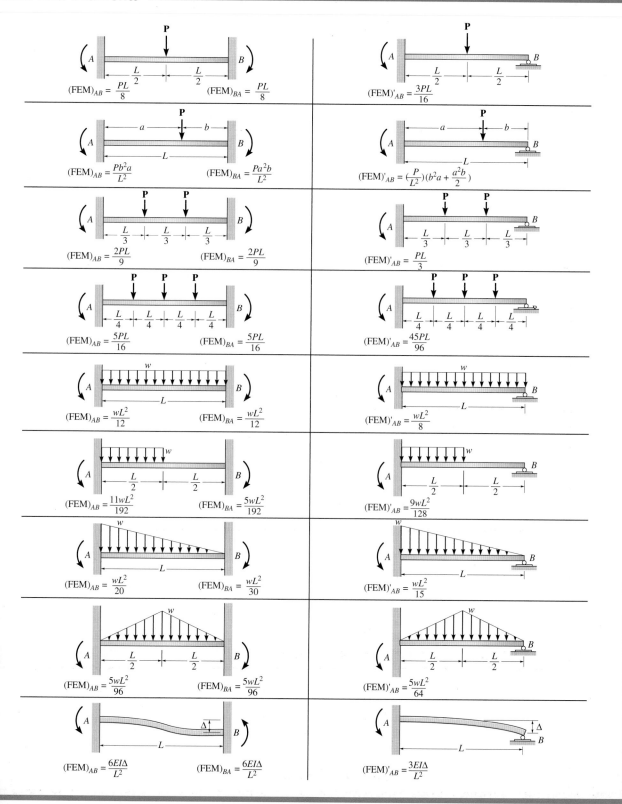

$(\text{FEM})_{AB} = \dfrac{PL}{8}$ $(\text{FEM})_{BA} = \dfrac{PL}{8}$

$(\text{FEM})'_{AB} = \dfrac{3PL}{16}$

$(\text{FEM})_{AB} = \dfrac{Pb^2a}{L^2}$ $(\text{FEM})_{BA} = \dfrac{Pa^2b}{L^2}$

$(\text{FEM})'_{AB} = \left(\dfrac{P}{L^2}\right)\left(b^2a + \dfrac{a^2b}{2}\right)$

$(\text{FEM})_{AB} = \dfrac{2PL}{9}$ $(\text{FEM})_{BA} = \dfrac{2PL}{9}$

$(\text{FEM})'_{AB} = \dfrac{PL}{3}$

$(\text{FEM})_{AB} = \dfrac{5PL}{16}$ $(\text{FEM})_{BA} = \dfrac{5PL}{16}$

$(\text{FEM})'_{AB} = \dfrac{45PL}{96}$

$(\text{FEM})_{AB} = \dfrac{wL^2}{12}$ $(\text{FEM})_{BA} = \dfrac{wL^2}{12}$

$(\text{FEM})'_{AB} = \dfrac{wL^2}{8}$

$(\text{FEM})_{AB} = \dfrac{11wL^2}{192}$ $(\text{FEM})_{BA} = \dfrac{5wL^2}{192}$

$(\text{FEM})'_{AB} = \dfrac{9wL^2}{128}$

$(\text{FEM})_{AB} = \dfrac{wL^2}{20}$ $(\text{FEM})_{BA} = \dfrac{wL^2}{30}$

$(\text{FEM})'_{AB} = \dfrac{wL^2}{15}$

$(\text{FEM})_{AB} = \dfrac{5wL^2}{96}$ $(\text{FEM})_{BA} = \dfrac{5wL^2}{96}$

$(\text{FEM})'_{AB} = \dfrac{5wL^2}{64}$

$(\text{FEM})_{AB} = \dfrac{6EI\Delta}{L^2}$ $(\text{FEM})_{BA} = \dfrac{6EI\Delta}{L^2}$

$(\text{FEM})'_{AB} = \dfrac{3EI\Delta}{L^2}$

ts noted that what they read led them to important discoveri
al construction or a performance, that intersections of rac
xuality are critical but difficult and complex, and that cultura
s contexts must be accounted for.

ndents identified particular moments that spurred their identifi
st issues or with feminist scholarship. These are autobiographi
olving a positive or negative experience with a paper accepted
anel or speaker attended, a significant moment in an associa-
e exclusion, or invisibility of women in conferences or events, or
ith a feminist colleague begun. For example, there were negative

few bonding moments with other women for me. Everywhere I looked
s were men and the professors were men... My dean told me to abandon
bout women in media management because I could more easily get tenure
genda setting research... I didn't like that either and didn't listen. It angered
anyone would suggest feminist work wasn't worthy or tenurable ... I realized
that the communication conventions I attended had rules about scholarship
inist work was not particularly well-received.

woman said:

first time I went to AEJMC was in 1979—when the association still had its
al meetings at university campuses. I distinctly recall going into the cafete-
at lunchtime—so this would be the best opportunity to see the "whole" of the
mbership, as opposed to smaller audiences at individual sessions—and being
eeted by a sea of white men in short-sleeved blue shirts, only distinguished, it
emed to me, by whether they had white hair or were bald. Frankly, my first ques-
on was why the average age skewed so high—but again, this was at a point when
he membership was very dominated by former/retired journalists. The very next
thought was, where are the women? At some party I met some women, but nearly
all of them were the wives of these men; they took tours during the day during the
convention, and then returned at night. And nearly every man I met asked me to
whom I was married, assuming that I was hanging around because I was married to
an AEJMC member.

Not everyone's experience with an association was negative:

I joined AEJMC in 1976 and have attended every convention since then (through
2006 except for the year 1977). In the early years I found support from exemplary individuals. . . I participated in the activities of the Commission on the Status of Women
and found it a congenial group. I must say I did not find much (if any) hostility from
male scholars in AEJMC. Perhaps by the time I was really involved in the organization (the 1980s) the groundwork for women's participation had been laid by others.

Women (and one man prominent in masculinity studies) outside of the academic discipline of communication comprised a second group significant to the
thinking and research of respondents. These are theorists from other disciplines
as well as public scholars and figures. Included in this group of eighty-four individuals (see Table 4.1) are feminist philosophers, feminist theologians, feminist
sociologists, women's rights advocates, and feminist activists from the nineteenth century through the twentieth century, and women journalists and other
writers. Those identified ranged from Betty Friedan (*Feminine Mystique*), Carol
Gilligan (*In a Different Voice*), bell hooks (*From Margin to Center*), Luce Irigaray
(*This Sex Which is Not One*), Thersesa de Lauretis (*Technologies of Gender*),
Patricia Hill Collins (*Black Feminist Thought*), Rosa Luxemburg (*Social Reform
or Revolution*), Batya Weinbaum (*Pictures of Patriarchy*), and Fatima Mernissi
(*Women's Rebellion and Islamic Memory*). Clearly, both lists are not exhaustive:
other feminist communication scholars will have names to add. As a step toward
constructing a genealogy (matrilineal rather than patrilineal) of ideas relevant to
feminist work in communication, a table of individuals outside of the discipline
of communication is included (see Table 4.1). It is one small corrective step to
Beniger's list of top theorists.

Table 4.1. A sampling of contributors to the history of ideas in communication

Individual authors and public figures outside of the academic field of communication identified
by a group of scholars as significant to their development as feminist communication scholars.

Abzug, Bella
Acker, Joan
Alexander, Shana
Anthony, Susan B.
Anzaldua, Gloria
Atwood, Margaret
Barrett, Michelle
Bartky, Sandra Lee
Benderly, Beryl
Benedict, Helen
Bosmajian, Hamida
Butler, Judith
Christ, Carol
Cixous, Helene
Cockburn, Cynthia
Collins, Patricia Hill
Daly, Mary

De Beauvoir, Simone
de Lauretis, Theresa
Dickinson, Emily
Donovan, Josephine
Duniway, Abigail Scott
Dworkin, Andrea
Eisenstein, Zillah
Eiserly, Riane
Elgin, Suzette Baden
Enlow, Cynthia
Faludi, Susan
Flax, Jane
Ferguson, Kathy
Fraser, Nancy
Friedan, Betty
Gherardi, Silvia
Gilligan, Carol
Gilman, Charlotte Perkins
Goldman, Emma
Gottfried, Heidi
Haraway, Donna
Harding, Sandra
hooks, bell
Irigaray, Luce
Johnson, Sonja
Kanter, Rosabeth Moss
Kimmel, Michael
King, Carole
Kristeva, Julia
Le Guin, Ursula
Lorde, Audre
Luxemburg, Rosa
Mani, Lata
MacKinnon, Catherine
Mernissi, Fatima
Montenegro, Sofia
Morgan, Robin
Morton, Nelle
Mulvey, Laura
Parsons, Elsie Clews
Piercy, Marge
Plath, Sylvia
Rich, Adrienne
Richardson, Laurel
Ruether, Rosemary Radford
Sandler, Berenice
Sarton, May

Scott, Joan
Simon, Carly
Spivak, Gayatri
Stanton, Elizabeth Cady
Starhawk
Steinem, Gloria
Taylor, Verta
Tong, Rosemary
Vance, Carole
Walker, Alice
Weedon, Chris
Weinbaum, Batya
Welty, Eudora
Willard, Frances
Williams, Judith
Wilson-Kastner, Patricia
Wollstonecraft, Mary
Woodhull, Victoria
Woolf, Virginia
Zita, Jacquelyn

Other responder
that gender is a so
class, gender, and s
global, and religiou
Second, respo
cation with femin
cal moments inv
or presented, a
tion's history, th
a relationship v
experiences:

There were
the schola
research a
with my
me that
early or
and fe

Respondents emphasized how the works they
fessional, and personal turning points for them. One

In 1982, when I was a student in a major university's mast
program, I somehow came across a copy of *Women in Med*
Book, edited by Maurine Beasley and Sheila Gibbons . . . I don
it, but I remember carrying it around like a Bible. With its shor
of women journalists and its radical "Principles of Feminist Jou
how a sacred text to me—in part because it was in every way dif
nalism program, where all the professors were male except one and
to take one of the required classes because I'd heard that the profes
his shoe and thrown it across the room at a female student. The truth
program. Journalism felt shallow and false to me . . . In some way I coul
the time, the book felt true to me. The problem was, I didn't know what

Another
The
ann
ria
me
gr
s
t

Work from other disciplines was equally influential. Feminist
have been included in respondents' undergraduate or graduate progi
ing them to find it on their own or rely on advice from role models in
plines on their campuses. One respondent said:

But in the 1980s when I was coming through graduate studies, I read widely in
eral fields and this really helped me situate feminist communication scholarship
the larger developing landscape of feminist scholarship.

NCA also had its forerunners at a time when feminist concerns were growing on campuses in every discipline. As one respondent described her experience:

> I gave my first convention paper at ICA in Montreal in 1974, on women's and men's speech. The man checking my luggage at the Montreal airport looked through some of the pages of a copy of my convention paper (clearly marked as a paper to be presented at ICA) and asked me to step into an office for further questioning. Eventually another man came in and asked me if I was going to try to sell my papers. There must be thousands of people coming through that airport most weeks on their way to a conference, so I've always wondered if the topic of the talk called forth this extra attention . . . Feminists were looking at all the communication troubles in many academic fields, so there were wonderful discussions across disciples. Since initially there was no one on campus to talk with (while there often was a lot of ridicule), I was delighted to meet feminists at these conferences.

Clearly the support of women in the associations—who organized panels, gave new scholars opportunities to review papers or present, chose helpful respondents, and encouraged research and paper submissions—were significant to nurturing feminist scholars and, thereby, feminist scholarship in the associations. More importantly, the creation of organizations and divisions devoted to research on women and communication and feminist scholarship provided spaces for the development of feminist scholarship. These included the six divisions and associations represented in the call for this research, as well as the much newer International Gender and Language Association (established in Berkeley in 1999) and other special workshops or conferences on gender. (One respondent noted that her first experience was in a 1994 workshop of the section "public, media and gender" of the Deutsche Gesellschaft für Publizistik und Kommunikationswissenschaft.)

> I remember it being ICA. It was a much more activist and politicized time. Members of FSIG [Feminist Scholarship Interest Group] were involved simultaneously in building the interest group into a division as well as in the broader task of getting the discipline, as represented by ICA, to take both feminist scholarship and women seriously. For example, committees would routinely be staffed only by men. Decisions made only by men. Editorial boards and scholarship committees only composed of men. On another level, any woman was treated as a feminist when there were [and still are] many women who are not only not feminist but are positively anti-feminist. Issues of men in feminism were also hot. Should we give a whole precious panel to men in feminism [when virtually all other divisions overrepresented men]? Also [I] remember it being far more ethnic- and race-sensitive than it is now. Women of color attended the FSIG meetings and felt as an integral part of the discussion. I met many of the long-lasting academic collaborators in the women of color communications community on those days. Many of the "white" feminists were more sensitive to these issues as well. It was a far more inclusive time than it is now . . .

FSD was my first and most important home, and without that framework and identity, I might not have survived these last 20 years (and I really mean this)... I was invited to be on a panel at AEJMC that was co-chaired by CSW; this proved to be another moment in which I realized there were more sisters out there who shared my intellectual passions . . .

Certainly OSCLG has always played a role for me. I felt welcomed, it was more informal than NCA, and lots more exchange of ideas.

AEJMC and ICA around 1987. I went to research sessions where I heard what I was experiencing and had experienced before entering academia.

I remember particularly participating in the Organization for Research on Women and Communication (ORWAC) in the mid-1970s . . . and the Women's Caucus of NCA (then SCA). At the time, both were amazing groups of support, rebellion, and fun because this was a time when the discipline was really just beginning to pay attention to women. The early Purple Fridays or Purple Saturdays, sponsored by the Women's Caucus, were successful and created community among women and visibility in the Association. These days, the Women's Caucus seems to have less of a reason for existing, and I frankly find their meetings tedious and largely irrelevant.

As this comment indicates, not all experiences with these groups are positive. There have been struggles within groups, there has been some dissatisfaction with the direction of groups at times, and some women have not felt their scholarship is welcomed if it does not meet certain criteria. The comments in the following text, reflecting recollections of feminist organizations, also raise these tensions.

ICA, Chicago, 1986 or so. Brenda Dervin was President that year, and although she doesn't identify as a feminist scholar, her scholarship and presence as a critical scholar in such a leadership role was a powerful statement about possibilities for women in the association and the potential of discourse. SCA, New Orleans, 1987 or so. I was on a student panel called something like, "Bright New Gems," and remember my paper based on the work of a feminist theologian being welcomed with smiles. I felt that there was a space where I could focus my attention on feminist issues without having to fight for a space in which or from which to do so; that had been created by the activist feminists who had been working just before me. Roanoke, 1991 or 2. First feminist conference. I loved the openness and encouragement of being a feminist and sharing with other feminist scholars; but also experienced the odd, confusing tensions/politics around [a] speaker . . . My first taste of in-group struggling among feminists.

IAMCR when I was a graduate student—but the feminist work most presented was rather case studies—I have been and still am more interested in theorists.

I sometimes feel like there is a feminist "yardstick" by which I am measured and up to which I fail to measure. While I belong to CSW [of AEJMC] and do research that affects women, I don't feel a part of a feminist community of scholars.

Third, themes emerged from respondents' answers to key strands or moments affecting feminist scholarship. There were disappointments as well. They identified as significant:

Periods of Feminist Social Activism:

> I guess I would return to the French Revolution and writing from that period for establishing general understanding of feminist communication patterns. Then perhaps the 1890s is a fertile writing period; with some great writing also in the 1920s and of course 1960s authors like Bella [Abzug] and Gloria [Steinem].

Theoretical Shifts and Research Trends in Scholarship:

> I suppose, for me, the inclusion of figures of importance were initially important. This is because I teach history of journalism and we're still in the "great man/woman mode." Undergraduate students seem to need to learn about important figures . . .

> One key point was when feminist scholars stopped merely adding women's names to journalism history, bringing women *Up From the Footnotes,* as the title of M. Marzolf's history read, and experimented with new questions, new issues, and new methods.

> That "Women Talk" was plural and socially constructed based on situations and contexts. As such, I've moved to more interest in masculinities and religious identity too at an intersection with gendered identities.

> In theoretical terms, how to conceptualize gender and its intersections with sexuality and race have been, and continue to be, the most pressing, important, interesting, and maddening work of communication feminists. Theoretically, we have gone from biology is not destiny, to women's (essentialist) communication, to differences among women (and strategic essentialism and strategic coalitions), to queer theory and transgender theory, to performance.

> Key trends: the disciplinarity and increasing internationality of feminist communication research; the present incorporation of race, ethnicity, sexuality and class into feminist communication concerns, the growing number of men who seem to be drawing on our work. Significant strands: gender representation in content, gender dimensions of political-economy of industries, the female audience, women's media activism, women and media policy (though too few deal with this or other macro issues).

> In the mid-80s our work became critical and/or radical. Shift in much of the scholarship was almost revolutionary and at the same time feminist activism resulted in great changes at AEJMC, ICA, etc. In the mid-90s the debates became more political economy vs. cultural studies in nature, or seems that was a major focus . . . Also dealing through the 90s with all posts—post-modernism, structuralism, colonialism. Even post-feminism. Now both in feminist studies and critical studies, the concern seems to be reviving the centrality of activism—and untangling that centrality from disagreements re theory and method.

I think [the] early eighties were really exciting and defining moments for the field. Due to much work (and not a little bit of personal cost) of many scholars, feminist scholarship really became a major and integral component of communications scholarship. I could envision myself as a scholar only through the development of feminist communications research. It was a heady time. We were really at the beginning of something that would change our discipline. We were a rather small group though with great desire to expand and include. Most memorable were the ICA preconference retreats where senior colleagues got to know junior scholars and graduate students. These retreats were vital to community formation. Stopping them has been a major disappointment. Another area that has flourished is sexuality and queer studies. That is important though it foregrounds whiteness and queerness. Masculinity studies I feel conflicted about—I realize it's important but yet. . .

A key trend I have observed in the literature is the linking of feminism to Marxist/socialist theory and the rejection of capitalism—with capitalism serving as the instrument of patriarchy. I also see this as a disappointment, in that, realistically, I don't see how such research can truly transform patriarchal society (as found in the capitalist U.S.) into a more egalitarian one . . .

The development of Feminist Standpoint Theory has been influential in my scholarship, and the development of distinctly feminist/womanist theory in our field suggests we are becoming seasoned. It is hard for me to separate out feminist trends, strands, etc., from the emergence of other critical theories and cultural studies. It was disappointing that women scholars of color felt the need to establish themselves as distinct from feminists because some felt they did not fit well with feminist commitments. I celebrated when Patricia Hill Collins was a keynote speaker at a recent NCA conference.

Regarding the German situation a time lag between German and American research becomes evident. The trends might be similar but not synchronized. (1) The main trend of the 1980s was a disappointment with the current situation of female academics and journalists. In this time a lot of research (mainly by young female academics and female students) tried to prove the inequality of opportunity and the existence of stereotypes in mass media that injure. (2) In the 1990s research pointed at differences among women and men, but also among women. (3) The linguistic and cultural turn arrived in the middle and end of the 1990s. Judith Butler's work was very influential. But there is still an imbalance between the deconstructivist approach which is broadly perceived as very inspiring on a theoretical level and the lack of empirical studies that actually try to "puzzle gender" . . . Whereas in the end of the 1980s and beginning 1990s (female) students were very interested in a feminist perspective, the interest has decreased dramatically within the last 10 years.

Conflict, Decline, and Disappointments:

It seems to me that those engaged in feminist theory did not fully understand the contributions of those more oriented to professional media careers. The reverse seemed true too. I thought that the solidarity and good feeling of the early days of the women's movement petered out in the 1980s and certainly by the 1990s with women "putting down" other women on alleged scholarly grounds. . . To a degree what happened to the feminist presence in these organizations paralleled what happened to the feminist movement outside journalism/mass communication circles. Feminism seemed to run out of steam during the Reagan years. Also, I think some women were tenured and promoted and then lost sight of the need to tenure and promote others.

I'm constantly disappointed by the dearth of gender theory and feminist theory in AEJMC feminist scholarship . . .

Obsession with analysis of representation, although I can see the purpose of it; lack of feminist communication and citizenship; lack of feminist international communication intervention. I detest the use of gender as a replacement for women.

For me it was the near death of theorizing gender in rhetoric. Communication scholars gravitated to Foucault, Butler and studies of sexuality in the 1990s. But those who are interested in rhetorical theory just stopped producing theory about it. The major trend in rhetoric has been reclamation of lost voices without the attendant theorizing of what that reclamation means. Given the ongoing and useful discussion of speaking for others that has been going on in postcolonial scholarship, I would hope similar issues might become more important in rhetorical studies, but I don't sense that they are. I think feminism in rhetoric ran into a moment of change and did not adapt so well . . .

[A disappointing] trend is the move away from engaged scholarship to totally intellectual pursuits—as if these can ever be separated from everyday issues . . . As feminist communications scholarship became more institutionalized in many communications locations—such as NCA, IAMCR, and ICA—it turned whiter and less focused on issues of class and labor. Institutionalization deradicalizes and it seems that we feminist communications scholars are no less impervious to this than other movements. If we look at leadership, editorial boards, scholarship committees, etc., these are overwhelmingly composed of white feminists, few of whom have any deep engagement with global, ethnic, or class issues. This is truly a disappointment.

Disappointments: I have quite a few of these. (1) Feminist communication scholars do not give enough credence to varied research. (2) The lack of theory-building. (3) The persistent dominance of research on women's representation in the media and the ignoring of structural issues that shape women's representation. (4) Sloppy methodology—simplistic research questions, lack of data gathering, little investigation but lots of assumptions and claims, long on descriptions and short on real analysis of the problem. (5) Dismissal of the emancipatory/social change aspect of critical theory (for those who do critical research) by feminist scholars. We forget that all critical theory has an inherent social change goal. (6) The few numbers of men who read, understand, care about or use feminist communication scholarship—though there are a few more these days, most just think it's irrelevant. (7) The persistent difficulty of getting our work into journals—this may be related to the realities of my #4 above, but it also suggests discrimination.

For me the hardest frustration is the constant clash with pop culture's notion of "gender wars and differences." What helped most was the contribution of African American feminist scholars' notion of intersectionality and the move to thinking about gender, race, class, and now global concerns. I feel like it is a constant fight to get beyond this focus on essentialized differences, particularly when I, too, am a product of this socialized assumption of heterosexual differences.

Turning Points in the Associations:

Having Brenda Dervin as President of ICA seemed significant to other feminists in that organization, as well as to me personally . . . It was significant when the FSIG of ICA became a division, the FSD, and when feminist journals were established.

The publication of FMS [*Feminist Media Studies*], and, most recently, getting ICA to adopt a new journal (the latter being about the best evidence of our scholarly influence I've seen).

In my (limited) world of experience, the most significant turning point in scholarship in communication occurred when the Feminist and Women's Studies Division of NCA was formed. At that point, it seems to me, the discipline could no longer deny the importance of feminism to the study of communication. Also, the division provided a much needed venue for sharing feminist scholarship.

I'm surprised that feminists in AEJMC, NCA, ICA, ORWAC, OSCLG, and other women's communication groups have not formed alliances to advance their common agendas.

Fourth, on the field's accounting of and for feminist scholarship, respondents had mixed emotions:

Our field is doing better. That is, there is more respect now for feminist scholarship than there was in the 1970s, for instance. Also, I don't sense anyone has to apologize for sifting through the theories connected to feminist scholarship. That's progress. Right? However, our work definitely is NOT mainstream. But perhaps that's a good thing.

I still don't think feminist scholarship has the respect that it should. A former Ph.D. advisee of mine, who has a tenure track appointment in a good university system, told me that she was expected to do more than others to get tenure because she wrote about women . . .

We [two respondents answering together] believe feminists have been quite successful in having an impact on the communication field. In fact, we think it's difficult to find essays in our journals these days that don't evidence a feminist sensibility. Even the number of programs slots allotted to the Women's Caucus and the Feminist and Women Studies divisions is evidence of the discipline's embrace of and acknowledgment of the importance of feminism. Sometimes, when we attend Women's Caucus meetings and they are still talking like we did 25 years ago about how women and feminist scholarship aren't a part of the discipline, we want to say, "I think we won. We should be doing something else now." It's difficult, however, for many to envision other possibilities besides a world where we continue to play the "ain't it awful" refrain. This is also why we think there hasn't been as much reconceptualizing of the field as one might hope: it's hard to figure out how to do that. But we'd like to see the discipline turn its attention to creating a world that incorporates the values we want rather than continuing to critique the world we don't want.

Women, and particularly feminists, are still "Other."

Feminist scholarship is still barely tolerated. If it weren't for the sheer numbers of women (and men) as members of those very powerful women's groups in AEJMC, NCA, and ICA, and others, that feminist scholarship would not be as publicly acceptable as it is today in communication and mass communication. But privately, we're still viewed as trouble-makers, which, of course, we must be.

I am not sure the object is clearly discernable anymore. For instance, some of the work I have written might be said to be feminist, but some might contest that it is not, only being influenced by feminism but not really feminist. So I think, first, a really good discussion of what is this category in the first place might be helpful. Undoubtedly feminism has had an enormous impact. Some of that can be traced with relatively simple measurements. Although I think some kind of quantitative account of publications, dissertations, curricular changes, etc., would be of interest, I am not sure that adequately captures the texture of feminism in the field. My sense is that feminism has been thoroughly acknowledged in rhetorical studies but that the significance of that is not really on anybody's mind in a serious way.

As members of the western academy we feminists of color have to learn the white feminist canon and continue on the diverse research in addition to expertise in the white feminist canon. Most white feminists do not engage with the non-white canon at all or if they do, they do so in a token level (i.e., the occasional reference to hooks, Anzaldúa, min-ha). Yet it is these same feminists who are in a position to pick and choose inclusivity wise.

Unfortunately, I see feminist scholarship still being treated in the communication and mass media fields as an aside, with no real incorporation, dare I say, infiltration, into research in general. I still get the sense that to identify either one's personal self and one's scholarly self as feminist, the "man-hater" label still exists. More needs to be done on how the personal is political; though it has become a cliché, this aspect of feminism needs more emphasis. Women today are still adhering to feminine ideals without question—with mass media fueling the idea that "everything is OK now." One more thought: I see feminist scholarship and the efforts of women scholars in general to be unorganized . . .

My answer refers to the situation in Germany. Feminist media studies are very elaborated on a theoretical level and this high theoretical level you find also in courses dealing with this field. But due to a still-existing marginalization of feminist media studies, the elaborated theoretical approaches in this field do not receive as much academic attention as they deserve . . . I would even argue that working in the field of feminist media studies can harm your academic career. In my opinion the prejudice that the main motivation of feminist scholars is rather personal affection than deep academic interest still exists.

I may be projecting my own impression onto the field, but I feel feminist scholarship works best when it is assertive and distinctive but appreciative and respectful of related voices in the discourse; by the way, it is then less easy to dismiss. It is distinctly off-putting when it becomes exclusionary and assumes malevolence. What is the point of a feminist perspective if not in relation to other perspectives?! It is at its best when encouraging silenced voices, even privileging them; at its worst when it becomes insular and exclusionary, and presumes other emphases in other parts of the field are patriarchally biased.

JMC [journalism and mass communication] programs have heavy female enrollments, still have a dearth of female professors and are losing status in the academy as professional media organizations scurry to figure out what's next.

It's a continuing battle. We have more journals now, and yes there has been more feminist scholarship in "mainstream" ones, but only because key leaders in our field . . . keep fighting. It never ends. Just when we make progress, journal editorship will change, and things go backward again. Also feminist scholars continue to endure horrible situations re. tenure and promotion. Ironically this happens as ACEJMC [Accrediting Council on Education in Journalism and Mass Communications] standards—again thanks to activism—increasingly emphasize diversity.

Finally, responses to the open-ended questions about what we should remember about feminist communication produced varied answers:

I think it is important to keep time lines so our collective memory is correct or at least visible.

It's been a long road and an uphill battle for many . . . Nevertheless, feminism and feminist scholars have made an impact on the field of journalism/mass communication education by teaching courses in women and media and using feminist-oriented textbooks. Standard 12 of the accreditation process reinforces the need to teach women and media.

I think one thing we should remember is that feminist scholarship in communication started long before the 1960s. It seems to me that there are many 19th Century figures who are almost forgotten, who did much to advance 20th Century feminist communication scholarship, whether or not that was their intention.

I wish the "wave" labels (first wave, second wave, third wave, etc.) were eliminated. More emphasis on the intertwining between women's "real" lives and feminist scholarship is needed. Who were/are the women doing this kind of work? How do they manage to do so and still compete with men in the same field?

There doesn't seem to be much theory—theory is what survives ephemeral case studies. We need large ethnographic studies, feminist policy studies. I am up for it if colleagues are.

I appreciate the sustained commitment and involvement of those in the field with a feminist perspective; and that the organizational existence persists despite early leaders moving on. This suggests successful establishment of an institutionalized presence of this perspective.

Feminism crosses the gender line to claim new territory where status is earned by hard work, intellectual prowess, and compassion. Although for centuries women were emotionally and financially shackled by the role of care-giver, one strength women offer the world is a deep-seated commitment to others as well as self—the realization that things, money, and power do not equate with righteousness. . .

[We should remember that] one's gender (masculine or feminine) is fluid and strategic and that we are standing on the shoulders of giants: Simone de Beauvoir, Luce Irigaray and so many others. We are "here" because they were "there."

While it's important to record and remember the history of women in communication, the "Who's Who" approach inevitably fails unless we do a roll call of every feminist communication scholar who ever presented a feminist paper, taught a feminist course, mentored a graduate student or junior faculty member, wrote a feminist book or chapter, published a feminist peer-review article, served on a board or committee, was elected to a position, worked behind the scenes to advance the cause, or stood up to cry "foul" when no one else would. Otherwise, we forget someone worth remembering . . . Communication is just too big and diverse, and there are too many women, feminist or otherwise, deserving of recognition. We should give out buttons that read, "I am a communication feminist, and my contribution is valuable."

[P]lease remember that feminist scholarship is not the same as learning to think about and practice feminism as a communication scholar. I think the latter is something that has not been addressed much whereas the former is something that has had some attention.

It wasn't easy for the forerunners. I know of battles and obstacles that they faced. It wasn't pretty—it wasn't clean. Many have had, and continue to have, ugly battles with administrations . . . It was also fun—when we got together it was a great community. . . I also want to reiterate that the gains are not permanent. That these gains have to be struggled for continuously, even for mainstream white feminists. For feminists of color, these are hard times . . .

Well, our history is still before us, really. While there is large body of feminist communication research "out there," we still have not institutionalized our work in the discipline the way it should be. I think our scholarship needs to be much stronger if we are going to do this, and our members must take bigger intellectual risks.

THE LESSONS OF FEMINIST HISTORIOGRAPHY

What can be learned from this review of the state and method of the histories in and of our field and of feminist scholarship? We know that conventional histories, whether from the dominant empirical tradition or cultural and critical alternatives, have neglected women; the gendered nature of communication, society, and the field; and the challenges of feminist scholarship. We know that feminist scholars have been retrieving women's communication, challenging research traditions and conceptual categories, and documenting their work in the field.

Now, from the contributions of a group of feminist communication scholars, we know more. We have the beginning of a matrilineage of people and ideas to correct the patrilineage propagated by other histories. It is clear there are significant twists, turns, and struggles forward and sometimes back in the development of theoretical and methodological lines of research, with significant moments attached to shifts in thinking about women (and their differences of culture, class, race, religion, sexuality, global position, and ideology) and about gender (in its complexity as a category of social organization and meaning). Relationships among women and the rise of spaces for feminist work in associations and separate organizations have had an enormous impact on the development of feminist scholarship, in comparison to the other recognized "schools" of thought and research. Although there is some optimism that feminist scholars have a clear field to do their work and to reconceptualize what that work is, others are not so optimistic about the activist and critical edge to feminist work that once seemed to characterize it. The historical work of gathering feminist experiences and narratives has its pitfalls. We run the risk of substituting the "great men" approach to history with a "great women" or "who's who" approach. On the other hand, we run the risk of reproducing the problem named by Virginia Woolf, "For most of history, anonymous was a woman." We are cautioned that feminism and feminist scholarship is elusive (the range of theoretical positions represented in responses emphasizes that fact), and that feminist practice in communication may be harder to achieve than feminist theory.

As we continue making history and making histories, feminist scholars might begin our transformation of the field by reconceptualizing history. Based on themes that emerged from the group of feminist colleagues who participated in the call for contributions, we might replace the standard "great men," "great events," and "great places" history-making in communication with histories of the ideas, relationships, organizing, struggles, and victories that characterize women's communication as well as the work of feminist scholars.

Histories of Ideas

These histories will want to account for the transdisciplinary nature of our work on women and gender and the ideas of those we draw on for insight and inspiration. These histories will look at not only academic scholarship but also public scholarship by activists and feminist public intellectuals who make their arguments through public discourse and raise public consciousness. These histories will trace insights about women, gender, social theory, race, class, colonialism, sexuality, oppression, resistance, and social change. They will recover women's ideas about communication, about women, and about gender and hold the field accountable for preserving them and incorporating them into our narratives.

Histories of Relationships

These histories will want to account for the place of women's mentors and role models, forerunners and co-runners, graduate programs and advisors, and friends and collaborators in women's communication and in feminist scholarship. Recognizing that great work—work that illuminates and emancipates—arises from the collective efforts of many, these histories will look for the contours and patterns of association that have developed and sustained feminist scholars and activists.

Histories of Organizing

It took risk-taking and sacrifice to establish the presence of women and the legitimacy of feminist work to the extent that they have been. These histories will document how the field was challenged and how associations were changed through the work of feminist activism and scholarship. Feminist activism will be explored further for ideas that are likely to challenge conventional ideas about women and about communication. Connections between feminist activism in public and in the academy and the flow of ideas between them will need to be traced. With their histories in front of them, the divisions and organizations who keep the fire of feminist work alive may find a new future in collaborations across subfields, methodologies, and theoretical strategies.

Histories of Struggles and Victories

What did we fight for, what did we win, what did we lose? Women's struggles and victories and those of feminist scholars need recounting and preserving, including those that are "internal" and those that are disciplinary and public. The past has not been a linear, progressive movement toward women's emancipation, toward understanding our differences, and toward the acceptance of feminist scholarship. Revival of activism, public and scholarly, should follow from what we have learned from the past. These histories will help us see our connections to the struggles of other groups and the strategies we need to create coalitions for change.

What is feminist history in the field of communication? It will be all of these things and more. With these histories, we could work toward that radical reconceptualizing of the field held out by Foss, Foss, and Griffin.[32] Will the field acknowledge our work? Only history will tell.

NOTES

1. See more responses in the pages ahead.
2. Robinson, "Monopolies of Knowledge."
3. See examples that challenge the knowledge and history of multiple disciplines that bear reviewing, in Kramarae and Spender, *The Knowledge Explosion*, and the particular recent example of philosophy, in Witt, "Feminist History of Philosophy."
4. Kerber, *Toward an Intellectual History of Women*; Lerner, *The Majority Finds Its Past*; Rosen, *The World Split Open*; Rowbotham, *Hidden from History*; Scott, *Gender and the Politics of History*; Smith, *The Gender of History*.
5. Smith, *The Gender of History*, 1.
6. See Hardt, "Shifting Paradigms."
7. Berry and Theobald, *Radical Mass Media Criticism*; Carter, "The Transformative Power of Cultural Criticism."
8. Peters and Simonson, *Mass Communication and American Social Thought*.
9. Beniger, "Who Are the Most Important Theorists of Communication?"
10. Kramarae and Treichler, "Words on a Feminist Dictionary."
11. See Katz and others, *Canonic Texts in Media Research*.
12. Loshitzky, "Afterthoughts on Mulvey."
13. Marzolf, *Up From the Footnotes*.
14. Lowery and DeFleur, *Milestones in Mass Communication Research*.
15. See Baran and Davis, *Mass Communication Theory*.
16. See exemplary work in journalism and mass communication such as Beasley and Gibbons, *Women in Media*; Marzolf, *Up from the Footnotes*; and an update to an early report on women in the field found in Rush, Oukrop, and Creedon, *Seeking Equity*.
17. Jansen, "The Future Is Not What It Used To Be."
18. Lueck, "Like Newsroom, Like Classroom."
19. See other feminist critiques, reviews, and challenges in the historical context of the field and the development of feminist scholarship such as Altman, "Conversing at the Margins"; Ardizzoni, "Feminist Contributions to Communication Studies"; Blair, Brown, and Baxter, "Disciplining the Feminine"; Brunsdon, "A Thief in the Night"; Campbell, *Man Cannot Speak for Her*; Franklin, Lury, and Stacey, "Feminism and Cultural Studies"; Gallagher, "Feminist Media Perspectives"; Kramarae, "Feminist Theories of Communication"; Rakow "The Field Reconsidered"; Rakow, "Feminist Studies: The Next Stage"; Steeves, "Feminist Theories and Media Studies"; Treichler and Wartella, "Interventions;" and Wackwitz and Rakow, "Feminist Communication Theory."
20. Valdivia and Projansky, "Feminism and/in Mass Media."
21. Foss, Foss, and Griffin, "Feminist Perspectives in Rhetorical Studies."
22. Ibid., 14–15.
23. See Rakow and Kramarae, "Introduction," 4.
24. Douglas, "*Personal Influence* and the Bracketing of Women's History."
25. Dicenzo,"Feminist Media and History."
26. Flores, "Gender With/out Borders"; Hegde, "A View from Elsewhere"; Rhodes, "The Visibility of Race and Media History;" and Valdivia, "Feminist Media Studies in a Global Setting."
27. Robinson, "Monopolies of Knowledge," 1.
28. Steiner, "The Uses of Autobiography."

29. My thanks to the officers and editors and board members of these groups who supported the project; facilitated access to their memberships through their e-mail listservs and bulletin boards.
30. On the other hand, one recipient complained about my decision to keep responses confidential, arguing that I would be taking credit for the work done by others. That certainly was not my intent. Perhaps others felt this way as well and did not respond for that reason.
31. The names of individuals within the field of communication have not been included because of the great risk of leaving out those who would be identified if more scholars were to respond.
32. See n. 21.

WORKS CITED

Altman, Karen E. "Conversing at the Margins: A Polemic, or Feminism and Communication Studies." *Journal of Communication* 11, no. 1 (1987): 116–17.
Ardizzoni, Michela. "Feminist Contributions to Communication Studies: Past and Present." *Journal of Communication Inquiry* 22, no. 3 (1998): 293–305.
Baran, Stanley J., and Dennis K. Davis. *Mass Communication Theory: Foundations, Ferment, and Future.* 4th ed. Belmont, CA: Thomson Wadsworth, 2006.
Beasley, Maurine, and Sheila Gibbons. *Women in Media: A Documentary Sourcebook.* Washington, DC: Women's Institute for Freedom of the Press, 1977.
Beniger, James R. "Who Are the Most Important Theorists of Communication?" *Communication Research* 17, no. 5 (1990): 698–715.
Berry, David, and John Theobald, eds. *Radical Mass Media Criticism: A Cultural Genealogy.* Tonawanda, NY: Black Rose Books, 2006.
Blair, Carole, Julie R. Brown, and Leslie A. Baxter. "Disciplining the Feminine." *Quarterly Journal of Speech* 80 (1994): 383–409.
Brunsdon, Charlotte. "A Thief in the Night: Stories of Feminism in the 1970s at CCCS." In *Stuart Hall: Critical Dialogues in Cultural Studies*, edited by David Morley and Kuan-Hsing Chen, 276–87. New York: Routledge, 1996.
Campbell, Karlyn Kohrs. *Man Cannot Speak for Her.* New York: Praeger, 1989.
Carter, Cynthia. "The Transformative Power of Cultural Criticism: bell hooks's Radical Media Analysis." In *Radical Mass Media Criticism: A Cultural Genealogy*, edited by David Berry and John Theobald, 212–33. Tonawanda, NY: Black Rose Books, 2006.
Dicenzo, Maria. "Feminist Media and History: A Response to James Curran." *Media History* 10, no. 1 (2004): 43–49.
Douglas, Susan J. "*Personal Influence* and the Bracketing of Women's History," In "Politics, Social Networks, and the History of Mass Communications Research: Rereading *Personal Influence*." edited by Peter Simonson, special issue. *Annals of the American Academy of Political and Social Sciences* 608 (2006): 1–50.
Flores, Lisa A. "Gender With/out Borders: Discursive Dynamics of Gender, Race, and Culture." In *The Sage Handbook of Gender and Communication*, edited by Bonnie J. Dow and Julia T. Wood, 379–98. Thousand Oaks, CA: Sage, 2006.
Foss, Karen A., Sonja K. Foss, and Cindy L. Griffin. "Feminist Perspectives in Rhetorical Studies: A History." In *Feminist Rhetorical Theories*, edited by Karen A. Foss, Sonja K. Foss, and Cindy L. Griffin, 14–32. Thousand Oaks, CA: Sage, 1999.
Franklin, Sarah, Celia Lury, and Jackie Stacey. "Feminism and Cultural Studies: Pasts, Presents, Futures." *Media, Culture and Society* 13 (1991): 171–92.

Gallagher, Margaret. "Feminist Media Perspectives." In *A Companion to Media Studies*, edited by Angharad N. Valdivia, 19–39. Malden, MA: Blackwell, 2003.

Hardt, Hanno. "Shifting Paradigms: Decentering the Discourse of Mass Communication Research." *Mass Communication & Society* 2, no. 3–4 (1999): 175–83.

Hegde, Radha S. "A View from Elsewhere: Locating Difference and the Politics of Representation from a Transnational Feminist Perspective." *Communication Theory* 8, no. 3 (1998): 272–97.

Jansen, Sue Curry. "'The Future Is Not What It Used To Be': Gender, History, and Communication Studies." *Communication Theory* 3, no. 2 (1993): 136–48.

Katz, Elihu, John Durham Peters, Tamar Liebes, and Avril Orloff, eds. *Canonic Texts in Media Research: Are There Any? How About These?* Cambridge, MA: Polity, 2003.

Kerber, Linda. *Toward an Intellectual History of Women.* Chapel Hill: University of North Carolina Press, 1997.

Kramarae, Cheris. "Feminist Theories of Communication." In *International Encyclopedia of Communications*, edited by Erik Barnouw, 156–60. New York: Oxford University Press, 1989.

Kramarae, Cheris, and Dale Spender, eds. *The Knowledge Explosion: Generations of Feminist Scholarship.* New York: Teachers College Press, 1992.

Kramarae, Cheris, and Paula A. Treichler. "Words on a Feminist Dictionary." In *A Feminist Dictionary*, edited by Cheris Kramarae and Paula A. Treichler, with assistance from Ann Russo, 1–22. Boston, MA: Pandora Press, 1985.

Lerner, Gerda. *The Majority Finds Its Past: Placing Women in History.* New York: Oxford University Press, 1979.

Loshitzky, Yosefa. "Afterthoughts on Mulvey's 'Visual Pleasure' in the Age of Cultural Studies." In *Canonic Texts in Media Research: Are There Any? How About These?*, edited by Elihu Katz, John Durham Peters, Tamar Liebes, and Avril Orloff, 248–59. Cambridge, MA: Polity, 2003.

Lowery, Shearon, and Melvin L. DeFleur. *Milestones in Mass Communication Research: Media Effects.* New York: Longman, 1983.

Lueck, Therese L. "Like Newsroom, Like Classroom: Women Journalism Educators Temper the Times." *American Journalism* 20, no. 1 (2004): 83–104.

Marzolf, Marion. *Up From the Footnotes: A History of Women Journalists.* New York: Hastings House, 1977.

Peters, John Durham and Peter Simonson, eds., Mass *Communication and American Social Thought: Key Texts, 1920–1968.* Lanham, MD: Rowman and Littlefield, 2004.

Rakow, Lana F. "Feminist Studies: The Next Stage." *Critical Studies in Mass Communication* 6, no. 2 (1989): 209–13.

———. "The Field Reconsidered." In *Women Making Meaning: New Feminist Directions in Communication*, edited by Lana F. Rakow, 3–17. New York: Routledge, 1992.

Rakow, Lana F., and Cheris Kramarae. Introduction to *The Revolution in Words: Righting Women, 1868–1871*, edited by Lana F. Rakow and Cheris Kramarae. New York: Routledge, 1990.

Rhodes, Jane. "The Visibility of Race and Media History." In *Gender, Race and Class in Media: A Text-Reader*, edited by Gail Dines and Jean M. Humez, 33–39. Thousand Oaks, CA: Sage, 1995.

Robinson, Gertrude. "Monopolies of Knowledge in Canadian Communication Studies: The Case of Feminist Approaches." *Canadian Journal of Communication* 23, no. 1 (1998): 65–72.

Rosen, Ruth. *The World Split Open: How the Modern Women's Movement Changed America.* New York: Viking, 2000.

Rowbotham, Sheila. *Hidden from History: 300 Years of Women's Oppression and the Fight against It.* London: Pluto Press, 1975.

Rush, Ramona, Carol E. Oukrop, and Pamela J. Creedon, eds. *Seeking Equity for Women in Journalism and Mass Communication*. Mahwah, NJ: Lawrence Erlbaum, 2004.

Scott, Joan Wallach. *Gender and the Politics of History*. Rev. ed. New York: Columbia University Press, 1999.

Smith, Bonnie G. *The Gender of History: Men, Women, and Historical Practice*. Cambridge, MA: Harvard University Press, 1998.

Steeves, H. Leslie. "Feminist Theories and Media Studies." *Critical Studies in Mass Communication* 4, no. 2 (1987): 95–135.

Steiner, Linda. "The Uses of Autobiography." In *American Cultural Studies*, edited by Catherine A. Warren and Mary Douglas Vavrus, 115–33. Urbana and Chicago: University of Illinois Press, 2002.

Treichler, Paula A., and Ellen Wartella. "Interventions: Feminist Theory and Communication Studies." *Communication* 9, no. 1 (1986): 2–18.

Valdivia, Angharad N. "Feminist Media Studies in a Global Setting: Beyond Binary Contradictions and into Multicultural Spectrums." In *Feminism, Multiculturalism, and the Media: Global Diversities*, edited by Angharad N. Valdivia, 7–29. Thousand Oaks, CA: Sage, 1995.

Valdivia, Angharad N., and Sarah Projansky. "Feminism and/in Mass Media." In *The Sage Handbook of Gender and Communication*, edited by Bonnie J. Dow and Julia T. Wood, 273–96. Thousand Oaks, CA: Sage, 2006.

Wackwitz, Laura A., and Lana F. Rakow. "Feminist Communication Theory: An Introduction." In *Feminist Communication Theory: Selections in Context*, edited by Lana F. Rakow and Laura A. Wackwitz, 1–10. Thousand Oaks, CA: Sage, 2005.

Witt, Charlotte. "Feminist History of Philosophy." *Stanford Encyclopedia of Philosophy*, edited by Edward N. Zalta. Fall 2006 ed. http://plato.stanford.edu/entries/feminism-femhist/.

II. INSTITUTIONAL HISTORIES

CHAPTER FIVE

Institutional Opportunities FOR Intellectual History IN Communication Studies

JOHN DURHAM PETERS

Here I take up the editors' invitation to reconsider my early essay "Institutional Sources of Intellectual Poverty in Communication Research" (1986).[1] That essay was both a history lesson about where communication research went wrong and a manifesto for an intellectually more cosmopolitan field. In many ways, it opened the line of inquiry that culminated in *Speaking into the Air* (1999).[2] The essay argued that communication research had sacrificed intellectual vitality on the altar of institutional autonomy. More specifically, communication research suffered from its nationalism: it sought to be intellectually self-sustaining in a world already abounding in rich resources around communication. The very idea of a field of communication was a contradictory project: how could one specialty claim a unique stewardship over something of such universal interest as communication? A stunning variety of twentieth-century intellectuals had tackled communication as one of the great problems of the age, almost all of them in happy ignorance of the academic field that claimed special expertise about that problem. This left communication research in the position, I argued, of an academic Taiwan—claiming to possess all of China while isolated on a small island. The solution to the perennial lamentation—why doesn't anyone understand what we do in communication studies?—lay, I suggested, in the alarming mismatch between institutional claim and intellectual delivery.

Rereading "Institutional Sources" for the first time in almost two decades, I see both reminders of my academic biography and opportunities for more nuanced narratives. Combining two uncomfortably presumptuous genres (the memoir and the wish list of research projects that need to be done some day) this chapter suggests that my experiences as a graduate student in the early 1980s can be read as symptoms of the larger postwar intellectual landscape around "communication," a landscape that is both more complicated and more focused than I realized when I wrote "Institutional Sources." The institutional and intellectual landscape today is different from that of 1986 in too many ways to count, but I would still ardently endorse the central operating assumption of "Institutional Sources." It is that intellectual history is about more than paternity suits, priority disputes, or property rights: its task is the enlarging of intellectual possibilities.

A CASE OF THE BENDS

I arrived at Stanford University in fall 1982 fresh from a master's degree in Speech Communication, as it was then known, as the University of Utah. I had found my way into the field of communication by lucky accident, one day prowling the University of Utah bookstore and discovering a class that assigned many of the books I had been wanting to read. I signed up at once. I had just completed a bachelor's in English at Utah with strong interests in language, linguistics, and folklore, and I had long sensed that literary studies gave me the tools I needed to understand the world in general but confined those tools to a canon of printed works. Instead of the "pop field" that my grandfather, a former university president, warned me about, I saw in communication studies ways and means to understand the things I had wondered about while living for two years as an LDS missionary in the Netherlands—cross-cultural mixing, questions of translation, the relation of language, and social order. My master's thesis at Utah on "The Sacred and Sociality" (1982), advised by Leonard Hawes, was an ambitious essay that brewed an enthusiastic but perhaps not very coherent potion from the writings of mostly Continental thinkers who, it seemed to me, were asking fundamental questions about communication. Structuralism, hermeneutics, critical theory, psychoanalysis, ethnomethodology, and conversation analysis were some of the bubbling new ideas that seemed poised to reveal the mysteries of how intelligent beings organize meanings, which is what I thought communication theory was about. At Utah, I had studied Jürgen Habermas's grand critical theory of communication, Jacques Lacan's dazzling rereading of Freud, and Paul Ricoeur's rethinking of the task of interpretation. Earlier, at Brigham Young University (BYU), I had learned about the structuralism of Roman Jakobson and Claude Lévi-Strauss and was thrilled by the project of a general linguistics that would account for culture in general. Gregory Bateson seemed to offer the exactly right kind of vision about why we should study communication, with interests that spanned philosophy and natural science, mental illness and otters, aesthetic form and cybernetics.

Arriving at Stanford, I liked the eucalyptus trees and the international neighbors but wondered at once what I had gotten myself into. Poking around the Quonset hut setting of the Institute for Communication Research, founded by Wilbur Schramm in the 1955–1956 academic year, and located on the periphery of campus as was then the fashion for research institutes at Stanford, I stepped into one of several remaining outposts of postwar social science as it had been pulled together in the late 1940s and 1950s. The spirit was applied, busy, collegial, and slapped together. Intelligence was quick and analytic, not erudite or philosophical. This was "social research for social problems," as Robert Hornik once characterized Schramm's project. There was as little ivy on the outer walls of Spruce and Cypress halls as there was literary talk; indeed, my class of 1982 was the last to be housed in Spruce and Cypress, architectural remnants of the "military-industrial research culture" that had flourished in much more than just the social sciences after the war.[3] The theory courses registered little of the intellectual quakes that had shaken social science in the 1970s (such as Marxism, cultural studies, or feminism).[4] The Institute also boasted the Franklin Fearing room, affectionately known by the graduate students as the Fearing and Loathing Room, which housed a collection of journals in social psychology from the 1920s to the 1950s donated by Professor Fearing. No one was exactly sure how the library of a University of California, Los Angeles (UCLA) social psychologist ended up in the Institute, but we were happy to use the room, more often for games of Boggle than for consultation of the aging print matter.[5] (The Fearing Room was demoted into a vestibule once the department moved to the Quadrangle on the main campus in winter 1983 and seems not to exist today.) The nth-year graduate students eagerly initiated us newcomers into a brave new world of heart disease prevention projects, data, children's television, international health campaigns in Belize and the Gambia, and hot tub parties. Their booklet, *Schramm for Beginners: A meta-comic book elucidating some of the finer points of communication theory, and featuring the adventures of Pfil and Pfyllis Pfd, those fearless defenders of truth, justice, and logical positivism,* was hilarious and may still be in samizdat circulation. It was all exciting and strange, socially pleasant, and intellectually alienating, and it posed a question for me that was existential before it was academic: how could two such diverse curricula as Utah and Stanford both belong to the same field? What did communication mean?

To be sure, I went to Stanford with my eyes open. In my letter of application, I grandly compared myself to Freud, who, alarmed at the fecundity of his imagination, went to work in a histological lab for three years. Michael Pacanowsky, a Stanford PhD who had taught the seminar that introduced me to communication studies, sent me off to Stanford with the mission of philosophizing mass communication. My plan was to go to the heart of the behavioral sciences beast and tame it from the inside out. (As a Korean classmate at Stanford told me, if you want to catch a lion, you must go to his den.) I thought I would get a rigorous training in technique and method in positivist social science and emerge with shiny skills that I could apply a ferocious philosophical rigor to criticizing. Instead, I learned just how improvised empirical social science is, at least in its more applied wings. Quantitative social science was less a towering Moloch than a swarm of termites. I was looking to be initiated into priesthood, with ascetic rules for operationalizing meanings or spiritual disciplines for cloud-free cognition; instead, we were invited into a guild, with rules of thumb, research designs, and remarkably inventive ways of finding the fish in an ocean of data. (Even so, the guild did sometimes talk as if it would rather be a priesthood; statistics was the closest we got to a secret doctrine).[6] Communication research at the Institute was an advanced form of puzzle solving, and Boggle was perhaps not so irrelevant after all for the cognitive skills we were supposed to be developing. Probably my biggest surprise was learning that good social science turned on good questions and creative ways of producing evidence. In retrospect, all the anti-positivist venom of the era seemed more a response to the official self-image of social science than to its grounded practice, although clearly there was much to criticize in the political abuses of cold war social science. The foes of quantitative social science probably took its visions of grandeur and quest for law-like generalizations more seriously than its practitioners do. It is probably not the first time in history that a story persuaded the critic more deeply than the exponent.

I don't think I had such a conciliatory vision of social science at the time. Rather, I often felt frustrated at the lack of high-level intellectual discussion in the Institute at a time when thinkers of all sorts were debating what was at stake in communication. The chief sin of positivist social science remains, as Habermas somewhere said, "The denial of self-reflection." There were no grand paradigm clashes at the Institute, although there was certainly some variety among the intellectual traditions of the four faculty there—political research and survey methods for Steve Chaffee, information science for Bill Paisley, social-psychological media effects research on children for Don Roberts, and rural sociology for Ev Rogers—but they were all quite insulated from the critiques of behaviorist social research that were de rigueur at the time in any dozen different theoretical perspectives. Basically, it wasn't that communication research at Stanford was indifferent to critique—it was indifferent to all ideas "grander" than the middle-range period. Indeed, the notion of "middle-range" theory and the focused interview method were the sole parts of Robert Merton's much richer legacy taught in the Institute.[7] Ideas were just another name for the crude early phase on the path to formulating testable hypotheses. Several of us doctoral students disobeyed such strictures, and whimsically following Wilbur Schramm's memory of a "hot air club" sponsored by Kurt Lewin at the University of Iowa, we created a weekly forum for crazy ideas, which resulted in some memorable debates. (Some of the faculty seemed secretly pleased with this short-lived rebellion. The fathers know that even Oedipality can bear fruit.)

The faculty members were often less interested in communication the topic than communication the field. I asked about the Palo Alto Group, led by Paul Watzlawick, which had become famous for its exploration, in the intellectual wake of Bateson, of the part played by the double bind and distorted communication in the familial genesis of schizophrenia.[8] No one on the faculty had any connection with them, even though they were in the same town and ostensibly studying the same thing. Quite greenly, I found this lack of communication about communication astonishing. The orientation was to one's peers in what Ev Rogers, building on an old trope, liked to call the "invisible college." How these peers were defined mystified me. Steve Chaffee—a complex and brilliant figure with a proverbial talent around a data-set, an uncanny memory for movie and baseball trivia, a passion for Friday afternoon volleyball, and mysteriously moody modes of interaction, sometimes morosely peering over his glasses and other times lighting up into comradely sunshine as he shared an insight or recounted a story—was clearly more interested in the latest work of minor figures who were his colleagues than of leading thinkers such as Foucault or Habermas, whose critical questions about empirical inquiry in social research, one got the distinct feeling from him, probably made them bad people. (Curiously enough, when Habermas visited the International Communication Association in 2006, he spoke on one of Steve's favorite topics: normative political theory and its relation to empirical political communication research.[9] I wished Steve had been alive to hear it.) Of course, I should have understood that orientation to peers rather than to remote grand thinkers is standard practice in academic professionalism, but in a field of study that claimed unique stewardship over the topic of communication, the ignorance of the wider ferment at the time was startling.

One answer for this combined existential and intellectual problem—what in the world did communication mean?—took shape in "Institutional Sources," which Chaffee with remarkable generosity published as the lead article in *Communication Research* in the last issue of his editorship. The essay's central metaphor of nationalism was inspired by Benedict Anderson's groundbreaking *Imagined Communities*. Another stimulus was Bill Paisley's valedictory study, written as he was leaving Stanford for a career in Silicon Valley. He demonstrated that communication scholars cited other social scientists but that other social scientists did not cite communication scholars in return. Communication as a field risked becoming, he ventured, an "oxbow"—that is, the island left when a river changes course and cuts across a bend, turning a once vital shoreline into a muddy pool of stagnant water.[10] In this, he offered a melancholy alternative to Schramm's metaphor of communication research as "a crossroads where many pass but few tarry."[11] Paisley's analysis of co-citation patterns struck me as exactly the same problem that dependency theorists had noted in the world system, in which the center siphons off the resources and talents of the periphery without providing any goods or recognition in return. Nationalist exclusivity was one response to the specter of extraterritorial domination, and I saw this response in the project of a field of communication. The problem was if you happen to live in a large rich country, nationalist exclusivity can provide a fine array of goods and services but it is a recipe for disaster in a small poor one.

The logical conclusion was not, I thought, to bemoan an asymmetrical balance of trade, but to rethink academic institutionalization. Unlike real nations, which educate, tax, and conscript their members, academic fields, I thought, should pursue truth free of realpolitik. It was a strategically outrageous act of idealism on my part to think that ideas rather than identities should be the key factor in academic life. (I still do.) Why define "communication scholar" by departmental residency rather than intellectual contribution? Who was to say that Adorno, Austin, Bateson, Benjamin, Chomsky, Foucault, Freud, Gadamer, Geertz, Habermas, Jakobson, Lacan, Ricoeur, Spivak, and Wittgenstein were not communication scholars? Of course, none of these people held passports in the state of communication research. (But neither did Franklin Fearing.) How had so many first-class thinkers found communication a good thing to think about in complete ignorance of the field dedicated to its study? Why did the world have Gregory Bateson to guide its thinking about communication when the field had Wilbur Schramm? The diminuendo was deafening. (Clearly, my professional socialization had failed somewhere.)

UNTAPPED POSTWAR WEALTH

In "Institutional Sources," I pointed to two large bodies of work that had not been sufficiently reckoned with intellectually. One was the pragmatist-progressive social thought associated with John Dewey and his peers and another was the postwar ferment around information theory. In the remainder of this chapter, I want to take up the postwar landscape in greater depth, having dealt with the pragmatist-progressive legacy at length elsewhere.

I now see that my experiences at Utah and Stanford represented two of the dominant wings of thinking about communication to emerge in the United

States after World War II. One of the war's many domesticated by-products was the concept of communication, though this term of course has a much longer intellectual pathway.[12] The explosion of interest in communication after the war was a classic peace dividend, as scientists and scholars of many stripes demobilized, bringing experiences of war—with its codes, interference, overload, "radio contact," "communication breakdown," radar, and cryptography—into civilian spheres. One wing was the intellectually ambitious interdisciplinary project we can roughly call cybernetics. Its university center was MIT, which boasted not only Norbert Wiener, but also Vannevar Bush, the ringleader of militarized science during the war, George A. Miller, one of the inventors of cognitive science, John von Neumann, the father of the computer, and linguist Roman Jakobson, among other luminaries. "Institutional Sources" touched on the explosion of interest in information theory in the 1950s, noting that even Wilbur Schramm had played a minor role in it, but the essay did not have a clear enough sense of its intellectual network or its global resonance. The other wing was the intellectually more modest tradition of social-psychological study of media effects associated with Schramm among many others.[13] My studies at Utah were guided by Leonard Hawes, whose teaching enacted Bateson's couplet of rigor and imagination, and whose thinking voraciously pursued the postwar project of a general communication theory; my years at Stanford caught the tail end of Schramm's project of administrative effects research. In "Institutional Sources," I answered an existential crisis of having lived in two estranged worlds of communication theory. W. H. Auden wrote that poets were hurt into poetry; the bends between Utah and Stanford hurt me into intellectual history.

However, the postwar range of thinking about communication was hardly represented by my narrow experience or "Institutional Sources." We still do not understand fully the intellectual buzz around the concept of communication from about 1946 to 1968. At least I don't. The postwar era is ripe for revisionist history, and a scholar armed with the tools of network analysis could tell us much about the links and cross-fertilizations among those who studied "communication" after the war.[14] In what remains, I want to sketch a research agenda for four main intellectual traditions, and one institutional problem, around communication since 1945. Each one would make an excellent dissertation topic, and I hope other scholars will take them up.

The Social Psychology of Media Effects

The key point about the effects tradition is that it was always shadowed by the broader and richer sources of sociology. Put eponymically, where there was Lazarsfeld there was also Merton. The best postwar social-psychological work on media effects was in dialogue, though sometimes sublimated, with broader sociological work on communication by thinkers such as Hannah Arendt, Daniel Bell, Warren Breed, Kenneth Burke, Erving Goffman, C. Wright Mills, and David Riesman. To this list we might add Habermas whose *Structural Transformation of the Public Sphere* (first published in German in 1962) launched his long career of importing Anglo-American thought into German civic and academic life. This book was not only a wrestle with the legacy of Frankfurt critical theory or a revisionist history of parliamentary democracy in Europe; it was also a dialogue with thinkers such as Arendt, Dewey, Lippmann, Riesman, Mills, and Katz and Lazarsfeld, all of whom who searched for the public in an age of mass communication. We might add many more postwar European thinkers to the list.

Such dialogue occurred in many countries after the war, and the role of the effects tradition as a postwar American export has barely been studied. We need an international intellectual history of communication research.[15] The global export of communication research has yet to be told in the way that CIA support for literary magazines and abstract expressionism via the Congress for Cultural Freedom has been.[16] In Germany and Japan, for instance, the American occupation clearly did not confine its reconstruction efforts to institutions or infrastructure: it encouraged liberalizing (anticommunist) intellectual reforms as well. No one less than Adorno, contrary to his American reputation as the scourge of positivism, argued in postwar Germany for the potentially emancipatory role of empirical social research, suggesting that public opinion surveys, for instance, would have been a counterweight against Nazism.[17] After his return to Germany in the late 1940s, Adorno worked hard to introduce empirical methods into a largely historical and philosophical German sociological tradition, and one of the first publications of his young assistant Habermas was an empirical survey of student attitudes.[18] That the political valence of empirical methods was not necessarily left-liberal is clear in another German importer of empirical research, Elisabeth Noelle-Neumann, whose Allensbach Institut spent five decades surveying public opinion for the conservative party in Germany. In Japan, one important venue was the Institute for Media and Communications Research at Keio University, which was founded in 1946 as a research bulwark against the media's perceived role in the bellicose fascism of the previous two decades. The Israel Institute of Applied Social Research was founded in 1947 by Louis Guttman, first as a branch of the Hagana, the paramilitary organization that was to become the Israeli Defense Forces with the founding of the state of Israel in 1948. The Institute later played a programmatic role in Israeli media policy and social research. In Poland, the Soviet Union, and Slovenia during the cold war, traditions of empirical research struggled for existence. Behind the so-called Iron Curtain, empirical research served as an alternative voice to dominant thinking, much as critical research was in the West in the same period. These stories are just starting to be told.[19]

Institutes for Communication Research were cropping up around the world in the late 1940s. All of them responded to the lessons of the war, and their intellectual charter owed something to the intuition that media were key agencies of mobilization in war and peace. When Wilbur Schramm formed the Institute of Communications Research at the University of Illinois in 1948, he was part of a postwar wave of American power that was establishing outposts of social research for social problems at home and abroad. What took place in Urbana or Palo Alto was part of a global expansion.

Cybernetics

The broad outlines of the interdisciplinary explosion of interest in cybernetics are well known. The development of the computer during World War II, and the wartime opportunity for cross-fertilization among diverse scientists, led to a remarkable postwar fascination with communication, information, systems, probability, noise, redundancy, entropy, interference, breakdown, feedback, homeostasis, and so on. One participant in the Macy Conferences on Cybernetics (1946–1953) caught some of the excitement: "the new synthesis heralded in [cybernetics and information theory] was destined to open new vistas on everything human and to help solve many of the disturbing open problems concerning man and humanity."[20] These conferences brought together anatomists, anthropologists, engineers, mathematicians, neurologists, physiologists, psychiatrists, and sociologists (including Lazarsfeld). The project was a "communication theory" that spanned logic and language, genes and proteins, computation and telecommunications, mental health and world politics, and much more all the way up to general systems. During the 1950s cybernetics boom, specialties involving communication blossomed. One of my favorites is the field of "Communications Biophysics."[21] Why it vanished and "communication studies" survived and flourished is unknown.

In "Institutional Sources," I had some inkling of the postwar explosion, but I didn't understand just how closely tied it was to French structuralism. Claude Shannon's mathematical theory of communication grew out of the economic imperative of the Bell system to maximize intelligibility on telephone lines while minimizing the complexity (and expense) of the auditory signal, and he always insisted, sometimes with disgust at the stranger adaptations of his work, that the theory was about signals, not significance. Roman Jakobson, probably the greatest linguist of the twentieth century, was fascinated by the very same problem: the production and recognition of "distinctive features" that serve as the building blocks ("phonemes") of intelligible speech. Jakobson collaborated with colleagues at MIT to produce *Preliminaries to Speech Analysis* (1952), a landmark in speech synthesis and psychoacoustics as well as a key advance in his own thinking.[22] Throughout the 1950s, he made "communication" a centerpiece of his work.[23] In the 1940s, his friend, the anthropologist Claude Lévi-Strauss, developed a theory of human cognition and culture that is essentially combinatoric. Humans are, as Lévi-Strauss famously said, *bricoleurs*, inveterate do-it-yourselfers who tinker myth and culture together. His direct tie to cybernetics and information is clear in a 1954 essay published in a UNESCO journal. Here he explains his discovery, aided by Jakobson, that kinship structures and language are both "rules of communication," citing Shannon and Weaver, Wiener, game theory, as well as Lazarsfeld.[24] As it happens, Lévi-Strauss and Shannon lived in the same New York City apartment building for several years in the early 1940s, and although there is no record of any face-to-face contact between the two men, we do know that gossip about one of his neighbors trying to build an "artificial brain" was a kind of trigger for Lévi-Strauss's evolving anthropology.[25]

Jacques Lacan, a highly influential psychoanalytic theorist and one of the leading lights of that motley band of French thinkers sometimes called (post)structuralists, also has clear ties to cybernetic thinking. Lacan was well informed about contemporary scientific developments, and arguably developed a "mathematics of the unconscious."[26] (His central terms—the real, the symbolic, and the imaginary—are, after all, mathematical categories.) Lacan made communication a central problem, and discussed Shannon in his epochal speech, "The Function and Field of Language" (1953): "what is redundant as far as information is concerned is precisely that which does duty as resonance in speech."[27] Lacan elsewhere made appreciative use of information theory.[28] So did Jean Hyppolite, whose work on Hegel was at the epicenter of mid-twentieth-century French thought. He deployed information theory to explicate Mallarmé's poetry in a 1958 essay, to ask whether machines can think in a 1961 Moscow address, and to ponder information and communication in a wide-ranging mathematically adorned essay in 1967.[29] Hyppolite was the honored teacher of Michel Foucault, who assumed his chair at the Collège de France in 1970. Clearly, Foucault was also in touch with the postwar ferment around communication. In a climactic moment in his famous chapter on the Panopticon in *Discipline and Punish*, he notes the sorry state of the inmate: "Il est vu, mais il ne voit pas; objet d'une information, jamais sujet dans une communication" ["He is seen, but he does not see; object of information, never a subject in communication"].[30] Here Foucault strikes an unusually Habermasian note, with "communication" suggesting some normative pathos, however much Foucault was ultimately critical of the metaphysical divide between subject and object. Foucault's later work on the dyadic play of "truth-games" in classical antiquity can be seen as meditations on communication. And his notion of governmentality has striking affinities with cybernetics (a word that comes, after all, from the Greek word for *governor*).[31] Postwar France saw an intense interest in communication theory writ large.

Though France is certainly not the only country of interest—we know that Wiener's *Cybernetics* was translated into Russian in the mid-1950s and was widely influential in the Soviet Union[32]—the key point is that the network of mid-century thinkers about communication is tighter and more international than I suspected in "Institutional Sources." Habermas had a clear link to the effects tradition, just as Foucault had links to the cybernetic tradition. What I thought were radical alternatives turn out to be variations on the family tree.

Psychiatry

One of the chief disciplinary frameworks for considering communication during the 1940s and 1950s was psychiatry. Cybernetics was a key nodal point. The leader of the Macy Conferences was Warren McCulloch, a psychiatrist, and Gregory Bateson was one of the most influential participants. In *Naven* (1936), Bateson developed his concept of schismogenesis, and cybernetics provided him a systems language for thinking about communication gone berserk. Together with physician Jurgen Ruesch, Bateson wrote a book widely neglected in communication studies, *Communication: The Social Matrix of Psychiatry* (1952), which is one of the most sustained twentieth-century meditations on communication, brimming with concepts made famous thanks to the Palo Alto Group such as the double bind, "you cannot not communicate," and metacommunication.[33] To be sure, communication had long been seen as a binary contrast to mental pathology, and many of the chief theorists of communication have also been theorists of madness (Deleuze, Foucault, Goffman, Habermas, Irigaray, Jaspers, and Lacan all come to mind).

How psychiatry lit upon communication at mid-century is another dissertation to be written, but the key figure in the United States would have to be Harry Stack Sullivan (1892–1949), who understood schizophrenia and other forms of mental deviance interpersonally, that is, as a breakdown in social relations. (An openly gay man for the standards of his age, he seems to have solved the problem of how to legitimize the homosexual family by adopting, rather than marrying, his lover.) Heavily indebted to Chicago sociology, and close friends with Harold Lasswell and Edward Sapir, he defined psychiatry as "the study of interpersonal relations." Toward the end of his life, he envisioned psychiatry as moving beyond the dyad to larger groups and societies, thus helping to prevent the insanity of mass bloodshed, a project that was widely shared after the war.[34] The journal he founded in 1938, *Psychiatry: Journal of the Biology and Pathology of Interpersonal Relations*, was one of the leading outlets for thinking about communication for the next three decades.[35] It featured work by anthropologists such as Ruth Benedict, Raymond Firth, Hildred Geertz, Margaret Mead, and Sapir;

sociologists such as Amitai Etzioni, Erving Goffman, Merton, Talcott Parsons, and Riesman; political scientists such as Karl Deutsch and Harold Lasswell; and many humanistic psychiatrists, psychologists, and clinicians. It included the groundbreaking essay on parasocial interaction by Donald Horton and Richard Wohl.[36] Of particular note are a semiotic essay by Walker Percy, a medical doctor later famous for his brooding novels, and a 1955 essay by Timothy Leary, later famous as an LSD guru.[37] In a somewhat technical piece on interpersonal communication, Leary builds on George Herbert Mead, Sapir, and Sullivan to demarcate different styles of relating. Indeed, interpersonal communication was a lifelong concern of Leary's, and he often mused that LSD expanded our ability to communicate "authentically." Like his LSD-dropping compatriot John Lilly, who set up a "Communication Research Institute" in the U.S. Virgin Islands to crack the dolphin code in 1961, Leary shows that an intellectual history of communication in the postwar era would have to venture into the far-out zones of drugs, the sea, aliens, and madness. Wilbur Schramm is not the whole story.

Psychiatry is one of the key sources of twentieth-century communication theory. Apart from systems theorists in interpersonal communication research (indeed, such as my Utah advisor Leonard Hawes), the mid-century boom of psychiatric work has had remarkably little influence on those with custody over the study of communication today. In part, this disappearance is more general: psychiatry's intellectual cachet plummeted in the last three decades of the twentieth century, thanks to the success of both the antipsychiatry movement of the 1960s and of psychopharmacology as the leading form of treatment for mental illness since the 1980s. At mid-century, mental illness was at the center of the intellectual agenda; today that problem has been vastly reconfigured, and a chapter in the history of theorizing about communication has been lost.

Literature Slouching toward Culture

The international rise of cultural studies around 1960, taking that term in the loosest sense, is one more crucial factor in the postwar mix. Narrators of the rise of British cultural studies often resort to a tale of British exceptionalism: unlike Germany, France, or the United States, sociology was a latecomer to British academic life, leaving departments of literature to incubate cultural studies as a compensation for the lack of social inquiry elsewhere. Although true of Britain, the tale is not at all exceptional: the global norm for the disciplinary matrix of studies of popular culture and communication has been literary studies. It is true in France, Canada, Scandinavia, Italy, Spain, Latin America, and Russia; even Wilbur Schramm, the great American founder, had a PhD in English. (I too came to communication studies from English, as did my Stanford advisor, Don Roberts.)

Obviously, the new remit of literary study owed to a world-historical shift in the place of literature. In an age of cinema and stereos, books no longer ruled the pantheon. Marshall McLuhan, a key figure in all this, noted the readiness of literary scholarship to tackle the new world:

> Insofar as literature is the study and training of perception, the electric age has complicated the literary lot a good deal. . . Yet the literary man is potentially in control of the strategies needed in the new sensory environment.[38]

This might be read as a mission statement for cultural studies: to redeploy literary modes of analysis for nonliterary objects. The key factor in the rise of cultural studies was the collaboration of literary and anthropological scholars.[39] In Canada, the University of Toronto's Centre for Culture and Technology rested on the twin pillars of Marshall McLuhan, a literary scholar, and Edmund Carpenter, an anthropologist. In England, one can point to both the joint work on literacy in the early 1960s by Ian Watt, a literary scholar, and Jack Goody, an anthropologist, and to the essential late 1950s texts of Richard Hoggart and Raymond Williams, trained literary scholars who saw culture both as a body of (artistic) artifacts and a whole (anthropological) way of life. Hoggart's founding of the Birmingham Centre for Contemporary Cultural Studies (CCCS) in 1964 is a key point in the story of cultural studies that leads to Stuart Hall. (Goody and McLuhan were both students of F. R. Leavis in Cambridge, and Williams was deeply influenced by him.) In France, the opening into culture was made by the sociologist Georges Friedmann, the semiologist Roland Barthes, and the sociologist and philosopher Edgar Morin, who together founded the Centre d'études des communications de masse [Center for the study of mass communications], often known as Cecmas, in 1960. In several striking ways, Cecmas is a French parallel to CCCS. The Frankfurt School, securely reinstalled in Germany around 1950, though already famous for work on the sociology of culture, got a fresh and controversial jolt of political engagement from Habermas's *Structural Transformation*, which Adorno refused as too political. There is a clear kinship among figures such as McLuhan, Carpenter, Williams, Hoggart, Goody, Watt, Barthes, Friedmann, Morin, and Habermas around 1960—in vision, method, and structure of feeling. With the exception of the Canadians, they all shared leftist politics. Another dissertation should be written on the international rise of cultural studies in the North Atlantic region, that is, literary-anthropological studies, circa 1960.

By the 1970s, fortunes had improved: McLuhan, British cultural studies, French semiology, and German critical theory were all internationally famous. Though the French and the British, unlike McLuhan and Habermas in their very different ways, claimed to have outgrown their erstwhile explicit engagement with "communication," all of these intellectual currents sprang from earlier efforts to sound the strange new world of industrial culture by combining textual and anthropological methods.

Institutional Problems: Undergraduates, Globalization

Almost all the intellectually generative institutions in twentieth century thought about communication were outside of university departments: the Frankfurt Institut für Sozialforschung; the Payne Fund Studies; the Bureau of Applied Social Research at Columbia; the wartime mobilization of the social sciences in the Office of Facts and Figures, the Office of War Information, and the Office of Strategic Services; the Washington School of Psychiatry; the Hutchins Commission (1947); UNESCO; the Macy Conferences on Cybernetics; the Centre for Culture and Technology; Schramm's Institutes at Illinois and Stanford; Cecmas; CCCS, and so on. The story of communication studies in the twentieth century can be crudely summarized as an institutional shift from parauniversity interdisciplinary postdoctoral research institutes to academic departments tasked with undergraduate education. How departments with some version of the word "communication" in their names cropped up in universities is, as I remarked in footnote 2, in "Institutional Sources," an untold story. The history of the field of communication studies has yet to be written, despite many good local studies about departments, universities, subfields, and misadventures among the family tree of speech, rhetoric, journalism, broadcasting, communication research, interpersonal and organizational communication, speech pathology, and theater.

Taking this history into an international context will be even more challenging. But one thing is clear: since 1989 or so, communication studies—or "communications" or "media and communications" or "culture and communication" among several other options—has exploded as a field worldwide. In China, Greece, Turkey, and Ukraine, to take instances with which I am familiar, new departments on such topics have sprouted up in the past fifteen years. "Media and Communications" was the largest undergraduate major in the United Kingdom in 2000, which led to no small stir among some political pundits. Each country and each department has its own story, and the closer you get the more jagged things appear. But some features stand out amid the recent global flowering. One is the recapitulation of the disciplinary diversity of the liberal arts within single departments. Communication departments often reconfigure the full range of social sciences and humanities within a single home. World-historical changes in the infrastructure of communications—technical, political, and economic—continue to call old humanistic and social science fields to new objects. For instance, the Department of Communication and Mass Media at the University of Athens, Greece, where I taught in 1998–1999 as a Fulbright Scholar, was founded in 1990. When I was there, the department's faculty included a mathematician, a semiotician, a political economist, a film scholar, a clinical psychologist, a linguist, a political scientist, a sociologist of culture, a statistician, and a scholar of arts administration, among others. Few were trained in communication but all were retooling to focus on it (whatever "it" is); the organizational rationale was topical rather than disciplinary.[40] In the United States, many programs prefer disciplinary to topical identity, though departments in many of the better universities such as Stanford, UCSD, NYU, and Northwestern seem to prefer to hire people with noncommunication doctorates. What does this all add up to? The best view of the interdisciplinary mix of communication studies is probably had by those least able to decipher it: undergraduate students. Each professor only contributes a few specialized ingredients to the soup, but the students drink the whole thing. In any case, we might think of communication studies as one of the first postmodern fields rather than as a stillborn modern one.[41] The global proliferation of departments and the continued power of American work as a global benchmark have radically changed the landscape since 1986.

CODA

"Institutional Sources" was quite negative about the intellectual possibilities of "communication" as a term of art, seeing it co-opted by institutional interests. *Speaking into the Air*, in contrast, argued that communication theory could be taken as an inquiry into the fundamental questions of human life and sought to demonstrate the endless potentials for weirdness and interest that that term can provide. Being precedes consciousness, and all the years of teaching in a Department of Communication Studies have perhaps softened my views. Or perhaps getting a job at the University of Iowa, a place that was wide open to intellectual exploration thanks to the leadership of Sam Becker among others, allowed for a reconciliation with communication as a term. We are, for better or worse, stuck with communication as an academic field. Fortunately, it can boast increasing good work as well as traditions of scholarship that can be enriched even more by serious digging.

As I see it now, the right strategy for the institutional field is to opt for what my late colleague Ken Cmiel liked to call "small-country cosmopolitanism." The provinces, whether in society or the university, are often the best soil for breeding cosmopolitanism. I invariably find people in Iowa City, where I live, better informed about what is happening in New York City than New Yorkers are about

Iowa City. Some New Yorkers can manage not to care about anything but what is happening in New York, since they assume anything in the world will sooner or later wash up on their doorstep; people in Iowa City know they are not the center of the universe and have to keep current with the rest of the world. In the same way, most of the world knows a lot about America but America does not know a lot about the rest of the world. Canadians perforce know a lot about the United States, but Americans rarely know much about Canada. Living in an empire can ruin cosmopolitan perception. I have met world-class literary scholars who are surprised to learn that communication studies exists. Such ignorance of other nations is a luxury that no small nation can afford. Those at the center often carry on in blissful ignorance of the periphery. Hegemony is epistemologically hazardous. Communication studies has the comparative advantage of marginality and has become, at least in part, a site for wide-ranging interest in social and cultural theory, and a gathering place for reconfiguring the humanities and social sciences, and maybe even the arts and sciences. Maybe in a global cultural economy, being an island like Taiwan isn't really so bad.[42]

NOTES

1. Peters, "Institutional Sources of Intellectual Poverty in Communication Research." See also the critique by Hernando Gonzalez ("The Evolution of Communication as a Field") and my response ("The Need for Theoretical Foundations").
2. Peters, *Speaking into the Air.*
3. The phrase is from Turner, *From Counterculture to Cyberculture.*
4. The Stanford curriculum was brilliantly parodied by Michael Pacanowsky, writing under the pseudonym of Murdock Pencil ("Salt Passage Research: The State of the Art").
5. Franklin Fearing, one of the founding editors of the *Hollywood Quarterly*, was noted at Stanford for his essay "Toward a Psychological Theory of Communication."
6. I discuss the priestly aspirations of social science in *Courting the Abyss*, chapter 5.
7. When the Department of Communication moved into a building with the Department of Sociology on the Stanford Quadrangle, a proposal was aired to bring Merton as a unifying speaker for the dedication of the building, but it fell through, I recall, for reasons of his health. *Social Theory and Social Structure* was in the Institute's book collection, and I read parts of it with recognition around 1984, thanks to Todd Gitlin's work.
8. Watzlawick, Beavin, and Jackson, *Pragmatics of Human Communication.* See also Wilder, "The Palo Alto Group."
9. Habermas, "Political Communication in Media Society."
10. Paisley, "Communication in the Communication Sciences."
11. Schramm, "The State of Communication Research: Comment."
12. On the importance of the 1940s, see Cmiel, "Cynicism, Evil, and the Discovery."
13. For an exciting revisionist account of postwar thinking about communication, see Schüttpelz, "Get the message through."

14. For two different examples of the power of network analysis in intellectual history, see Turner, *From Counterculture to Cyberculture*; and Collins, *The Sociology of Philosophies*.
15. For one start on this project, see Simonson and Peters, "Communication and Media Studies, History to 1968."
16. See, for instance, Saunders, *The Cultural Cold War*.
17. Adorno, "Zur gegenwärtigen Stellung."
18. Habermas and others, *Student und Politik*.
19. See, for instance, Shlapentokh, *The Politics of Sociology in the Soviet Union*; Splichal, "Indigenization versus Ideologization"; and Vihalemm, "Development of Media Research in Estonia."
20. Yehoshua Bar-Hillel, quoted in Heims, *The Cybernetics Group*, 28.
21. On the jacket of Rosenblith, *Sensory Communication*, Rosenblith is listed as "Professor of Communications Biophysics" at MIT.
22. Jakobson, Fant, and Halle, *Preliminaries to Speech Analysis*.
23. See, for example, Jakobson, "Linguistics and Communication Theory." On Jakobson and communication more generally, see Schüttpelz, "Von der Kommunikation zu den Medien." Kay, *Who Wrote the Book of Life?* chapter 7, argues Jakobson to be a key architect of the notion that DNA is a linguistic code.
24. Lévi-Strauss, "Introduction: The Mathematics of Man," 585.
25. Hörl, *Die heiligen Kanäle*, 231–32.
26. Bitsch, *"Always crashing in the same car."*
27. Lacan, *Écrits: A Selection*, 86.
28. Probably the fullest and strangest fruit of the marriage of Lacanian psychoanalysis and cybernetics is Wilden, *System and Structure*.
29. Hyppolite, "Le 'coup de dés' de Stéphane Mallarmé et le message," 877–84; "La machine et la pensée," 891–919; and "Information et communication," 928–71, in Hyppolite, *Figures de la pensée philosophique*.
30. Foucault, *Surveiller et punir*, 202.
31. See Combes, "L'hypothèse du bio-pouvoir."
32. Benjamin Peters, "The Soviet Translation of Norbert Wiener."
33. Ruesch and Bateson, *Communication*. Ruesch is another neglected figure. No less than 16 pieces have "communication" in their title in his collection, *Semiotic Approaches to Human Relations*.
34. See Heims, *The Cybernetics Group*, chapters, 6–7.
35. For one valuable start on this context, see Yves Winkin's forthcoming work on Erving Goffman.
36. Horton and Wohl, "Mass Communication and Para-Social Interaction."
37. Percy, "The Symbolic Structure of Interpersonal Process"; and Leary, "Theory and Measurement Methodology."
38. McLuhan to Michael Wolff, July 4, 1964, in *Letters of Marshall McLuhan*, 304.
39. Schüttpelz, "Von der Kommunikation zu den Medien," 526–28.
40. If the word "organizational" can be used in Greece.
41. John Durham Peters, "Genealogical Notes on 'The Field.'"
42. I would like to acknowledge helpful commentary by David Park, Benjamin Peters, Peter Simonson, Fred Turner, and Yves Winkin, and only wish I could have done all they recommended. An earlier version was delivered at Carleton University in Ottawa in March 2007, and I thank Michael Dorland, Jason Hannan, and John Shiga and their colleagues for the warm invitation.

WORKS CITED

Adorno, T. W. "Zur gegenwärtigen Stellung der empirischen Sozialforschung in Deutschland." In *Soziologische Schriften*, vol. 1, 478–93. Frankfurt: Suhrkamp, 1972. First published 1952.

Anderson, Benedict R. *Imagined Communities: Reflections on the Origin and Spread of Nationalism.* London: Verso, 1983.

Bateson, Gregory. *Naven, a Survey of the Problems Suggested by a Composite Picture of the Culture of a New Guinea Tribe Drawn from Three Points of View.* Cambridge, MA: University Press, 1936.

Bitsch, Annette. *"Always crashing in the same car": Jacques Lacans Mathematik des Unbewussten.* Weimar: Verlag und Datenbank für Geisteswissenschaften, 2001.

Cmiel, Kenneth. "On Cynicism, Evil, and the Discovery of Communication in the 1940s." *Journal of Communication* 46 (1996): 88–107.

Collins, Randall. *The Sociology of Philosophies: A Global Theory of Intellectual Change.* Cambridge, MA: Harvard University Press, 1998.

Combes, Muriel. "L'hypothèse du bio-pouvoir: Entre polémique et cybernétique." *Multitudes*, February 19, 2006. http://multitudes.samizdat.net/L-hypothese-du-bio-pouvoir-entre.html (accessed September 15, 2006).

Fearing, Franklin. "Toward a Psychological Theory of Communication." *Journal of Personality* 22 (1953): 71–88.

Foucault, Michel. *Surveiller et punir: Naissance de la prison.* Paris: Gallimard, 1975.

Gonzalez, Hernando. "The Evolution of Communication as a Field." *Communication Research* 15, no. 3 (1988): 302–08.

Habermas, Jürgen. *The Structural Transformation of the Public Sphere: An Inquiry into a Category of Bourgeois Society.* Cambridge, MA: MIT Press, 1989. First published in German in 1962.

———. "Political Communication in Media Society: Does Democracy Still Enjoy an Epistemic Dimension? The Impact of Normative Theory on Empirical Research." *Communication Theory* 16 (2006): 411–26.

Habermas, Jürgen, and others. *Student und Politik: Eine soziologische Untersuchung zum politischen Bewusstsein Frankfurter Studenten.* Neuwied: Luchterhand, 1961.

Heims, Steve Joshua. *The Cybernetics Group.* Cambridge, MA: MIT Press, 1991.

Hörl, Erich. *Die heiligen Kanäle: Über die archaische Illusion der Kommunikation.* Zürich-Berlin: Diaphanes, 2005.

Horton, Donald, and R. Richard Wohl. "Mass Communication and Para-Social Interaction." *Psychiatry* 18 (1956): 215–29.

Hyppolite, Jean. *Figures de la pensée philosophique: Écrits de Jean Hyppolite (1931–1968).* Vol. 2. Paris: Presses universitaires, 1971.

Jakobson, Roman. "Linguistics and Communication Theory." In *On Language*, edited by Linda R. Waugh and Monique Monville-Burston, 489–97. Cambridge, MA: Harvard University Press, 1990. First published in 1960.

Jakobson, Roman, Gunnar M. Fant, and Morris Halle. *Preliminaries to Speech Analysis: The Distinctive Features and their Correlates.* Acoustics Laboratory, MIT: Technical Report no. 13, 1952.

Kay, Lily E. *Who Wrote the Book of Life? A History of the Genetic Code.* Stanford, CA: Stanford University Press, 2000.

Lacan, Jacques. *Écrits: A Selection.* Translated by Alan Sheridan. New York: Norton, 1977.

Leary, Timothy. "The Theory and Measurement Methodology of Interpersonal Communication." *Psychiatry* 18 (1955): 147–60.

Lévi-Strauss, Claude. "Introduction: The Mathematics of Man." *International Social Science Bulletin* 6 (1954): 581–90.

Merton, Robert K. *Social Theory and Social Structure.* Enl. ed. New York: Free Press, 1968.

Molinaro, Matie, Corinne McLuhan, and William Toye, eds. *Letters of Marshall McLuhan.* Toronto: Oxford University Press, 1987.

Paisley, William. "Communication in the Communication Sciences." In *Progress in the Communication Sciences,* edited by Brenda Dervin and M. J. Voigt, vol. 5, 1–43. Norwood, NJ: Ablex, 1984.

Pencil, Murdock [Michael Pacanowsky]. "Salt Passage Research: The State of the Art." *Journal of Communication* 26 (1976): 31–36.

Percy, Walker. "The Symbolic Structure of Interpersonal Process." *Psychiatry* 24 (1961): 39–52.

Peters, Benjamin. "Traduttore, Traditore: The Soviet Translation of Norbert Wiener's Early Cybernetics." Paper presented at the annual meeting of the International Communication Association, Dresden, Germany, June 2006.

Peters, John Durham. "Institutional Sources of Intellectual Poverty in Communication Research." *Communication Research* 13, no. 4 (1986): 527–59.

———. "The Need for Theoretical Foundations: Reply to Gonzalez." *Communication Research* 15, no. 3 (1988): 309–317.

———. "Genealogical Notes on 'The Field.'" *Journal of Communication* 43 (1993): 132–39.

———. *Speaking into the Air: A History of the Idea of Communication.* Chicago: University of Chicago Press, 1999.

———. *Courting the Abyss.* Chicago: University of Chicago Press, 2005.

Rosenblith, Walter A., ed. *Sensory Communication: Contributions to the Symposium on Principles of Sensory Communication, July 19–August 1, Endicott House, MIT.* Cambridge, MA: MIT Press, 1961.

Ruesch, Jurgen. *Semiotic Approaches to Human Relations.* The Hague, the Netherlands: Mouton, 1972.

Ruesch, Jurgen, and Gregory Bateson. *Communication: The Social Matrix of Psychiatry.* New York: Norton, 1951.

Saunders, Frances Stonor. *The Cultural Cold War: The CIA and the World of Arts and Letters.* New York: New Press, 2000.

Schramm, Wilbur. "The State of Communication Research: Comment." *Public Opinion Quarterly* 23 (1959): 6–9.

Schüttpelz, Erhard. "'Get the message through': Von der Kanaltheorie der Kommunikation zur Botschaft des Mediums: Ein Telegramm aus der nordatlantischen Nachkriegszeit." In *Medienkultur der 50er Jahre,* edited by I. Schneider and P. M. Spangenberg, 51–76. Wiesbaden, Germany: Westdeutscher Verlag, 2002.

———. "Von der Kommunikation zu den Medien: In Krieg und Frieden (1943–1960)." In *Gelehrte Kommunikation,* edited by Jürgen Föhrmann, 483–551. Vienna, Austria: Böhlau Verlag, 2005.

Shlapentokh, Vladimir. *The Politics of Sociology in the Soviet Union.* Boulder, CO: Westview, 1987.

Simonson, Peter, and John Durham Peters. "Communication and Media Studies, History to 1968." In *International Encyclopedia of Communication,* edited by Wolfgang Donsbach. Oxford: Blackwell, forthcoming.

Splichal, Slavko. "Indigenization versus Ideologization: Communication Science on the Periphery." *European Journal of Communication* 4 (1989): 329–59.

Turner, Fred. *From Counterculture to Cyberculture.* Chicago: University of Chicago Press, 2006.

Vihalemm, Peeter. "Development of Media Research in Estonia." *Nordicom Review* 22 (2001): 79–92.

Watzlawick, Paul, Janet Helmick Beavin, and Don D. Jackson. *Pragmatics of Human Communication: A Study of Interactional Patterns, Pathologies, and Paradoxes.* New York: Norton, 1967.

Wiener, Norbert. *Cybernetics.* New York: J. Wiley, 1948.

Wilden, Anthony. *System and Structure: Essays in Communication and Exchange.* London: Tavistock, 1972.

Wilder, Carol. "The Palo Alto Group: Difficulties and Directions of the Interactional View for Human Communication Research." *Human Communication Research* 5 (1979): 171–86.

CHAPTER SIX

"Communication": FROM Concept TO Field TO Discipline

J. MICHAEL SPROULE

By the mid-1990s, the term "communication" had so much supplanted that of "speech" as to make one suppose that the former label finally had become a rubric under which an entire academic discipline confidently could deploy itself. Yet from a broader perspective on the study of human symbolic interplay, the titles of relevant scholarly societies still suggested, more than a decade later, that "journalism," "broadcasting," and "rhetoric" could challenge communication's claims as a universal disciplinary designator. In each case, however, the contending terminologies ultimately paid fealty to the master concept of communication as indicated by certain elaborations or substitutions—for example, the marrying of "mass communication" to journalism, the parallel locution of "telecommunications" for broadcasting, and the frequent elaboration of rhetoric by the phrase "and communication."

The seventy-year expansion of "communication" as designator of concept, field, and discipline represents far more than a superficial semantic overlay. The replacement of "speech," "journalism," "radio/TV/film," "broadcasting," and "rhetoric" by "communication" constitutes the visible sign of an underlying struggle as to whether the latter term should be considered in an exclusionist or an inclusionist light. What follows is a history of "communication" in the context of pressures to define the term narrowly as an interdisciplinary field of study or broadly as an institutionalized academic discipline.

COMMUNICATION IN THE EXCLUSIONARY MODE

Although the term "communication" was occasionally used as a significant conceptual locution before World War II, it was the war that brought the term into common academic parlance as a designator for a niche of social science scholarship narrowly drawn as administratively focused, quantitatively conducted, and located in high-level academic and other institutions.

Because the original bound volumes of *Psychological Abstracts* indexed journals of psychology, sociology, political science, market research, public opinion, and education, they give us a broad indication of how the term "communication" emerged between the mid-1920s and the mid-1950s in relation to such other designators as "attitude," "public opinion," and "propaganda." Here we find "communication" rarely used during the 1930s—and three of the five articles so indexed dealt with the communication of insects or birds. Issues of *Public Opinion Quarterly* show that, before the early 1950s, media and opinion research was infrequently deployed under the banner of "communication process." In academic parlance, "public opinion" remained more attached to "propaganda" than to "communication" until the rapid decline of the propaganda label in social science of the 1950s. Although we find "communication" (along with "transportation") employed as a principal concept in what became known as the Chicago School of social science, the overarching rubric of research was that of "urban studies."[1]

Communication study was afoot in the 1930s, but it was not until 1942 that the bound indices of *Psychological Abstracts* regularly included "communication" as an entry, and the total of such entries between 1942 and 1947 was but twenty-two articles. Yet by 1948, the annual number had risen to sixteen, thence to thirty-two in 1949, and forty-eight in 1950. Yearly averages thereafter were between fifty and one hundred articles in the 1950s, increasing typically to the upper 100s in the 1960s and 1970s, indicating the growth and stability of the term as a major focal concept in social science research.

Accounting for the rise of communication study was the desire of academicians to be of practical (and politically noncontroversial) service to a nation confronted by fascism (and, later, communism). An important component was wartime work carried out under the explicit rubric of communication. The turn to communication as master term for studying social influence was characteristic of two particularly important research programs, those of Harold D. Lasswell in the Library of Congress and Carl I. Hovland for the U.S. Army.

Lasswell's research program originated in his having been a regular participant since 1939 in an informal working group, sponsored by the Rockefeller Foundation, that aimed to refine new approaches to knowledge in the promising area of communication, with the further objective of bringing this information to the attention of relevant government administrators. In 1940, Lasswell was invited to apply for a large grant from the foundation to set up an organization, administratively housed in the Library of Congress, that would pursue two objectives. First, Lasswell's Experimental Division for the Study of Wartime Communications would refine quantitative content analysis. Second, the Lasswell group would directly apply this method, on the one hand, to the immediate analysis of organizations (for the benefit of the Department of Justice) and, on the other, to the training of personnel for assignments in propaganda analysis and intelligence.[2] The practical and administrative contributions of Lasswell's group included a comprehensive analysis of worldwide media (the World Attention Survey).

Lasswell's turn to the rubric of communication was directly related to the confluence of wartime conditions whereby grantors, administrators, and academicians all sought to distance their work (the study of "communication") from that of the enemy (whose output was "propaganda"). Here consultant Lasswell continued to function as an academician building up "basic theory in the field of communication." Invocation of the politically neutral rubric of communication permitted foundations, government agencies, and individual scholars to study domestic symbolic output absent the inconvenient reminder—inherent to the term "propaganda"—that official messages harbored deeper persuasive purposes. As World War II segued into the cold war, social scientists increasingly took Lasswell's lead such that, by 1952, *Psychological Abstracts* reported but six studies of "propaganda" as compared to thirty-seven of "communication," a reversal of the situation in the early 1940s. Five of these six propaganda studies related not to domestic U.S. social influence but to enemy communication or the international competition of ideologies.[3]

Where the Rockefeller Foundation supplied the key impetus for Lasswell's program, the Hovland line of communication research began with the Army's interest in understanding and motivating the American soldier. Within the Army's overall Research Branch, headed on the technical side by Chicago sociologist Samuel Stouffer, the Survey Section produced a regular series of reports on the attitudes of GIs, for example soldiers' beliefs and opinions about such general matters as combat and perceived opportunities for advancement together with such particular questions as the wartime roles of big business and America's several military allies. Later published as *The American Soldier* (two volumes), both the findings and the quantitative approach of the Survey Section had a significant impact on social science, generally, by illustrating how extensive quantitative data collection could build up a systematic body of social knowledge relating to such concepts as "attitude," "adjustment," and "satisfaction."[4]

Although the Survey Section diffused quantitative survey research broadly, the two *American Soldier* volumes called less attention to communication as a new and coalescing field than did related work by the Experimental Section, headed by Carl I. Hovland, Yale psychologist. Where the term "communications" is indexed but once (and as an atheoretical label) in the 1250 pages of the two survey volumes, the concept was central to Hovland's experimental researchers as chronicled in *Experiments on Mass Communication* (volume 3 of the series reporting findings of the Research Branch). Hovland's continued pursuit of quantitative social influence studies was reflected in succeeding books, the first of which, *Communication and Persuasion*, appeared in 1953.[5]

Postwar social science—notably, interdisciplinary communication research—was greatly influenced by the multivolume *Studies in Social Psychology in World War II* and other books and articles that presented the major wartime quantitative research programs.[6] In contrast, wartime work by professors of speech, both those who joined industrial and government agencies and those who remained on campus, was notable for lack of postwar public visibility and for minimal theoretical-methodological impact on the discipline. War-related articles by speech faculty focused on how they employed their existing approaches and competencies to contribute to officer training, industrial training and discussion, civil defense, volunteer organizations, adult education, and public information. In a review of what such undertakings portended for postwar departments of speech, the emphasis was on teaching and service—rather than Stouffer-Hovland style innovations in theory or methods. For speech faculty, prospects for cooperation with English in basic education loomed large in this anticipated future. Paul Bagwell's review of work going on at Michigan State College included hints that what speech and English shared was a common interest in teaching something broader than their familiar composition and public speaking courses—something called communication.[7] (Here it should be noted that speech was not the only academic field to produce wartime articles that laid out routine application- and service-type opportunities for scholars. Such could be found in sociology and psychology as well.)[8]

In contrast to the minimal impact of wartime work by speech faculty, leading lights in certain disciplines affiliated with the Social Science Research Council (SSRC)—notably, psychology, sociology, and political science—continued to confer about their common interest in studies of communication. The 1954 "Conference on Research on Public Communication" represented a key moment in this effort to assess what might be the defining features, overall parameters, and best means to advance communication study. In this round table of sixteen prominent scholars, the objectives of laying out communication's features and parameters were most succinctly reflected in Bernard Berelson's summary of six lines of ongoing research. As reflected in the minutes of the meeting, Berelson gave most weight to four lines of endeavor: (1) "the Lasswell tradition," (2) "the [Paul] Lazarsfeld tradition of empirical field studies," (3) the tradition of group dynamics, notably in the work of Kurt Lewin, Robert Bales, and Leon Festinger, and (4) "the Hovland psychological tradition" with its focus on experimentally induced communication effects. Although Berelson also credited Robert Leigh's work on public policy and Harold Innis's historical studies, these latter two bodies of work orbited quite distantly from his favored foursome, as became further clear in Berelson's 1959 *Public Opinion Quarterly* article, which reiterated the first four as primary and presented the other two (public policy and history) as "too value laden" and "unscientific" to be proper exemplars of the field.[9]

Although the work cited by conferees was being undertaken in psychology, sociology, and political science (all SSRC-constituent disciplines), schools of journalism (and media research centers) were at several points highlighted in the 1954 conference as promising institutional homes for pursuing interdisciplinary communication research and for building links between academicians and professionals. Carried to its logical conclusion, this institutional line of thought might well have pointed to the designation of communication as a discipline more than field. Yet the surface consensus of the time, as reflected in comments following Berelson's 1959 article, was to regard communication less as a coalescing discipline and more as an interdisciplinary field plagued by thorny boundary issues. Notable here were Wilbur Schramm's explicit stipulation of "field" over "discipline" and Raymond Bauer's reference to his Stanford colleague, Schramm, as a scholar of journalism (the discipline) who was interested in a cybernetic model of communication (the field).[10]

The lingering postwar impetus to consider communication as a narrow boundary region flowed naturally from the field's origins in a time when key researchers amassed quantified data under official government or grantor auspices with the objective to help solve the practical problems of agency managers. As previously explained, Lasswell's and Hovland's research fit this description; but the three characteristics of exemplary communication inquiry also applied to the wartime work of Lazarsfeld, whose Columbia University research bureau both enjoyed Rockefeller Foundation support and contributed to Stouffer's Research Branch, and to Lewin, who worked under the auspices of the Iowa Child Welfare Research Station and also contributed to the Committee on National Morale.

The role of the research-sponsoring and research-supporting agencies was crucial in reinforcing the presumption that communication study should be narrowly conceived. Grantor support from the major philanthropies pointed in the direction of priming the pump to accelerate innovation in theory and methodology in a politically noncontroversial context. Representative was the focus of the Rockefeller Foundation round tables on "new methods of acquiring knowledge"— that is, research methodologies and hypotheses—that informed public policy.[11] By supporting selected key scholars at major institutions, the Rockefeller and other foundations increased the chances that their seed money would sprout significant results nationally. In addition, the continuing role of the SSRC as intermediary between grantors and researchers assured that the work would be highly quantitative inasmuch as a key animating purpose of the Council remained that of redefining social science in the image of a pure science of rigorous method in which value questions were eschewed.[12]

Out of World War II, therefore, emerged a field of communication defined sharply according to what was seen as archetypal work. To the extent that leading researchers viewed communication study as an oasis (Schramm's metaphor[13]) occasionally visited by representatives of social science disciplines, it was appropriate to approach the term with the exclusionary objective of properly delimiting a shared area of study. Accordingly, focus was on the paradigmatic work occurring in six or so high-level locations where researchers undertook to collect quantified survey or experimental data to inform pressing administrative problems.

COMMUNICATION IN THE INCLUSIONARY MODE

Minutes of the 1954 SSRC conclave suggest that the sixteen conferees remained unaware that their pronouncements perhaps demanded too much of a borderline field of study. Could communication, considered as a mere oasis for interdisciplinary dalliance, successfully bring together studies of institutions and channels, on the one hand, and individual responses and psyches, on the other hand, while simultaneously integrating practitioners and academicians in furtherance of research, training, and service to society? Reflecting the implicit disciplinary pretensions of communication study, this question set the stage for an inclusionary enterprise conducted in several academic venues by which communication came to be regarded more as discipline than field. Here we may draw analogy to the contemporary condition of women's studies and American studies. Within these two areas, which began as interdisciplinary fields supported by affiliated scholars in centers, many program directors now seek departmental status for their units. Acceptance as a separate department is a historical precursor to disciplinary recognition.[14]

Had it not been for the longstanding mission of professional training, journalism schools and departments might have rapidly become *the* recognized disciplinary venue for communication research during the time when many specialists from psychology, sociology, and political science were departing homeward. Yet although journalism units were included in the nexus of narrowly conceived communication study by the 1954 conferees, Wilbur Schramm noted the problem that only about a third of accredited schools or departments were active either as "mature" contributors to or as "intelligent" consumers of cutting-edge social science scholarship in communication. Given that education in journalism had begun with a strong practical orientation (to which was added a close link to the humanistic discipline of English), it is not surprising that no articles dealing with "communication theory" appeared in the first twenty-one volumes of *Journalism Quarterly*. Yet, by the 1970s, around ten such articles were appearing yearly, and in 1976, the Association for Education in Journalism (AEJ) broadened its masthead to AEJMC (appending the phrase "and Mass Communication"). Accounting for these changes was the emergence of social science research units in J-schools, sometimes as semi-independent communication institutes.[15]

The situation of bona fide participation in communication research proved to be initially more tenuous in departments affiliated with another discipline-defining organization, the Speech Association of America (SAA). Here, as Herman Cohen later observed, scholars in the postwar period, having "slumbered through the early days of the [social scientific] revolution" now "awoke to find that social scientists had undertaken systematic studies of human communication"—studies that resonated well with speech research but were entirely separate from it. Because speech scholars had been uninvolved in the Lasswell, Hovland, and other wartime research enterprises, the postwar outpouring of communication-based findings raised the possibility that a generation of progress recorded in speech journals, and conducted under such rubrics as "rhetoric," "public address," "general speech," and "discussion and debate," might be supplanted. Seemingly a new enterprise was afoot that appeared, on the one hand, entirely oblivious to the speech discipline and, on the other, so self confident as to implicitly denigrate what Cohen described as the "tranquil, if obscure" world of the SAA.[16]

During the 1950s, some speech researchers responded to the Lasswell-Lazarsfeld-Hovland-Lewin juggernaut by appropriating certain relevant terminologies and by augmented attention to quantitative methodologies. The SAA's community of scholars lacked the AEJ's Wilbur Schramm, who not only acted as an evangelist for social-scientific communication research but also seized entrepreneurial opportunities to bring forth research institutes at Illinois and Stanford. Speech scholars digested the wartime work both by forming sympathetic interest groups within the SAA and by publishing quantitative studies in mainstream speech journals. This gradualist undertaking mounted throughout the 1950s and 1960s culminating in a 1968 New Orleans special conference that recommended that greater attention be given to the "field" of *scientific* speech-communication studies within the general disciplinary area of "spoken symbolic interaction."[17]

The New Orleans recommendations included the idea of focusing the instructional program on scientific studies more than on "principles derived from ancient and modern rhetorical studies"[18]; but such an enthusiasm for science did not overly concern the rhetorical and historical scholars present at New Orleans. These nonbehaviorists seemed to regard science as a welcome addition that permitted speech scholars to better engage a wider study of communication then ongoing in many areas of the contemporary campus. Perhaps they also shared Malcolm MacLean's sentiments that scientific communication study would chiefly augment existing scholarly associations in a context in which, "in both speech and journalism, teachers make little use of research."[19] But the qualitatively oriented scholars did look ahead to their own special conferences.

While members of the SAA continued to arrange their minds better to accommodate the master term of communication, others within the SAA had earlier come to believe that creating a separate but affiliated organization might better promote the establishment of communication as a rubric for integrating the entire range of human symbolic action in both research and teaching. To this end, the National Society for the Study of Communication (NSSC) was organized at the 1949 SAA convention with the objective of "bring[ing] greater unity into the whole field of speech education by showing the interrelations of the basic speech and language communication behavior to the total interpersonal communication." The NSSC's marrying of organizational independence and SAA affiliation seemed a combination most likely to overcome inertia in an existing organization where new approaches might be otherwise dismissed as "esoteric" or "far-out."[20]

Foreshadowing in several ways the approach later endorsed in the SAA's New Orleans conference, the NSSC's mission of 1949–1950 became that of attuning the parent SAA to "the emerging new field of communication" that integrated the whole spectrum of research—from face-to-face interactions to those located in mediated settings, from aesthetics to practical communication in industry, from specific channels to intercultural exchange—and including, on the teaching side, all the clinical skills of reading, writing, listening, and speaking.[21] One important source of this broad objective was the legacy of wartime communication research that, as Ralph G. Nichols (the NSSC's second president) explained, was taking place in "communication centers" located at such places as Illinois, North Carolina, and Iowa. Here the NSSC's unique contribution would be to organize study and research committees to stimulate work in such areas as methodology, industrial communication, intercultural relations, and communication in families.[22]

Yet the mission of the NSSC responded not only to perceived research imperatives but also to closely related needs in the development of teaching programs in colleges and in K-12 settings. As Nichols explained, the NSSC's mission to unify listening, speaking, reading, and writing related to a postwar teaching approach that married these areas through interdisciplinary pedagogical cooperation between and among professors of speech, English, and related specialties. Here the NSSC (affiliated with SAA) represented the culmination of efforts on the speech side while the English discipline also contributed a new organization called the College Conference on Composition and Communication (affiliated with the National Council of Teachers of English).[23]

While the term "communication" became ever more familiar in the academy during the 1950s and 1960s, there remained a dichotomy in how broadly the term was applied within both the SAA and NSSC. Following the SAA's 1968 special conference, the terms "speech" and "communication" became melded not only in the new name of that sponsoring organization (i.e., Speech Communication Association—SCA) but also, increasingly, in the designations of interest groups and in terminologies used for research and teaching. Here "communication" assumed a clear disciplinary mission, although sharing the stage with the older term of "speech." The implicitly (and sometimes explicitly) hyphenated locution of "speech communication" suggested that "communication" enjoyed only an ambiguous status in the newly named association. However, in actual outcome, the chief effect of the terminological merger was to widen the mission of all members of the SCA—scientists, humanists, performance specialists, and professionals—in what became a gradual evolution away from "speech" and toward "communication" as the association's disciplinary masthead.

If the SCA struggled to reconcile the disciplinary connotations of "speech" and "communication," the NSSC continued to reflect the old field/discipline tension that, from the first, had been associated with the Lasswell-Lazarsfeld-Hovland-Lewin paradigm. Although the term "communication" clearly held sway in the NSSC, the kind of communication research undertaken under its auspices began to narrow somewhat from that described in the organization's first pronouncements. As the NSSC deployed itself away from interdisciplinarity, it simultaneously defined communication study in the restricted wartime mode such that (1) communication meant research more than teaching and (2) the preferred research should reflect exemplary quantitative-administrative programs. To these ends, the NSSC rechristened itself in 1969 as the International Communication Association (ICA), not only abandoning its former approach of jointly meeting with the SAA/SCA, but actively competing with the progenitor organization for recognition as the premier worldwide representative of an entire discipline.

Summarizing research trends in the ICA of 1963–1983, Everett Rogers noted that "over the past twenty years or so, the statistical analysis of quantitative data has predominated." He added that, with research rigor becoming synonymous with experimental and survey data collection, "this limited approach was preferred as safer for younger scholars." In earlier remarks, Rogers had coupled the ICA's preference for "empirical" over "critical" research with a prediction that the organization would see a growing methodological flexibility based on awareness of work being conducted in European and Latin American venues.[24] Notwithstanding Rogers's prognostication of future catholicity, the ICA's divisions throughout the preceding twenty-year period had prided themselves in an overall unity of theoretical-methodological approach such that, at the 1980 Acapulco convention, I recall joining with a small band of participants who, meeting somewhat in the mode of refugees, aimed to further a new Philosophy of Communication division. Debuting in 1981, this unit numbered a mere twenty members—one tenth the size of the next smallest ICA division, and one-thirtieth the size of the largest.[25]

At that same Acapulco convention, Wilbur Schramm delivered his well-known encomium to the "four fathers" of communication study—Lasswell, Lewin, Lazarsfeld, and Hovland—who, he argued, commonly illustrated the pattern of disciplinary specialists entering for a time the field of communication to solve particular real-world problems by establishing or working through institutes or centers and amassing quantified data. Here Schramm transmuted Berelson's idealized view of four exemplary research lines into an ostensible history of something that Schramm continued to characterize as a working area for visitors. Even at the beginning of the 1980s, Schramm's "four fathers" thesis was less a history and more a reiteration of Berelsonian assumptions that, increasingly, did not seem to hold for communication study outside of a few selected locales. But the selective memory implicit in Schramm's "four fathers" interpretation served as useful counterpoint to his approach of creating enclaves of scientific communication study that, when knit together, might culminate in departments exclusively devoted to this approach.[26] Here Kuhn observes that a paradigm shift typically involves "a drastic distortion in the scientist's perception of his discipline's past" such that history becomes but a straight line to a discipline's present or, we may add, to its idealized immediate future.[27]

Despite many a turn and twist, by the 1990s, communication ever more appeared to function as a true discipline that (1) was centered in its own departments, schools and colleges in which (2) a characteristic collection of teaching and research specialties were pursued in ways that (3) exhibited every bit of the unity-in-diversity combination of psychology, sociology, or political science in the context of (4) vastly growing numbers of majors whose needs were met by (5) a goodly supply of teacher/scholar recruits mustered from expanding communication graduate programs. In view of the spectrum of approaches long present in the SAA/SCA, combined with the ICA's growing recognition (in the aftermath of the Berlin and Acapulco conferences) of a philosophical/methodological "ferment in the field," it was more the case that only a minority of the discipline might view the "four fathers" as anything like paternal forbearers—with a greater majority regarding them more as distant uncles in what was becoming a large family of many branches.

THE FUTURE OF COMMUNICATION

Beginning with the 1990s and working back from the 1930s to the 1980s, this review of the term "communication" has now come full circle. In the United States, the idea of communication as a disciplinary reference point had gained strength as exemplified by developments that led to the redesignation of the NSSC as ICA in 1969, the AEJ as AEJMC in 1976, and the SAA/SCA as NCA in 1997. These nomenclatural changes were visible signs of a growing consensus that communication was the most suitable disciplinary designator for the expanding missions and growing numbers seen in affiliated departments, schools, and colleges.

Yet the continuing interplay of exclusionary versus inclusionary senses of "communication" may be observed in efforts to map out the history of a master term that has metamorphosed from concept to field to discipline. Here one finds a surprising lack of consensus in historical works endeavoring to chronicle the origins of the discipline. To take a salient example, we may compare how the foundations of communication research and teaching are laid out in two histories, one that reviews 1914–1945 by a former president of the NCA (Herman Cohen) and another that documents the period 1830–1990 by a past ICA president (Everett Rogers). Juxtaposing Rogers's 66 "Principal Figures in the History of Communication Study" to Cohen's author index of approximately 270 individuals, we find an overlap of but five persons: David Berlo, John Dewey, Sigmund Freud, Walter Lippmann, and Herbert Spencer. Granting that these five have appeared, and will continue to appear, in the endnotes of communication scholarship, few would describe this list of five as reflecting any coherent history of the departments, associations, and journals of the communication discipline.[28]

As further evidence that we have far to travel in working out the historical parameters of communication study, I consulted the names index of my own *Propaganda and Democracy*,[29] and pulled out more than twenty individuals, listed neither by Cohen nor Rogers, whose work I would consider as pivotal in setting the conceptual foundations of communication study in the twentieth century. Dividing this roster into categories of contribution, the group includes:

1. Influential communication practitioners: Edward L. Bernays (key founder, author, and promoter of professional public relations); and Ivy L. Lee (major founder of public relations).
2. Muckrakers and others who developed the political criticism of media: Ray Stannard Baker (author of such muckraking studies as "How Railroads Make Public Opinion"); Will Irwin (muckraker cited by Robert Park as one of the best historians of the press); A. J. Liebling (press critic for the *New Yorker*); George Seldes (dean of American popular media criticism);

I. F. Stone (independent press critic); and Upton Sinclair (author of two major muckraking works on propaganda in education—in addition to his notable *The Jungle*).

3. Major contributors to academic and popular propaganda analysis: Edgar Dale (educator, propaganda critic, and researcher on readability); Clyde R. Miller (founder of the Institute for Propaganda Analysis); Peter Odegard (leading political scientist contributor to propaganda study); John W. Studebaker (promoter of public discussion forums while serving as U.S. Commissioner of Education).

4. Prominent semanticists: Alfred Korzybski (major founder of general semantics); and S. I. Hayakawa (English professor who greatly popularized semantics).

5. Important sociocultural critics: Jacques Ellul (French author of a key trilogy of works on propaganda, society, and media); Alfred McClung Lee (media sociologist and critic); Gilbert Seldes (major author, first dean of the Annenberg School for Communication); and Stephen Toulmin (theorist of argumentation and science).

6. Major contributors to the study and practice of critical thinking: Stuart Chase (author of popular works on the influence of language on thought); and Goodwin Watson (developer of scales to measure critical thinking).

7. Significant contributors to opinion research: Archibald Crossley (early developer of broadcast ascertainment research); and Elmo Roper (early pollster).

The chief lesson that we may take away from Cohen's, Rogers's, and my own lists of the founders of communication study is a humble recognition that much work remains to solidify the parameters of communication as both academic and social enterprise and as manifested in teaching, research, and service to society. But if the full story of communication has yet to be written, we may at least depend on its having safely metamorphosed from loose concept in the 1930s to borderline field in the 1950s to full-fledged discipline at the beginning of the twenty-first century.

If communication has attained disciplinary status through a successful coalescing process, it does not follow that everyone would endorse either the route or the result. The decades-long process of harmonizing the humanistic, social scientific, performance, and professional elements of communication study remains incomplete as evidenced both by the great variety of ways that departments, schools, and colleges are organized and by the several somewhat overlapping national associations. Some would argue that the resulting synthesis—what I would call a *discipline for all seasons* and what Frank Macke terms a "human science"—lacks the theoretical elegance of the pure humanities or the methodological rigor of quantitative social science. To this, some would add the argument that the discipline's familiar performance courses and professional training both detract from a proper deference to the imperative of research. Here it is relevant to note that a focus on research rigor to secure external funding is not without contestation, as evidenced by criticism in the 1940s that the government-supported Stouffer-Hovland program reflected a social-engineering approach inconsistent with the needs of human beings. Contemporary communication historians such as Christopher Simpson similarly question whether the four fathers and their associates might have served as minions of such questionable funding sources as the Central Intelligence Agency.[30]

Clearly, the final verdict on communication's circuitous route to disciplinary catholicity has yet to be written. Yet despite varieties of selective memory, and continuing disputes over purposes, concepts, and methods, the discipline variously succeeds in synthesizing its humanistic, social-scientific, performative, and professional strands. At the same time, communication—both as popular objective and as academic discipline—grows in size, recognition, and influence.

NOTES

1. Albig, "Two Decades of Opinion Study," 16; Doob, *Public Opinion and Propaganda*, iii–iv; and Doob, *Public Opinion and Propaganda*, 2nd ed., vi; and Park, Burgess, McKenzie, *The City*.
2. Lasswell, "Affidavit," October 23, 1951, in file "Army-Navy-Air Force Personnel Security Board," Papers of Harold D. Lasswell, Yale University Library.
3. Sproule, "Propaganda Studies;" and Lasswell to Archibald MacLeish, August 25, 1941, Lasswell Papers, 2.
4. Stouffer et al., *The American Soldier*.
5. Hovland, Lumsdaine, and Sheffield, *Experiments on Mass Communication*; and Hovland, Janis, and Kelley, *Communication and Persuasion*.
6. See, for example, Lasswell and Leites, *Language of Politics*; and Merton and Lazarsfeld, *Continuities in Social Research*.
7. Dickens, "Discussion Method"; Balduf, "How Departments of Speech"; Westfall, "What Speech Teachers"; Cathcart, "War Responsibilities"; Wallace, "A Glance Ahead"; and Bagwell, "A Composite Course." As illustration of the postwar public face of quantified social science, we may turn to Chase, *Proper Study of Mankind*.
8. See, for example, Bateson and Mead, "Principles of Morale Building"; and Watkins, "Further Opportunities."
9. "Conference on Research on Public Communication," May 31–June 1, 1954, Box 8, Folder 41, Papers of Carl I. Hovland, Yale University Library; and Berelson, "State of Communication Research." Four other lines were identified by the SSRC conferees in discussion of Berelson's list, to wit the psychiatric approach of Jurgen Ruesch and Gregory Bateson, the information theory characteristic of Warren Weaver and Claude Shannon, psycholinguistic studies such as by Charles Osgood, and "role theory."

10. Schramm and Bauer quoted in Berelson, "State of Communication Research," 8, 14. It should be noted that, although Schramm founded communication institutes at Illinois and Stanford, his PhD was taken in the field of English.
11. "Memorandum on Communications Conference," January 18, 1941, Papers of Harold D. Lasswell, Yale University Library.
12. Higham, "Schism in American Scholarship."
13. Schramm, "Beginnings of Communication Study."
14. In the history of American higher education, the term "discipline" tends to be associated with a field of study that has gained separate departmental status in universities signaling that its collection of subspecialties has become recognized as permanent. See Veysey, *Emergence of the American University*, 320–24.
15. "Conference on Research on Public Communication," Carl I. Hovland Papers, Yale University Library; Sutton, *Education for Journalism*; Schramm, "Master Teachers," 130–33; and Schramm, *Beginnings of Communication Study*, 155–77.
16. See, respectively, Cohen, *The History of Speech Communication*, 323; and Cohen, "The Development of Research," 293. Here it is relevant to point out that Lasswell, the most comprehensive and catholic of the communication researchers, had come across public speaking textbooks by the mid-1930s but did not discover classical rhetoric, one of the key theoretical underpinnings of the speech discipline's basic course, until the later 1940s; Lasswell, Casey, and Smith, *Propaganda and Promotional Activities*; and Lasswell and Leites, *Language of Politics*, 3–5.
17. Kibler and Barker, *Conceptual Frontiers in Speech-Communication*, 18.
18. Ibid., 27.
19. MacLean, "Communication Research," 204.
20. Weaver, "A History of ICA," 608, 609.
21. Weaver, "A History of ICA," 609–10. Also see Elwood Murray, in Bagwell, "The Gilman Plan," 334–35; and Nichols, "Material for Courses," 464–67.
22. Nichols, "Material for Courses," 468; and Weaver, "A History of ICA," 610.
23. Nichols, "Material for Courses," 466–68.
24. Rogers in Rogers and Chaffee, "Communication as an Academic Discipline," 27; and Rogers, "President's Column," 2.
25. "Two Committees Study ICA Membership," ICA News, 1.
26. Schramm, "Beginnings of Communication Study," 73–82.
27. Kuhn, *Structure of Scientific Revolutions*, 157–58, 166–68 (quotation, 167).
28. Cohen, *History of Speech Communication*, 345–49; and Rogers, *History of Communication Study*, 496–502.
29. Sproule, *Propaganda and Democracy*, 311–17.
30. Macke, "Communication Left Speechless"; Burgoon, "Instruction about Communication," 305–8; Lynd, "The Science of Inhuman Relations," 22–25; Lee, Review, 173–75; and Simpson, *Science of Coercion*, 81.

WORKS CITED

Albig, William. "Two Decades of Opinion Study: 1936–1956." *Public Opinion Quarterly* 21 (1957): 14–22.
Bagwell, Paul D. "A Composite Course in Writing and Speaking." *Quarterly Journal of Speech* 31 (1945): 79–87.

————. "The Gilman Plan for the Reorganization of the Speech Association of America: A Symposium." *Quarterly Journal of Speech* 38 (1952): 331–42.

Balduf, Emery W. "How Departments of Speech Can Cooperate with Government in the War Effort." *Quarterly Journal of Speech* 29 (1943): 271–76.

Bateson, Gregory, and Margaret Mead. "Principles of Morale Building." *Journal of Educational Sociology* 15 (1941): 206–20.

Berelson, Bernard. "The State of Communication Research." *Public Opinion Quarterly* 23 (1959): 1–17.

Burgoon, Michael. "Instruction about Communication: On Divorcing Dame Speech." *Communication Education* 38 (1989): 305–8.

Carl I. Hovland Papers. Yale University Library. New Haven, CT: Yale University.

Cathcart, Raymond. "War Responsibilities of the Speech Correctionist." *Quarterly Journal of Speech* 29 (1943): 137–40.

Chase, Stuart. *The Proper Study of Mankind.* New York: Harper, 1948.

Cohen, Herman. "The Development of Research in Speech Communication: A Historical Perspective." In *Speech Communication in the Twentieth Century*, edited by Thomas Benson, 282–98. Carbondale: Southern Illinois University Press, 1985.

————. *The History of Speech Communication: The Emergence of a Discipline, 1914–1945.* Annandale, VA: Speech Communication Association, 1994.

Dickens, Milton. "Discussion Method in War Industry." *Quarterly Journal of Speech* 31 (1945): 144–50.

Doob, Leonard W. *Public Opinion and Propaganda.* New York: Henry Holt, 1948.

————. *Public Opinion and Propaganda.* 2nd ed. Hamden, CT: Archon, 1966.

Harold D. Lasswell Papers. Yale University Library. New Haven, CT: Yale University.

Higham, John. "The Schism in American Scholarship." *American Historical Review* 72 (1966): 1–21.

Hovland, Carl I., Arthur A. Lumsdaine, and Fred D. Sheffield. *Experiments on Mass Communication: Studies in Social Psychology in World War II*, vol. 3. Princeton, NJ: Princeton University Press, 1949.

Hovland, Carl I., Irving L. Janis, and Harold H. Kelley. *Communication and Persuasion.* New Haven, CT: Yale University Press, 1953.

Kibler, Robert J., and Larry L. Barker, eds. *Conceptual Frontiers in Speech-Communication.* New York: Speech Association of America, 1969.

Kuhn, Thomas S. *The Structure of Scientific Revolutions.* 2nd ed. Chicago: University of Chicago Press, 1970.

Lasswell, Harold D., and Nathan Leites. *Language of Politics.* Cambridge, MA: MIT Press, 1965.

Lasswell, Harold D., Ralph D. Casey, and Bruce L. Smith, eds. *Propaganda and Promotional Activities: An Annotated Bibliography.* Minneapolis: University of Minnesota Press, 1935.

Lee, Alfred McClung. Review of *The American Soldier*, by Samuel A. Stouffer et al. *Annals of the American Academy of Political and Social Science* 265 (September 1949): 173–75.

Lynd, Robert S. "The Science of Inhuman Relations." *New Republic* 121 (August 29, 1949): 22–25.

Macke, Frank J. "Communication Left Speechless: A Critical Examination of the Evolution of Speech Communication as an Academic Discipline." *Communication Education* 40 (1991): 125–43.

MacLean, Malcolm S. "Communication Research: The Tie that Binds—But Loosely." In *Conceptual Frontiers in Speech Communication*, edited by Robert J. Kibler and Larry L. Barker, 204–5. New York: Speech Association of America, 1969.

Merton, Robert K., and Paul F. Lazarsfeld, eds. *Continuities in Social Research: Studies in the Scope and Method of "The American Soldier."* Glencoe, IL: Free Press, 1950.

Nichols, Ralph G. "Material for Courses in Communication." *Quarterly Journal of Speech* 38 (1952): 464–67.

Park, Robert E., Ernest W. Burgess, and Roderick D. McKenzie. *The City.* Chicago: University of Chicago Press, 1967.

Rogers, Everett M. *A History of Communication Study: A Biographical Approach.* New York: Free Press, 1994.

———. "President's Column." *International Communication Association News* 8 (Fall 1980): 2 and (Winter 1980): 2.

Rogers, Everett M., and Steven H. Chaffee. "Communication as an Academic Discipline: A Dialogue." *Journal of Communication* 33, 3 (1983): 18–30.

Schramm, Wilbur. *The Beginnings of Communication Study in America: A Personal Memoir*, edited by Steven H. Chaffee and Everett M. Rogers. Thousand Oaks, CA: Sage, 1997.

———. "The Beginnings of Communication Study in the United States." In *Communication Yearbook* 4, edited by Dan Nimmo, 73–82. New Brunswick, NJ: Transaction Books, 1980.

———. "The Master Teachers." In *American Communication Research: The Remembered History*, edited by Everette E. Dennis and Ellen Wartella, 123–133. Mahwah, NJ: Lawrence Erlbaum, 1996.

Simpson, Christopher. *Science of Coercion: Communication Research and Psychological Warfare, 1945–1960.* New York: Oxford University Press, 1994.

Sproule, J. Michael. *Propaganda and Democracy: The American Experience of Media and Mass Persuasion.* New York: Cambridge University Press, 1997.

———. "Propaganda Studies in American Social Science: The Rise and Fall of the Critical Paradigm." *Quarterly Journal of Speech* 73 (1987): 60–78.

Stouffer, Samuel A., Edward A. Suchman, Leland C. DeVinney, Shirley A. Star, Robin M. Williams, Jr., Arthur A. Lumsdaine, Marion Harper Lumsdaine, M. Brewster Smith, Irving L. Janis, and Leonard S. Cottrell, Jr. *The American Soldier. Studies in Social Psychology in World War II*, vols. 1 and 2. Princeton, NJ: Princeton University Press, 1949.

Sutton, Albert A. *Education for Journalism in the United States from Its Beginning to 1940.* Evanston, IL: Northwestern University Press, 1941.

"Two Committees Study ICA Membership." *International Communication Association News* 8 (Winter 1980): 1.

Veysey, Laurence R. *The Emergence of the American University.* Chicago: University of Chicago Press, 1965.

Wallace, Karl R. "A Glance Ahead at the Field of Speech." *Quarterly Journal of Speech* 30 (1944): 383–87.

Watkins, John G. "Further Opportunities for Applied Psychologists in Offensive Warfare." *Journal of Consulting Psychology* 7 (1943): 135–41.

Weaver, Carl H. "A History of the International Communication Association." In *Communication Yearbook I*, edited by Brent D. Ruben, 607–618. New Brunswick, NJ: Transaction Books, 1977.

Westfall, Alfred. "What Speech Teachers May Do to Help Win the War." *Quarterly Journal of Speech* 29 (1943): 5–9.

Opportunity Structures AND THE Creation OF Knowledge: Paul Lazarsfeld AND THE Politics OF Research

DAVID E. MORRISON

It is easy, possibly more so than in other intellectual areas, for communications research to fail to appreciate the past, that is, to fail to pay recognition to past efforts that have helped produce the intellectual present. Perhaps this tendency can be attributed to the speed of change in communications technology that pins the student of communications to the present in a way not so common in other fields. Indeed, the creation of knowledge appears as a set of responses to changes in the technology of communications, or as a response to social problems that are linked to changes in communications. In the process, the wider context of research, the politics of the setting, especially the historical political setting, is overlooked. The purpose of this chapter, therefore, is to locate the production of knowledge within the political situation of its production.

LOST BEGINNINGS

In the natural history of the field, one might expect that the first major study of television would have been in America, and not England. However, this was not the case. Hilda Himmelweit's study *Television and the Child*[1] was produced in England in 1958, and not until three years later, in 1961, did Wilbur

Schramm's American study, *Television in the Lives of Our Children*,[2] appear. This paucity of television research in the United States did not go unnoticed. In his UNESCO report on American mass communication research covering the decade 1945–1955, Charles Wright reported:

> In reviewing the recent history, one is struck by the absence of material on what is perhaps the most salient development of the mass media in the United States during this period; that is, television. There has been no major study of the new medium to date.[3]

This chapter explains why there was no major study of the kind clearly expected by Wright. Paul Lazarsfeld is the central figure in this story. Lazarsfeld had been appointed in 1952 by the Ford Foundation to be head of their Advisory Committee on Television. The expectation was that the report and working papers of this committee would put television research on the map, just as Lazarsfeld's Rockefeller Foundation–supported research in the 1930s and early 1940s had placed radio research on the map. The Ford Foundation's effort to establish television research, however, was a complete failure: the foundation withdrew support when it came under attack by right-wing voices.

This episode in the history of communications research is a story of failure, but I wish also to examine a success story, namely, Paul Lazarsfeld's establishment of the Österreichische Wirtschaftspychologische *Forschungsstelle*, in Vienna, in 1925. It was the first research center of its kind in the world.[4] The importance of the *Forschungsstelle* in the history of higher education is that it broke with existing institutional arrangements for the production of knowledge. Its importance to communications research, however, is that it became the model for Lazarsfeld's establishment of the Bureau of Applied Social Research (BASR) at Columbia University in New York in the late 1930s—the "Bureau" became the center of mass communications research. I will now look at these twinned stories of success and failure, and I will do so through an examination of the political settings that contributed to success and failure, and by drawing on the notion of opportunity structures. Robert Merton, in his classic article "Opportunity Structure: The Emergence, Diffusion, and Differentiation of a Sociological Concept," notes, "Opportunity structure designates the scale and distributions of conditions that provide various probabilities for acting individuals and groups to achieve specific outcomes." He adds, "From time to time, opportunity structures expand or contract, as do segments of that structure."[5] It is essential, however, for the explication of events that the notion of opportunity structure is related to the part that ideas themselves play in the location of the individual within social structures, which then assists or restricts the achievement of goals. That is, I draw attention to "political climates" as forming opportunity structures, and how sets of political sentiments in Austria offered certain opening and closing of opportunities, and in America.

WORLDS APART

In his *History of Communication Study*,[6] Everett Rogers takes, as he subtitles the book, "a biographical approach" to understanding the development of the field. He is right to do so. Understanding the biographies of those who have shaped the field assists us in our efforts to contextualize how the field was developed. To read the recently republished *Personal Influence*[7] is to hold more than a tract of communications history. It is also to hold a story of one individual, Paul Lazarsfeld, who happened to light on communications research as the vehicle by which he would, upon exile from Austria, establish himself in America. To understand the history of communications research, it is essential to comprehend a history that rests outside the field of communications inquiry itself.

Lazarsfeld was born in Vienna in 1901 into an intellectual, middle-class Jewish family. Years later, in the United States, when referring to the Austria where he was born and where he grew to early manhood, he described it as the "old world." For him, this use of the term "old" referred not to distance in time, but to a world to which he had belonged but which had passed into memory.

With the passing of time all places change, often quite dramatically, but usually they still have the imprint of their former self. Robert Merton, for example— Lazarsfeld's colleague and close friend at the BASR—was born in 1910 to East European Jewish immigrants in a South Philadelphia slum. Of course, the area where Merton grew up, and more particularly the way of life of that area, now exists only as historical record; nevertheless, Philadelphia is still Philadelphia in a fashion that cannot be said of Lazarsfeld's Vienna. It was its position as the glittering administrative capital of a disintegrating empire, overseen by a ruling elite of nobility and higher bureaucracy, more tightly knit than anywhere else in Imperial Europe, that gave Vienna a specialized culture. When the end came, it came suddenly, leaving little trace other than the grand architecture to symbolize its once great imperial power. With the establishment of the First Republic by politically radicalized soldiers following defeat in the World War I, the old order was swept away, little more than memory.

Lazarsfeld's interwar Vienna was a site of conflict, and this conflict shaped Lazarsfeld's life and intellectual trajectory. Historian Charles Gulick states that, "to a degree unique in the history of nations the story of Austria between the world wars, particularly between that of 1918–1934, was that of the fight for and against the aspirations and achievements of the working-class movement." Put in stark political terms, "this struggle was that between democracy on the one side and reaction and Fascism on the other."[8] Lazarsfeld was part of that struggle. Indeed, his life straddles the glitter of the Hapsburg Empire and the struggle for socialism of the Social Democratic Party.

Although it is not difficult to connect Robert K. Merton's early life in the mean poverty of Philadelphia to his sociological work, it is relatively challenging to connect Lazarsfeld's work with his early life. Lazarsfeld, after all, was a man who was accustomed to seeing the grand emperor Franz Josef, and who had Friedrich Adler living with his family at the time he assassinated the Austrian Prime Minister, Count Stürgkh. Yet, if we do not understand that world, it is almost impossible to chart the path of Lazarsfeld's intellectual career.

MATHS AND MURDER

Adler had competed with Einstein in 1908 for a professorship in physics at Zurich University. While awaiting execution for the killing of Count Stürgkh, Adler continued his work on relativity theory. The young Lazarsfeld would visit Adler in prison and help to type up smuggled-out manuscripts. To the sixteen-year-old Lazarsfeld, Adler was a hero, or, as he described him to me, "a glorious murderer." In many ways Adler epitomized the world within which Lazarsfeld grew up—a peculiar combination of political dedication and academic scholarship. Indeed, Lazarsfeld knew all the great Austro-Marxist thinkers who led the Party. "These people," he said, "were all, so to say, my uncles in some way."[9] Adler even wrote to him from his cell to encourage Lazarsfeld, as when he wrote, "Dear Paul, I'm glad that you are doing well in mathematics. Whatever you do later mathematics will always be useful to you."[10]

No doubt the encouragement of a "glorious murderer" assisted his interest in mathematics, but really, it was the place of mathematics in the socialist intellectual circles of Vienna to which Lazarsfeld belonged that was key to his interest. As he would later put it, "the prestige of mathematics among those charismatic Austro Marxists like Bauer—that certainly played a role."[11] Even so, his political interests did interfere with his studies. As he said: "One term I was enrolled in a social science course and then the next term I was doing *Staatswissenschaft*, and then sociology . . . enrolled in mathematics and so on. But I didn't do much work in any as I was so politically active."[12] Lazarsfeld's recollection was supported by one of his fellow students, Berta Karlick, who recalled that Lazarsfeld "was more interested in social problems than in studying physics and mathematics."[13] He did graduate, however, with a doctorate in applied mathematics in 1925, which took him, not surprisingly, a year longer than usual to complete. Despite his motivation and gifts, a career in the social sciences—which would have allowed him to combine his interests in social problems and mathematics—was out of the question at the University of Vienna.

Anti-Semitism at the university had a long history. Indeed, it had long been a feature of the Empire. In fact, if one had had to predict the most likely territory for the systematic destruction of the Jews, it would have been in the regions of the Austro-Hungarian Empire, not in developed, industrialized Germany. After World War I, and the collapse of the Empire, the anti-Semitism that had been a permanent feature of social and political life shifted in tone and style. Anti-Semitism became fused with attacks on socialism and liberalism. The result was a radical anti-Semitism and a growing acceptance of physical violence. Although violence toward Jewish students at the university came from other students, such violence was looked upon with sympathetic eyes by many of the teaching staff,[14] including the powerful Othmar Spann, professor of economics, and—owing to the structure of the university—a man who dominated the social sciences, including sociology.

It is essential to grasp that students at the university in Lazarsfeld's time needed the sponsorship of the professoriate to advance. This support was particularly necessary for *habilitieren*, a process that involved producing and defending a body of work beyond the doctorate. Without *Habilitation*, one could not become a *Dozent* (Lecturer), and Spann was not one to approve the *Habilitation* of Jews. In 1925, for example, the same year that Lazarsfeld earned his doctorate, at a meeting that included Czermac, the minister for education, Spann attacked his fellow professor of economics, Mayer, for supporting the *Habilitation* of two Jews, Wiesser and Schlesinger, calling the Jew the enemy, but using the term *Ungeraden*—meaning odd or nonconformist as a cover word (*Geheimwort*) for Jew. He also turned on Mayer for his own liberal teaching. Indeed, the term liberal was often used as a *Deckbild* (cover picture) for Jew.[15] For Spann, and the group of professors known as the "Spann Circle,"[16] the Jew and the liberal were rolled into one abhorrent whole.

If belonging to one of Vienna's best-known Jewish socialist families had not been sufficient to block Lazarsfeld's acceptance by the university, his research alone would have created difficulties. Empirical social research was anathema to Spann. He considered such research individualistic—"the most severe condemnation in [Spann's] vocabulary."[17] For both intellectual and political reasons, Spann rejected fact-finding, in favor of the theoretical search for the rightful, or organic, structure of society. His idea of the Corporate State[18] influenced paramilitary right-wing groups such as the Stahlheim in Germany and the Heimwehr in Austria. It is hard to imagine anyone more likely than Lazarsfeld to capture in a single individual all that Spann stood against: he was Jewish, he was a socialist, and he was committed to empirical social research.

However, even without such objections, if Lazarsfeld *had* been appointed to a position, his type of expensive empirical social research would have been impossible to finance in a university that was bankrupt, as indeed was the whole of post–World War I Austria.[19] Lazarsfeld was very aware that, "In order to do empirical studies you needed machinery, and you needed money."[20] A solution to this was the establishment of the *Forschungsstelle,* initially, as Gertrud Wagner, one of the early staff, said, "in just two rooms in the flat of a friend." She continued, "We started with nothing. Paul started to get orders from firms, and we started the market research. He got a few coworkers to help—actually we were friends of his."[21] She should have added "socialist" to "friends," since as Lazarsfeld, in discussing the establishment of the *Forschungsstelle,* commented, "it enabled this whole defeated socialist group," meaning his friends, to transfer to "a new activity which was close enough to social reality, and had some academic glamour."[22]

By "defeated socialist group," Lazarsfeld was referring to what he called "the failed revolution," by which he meant the refusal, or inability, of the Social Democratic Party to push through a complete economic transformation of society following the establishment of the First Republic. In many ways the Party was suspended between success and failure, but at least in Vienna, as Henry Brailsford[23] observed, socialism was "lived as a creed that covers the whole of life," as indeed it did for Lazarsfeld. His life was lived within and bounded by the dreams, hopes, and ambitions of the Social Democratic Party.

In talking about their Vienna years, Marie Jahoda, his first wife, who was very active in the Party herself, said, "You know, all the time Paul was in Vienna and was director of the *Forschungsstelle* it was still a time that he was very conscious that if only he could, he would be in politics rather than social research." She added, "Paul was very political . . . he had great political ambitions . . . and I think the great dream of his life would have been to be Foreign Minister for a socialist Austria." Given the course of European history, this could never have been, but at the more immediate level, Jahoda made the following observation:

> You know, Paul was so obviously Jewish and he just didn't have a chance in the political party. Other young people did make a spectacular career in the Social Democratic Party, but for Paul it was difficult because he was so very intelligent that nothing on the second level would have suited him, and the fear of general reaction to (another) Jewish dominant figure in the Party was strong.[24]

Elizabeth Schilder, a Party activist and friend of the Lazarsfelds, confirmed this, stating that there was a "sensitivity in the Party at the Jewish people" and that "Paul by his exterior was a type of Jew, his movements were very Jewish. An intellectual Jew." She concluded that "it is possible that he had the feeling not to be . . . that he had not a chance to be a leader of the Party by his Jewishness."[25] Thus, while Lazarsfeld presents the establishment of the *Forschungsstelle* as a response to "the failed revolution" that then allowed a sense of social engagement for, as he calls it, "this whole defeated socialist group," this overlooks Lazarsfeld's own individual situation. He was imprisoned by failure. His attempt to break out of this prison was to strive to create an activity, namely the *Forschungsstelle*, that could nourish him intellectually, whilst also affording him the opportunity for some kind of security.

The brutal fact is that Lazarsfeld suffered the pain of being a Jew in a Europe where Jews were not wanted. More than anything the establishment of the *Forschungsstelle* is to be viewed as the response of one individual to the curtailment and constriction of the life world of the European Jew. It was created out of failure—the failure of the socialist revolution, the failure to find a desired position within the Social Democratic Party owing to anti-Semitism, and the failure to secure a teaching position at the university for the same reason. Indeed, when it came to promotion at the Psychology Institut, where Lazarsfeld held a very lowly position, Karl Bühler, the main professor, and who also served on the board of the *Forschungsstelle*, was fearful to put him forward, especially in light of the fact that his wife, Charlotte, an assistant professor at the Institut, was half Jewish. They themselves had to flee a few years later, but by way of compensation for being unable to promote him, Bühler put Lazarsfeld forward for a Rockefeller Fellowship to America.

AMERICA: AN ALTERNATIVE CASE

Insecurity is not a condition conducive to creativity, and in terms of success in academia following emigration, much depends on the immigrant's discipline, international reputation, age, and the openness or closure of the host academic culture to which the immigrant attempts to enter. The Bühlers', for example, is a sad tale: their Institut in Vienna was internationally renowned, but they never found acceptance in America by way of position. They were too old.

Lazarsfeld did not belong to any school of thought that would give him a ready-made home, and he lacked an international reputation. Still, in many ways, with his empirical research background, he was an American before he even arrived. In addition, his deep knowledge of European theoretical and philosophical thought gave him an advantage over his empirical American cousins. When he arrived in 1933, however, it was a time when America was, as he said, "at the depth of the Depression still, and the depth of the misery of the American University."[26] Thus with the ending of his fellowship in 1935, by which time return home to Austria was no longer "advisable," he was faced with the difficult task of finding a new job. In his memoir, he recounts that "it never occurred to me to aspire to a major university job. I took it for granted that I would have to make some move similar to the creation of the Vienna Research Centre if I wanted to find a place for myself in the United States."[27]

This is not Lazarsfeld "creating" his biography. He says as much in a letter to Hadley Cantril of Princeton University at the time he accepted the directorship of the Rockefeller-funded Princeton Radio Research Project that was to launch his career in America: "I feel strongly that I don't want to go ahead alone, that I want to stand for an institution and try to build up an institution which is able and willing to stand for me."[28] The letter was an attempt by Lazarsfeld to have the grant transferred from Princeton to the University of Newark where he had managed to establish a small research center. Although Newark was hardly a prestigious university, and later folded, at the time he was offered the directorship of the project he had high hopes for the university and by extension his associated research center.

Lazarsfeld arrived in America as the quantitative wing of sociology was gaining influence. His empirical expertise gave him an obvious advantage over his erstwhile European colleagues schooled with a veneration of theory. Perhaps the most important factor helping him to establish himself was his coming to the attention of the Rockefeller Foundation just as it was making a massive move into communications research.[29] Although the *Forschungsstelle* had undertaken some media research, Lazarsfeld had no or little interest in communications research as such. The offer to become director of the Princeton Radio Research Project gave him the perfect opportunity to pursue his real passion: methodology. He was very open about this: "I had no interest whatsoever in mass communications—I mean everything is interesting to a methodologist."[30] When this was put to John Marshall of the Rockefeller Foundation, Marshall agreed: "I suspect Paul's right. He perhaps says it inexactly. He was interested primarily in methodological opportunities which research in the media offered."[31] This interest in methods did not bother the foundation; indeed, in many ways it was a plus, or at least his actual methodological expertise was. For example, the records reveal that, given the newness of the enterprise, concern was expressed within the foundation about the lack of methods with which to measure the effectiveness and role of radio on people's lives, a consequence of which was that "of the four years the proposed study will require, the first will be devoted to developing and testing techniques."[32] The newness of what the foundation was attempting, and the methodological groundwork required, is well captured by Lazarsfeld's account of an encounter:

> I remember one day, a friend of mine, in 1937 or so, introduced me to a group of colleagues and said, "this is a European colleague who is the up-most authority on

communications research," and he saw that no-one was especially impressed, so he wanted to press the point and said, "as a matter of fact, he is the only one who works in this field."[33]

The newness of the field also meant, as Lazarsfeld stressed in explaining how he was offered the directorship, "that no experienced American professor wanted to give up a regular position to run a new thing like a project. You see it was a completely new field."[34]

MEDIA INDUSTRY AND OPPORTUNITIES

In terms of opportunity structures, the newness of communications research provided Lazarsfeld with an authority to act that he would not have possessed in an established field of research. Lazarsfeld experienced little difficulty getting the Radio Project—then under the auspices of Princeton University, where Hadley Cantril was professor and recipient of the grant—moved to Columbia University in New York.[35] It found a happy home at Columbia, or at least more so than at Princeton, but the real advantage of having a base at Columbia was that New York had become the capital of the emerging communications, advertising, and national marketing industries. For Lazarsfeld, New York was the place to be, and hence, as Everett Rogers notes, the BASR became the "institutional birthplace of mass communications research."[36] Indeed, viewing Lazarsfeld's success in America, I would like to suggest an axiom for the sociology of knowledge: individual talent makes a difference in the development of ideas into an organized body of knowledge and the institutionalization of that knowledge into a recognized field or discipline, but structure provides the opportunity for the transformation of that talent into the process of development.

It should not be assumed, however, that Lazarsfeld saw himself as creating a new area of intellectual activity. His entry to the field was no more than an agreement to temporarily direct a research project; his expectation was, following advice received from Cantril, that after the publication of "a couple of first class studies," he would look for work in some other area. (As Cantril put it in urging Lazarsfeld to accept the post, "you could really land something you liked when this was up.")[37]

What Lazarsfeld did not appreciate at the time was that the opportunity structures available to him in mass communications research were more favorable than in any other potential area. Here was a field of research that had attached to it a rich and growing industry. Given Lazarsfeld's lack of substantive interest in media questions, the financial resources offered by the industry probably helped to keep him in the field longer than otherwise would have been the case. Indeed, "one of the reasons," according to Jefferson Pooley, "that he withdrew from media research in the late 1940s was that the networks had by then moved nearly all their research in-house."[38]

The "Bureau's 'golden era' was between the later 1940s through to the mid-1960s, gaining strategic position within sociology, especially through its doctorate program."[39] Many of its students, including James Coleman, Peter Blau, Peter Rossi, Seymour Martin Lipset, Alice Rossi, David Sills, Charles Glock, Morris Rosenberg, and Elihu Katz, became leaders in their respective fields, extending the influence of the Bureau well beyond its original institutional base. Lazarsfeld had established the Bureau out of communications research, and had established himself in the process; when he stepped down from the directorship in 1950, he had already moved away from communications research. Yet, in 1952, the opportunity arose for Lazarsfeld to occupy once more the center stage of communications research, as he had in the 1940s.

SOCIAL STRUCTURE, AGENCY, AND EXPLANATION

In the winter of 1951–1952, the Ford Foundation began to discuss the feasibility of studying the impact of television on American society. On advice from Frank Stanton, the foundation offered Lazarsfeld the chair of an "Advisory Committee on Television." This he accepted. The remit of the committee was to develop proposals for something akin to a commission on television, with a million-dollar budget commensurate with the task. The committee first met in New York in late August 1952, and reported in the summer of 1953. According to Lazarsfeld, "there were eight or ten monographs written around it," but "one day, out of the blue, it was all ended." No action was taken on the recommendations made in the report, and no explanation was given as to why it was decided not to proceed with the plan to study television. The abruptness of the end of the project, and the total silence that followed, prompted Lazarsfeld to say, "it's a very funny episode."[40]

The episode may have been "funny," but Lazarsfeld did have an explanation for the project's termination; indeed, he had "a definite theory about it." "I have no evidence really," he admitted, then laid out his theory:

> There was one man on my committee who was very famous for many reasons but also completely erratic—that was [Beardsley] Ruml. I think Ruml took a definite dislike to me. I always had a theory that Ruml torpedoed the whole project. There was a Vice-President, [William] McPeak. Ruml had great influence on McPeak, and

McPeak had no reason to like me either. I had hardly any contact with him. But between these two men it was killed.[41]

His theory is interesting, not so much for what it says about the episode, but for what it says about Lazarsfeld. His explanation—that the project was killed because of Ruml's dislike, and that Ruml had the ear of the vice president—shows an absolute failure, perhaps even refusal, to credit social structures as organizational points of social action, at least when explaining his own experience. His turn to personalized explanation is appropriately grounded in features of his peculiar scholarly life.

While developing a new institutional setting for the organization of knowledge—the research institute—Lazarsfeld was also developing a new academic role: that of the entrepreneurial scholar. His academic life, in response to this new role, involved a great deal of wheeling and dealing. This, when added to the experience of uncertainty and political maneuvering of his Vienna days, led to a belief in the effectiveness of personal intervention in altering social arrangements. Indeed, what was notable in his recounting of his life was an absolute lack of appeal to sociological explanation for understanding the world around him. What he demonstrated was a belief in his ability to penetrate the motives and character of those he had to deal with. The individual component of action took precedence over social explanations of outcomes. In this light, it is interesting that his substantive intellectual interest was the empirical study of action, in particular the moment of decision making, the study of how an individual arrives at a choice of, say, one election candidate over another, or one brand of coffee over another. In terms of the empirical study of action, consumer choice and voting choice were methodological equivalents, but this attention to the individual and to motivation tended to frame his sociology with little reference to social structure.[42] If we pull together his belief in the importance of the individual in producing effects and his (perhaps resultant) interest in the empirical study of action, the scene that emerges is one of a biographical experience that pushes attention away from explanations that originate in social structure. This individual volitional outlook helps to explain his comment to Merton that he had difficulty understanding the very idea of "social structure."[43]

OPPORTUNITY STRUCTURE
AND PHILANTHROPIC FOUNDATIONS

The term "opportunity structure," as used here, refers not simply to organizational arrangements, legal provisions, and the like but to *ideas* as part of such structures.

Lazarsfeld's sociology stood in perfect alignment with the definition of knowledge long embraced by the major foundations. Just as sociology had legitimated itself as an intellectual discipline by its claims to objectivity,[44] so too had the philanthropic foundations turned to the idea of scientific objectivity as a mantra of performance in the pursuit of legitimacy.[45]

Formed from the fortunes of the industrial and financial barons of the late nineteenth and early twentieth century, the major foundations faced the charge of "tainted money"—that they were formed to further the economic and industrial interests of their donors.[46] As early as 1915, the Rockefeller Foundation was subject to a Senate inquiry, which delivered the damning indictment that its research in the area of industrial relations "has not, as is claimed, either a scientific base or a social base, but organized to promote the industrial interests of Rockefeller."[47] The result was that the trustees of the foundation "were given an uncomfortable lesson about the hazards of becoming involved in social and economic issues."[48] The Rockefeller Foundation stayed clear of support for the social sciences until the 1920s. After this, the increasing acceptance of the benefit of knowledge gained from the social sciences, "the war gift to industry"[49] of World War I, and the enhanced status during the New Deal of social scientists acting as advisers to the pragmatic Roosevelt Administration, made entry to the social sciences before and after World War II relatively noncontroversial. In short, sociology, by stressing the objectivity of its activity, and the foundations, by their support in fostering the empirical quantification wing of sociology and its stress on the objective, came together to present a particular view of knowledge that found social and political acceptance. It was within such a climate that the Rockefeller Foundation could come to the support of radio research in the late 1930s and early 1940s. No longer was it held that the foundation operated to protect the business interests of the donors against the interests and welfare of the American people. They were now more often cast as progressive in their attempt to understand social association and human problems.

In terms of how ideas can shape opportunity structures, the difference for someone such as Lazarsfeld between his accepting the directorship of the Radio Research Project and his acceptance of the chair of the Ford Foundation's Television Advisory Committee in 1952 was that the movement of ideas saw a restriction of opportunity for the establishment of new areas of research. Whereas the foundations had won the trust of the public, or public voices, by the late 1930s to give confidence to the Rockefeller Foundation sufficient to venture a study of the role of radio in American life, that trust had been undermined in the course of the cold war politics of the 1950s to sap the confidence of the Ford Foundation to study the role of television in American life. In general, the social sciences had come to be seen as politically and socially subversive, and hence by giving funds to support such research, the foundations were similarly seen as politically suspect. Given the importance of the media to the political process, for the foundation to propose a study of television and its part in American life was bound to make it the object of political interest and engulf it in controversy.

FORD AND FUNDING

If the early 1950s was not a propitious time for any foundation to be entering into studies of controversial social issues, the Ford Foundation was the most exposed to charges of support for communism and, given its source of funding, the most vulnerable to pressure.

The foundation, although established in 1936, did not support the social sciences until 1950. In entering the field, the trustees observed that "the problems and opportunities of our time arise out of man's relation to man . . . rather than his relations to the physical world."[50] Accordingly, five programs of research were mapped out for attention, the fifth becoming known as the Behavioral Sciences Division, a term created by Bernard Berelson to describe its operations.

Berelson had been appointed to the foundation in the summer of 1951 from the deanship of the Graduate Library School at the University of Chicago to develop plans for social scientific research. These plans were completed by fall 1951 and accepted by the trustees in February 1952, with Berelson named director of the new Behavioral Sciences Division to carry out the plan, part of which was to establish the Center for Advanced Study in the Behavioral Sciences at Palo Alto—the actual structure of the center being accepted by the board of trustees on February 13, 1954.[51] Part of the remit of the Palo Alto center was to "contribute directly to the further development and integration of the behavioral sciences."[52]

The fact that the Ford Foundation became a major force in the 1950s in the financing and support of social research would, as Peter Seybold[53] documents, make it a key force in assisting the rise of behavioralism in political science. This would on its own have made it an obvious target for the conservative forces opposed to the social sciences, but the gift of $15 million for the establishment of the liberal Fund for the Republic marked the foundation out for special attention. Established in 1953, the Fund for the Republic had in fact been conceived some eighteen months earlier with the actual founding postponed because of the presidential election of November 1952. The purpose of the fund was to support work in the defense of civil liberties. The fund sponsored Stouffer's communism studies,[54] Lazarsfeld and Thielens's study, *The Academic Mind*,[55] examining the influence of McCarthyism on higher education, and at one point even intended to study the American Legion, criticism of which worried the foundation.[56]

The Fund for the Republic was headed by the outspoken Robert Hutchins, who had moved over from Ford to take charge. Even though the fund was independent from Ford, given the source of money and the "loan" of personnel, the foundation was obliged to assume parental responsibility for its wayward child. The actions of the fund repeatedly stung its right-wing critics to fury, but attacks on the Ford Foundation began even before the fund was established. The opening salvoes were fired in 1951 when the *Chicago Tribune* ran a news story under the headline: "Leftist Slant Begins to Show in Ford."[57] This was an allusion to Paul Hoffman, head of the foundation, who, when administrator of the Marshall Fund to reconstruct Europe after the destruction of war, had charged the *Tribune*, "given away 10 billion dollars to foreign countries."[58] This theme was taken up by Hearst columnists such as Westbrook Peglar, George Sokolsky, and Fulton Lewis, Jr. Hoffman was described by Peglar as "a hoax without rival in the history of mankind," and the Marshall Plan was mocked as "the fabulous Roosevelt-Truman overseas squanderbund."[59]

Fulton Lewis, Jr., had serious political influence in Washington. His radio broadcasts reached more than sixteen million people each weekday evening, and many watched him on more than fifty television stations, while millions read his syndicated columns in the Hearst press.[60] He repeatedly used his position to call upon his audience to mount pressure for various causes. He was credited with helping to spark over a dozen congressional inquiries.[61] Into this witch's brew were added voices from groups such as the right-wing Constitutional Educational League, who produced pamphlets linking the Ford Motor Company with communism. Its argument was that by owning a Ford, the driver assisted communism because profits from the company went to the Ford Foundation, which in turn funded left-wing academics.[62]

Neither the shrillness of these attacks, nor the audience reach, should not be underestimated. They helped to create a mood oppositional to the open support for research. Despite Henry Ford's claim to the contrary, such attacks seriously worried the foundation. Henry Ford is quoted as saying: "The dealers send us in letters from customers accusing the foundation of being communist and warning that they'll never buy another Ford. But I don't bother much with that sort of mail. Why should I?"[63] But he *was* bothered. Evidently, Ford showed letters from dealers and customers to his speechwriter, to whom he "expressed deep concern" over their content.[64]

Hoffman's account of his time at the foundation, spent having to placate Henry Ford following media attacks on the foundation, lends further support to views that Ford was affected: "I told Ford that I wanted to experiment" and that "change always means trouble," but "every time we got a dozen letters to something we had done . . . a radio show of an overseas program or what not . . . I'd have to spend hours reassuring the board."[65] Matters only got worse once the Fund for the Republic was up and running. The attacks were such that they led Henry Ford to go so far as to put pressure on the marketing researcher, Elmo Roper, to step down from its board and support the removal of Hutchins as its president. Roper knew the pressure that Ford was facing from officials at the motor company when "hints were given that his own business relations might be cut" if he did not comply. "At one point, Ford himself had asked Roper to leave the fund's board."[66]

WHY ATTACK THE SUPPORT OF KNOWLEDGE?

Two congressional inquiries were launched to examine the functioning of the foundations: the Cox Commission of 1952, and then the Reece Commission at the height of the McCarthy "reign" in 1953. The Reece Commission was the most incendiary in its attacks, charging the foundations with communist infiltration, and malfeasance in allocating grants to communists or to communist sympathizers. The Rockefeller, Carnegie, and Russell Sage foundations, for example, all came in for attack for their initial sponsorship of the *Encyclopedia of the Social Sciences*, which was seen as left wing. The Ford Foundation was singled out for "its dubious staff." Bernard Berelson in particular came in for special mention in this regard,[67] which is perhaps not too surprising given what has already been noted about his central role in developing the social sciences for the foundation; indeed, the Palo Alto center, the brainchild of Berelson, drew special criticism from Reece.[68] According to Berelson, Hans Zeisel, who ran the *Forschungsstelle* after Lazarsfeld left for America, and co-author with Lazarsfeld and Jahoda of *Die Arbietslosen von Marienthal*, became concerned by the attacks on Berelson led by an Austrian émigré who had been a comrade of Lazarsfeld and Zeisel in the Social Democratic Party. Zeisel, for some reason and somehow, was in possession of this person's old party card, and sent it to Berelson with the note to use if necessary "to expose the son of a bitch."[69] Unsavory times promote unsavory people, but the question remains, why such objection to social research?

Before examining the refusal of the Ford Foundation to support Lazarsfeld's efforts in establishing television research, it is essential to understand why the social sciences were anathema to the Reece Committee, and more, why the defense of objectivity failed to work during this period.

The complexities of the McCarthy period and the conservative turn against the liberal establishment cannot be explored here,[70] nor can the partisan politics of the period with the attempt of the social sciences to advance its position by following and hanging onto the coattails of the natural sciences, so well documented by Mark Solovey.[71] It is only possible to point to some features that carry importance not simply for the failure of television research at this time, but also for Lazarsfeld in the sense of the bearing his empirical operations had for political objection: his fact gathering, so often seen as conservative, was not considered safe by those holding power.

It was Norman Dodd, the research director of the Reece Committee, who drew attention to the "dubious" Berelson. What Dodd objected to, and what was captured by Berelson's unwavering support for the "behavioral sciences," was fact gathering. This was not some epistemological dispute, but rather the failure of the social sciences, as Dodd saw it, to operate in support of "traditional American values."[72] This raises the further question, therefore, of why "fact gathering" would be seen as antithetical to American values.

What Dodd means by "American values" is not spelled out for the simple reason that there is a presumption of no requirement to do so. Such terms refer to an approved "way of life." That "way of life," in the context of the time, is ensured by democratic government. The value or benefit of democracy, however, in a testable rather than political philosophical sense, is in pragmatic fashion through appeal to what America has to offer in terms of material achievement, standard of living, and technical advancement. Viewed this way, the American way of life is a story of success that requires no examination; hence empirical studies that produce facts counter to the story of "the American way of life" are, by reason, written to undermine American values. It is but a short step from this to see empirical social scientists as divergent from American values. In the context of the cold war, this seeming divergence was understood to mean that empirical social scientists shared the values of those who would seek to destroy America. The enemies within, communists and communist sympathizers, were being supported by the foundations. Social scientists were seen not as researchers, but as excavators of facts, digging away at the foundations of American society. An illustration here will help.

A decade previously, the Carnegie Corporation entered the choppy waters of race relations by funding Gunnar Myrdal's classic study, *An American Dilemma*.[73] To read the foreword to the book, written by Frederick Keppel, Carnegie's president, is to witness a dreadfully worried expectation of what might await the corporation for daring to examine the facts of race in America. In the foreword, Keppel attempts to head off criticism by making the case that so long as the corporation "makes facts available and lets them speak for themselves . . . studies of this kind provide a wholly proper . . . use of their funds."[74]

The interest that informs the use of terminology such as making "facts available" that "speak for themselves" is to seek shelter behind the defense of objectivity in the legitimation of research practice. Even in 1944, when the foundations were enjoying public confidence, the social sciences had the capacity to mire them in controversy. A decade later, at the time of the Reece Committee, no such defense was possible. What was demanded was theorizing that would offer values consistent with the "American way," not facts that might tarnish the projection of an ideal, or expose structural faults that could challenge the narrative construction of American society.

In consideration of the political controversy that attended these studies, it is a mistake of leftist critics to assume that empirical social inquiry, the "abstracted empiricism" of C. Wright Mills's attack on Lazarsfeld,[75] is conservative and supportive of existing social arrangements. This supposed conservatism is in large measure determined by the use to which the information is put, or by the nature of the society from which the findings are drawn and in turn refer. Facts can be destabilizing. They do not necessarily reproduce, as Adorno would have it in his attack on American empirical methods,[76] the very conditions that produced the facts. The delivery of facts, especially in the modern Western world, has a power of acceptance capable, so to speak, of making an argument of their own. To that extent, Keppel's defense of Myrdal's study in asking for the facts be allowed to "speak for themselves," as if somehow the facts were neutral, is simply wrong and unworkable as a protective device. The facts of Myrdal's study, once delivered into the public domain, become alive with the values of the environment to which they are introduced. The second matter, at least in so far as the congressional inquiries were concerned, is that research should not be allowed to wander at will unguided by approved political purpose. In so far as Lazarsfeld is concerned, there is a connecting spirit here with his days in Vienna and Othmar Spann's idea of sociology—namely, that research should be at the service of values. It is piquant that for all the attacks on Lazarsfeld's empirical sociology, among other things for the intrinsic conservatism of research so wedded to facts, his empirical style of research should have been found so offensive, indeed, dangerous, first to fascist circles in Austria, and then right-wing anti-intellectual groups in America.

THE DEATH OF TELEVISION RESEARCH

The decision to refuse to support research on television must be set not just within the political climate faced by all foundations, but must be considered, too, in terms of the specific context of the Ford Foundation. While it is true, as Seybold points out, that an inventory of areas in need of support by the foundation had identified communications and the need to understand the formation of public opinion,[77] the fact was that from the very outset of the establishment of Lazarsfeld's television committee, Henry Ford had expressed misgivings about the political wisdom of conducting such work in this climate. His unease probably stemmed from his experiences with Frank Stanton at CBS. When Ford set up the committee, he phoned Stanton to ask if he would participate. Stanton agreed, and then found himself the object of criticism by others in the industry. All of this got back to Ford. What is interesting here is that, as the papers surrounding the decision to enter television research show, the foundation initially wished to exclude members of the industry from serving on the advisory committee on the grounds of the industry's "vested interest."[78] This sounds like the decision of a confident person, someone such as Hoffman. However, the pervasive nervousness of the period resulted in a volte-face and the decision that the research could not go ahead without the involvement of the industry. This, no doubt, led to Ford's fateful phone call to Stanton. This would also account for Lazarsfeld's letter to Vice President McPeak, which, in the context, was clearly meant to reassure the foundation that matters were running smoothly. In the letter, Lazarsfeld assured McPeak that he was "very eager indeed to talk with you about the progress of the preparatory committee on the TV Commission," and stated that "we have made good progress intellectually as well as in our effort to secure the support of the industry."[79] When it was put to Lazarsfeld that the foundation had insisted on industry members joining the board and that they were responsive to the idea, he replied, "Of course. Look, I had all sorts . . . I remember big station managers from the Midwest, and I was fairly skilled to reconcile them. The station representative liked the idea, at least they never created any trouble."[80]

They did not need to cause trouble to have the project dashed. No amount of personal skill on Lazarsfeld's part could turn back the forces of reaction massing in the wider world against the liberal conception of scholarship of which the Reece Committee was a part. The foundation was certainly sensitive to all the criticism, as Lazarsfeld revealed in planned testimony before the Kefauver Committee on Television and Juvenile Delinquency.

In the spring of 1955, Lazarsfeld was called as an expert witness to the Kefauver Committee. Much like Charles Wright in his UNESCO report, the committee was puzzled as to why there had been no research on television. Lazarsfeld answered:

> Unfortunately, the chance for such a turn of affairs [foundation support] are limited at this moment, because of the kind of criticism which has been levelled against foundations in recent years. A Congressional Committee [Reece] has criticised the foundation boards for certain actions in other areas. The boards are frequently cautious in making funds available for new areas of study. When radio appeared on the scene, the Rockefeller Foundation was still quite willing to finance large studies on the effects the new medium might have on American life. Now that television is here,

with presumably more intensive effects, no foundation has seen fit to sponsor the necessary research.[81]

This is clear enough, but it was not the full testimony. Part of the original document submitted by Lazarsfeld is missing. The missing part reads:

> Just as our Committee [the Ford Advisory Committee on Television] submitted a detailed plan endorsed by the industry as well as by critical reform groups, the attacks on foundations began and the sponsoring organisation [Ford] decided to drop the whole matter.[82]

Who interfered with Lazarsfeld's evidence, and how they did it, is difficult to say. Certainly, though, it was the foundation's doing, and more than likely involved the hand of McPeak. In examining the existing correspondence, it is clear that McPeak had got wind of Lazarsfeld's testimony and had tried to pressure him to detract or alter parts of what he was to say. For example, in a letter to McPeak, Lazarsfeld writes, "Your letter came too late for changes to be made in the Congressional record." In the same letter, he uses the opportunity to try to discover what had happened to the proposed research: "I am, of course, very eager to avoid any misunderstanding as to what happened with the Citizen's Committee on Television."[83] He then added, "I have never had any communications from the Foundation as to what disposition was made of our proposal." He concludes the letter with an arch, slightly aggrieved offer: "Let me assure you that I will be equally cooperative in adjusting to your way of looking at the matter once you have explained it to me."[84]

This lack of communication by the foundation had evidently proved an embarrassment to Lazarsfeld when inquiries were made of him by board members. To protect himself, and the foundation, he blamed the Reece inquiry:

> The most reasonable interpretation was that the matter had got forgotten in the turn-out created by the Reece Committee. This formulation saved me from embarrassment with my colleagues who had worked so hard on this assignment, and I thought it also saved the foundation from the reproach of being discourteous to this distinguished group of men.[85]

Considering the amount of work Lazarsfeld had done on behalf of the foundation, and the lack of respect shown by the foundation toward him and his board, this is a remarkably restrained and polite letter, but it was followed a few months later with a much pricklier missive, which stated:

> At the time I rendered my report every member of the group got a thank-you letter and was told in due time we would learn of the disposition the foundation made of

our recommendations. Since then, neither I nor any other member of the group as far as I know has heard from you. We of course took it for granted that the foundation had decided to drop the matter but because we were left without any information most of us developed the theory that the Reece episode accounted for it all . . . including the silence.[86]

He finished the letter demanding that the matter be cleared up, as he puts it, "laying the ghost of the old proposal." No explanation was forthcoming. All that Lazarsfeld received was the repeat notice that the project had been refused funds.

COMMUNICATIONS RESEARCH AND ACTION

Clearly, Lazarsfeld recognized the nature of the times in which his report from the advisory committee had to find its way. Speaking of this period he said, "the atmosphere was very unpleasant," and added, "this Reece [Committee] was really just like McCarthy."[87] In my later interviews with Lazarsfeld, he freely discussed the Reece Committee and his statements to the Kefauver panel. Still, it was in the course of our interview that he presented his "theory" about how Ruml and McPeak "torpedoed" the project. Despite his appreciation for the political climate surrounding foundations, he still gave primacy to individual action and decisions in explaining the course of events.

As previously discussed, such explanatory "primacy" is revealing in terms of Lazarsfeld and his intellectual orientation toward individual action. But the broader import of the quashed Ford television project is the light the whole episode sheds on the politics of research. These politics are crucial components of *opportunity structure*—not in the determinant sense of politics as a "force" shaping outcomes, but politics, instead, as fluid states faced by, and worked upon, by actors. The advantage of this framework for historical understanding is evident through Lazarsfeld's experiences in both Austria and America. In Austria, despite restricted opportunity structures formed by the politics of the time, the restrictiveness that Lazarsfeld faced was directly, in one sense, on him. Thus, aspirations that were blocked in one direction did not mean that he could not create other avenues of expression for himself. In America in the 1950s, however, the obstacles he faced in pursuit of the establishment of television research were not personal, but came as a result of the restriction of opportunities faced by foundations. Because of this, Lazarsfeld could not turn a restricting situation into one of opportunity, as he had done in Austria. The block to success was not focused directly on him, but on an institution over which he had no control, but yet was totally dependent upon if success was to be attained.

NOTES

1. Himmelweit, Oppenheim, and Vince, *Television and the Child.*
2. Schramm, Lyle, and Parker, *Television in the Lives of Our Children.*
3. Wright, "Sociology of Mass Communication," 83.
4. A translation would be the Austrian Economic and Psychological Research Center. It was a cross between a market research company and an academic department, the idea being that money from its market research operations would fund its social research.
5. Merton, "Opportunity Structure," 25.
6. Rogers, *History of Communication Study.*
7. Katz and Lazarsfeld, *Personal Influence.*
8. Gulick, *Austria,* 1:1.
9. Lazarsfeld, personal interview, June 15, 1973.
10. Lazarsfeld, personal interview, May 25, 1973.
11. Lazarsfeld, personal interview, May 19, 1973.
12. Lazarsfeld, personal interview, May 25, 1973.
13. Karlick, personal interview, August 9, 1973.
14. Pulzer, *Rise of Anti-Semitism,* 308.
15. Siegert, "Numerus Juden Raus," 35.
16. Landheer, "Universalistic Theory of Society."
17. Konig, "Germany," 785.
18. Jedlicka, "Austrian Heimwehr," 137.
19. Bullock, *Austria,* 105.
20. Lazarsfeld, personal interview, May 25, 1973.
21. Wagner, personal interview, October 9, 1973.
22. Lazarsfeld, personal interview, May 25, 1973.
23. Brailsford in Braunthal, *Search of the Millennium,* 8.
24. Jahoda, personal interview, September 26, 1973.
25. Schilder, personal interview, June 28, 1974.
26. Lazarsfeld, personal interview, May 15, 1973.
27. Lazarsfeld, "An Episode," 301.
28. Lazarsfeld letter to Cantril, August 8, 1937. Lazarsfeld Papers, Columbia University Library.
29. Gary, "Communication Research."
30. Lazarsfeld, personal interview, June 15, 1973.
31. Marshall, personal interview, July 6, 1973.
32. Marshall, Inter-Office Memo, May 21, 1937. Marshall Papers, Rockefeller Foundation Archives.
33. Lazarsfeld, personal interview, May 25, 1973.
34. Ibid.
35. See Morrison, *Search for a Method,* 70–77 for details of the move from Princeton to Columbia and the internal difficulties the Project faced at Princeton, so much so that Cantril was pleased to let it go. Marshall was so tired of university politics and academic egos that he wondered whether a university was the most appropriate place for the continuation of the Project. Due to the support of Robert Lynd the Project was well received by Columbia, although the full integration of the Bureau into the University, of which the Project formed the base, took time and was not without its difficulties.

36. Rogers, *History of Communication Study*, 290.
37. Cantril letter to Lazarsfeld, August 9, 1937. Lazarsfeld Papers, Columbia University Library.
38. Pooley, "An Accident of Memory," 269.
39. Rogers, *History of Communication Study*, 293.
40. Lazarsfeld, personal interview, June 15, 1973.
41. Ibid.
42. Coleman, "Paul F. Lazarsfeld," 165; and Rogers, *History of Communication Study*, 309.
43. Lipset, "Paul F. Lazarsfeld," 263.
44. Gouldner, *For Sociology*, 10.
45. Hollis, *Philanthropic Foundations*, 255; and Berelson, "Place of Foundations," 8.
46. Macdonald, *Ford Foundation*, 23.
47. U.S. Congress, Senate. Commission on Industrial Relations, 83.
48. Neilson, *Big Foundations*, 53–54.
49. Lyons, *Uneasy Partnership*, 31.
50. Ford Foundation, "Trustees Report."
51. See Ralph Tyler, Program Report to the Trustees. June 23, 1954. Ford Foundation Archives, New York.—Tyler, formerly dean of social sciences at University of Chicago, was the first director of the Center for Advanced Study in the Behavioral Sciences.
52. Docket July 15–16. 1952. Ford Foundation Archives, New York.
53. Seybold, "Ford Foundation," 269–303.
54. Stouffer, *Communism, Conformity and Civil Liberties*.
55. Lazarsfeld and Thielens, *Academic Mind.*
56. Reeves, *Freedom and the Foundation*, 138.
57. Macdonald, *Ford Foundation*, 25.
58. Ibid., 25.
59. Reeves, *Freedom and the Foundation*, 15.
60. Ibid., 124.
61. Macdonald, *Ford Foundation*, 25.
62. Ibid., 27.
63. Ibid.
64. Reeves, *Freedom and the Foundation*, 15.
65. Ibid.
66. Ibid., 177.
67. U.S. Congress. House of Representatives, 36.
68. Ibid., 36.
69. Berelson, personal interview, July 12, 1973.
70. See Caute, *Great Fear.*
71. Solovey, "Riding Natural Scientists' Coattails."
72. Dodd's report to Reece. U.S. Congress, House of Representatives.
73. Myrdal, *An American Dilemma.*
74. Keppel in Myrdal, *An American Dilemma*, v.
75. Summers, "Perpetual Revelations."
76. Adorno, "Scientific Experiences."
77. Seybold, "Ford Foundation," 280.
78. Discussion Document, Advisory Committee on Television, Ford Foundation Archives, New York.

79. Lazarsfeld letter to McPeak. June 15, 1953. Ford Foundation Archives, New York.
80. Lazarsfeld, personal interview, June 15, 1973.
81. U.S. Congress, Senate. Juvenile Delinquency, 54.
82. This missing part to Lazarsfeld's testimony was discovered in the Ford Foundation Archives, New York.—Files P.A. 15–16.
83. Lazarsfeld, when interviewed, called the Advisory Committee on Television "the so-called Citizen's Committee."
84. Lazarsfeld letter to McPeak. June 15, 1955. Ford Foundation Archives, New York.
85. Ibid.
86. Lazarsfeld letter to McPeak. September 27, 1955. Ford Foundation Archives, New York.
87. Lazarsfeld, personal interview, June 15, 1973.

WORKS CITED

Adorno, Theodor. "Scientific Experiences of a European Scholar in America." In *The Intellectual Migration: Europe and America, 1930–1960,* edited by Donald Fleming and Bernard Bailyn, 338–370. Cambridge, MA: Harvard University Press, 1969.

Berelson, Bernard. "The Place of Foundations." Paper presented at American Sociological Association meeting, New York, 1960.

Braunthal, Julius. *Search of the Millennium,* with an introduction by Henry Brailsford. London: Gollanz, 1945.

Bullock, Malcolm. *Austria, 1918–38: A Study in Failure.* London: Macmillan, 1939.

Caute, David. *The Great Fear: The Anti-Communist Purge under Truman and Eisenhower.* New York: Simon Schuster, 1978.

Coleman, James S. "Paul F. Lazarsfeld: The Substance and Style of His Work." In *Sociological Traditions from Generation to Generation: Glimpses of the American Experience,* edited by Robert K. Merton and Matilda W. Riley, 153–74. Norwood, NJ: Ablex, 1980.

Ford Foundation. "Trustees Report." Detroit, MI: Ford Foundation, 1950.

Ford Foundation Archives. New York.

Gary, Brett. "Communication Research, the Rockefeller Foundation, and Mobilization for the War on Words." *Journal of Communication* 46, no. 3 (1996): 124–47.

Gouldner, Alvin W. *For Sociology: Renewal and Critique in Sociology Today.* London: Allen Lane, 1973.

Gulick, Charles A. *Austria: From Hapsburg to Hitler.* 2 vols. Berkeley: University of California Press, 1948.

Himmelweit, Hilde T., A. N. Oppenheim, and Pamela Vince. *Television and the Child: An Empirical Study of the Effect of Television on the Young.* London: Oxford University Press for the Nuffield Foundation, 1958.

Hollis, Ernest V. *Philanthropic Foundations and Higher Education.* New York: Columbia University Press, 1938.

Jedlicka, Ludwig. "The Austrian Heimwehr." *Journal of Contemporary History* 1, no. 1 (1966): 127–44.

John Marshall Papers. Rockefeller Foundation Archives. Sleepy Hallow, New York.

Katz, Elihu, and Paul F. Lazarsfeld. *Personal Influence: The Part Played by People in the Flow of Mass Communication.* Glencoe, IL: Free Press, 1955.

Katz, Elihu, and Paul F. Lazarsfeld. *Personal Influence: The Part Played by People in the Flow of Mass Communication*. New Brunswick, NJ: Transaction, 2006.

Konig, Rene. "Germany." In *Contemporary Sociology*, edited by Joseph Slabey Roucek, 779–806. London: Peter Owen, 1959.

Landheer, Barth. "The Universalistic Theory of Society of Othmar Spann and his School." In *An Introduction to the History of Sociology*, edited Harry E. Barnes, 385–99. Chicago: University of Chicago Press, 1958.

Lazarsfeld, Paul F. "An Episode in the History of Social Research." In *The Intellectual Migration: Europe and America, 1930–1960*, edited by Donald Fleming and Bernard Bailyn, 270–337. Cambridge, MA: Harvard University Press, 1969.

Lazarsfeld, Paul F., and Wager Thielens, Jr. *The Academic Mind: Social Scientists in Crisis*. Glencoe, IL: Free Press, 1958.

Lipset, Seymour Martin. "Paul F. Lazarsfeld of Columbia: A Great Methodologist and Teacher." In *Paul Lazarsfeld (1901–1976): La Sociologie de Vienne à New York*, edited by Jacques Lautman and Bernard-Pierre Lécuyer, 255–70. Paris: L'Harmattan, 1998.

Lyons, Gene M. *The Uneasy Partnership: Social Science and the Federal Government in the Twentieth Century*. New York: Harcourt, Brace and Co., 1969.

Macdonald, Dwight. *The Ford Foundation: The Men and the Millions*. New York: Reynal, 1956.

Merton, Robert K. "Opportunity Structure: The Emergence, Diffusion, and Differentiation of a Sociological Concept, 1930s–1950s." *Advances in Criminological Theory* 6 (1995): 3–78.

Morrison, David E. *The Search for a Method: Focus Groups and the Development of Mass Communication Research*. London: University of Luton Press, 1998.

Myrdal, Gunnar. *An American Dilemma: The Negro Problems and Modern Democracy*, with an introduction by Frederick L. Keppel. New York: Harper & Row, 1944.

Neilson, Waldemar A. *The Big Foundations*. New York: Columbia University Press, 1972.

Paul F. Lazarsfeld Papers. Columbia University Library, New York.

Pooley, Jefferson. "An Accident of Memory: Edward Shils, Paul Lazarsfeld and the History of American Mass Communication Research." PhD diss., Columbia University, 2006.

Pulzer, Peter G. J. *The Rise of Anti-Semitism in Germany and Austria*. New York: Wiley and Sons, Inc., 1964.

Reeves, Thomas C. *Freedom and the Foundation: The Fund for the Republic in the Era of McCarthyism*. New York: A. Knopf, 1969.

Rogers, Everett M. *A History of Communication Study: A Biographical Approach*. New York: Free Press, 1994.

Schramm, Wilbur, Jack Lyle, and Edwin B. Parker. *Television in the Lives of Our Children*. Stanford, CA: Stanford University Press.

Seybold, Peter J. "The Ford Foundation and the Triumph of Behavioralism in American Political Science." In *Philanthropy and Cultural Imperialism: The Foundations at Home and Abroad*, edited by Robert F. Arnove, 269–303. Boston: G. K. Hall & Co., 1980.

Siegert, Michael. "Numerus Juden Raus." *Neues Forum* 21 (June 1974): 35–37.

Solovey, Mark. "Riding Natural Scientists' Coattails onto the Endless Frontier: The SSRC and the Quest for Scientific Legitimacy." *Journal of the History of the Behavioral Sciences* 40, no. 4 (2004): 393–422.

Stouffer, Samuel. *Communism, Conformity and Civil Liberties: A Cross-Section of the Nation Speaks its Mind*. New York: Doubleday, 1955.

Summers, John H. "Perpetual Revelations: C. Wright Mills and Paul Lazarsfeld," In "Politics, Social Networks, and the History of Mass Communications Research: Rereading *Personal Influence*." edited by Peter Simonson, special issue. *The Annals of the American Academy of Political and Social Science* 608 (2006): 25–40.

U.S. Congress. House of Representatives. Report of Special Committee to Congress on Tax-Exempt Foundations and Comparable Organizations. 83rd Congress, 2nd sess., 1954.

U.S. Congress. Senate. Commission on Industrial Relations. 64th Congress, 1st sess., 1916.

U.S. Congress. Senate. The Sub-Committee to Investigate Juvenile Delinquency. 83rd Congress, 2nd sess., 1954. 1 Senate Resolution 89, Catalogue No. 3151–3154.

Wright, Charles R. "Sociology of Mass Communication, 1945–1955." In *Sociology in the United States of America: A Trend Report,* edited by Hans Lennart Zetterberg, 78-83. Paris: UNESCO, 1956.

CHAPTER EIGHT

How Does A Discipline Become Institutionalized?

VEIKKO PIETILÄ

LESSONS FROM GERMANY AND FINLAND

As James Carey has remarked, the "originating impulses" behind research concerning mass communication were European rather than American: although it was "not until the twentieth century that journalism and the mass media became subjects of scholarship in the United States in any significant way," the press had captured the European scholarly mind much earlier than this.[1] German scholars in particular had found the press to be "a problematic institution" and worthy of scholarly attention at a time when American scholars still believed mass communication to be, in Carey's words, "an invisible hand leading the will of individuals to the maximization of the social good."[2]

As Hanno Hardt's[3] excellent survey shows, nineteenth-century German theories examined the press as an organic part of society, connected to politics, economics, and culture. The press was viewed variously: as a nervous system linking the parts of society to one another, as a transmission belt of opinions between the power holders and the citizens, or as a peculiar economic institution in which its private pursuit of profit and its public obligations coexisted in a state of uneasy tension.

These early attempts to understand the press prompt questions concerning the institutionalization of scholarly work. Does the attention paid to the press mean that its study had already become institutionalized as a discipline in nineteenth-century Germany? More generally, what are the criteria that allow one to conclude that the study of a subject has reached that stage? Any answer will, perforce, require attention to the meaning of institutionalization. Institutions have been defined as "stable social arrangements"[4] or "established patterns of behavior"[5]—activities, that is, that are somehow fixed, arranged in an orderly fashion, or regulated by a set of rules. A shared scholarly interest in a subject (such as the press) does not suffice to constitute an institutionalized discipline. At most, such interest serves as a starting point of a process that ends in a discipline that is thoroughly institutionalized, both cognitively and socially. A discipline is institutionalized *cognitively* to the extent that it rests on commonly accepted principles and rules. It is institutionalized *socially* to the extent that it is exercised within a socially organized context.[6]

Between the starting and endpoints of this process, there are many intermediate stages, though no defined path. Numerous factors—from within and without the scholarship—conspire to shape whether or not such a process even begins, and then what the outcome looks like. Certainly, many fields proceed through a cognitive phase first, and only later cohere socially. That is, a discipline often assumes the shape of an informal "invisible college"[7]—groups of scholars studying and theorizing informally the same subject—before it takes on a distinct, socially institutionalized form. An example is German sociology, which started as an intellectual enterprise during the nineteenth century before there were any stable, socially organized contexts for it.[8] At other times, the social phase may *precede* cognitive institutionalization. For example, the social institutionalization of education in some practical subject may call subsequently for a systematic inquiry into that subject. In such a case, the prior social institutionalization may largely fix the cognitive framework within which the subject is approached.

In what follows, I elucidate this latter route to disciplinary institutionalization by tracing the development of mass communication research in both Germany and Finland. In particular, I take up the German newspaper science (*Zeitungswissenschaft*) as a case study of the tensions—internal and external—which shape the institutionalization process. I then turn to Finnish mass communication research as a parallel case, influenced initially by the German scholarship.

THE ESTABLISHMENT OF THE GERMAN NEWSPAPER SCIENCE

Though the press was the subject of theoretical speculation in nineteenth-century Germany, the theorizing had no identity of its own. It comprised only a part—and a subordinate one at that—of a far broader theoretical approach to society as a whole. That is why the press was examined "in a synoptic rather than a disciplinary frame."[9] In a way, the press was approached sociologically, although there was at the time no socially institutionalized sociology.[10] There was, however, a sociological habit of thinking that crossed the boundaries of the established social sciences such as political economy, political science, history, and jurisprudence—branches represented by Karl Knies, Albert Schäffle, Karl Bücher, and some others who at that time also theorized the press in their treatises.[11]

This broad societal view represents one possible paradigm on which a discipline of the press and other mass media might have been erected. In fact, when the need for such a discipline made itself felt—very prominently so after World War I—some stressed the necessity to study the social role of the press, in particular its function as an instrument of public opinion. Such suggestions resonated with the views of the early press theorists, above all Ferdinand Tönnies and Max Weber.[12] The claim that the influence of the press on public opinion ought to be subjected to a systematic scientific scrutiny was fueled by the widespread belief that the German press had lost the wartime propaganda battle, and that this was one of the main causes of the Axis Powers' defeat. One of the early representatives of the *Zeitungswissenschaft* (newspaper science), Johannes Kleinpaul, summarized this common view:

> The experience of the World War and the time immediately afterward opened the eyes of many to see what a power the press has, if it is led adequately, and what a danger it is, both in domestic and foreign political sense, if it is conducted badly. This gave the first impetus to the scientific study of the press as an important vehicle of public opinion.[13]

As a result, a range of university institutes devoted to the study of the press emerged in the years after the war. Yet—and *pace* Kleinpaul—the field did not become institutionalized as a "scientific study of the press as an important vehicle of public opinion." Hans Bohrmann's statement that public opinion as an object of study was "with a light heart pushed into the margins and left to other disciplines"[14] may be somewhat exaggerated but, on the whole, public opinion as such remained a secondary issue, as did other topics with a wider, societal scope. Thus, the earlier theories of the press—with their emphasis on broad societal understanding—did not enter into the newspaper science's grounding view of its subject.

Neither did the newspaper science take due notice of the remarkable proposition for a sociological study of the newspaper that Max Weber had made in his address to the first congress of the German Sociological Association in 1910.[15] One of the objectives Weber set for this study was to understand the role of the press in the formation of public opinion. Methodologically, Weber emphasized an empirical approach, and suggested that scholars start "with scissors and compasses to measure the quantitative changes of newspaper content," and "from these quantitative analyses" scholarship could then move "to qualitative ones."[16]

The proposed study, however, was never conducted, for various reasons.[17] Had the newspaper scientists taken the proposal seriously, it might have pushed them in the direction of an empirical, social scientific agenda, but this did not happen.

The possibilities opened up by earlier thinking for the coming postwar newspaper science were not, then, seized. Instead of focusing on the press in a broad societal context, the emerging newspaper science concentrated narrowly on the newspaper as a phenomenon *sui generis*. Instead of pursuing empirical research, the newspaper science was mostly content with a conceptual approach, with a focus on slightly varying definitions, categorizations, and classifications. This prompts the question: why did the newspaper science assume this form, given the other options available to it? Why wasn't Weber's call heeded?

SOCIAL INSTITUTIONALIZATION AS A DETERMINANT OF COGNITIVE INSTITUTIONAL FORM

The particular shape that newspaper science assumed was largely determined by the fact that the forces driving its institutionalization were practical, rather than intellectual. These practical exigencies, moreover, came from quarters outside of academe. The new discipline was closely tied up, in particular, with the felt need to provide journalists with professional training. In response to the transformation of the German press in the late nineteenth century, many in the industry claimed that untrained journalists were a poor match for the new mass press.[18] This situation in turn prompted suggestions that a university-level education might solve the problem.

The impulse to forge a connection between scholarship and the press was not terribly new. As early as the late seventeenth century, German universities sporadically arranged lectures on the press.[19] By the late nineteenth century, this sort of lecturing had become more routine.[20] The lecturers delivered mainly general facts and information about the press. Here were the early roots of the postwar partnership between the academy and the press, though still "far away from any systematic discipline."[21]

A dramatic step in the institutionalization of journalism training was taken in 1916 with the establishment of the Institut für Zeitungskunde at the University of Leipzig, by Karl Bücher, professor of political economy and one of the early press theorists. From the beginning, the institute's mission was far more educational than scientific. The very idea for an institute of journalism education, as well as the institute's early funds, came from a newspaper publisher.[22] Bücher himself, moreover, "strongly believed in a sound education of journalists."[23] He was just as strongly against an independent, scholarly discipline focused on the press. He thought such a field unnecessary, noting at the time that the press was already being studied within existing disciplines.[24]

A quite different understanding of press scholarship came from academics who held that the study of the press involved an obligation to carry out research, in addition to education. In response to this view, Bücher's institute announced in 1917 that its activity was "directed to the scientific study of the newspaper from all viewpoints in like manner as to the training of an academically educated offspring of journalists."[25] To talk of a "scientific study of the newspaper" does not in itself make a claim for a specific discipline—most obviously it only meant the study of the newspaper from the viewpoints of the established sciences—but nonetheless the phrase implies that such a discipline might be possible. As other university institutes for the education of journalists proliferated, this scientific vision of the study of the press began to build momentum. Most of these institutes began to call themselves institutes for newspaper science or for the "science of public communication" (*publizistische Wissenschaft*).[26] Thus, newspaper science became socially institutionalized, and it began to develop a degree of cognitive institutionalization.

That the gaze of the discipline was mostly concentrated on the press as isolated from broader societal contexts was mostly owing to the outside pressures that prompted the field's emergence. For Bohrmann, the most important motivation behind the field's quest for scientific status was the desire of the publishers' and journalists' organizations for raising the low esteem of journalistic work and business through the prestige of an academic science.[27] He argues further that it was, above all, the necessity to procure the continuation of the institutes' outside financing that led the field dogmatically to strive for an academic scientific status of its own.[28] Certainly, the fact that many of the institutes received financial support from these news businesses indicates the outside interests at work in the formation of the discipline. The emerging discipline found itself in a paradoxical situation: being dependent on nonacademic circles, it strove, for an independent academic status.

With these outside pressures solidly in place, the newspaper science developed, according to Lutz Hachmeister, into a field that "oscillated quite helplessly" between practical know-how and "scientificism" consisting predominantly of "an extensive use of definitions and classifications."[29] Some other disciplinary historians have taken a less critical stance regarding the nonacademic influences on the discipline's formation. Rüdiger vom Bruch, for one, argues that what was at stake was a "scientification" of a field oriented for professional practice.[30] The outside circles naturally expected the discipline to contribute to the training of journalists along with its research activity, but one may wonder how far the field with its meager output of conceptual treatises and historical studies was able to fulfill this expectation.

As stated above, the field strove for an acknowledged status of an independent, full-fledged academic science. However, it is in general not easy for a newcomer in the academic milieu to achieve this status.[31] The situation of the newspaper science was even more difficult, because its aspiration was severely handicapped by its dependency on outside interests. Such a dependency was everything but a merit in the eyes of those days' academic circles. In his sketch of the field's history, vom Bruch states that these circles felt a deep distrust of a commercialized press that "showed no scientific dignity."[32] According to Wilmont Haacke, the critics argued, in more general terms, that "an object whatsoever taken from today's everyday life does not qualify to constitute a science tomorrow."[33]

At the same time, the newspaper scientists' aspirations for a science of their own were opposed by the argument that the established disciplines already studied the press. For example, after having been for a long time regarded needless as a science of its own by the established sciences, sociology, having attained its academic foothold, turned against the newspaper science with the same argument. As Dirk Käsler[34] explains, the renowned sociologist Ferdinand Tönnies scoffed at the newspaper science's right to exist by remarking mockingly that although ornithology is necessary, an independent science of the hen or the duck is not. The implication, of course, was that the study of the newspaper belongs to the domain of sociology.

These hostile arguments naturally pushed newspaper scholars to seek out arguments they could use to make a case on behalf of the field's independence. They had to assure the academic world that the field merited the status of a genuine science. To be recognized as a genuine science in the Germany of the time, the most important thing was, according to Otto Groth,[35] an object of study researched by no other field—similar perhaps to Magali Sarfatti Larson's notion of a "monopoly of competence."[36] Not surprisingly, the newspaper scientists argued for the scientific character of their field with the claim that they had such an object.

QUEST FOR INDEPENDENCE AND DISCUSSION AROUND THE FIELD'S SUBSTANCE

The fact that the press was already studied within other disciplines—its development in history, its economic features in economics, and its legal aspects in jurisprudence—put the newspaper scholars in a thorny situation. How might they prove that the press was *their* legitimate object of study—a topic researched by no other discipline? Scholars used varying strategies to address this dilemma. One strategy was to argue, like Erich Everth,[37] that even if the existing disciplines had examined the newspaper, the scant attention they paid in no way corresponded to its social importance. An independent newspaper science was necessary, therefore, to remedy this neglect. A second strategy was to stress that other disciplines had no competence to study the newspaper in and of itself. Utilizing this strategy, Emil Dovifat wrote that other disciplines

> study the newspaper only because it mirrors life on realms in which they are interested. The newspaper science, for its part, is not interested in mirror images but in the mirror itself and the laws of mirroring.[38]

For this reason, Dovifat rejected the idea of handing the newspaper over to "different sciences and their methods." "Our discipline," he proclaimed, "is as independent as its object of study. Distinct and autonomous. A typical phenomenon sui generis."[39]

A third strategy, represented for example by Karl d'Ester, was to assert that—because "the newspaper can be studied from different angles"—the newspaper science is not a new discipline per se, but instead "a combination of existing disciplines."[40] As a composite, d'Ester reasoned, the field is not subordinate to any of its constituent disciplines, but is rightfully autonomous.[41]

Although all of these strategies were used to champion the newspaper science's independence, they rest on divergent conceptions of the field's nature and identity. Such disagreement implies, at the very least, that the degree of the field's cognitive institutionalization was not particularly high. In light of these competing strategies and self-definitions, it is clear that the field was *not* united by common theoretical principles and views that, in a cognitively institutionalized discipline, enable "researchers to study quite specific problems in a fairly standardized manner."[42] Such a situation, to be fair, is quite characteristic of the social sciences in general.

Within the newspaper science, disagreements surfaced most clearly in the debate over its proper object domain. Some scholars, for example, began to complain that an exclusive focus on the newspaper would restrict the field unnecessarily. Karl Jaeger, for one, emphasized that the object of study should be the communicated message, and that the field, as a result, should "be extended to all forms that the message can assume, to conversation, letter, document, poster, leaflet, newspaper, magazine, almanac, book."[43] This would mean, of course, that newspaper science would no longer stand as an adequate name for the field. Jaeger proposed that the field be referred to as the science of public communication. Hans Traub agreed that the field's object domain should be broadened to cover all means of the "anonymous, public, mental intercourse of human beings,"[44] but he saw no need to change the field's name because he understood all of these means of interaction to be newspaper-like.

Dovifat, for his part, was skeptical of efforts to broaden the object domain of this incipient field. For him, a science of public communication would necessarily be "a science whose subject matter would stretch to infinity."[45] He thought that it would be best, at least in the initial stage, to fix "one's gaze on a very restricted area"—that is, on the newspaper, in which "the whole spectrum of public life assumes forms which can be observed exactly."[46] Walter Schöne, too, was critical of Jaeger's suggestion for a broadening of the field. For him, it was a mistake to subsume "conversation, letter or book" within the field's object domain, for only those message forms that "contain ingredients typical of the newspaper" were relevant to this young specialty.[47] Still, he left open the question of whether or not the object domain should be extended to cover such newspaper-like forms.

In addition to these fundamental issues relating to the nascent field's definition, there were other major points of contention. For example, the essential characteristics of the newspaper—traits that distinguish the medium from its close relatives—could not be agreed upon.[48] Scholars also clashed over basic issues such as the appropriate questions to research and the methods to pursue them. Many of the early representatives of the field had pursued studies in the humanities, particularly in history, and they preferred questions that might be tackled with historical or philological methods, while some younger scholars, with more sociological backgrounds, began to pose questions demanding the application of empirical methods of social research.[49] Despite the younger generation's push, research output in this initial stage consisted mainly of writings in the history of the press, in addition to treatises attempting to clarify the conceptual basis for the field.

It has been estimated that the "scientific yield" of these efforts "remained scanty."[50] There is an obvious reason for this: when a field is institutionalized socially to serve the needs of an outside industry, this does not bode well for the scientific success of that field.[51] Moreover, most of the field's institutes were too small to be scientifically effective. Indeed, the best research in newspaper science, in its initial phase, was conducted outside the universities. Otto Groth's massive, four-volume opus *Die Zeitung*, published between 1928 and 1930, is a case in point. Groth longed for a university career but failed in this and remained separate from the university as an independent scholar, a *Privatgelehrte*, without any institutional position in the field. His output was nevertheless impressive. In addition to *Die Zeitung*, he also authored a history of theoretical approaches to the German press, published in 1948, and a seven-volume treatment of the newspaper as a social and cultural institution, published between 1960 and 1972. He accomplished all of this while working first as a professional journalist and then as a pensioner after the Nazi authorities had forced him to retire.

THE NEWSPAPER SCIENCE DURING
AND AFTER THE NAZI ERA

After the Nazis took over in 1933, academic disciplines were recast to fit the objectives of the new rulers. Jewish scholars, and those otherwise regarded as suspicious, were dismissed, and their syllabi were revised. A number of scholars emigrated to the United States, among them individuals—such as Paul F. Lazarsfeld from Austria and some members of the Frankfurt School from Germany—who deeply influenced the development of American mass communication research. Although many newspaper scientists were also persecuted, none of the leading names left Germany—this explains, in part, why the newspaper science has remained poorly known outside the German-speaking sphere.

The Nazi recasting changed the outside, practical needs that the newspaper science had served since its birth. Now the science was asked to train journalists in National Socialist spirit to contribute to the advancement of the new rulers' political ends.[52] This rendered the field formally quite homogeneous. An aspect of this formal homogenization was that the discipline was now bestowed the long desired status of an academic science in all institutes. Still, this formal homogenization did not render the field cognitively unanimous. On the contrary, old disagreements became even more pronounced so that, during the Nazi era, "the discipline split into two hostile camps."[53] One camp demanded that the field concentrate solely on the newspaper. An opposing camp demanded an extension of the field's domain into radio and film; the argument was to transform the field from a newspaper science into a science of public communication.[54] In addition, the latter camp was more inclined to social research with empirical methods while the members of the former cluster looked askance at this kind of research.[55] Eventually, the latter camp—consisting largely of younger scholars with clear Nazi sympathies—seemed to get the upper hand.[56]

After World War II, the field had to reinstitutionalize itself thoroughly. The careers of scholars who had open Nazi sympathies, of course, were over. One inadvertent consequence was that the younger generation's emphasis on empirical social research fell into disrepute. Without this pairing of Nazi sympathies and empirical methods, the field might have taken a different postwar course—one similar to American mass communication research. Instead, the field reverted, for the most part, to its pre-Nazi cognitive form, with, however, one conspicuous change: most of the field's representatives were now ready to extend its domain beyond the newspaper, to other media of public communication. As a result, the discipline adopted the science of public communication moniker.[57] Some scholars, to be sure, resisted the extension. Groth, for one, claimed that this expansion would ruin the field by compelling it to cover wholly alien subjects, such as the theater or even transport services.[58]

The phases of the newspaper science, described above, provide a telling example of a field whose prior social institutionalization for serving practical needs has determined its cognitive institutional form as a discipline. The obligation to train journalists placed the newspaper at the nexus of disciplinary thinking and research, and the newspaper retained a dominant position long after the extension of the field because it remained the focal object of study. The reconstitution of the discipline on the pre-Nazi basis is a fact that helps to explain this inertia. In addition, the field's early, newspaper-based approaches served as models for inquiries into other media. This kept the field quite media-centric. However, a major transformation in this respect was already drawing near. I explore this transformation via a detour through the Finnish case.

THE INITIAL PHASES OF FINNISH MASS COMMUNICATION RESEARCH

The German newspaper science had little visible impact on the international development of mass communication research. Nevertheless, it affected quite strongly the cognitive form that the Finnish media research adopted at its outset, most obviously because Finland's research, like Germany's, received its initial impetus from the perceived need to professionalize the training of journalists. This training was institutionalized in 1925 through the establishment of a journalism curriculum at the newly established School of Social Sciences in Helsinki. In contrast to the German case, however, the newspaper industry was not involved. The incentive for the curriculum—as for the School as a whole—came from educated circles who, traumatized by the Civil War in 1918, argued that civic affairs knowledge among citizens was a crucial bulwark against further social unrest. For this reason, too, the School of Social Sciences became "a kind of 'free university', as its students were not required to have matriculated."[59]

The School's curriculum, in general, consisted mainly of scholarship from the vantage points of political economy, constitutional and governmental law, history, political science, and social policy. Training in journalism, however, consisted mostly of practical exercises. This left little room for a cognitive institutionalization of a field, let alone for a bona fide discipline of media research. Consequently, there emerged no theoretical or conceptual work, nor any systematic research, despite the fact that the School's statutes from the early 1930s obliged it to "to promote scientific research within social sciences."[60] The scattered studies that were produced, mostly by students completing theses, tended to cover the history of the Finnish press and its personalities.

Steps toward a cognitive institutionalization were taken, first, after the School established a chair in *sanomalehtioppi* (approximately, "newspaper studies") in 1947. There existed, though, clashing conceptions regarding the proper makeup of this academic discipline. As a result, it took nine full years to fill the post. The first professor, Eino Suova, resembled the early scholars of the German newspaper science in that he, too, had studied in the humanities and had a long career as a practicing journalist. Moreover, he had been responsible for the training of journalists at the School for several years before his appointment. It is no surprise, then, that when he wrote in 1956 an article covering the principles of the discipline, he called it *sanomalehtitiede* ("newspaper science"), and defined it in quite practical terms as "a science investigating the working methods of the newspaper, the composition of its content, and its effects."[61] Here, the German accents are unmistakable.

The cognitive form, then, that Suova gave to the infant discipline rested on a foundation of sorts—but a foundation that was to disintegrate all too soon. Suova died only a few years after his nomination in 1956. He did not, as a result, have sufficient time to inaugurate a legacy through the education of loyal disciples. There was, moreover, a major transformation in the Finnish social sciences already underway during Suova's last years. Many young Finnish scholars had studied in the United States during the 1950s, and upon return, they brought with them the tenets of empirical social science research so prevalent in the postwar period. In rapid fashion, the turn to the American behavioral sciences spread across the whole complex of the Finnish social sciences.

Sanomalehtioppi was caught up in this same turn to imported U.S. methods. One striking consequence of this shift was the discipline's adoption of a new name, *lehdistö- ja tiedotusoppi* (approximately, "press and communication studies"). Raino Vehmas, a journalist and sociologist who succeeded Suova in 1962, institutionalized the discipline, in cognitive terms, on this new basis. For him, the field's purpose was "to create a whole and tested structure of knowledge as to how the individual behaves when he or she obtains information about his or her surroundings and when he or she sends or transmits messages to other individuals."[62] This approach, he argued, was more fruitful than the "cumbersome" newspaper science of the German style, which, for him, remained "at the stage of closet scholarship."[63]

Along with this transformation of the discipline's cognitive form came an attendant alteration in the field's *institutional* form. Hitherto its social organization had been anchored in the training of journalists, in line with Suova's practical view of the field. But in its new cognitive suit, the discipline could be harnessed to conduct contract research for interested nonacademic partners. The most important partner, in this vein, was the Finnish Broadcasting Company, Finland's public service broadcasting institution. In need of "information on the role of radio, television, and other media in people's lives,"[64] the company launched a major audience research project in 1965 that was carried out at the University of Tampere (the former School of Social Sciences)[65] and three other universities. This and other, similar projects strongly reinforced the discipline's new cognitive orientation, and pushed it further away from the newspaper science as formerly conceived and practiced.

It did not take long for *this* cognitive form to be challenged by new ones. At the end of the 1960s, the field began to split into different camps. Consequently, the degree of its cognitive institutionalization declined significantly. Or, perhaps more accurately, it became cognitively institutionalized in multiple, competing ways. In any case, the diffusion of the field in Finland reflected *international* processes of diversification.[66] This is a story for a different time. I conclude this chapter with a return to Germany.

THE MARGINALIZATION OF THE NEWSPAPER SCIENCE IN GERMANY

In Germany, too, the newspaper science (or the science of public communication) began to lose its position in the early 1960s. The influences coming from the United States were, again, decisive. As Carey has stated, there emerged, in the 1940s and 1950s, "a kind of intellectual Marshall plan" that led "to the widespread exportation of the American scholarship" to Europe.[67] The Germans, however, were not as ready as the Finns to welcome this scholarly aid. In particular, the older generation of social scientists had very mixed feelings toward American behavioral science and empirical social research. Within sociology, for example, it was the younger generation who first embraced the American approaches to their own discipline.[68]

This same pattern of generational conflict took hold in German mass communication research. After its postwar reinstitutionalization, the field reclaimed its traditional cognitive form, inherited from the pre-Nazi era, through the 1960s. Although the field's purpose was subject to frequent debate throughout these years, the debates did not transgress the principles of this traditional cognitive form. During this period, however, a younger generation emerged, whose members began to familiarize themselves with American mass communication research. One conclusion they drew from this approach was that the field's proper object of study is not definable by starting, in the traditional manner, with the newspaper or even with the media in general, but, instead, by looking to *communication* in society.[69] But what was more important was the severe criticism to which the younger generation subjected the basic principles of the older generation's traditional cognitive form.

From the point of view of the younger scholars, the traditional principles seemed totally "estranged from reality."[70] In the old view, it was "much more 'scientific' to speculate about facts than to investigate them with appropriate methods."[71] The traditional approach's body of theory, younger scholars charged, was nothing but "a catalogue of 'basic concepts'" that could not rise to the task assigned to it—that is, to grasp "the essence and influence of the different forms of the press."[72] Moreover, as Hachmeister shows, for the representatives of the new cognitive form, it did not make sense to brood over "what would constitute the essence of a medium or a process of public communication."[73] Younger scholars also reproached some of the older representatives for having a normative slant in their thinking, and demanded that the discipline should become value-free.[74] In sum, then, the traditional cognitive form was seen to be "nothing but ideology concerning either the publicist as a personality or journalistic profession."[75]

The new cognitive form pushed aside the field's traditional orientation, but did not annihilate it altogether.[76] This is an important difference between the situations in Germany and Finland, where the old newspaper science was fully eclipsed. Still, the German and Finnish cases have many parallels: the German field's new cognitive form became embedded in a new kind of organized social research, in which the field conducted studies commissioned by outside, interested partners.[77] This new reliance on contract research reduced the field's reliance on journalism education as the basis for the field's institutional shape. As in Finland, the unchallenged reign of this Americanized, empirical outlook did not last long. From the late 1960s on, the German field began to splinter into different camps. Like the Finnish case, the field's growing diversification was part of a much broader international process.

CONCLUDING REMARKS

As I have asserted, the fact that the newspaper science was institutionalized socially in the context of journalism education had a tremendous influence upon its subsequent cognitive form. In this respect, its cognitive and social institutionalization were linked, but they were discrepant in an important respect. The field became, as we have seen, rather highly institutionalized in social terms—it attained, at least formally, the status of an academic discipline. But its degree of cognitive institutionalization remained low, mostly because its representatives failed to come to a meaningful consensus concerning its nature and domain.

Was this low level of cognitive institutionalization a result of the preceding social institutionalization? Maybe. On the other hand, a social institutionalization along professional training lines might as well have led to a high-level, even if very restricted, cognitive institutionalization by providing a homogeneous basis, on which the object of study might have been analyzed in a fairly standardized manner. It is difficult to say why this did not happen. The fact is that the social sciences often display a low level of overall cognitive institutionalization with their many different—even conflicting—traditions of thought.

A similar discrepancy between social and cognitive institutionalization that was observed, particularly in the German case, has characterized the whole international field of mass communication research since. Academic institutes in different parts of the world have guaranteed the field's survival as a social institution, but at the same time it has become, cognitively, more dispersed than ever. This dispersal has occurred despite the fact that there have been periods when a given cognitive form has seized a hegemonic position, as with the spread of American behavioral mass communication research in the 1950s and 1960s. These centrifugal tendencies have, now and then, prompted the field's representatives to demand that the field be rendered more coherent or, in other words, institutionalized cognitively on a higher level.[78]

A high degree of cognitive institutionalization is, however, a two-edged sword. While a discipline cohering around a paradigm or a set of theoretical views might be effective in doing "normal science," as Thomas S. Kuhn[79] would put it, such a field would, at the same time, prove intolerant toward new ideas that do not fit the paradigm. A dispersed discipline is more tolerant in this respect. Thus, if international mass communication research nowadays consists of "a number of isolated frog ponds with no friendly croaking between the ponds," as Karl-Erik Rosengren[80] has pointedly maintained, this is not an altogether unhappy situation.

NOTES

1. Carey, foreword, 11.
2. Ibid., 11–12.
3. Hardt, *Social Theories of the Press*.
4. Bryant, "Institutions and Total Institutions," 686.
5. Lawson and Garred, *Dictionary of Sociology*, 123.
6. See Whitley, "Cognitive and Social Institutionalization."
7. See, for example, Crane, *Invisible Colleges*.

8. See, for example, Lepenies, *Between Literature and Science.*
9. Carey, foreword, 13.
10. See Lepenies, *Between Literature and Science.*
11. See Hardt, *Social Theories of the Press.*
12. Ibid., 138–86.
13. Kleinpaul, *Zeitungskunde*, 1–2.
14. Bohrmann, "Grenzüberschreitung?" 105.
15. Weber, "Deutschen Soziologentage."
16. Ibid., 441.
17. See, for example, Hardt, *Social Theories of the Press*, 183–4.
18. Bruch, "Zeitungswissenschaft zwischen Historie," 2.
19. See Groth, *Geschichte der deutschen Zeitungswissenschaft*, 25–26, 39–40; and Jaeger, *Von der Zeitungskunde zur publizistischen Wissenschaft*, 3–4.
20. See Bruch, "Zeitungswissenschaft zwischenHistorie"; Bruch, "Einleitung;" and Maoro, *Zeitungswissenschaft in Westfalen 1914–1945*, 23–34.
21. Klose, *Zeitungswissenschaft in Köln*, 15.
22. See, for example, Straetz, "Institut für Zeitungskunde," 75.
23. Hardt, *Social Theories of the Press*, 118.
24. Ibid., 119; and Straetz, "Institut für Zeitungskunde in Leipzig," 79–80.
25. Quoted in Straetz, "Institut für Zeitungskunde in Leipzig," 80.
26. Bruch, "Einleitung," 10.
27. Bohrmann, "Grenzüberschreitung?" 95–96; see also Maoro, *Zeitungswissenschaft in Westfalen*, 151–67.
28. Bohrmann, "Grenzüberschreitung?" 104.
29. Hachmeister, *Theoretische Publizistik*, 30, 94.
30. Bruch, "Einleitung," 10, 13–14.
31. A new field's quest for the status of an academically institutionalized discipline is often met with hostility from the already-institutionalized disciplines. For example, as soon as German sociology "had advanced its claim to be a self-sufficient discipline," it saw itself confronted "by the ill will of the established disciplines." See Lepenies, *Between Literature and Science*, 6. For the opponents to the new field, there was no need for sociology as a special discipline, because questions claimed by sociologists to be their own were already studied by other disciplines.
32. Bruch, "Zeitungswissenschaft zwischen Historie," 580.
33. Haacke, *Publizistik und Gesellschaft*, 29.
34. Käsler, "Streit um die Bestimmung der Soziologie," 234.
35. Groth, *Geschichte der deutschen Zeitungswissenschaft*, 5.
36. Larson, *The Rise of Professionalism*, 208.
37. Everth, *Zeitungskunde und Universität*, 1–7.
38. Dovifat, *Wege und Ziele der Zeitungswissenschaftlichen Arbeit*, 10.
39. Ibid., 11.
40. d'Ester, *Zeitungswesen*, 124, 129.
41. Views depicting mass communication research as a kind of crossroads have been presented later, too, most notably by Wilbur Schramm in the United States. See Schramm, "The State of Communication Research: Comment," 6–9; and Schramm, "Communication Research in the United States," 1–2.
42. Whitley, *The Intellectual and Social Organization*, xxxv.
43. Jaeger, *Von der Zeitungskunde*, 67.

44. Traub, *Grundbegriffe des Zeitungswesens*, 180.
45. Dovifat, *Wege und Ziele*, 7.
46. Ibid., 8.
47. Schöne, *Die Zeitung und ihre Wissenschaft*, 25.
48. See, for example, Bömer, *Bibliographisches Handbuch der Zeitungswissenschaft*, 10–11; Groth, *Zeitung*, 22–90; and Groth, *unerkannte Kulturmacht*, 258–343.
49. See Straetz, "Institut für Zeitungskunde in Leipzig," 89–90.
50. Bohrmann and Sülzer, "Massenkommunikationsforschung in der BRD," 86.
51. In general, a one-sided social institutionalization is not conducive to intellectual progress. In his analysis of the U.S. situation, John Durham Peters maintains that the field's administrative, not conceptual, orientation there has led it into intellectual poverty. See Peters, "Institutional Sources of Intellectual Poverty." In a certain sense, the German case is parallel to this.
52. See, for example, Benedikt, "Berliner Institut," 119–20.
53. Hachmeister, *Theoretische Publizistik*, 55.
54. Ibid., 54–61; see also Boguschewsky-Kube, *Theoriestreit zwischen Publizistik*, 53–58.
55. See Straetz, "Institut für Zeitungskunde in Leipzig," 91–95.
56. Hachmeister, *Theoretische Publizistik*, 58–60; and Maoro, *Zeitungswissenchaft in Westfalen*, 399–410.
57. Bohrmann, "Grenzüberschreitung?" 109–12.
58. Groth, *Geschichte der deutschen Zeitungswissenschaft*, 254–55.
59. Rasila, *Yhteiskunnallinen korkeakoulu 1925–1966*, 301.
60. Ibid., 135.
61. Suova, "Sanomalehtitiede valinkauhassa," 5.
62. Vehmas, "Lehdistö- ja tiedotusopin tavoitteista," 463.
63. Ibid., 466.
64. Pietilä, Malmberg, and Nordenstreng, "Theoretical Convergences and Contrasts," 69.
65. This part was carried out at the University's Research Institute for Social Sciences. The Institute was established in 1945 as a semi-autonomous department within the School of Social Sciences for doing empirical social research. In a certain sense it resembled the famous Bureau of Applied Social Research in New York. Research initiatives came mainly from public organs, although there was room for the staff's own initiative, too. The first research project was a study on the social adjustment of Carelian people, evacuated to Finland from the areas surrendered to the Soviet Union in 1944, to their new places of residence. The study was funded by the Rockefeller Foundation. It was this financing that initially made the Institute's establishment possible.
66. I have made a brief sketch of this international diversification process in another context. See Pietilä, *On the Highway of Mass Communication Studies*, 179–82. See also the "Ferment in the Field" issue of the *Journal of Communication*. The turmoil, begun with the critique of positivism at the end of the 1960s, broke quickly through all social sciences. For example, as Levine states, "the eruptions of the late 1960s shattered" the vision of a unified sociology "once and for all (*Visions of the Sociological Tradition*, 279). Since then sociology has been "highly diverse and fragmented" and there have been insurmountable "difficulties in finding the core of sociology." See Boje and Svallfors, introduction to *The New Millennium*, 3.
67. Carey, foreword, 11.
68. See, for example, Tenbruck, "Deutsche Soziologie im internationalen Kontext."
69. See, for example, Dröge and Lerg, "Kritik der Kommunikationswissenschaft," 254–55.
70. Silbermann, "Schwächen und Marotten der Massenmedienforschung," 4.
71. Eberhard, "Thesen zur Publizistikwissenschaft," 261.
72. Dröge and Lerg, "Kritik der Kommunikationswissenschaft," 252.

73. Hachmeister, *Theoretische Publizistik*, 264.
74. See, for example, Eberhard, "Thesen zur Publizistikwissenschaft"; and Dröge and Lerg, "Kritik der Kommunikationswissenschaft."
75. Dröge and Lerg, "Kritik der Kommunikationswissenschaft," 252.
76. Bohrmann and Sülzer, "Massenkommunikationsforschung in der BRD," 99–100.
77. Ibid., 95–97.
78. See, for example, Pietilä, *On the Highway of Mass Communication Studies*, 321–25.
79. Kuhn, *Structure of Scientific Revolutions*.
80. Rosengren, "From Fields to Frog Ponds," 9.

WORKS CITED

Benedikt, Klaus-Ulrich. "Das Berliner Institut für Zeitungskunde/Zeitungswissenschaft." In *Von der Zeitungskunde zur Publizistik*, edited by Rüdiger vom Bruch and Otto B. Roegele, 105–41. Frankfurt am Main: Haag + Herchen Verlag, 1986.

Boguschewsky-Kube, Sigrid. *Der Theoriestreit zwischen Publizistik und Zeitungswissenschaft*. Munich: tuduv-Verlag, 1990.

Bohrmann, Hans. "Grenzüberschreitung? Zur Beziehung von Soziologie und Zeitungswissenschaft 1900–1960." In *Ordnung und Theorie*, edited by Sven Papcke, 93–112. Darmstadt: Wissenschaftliche Buchgesellschaft, 1986.

Bohrmann, Hans, and Rolf Sülzer. "Massenkommunikationsforschung in der BRD." In *Gesellschaftliche Kommunikation und Information*, edited by Jörg Aufermann, Hans Bohrmann, and Rolf Sülzer, Band 1, 83–120. Frankfurt am Main: Fischer Athenäum, 1973.

Boje, Thomas P., and Stefan Svallfors. Introduction to *The New Millennium: Essays on the Current State in Sociology*, edited by Thomas P. Boje and Stefan Svallfors, 1–8. Umeå: Umeå University, 2000.

Börner, Karl. *Bibliographisches Handbuch der Zeitungswissenschaft*. Leipzig: Otto Harassowitz, 1929.

Bruch, Rüdiger vom. "Zeitungswissenschaft zwischen Historie und Nationalökonomie." *Publizistik* 25, no. 4 (1980): 579–600.

———. "Einleitung." In *Von der Zeitungskunde zur Publizistik*, edited by Rüdiger vom Bruch and Otto B. Roegele, 1–30. Frankfurt am Main: Haag + Herchen Verlag, 1986.

Bryant, Robert D. "Institutions and Total Institutions." In *International Encyclopedia of Sociology*, edited by Frank N. Magill, vol. 1, 686–89. Chicago: Fitzroy Dearborn Publishers, 1995.

Carey, James W. "Foreword to *Social Theories of the Press: Early German & American Perspectives*," by Hanno Hardt, 9–14. Beverly Hills, CA: Sage, 1979.

Crane, Diana. *Invisible Colleges: Diffusion of Knowledge in Scientific Communities*. Chicago: University of Chicago Press, 1972.

d'Ester, Karl. *Zeitungswesen*. Breslau: Ferdinand Hirt, 1928.

Dovifat, Emil. *Wege und Ziele der Zeitungswissenschaftlichen Arbeit*. Berlin: Walter de Gruyter, 1929.

Dröge, Franz W., and Winfried B. Lerg. "Kritik der Kommunikationswissenschaft." *Publizistik* 10, no. 3 (1965): 251–66.

Eberhard, Fritz. "Thesen zur Publizistikwissenschaft." *Publizistik* 6, no. 5–6 (1961): 259–66.

Everth, Erich. *Zeitungskunde und Universität*. Jena: Verlag von Gustav Fischer, 1927.

Gerbner, George and Marsha Siefert, eds. "Ferment in the Field." Special issue, *Journal of Communication* 33, no. 3 (1983)."

Groth, Otto. *Die Geschichte der deutschen Zeitungswissenschaft*. Munich: Weinmayer, 1948.

————. *Die unerkannte Kulturmacht*, Band 1. Berlin: Walter de Gruyter, 1960.

————. *Die Zeitung*, Band 1. Mannheim: J. Bensheimer, 1928.

Haacke, Wilmont. *Publizistik und Gesellschaft*. Stuttgart: K. F. Koehlers Verlag, 1970.

Hachmeister, Lutz. *Theoretische Publizistik*. Berlin: Volker Spiess, 1987.

Hardt, Hanno. *Social Theories of the Press: Early German & American Perspectives*. Beverly Hills, CA: Sage, 1979.

Jaeger, Karl. *Von der Zeitungskunde zur publizistischen Wissenschaft*. Jena: Verlag von Gustav Fischer, 1926.

Käsler, Dirk. "Der Streit um die Bestimmung der Soziologie auf den deutschen Soziologentagen 1910 bis 1930." In *Soziologie in Deutschland und Österreich 1918–1945*, edited by M. Reiner Lepsius, 159–99. Opladen: Westdeutscher Verlag, 1981.

Kleinpaul, Johannnes. *Zeitungskunde*. Leipzig: Verlag Julius Mäser, 1927.

Klose, Hans-Georg. *Zeitungswissenschaft in Köln*. Dortmunder Beiträge zur Zeitungsforschung, Band 45. Munich: K. G. Saur, 1989.

Kuhn, Thomas S. *The Structure of Scientific Revolutions*. 2nd ed. Chicago: University of Chicago Press, 1970.

Larson, Magali S. *The Rise of Professionalism: A Sociological Analysis*. Berkeley, CA: University of California Press, 1977.

Lawson, Tony, and Joan Garred. *Dictionary of Sociology*. Chicago: Fitzroy Dearborn Publishers, 2001.

Lepenies, Wolf. *Between Literature and Science: The Rise of Sociology*. Translated by R. J. Hollingdale. Cambridge: Cambridge University Press, 1988.

Levine, Donald N. *Visions of the Sociological Tradition*. Chicago: University of Chicago Press, 1995.

Maoro, Bettina. *Die Zeitungswissenschaft in Westfalen 1914–1945*. Dortmunder Beiträge zur Zeitungsforschung, Band 43. Munich: K. G. Saur, 1987.

Peters, John Durham. "Institutional Sources of Intellectual Poverty in Communication Research." *Communication Research* 13, no. 4 (1986): 527–59.

Pietilä, Veikko. *On the Highway of Mass Communication Studies*. Cresskill, NJ: Hampton Press, 2005.

Pietilä, Veikko, Tarmo Malmberg, and Kaarle Nordenstreng. "Theoretical Convergences and Contrasts: A View from Finland." *European Journal of Communication* 5, no. 2–3 (1990): 165–85.

Rasila, Viljo. *Yhteiskunnallinen korkeakoulu 1925–1966* [The School of Social Sciences 1925–1966]. Acta Universitatis Tamperensis, ser. B, vol. 7. Vammala: Vammalan kirjapaino, 1973.

Rosengren, Karl-Erik. "From Field to Frog Ponds." *Journal of Communication* 43, no. 3 (1993): 6–17.

Schöne, Walter. *Die Zeitung und ihre Wissenschaft*. Leipzig: Verlag Heinrich F. A. Timm, 1928.

Schramm, Wilbur. "Communication Research in the United States." In *The Science of Human Communication*, edited by Wilbur Schramm, 1–16. New York: Basic Books, 1963.

————. "The State of Communication Research: Comment." *Public Opinion Quarterly* 23, no. 1 (1959): 6–9.

Silbermann, Alphons. "Schwächen und Marotten der Massenmedienforschung." In *Publizistik als Gesellschaftswissenschaft*, edited by Hansjürgen Koschwitz and Günter Pötter, 3–18. Konstanz: Konstanz Universitätsverlag, 1973.

Straetz, Sylvia. "Das Institut für Zeitungskunde in Leipzig bis 1945." In *Von der Zeitungskunde zur Publizistik*, edited by Rüdiger vom Bruch and Otto B. Roegele, 75–103. Frankfurt am Main: Haag + Herchen Verlag, 1986.

Suova, Eino. "Sanomalehtitiede valinkauhassa" [Newspaper science in the casting ladle]. *Tiedotustutkimus* 5, no. 4 (1982): 24–37. First published in 1956 as a separate article, Helsinki: KK:n kirjapaino.

Tenbruck, Friedrich H. "Deutsche Soziologie im internationalen Kontext." In *Deutsche Soziologie seit 1945*, edited by Günther Lüschen, 71–107. Opladen: Westdeutscher Verlag, 1979.

Traub, Hans. *Grundbegriffe des Zeitungswesens*. Stuttgart: C. E. Poeschel Verlag, 1933.

Vehmas, Raino. "Lehdistö-ja tiedotusopin tavoitteista" [On the goals of press and communication studies]. *Suomalainen Suomi* 32, no. 8 (1964): 462–68.

Weber, Max. "Rede auf dem ersten Deutschen Soziologentage in Frankfurt 1910." In *Gesammelte Aufsätze zur Soziologie und Sozialpolitik*, by Max Weber, 431–49. Tübingen: J. C. B. Mohr, Paul Siebeck, 1924.

Whitley, Richard. "Cognitive and Social Institutionalization of Scientific Specialities and Research Areas." In *Social Processes of Scientific Development*, edited by Richard Whitley, 69–95. London: Routledge & Kegan Paul, 1974.

———. *The Intellectual and Social Organization of the Sciences*. 2nd ed. Oxford: Oxford University Press, 2000.

Institutional Networking: THE Story OF THE International Association FOR Media AND Communication Research (IAMCR)

KAARLE NORDENSTRENG

Research in any field is both a source and an outcome of the institutions that psurround it. In communication research, the most obvious institutions are academic programs and the media industry's own research centers. These institutions are typically national and are firmly rooted in the respective political, economic, and cultural conditions. Beyond them are international institutions, which come into play once similar establishments have emerged in several countries. These international institutions are relatively weak, but they both reflect and shape the national landscapes. What follows is a story of the most central international institution in media research, the International Association for Media and Communication Research (IAMCR), whose fifty-year history provides a panorama of how research in the field has developed and how it has been influenced by international contacts and cooperation.[1]

Looking at histories of the emerging field of mass communication[2]—in continental Europe from the late seventeenth century onward and in the United States from the early nineteenth century onward—leads one to notice how little and how late international institutions have played a role in shaping communication research. Although the roots of the field go back to the classics of sociology and political science, it is only in the twentieth century that we can find any systematic international networking of research, built through particular structures such as international meetings or associations among relevant scholars. Journalists and other "press people" had their first international congress in 1894, followed by their own international association(s) in the first half of the twentieth century.[3] Global media policies began to take shape in the League of Nations in the 1920s[4]—at a time when communication research was not only established but already being divided into various traditions. But communication research remained conspicuously remiss on its own international platforms and structures until the end of World War II.

PREHISTORY OF IAMCR AND ITS LESSONS

The history of IAMCR begins in the first years of the United Nations Educational, Scientific, and Cultural Organization (UNESCO), formed in the aftermath of World War II. In 1946, UNESCO proposed an "International Institute of the Press and Information, designed to promote the training of journalists and the study of press problems throughout the world."[5] This initiative was marked by the idealism that had inspired the founding of the United Nations (UN) itself.

At this time, the mass media included mainly the press, radio, and cinema— as television was still at an experimental stage. Given their role during the war, the mass media were being recognized as an important factor in many fields, including international relations. Accordingly, one of the first special conferences organized by the UN in April 1948 was devoted to the freedom of information. This was where the famous Article 19 on Freedom of Expression and Information was drafted as part of the Universal Declaration of Human Rights, adopted by the UN General Assembly in December of the same year. Two founders and future presidents of IAMCR—Fernand Terrou and Jacques Bourquin—were actively involved in drafting Article 19 during the UN Conference on Freedom of Expression and Information.

A decade passed, however, before the IAMCR was established. One reason for this slow progress was the rapid deterioration of East-West relations and the onset of the cold war in the late 1940s. Issues related to the role of public opinion and the media were of concern to domestic politics and became increasingly important to international relations, not least with respect to "the ideologies of freedom." In addition, the International Press Institute (IPI) was established in 1951 as an international association of newspaper editors and publishers in the Western countries, representing the "free world," as opposed to the "Communist world." At this stage, UNESCO refrained from promoting the establishment of a separate research association, anticipating that the IPI would meet this need when it undertook, for example, a content analysis of the international news flows.

However, the limitations of the IPI's geopolitical and thematic base soon became obvious. UNESCO realized that in addition to press freedom there were other issues in the growing field of mass communication, particularly relating to journalism education, that would benefit from internationally coordinated activity by a separate organization. In 1952, UNESCO returned to this topic, setting out two lines of activity: establishing training centers for journalists and founding an international organization for the promotion of scientific research on mass communication.

At this time the UNESCO Secretariat established a "Clearing House" within its Department of Mass Communication, which was charged "to collect, analyse and disseminate information on press, film, radio and television, pointing out their use for educational, cultural and scientific purpose," as stated in the standing preface of its publication series "Reports and Papers on Mass Communication." The first twenty issues of this series, published between 1952 and 1956, covered topics related mainly to film, television, and newsprint, but its December 1956 issue was titled, "Current Mass Communication Research—I."[6] This volume included a register of ongoing research projects and a bibliography of books and articles published since early 1955, each divided into eight topics relating to mass communication, including history, economic and legal aspects, government information and propaganda services, advertising and public relations, psychological and sociological studies on mass communication and public opinion, and the pedagogical and cultural role of mass communication. The mass communication research in progress included a list of nearly 400 projects in fourteen countries, while the bibliography listed some 800 publications in twenty-five countries. This impressive research overview was compiled with the aid of a questionnaire sent to thirty-two selected institutions in nineteen countries. The data gathering was assisted by national clearinghouses established in France, Japan, and the United States. The process encouraged the establishment of clearinghouses in other countries, beginning with West Germany and Italy.

The year 1956 was crucial for developments under the aegis of UNESCO. In April, a meeting of experts on the professional training of journalists was held at the UNESCO headquarters in Paris. This meeting of forty professors and other media experts, with accompanying documents and resolutions, demonstrated that there indeed existed a dynamic field of research and training in need of international coordination. A list of establishments for professional training of journalists included a hundred institutes from the United States alone, and nearly a hundred more from some thirty other countries. In November to December of that year, the General Conference of UNESCO (held in New Delhi) adopted a resolution "to promote the coordination of activities of national research institutes in the field of mass communication in particular by encouraging the establishment of an international association of such institutes."

Immediately after the General Conference, a colloquium was held in Strasbourg, where the International Centre for Higher Education in Journalism had been established. It was on this occasion, in December 1956, that a preparatory group called the "Interim Committee" (Comité Intérimaire) was formed by four dedicated colleagues: Fernand Terrou (director of the French Institute of the Press and president of the French Association for Communication Sciences), Mieczyslav Kafel (director of the Institute of Journalism at the University of Warsaw), Marcel Stijns (editor-in-chief of the Belgian journal *Het Laatste Nieuws* and vice president of the International Federation of Journalists), and David Manning White (professor of journalism at Boston University and chairman of the Council on Research of the Association for Education in Journalism [AEJ]).

Terrou chaired the committee. He invited Jacques Kayser, director of research at the French Institute of the Press, to serve as its executive secretary. Jacques Bourquin was not a member of the committee, but as representative for the French-speaking press in Switzerland, he lobbied strongly for IAMCR within the International Federation of Newspaper Publishers (FIEJ)—the predecessor of today's World Association of Newspapers (WAN). UNESCO did not favor Bourquin's inclusion in the Interim Committee lest it appear that French or Francophone interests were overrepresented. A hidden and perhaps more relevant reason was that Bourquin had taken sides in an earlier dispute within UNESCO against the then-director of the Department of Mass Communication. This is a classic example of how personal factors can intervene in institutional history: Bourquin was a decisive player in rallying the media industry behind IAMCR, while UNESCO excluded him from the Interim Committee—most likely because of personality conflicts in the past.[7]

The tasks to be carried out by the new Association were now foreseen to include not only general promotion of international contacts within the field but also specific clearinghouse functions, such as the production of bibliographies and lists of institutions as had been issued in UNESCO's inventory. The committee prepared a draft constitution and sent two circular letters out to potential participants. It convened the founding conference in December—after the IPI held its conference in Asia (Colombo) in November.

In summary, once mass communication, like other fields of socioeconomic activity, had reached a certain level of importance and specialization in society, this led to an institutionalization of the field, both nationally and internationally.

Accordingly, IAMCR grew out of a rapidly developing media field, particularly with respect to journalism, which created its own branch of institutional interests and a need for professional education as well as for scientific research. As Terrou wrote in *Etudes de Presse*, the periodical of the French Institute of the Press, in 1956, "The professional training of journalists and the science of communication are the agenda of the day," and added, "This is very good for the freedom of information." For Terrou, as for Bourquin, IAMCR represented not only a technical project to promote training and research, but also an ideological project to serve a broader cause aimed at fostering peace and freedom in an international order.

In terms of its focus, IAMCR initially concentrated first and foremost on journalism and mass communication—rather than, for example, on speech communication (which had a long academic tradition in the United States), or on telecommunication (which at the time remained largely a technical subject). The actors involved were predominantly academics, with a strong presence of print journalists and others from the media industry.

The springboard for IAMCR was a combination of training needs and the growth of research in mass communication. As has not been the case in other fields, the emergence of a scientific association proceeded—on national and international levels—according to the demands of not just academic research but also of nonacademic professional training. From the beginning, mass communication research has been inseparable from the training of communicators, especially journalists. Contrast this with, say, political science, which has played only a very small role in the training of politicians. However, although training was crucial for ensuring that the research interests received international recognition, at least in getting the association started, training and research would need to be separated eventually.

Geopolitically, IAMCR had a broad—even global—base, with institutions and individuals from all continents affiliated with it. There is no doubt that the initiative to create IAMCR was dominated by Europeans, particularly the French, but colleagues from countries such as Brazil, Peru, Uruguay, Egypt, Israel, India, Indonesia, Japan, Australia, the United States, and Canada were also involved. The new Eastern Europe, behind the so-called Iron Curtain, was represented by leading academics from Poland, Czechoslovakia, and the Soviet Union, making the IAMCR configuration more balanced than the IPI or the two international associations of professional journalists, the International Organization of Journalists (IOJ, representing mainly the East and the South), and the International Federation of Journalists (IFJ, representing mainly the West). Accordingly, IAMCR was not a cold war project. On the contrary, it was founded on ecumenical soil crossing both East-West and North-South divides.

HISTORY OF IAMCR

IAMCR's history spans five decades, with four stages of development: (1) the foundation, 1957–1964; (2) a period of consolidation, 1964–1972; (3) the years of growth, 1972–1990; and, finally, (4) the period of challenges, since 1990. It has held twenty-five biennial conferences and has had nine presidents, reflecting its global profile (see Table 9.1).

Table 9.1. IAMCR Meetings and Presidents

1957	Paris (France)	Fernand Terrou (France)
1959	Milan (Italy)	Raymond B. Nixon (United States of America)
1961	Vevey (Switzerland)	
1964	Vienna (Austria)	Jacques Bourquin (Switzerland)
1966	Herceg Novi (Yugoslavia)	
1968	Pamplona (Spain)	
1970	Konstanz (West Germany)	
1972	Buenos Aires (Argentina)	James D. Halloran (United Kingdom)
1974	Leipzig (East Germany)	
1976	Leicester (United Kingdom)	
1978	Warsaw (Poland)	
1980	Caracas (Venezuela)	
1982	Paris (France)	
1984	Prague (Czechoslovakia)	
1986	New Delhi (India)	
1988	Barcelona (Spain)	Cees Hamelink (Netherlands)
1990	Bled (Yugoslavia)	
1992	Guaruja (Brazil)	Hamid Mowlana (United States)
1994	Seoul (South Korea)	
1996	Sydney (Australia)	Manuel Pares i Maicas (Spain)
1998	Glasgow (Scotland)	
2000	Singapore	Frank Morgan (Australia)
2002	Barcelona (Spain)	
2004	Porto Alegre (Brazil)	Robin Mansell (United Kingdom)
2006	Cairo (Egypt)	

Source: Hamelink and Nordenstreng, "A Short History of IAMCR," 24–26.

The Foundation, 1957–1964

The "constitutive conference" (as it was called following the French terminology) was held at UNESCO headquarters in Paris on December 18 and 19, 1957. According to UNESCO's press release,

> Fifty experts on information media, from 15 countries, have just completed in a two-day session at UNESCO House, Paris, the task of establishing the International Association for Mass Communication Research. Created with the co-operation of UNESCO, the new association, which is independent, has its headquarters in Paris, in the offices of the Institut Français de Presse of the University of Paris, 27 rue St. Guillaume. Its function is the promotion throughout the world of the development of research on problems related to press, radio, television and films. The association's membership list includes about 200 names of institutes, educational establishments and individuals. Educators in journalism are the most numerous on the individual list of educators and sociologists.[8]

The first general assembly after the founding conference was held in Milan, Italy, in October 1959. At this general assembly, IAMCR named its first American president, Raymond Nixon, while the French founding president, Fernand Terrou, became secretary general. Most of those elected as officers in Paris in 1957 remained in office in Milan in 1959, although some changed positions.

The 1959 general assembly, in Milan, modified the organizational structure of IAMCR with a rotating presidency and a permanent secretariat with the posts of secretary general and deputy secretary general. Four permanent "sections" were also established, for historical research, legal and political research, psychological and sociological research (headed by Wilbur Schramm from Stanford University), and economic and technical research.

It is interesting to read President Terrou's circular letter of 1958, because there he invites members to consider joining not only the sections devoted to history and to law, but also sections for audiovisual media and for problems of media and children. Obviously there was not enough interest in audiovisual media and in the specific problem of media influence on children—the latter topic falling under the section for psychological and sociological research.

Apart from the work of the sections, IAMCR mobilized scholarly attention to topical issues through separate colloquia and thematic sessions at its biennial conferences. The first major thematic undertaking was an international colloquium on the professional secrecy of journalists, held in Strasbourg in October 1958. This was followed by a study on the same topic commissioned by UNESCO and issued jointly by IAMCR and IPI in 1959. Another early project that UNESCO invited IAMCR to contribute was the preparation of a selective bibliography on the influence of cinema on children. IAMCR also began to collect and publish general bibliographies on mass communication research—something that was foreseen as one of its main tasks at the founding conference. Although the harvest of these inventories was not as abundant as that gathered by UNESCO's Clearing House in 1956, the first *IAMCR Bulletin* served as a channel to share bibliographical data.

The foundation stage continued through the second general assembly (in Vevey, Switzerland, June 1961), and the third (in Vienna, June 1964). President Nixon introduced the transfer of the secretariat from Paris to Amsterdam, with Maarten Rooy as secretary general, *Gazette* as the official journal of the association, and the *IAMCR Bulletin* as its supplement. At this stage, there was clear friction between the French and the Americans—UNESCO siding with the Americans rather than with the French—but formally the association was functioning normally. Nixon was succeeded in Vienna by Jacques Bourquin as president, leading to a new era for IAMCR.

Consolidation, 1964–1972

After Vienna, the secretariat was transferred from Amsterdam to President Bourquin's office in Lausanne, Switzerland, where he began to issue presidential letters. During his eight-year presidency, twenty-six letters were issued in English and French as mimeographed copies reporting on the organization's news. Bourquin wrote these himself, though, as director of the association for Switzerland's French-language press, he had a personal assistant for translations, mailing, and other such duties. The letters evoke an active international association but in a manner that resembles a familiar cottage industry rather than an official world body.

In this new Lausanne-based IAMCR, *Gazette* had only a nominal relation to the association, and its supplement, the *Bulletin*, gradually disappeared, owing to lack of funding. Formally, Rooy was appointed as an officer in charge of publications, while Terrou was listed as director of research. In practice, the clearinghouse function of IAMCR was limited to occasional lists of studies and publications by the members, distributed with the presidential letters. Yet some thematic studies were carried out under UNESCO contracts, including a comparison of the statutes of radio and television companies.

The main activity of IAMCR at this stage was the biennial conferences and other symposia; these were fairly well attended and served as important platforms for academic and political exchanges. Through these live occasions, the membership grew slowly but surely and kept its balance, especially between Eastern and Western Europe.

The 1966 conference, held in Herceg Novi, Yugoslavia, was a milestone in IAMCR's history. More than seventy participants from seventeen countries of Europe, Asia, and the United States discussed, among other things, the topic of mass media and national development. The development theme was introduced by Gerhard Maletzke of Germany, while Lakshman Rao from India was present as UNESCO representative. This was the first time that Finns attended an IAMCR conference, and they quickly became active members—after Bourquin's invitation to the research office of the Finnish Broadcasting Company (headed by Kaarle Nordenstreng). Many Americans, including Alex Edelstein and Herbert Schiller, as well as Walery Pisarek from Poland and Yassen Zassoursky from the Union of Soviet Socialist Republics (USSR), also joined the association at this time. George Gerbner of the United States was there as well; he had been an active member since before 1966.

Additional sections were established in Herceg Novi, for terminology and methodology, for professional training, as well as for marketing and advertising. The latter section was led by an American, Leo Bogart, who proposed, in a letter to the general assembly, a merger with the World Association for Public Opinion Research (WAPOR). The assembly invited Bogart to seek close cooperation—and eventually a merger—between his section and WAPOR. Participants envisioned a similar merger between the history section and the International Council of Historical Sciences. There seems to have been little follow-up on these initiatives.

In 1968 IAMCR held its general assembly in Pamplona, Spain—under Franco's regime. Paradoxically, this venue replaced Oxford in the United Kingdom, where visas could not be guaranteed for representatives from the East European "Communist countries." Thus, in this turbulent year 1968, there were red flags at the University of Navarra marking the presence of delegates such as Emil Dusiska from East Germany's main School of Journalism, Karl Marx University in Leipzig. Politically, this was an indication of "liberal" tendencies in Spanish society in the sensitive field of mass media, and served as encouragement for the radical elements among the students and faculty.[9]

In Pamplona, Zassoursky was elected vice president—the first Russian in the leadership, although his predecessor as dean of the faculty of journalism at Moscow State University, Evgeniy Khudyakov, had been involved in the preparatory process. Irena Tetelowska of Poland became head of a new section on bibliography, and this director of the Press Research Centre in Cracow was the first woman to hold a leading position in the association. Another section was established for research on mass media and international understanding, after a big international symposium on this topic held in Ljubljana jointly with the Yugoslav IAMCR members on September 3–6, 1968—to celebrate the twentieth anniversary of the Universal Declaration of Human Rights. Ironically, this was just after the Prague Spring was crushed.

The 1970 general assembly was held in Konstanz, West Germany—after Moscow had been considered as a venue in connection with a world conference of historians. Here IAMCR adopted an extensive set of recommendations to the UN and its member states on various aspects of mass communication, notably freedom of information, the cultural integrity of nations, and the use of satellites. Those active in drafting the recommendations included Bogdan Osolnik of Yugoslavia and Dallas Smythe of Canada. A separate resolution was adopted in support of journalists on dangerous missions and another in support of the efforts to set up a UN University, with the expressed hope that "the science of mass communication" would be accorded the recognition appropriate to its importance. This conference, incidentally, was the first attended by James Halloran, and he was immediately elected vice president.

At this time—from the late 1960s to the early 1970s—significant developments took place in the association. First, mass communication research grew everywhere along with the expansion of the media themselves, especially television. New university programs were established and national committees were appointed to highlight the field, leading to new institutions like Halloran's Centre for Mass Communication Research at the University of Leicester. Second, the scientific and political orientation of communication research was diversified by the entry of critical, "anti-positivist" scholars—among them many who were active in IAMCR, notably Smythe from Canada and Schiller from the United States—and younger scholars, such as Robin Cheesman, Nicholas Garnham, and Armand Mattelart from Western Europe. Third, UNESCO reserved a more prominent role for itself after 1968—not by chance but as a consequence of the significance of developments on the national level.

The UNESCO initiative is a story in itself—its explicit policy orientation bringing it close to the critical generation of scholars, and its global resonance connecting it to the Non-Aligned Movement of the developing world (with Yugoslavia as one of its leading members). The General Conference of UNESCO had adopted, in November 1968, a new strategy for promoting communication research and policy, and authorized the director-general,

> in cooperation with appropriate international and national organizations, governmental and non-governmental, to undertake a long-term programme of research on technological progress in means of communication and to promote study on the role and effects of mass communication in modern society.[10]

One of the first activities taken up by UNESCO's Department of Mass Communication, with Pierre Navaux as its director, was to commission from

Halloran (as director of the Leicester Centre) a working paper on mass media and society and to convene a meeting of experts on the topic in Montreal in June 1969. Halloran was consulted about whom to invite and many on his list were active in IAMCR (including Bourquin, Edelstein, Maletzke, Nordenstreng, and Smythe).

Halloran's working paper for Montreal and the final report of the meeting were issued as a publication by UNESCO (in English, French, and Spanish).[11] Taken together, this event and the publication stand as a milestone in the history of mass communication research. They were followed by UNESCO's international panel of consultants on communication research, which included Halloran, Nordenstreng, Pisarek, Smythe, and others from the association. The panel was first convened in April 1971 to prepare *Proposals for an International Programme of Communication Research*—another landmark document (known by its code COM/MD/20).[12]

Growth, 1972–1990

Buenos Aires, in 1972, was more than just the first conference that IAMCR held in Latin America. It also signaled a new era of cooperation with UNESCO—at a time when mass communication research was going through what might be called a "social turn," the stage when a young field becomes conscious of itself and actively involved in social policy. UNESCO's role in Buenos Aires was crucial, as it paid the travel costs of the twelve members of its panel on communication research (which held its second meeting there on the eve of the conference). However, many others found the means to travel to Argentina at the time, including Dusiska and Schiller.

According to Bourquin's last presidential letter, the Buenos Aires conference, under the theme "Communication and Development," was attended by some fifty IAMCR members in addition to "an important South American and Argentinean participation." Elections resulted in Halloran as president and Dusiska as secretary general. Nordenstreng and Schiller were elected vice presidents, in addition to four others who had held office in the earlier years. A sign of the times was the establishment of a new section for research on media and developing countries. Alfred Opubor of Nigeria was elected head of this section, while Annette Suffert of France was appointed head of another new section on television studies.

From 1973, we can follow IAMCR developments in Halloran's presidential letters, issued from the Leicester Centre, where the secretariat was effectively moved from Bourquin's office in Lausanne. The association's bank account remained in Lausanne, however, for a few more years, so that membership fees were paid in Swiss francs. Halloran's mimeographed letters followed the same familiar tone established by Bourquin, and they became longer and longer, reporting in detail not only the organization's events and plans but also whom he had met and who had contacted him as the association president. This networking established "Jim Halloran" as a man known by hundreds of colleagues around the world and made Leicester a focal point in the field, with Peggy Gray as the president's right hand in administrative matters.

Secretary General Dusiska, at his Leipzig office, remained somewhat in the background but cooperated effectively with Halloran. In Leipzig, at the end of May 1973, Dusiska hosted the first meeting of the executive committee during Halloran's tenure. At the meeting, a considerable debate ensued, involving the association's past, present, and future. On specific matters, "it was decided to give priority to the question of publications and investigate the possibility of launching a Journal for the Association and/or establishing a co-operative working relationship with new or existing Journals."[13] The executive committee welcomed the offer to organize the next biennial conference in Leipzig, employing the large institutional resources that Secretary General Dusiska had at the Karl Marx University with the backing of the East German authorities. With Leipzig, the association's tradition of holding successive conferences in the Eastern, Southern, and Western hemispheres was established.

The Leipzig conference, held in September 1974, addressed the general theme of "Mass Communication and Social Consciousness in a Changing World," with four subthemes cutting across this grander perspective: economics, participation, socialization, and developing nations. More than sixty papers were presented, and all the papers together with the keynote addresses were printed by the Leipzig host in a multilingual book of two volumes. The conference had a record attendance of 250 delegates from thirty-one countries. They included, again, the UNESCO panel members who were scheduled to meet before the conference, which "meant that several prominent communication researchers—from places as far afield as Colombia and Singapore, Canada and Lebanon—were able to attend our proceedings because UNESCO met their not inconsiderable traveling expenses," as reported in a presidential letter. The close cooperation with UNESCO was handled by John Willings, the acting chief of the Division of Communication Research and Planning, which had been established within the Sector of Communication (under Pierre Navaux).

President Halloran praised this conference in his "Dear Friends and Colleagues" letter of December 1974:

> To me, one of the most encouraging features was to be found in the number of new faces at Leipzig. For an Association like ours it is good to have the continued support and loyalty of old friends, but it is also absolutely essential to attract and keep the interest of new and younger researchers. The future depends on this.

He also noted that the conference

> offered many of the participants their first opportunity for discussing research policies, aims, theories, methods, results, and the application of results with fellow researchers whose basic assumptions, aims, purposes policies, strategies, and social and political environments are quite different from their own.

On the whole, Halloran could be satisfied with the first two years of his presidency. As he put it at the time:

> the signs augured well for our future progress. Membership was increasing, enquiries were coming in from all over the world, an effective co-operative working relationship had been established with UNESCO at several levels, and possibilities for co-operation were being explored with other international bodies such as the International Communication Association, the Association for Education in Journalism, and the International Sociological Association.

In the area of publications, Halloran reported that,

> for the time being, it had been decided not to proceed with the idea of regularly publishing an official journal of the Association. Fortunately, one or two related journals have agreed to carry information about our plans and activities, and these served us well in the past year. We now learn that UNESCO is willing to subsidize the publication of an IAMCR Monograph which, in addition to containing articles and an extended bibliography on a selected theme appropriate to our field of interest, will include two other sections.

The Monograph became a book of 130 pages, published on the eve of the next conference.[14] It contained two substantive articles on media and socialization (by a British and a Soviet author), with an extensive international bibliography on the topic, compiled by Pisarek. In addition, the book included Halloran's profile of IAMCR and Nordenstreng's presentation of an emerging global system of documentation and information centers for mass communication research.

The general assembly in Leipzig left the officers elected in Buenos Aires in place, but voted to appoint a committee to prepare a proposal to overhaul the association's statutes for the next conference, in two years' time. The conference, it was decided, would take place in Leicester, with a title lifted from UNESCO's International Program: "Mass Media and Man's View of Society." In the general assembly debate, Gerbner pointed out "the sexist interpretation of 'Man's View,'" and this reminder led to the inclusion of Elisabeth Noelle-Neumann as additional member of the conference planning committee, comprised of Halloran, Dusiska, Nordenstreng, and Schiller. This was the first time that gender issues were recorded in IAMCR proceedings.

The Leicester conference was held in late August and early September 1976. In the words of its report prepared by a team of four rapporteurs (Alice Bunzlova of Czechoslovakia, Michael Gurevitch of the United States, Hans Kepplinger of the Federal Republic of Germany [FRG], and Robin McCron of the United Kingdom), the event

> was attended by more than three hundred people from forty different countries. This made it the best ever attended meeting in the history of the Association … The conference was supported by a grant from UNESCO which inter alia made it possible for young scholars and members from the third world to be better represented than had been the case in the past.[15]

The program was divided into four main themes: (1) the state-of-the-art in communication research, addressed by Lothar Bisky (German Democratic Republic), George Gerbner (United States), and Peter Golding (United Kingdom); (2) structures and contexts of media production, addressed by Stuart Hall (Jamaica/ United Kingdom), Michael Tracey (United Kingdom), and John Pollock (United States); (3) media influence, addressed by Neville Jayaweera (Sri Lanka/World Association for Christian Communication), N. Mansurov (USSR), and Elisabeth Noelle-Neumann (FRG); and (4) media and international understanding, addressed by Luis Beltran (Colombia/Canada), Phil Harris (United Kingdom), Al Hester (United States), and Frank Ugboajah (Nigeria). Discussants included Jay Blumler, Theodore Glasser, Jan Ekecrantz, Cees Hamelink, Elihu Katz, Ramona Rush, and Percy Tannenbaum.

In addition to the plenary sessions, workshops were convened around specific topics and projects, including "Cultural Indicators," "Women and Mass Media," "Media/Cultural Imperialism," "Ethical Problems in Mass Communication", and "Communication and Community" (an eight-nation project contracted by UNESCO to IAMCR). These as well as the section meetings facilitated the presentation of dozens of individual papers. Most of the proceedings raised highly topical issues, making the Leicester conference a timely platform for intellectual exchanges. Indeed, the conference offered a panorama of mass communication research that no historian of the field should neglect. The conference was also remarkable because of the adoption of resolutions on the need for international communication policies in the service of democratic development, and for the support for the universal right to communicate.

Leicester 1976 was a landmark in the organizational history of IAMCR, since the statutes were revised, following recommendations by the committee appointed in Leipzig, and elections held according to the new structure, whereby the executive board was to be supervised by a large International Council that included both ordinary members and the section heads. The composition of the executive board remained more or less the same, while the International Council was internationally worldwide in its membership. The slate for elections was adopted by the general assembly, as proposed by Halloran. His proposal was prepared behind the scenes during the conference itself, by Nordenstreng, who tried to achieve a balanced representation not only in terms of geopolitics but also of scholarly generations as well as gender. Accordingly, Nordenstreng brought to the list seven women, including Nelly de Camargo of Brazil, Anita Werner of Norway, and Gertrude Robinson of Canada. Robinson's entry pushed out Dallas Smythe, who had been a bureau member since 1970.

After Leicester, IAMCR published another book with UNESCO's support.[16] Halloran and his assistants reviewed the main themes of the Leicester conference, and relevant bibliographies were compiled by Pisarek in Cracow and documentary workers in other regional centers of communication research.

The next conference was held in Warsaw, in September 1978, with the general theme "Mass Media and Culture." Its attendance once more surpassed the preceding conferences: nearly 500 participants from thirty-eight countries. The association's membership had already grown to nearly a thousand, from more than fifty countries. While the membership kept growing, special interest groups became more organized. One of these was a Marxist or "materialist theory" approach to communication research, which was first convened as an informal group at the Leicester conference and was approved in the Warsaw general assembly, after heated debate, as a section called "Political Economy." The section on professional education convened a special session jointly with the American AEJ and the IOJ, with Nordenstreng as its president since the Leicester conference.

As before, this conference also led to a UNESCO-supported book.[17] It covered the four thematic aspects of the conference: ideologies, theories, and methodologies of mass media and culture; structure, content, and influence of national cultures; political, economic, and technological factors of cross-cultural and international communication; and content, values, and effects of cross-cultural and international communication. The four reviews were written by two Polish and two British scholars, followed by a comprehensive bibliography compiled with the assistance of the UNESCO-related International Network of Information and Communication Centres on Communication Research (COMNET).

IAMCR returned to Latin America in August 1980, eight years after Buenos Aires, to Caracas, Venezuela, for a conference titled "New Structures of International Communication." This was another successful event, although not very many participants from other continents could afford the trip. Among those present, in addition to most of the officers, were Nelly de Camargo of Brazil and Cees Hamelink of the Netherlands—both of whom were elected vice president in Caracas. A highlight of the conference was an unscheduled debate between Ithiel de Sola Pool and Herbert Schiller on the topic of media technology and ideology. An offshoot of the Caracas conference was a critical examination of the draft report of the International Commission for the Study on Communication Problems. The draft, known as the MacBride Report for the commission's chair, Sean MacBride, had just been issued and closely read by several IAMCR activists, and led to a collection of essays.[18]

By this time, the UNESCO panel of consultants on communication research had ended its term and could no longer meet parallel to IAMCR, thus ending UNESCO's indirect subsidy. UNESCO's support to thematic publications was also discontinued, a development parallel to its declining support to COMNET. This was due to changing priorities in UNESCO's communication program that, in the late 1970s, was increasingly concerned with the MacBride Commission.[19] Several IAMCR members, including President Halloran and Vice President Zassoursky, contributed to the commission's work through the association's secretariat and series of background papers, but this work bypassed IAMCR as an institution.

Nevertheless, UNESCO did contract with IAMCR to carry out a major study on foreign news.[20] Also, the section on professional education (headed by Zassoursky and later by Nordenstreng) mobilized, together with the AEJ, IOJ, WACC, and the regional sister associations in Africa, Asia, and Latin America, a project for the promotion of textbooks in journalism education in the developing world. This project received a major grant from UNESCO's new International Programme for the Development of Communication (IPDC)—thanks to the efficient lobbying of Alfred Opubor, who represented Nigeria on the IPDC board. Later, in the 1990s, the project continued with support from the Finnish International Development Agency (FINNIDA).

Caracas was followed by conferences in Paris (1982), Prague (1984), Delhi (1986), and Barcelona (1988). In Barcelona, the association's statutes were modified again, to restrict the presidency to a single term: two years as president-elect, followed by four years as president, and then two more years as "past president." Halloran had been reelected three times since Buenos Aires in 1972, making him the longest-serving president of the association, with eighteen years of service (with the final two alongside President-Elect Cees Hamelink). The Iranian-American Hamid Mowlana became vice president, together with K. E. Eapen of India. The revised statutes no longer called for a long list of vice presidents (twelve had been elected in Prague); the cap was now placed at five, to be followed by a limit of two after the Barcelona conference. Halloran's term ended formally in Bled in 1990, after which he became "honorary president," along with his predecessors Bourquin, Nixon, and Terrou before him—Bourquin actively continuing this role throughout Halloran's presidency.

The conferences and other activities of IAMCR in the 1980s continued to be broadly based and successful, with Barcelona in 1988 as the highlight—attended by more than 600 participants from forty-six countries. The eight sections and thirty-two ad hoc working groups organized more than seventy panels in all, with more than 250 papers presented. Adding to this record attendance, Halloran could proudly announce the latest membership figures: 1850 members from sixty countries.

Yet the rapid growth and dynamism of the previous decade, partly stimulated by UNESCO's financial assistance, was no longer present. IAMCR continued its established forms of formal activity, without introducing a newsletter, or its own journal. After Gerbner became editor of the *Journal of Communication*, published by the Annenberg School for Communication in Philadelphia, he suggested that it become an IAMCR journal, but the proposal was rejected by the International Council, mainly on financial grounds but also because of hesitancy about being tied to only one journal—and an American one at that. Meanwhile, other associations in different regions mobilized researchers closer to home, including the African Council for Communication Education (ACCE), the Asian Media Information and Communication Centre (AMIC), and the Asociación Latinoamericana de Investigadores de la Comunicación (ALAIC).

By 1987 there was already a commonly held sentiment within the leading IAMCR bodies that the association was approaching a state of stagnation and that it was time for a change of generation—as well as a change in the presidency. This message was first presented to Halloran in a letter by the president of the Finnish Association of Mass Communication Research during a meeting of the executive board in Tampere in August, which suggested two candidates for a replacement: Hamelink and Mowlana.

The elections in Barcelona 1988 were historic in the sense that there was an open election for the International Council, with each position having a male as well as a female candidate—a process proposed by Gerbner. As Robinson reported in the next IAMCR conference in Barcelona in 2002, it was after 1988 that

> female members began to penetrate what until then had been the top management 'glass ceiling' in our organization, thanks in part to a more egalitarian attitude on the part of our male colleagues and pressure from the newly formed Women's Network.[21]

But gender did not just surface in IAMCR management; it also attracted scholars doing research on media and gender, leading to a section headed by Madeleine Kleberg of Sweden. A new section was also established in media education, headed by Birgitte Tufte of Denmark.

IAMCR was a close witness to the "collapse of Communism" in Eastern Europe from 1989 to 1991, first in August 1989 in Budapest, where Tamás Szecskö hosted a meeting of the International Council during the days when the first East Germans escaped to the West via their embassy in Budapest—a prelude to the fall of the Berlin Wall in November 1989. The next IAMCR conference was held in Bled, Slovenia, which in August 1990 was in a state of violent conflict, leading to its secession from Yugoslavia. Yet the resort town at Lake Bled hosted a peaceful and professionally efficient meeting with the theme "Developments in Communication and Democracy," paving the way to the new millennium.

CHALLENGES AND LESSONS

Here I conclude with reflections on the major challenges faced by the IAMCR since 1990 and the lessons to be learned from its five-decade history. The list of conferences after 1990 is quite long, since "off-year" meetings were held at sites offering to host IAMCR between the regular biennial events.[22] Starting with Istanbul in 1991, these conferences were intended to be smaller and to combine a meeting of the International Council with selected plenary sessions and a number of section meetings. In practice, they were often quite large and rich in their scholarly content—for example, Dublin in 1993 with Anthony Giddens as the keynote speaker, and Oaxaca in 1997 with a celebration of IAMCR's fortieth anniversary—so that there was little difference between these meetings and the regular biennial conferences. There was, clearly, a demand for international platforms to be catered and more than enough enthusiastic hosts. On the other hand, the rising costs of international travel and improved Internet facilities for maintaining virtual contacts depressed to some extent the spontaneous interest in using IAMCR conferences for networking.

These developments created a challenge that led to a variety of proposals: to orient the association toward virtual networking based on special interests mobilized by the sections; to focus on regional meetings in the off-years; and to convene the main conferences less frequently than every other year—perhaps every fourth or fifth year as many other scientific world congresses have done. However, no consensus emerged, and the association continued more or less as before. For rank-and-file members, the core activities were organized by the sections, which grew in number despite attempts by a Section Review Committee to establish a logic that would avoid proliferation. Working groups were introduced as a subcategory of sections, gradually leading to a total of thirty sections and working groups.[23]

There continues to be a fundamental challenge, one that has followed IAMCR throughout its history, created by the tension between special disciplinary approaches in the field (history, law, etc.) and a more general interest in interdisciplinary areas of research (development, new technology, etc.). This poses a dilemma that cannot easily be resolved through organizational arrangements. This was recognized by Halloran and his predecessors and, in consequence, they were quite open to different initiatives and concerned with achieving a truly international representation. As Terrou used to say, no scientific progress was possible without extensive international collaboration.

Maintaining a worldwide association—first in terms of its East-West balance and later its North-South balance—has been a challenge throughout the history of IAMCR. There have also been periods of friction between different regional interests within the Western world—Spanish-speaking versus Anglophone regions, Europe versus North America—but these conflicts never overtook a common interest in a global platform. In fact, it can be argued that geopolitics has not been an obstacle so much as a positive factor that has made IAMCR both internationally representative and intellectually stimulating. If there have been obstacles throughout our history—as in all human organizations—they are to be found in personalities and their "chemistry" rather than in scholarly traditions as such.

The changing nature of mass communication itself in the era of new media and digitalization has also presented many challenges. A manifestation of this was the decision to change the Association's name: "Mass" was replaced by "Media" at the general assembly in Sydney in 1996. The proposal was made by Wolfgang Kleinwächter, then head of the legal section, and it was approved without discussion. The name change was smooth because there was no need to revise the English acronym IAMCR. Moreover, the other language versions of the name remained unchanged, as "mass" was absent from both the French Association Internationale des Études et Recherches sur l'Information et la Communication (AIERI) and the Spanish Asociación Internacional de Estudios en Comunicación Social (AIECS).

The change was a natural step reflecting a general trend since the 1990s to do away with "mass" as the distinctive feature of the field and instead elevate "media" as its central designator. Accordingly, many academic programs and institutions adopted "media and communication" in their names. On the other hand, "mass communication" has not become totally anachronistic either, retaining its status as a valid label for the field in individual institutions[24] as well as associations.[25] One should recall that five decades ago the concept of mass communication was quite modern, and in the 1940s it was even written into UNESCO's constitution, according to which the organization was charged to

[c]ollaborate in the work of advancing the mutual knowledge and understanding of peoples, through all means of mass communication and to that end recommend such

international agreements as may be necessary to promote the free flow of ideas by word and image (Article 2a).[26]

The association has continuously faced new challenges, both institutionally and substantively. The institutional challenge was met by expanded relationships with the UN system beyond UNESCO, to include the Economics and Social Council (ECOSOC) and the World Intellectual Property Organization (WIPO), as well as NGOs, including the IFJ, which after the cold war embraced most of the former IOJ. Perhaps the most significant challenge was presented by the particularly topical debates in the field of human rights in platforms such as the Organization of Security and Co-operation in Europe (OSCE, Helsinki 1992), the World Conference of Human Rights (Vienna 1993), and the World Summit on the Information Society (WSIS, Geneva 2003 and Tunis 2005). A focus on communication rights meant a return to IAMCR's roots, when its founders had participated in the drafting of Article 19 of the Universal Declaration of Human Rights. IAMCR members offered a critical-analytical approach, seeking to achieve a balance between those with proprietary interests in employing the concept of human rights as an instrument to justify globalization for commercial purposes, and those with democratic interests, championing fundamental freedoms in pursuit of enhancing civil society and its media in the post–cold war world.

These challenges were created in part by UNESCO's move away from the so-called New World Information and Communication Order (NWICO), its abandonment of the MacBride Commission's approach, and its acceptance of a neoliberal solution to communication problems.[27] Personality changes in UNESCO's communication sector led to less emphasis on the scientific tradition represented by IAMCR's activities. The most obvious change occurred in the early 1990s when Alain Modoux (former head of UNESCO's press relations) became the director of the Communication Sector, succeeding Alan Hancock, the media research and policy expert. Until Hancock, the sector had always been quite sympathetic to communication research, and regarded IAMCR as its prime representative. With Modoux, IAMCR lost its special status at UNESCO, which had begun looking for partners among media proprietors and public relations circles. This was also reflected in the process of reorganizing the NGOs associated with UNESCO, whereby IAMCR is nowadays found among a dozen media-related NGOs, replacing its earlier status as one of three media NGOs and the only one representing research.[28]

As part of this reorientation, UNESCO helped to create a new body called ORBICOM, the International Network of UNESCO Chairs in Communication. It was established in 1994 around the so-called UNESCO Chairs in Communication, which were endowed partly with UNESCO funding in several universities, particularly in developing and former socialist countries. However, in practice, most of its membership was made up of academic and industrial representatives from the West. Formally, this new body was not directed against IAMCR, but in practice it contributed to a confusing and divisive state in the international field of communication research.

The global landscape of communication associations was further confused by the strengthening of regional research associations, although most of them have friendly and even formalized relationships with IAMCR. The first of these was AMIC, the Asian Media Information and Communication Centre, established in 1971, and the most recent is ECREA, the European Communication Research and Education Association, established in 2005. Also, several international organizations of a special thematic nature have entered the field, including the Association for Cultural Studies (ACS) and the Association for Internet Researchers (AoIR).

A particular challenge was posed by ICA, the International Communication Association, which in the 1990s began a process aimed at internationalizing its membership beyond its predominantly North American base. At the same time, Klaus Krippendorff of the Annenberg School for Communication at the University of Pennsylvania, with David Mitchell from the University of Calgary in Canada, launched IFCA, the International Federation of Communication Associations, aiming to establish a common platform for different research-oriented associations in the field. IAMCR was lukewarm to this initiative, which was more or less the same as its own original mission, but it did not oppose the idea as such. Other international and national associations were not too enthusiastic, either, to join a common global platform for the sharing of research in the field. After some initial enthusiasm, this voluntary club has remained more or less a paper tiger.

By the turn of the millennium, the international landscape of media and communication research organizations is quite abundant and diverse. What does this suggest about IAMCR?

On the one hand, we may say that the mission of IAMCR has been successfully carried out, as the field has expanded not only in terms of national institutions but also of international networks. On the other hand, we may ask to what extent this historical success story is attributable to IAMCR—has it happened perhaps in spite of, rather than because of, IAMCR influence?

The present author's answer to this question is quite cautious and even cynical: IAMCR cannot be celebrated as a decisive factor in internationalizing the field—the association has followed rather than driven the development. Still, IAMCR has played a vital part in mobilizing the international dimension of the field, especially in the earlier decades. It is unlikely that any other body could have more effectively promoted international networking in a field so deeply rooted in national conditions of politics, economy, and culture. This is a research problem for the sociology of science, but in any case we may conclude that an overall lesson of the IAMCR history is a paradox—an irony—of the fact that its relative importance seems to have decreased the more the field has developed in general and become internationalized in particular.

NOTES

1. The present article is based on the author's joint project with Cees Hamelink on the history of the IAMCR. Its first manifestations are a booklet published for the 50th anniversary conference, *IAMCR in Retrospect* (by Hamelink and Nordenstreng); and an entry, "IAMCR," in the forthcoming *International Encyclopedia of Communication* (by Hamelink). The project will culminate in a book on the association in the context of the history of the field.
2. See, for example, Hardt, *Social Theories of the Press*; and Pietilä, *On the Highway of Mass Communication Studies*.
3. See Kubka and Nordenstreng, *Useful Recollections*.
4. Nordenstreng, "The Forgotten History of Global Communication."
5. Hamelink and Nordenstreng, "A Short History of IAMCR," 6.
6. UNESCO, *Current Mass Communication Research – I*.
7. Santoro, *La liberté de l'information*.
8. UNESCO, "International Association Established."
9. Barrera, "The IAMCR International Conference."
10. UNESCO, *Mass Media in Society*, 3.
11. UNESCO, *Mass Media in Society*.
12. For a more detailed history of the panel and its proposals, see Nordenstreng, "The UNESCO Expert Panel."
13. Hamelink and Nordenstreng, "A Shorth History of IAMCR," 16.
14. Halloran, *Mass Media and Socialization*.
15. Hamelink and Nordenstreng, "A Shorth History of IAMCR," 17.
16. IAMCR, *Mass Media and Man's View of Society*.
17. IAMCR, *Mass Media and National Cultures*.
18. Hamelink, *Communication in the Eighties*.
19. MacBride and others, *Many Voices, One World*.
20. Sreberny-Mohammadi and others, *Foreign News in the Media*.
21. Robinson, "IAMCR Then and Now."
22. For a full list of the conferences, see Hamelink and Nordenstreng, "A Short History of IAMCR," 24–26.
23. These are presented in the IAMCR Web site: http://www.iamcr.org.
24. See, for example, that of the present author: http://www.uta.fi/jour/index1.html.
25. See, for example, that of the U.S. journalism-based academics http://www.aejmc.org.
26. See http://www.unesco.org.
27. For a review of these changes, see, for example, Mansell and Nordenstreng, "Great Media and Communication Debates."
28. Nordenstreng, "IAMCR as NGO."

WORKS CITED

Barrera, Carlos. "The IAMCR International Conference at Navarra (Spain) in April 1968: A Little History." Paper presented at the IAMCR conference, Paris, July 23–25, 2007.

Halloran, James D., ed. *Mass Media and Socialization: International Bibliography and Different Perspectives.* Leeds: IAMCR, 1976.

Hamelink, Cees, ed. *Communication in the Eighties: A Reader on the "MacBride Report".* Rome: IDOC, 1980. Reprinted in *Mass Communication Yearbook*, edited by D. Charles Whitney, Ellen Wartella, and Sven Windahl, vol. 3, 236–87. Beverly Hills, CA: Sage, 1982.

———. "International Association for Media and Communication Research (IAMCR)." In *International Encyclopedia of Communication*, edited by Wolfgang Donsbach. Malden, MA: Blackwell Publishing, forthcoming.

Hamelink, Cees, and Kaarle Nordenstreng. "A Short History of IAMCR." In *IAMCR in Retrospect 1957–2007, L'AIERI en retrospective 1957–2007*, 5–26. Paris: Maison des Sciences de l'Homme, Paris Nord Université Paris 13, 2007.

Hardt, Hanno. *Social Theories of the Press: Constituents of Communication Research, 1840s to 1920s.* 2nd ed. Lanham, MD: Rowman & Littlefield, 2001.

IAMCR. *Mass Media and Man's View of Society: A Conference Report and International Bibliography.* Leicester, UK: IAMCR, 1978.

IAMCR. *Mass Media and National Cultures: A Conference Report and International Bibliography.* Leicester, UK: IAMCR, 1980.

Kubka, Jiri, and Kaarle Nordenstreng. *Useful Recollections: Excursion into the History of International Movement of Journalists I.* Prague: International Organization of Journalists, 1986.

MacBride, Sean, Elie Abel, Hubert Beuve-Méry, and others. *Many Voices, One World: Towards a New More Just and More Efficient World Information and Communication Order.* Paris: UNESCO, 1980.

Mansell, Robin, and Kaarle Nordenstreng. "Great Media and Communication Debates: WSIS and the MacBride Report." *Information Technologies and International Development* 3, no. 4 (2006): 15–26.

Nordenstreng, Kaarle. "The UNESCO Expert Panel with the Benefit of Hindsight." In *Mass Communication Research: On Problems and Policies*, edited by Cees Hamelink and Olga Linne, 3–19. Norwood, NJ: Ablex, 1994.

———. "The Forgotten History of Global Communication Negotiations at the League of Nations." In *The World Summit on the Information Society: Moving from the Past into the Future*, edited by Daniel Stauffacher and Wolfgang Kleinwächter, 119–124. New York: United Nations (Information and Communication Technologies Task Force, Series 8), 2005.

———. "IAMCR as NGO." *IAMCR Newsletter* 17, no. 1 (2007): 2.

———. "Myths About Press Freedom." *Brazilian Journalism Research* 3, no. 1 (2007): 15–30.

Pietilä, Veikko. *On the Highway of Mass Communication Studies.* Cresskill, NJ: Hampton Press, 2005.

Robinson, Gertrude. "IAMCR Then and Now: Lessons from Gender Research." Paper presented at the IAMCR conference, Barcelona, July 21–25, 2002.

Santoro, Jean-Louis. *La liberté de l'Information: Logiques institutionelles et logiques professionelles au plan international (1947–1972).* Bordeaux: Universite Michel de Motaigne, Bordeaux III, 1991.

Sreberny-Mohammadi, Annabelle, Kaarle Nordenstreng, Robert Stevenson, and Frank Ugboajah, eds. *Foreign News in the Media: International Reporting in 32 Countries.* Reports and Papers on Mass Communication, No 93. Paris: UNESCO, 1985.

UNESCO, *Current Mass Communicatiopon Research – I.* Reports and Papers on Mass Communication, No 21. Paris: UNESCO, 1956.

———. "International Association for Mass Communication Research Established." Paris: UNESCO Press release (no. 1727), December 23, 1957.

———. *Mass Media in Society: The Need of Research.* Reports and Papers on Mass Communication, No 59. Paris: UNESCO, 1970.

III. PEOPLE AND PLACES
IN THE HISTORY OF THE FIELD

CHAPTER TEN

The Two-Step Flow vs. *The Lonely Crowd*: Conformity AND THE Media IN THE 1950s

DAVID W. PARK

Similar to many scholarly fields, mass communication research frequently traces its own history through familiar milestones[1] that construct important moments of stability or change in the history of the field. As milestones go, Elihu Katz and Paul Lazarsfeld's 1955 book *Personal Influence* easily qualifies as a formative moment in the study of mass communication. Through the work of critics as well as proponents, who together collude in locating *Personal Influence* as a crucial moment of crystallization, the book has been granted a unique place within the grand narrative of mass communication research in the United States. We understand *Personal Influence* in terms of how it diverged from what preceded it and how it laid a new foundation.

With our own (functional, convenient) disciplinary blinders on, we locate *Personal Influence* in the context of other major works of mass communication research. The "milestone" approach to *Personal Influence* leads us to conceptualize the book only within the context of mass communication research, severing its connections to the rest of the social sciences and to intellectual thought more broadly. Its position within what we have defined as our field has occluded its rich, complex, and controversial place in the intellectual milieu of its time. In this chapter, I hope to reconsider *Personal Influence* by taking it out of the ready-made context of our field. When *Personal Influence* is placed in the context of contemporaneous ideas from outside the world of mass communication research, we can gain a deeper perspective on what is actually put forward in the book and understand more about how our field relates to other traditions, ideas, and disciplines.

The model of social behavior that is advanced in *Personal Influence* lines up in some curious ways with David Riesman's ideas in his 1950 book *The Lonely Crowd* (written with Nathan Glazer and Reuel Denney), and in Riesman's other writing. The kind of perspective-by-incongruity offered in this comparison shows us how *Personal Influence* involves a very particular vision of social relations and social networks, one that stands in marked contrast to other sociological and psychological understandings in the 1950s. In short, *Personal Influence*'s approaches to conformity and to social relations represent a striking inversion of some of the key ideas in the social sciences of the time. To understand *Personal Influence* merely in the presentist, Whiggish, intradisciplinary terms that we have in the past is to miss out on some of what makes the book—and perhaps, by extension, our field—so fascinating.

I assert that the ideas in *Personal Influence* can be placed in the context of David Riesman's ideas concerning conformity, as detailed in *The Lonely Crowd*, 1952's *Faces in the Crowd* (written with Nathan Glazer), and 1954's *Individualism Reconsidered*. I argue that *Personal Influence* is to a great degree a book about conformity, and that the treatment of conformity that Katz and Lazarsfeld use differs dramatically from Riesman's evocation of a society where an ever-present urge to conform to intragroup pressures threatens to stamp out individuality. *Personal Influence* calls our attention to how ingroup conformity insulates us from the persuasive effects of the media. The tonal contrast between these two classic works of 1950s social science is striking. At the same time that Riesman was warning us of the dangers of conformity, Katz and Lazarsfeld assured us that conformity was a crucial bulwark against feared direct effects of the media.

CONFORMITY IN *PERSONAL INFLUENCE*

Though certainly not a preordained milestone, *Personal Influence* has for years functioned as the centerpiece in the story that media studies scholars tell themselves about the field. The most important moment in placing *Personal Influence* at the center of media studies comes with Todd Gitlin's oft-cited "Media Sociology: The Dominant Paradigm." In this essay, Gitlin critiques *Personal Influence* by way of holding it accountable for the sins of a field that, he fears, has lost touch with any sense of social or political purpose. As Simonson puts it, Gitlin's essay "had the effect of raising the status of *Personal Influence* even as it was attempting to destroy it."[2] The fiftieth anniversary of the publication of *Personal Influence* has provided contemporary scholars with the opportunity to revisit the book's unique story and influence on the field. A recent special issue of the *Annals of the American Academy of Political and Social Science* (culled from papers presented at a conference at Columbia University in 2005) was dedicated to such a reappraisal. At this point, it is difficult to find a work in communication studies that has been more rigorously fixed through secondary scholarship as a definitive component of the makeup of the field.

The idea of the two-step flow of mass communication has become such a convenient shorthand for *Personal Influence* that many of the intellectual currents at work in the book have been retrospectively (if accidentally) downplayed, and subsequently neglected. The idea of the two-step flow is, of course, that "ideas, often, seem to flow *from* radio and print *to* opinion leaders and *from them* to the less active sections of the population."[3] It would certainly be possible to interpret this as a sobering reassertion of the powerful effects of the mass media, with the opinion leaders acting as little more than midwives in an all-but-direct (two is just one more than one, after all) process of domination via mass communication.[4] Jefferson Pooley calls our attention to how *Personal Influence* tells a much more optimistic story. Drawing on ideas that find their roots in the writings of Edward Shils, Katz and Lazarsfeld emphasized the social network as a source of resistance to the modern media. As Pooley describes, the

> measured finding of *limited* effect…has its roots in an image of society that recognizes the endurance of primary ties and a rich associational life. The wrongness of the interwar media analysts, in other words, rested on their warped view of modern life: "Their image, first of all, was of an atomistic mass of millions of readers, listeners and movie-goers prepared to receive the Message."[5]

Thus plotted, *Personal Influence* became a story about the strength of community ties.

When contrasted with theories that outlined a more powerful role for the media, the two-step flow offered an attractive narrative for Katz and Lazarsfeld. The limited effects story was embraced in part "because of the scholarly support it lent to the public intellectual defense of American popular culture, in the context of an evolving cold war liberalism."[6] In addition, of course, the new field of communication "retained the story line as a usable, and teachable, past."[7] Framed as a story of "*Gemeinschaft* after all," *Personal Influence* struck many as very good news about the strength of primary ties. Lazarsfeld and Robert K. Merton had previously taken note of the "curiously girlish attitude toward anyone who might be doing propaganda. 'Don't let that man come near. Don't let him tempt me, because if he does, I'm sure to fall'."[8] *Personal Influence*'s conclusions about mass communication reassure us that we are insulated from propagandists' nefarious designs, making this "girlish attitude" all the more laughable.

Though it is rarely remembered as such, *Personal Influence* is, to a great extent, a book about conformity. Conformity is treated in the book as a positive thing, linked to an inoculation against media manipulation. Katz and Lazarsfeld address conformity through reference to social networks. Attached very closely to what Katz and Lazarsfeld describe as the "'rediscovery' of the primary group"[9] in the social sciences is a concern for how communication works within social networks. They describe "two characteristics of small, intimate groups—(1) person-to-person sharing of opinions and attitudes (which we shall often refer to as 'group norms') and (2) person-to-person *communications networks*."[10] It takes little time for them to move to a discussion of "the benefits of conformity"[11] in which they review in quick succession much of the most important social psychological research on conformity. They emphasize the potential for groups to provide individuals with "frames of reference." This is itself placed within a frame of reference provided by social psychological studies of conformity by Theodore Newcomb, Edward Shils, Leon Festinger, Harry Stack Sullivan, Muzafer Sherif, and even Franklin Giddings.[12] Particular attention is drawn to Lewin and Grabbe's contention that "what exists as 'reality' for the individual is to a high degree determined by what is socially accepted as reality."[13]

At the end of Katz and Lazarsfeld's review of the idea of conformity is a section titled "interdependent individuals demand conformity of each other," in which Katz and Lazarsfeld summarize the "instrumental value—the 'benefits'—of conformity."[14] This takes the form of a list of "reasons why group members demand conformity of each other." In this list, they note that,

> First of all, individuals do not like to find their associates departing from a traditional way of "seeing" something...Secondly, groups like to preserve their identities, and one of the chief ways a group can make its boundary lines clear is by the requirement of uniform behavior on the parts of its members...Third, and most important perhaps, is the fact that groups, like individuals, have goals; and group goals often cannot be achieved without consensus.[15]

It is telling that Katz and Lazarsfeld finish this chapter (alliteratively titled "Norms and Networks in the Process of Persuasion") with "a word about 'conformity'," wherein they note that their "discussion of the consequences of deviation and the 'benefits' of conformity, etc., is on the level of agreement or divergence of opinion among intimately interacting individuals."[16] Katz and Lazarsfeld were, of course, perfectly aware of the novelty of asserting any kind of proconformity message. They assiduously used quotation marks around the term "benefits" when describing the "'benefits' of conformity," and they frequently offered asides and sidebars that called attention to their own special version of conformity. "Obviously," they assure the reader, "this is quite a different level from the one on which current political discussions concerning conformity and orthodoxy in American thought are taking place."[17]

Conformity, as it is dealt with in *Personal Influence*, is closely linked to interpersonal communication, which Katz and Lazarsfeld address under the heading of "gregariousness." As others have pointed out,[18] *Personal Influence* is a book that locates its subjects as being in near-constant conversation with each other. Gregariousness, "a measure of the extent of a woman's social contacts,"[19] is operationalized as a measure of size of social network ("How many people are there with whom you are friendly and talk with fairly often who are not and never have been your neighbor?") and of membership in organizations ("What organizations, clubs or discussion groups do you belong to?").[20] Thus operationalized, gregariousness in turn becomes a crucial measure across the four areas of research in *Personal Influence*: marketing leadership, fashion leadership, public affairs leadership, and movie leadership. Gregariousness becomes the factor that sets the two-step flow into motion. Social contact—frequent, multifarious, often friendly—is situated via gregariousness as the linchpin of the two-step flow. In this context, Katz and Lazarsfeld held up conformity as a helpful mechanism for smoothing over the day-to-day frictions, allowing small groups to arrive at functional consensus, and, crucially, protecting individuals from the influence of mass media.

OPINION LEADERS IN A LONELY CROWD

Much of *Personal Influence* rests on Katz and Lazarsfeld's ideas concerning conformity. As has already been established, they made clear that their version of conformity was something special, set apart. It would be hard to find a more obvious exemplar of Katz and Lazarsfeld's "current political discussions concerning conformity and orthodoxy in American thought" than David Riesman's *The Lonely Crowd*. Much like *Personal Influence*, *The Lonely Crowd* has a long history. Originally published in 1950, by Yale University Press, the book was released as a paperback in 1953. The mid-1950s found Riesman dealing with the notoriety granted him by what had become one of the most talked-about books of the 1950s and the best-selling sociology book ever.[21] When *Personal Influence* was published in 1955, it entered a field of discourse that Riesman had played a large role in establishing Riesman's influence on *Personal Influence* was not merely indirect; he was one of the prepublication reviewers of the manuscript for *Personal Influence*,[22] and he is cited on three occasions.

Still, when we look at Riesman's relatively downcast picture of conformity, we find a reverse of the picture of conformity found in *Personal Influence*. Understanding Riesman's ideas of conformity requires a brief review of the threefold characterological typology he devised in *The Lonely Crowd*. This typology was comprised of: the tradition-directed type, the inner-directed type, and the other-directed type. Tradition-directed types learn to adapt, not to innovate. Tradition tells them exactly how they relate to society. Riesman compares their situation to that of molecules that have been cooled into slowness, in a sense frozen by their social framework.[23] Inner-directed types are generated by societies on the move, in transition. They are given greater ranges of choice, and are, in a sense, fit with inner gyroscopes that make them stable even in times of extreme change. For these types of individuals, "the source of direction for the individual is 'inner' in the sense that it is implanted early in life by the elders and directed toward generalized but nonetheless inescapably destined goals."[24]

The last of the three types that Riesman discusses—the other-directed—represented what he took to be the central form of character in American life during the mid-twentieth century. As he put it, "[w]hat is common to all other-directed is that their contemporaries are the source of direction for the individual—either those known to him or those with whom he is indirectly acquainted, through friends and through the mass media."[25] The other-directed type develops

> the elaborate equipment needed to attend to such messages and occasionally to participate in their circulation. As against guilt-and-shame controls, though of course these survive, one prime psychological lever of the other-directed person is a diffuse anxiety. This control equipment, instead of being like a gyroscope, is like a radar.[26]

Riesman explained that the "middle-class urban American of today, the 'other-directed,' is…in a characterological sense more the product of his peers—that is, in sociological terms, his 'peer-groups,' the other kids at school or in the block."[27] The other-directed individual has goals that shift with the peer groups in which he or she functions, and maintains these goals "through an exceptional sensitivity to the actions and wishes of others."[28] This characterological trait is instilled very early in life, as the "other-directed child is able to move among new associates with an almost automatic adjustment to the subtlest insignia of status."[29]

In terms of substance, the portrait Riesman paints of social commerce in the United States is quite similar to that offered by Katz and Lazarsfeld in *Personal Influence*. In both books, we find a vision of individuals engaged in social networks, talking to each other, sharing opinions, disagreeing occasionally, but almost constantly monitoring themselves and those around them, seeking agreement within their interpersonal networks.

Riesman's evaluation of conformity is the diametric opposite of Katz and Lazarsfeld's. Without oversimplifying Riesman's ideas, there is much that sets Riesman's other-direction in opposition to the vision of conformity that Katz and Lazarsfeld present. Riesman wrote for years about other-direction in American character. It was his most frequently returned-to and widely discussed idea. It is easy to read a narrative of decline/critique into this idea, whereby the once-great United States of America, founded and built on inner-direction, is now in danger of slipping because of the conformity of its citizens. In *Individualism Reconsidered*, Riesman arrived at the

> tentative conclusion that the feeling of helplessness of modern man results from both the vastly enhanced power of the social group and the incorporation of its authority into his very character...the point is that the individual is psychologically dependent on others for clues to the meaning of life. He thus fails to resist authority or fears to exercise freedom of choice even where he might safely do so.[30]

Riesman noted that "[i]n adult life," the other-directed person "continues to respond to [his or her] peers, not only with overt conformity, as do people in all times and places, but also in a deeper sense, in the very quality of his feeling."[31] Conformity, here, is something quite different from the relatively optimistic assessment of conformity found in *Personal Influence*. This is a critical approach to conformity.

The sense that other-direction is a bad thing is largely supported by a close read of *The Lonely Crowd*. As Riesman put it, in the context of modern production, "the product [now] in demand is neither a staple nor machine; it is a personality."[32] In the other-directed world, Riesman asserts, the individual succumbs to within-social-network conformity, to the extent that

> [a]pproval itself, irrespective of content, becomes almost the only unequivocal good ... : one makes good when one is approved of. Thus all power, not merely some power, is in the hands of the actual or imaginary approving group, and the child learns from his parents' reactions to him that nothing in his character, no possession he owns, no inheritance of name or talent, no work he has done is valued for itself but only for its effect on others.[33]

Industry becomes largely retooled around these goals, as "personnel directors today are weeding out of commerce and industry the lone wolf who is not cooperative, no matter what his gifts."[34] Office place personalization is "false, even where it is not intentionally exploitative, because of its compulsory character: like the antagonistic cooperation of which it forms a part, it is a manipulative and self-manipulative mandate for those in the white-collar ranks and above."[35]

Riesman brings out the tragic elements in the world of the other-directed, especially at the end of *The Lonely Crowd*, which he concludes on a sour note. He writes that

> of one thing I am quite sure: the enormous potentialities for diversity in nature's bounty and men's capacity to differentiate their experience can become valued by the individual himself, so that he will not be tempted and coerced into adjustment or, failing adjustment, into anomie. The idea that men are created free and equal is both true and misleading: men are created different; they lose their social freedom and their individual autonomy in seeking to become like each other.[36]

Clearly, Riesman's vision of conformity shared much with that of Katz and Lazarsfeld. From both points of view, we see the same "rediscovery" of the primary group, and a similar emphasis on the importance of conformity within these groups. Just as clearly, Riesman's approach to conformity is suggestive of conformity's negative potential. At the end of *The Lonely Crowd*, the reader was left with an unsettling picture of how conformity threatened to undo individual liberty and spontaneity. Though it came at us with a smile on its face, this type of conformity threatened to undo us.

Riesman occasionally rejected the idea that other-direction was an entirely bad thing. He struck a moderate pose, asking his readers to "skeptically question the demands for greater social participation and belongingness among the group-minded while, on another front, opposing the claims of those who for outworn reasons cling to individualism as a (largely economic) shibboleth."[37] Much later, Riesman would say that "the most important misinterpretation [of *The Lonely Crowd*] was the assumption that other-direction was wholly negative."[38] These reservations notwithstanding, Riesman's ideas were certainly received as a critical dispatch on the state of American society, in part owing to the environment of ideas concerning conformity at the time, and also to Riesman's not-so-ambiguous statements concerning the position of the other-directed individuals he saw around him.

Another major illuminating similarity between Riesman's ideas in *The Lonely Crowd* and the analysis of the media presented in *Personal Influence* concerns the emphasis that both books place on the habits of consumption. The notion of consumption has been a major part of the secondary literature on *Personal Influence*, as those who reappraise the book have often found it significant that it is focused so much on the decision-making apparatus attached to questions of consumption. Frequently, these retrospectives point to the funding of the Decatur project, by Macfadden Publications, which was interested in marketing their *True Story* magazine.[39] Todd Gitlin was particularly thorough in his treatment of the marketing focus in *Personal Influence*. Gitlin demonstrated that marketing goals occupied a central place in the assumptions that informed the book's research, making clear that, in the four "issue-areas," three were tied directly to consumption decisions, and the final area (politics) was treated largely as a form of consumer behavior. From this, Gitlin concludes that "the blithe assumption of the commensurability of buying and politics, never explicitly justified, never opened up to question, hung over the entire argument of *Personal Influence* like an ideological smog."[40]

Riesman has been credited as one of the definitive theorists of consumption.[41] Certainly, this reputation was solidified by the time Riesman wrote his appreciative intellectual biography, 1953's *Thorstein Veblen: A Critical Interpretation*. Riesman's writings on consumption, consonant with the general gist of *Personal Influence*, portray a picture of Americans as gregarious traders of opinion and information whose interactions (within social networks) quite explicitly place politics within the realm of consumption. Riesman describes how children become "gifted as consumers long before they have a decisive say themselves."[42] For Riesman, this tendency to treat everything with the radar of the other-directed character, always on the lookout for more options, informs how friendships are established. Friendships become "the greatest of all consumables; the peer-group is itself a main object of consumption, its own main competition in taste."[43] He traces the concern for identity through consumption to music,[44] food,[45] literature,[46] and much else. Riesman's updating of Veblen—and his critical understanding of this conformity—is quite clear here, as when Riesman points out that the Veblenian model is "obsolete today," because "we are all of us—that is, almost all—members of the leisure class, and face its problems."[47] For Riesman, the notion of other-direction shifts the entire notion of consumption. The conspicuous consumer "is seeking to fit into a role demanded of him by his station, or hoped-for station," while the "other-directed consumer seeks experiences rather than things and yearns to be guided by others rather than to dazzle them with display."[48] The reader would be right to detect a note of gloom here, as

> the demand on the other-directeds that every sphere of life may be attended by the responsibility for "making a group mood" (by manipulating one's own mood and often by seeing to it that everyone gets equally tight) constantly interferes with spontaneous sociability, not to mention privacy, in play.[49]

Once again, the gregariousness that Katz and Lazarsfeld describe as a saving grace of modern society rears a more vicious head in Riesman's writing.

A similarly critical stance is apparent when Riesman applies this gaze to politics. Riesman was very much concerned with (as one section in *The Lonely Crowd* put it) "politics as an object of consumption."[50] What Katz and Lazarsfeld treated implicitly (that politics was a kind of consumption) Riesman stated baldly. And, in keeping with other divergences in tone, Riesman's understanding of this link between consumption, conformity, and politics is particularly sobering. Among other-directeds, political opinions "resemble ... the peer-group exchange of consumption preferences, though unlike the latter, the preferences are seldom taken into the political market and translated into purchases of political commodities."[51] Other-directeds'

> loyalty is at large, ready to be captured by any movement that can undercut their frequent cynicism or exploit it. In all these ways they place hardly any barriers, even those of their own tastes and feelings, between themselves and the politically organized community. The only barrier is their apathy.[52]

Riesman observed that, "when manifested in a political style," other-direction "intensifies the tendencies in political life that encourage still more other-direction."[53]

The workings of political conformity—and of political consumerism—are perhaps nowhere more obvious that in Riesman's use of the term "inside-dopester," a neologism he created to get at the intersection of politics, consumption, and conformity. The inside-dopester, explained Riesman,

> tends to know a great deal about what other people are doing and thinking in the important or "great-issue" spheres of life; he is politically cosmopolitan rather than parochial. If he cannot change the others who dominate his political attention, his characterological drive leads him to manipulate himself in order not to change the others but to resemble them. He will go to great lengths to keep from looking and feeling like the uninformed outsider.[54]

That this characterological quirk could spell trouble in the political arena is not lost on Riesman. He notes that the nineteenth century had jobs to be done, and "character types to do them," but that "today this unity has broken down." He continues: "The tasks still get done—astonishingly well, all things considered—but they get done primarily by the older, inner-directed type, while the newer type watches with anxiety, or tacit agreement or, sometimes, amusement." In the end, the people's "feeling of control and competence seems to diminish; people no longer feel confident that they can affect their destiny, in anything that matters, by political action, individual or collective."[55] This rather sobering conclusion shows Riesman once again sharing some of the same assumptions (so witheringly critiqued by Gitlin) of Katz and Lazarsfeld, while offering a very different assessment of then-contemporary social processes and their meaning to democracy.

In his informative attempt at an "exogenous"[56] approach to *Personal Influence*, Paddy Scannell lays out the similarities between Katz and Lazarsfeld's book and *The Lonely Crowd*. Scannell asserts that "the implicit characterization of the women in the Decatur study is in perfect accord with Riesman's other-directed typology."[57] This impels Scannell to conclude that

> what Riesman perceives and what Katz and Lazarsfeld so convincingly "prove" is the emergence into history of something that is now recognized and identified as "everyday life." The discovery of "people" in the Decatur study is, in a larger sense, the discovery of everyday life.[58]

This is right on. However, in his careful assessment of the similarities between *The Lonely Crowd* and *Personal Influence*, Scannell overlooks the differences regarding how conformity is framed in the two books. Katz and Lazarsfeld treated conformity with a counterintuitive and upbeat touch, parts of which would later become institutionalized into mainstream communication research, minus the self-awareness. Riesman approached conformity in a manner largely in keeping with a broader conformity trope running through 1950s intellectual culture.

CONFORMITY'S BROADER CONTEXT

Riesman was not the only intellectual of the time to have dealt with conformity in such detail. His ideas shared a great deal with many other scholars and intellectuals of the time. One of the most important and widespread ideas of the time— and one which Riesman himself incorporated to no small degree—was the notion that "conformity" or "adjustment" had become a social problem. Many writers addressed conformity in a manner roughly consonant with Riesman's analysis. Authors such as William H. Whyte, Robert and Helen Lynd, and C. Wright Mills all constructed narratives of middle-class America that emphasized the oft-stifling tendencies of American community. It is also significant that psychologists such as Stanley Milgram, Solomon Asch, and Muzafer Sherif had established the concept of conformity as a significant variable in social psychology.[59] Popular authors such as Sloan Wilson—whose 1955 book *Man in the Grey Flannel Suit* offered a critical jab at corporate conformity—also contributed to this broadly discussed issue of conformity.

We find an even more proximal (to Riesman) sense of the mid-1950s critique of conformity by looking at the popular ideas advanced by Erich Fromm and Robert Lindner. Fromm is particularly important here. Fromm's most popular book in the United States—*The Sane Society*—was published in 1955, the same year as *Personal Influence*. Fromm was also David Riesman's friend and therapist, and as such, he had an influence on Riesman's ideas. In a footnote, Riesman himself noted that his concept of the other-directed personality was "stimulated by, and developed from, Erich Fromm's discussion of the 'marketing orientation' in *Man for Himself*,"[60] and this represents only a portion of Fromm's influence on Riesman.

Fromm's ideas concerning conformity were central to his own considerable oeuvre. He concerned himself with "internal factors blocking the full realization of the freedom of personality."[61] With this normative background, Fromm engaged in a critique of how conformity prevented individuals from reaching this kind of freedom. "To be different, to find himself in a minority," he opined, "are the dangers which threaten [an individual's] sense of security; hence a craving for conformity produces in turn a continuously operating, though hidden, sense of insecurity."[62] In time, the

> conformity pattern develops a new morality, a new kind of super-ego. But the new morality is not the conscience of the humanistic tradition nor is the new super-ego made in the image of an authoritarian father. Virtue is to be adjusted and to be like the rest. Vice, to be different.[63]

Meanwhile, the tendency to treat one's own social network and individual identity as something to be traded—a tendency that was particularly important in Riesman's work—"is rooted in the experience of oneself as a commodity and of one's value as exchange value."[64] This perspective on conformity, with its emphasis on the constantly concerned individual who will do almost anything to avoid sticking out, and who is particularly conscious about the importance of one's social network—made no small impact on Riesman, and a major component of the conformity trope of the mid-1950s. As Neil McLaughlin has pointed out, "[w]hen Riesman met Fromm, American social criticism and pragmatism met the grand tradition of European critical theory."[65] Suspicion of conformity was a central theme in this transcontinental meeting.

Robert Lindner was another influential writer in the mid-1950s, and like Riesman, Lindner's take on conformity was shaped to a great degree by Erich Fromm. Like Fromm, Lindner was a lay analyst, and his ideas were exemplars of mid-1950s concerns about conformity. Lindner attempted in his writings to build a "case against adjustment,"[66] using the word "adjustment," as did Fromm, as a synonym for the conformity in modern man. He lamented that "[i]n every area of our life today individuals are forced to conform. They are commanded to distort their personalities, sacrifice personal freedoms and fit themselves into a prevailing pattern."[67] Lindner felt strongly that this pressure to conform is particularly strong within specific social networks, as individuals whose instincts to rebel were thwarted, and turned into what he called "negative rebellion,"[68] in which the desire for individual expression is channeled into further conformity. This conformity found its truest expression in hyper-conformist subgroups, be they delinquent gangs, fascists, or communists. Lindner's ideas, printed in popular periodicals and books, received considerable discussion and attention in the mid-1950s.

The examples of Lindner and Fromm show us that even if Riesman had not intended *The Lonely Crowd* and its sequels to serve as harbingers of a conformity-initiated doom, his ideas were still contextualized largely within this broader notion of a powerful American impulse to conform. Fromm's pessimism about conformity no doubt influenced Riesman; he was one of the authors Riesman was most likely to cite, and his influence can be found quite easily in Riesman's concerns about individuals who have become "psychologically dependent on others." In light of the quite widespread notions of conformity in the 1950s, it is likely that this was how the audience of the 1950s read *The Lonely Crowd*. This makes *Personal Influence*'s upbeat version of conformity all the more unusual for the time. To the extent that the field of mass communication picked up where *Personal Influence* led it, the field was led away from the mainstream understanding of conformity. Mainstream communication research has been based in part on a most unorthodox approach.

CONCLUSION

What do we get from this juxtaposition of *Personal Influence* with other takes on conformity in the 1950s? A first point worth bringing out here is the easy-to-forget role of conformity in *Personal Influence*. Katz and Lazarsfeld placed conformity front and center in the theoretical discussion in the book. This notion of conformity was then made dynamic through the operation of the variable "gregariousness." Because we often think of *Personal Influence* as a book that cleared the way for minimal effects, emphasized what we have come to know as the "active audience," and reassured us of the viability of *Gemeinschaft*-like functioning even in the world of electronic media, we often overlook the role played by conformity in the book. The conclusions that Katz and Lazarsfeld offer, with their trademark refusal to engage in jeremiad, absolutely depend on this notion of the benefits of conformity. They tell us that it is our own need for conformity in our social networks that keeps us from being affected directly by the media. This might strike anyone as at least unsettling, were they not operating from the frame that Katz and Lazarsfeld provide, wherein direct effects are linked to dysfunction and to totalitarianism/fascism, while indirect effects are linked to community and democracy.

A second point here is that juxtapositions like this offer us a way of understanding the place that the field of communication occupies in the system of the social sciences. The study of mass communication, though still young (even as the term "mass communication" begins to appear utterly pre-Cambrian), has claimed its own particular intellectual tradition. John Durham Peters compared the field's understanding of itself to the ideology of a nation-state, emphasizing the degree to which communication has played up the idea that it has some unique claim—what Magali Sarfatti Larson calls a "monopoly of competence"[69]—to the study of mass communication.[70] This nationalistic bent in the field's understanding of itself has resulted in a tendency to treat as irrelevant the ideas that come from "outside" this very particularized tradition. So, we wind up with a milestone-oriented tunnel vision of *Personal Influence*, which places the book on a Whiggish, temporal continuum (often between Hovland's "Why We Fight" studies and Everett Rogers's diffusion of innovation research), instead of considering the broader, synchronic intellectual field. I propose that a fuller understanding of the study of mass communication will require more of this synchronic, inclusive work, as we realize that the study of communication can be enriched by letting down our nationalistic guard and admitting that there are important connections between the ideas "inside" and "outside" the field. The cultural studies projects (whether British- or American-flavored) have outlined strategies for more interdisciplinary studies of communication, but these have to a great extent become as institutionalized as any other subfield in communication. To understand the idea of conformity in *Personal Influence* and *The Lonely Crowd* (and other works of the time) is only a small application of this ideal of moving away from the bounded, intramural sense of communication study.

My third and final point is that this contrast between understandings of intranetwork conformity demonstrates an enduring component of how mass communication research has conceptualized democracy and the public sphere. By emphasizing the benefits of conformity and the part played by gregariousness in mass communication processes, Katz and Lazarsfeld assured us that interpersonal communication—nestled in the rediscovered primary group—defused the threat of direct effects. Michael Schudson[71] has argued that a similar approach to face-to-face communication has profoundly shaped theories in communication. He shows how the ideas of Jürgen Habermas, John Dewey, and James Carey, among others, have involved the idea that democracy is driven largely by face-to-face communication, and that a "romance-of-conversation"[72] has installed itself quite securely in the field. I assert that we find a very similar romance at work in *Personal Influence*, in which the conversational space of interpersonal networks is located as a site of gregarious commonality, guarding individuals from the depredations of mass communication. This approach to interpersonal networks is brought into stark relief when contrasted with Riesman's vision of conformity, where negative effects spring from interpersonal networks. Riesman's problem is Katz and Lazarsfeld's solution. Of course, this does not mean that either set of ideas is wrong. However, it does highlight the important role that *Personal Influence* played in marshaling the forces of conversation on the side of democracy, even during a time when social networks were being conceptualized as something most undemocratic. When Riesman's notion of other-direction is applied to the emerging school of "limited effects" and its intellectual progeny, we can envision a more pessimistic understanding of what limited effects might be. Riesman's sobering ideas about conformity in social networks lead us to face the possibility that audience activity—including what is often called agency—may be phenomena as troubling as any direct effect of the media. Riesman reminds us that threats to individual autonomy and freedom can come not only from the media, but also from our friends and neighbors.

NOTES

1. The classic version of this is to be found in Lowery and DeFleur's *Milestones*.
2. Simonson, introduction, 7.
3. Katz and Lazarsfeld, *Personal Influence*, 32.
4. John Summers has suggested that the initial research leader in the Decatur project—C. Wright Mills—indeed may have reached a very similar conclusion about the Decatur data, perhaps contributing to his fall-out with Lazarsfeld. See Summers, "Perpetual Revelations," 33.
5. Pooley, "Fifteen Pages," 133.
6. Ibid., 134.
7. Ibid.
8. Lazarsfeld and Merton, "Mass Communication," 95.
9. Katz and Lazarsfeld, *Personal Influence*, 34.
10. Ibid., 45.
11. Ibid., 50.
12. Ibid., 48–65.
13. Ibid., 54.
14. Ibid., 62.
15. Ibid.
16. Ibid., 65.
17. Ibid.
18. In particular, see Peters, "Part Played by Gentiles."
19. Katz and Lazarsfeld, *Personal Influence*, 227.
20. Ibid., 227.
21. Gans, "Bestsellers by Sociologists."
22. Scannell, *"Personal Influence,"* 119.
23. Riesman, *The Lonely Crowd*, 10–12.
24. Ibid., 15.
25. Ibid., 22.
26. Ibid., 26.

27. Ibid., v.
28. Ibid., 22.
29. Ibid., 71.
30. Riesman, *Individualism Reconsidered*, 106.
31. Riesman, *The Lonely Crowd*, v.
32. Ibid., 46.
33. Ibid., 49.
34. Ibid., 135.
35. Ibid., 312.
36. Ibid., 373.
37. Riesman, *Individualism Reconsidered*, 32.
38. Riesman, personal communication, March 20, 1999.
39. *Long Road to Decatur.*
40. Gitlin, "Media Sociology," 215.
41. Lunn, "Beyond 'Mass Man.'"
42. Riesman, *The Lonely Crowd*, 74.
43. Ibid., 82.
44. Riesman, *Individualism Reconsidered*, 192.
45. Riesman, *The Lonely Crowd*, 81.
46. Riesman, *Individualism Reconsidered*, 264.
47. Ibid., 203.
48. Riesman, *The Lonely Crowd*, 122.
49. Ibid., 343.
50. Ibid., 212.
51. Ibid., 189.
52. Ibid., 190.
53. Ibid., 177.
54. Ibid., 200.
55. Riesman, *Faces in the Crowd*, 33.
56. Scannell, *"Personal Influence,"* 15.
57. Ibid., 120.
58. Ibid., 121.
59. It is worth pointing out that *Personal Influence* cited both Asch and Sherif in its discussion of conformity.
60. Riesman, *The Lonely Crowd*, 23.
61. Fromm, *Escape From Freedom*, 104.
62. Fromm, "What Is Happiness?" 44.
63. Fromm, *The Sane Society*, 158.
64. Fromm, *Man for Himself*, 67.
65. McLaughlin, "Critical Theory Meets America," 21.
66. Lindner, *Prescription for Rebellion*, 149.
67. Lindner, "Raise Your Child," 102.
68. Lindner, *Must You Conform?* 89.
69. Larson, *The Rise of Professionalism*, 208.
70. Peters, "Institutional Sources."
71. Here it is worth pointing out that Schudson studied with David Riesman at Harvard.
72. Schudson, "Why Conversation Is Not," 307.

WORKS CITED

Fromm, Erich. *Escape From Freedom.* New York: Rinehart, 1941.

———. *Man for Himself.* New York: Rinehart, 1947.

———. *The Sane Society.* New York: Holt, Rinehart, and Winston, 1955.

———. "What Is Happiness?" *Science Digest* 39 (March 1956): 43–47.

Gans, Herbert. "Bestsellers by Sociologists: An Exploratory Study." In *Required Reading: Sociology's Most Influential Books,* edited by Dan Clawson, 19–27. Amherst: University of Massachusetts Press, 1998.

Gitlin, Todd. "Media Sociology: The Dominant Paradigm." *Theory and Society* 6, no. 1 (1978): 205–53.

Katz, Elihu, and Paul F. Lazarsfeld. *Personal Influence: The Part Played by People in the Flow of Mass Communication.* Glencoe, IL: Free Press, 1955.

Larson, Magali. *The Rise of Professionalism: A Sociological Analysis.* Berkeley: University of California Press, 1977.

Lazarsfeld, Paul F., and Robert K. Merton. "Mass Communication, Popular Taste, and Organized Social Action." In *The Communication of Ideas,* edited by Lyman Bryson, 95–118. New York: Harper, 1948.

Lindner, Robert. *Must You Conform?* New York: Rinehart, 1956.

———. *Prescription for Rebellion.* London: Victor Gallancz, 1953.

———. "Raise Your Child to Be a Rebel." *McCall's* (February 1956): 31, 100, 102, 104.

The Long Road to Decatur: A History of Personal Influence. DVD. Directed by Glenda Balas. Albuquerque: University of New Mexico, 2007.

Lowery, Shearon, and Melvin DeFleur. *Milestones in Mass Communication Research: Media Effects.* New York: Longman, 1983.

Lunn, Eugene. "Beyond 'Mass Man': *The Lonely Crowd,* the *Uses of Literacy,* and the Postwar Era." *Theory and Society* 19, no. 1 (February 1990): 63–86.

McLaughlin, Neil. "Critical Theory Meets America: Riesman, Fromm, and *The Lonely Crowd.*" *American Sociologist* 32, no. 1 (Spring 2001): 5–26.

Peters, John Durham. "Institutional Sources of Intellectual Poverty in Communication Research." *Communication Research* 13, no. 4 (October 1986): 527–59.

———. "The Part Played by Gentiles in the Flow of Mass Communications: On the Ethnic Utopia of *Personal Influence.*" In "Politics, Social Networks, and the History of Mass Communications Research: Rereading *Personal Influence,*" edited by Peter Simonson, special issue. *Annals of the American Academy of Political and Social Science* 608 (2006): 97–114.

Pooley, Jefferson. "Fifteen Pages That Shook the Field: *Personal Influence,* Edward Shils, and the Remembered History of Mass Communication Research." In "Politics, Social Networks, and the History of Mass Communications Research: Rereading *Personal Influence,*" edited by Peter Simonson, special issue. *Annals of the American Academy of Political and Social Science* 608 (2006): 130–56.

Riesman, David. *Individualism Reconsidered, and Other Essays.* Glencoe, IL: Free Press, 1954.

Riesman, David, with Nathan Glazer. *Faces in the Crowd: Individual Studies in Character and Politics.* New Haven, CT: Yale University Press, 1952.

Riesman, David, with Nathan Glazer and Reuel Denney. *The Lonely Crowd.* New Haven, CT: Yale University Press, 1950.

Scannell, Paddy. "*Personal Influence* and the End of the Masses." In "Politics, Social Networks, and the History of Mass Communications Research: Rereading *Personal Influence,*" edited by Peter

Simonson, special issue. *Annals of the American Academy of Political and Social Science* 68 (2006): 115–29.

Schudson, Michael. "Why Conversation is Not the Soul of Democracy." *Critical Studies in Mass Communication* 14, no. 4 (December 1997): 297–309.

Simonson, Peter. "Introduction." In "Politics, Social Networks, and the History of Mass Communications Research: Rereading *Personal Influence*," edited by Peter Simonson, special issue. *Annals of the American Academy of Political and Social Science* 608 (November 2006): 6–24.

Summers, John H. "Perpetual Revelations: C. Wright Mills and Paul Lazarsfeld." In "Politics, Social Networks, and the History of Mass Communications Research: Rereading *Personal Influence*," edited by Peter Simonson, special issue. *Annals of the American Academy of Political and Social Science* 608 (November 2006): 6–24.

CHAPTER ELEVEN

Making Sense OF Social Change: Studying Media AND Culture IN 1960s Britain

WENDY WORRALL REDAL

We have a thousand problems to resolve in this country, but the essential one is this: What kind of Britain do we want? What ideal are we going to set ourselves in the re-ordering of society? What truths do *we* hold to be self-evident?

—LINDSAY ANDERSON, FILMMAKER, 1957[1]

Culture is ordinary: that is where we must start . . . The making of a society is the finding of common meanings and directions, and its growth is an active debate and amendment under the pressures of experience, contact, and discovery, writing themselves into the land.

—RAYMOND WILLIAMS, "CULTURE IS ORDINARY," 1958[2]

In the two decades following World War II, Britain experienced a period of palpable social change that forced an engagement on the part of its socialist intellectuals with profound political questions. Confronting an unexpected world of consumer capitalism, a relatively affluent working class inside a new welfare state, and the pervasiveness of the mass media, which destabilized class lines further, they questioned socialism's future as such shifts undermined traditional Marxist assumptions about the working class as the catalyst for revolution, alongside capitalism's inevitable demise. Labour's postwar triumph, for all it initially accomplished, had stagnated, mired in bureaucracy and a retrenched conservatism in the face of an increasingly chilly cold war, opening the door to a Tory return under Churchill in

1951. Outside the superpower blocs the colonial world was fragmenting, spawning a diaspora that would reconfigure the face of Britain's populace. In the meantime, the nuclear threat heightened. Palliated by the availability of a wide array of consumer goods, much of the working class—and the British public generally—seemed largely content to avoid politics entirely.

Undoubtedly many people's lives were "better," especially in a material sense, with the gains made by social democracy in the postwar years. Yet a disillusioned sector of the Left knew that what had been attained was not a communitarian ideal in which a broad, humane ethics prevailed. It was merely a victory of sorts for a degree of redistribution within capitalism, beset with compromises and contradictions. The dream of a complete transformation in social values remained a dream deferred.

Such upheavals in the social terrain demanded analysis by Left thinkers who sought to come to grips with such unanticipated changes. They realized that any future for socialism would have to recognize the political dimensions of the cultural sphere, the realm of beliefs, values, and deeply ingrained practices that comprised the context in which political identity was formed. Understanding the role that culture played in politics, how, at its heart, it *was* politics, was crucial. Such an expanded notion demanded a connection with the lived experience of people's everyday lives, a site in which political struggles were played out as fervently as they were in party campaigns and general elections, if not as consciously.

A young Stuart Hall, recently graduated from Oxford[3] and teaching in secondary education in London when he took the helm of the *New Left Review*[4] (*NLR*), wrote in its inaugural editorial in 1960:

> We are convinced that politics, too narrowly conceived, has been a main cause of the decline of socialism in this country and one of the reasons for the disaffection from socialist ideas of young people in particular. The humanist strengths of socialism which are the foundations for a genuinely popular socialist movement—must be developed in cultural and social terms as well as in economic and political . . . The purpose of discussing the cinema or teen-age culture in *NLR* is not to show that, in some modish way, we are keeping up with the times. These are directly relevant to the imaginative resistances of people who live within capitalism . . . The task of socialism is to meet people where they *are*, where they are touched, bitten, moved, frustrated, nauseated—to develop discontent and, at the same time, to give the socialist movement some *direct* sense of the times and ways in which we live.[5]

Where the people often were, of course, was watching television, going to the cinema, listening to popular music, reading mass-market paperbacks and the tabloid press, and consuming an ever greater sweep of affordable products pushed by advertisers in a commercialized media sphere. How might their imaginations be captivated through popular art and culture in a mode that would engage them and inspire them to want more for their lives than simply the current top-hit record or the latest fashion from Marks & Spencer? What were the forces of paralysis at work that inhibited people from realizing their capacity to effect the changes they desperately wanted and needed?

Such issues animated an emergent New Left who were tackling analyses of capitalism, class, and cultural institutions, including education, mass communication, and the arts, conversing in coffee houses and salons and clubs and the pages of their movement's journals. Integral to their project was an engagement with the channels and products of the media, which were commanding concern more broadly across British society. By the mid-1960s, numerous official forays into media studies would be under way, launched not only by Left intellectuals associated with an approach that came to be known as cultural studies, but mainstream educators and government initiatives as well.

This chapter revisits Britain in the early 1960s, focusing on specific ventures that laid the groundwork for a flourishing and diverse body of media research by the 1970s. Its intention is not only to recover empirical detail about the historical circumstances within which formal media research endeavors arose—detail often lost or ignored in typically cursory accounts of this period—to better illuminate the motivations for such pursuits, but also to reveal the diversity of avenues via which investigations into the mass media were undertaken. Oversimplified historical accounts often perpetuate assumptions that "British cultural studies" is synonymous with British media studies, at least in its early incarnation. And while the set of approaches that came to be known as British cultural studies was (and is, however problematic the moniker has become) unquestionably an influential paradigm in thinking about the relationship between media and society, it was certainly not the only set of ideas and methods that shaped fledgling efforts at media research in the later postwar years. This chapter seeks to tell the fuller story, while recognizing that this account, too, is inevitably a limited overview.

If there was a unifying factor in 1960, it was the emerging recognition that the mass media were not only a legitimate focus of academic inquiry but also an essential one. That awareness would translate into a growing array of formal avenues for media studies in Britain in the decade to come, animated by a shared set of questions that invoked a moral purpose for scholarship: how can the role of culture be understood in the quest for a better world? How can it be influenced in the service of such a goal? How can intellectual practice, at its heart, also be political practice?

Such questions were not new in 1960, however. They had been percolating for years in adult education, inside the Workers Education Association and extramural studies programs in which Richard Hoggart, Raymond Williams, and James Halloran, who would become influential figures in British media research, figured prominently. E. P. Thompson was also deeply involved in adult education, and while he was never directly involved in media studies, his enormous presence in the New Left, and his ideas about the relationship between culture and power—specifically the role of human agency and the importance of the imagination in making a better world—would contribute significantly to new ways of thinking about the instruments of mass communication so pervasive in postwar society. Using unconventional methods foreign to the academic establishment, these tutors working outside the educational establishment were relating the study of cultural products to the wider world in which they were produced and encountered, a necessary act of making their courses relevant to the experiences of the adult pupils whom they served. It was a "shift of perspective about the teaching of arts and literature and their relation to history and contemporary society [that] began in Adult Education," according to Williams.[6]

Too often, however, cursory histories of British media studies subsume early media research under the cultural studies umbrella while simultaneously placing cultural studies' genesis in a troika of "founding texts." Traditionally included are Richard Hoggart's *The Uses of Literacy* (1957), Raymond Williams's *Culture and Society* (1958), and E. P. Thompson's *The Making of the English Working Class* (1963), which tend to be lumped together as the initial "culturalist" strand, as opposed to the "structuralist" strand of cultural studies that would emerge in the 1970s under the influence of continental theory, the latter questioning what were by then regarded as too-simplistic notions of individual agency in the creation of cultural forms and practices.[7] While continuities between the three texts are often stressed, few accounts bother to track the substantial divergences between the beliefs and approaches of British media studies pioneers and the respective intellectual circles they inhabited, differences that were in many cases as significant as common threads. Thompson, for instance, would never have aligned himself with "cultural studies," according to his widow, Dorothy Thompson,[8] nor considered himself—a socialist historian—an important figure in the study of mass communication. Although he did contribute groundbreaking intellectual work to the rethinking of "culture" in mid-twentieth-century Britain, he was a vocal critic of Williams and other New Left peers who came to be affiliated with cultural studies, regarding their work as lacking sufficient recognition of historical agency and power relations inhering in actual human beings and their decisions.[9]

The tendency to place origins in seminal texts also hides other developments integral to the institutionalization of media studies in Britain. For example, a host of governmental inquiries in the early 1960s provided an impetus and funding for more methodical analysis of the mass media, largely employing empirical methods from within the social sciences grounded in a behaviorist paradigm. Media research was also occurring within sociology, often concerned with the effects of television and incorporating an institutional focus, breaking ground for approaches that would later coalesce around political economy. However, the emphasis in many histories of British media research is on continuities and breaks within literary studies, under the umbrella of which cultural studies initially emerges. While this chapter attempts to map a more thorough picture of British media research in the 1960s, clearly, "rethinking culture" was central to the overall enterprise. This focus was the motivating force behind the emerging study of the media and the growing institutionalization of such practices within the academic realm.

Although the term "cultural studies" came to be applied by observers to the kind of work Williams was doing at Cambridge on film, for example, after he accepted an internal post there in 1960, its use by Hoggart was deliberately chosen. A year after the enthusiastic reception for *The Uses of Literacy*, Hoggart was offered a post as Senior Lecturer in English at the University of Leicester, where he began teaching in 1959. As an extramural tutor his initial hire was a long shot, yet the enormous success of the book proved decisive. Established now inside a university, Hoggart was offered a professorship at Birmingham just eighteen months later, which he accepted because they agreed to let him set up a postgraduate research center to continue the kind of work he had been doing with *The Uses of Literacy*. In 1964, he established the Centre for Contemporary Cultural Studies (CCCS), thus naming and institutionalizing the practice of cultural studies for the first time in any official sense. Hoggart implies that the response by his department was lukewarm yet officially supportive.[10] Yet Hall remembers him saying that the idea was met with "disbelief and dismay. Having appointed him, they couldn't say he couldn't do it; but they certainly did say they weren't going to give him any money with which to do it."[11] That was enough to persuade Hoggart to take the job. The details would be sorted out later. Here at the larger and more established "redbrick" university, Hoggart believed there was a chance to put his new project on the map. It would, however, remain at the margins of that map for a long while.

Hoggart was compelled to understand the significance and consequences of the changes affecting social and cultural life in Britain, many of which he did not regard positively, particularly the commercialism that had come to pervade cultural production and the channels of public communication. In a 1961 essay on "Changes in Working-Class Life,"[12] he attempted to catalogue the transformations he was observing, and the deeper questions they raised. What kind of relationships with neighbors does one have in an impersonal suburban housing estate, for example? And what is to become of close-knit extended families and the support network they constitute when relatives no longer live around the corner in the next street? What does "family" mean in a society in which new geographic mobility is seen as the ticket to greater social mobility? And how is solidarity among workers being restructured, as old, familiar patterns of working relationships are lost within more complex, elaborate "professionally 'lubricated'" approaches to industrial management? "As the changes increase," Hoggart argued, "the style of life has to alter or absorb."[13]

> It is difficult to live in a society which is changing so quickly, and perhaps especially difficult if much in your way of life was previously decided by the customs of the group, if the range of deliberate, planned, conscious, individual decisions was limited by custom as well as by lack of cash. Here a whole new feature of British life comes in strongly: mass communications.[14]

Mass communications are identified as much more than idle pastimes. They are, rather, significant forces that are "helping people to come to terms with, to live with, their new world and new opportunities."[15] Cut loose from their traditional moorings, people were turning to the sleek appeal of the media to help them define a new style of life, Hoggart contends, as the television replaced the hearth as the locus of family engagement. He is wary of their vulnerability, their susceptibility to sophisticated new modes of communication in the service of commercial, rather than public, interests:

> The new men, the classless men—the advertisers, their artists and copywriters, television personalities, magazine columnists—are creating an image of a new way of life which does not belong to the old class-defined society of the past. *All these are socially educative forces.* Their aims are not usually disinterested and we may suspect the sort of life they promote. But we must recognize that they are social teachers, are replying to an unspoken but powerful need for directions within a more open society.[16]

Yet Hoggart reiterates the point he made in *The Uses of Literacy*, that people can resist change, as well as assimilate, modify, and adapt things toward more traditional ends. Understanding contemporary culture would provide means to promote such resistance.

The shifts Hoggart charts were in the works at the close of the war, yet their scale and scope seemed to be accelerating in the 1960s, spurred on by the pervasive presence of the media. Much of the original impetus for cultural studies was "trying to come to terms with the fluidity and the undermining impact of the mass media," according to Hall, to examine their role in the "manifest break-up of traditional culture, especially traditional class cultures,"[17] which so concerned Hoggart.

Hoggart saw mass communication contributing to a narrowing and stratifying of the social order. In an era of growing concentration in ownership, it is most advantageous to "weld" consumers into large, coherent blocks, he explains, and then to offer up a product designed to appeal to that mass audience. He considers the press, observing that while more Britons are reading more newspapers in the early 1960s, there are fewer different newspapers published than there were twenty years prior. Such a situation is more profitable to the owners of the newspapers that remain, while larger, more concentrated markets are more appealing to advertisers as well. The same homogenizing phenomenon is at work in other media, including popular music and television programming. Embedded in such tendencies, Hoggart suggests, is an implicit disdain for the working class, a perception of 'the masses' as too ignorant or too lacking in sensibility to be able to appreciate truly creative or provocative or challenging cultural fare that attempts to move beyond "least common denominator" palatability.

At the heart of Hoggart's agenda—and that of the wider New Left—was a mission to affirm the subjectivity of ordinary people and to develop their critical faculties to equip them to resist the insidious changes altering social life, especially as perpetuated by the mass media. One hears echoes of the same contentions made earlier by other adult education tutors, and fainter echoes of a long line of English voices insisting upon education as a vital means of liberating people from forces that constrain and dehumanize them. Only in this case, it is not "culture" that will save society, but "culture"—especially popular culture—that must be critiqued: weighed, measured, informed distinctions made. As Leavis's project was devoted to training in discrimination, so, too, were early efforts in media studies. A crucial difference, however, was that by the 1960s, a number of Leavis's heirs were prepared to embrace the popular in order to make distinctions *within* it. This would be the case in the Centre's first years in Birmingham, as well as in other pioneering ventures in media studies taking place before the founding of the center in 1964.

The call for more systematic study of the media, especially television, was growing. By 1960, television had become a fixture in everyday British life, and many people shared a sense that its ability to shape and effect change was profound. The commercially funded Independent Television Network (ITV) arrived in 1955 as a competitor to the BBC, sparking concern over its ability to spread "mass culture" farther and wider than ever before. It was not just a few renegade academics and left-wing intellectuals who were interested in the power wielded by TV and other dominant forms of popular media. Politicians, too, sought to come to terms with the new culture of communications.

With the BBC's charter soon up for renewal and plans for a third channel in the works, Parliament convened a Royal Commission on Broadcasting in 1960 to set the direction for the future. The committee met under the leadership of industrialist Sir Harry Pilkington, and Hoggart was a member. Hoggart has been credited (or disparaged, depending upon one's point of view) with influencing the tone of the final report, which upheld traditional public service principles, seen by some as too elitist, while criticizing the tendency in independent programming especially to pander to "mass taste" through cheap, crass, or simply bland entertainment.

One function of the Pilkington Report, issued in 1962, was "a questioning of the cultural influence of television in general," Robert Hewison has observed. In this regard, the report

> came to two important conclusions. One was that "entertainment" was not the problem, but trivialization, which could happen at any level. The second was that to conceive of mass audiences at all was patronizing and arrogant[18]. . . "Giving the public what it wants" was almost an underestimation of any individual's taste, and a reinforcement of social and cultural stratification.[19]

What the committee did not address was any systematic inquiry into the actual effects of television. Describing it as a "missed opportunity," James Halloran criticized the report for making assumptions about television's effects that were based more on prevailing opinion that any compelling evidence or first-hand research.[20]

While the Pilkington Report set agreed-upon directions for the future of broadcasting, it did not ameliorate governmental concerns over the power of the pervasive new medium. The Home Secretary thus created the Television Research Committee (TRC) in July 1963 to facilitate the study of television's effects, especially on the moral values and attitudes of young people, who were regarded as particularly vulnerable to its coarser appeals and manipulative techniques. The committee recommended that the government, through the Social Sciences Research Council and the University Grants Committee, provide long-term support for mass communications research, focusing on the provision of resources to strengthen the institutional efforts that were beginning in that vein. The emphasis was on using the methods of social science to make an empirical inquiry into the subject.

The main beneficiary was the University of Leicester, where the Centre for Mass Communication Research (CMCR) would be established under the leadership of TRC secretary Halloran in 1966. Halloran, a sociologist who had been teaching on the media in Leicester's adult education department, was the obvious choice to run the new center. Whereas the TRC's goals were to assess the condition of existing mass communication research within the social sciences and to recommend courses for future studies, the Leicester Centre was to carry out a research program investigating both the communication process and media institutions.[21] At roughly the same time, the Television Research Unit had been established at the University of Leeds under the sponsorship of Granada Television and was also undertaking social scientific inquiry into the medium, focusing especially on television's influence on the political process.

Such early efforts, underway almost coterminously with Hoggart's founding of the Centre, were very different in tone and emphasis from what was taking place at Birmingham. Grounded in sociology or psychology, and shaped by the government's concerns with identifying behavioral effects, there was little context for a wider consideration of the cultural milieu within which communication took place or in which media institutions were situated. The language employed by social scientists traditionally did not include discussions of judgment and value, seen as the province of

> the moralizing literati, artists and educators who, judging from their comments, often feel that social scientists are so preoccupied with research techniques and method-ological devices that their works lack immediate social relevance and that they suffer further because they are unrelated to the general intellectual discussion of mass culture on the one hand and its historical development on the other. The social scientists reply to this by questioning the whole nature of the evidence produced by these writers and by criticizing the undisciplined nature of the generalizations, interpretations and speculations which abound in this field.[22]

The summary of the debate is Halloran's, who to his credit argued for a complementary combination of both approaches for the fullest understanding of the media in society, and saw such potential early on for Hoggart's project at Birmingham.[23] He would come to know Hoggart well across the Midlands in the 1960s, with whom he agreed that many of the important issues could not be considered solely inside a social scientific framework. Included were such questions as these:

> What is high culture? Is the popularization of genuine art and intellectual systems possible? Can genuine folk culture exist in mass society? Are divisions by 'height of brow' of any validity? Is a widely diffused common culture possible in our democratic society? Who decides what is good and bad? What standards are to be used?[24]

Such debates are beset by conflict, disagreement, and confusion, Halloran contends, because the terms are so fluid. Without the ability to measure anything systematically, "how can we possibly arrive at any conclusion about the alleged corruptive effects of mass culture, or high culture?" he asks. Yet to criticize such an approach is not to diminish the importance of its role. Cultural critics often "point the way," says Halloran, suggesting lines of investigation and prodding social scientists into activity, while acting as the "social conscience" of society.[25]

In many respects the language of the Pilkington Report played such a role. It was largely in tandem with many of the assumptions of intellectuals who were rethinking culture at the turn of the decade, hanging on to vestiges of "Leavisism" in wanting to promote "the best for the most," yet standing up for popular culture to a degree that hadn't been seen before. What mattered was not so much the question of whether or not a given cultural product was "good" in some abstract and objective sense or in comparison to other cultural products on an assumed hierarchy, but was it "good of its *kind*?" Popular entertainment programming ought not be dismissed out of hand, went the argument, but it should be creative, insightful, provocative, even audacious—that was certainly the view of Sir Hugh Greene, who had been installed as the new director-general of the BBC in 1960. A similar view was held by Sir Allen Lane, who headed Penguin Books, with a mission of making quality work available to the broad public at an affordable price. When *The Uses of Literacy* came out as a Penguin paperback in 1958, for example, it was then that the book gained widespread attention and influence.

Much popular culture, however, did not meet the standards held by Greene or Lane. A growing anxiety existed over the nature and style of the messages being promoted via most avenues of commercial culture. The impact of mass media on adolescents was a widely shared concern, particularly among educators. In 1960 the National Union of Teachers held a conference on "Popular Culture and Personal Responsibility" to air those fears and criticisms. Among the participants were Raymond Williams, recently appointed at Cambridge after a long stint in extramural adult education, and Stuart Hall, who was then teaching in a secondary modern school in London. The papers from the conference were published in 1964 in a rather despairing volume called *Discrimination and Popular Culture*, edited by Leavis's colleague Denys Thompson.[26]

Williams has identified his 1962 book *Communications*[27] as a response to the conference. His view of the media was not as dismissive as many participants'; rather, he argued for the potentiality of communications as a force for positive change. It wasn't the media technologies that were inherently the problem but the uses to which they were put, which were generally determined by the structure of ownership and control. Williams issued a call for a complete end to commercial television in order to utilize fully the medium in the service of public, rather than corporate interests. It was necessary to analyze the media critically, he insisted, including their institutional formation, if society was to employ them toward democratic ends. *Communications* was Williams's initial contribution to that goal, offering a textbook model for further academic inquiry.

While Williams was beginning his university career at Cambridge and Hoggart his own at Leicester, both in traditional English studies positions, Hall was formally teaching media studies at Chelsea College in London, having accepted a position there following his departure from the *NLR* at the close of 1961.

By then the New Left as a movement, and the journal that bore its name, were in decline. Many of the clubs were no longer meeting regularly, and the journal's circulation fell steadily amid dissension within New Left ranks and criticism of Hall's performance.[28] Under such pressure, a disparaged Hall resigned his editorship. Sensing he had blown a remarkable opportunity, Hall retreated in despair, though in choosing to accept an offer to teach film and media to college students in a program of liberal studies, he may not have sensed then the importance of the pioneering role he would play in a new field.

In 1962, Hall joined the staff of Chelsea College of Science and Technology, now part of London University. "They hired people to teach scientists what were called 'complementary studies,'" he explained. "I think I was the first person they'd ever hired to teach film and mass media. Very, very early on."[29] The position fit perfectly with his own long-standing interest in the cinema, and he had also developed a keen interest in television during the period of the New Left.[30] Through the movement, he had developed a close friendship with Paddy Whannel, head of the education department at the British Film Institute[31] and a regular contributor to *Universities and Left Review* (*ULR*) and *New Left Review* (*NLR*). Whannel had been a key contact for Hall in his earliest efforts to teach students about film. When they met, in the late 1950s,

> there was no teaching film anywhere in the country, but the education department used to mount lectures and extramural courses in film that you could take around and discuss, so I used to do that. People who knew me, through the New Left and through that interest in popular culture and television, hired me to teach in that area.[32]

In such a manner Hall was offered the job at Chelsea.

Out of this period, Hall and Whannel produced a book called *The Popular Arts*.[33] Geared toward secondary school teachers, it served as a bridge between Leavisite principles of literary criticism and a more open-minded engagement with contemporary popular culture on its own terms. The book assessed the growing scope of the communications media, noting how much of their output was directed toward young audiences, helping to create what was widely regarded as a distinct "youth culture." Their concerns reflected their proposition that the media

> provide young people with information and ideas about the society into which they are maturing. They can test few of these descriptions and interpretations against their own experience. At the deeper level, the use of the media to provide imaginative experiences through various forms of art and entertainment has a modifying impact upon young people's attitudes and values.[34]

What is crucial about such changes, they argue, is that "they cannot be held apart from education."[35] If part of a teacher's task is to help students understand the world in which they live, then teachers had better come to closer terms with the way in which the media were transforming that world. Hall and Whannel asserted that the media deserved formal study in the schools, calling for scrutiny of their "organization, content and impact."[36]

While they praised the 1960 National Union of Teachers conference for addressing these issues, they found problematic the fact that the resolution passed seemed to place the entire onus on the media providers. In true Leavisite fashion, the resolution had called for "a determined effort. . .to counteract the debasement of standards which results from the misuse of press, radio, cinema and television."[37] No attention was paid to what the "right" uses might have been. Yet all-out protection of young people from the insidious incursions of the media was not the answer, contended Hall and Whannel: teachers had a role to play, to help students learn to discriminate. The role of education ought to be an *active* one. It would not work to cordon young people off from the "menacing forces" of the media, shoring them up with exposure only to high art, literature, music, and drama. What had to be recognized was that good *could* be found among the popular. "'Pop' music," for example,

> is seen here as universally the opposite of—implicitly the enemy of—"good" music. There is no recognition that popular music is of different kinds which vary in achievement and aim, or that each may have its own standards. "Pop" here may cover anything from jazz to dance music: "good" music anything from the light classics to Bach.[38]

What matters in the end are "critical judgments of individual pieces of work."[39] "Surely the value of art and entertainment is that it does affect us?" the authors ask, rhetorically.

> [But] it is the quality of the effect that matters . . . We should be asking of these new means of communications not what harm they might be doing, but how they can be used more creatively. We should be seeking to train a more demanding audience.[40]

One sees how rooted the authors are in Leavis's language, how much in keeping with the tenets of literary criticism is their emphasis upon the audience's *response,* rather than on the production of messages and the conditions under which they are made. Yet the departure from Leavis is apparent as well. What is to be trained is not "taste" but a critical response, such that a student could find much in contemporary jazz praiseworthy while remaining able to critique the vapidness of many pop lyrics. Where *The Popular Arts* is noteworthy within the history of media studies is in its willingness to extend legitimacy to forms of popular expression, regarding them as not only worthy of but in need of criticism.

Hall takes the same approach in an early book on teaching film, published by the British Film Institute in 1964,[41] in which he explains the purpose and practice of teaching film as part of a liberal studies "enrichment" curriculum for science and technology students at Chelsea College, most of whom dismissed the cinema as not worthy of serious study. Students tended to judge films by their subject matter, said Hall, rather than to appreciate "the distinctive character of the cinema as an art form."[42] Occasionally, more sophisticated students would consider only avant-garde films "good," still "holding to a set of high-brow or advanced tastes."[43] To break down that false dichotomy, that good = "high-brow" and bad = "low-brow," or popular, had to be among the film studies teacher's main goals.

Rather, Hall argued, "the great power of the cinema lies in the fact that it has managed to hold a bridge between high art and authentic popular culture."[44] The aim of a course on cinema and popular culture was thus

> to break the false connection between quality and taste, and to develop some critical language by means of which the qualities of cinema as a *popular* art can be discriminated from the great welter of rubbish . . . In each case one has to suggest the ways in which the films establish their quality *within* the accepted popular conventions.[45]

In such a manner, for instance, could one set apart the uniqueness of John Ford's *My Darling Clementine* from a plethora of vulgar and predictable westerns, while neither dismissing nor exonerating the genre as a whole. The value of the popular art of the film, said Hall, "is not to teach us anything new about human experience, but to recreate what we already know—yet with a particular force and intensity." It was Ford's particular uses of certain received conventions of the western that set the film apart, Hall argued, giving it greater "life and quality" than other, less creative approaches.[46] Such distinctions were the sort of judgments Hall sought to foster, encouraging his students to confront popular culture critically, yet on its own terms.

At the same time, Hoggart espoused a similar pedagogical approach in his new post at Birmingham. Eighteen months before the formal establishment of his Centre, Hoggart gave his inaugural lecture as a professor. The address, titled "Schools of English and Contemporary Society,"[47] made the case for widening the boundaries of English as it was taught at universities. Hoggart argued for a stronger connection between scholars of English and the imaginative life of contemporary society, calling for a bold effort to tackle "interconnections between history, politics and the aesthetics of popular taste," an approach that "may be provisionally called Literature and Contemporary Cultural Studies."[48] It comprised three fields of analysis: historical and philosophical, sociological, and literary critical, blending, and going beyond traditional disciplinary predecessors. In this regard, cultural studies would break new ground. Yet one can see from the outset the foundation for cultural studies' heavier emphasis upon texts, rather than the other elements that were part of Hoggart's proposed arena of inquiry.

Whereas Halloran argued for the primacy of the sociological in approaches to the media and culture, Hoggart (not surprisingly, given his training in English) placed the weight on language and its uses. Resident within texts lay meaning and value, and elucidating those was foremost in Hoggart's approach. He proclaimed the need for a more developed critical practice of film, television, and radio; popular fiction, including crime, westerns, romance, science fiction, and detective stories; the press; strip cartoons; the language of advertising and public relations; and popular music and songs. Although he knew that many would dismiss such forms as "beneath serious critical consideration," he insisted that the starting point had to be at the very least "a little more humility about what audiences actually take from unpromising material."[49]

What was essential, Hoggart argued, is recognition of the *meaningfulness* of popular art in people's lives. It is not possible to speak of the effects of such arts "as confidently as some social scientists do unless we have a close sense of their imaginative working." The task of interpreting those meanings was to be the heart of cultural studies. To gain a sense of how cultural products "worked" in the minds of active subjects would be the foremost task of the graduate students at the new research center.

The center's locus within the Arts Faculty at Birmingham suggested the orientations that would emerge in its work, in contrast to the empirical model of social science research that would drive the agenda at Leicester. The divergent emphases and colliding approaches in media studies in Britain are evident from the field's inception; they would come to define the disparate, often competing, approaches to media research by the 1970s as exemplified in the "cultural studies" versus "political economy" split.[50]

With funding from Allen Lane at Penguin Books, which had published *The Uses of Literacy* in paperback, plus small grants from Chatto & Windus, the original publisher of *Uses,* and *The Observer,* Hoggart was able to hire a full-time secretary and a senior research fellow. In 1964, as Hall was writing of his own forays in teaching popular culture, Hoggart invited him to join him at Birmingham as the center's research director, after considering several candidates:

> There were some very good people who came for the job, one of whom was Peter Davis who became a great Shakespeare scholar. David Lodge came to me and said, "May I have the job," because he'd written three novels and they hadn't taken off. But I said, "David, you'd be making quite the wrong move, coming to the center—you want to write novels." And he said, "You're right." And within a few years he had written *Changing Places* which propelled him into the stratosphere. That was the sort of level of people coming, but I knew that if I could, I wanted Stuart. So I got him on the phone and said, "Would you be willing?"[51]

Hoggart was well acquainted with Hall and his work through the New Left. It was Hall who had first introduced Hoggart and Williams in connection with a *ULR* discussion on working-class communities. Hoggart had also read Hall's work in *ULR* and *NLR*, much of which focused on cultural analysis, including an enthusiastic review of *The Uses of Literacy*. Furthermore, Hoggart was impressed with *The Popular Arts*, which had just been published.

While their common interests were central, Hoggart said Hall's "luminous intelligence" made him the preferred candidate. "I knew he could give something to the Centre which I couldn't, because I'm not a theorist."[52] Hall suggests that Hoggart's sense that he hired him to bring the missing theoretical perspective to the center's work may be a view couched in hindsight. "I think that's what he knows now," said Hall, "because that is how I've evolved."[53] Hoggart didn't really recognize that there were theoretical lacunae in his own work at the time, Hall contends:

> We got on very well together, I loved the book [*Uses*]; he knew that. I was interested in some things that were complementary to him, but. . .I think he hired me to set his project going. Because he had to go on being head of the English department, teaching his courses, so he really needed someone to get graduate students and define it. He didn't have any notion of what he was doing. He thought we'd take three or four PhD students and instead of their editing a literary text, they might be writing the history of television, or writing about popular culture. I mean, that's as far, I think, as it went . . . We had the money, so 'go out and do it.'[54]

One of the first tasks was what to call the place. Hall remembers that discussion well:

> We'd thought of calling it 'the Institute,' but then we decided there was nothing institutional about it! It was the least institutional space there ever was. We had two rooms in a quonset hut underneath the Birmingham clock tower, which were destined to be removed, pulled down by some bulldozer any minute, so we didn't seem very institutional. So I said, "How about a Centre?"[55]

With that agreed upon, Hoggart then had to "negotiate his way through the university":

> "Cultural Studies" we could not be called, because after all, there was a Classics department, and they were doing culture; there was an Art department, they were doing culture; Literature, they were doing culture. We had to be the Centre for *Contemporary* Cultural Studies—a place in the ephemera.[56]

The ambiguous name wasn't enough to satisfy entrenched disciplinary interests. Hoggart received a firm letter from the head of the Sociology department warning him not to "pretend to be doing sociology," Hall recalls. The letter reminded Hoggart that there was "nothing systematic" about his method and thus, no claims could be made that the center's work would have any connection to sociology whatsoever. Hall remembers amusedly that there was a response something along the lines of, "Okay, we won't say a word about this society—'we look at texts!' We negotiated our way."[57] So, in an academic "no man's land," marginalized both physically and in the university's disciplinary structure, the work of the CCCS began. It was that very marginality, however, that permitted challenges to existing modes of thinking and doing which would come to define the center's most significant intellectual contributions during its heyday in the 1970s.

Other accounts[58] exist of the center's subsequent growth and development, offering specific discussions of the academic explorations that took place at Birmingham, particularly the center's evolution toward a central concern with embedding cultural forms and interpretive methods in a wider theoretical context, "a serious theorization of the concept of culture" that could not be found anywhere else at that time in the social sciences or the humanities, according to Hall.[59] Before the 1970s, however, most of the center's work remained with the analysis of popular media texts.

It was in the area of media studies that the Centre received a grant for its first funded research project, shortly after Hall assumed the directorship in 1969 following Hoggart's departure for Paris to serve as an assistant director-general for UNESCO. The Rowntree Memorial Trust provided the money for an analysis of the discourse of popular journalism, a comparative study of the *Daily Mirror* and the *Daily Express*. The venture was kind of an "experiment" in the eyes of the foundation, Hall notes, a "rival bid" to the sort of work that Hilda Himmelweit at the London School of Economics had received funding for to examine popular media from a psychological point of view. When Himmelweit's project was announced, according to Hall, he approached Rowntree to fund a study that would look at questions of meaning missing from psychological research. He argued for the importance of such a perspective, and got the support.[60]

While Hall admits that much of the thinking that went into the project initially was "rather primitive," they soon recognized that "you couldn't explain what was in the text until you look at the context." Such a recognition forced an engagement with the historical context, said Hall, pointing up the need "to understand a whole sociological framework," because these were "class voices" that were speaking:

> What is a class voice? You don't recognize a class voice just by reading it. You recognize a class voice because you relate what's in the text to some larger thing. That's what the center was playing around with: how to make this relationship between the

literary analysis of the. . .discourse to the larger sociohistorical context. That's the problematic of the center from the very beginning.[61]

While that early study, published as *Paper Voices*,[62] would now seem "naive," says Hall, it was immensely exciting at the time because of the new ground it broke, the connections it revealed between a cultural text and its wider social conditions of production. It was the *politics* of culture, unveiled. Such relationships were far more complex than many early media researchers were acknowledging at the time, contends Hall. Those recognitions fueled the desire of CCCS researchers to stake out a terrain for the contributions of cultural studies that they saw lacking in other endeavors. One can thus see the tensions early on between the style and substance of the different approaches to researching mass communication underway in Britain at the time. Hall cites the CCCS's interactions with the Leicester Centre as a case in point.

He describes the relationship between the two research centers as a "friendly rivalry," their divergences born of their attachments to somewhat "different problematics and theoretical orientations" within the same field of inquiry. The CMCR was also

> stimulated by certain debates about the impact and effect of the big spread of mass culture and mass media after the War in Britain . . . Halloran was interested in mapping social change through the media just as we were . . . But it took a much more conventional, sociological analysis, political economy of the media, survey of effects kind of approach.[63]

While the CMCR was doing traditional content analysis early on, Hall explains, the CCCS was critiquing it. Given the extensive funding of the CMCR by broadcasting organizations, however, it was logical that they would be interested in "audience studies and effects studies of a traditional kind," observes Hall. Yet there was at Leicester a spectrum of approaches which to some degree overlapped with certain views more commonly voiced at Birmingham. Hall cites Peter Golding and Graham Murdock in particular as closer to a similar "center of gravity," rather than strict empirical researchers, making possible a fruitful debate with them.[64]

The two centers were engaged in a lively relationship in which they exchanged speakers and occasionally attended one another's seminars. When Hoggart left and Birmingham considered closing the CCCS, Hall notes that Halloran "played an extremely important role in writing to the Vice Chancellor and saying, 'You can't possibly close it down' . . . So we worked alongside them, aware of a debate and discussion between us."[65] Hall cites what would become his landmark article on "Encoding/Decoding"[66] as an example of that awareness. The article originated as a contribution to a national conference hosted by the CMCR and was a direct engagement with content analysis, a "specific intervention."[67] His argument held that content analysis could not provide the texture of meaning necessary for a complete understanding of a media text.

The debate, at heart, was with forms of positivist methodology; Hall criticized the way in which content analysis "measures frequency but not necessarily significance." Influenced at this stage by an early introduction to French semiotics, Hall used an example in the piece that argued,

> if you see 85 VWs and one Peugeot, the one Peugeot is important. But not because it's one, because it's different. It's the relations of difference that *mean* something. A thousand dead Indians is not as important as the death of the hero's sidekick. . .We weren't saying, "Don't ask any questions of the audience," but we were saying, "Ask questions of the audience about the meaning of the text."[68]

To define the meaning of a text fully, one could not ignore its latent meaning:

> You can't just say, "You mention the Soviet Union 85 times"; you have to ask them about the Soviet Union! Because it may be more important that they didn't mention China. That absence may be the key thing you want to ask them about. . . Sociological survey methods tell you some things. We weren't even opposed to counting—counting tells you *some things*. But it can't tell you some other things. What we were critical of was thinking that what the methodology told you most easily was what you needed to figure out.[69]

What Hall was objecting to—and what cultural studies as a project had been born objecting to—was any kind of reductionist approach that failed to account for the intricate complexities, significant nuances and multitextured richness of the cultural sphere. It was only through interpretation that culture as a web of relations and determinations could be adequately encountered. That proposition, above all, was the central tenet and most important intellectual contribution made by cultural studies in the initial conceptualizations of the project at Birmingham. Reading the media for meaning, and legitimizing the meaning-making capacities of audiences, were the dominant influences of the cultural studies approach to media research in 1960s Britain and beyond.

The problem with an emphasis on interpretation, however, regardless of whatever corrective it provided, was that by itself it was insufficient to change the balance of power in British life, according to its critics. Although the life of the imagination must be fought for, as Hoggart asserted, that fight did not necessarily take place on the same battlefield on which material struggles ensued. Out of the sociological tradition, with its emphases on institutions and ownership and their power to influence behavior, political-economic approaches took root, with critical theory eventually pushing analyses beyond empirical observation and into the realm of judgment and values, albeit in a vein quite different from cultural studies.

I have argued that new modes of conceptualizing culture and its relation to the social, especially the political, were a crucial impetus for the rise of media studies in Britain. An interest in the instruments and processes of mass communication did not emerge as a subject of inquiry for its own sake, or within an isolated behaviorist effects paradigm, but rather as part of a wider discourse on the distribution of social power. How the mass media could serve, rather than harm, the public was the overarching concern, and early forays into media studies in Britain, even in their fledgling diversity, were arguably united in their focus as an emancipatory project. Only in engaging ideas—in the form of cultural texts—and the conditions in which those texts are produced in the first place, could the possibility exist to tap people's imaginative faculties, to affect the circumstances in which democratic potential could emerge in an era framed by unprecedented change and complexity. It was that awareness, and that faith, that spirited a generation of idealistic British scholars to inaugurate a host of formal investigations into the media as part of that quest.

NOTES

1. Anderson, "Get Out and Push!"
2. Williams, "Culture is Ordinary."
3. Hall came to Oxford from his native Jamaica on a Rhodes scholarship, commencing studies in 1951.
4. The journal was created through an amalgamation of the two primary 1950s New Left publications, *The New Reasoner* and *Universities and Left Review,* when they decided to pool their resources and influence in 1960.
5. Stuart Hall, "Editorial."
6. Williams, "Adult Education and Social Change," 256.
7. Hall, "Cultural Studies: Two Paradigms."
8. Personal telephone conversation, October 12, 1995.
9. For further discussion of such erroneous conflation, see Jones, "The Myth of 'Raymond Hoggart.'"
10. Hoggart, *A Measured Life,* 77.
11. Hall, "The Emergence of Cultural Studies."
12. Hoggart, *Speaking to Each Other, Vol. I.*
13. Ibid., 50.
14. Ibid.
15. Ibid.
16. Ibid. (Emphasis mine: Hoggart's Centre for Contemporary Cultural Studies will take as a central goal to understand just how the media function as such "socially educative forces," what their specific role is in contributing to the social changes under scrutiny.)
17. Hall, "The Emergence of Cultural Studies," 12.

18. Williams, of course, had written that "There are in fact no masses, only ways of seeing people as masses," in "Culture is Ordinary."

19. Hewison, *Culture and Consensus*, 118–19.

20. Halloran, *The Effects of Mass Communication*.

21. Halloran and Gurevitch, *Broadcaster/Researcher Co-Operation*.

22. Halloran, *The Effects of Mass Communication*, 40.

23. "An integration of approaches is obviously indicated," writes Halloran. "It is felt that the philosophers, the social scientists and the literary critics must get together and work together and it seems possible that this will be attempted at Birmingham University" (Halloran, *The Effects of Mass Communication*, 42).

24. Halloran, *The Effects of Mass Communication*, 42.

25. Ibid., 41–2.

26. Thompson, *Discrimination and Popular Culture*; and Thompson and Leavis, *Culture and Environment*.

27. Williams, *Communications*.

28. Dworkin, *Cultural Marxism*.

29. Hall, personal interview, July 4, 1995.

30. Ibid.

31. The department was originally instituted as the Film Appreciation Department, in response to an increasing demand for lectures and courses on the subject. It followed the earlier creation in 1943 of the BFI's Annual Summer School, established by Ernest Lindgren, who had established a reputation as a lecturer on film in the 1930s and 1940s at the City Literary Institute in London. A collection of lectures he gave to the Summer School was eventually published in 1948 under the title *The Art of Film*, one of the first "textbooks" in cinema studies. Stanley Reed was the first head of the Film Appreciation Department and had as his goal not merely the provision of lecture services by the BFI, but also to encourage colleges and schools to develop their own courses in teaching film. He sought to stimulate this by publishing pamphlets on methods of film education and by providing schools with teaching materials, including excerpts from feature films (Whannel and Harcourt, *Studies in the Teaching of Film*).

32. Hall, personal interview, July 4, 1995.

33. Hall and Whannel, *The Popular Arts*.

34. Ibid.

35. Ibid., 21.

36. Ibid.

37. Resolution of the conference, quoted in Hall and Whannel, *The Popular Arts*, 23.

38. Hall and Whannel, *The Popular Arts*, 25.

39. Ibid.

40. Ibid., 34–5, 37.

41. Hall, "Liberal Studies."

42. Ibid., 26.

43. Ibid.

44. Ibid.

45. Ibid.

46. Ibid., 27.

47. Hoggart, *Speaking to Each Other, Vol. II*.

48. Ibid., 254.

49. Ibid., 258.

50. For a discussion of the characteristics of and tensions between these two dominant strands in British media research, see the 1995 "Colloquy" on cultural studies versus political economy in *Critical Studies in Mass Communication,* in particular Garnham, "Political Economy and Cultural Studies."

51. Hoggart, personal interview, July 5, 1995.

52. Ibid.

53. Hall, personal interview, July 4, 1995.

54. Ibid.

55. Ibid.

56. Ibid.

57. Ibid. Hall also pointed out the irony in that exchange, in that the CCCS ultimately became the Department of Cultural Studies in the early 1990s and was absorbed into the Social Sciences Faculty.

58. Hall, "The Emergence of Cultural Studies"; Hall, "Cultural Studies and the Centre"; Hall, "Introduction to Media Studies at the Centre"; Green, "The Centre for Contemporary Cultural Studies"; Grossberg, "The Formations of Cultural Studies"; Schulman, "Conditions of Their Own Making;" and Redal, "Imaginative Resistance."

59. Hall, "The Emergence of Cultural Studies," 15.

60. Hall, personal interview, October 19, 1995.

61. Ibid.

62. Smith, Blackwell, and Immirzi, *Paper Voices.*

63. Hall, personal interview, October 19, 1995.

64. Ibid.

65. Ibid.

66. Hall, "Encoding/Decoding."

67. Hall, personal interview, October 19, 1995.

68. Ibid.

69. Ibid.

WORKS CITED

Anderson, Lindsay. "Get Out and Push!" In *Declaration,* edited by Tom Maschler, 153–178. London: MacGibbon & Kee, 1957.

Dworkin, Dennis. *Cultural Marxism in Postwar Britain.* Durham, NC: Duke University Press, 1997.

Garnham, Nicholas. "Political Economy and Cultural Studies: Reconciliation or Divorce?" *Critical Studies in Mass Communication* 12, no. 1 (March 1995): 62–71.

Green, Michael. "The Centre for Contemporary Cultural Studies." In *Re-Reading English,* edited by Peter Widdowson, 77–90. London and New York: Methuen, 1982

Grossberg, Lawrence. "The Formations of Cultural Studies: An American in Birmingham." In *Relocating Cultural Studies,* edited by Valda Blundell, John Shepherd, and Ian Taylor, 21–66. London: Routledge, 1993.

Hall, Stuart. "Cultural Studies: Two Paradigms." *Media, Culture and Society* 2 (1980): 57–72.

———. "Cultural Studies and the Centre: Some Problematics and Problems." In *Culture, Media, Language: Working Papers in Cultural Studies, 1972–79,* edited by Stuart Hall, Dorothy Hobson, Andrew Lowe, and Paul Willis, 15–47. London: Hutchinson & Co., 1980.

————. "Editorial." *New Left Review*, 1, no. 1 (1960): 1–3.

————. "The Emergence of Cultural Studies and the Crisis of the Humanities," *October* 53 (Summer 1990): 11–23.

————. "Encoding/Decoding." In *Culture, Media, Language: Working Papers in Cultural Studies, 1972–79*, edited by Stuart Hall, Dorothy Hobson, Andy Lowe, and Paul Willis, 128–138. London: Hutchinson & Co., 1980.

————. "Introduction to Media Studies at the Centre." In *Culture, Media, Language: Working Papers in Cultural Studies, 1972–79*, edited by Stuart Hall, Dorothy Hobson, Andrew Lowe, and Paul Willis, 117–121. London: Hutchinson & Co., 1980.

————. "Liberal Studies." In *Studies in the Teaching of Film in Formal Education*, edited by Paddy Whannel and Peter Harcourt, 10–27. London: British Film Institute Education Department, 1964.

Hall, Stuart, and Paddy Whannel. *The Popular Arts*. London: Hutchinson Educational, 1964.

Halloran, James D. *The Effects of Mass Communication with Special Reference to Television*. Leicester, UK: Leicester University Press, 1964.

Halloran, James D., and Michael Gurevitch, eds. *Broadcaster/Researcher Co-Operation in Mass Communication Research*. Leeds, UK: J.A. Kavanagh & Sons Ltd., 1971.

Hewison, Robert. *Culture and Consensus: England, Art and Politics Since 1940*. London: Methuen, 1995.

Hoggart, Richard. *A Measured Life, Vol. III: An Imagined Life*. New Brunswick, NJ: Transaction Publishers, 1994.

————. *Speaking to Each Other, Essays by Richard Hoggart, Vol. I: About Society*. London: Chatto & Windus, 1970.

————. *Speaking to Each Other, Essays by Richard Hoggart, Vol. II: About Literature*. London: Chatto & Windus, 1970.

Jones, Paul. "The Myth of 'Raymond Hoggart': On 'Founding Fathers' and Cultural Policy." *Cultural Studies*, 8, no. 3 (October 1994): 394–416.

Lindgren, Ernest. *The Art of Film*. London: Allen & Unwin, 1948.

Redal, Wendy Worrall. "Imaginative Resistance: The Roots of Cultural Studies as Political Practice in Britain." PhD diss., University of Colorado, 1997.

Schulman, Norma. "Conditions of Their Own Making: An Intellectual History of the Centre for Contemporary Cultural Studies at the University of Birmingham." *Canadian Journal of Communication* 18 (1993): 51–73.

Smith, A. C. H., T. Blackwell, and E. Immirzi. *Paper Voices*, with an introduction by Stuart Hall. London: Chatto & Windus, 1975.

Thompson, Denys, ed. *Discrimination and Popular Culture*. Baltimore, MD: Penguin Books, 1964.

Thompson, Denys, and F. R. Leavis. *Culture and Environment: The Training of Critical Awareness*. London: Chatto & Windus, 1933.

Whannel, Paddy, and Peter Harcourt. *Studies in the Teaching of Film in Formal Education*. London: British Film Institute Education Department, 1964.

Williams, Raymond. "Adult Education and Social Change." In *Border Country: Raymond Williams in Adult Education*, edited by J. McIlroy and S. Westwood, 255–264. Leicester, UK: National Institute of Continuing Adult Education, 1993.

————. *Communications*, 3rd ed. Harmondsworth, UK: Penguin Books, 1976.

————. "Culture is Ordinary." In *Conviction*, edited by Norman Mackenzie, 74–92. London: MacGibbon & Kee, 1958.

Writing Figures INTO THE Field: William McPhee AND THE Parts Played BY People IN Our Histories OF Media Research

PETER SIMONSON

There are many ways to write about the history of media research. We can focus upon governing paradigms, institutional settings, ideological commitments, intellectual lineages, animating problems, operative methodologies, key texts, or structuring contexts. As we move toward fuller and more adequate histories of media research, we need to pursue all of these avenues and more, and do so from a variety of theoretical and political perspectives. Kenneth Burke observed long ago that every way of seeing is also a way of not seeing, and each carries its own trained incapacities.[1] One way to counterbalance the limits of any one approach is to cultivate a kind of ecumenical pluralism at the collective and disciplinary level, which means making room for multiple and competing perspectives and histories. I take this volume as an important step forward in that direction.

For my contribution, I want to make a case for history that attends carefully to individual people, and contribute modestly to that project by attending to a person who hasn't figured into histories of the field thus far, William N. ("Wild Bill") McPhee (1921–1998). It is a two-part effort, which begins at the historiographical and conceptual level and then seeks specificity through McPhee, whom I try to write into greater collective awareness as a way to make a substantive contribution to the history of the field. Historiographically, I lay out roles that individuals can play in our histories, conceiving them as rhetorical figures of a sort that function in multiple ways, five of which I briefly describe. I then go on to illustrate those roles through McPhee, best known as third author of *Voting*,[2] the second of Paul Lazarsfeld's election studies, based on data collected in Elmira, New York, during the presidential campaign of 1948. McPhee, it turns out, was a fascinating figure whose life traversed a number of things consequential in the history of media research during its formative era from the 1940s through the 1960s. After sketching a number of these, I use his life to illustrate some of "the parts played by people" in the field's histories and other collective memories.[3]

THE PARTS PLAYED BY PEOPLE:
FIGURING SOME FUNCTIONS FOR INDIVIDUALS

Because individuals have fallen from favor as loci of social and historical analysis, my move to feature them requires some explanation. First let me say that I intend the focus on individuals as a supplement to social and institutional history, not as a replacement. The same goes for those histories that focus upon broader ideological and cultural patterns, and other structuring mechanisms. In this regard, I come to the individual through a different moment and orientation than earlier historical understandings of the field that also featured individuals—those that run from Bernard Berelson[4] through Wilbur Schramm[5] and Everett Rogers's[6] That historiographical line began with Berelson's "The State of Communication Research,"[7] which identified Harold Lasswell, Carl Hovland, Kurt Lewin, and Lazarsfeld as representatives and innovators of the four major approaches to communication research (the political, small-group, experimental, and sample-survey approaches, respectively). Schramm incorporated these four figures into the noteworthy but overlooked first chapter of his edited *Science of Human Communication*, "Communication Research in the United States,"[8] and he maintained the schema through multiple iterations, culminating in the book manuscript that Rogers and Steven Chaffee finished after Schramm's death.[9] By then, Rogers[10] had published his own "biographical approach" to the history of communication study, which made Schramm the founder of the field and then built out from the classic foursome by adding chapters on six other individuals and the Chicago School. The historical and conceptual shortcomings of that book might justifiably make one gun-shy about moving in the direction of biography as a vehicle for writing histories of media research. But despite the simplifications, errors, and naïveté, Rogers's book actually directs us toward something important: the individual.

The individual shows signs of making a comeback, as a recent article in *Social Science History* by Andrew Abbott[11] indicates. Operating at a sophisticated theoretical level, Abbott, the University of Chicago sociologist and historian, argues that individuals should be reinstated as an important force in history, not as a revival of some kind of "great-man" (or "great-woman") theory, but rather based on what he calls their "historicality," or continuity over time.[12] There are multiple dimensions to this continuity, he suggests. One is biological—individuals have bodies that are in some sense physically continuous, carrying forward records of past behavior (e.g., of occupation, exercise, drug abuse, disease, exposure to chemicals). A second dimension of continuity is memorial, for individuals carry "a huge mass of historical experience, written quite literally in and on their bodies," and thus (unlike other more dispersed forms of memory) concentrated in one particular locale. Third, continuity exists in the records preserved for an individual—legal and other public or semipublic records, along with other preserved texts that document aspects of a historical life unfolding over time. Finally, "substantive historicality," or the "mass of personal attributes and experience carried forward through time," both refracts the meaning of events for a cohort and helps determine the shape of the society as a whole.[13] Abbott puts his insights in service of history on the larger scale, where the historicality of individuals en masse becomes "an enormous reservoir of continuity with the past" and "a central force in determining most historical processes."[14] I'd like to take the individual in somewhat different directions.

Extending Abbott, I would suggest that the historical continuity of individual bodies and the records that index them are basic and fundamental resources to draw upon as we write and rewrite histories of media research. They operate on all the levels of historicity that Abbott identifies. As biological beings, individuals traverse time and space, and through their persistence link those things in the narrative of a life. As memorial beings, they record the past in both retrievable and inaccessible ways, and thus offer testimony to the particularities of earlier days. As record-producing beings, they leave behind textual trails that articulate themselves with social structures and, in the case of intellectuals, express thoughts about aspects of the world around them. Each kind of historicality points toward a particular kind of work we need to do. The pasts of the memorial beings and their memories and encoded experiences need to be recorded and heard, which we can do through oral history interviews. The texts of the record-producing beings need to be read and interpreted, which we can do through reading published and unpublished documents. The narratives of all of these beings require words to be told, which we do in our writings and lectures. Each of these things is in some measure rhetorical, I might add—the dia-logos of interviews, the reading and interpretation of texts, the spinning of narratives.

I'd like to attend to the narrative dimension here, suggest that individuals play a number of roles in broader histories of media research, and show that those roles can be understood rhetorically. More precisely, I would suggest that individuals function as what the Greeks and Romans called figures of thought (Gk. *schemata dianoias*; L. *figurae sententiarum*)—figures that both organize conceptions of the field and "add force and charm" to the stories we tell about its past, as Quintilian said of rhetorical figures.[15] In other words, individuals help shape our (substantive) historical thinking and inquiry, and they (stylistically) add narrative appeal and persuasive force to the histories we compose and communicate—which, I quickly add, are two mutually influencing processes. With partial assistance from Quintilian's classic catalog of Latin figures, I'll lay out five of the many ideal-typical roles that individuals can play in our historical compositions, which I will call *thread end, agent, representative figure, model,* and *conversationalist.* Let me give a condensed account of each, before working them out through McPhee.

The first is individual as *thread end,* a figure drawn from Confucian moral philosophy and ontology, which I bend here to very different purposes.[16] I use this term to mean that individuals sit metaphorically at the end of a thread that is woven into the broader social, cultural, institutional, and geographical fabrics of their historical moments. As a thread end, any given individual can serve as a point of entry into that fabric. By following the thread of that particular life, we gain entry into the social practices, cultural understandings, institutional formations, geographical locations, and broader historical events that structure the lives that traverse them. Theoretically speaking, the life of any particular individual provides the starting point for telling the whole of a story about media research in any given era. As thread ends, individuals can serve to guide both the discovery of historical truths and the construction of narratives about them. In their capacities as thread ends, individuals perform both substantive and stylistic historiographical functions. Their historicality (biological, memorial, and textual) makes them useful devices for figuring continuity, change, constraint, and possibility in the field over time. The lived drama and perspective of a single life, meanwhile, makes it an ideal device to shape and enliven the broader histories we compose.

In the course of following out the thread of a particular life, one discovers moments when that individual functioned as an *agent,* by which I mean his or her decisions and actions became socially consequential. Agency, as I use the term, is agnostic about motives, intentions, and ultimate causality—those entities called into question by critiques of the classic bourgeois subject—and it recognizes that consequences fall along a wide spectrum of significance. To say that individuals function as agents is to say their actions have consequences in the social or material world beyond them. Sometimes these actions maintain the status quo, and other times they help bring about change. In large measure, the consequences of individual agency are utterly trivial, even in the individual's own life. On rare occasions and for those situated in favorable social and institutional contexts, agency counts, and historical inquiry will try to determine how. For those who have worked in the field of media research, agency is most often exercised through publication, teaching, communicative collaborations, and the formation of institutes and professional organizations. These actions are never done in solitude, or without considerable structural constraint, but individuals sometimes make a difference in outcomes, and thereby function as agents. In this capacity, they deserve historical credit, as well as the praise or criticism that accompanies the judgments we often make about past actions of consequence and social significance.

In addition to thread ends and agents, individuals also function historically as *representative figures*. As Emerson said, "All mythology opens with demigods," and the mythologies of media research are no different. The most common mythologies feature demigods such as Paul Lazarsfeld, Theodor Adorno, Wilbur Schramm, John Dewey, Harold Innis, Marshall McLuhan, Dallas Smythe, Raymond Williams, and Stuart Hall. Others fill in the pantheon as figures worshipped or despised—Charles Cooley, Robert Park, Walter Lippmann, Queenie Leavis, Harold Lasswell, Helen McGill Hughes, Leo Lowenthal, Herta Herzog, Roland Barthes, Richard Hoggart, C. Wright Mills, Jürgen Habermas, Elihu Katz, and James Carey. As Emerson observed in the gendered language of his day, "Men have a pictorial or representative quality."[17] They are representative, "first, of things, and secondly, of ideas."[18] The demigods of our disciplinary histories represent "things and ideas" such as critical and administrative research, Chicago School progressivism, propaganda analysis, audience gratification studies, the focused interview, Toronto-based medium theory, positivism, literary analysis, semiotics, cultural studies, and so on. As Charles Cooley wrote, folding Emerson into the evolving argot of American sociology, fame exists "for our present use and not to perpetuate a dead past."[19] Public recognition of the demigods has a functional value by making individuals into the symbolic representatives of sentiments, ideas, and groups that continue to have purchase in the present.[20] As representative figures, historical individuals stand as synecdochical condensations of the ideas, methods, research styles, institutions, and broader identities of which they were part, and which in some way resonate in our own time.

Representative figures can easily slip over into becoming *models* as well, the fourth of the roles that individual people can play in our histories. The difference between the two is subtle but important, captured by the distinction between spectator and participant perspectives. When we talk of individuals as representative figures, we are focused on their historical roles in dramatic narratives of the past. As Emerson and Cooley both recognized, the play of representative figures from the past is given meaning by those of us writing and reading about the past in the present. We participate in these past dramas, but as spectators bearing witness from a distance. When individuals function as models, in contrast, they serve to guide our own productions in the present. Put in a rhetorical vocabulary, past individuals become subjects for *imitatio,* that creative appropriation of past substance or style poorly captured by the English word *imitation.*[21] Put in sociological terms, those individuals function as role models upon whom we draw to produce the lives that each of us creates. Models can be of the positive or negative sort—those we find ourselves appalled by can teach as much as those we admire, inviting a variation on what Kenneth Burke[22] called perspective by incongruity. Individuals can provide models in a number of different ways, from the overall trajectories of their lives or professional careers to the intellectual styles they construct, the lasting artifacts they produce, the methodologies they pioneer or practice, the political battles they fight, the moral commitments they maintain, the social identities they enact, and even the geographical places they call home. As models, past media researchers provide precedents creatively to reenact or watchfully to avoid. Careful consideration of them can invite a reflexive self-awareness in us. It can open up dialectical awareness about our own structuring contexts and the trained incapacities in our own ways of work.

Finally, as a fifth analytically distinct role, I suggest that individuals written into the field become recognized members of the broader conversation about media. To frame things more actively, they can become *conversationalists*—or in certain cases, guerrilla *conversationalistas* who challenge the terms of the discussion and fight for causes not featured on contemporary discursive agendas. Sometimes individuals contribute their *logoi* to the conversation but are "obliterated by incorporation," to borrow Robert K. Merton's useful phrase.[23] That is, we speak the words and concepts of past figures, or mobilize methodologies they pioneered, without recognizing their source in the individual lives that produced them. We run focus groups, for instance, without realizing that Merton and Patricia Kendall codified the methodology—or that Herta Herzog lay down the techniques before them, but was never properly credited with doing so. We speak of audience "gratifications," and in so doing channel Herzog's concept and keep it part of the conversation. But for the most part Herzog herself is obliterated in the process, at least insofar as we do not return to make contact with the texts she left for us, or learn about the life of which they were part, or cite her pioneering work in the field. There is a transhistorical conversation about media (a polyvocal tradition), which we energize and expand through dialogue with the living and dead we make part of it (i.e., through dialectical engagement). While there will always be limits to the number of conversationalists we can engage, the field as a whole is richer and more intellectually powerful if we expand the available figures and make the conversation more cosmopolitan. Furthermore, each of its localized dialects, marked by a preferred vocabulary and range of concerns, should be open to *conversationalistas* who might crash in and provide correctives to the inescapable parochialisms every cultural scene perpetuates.

Thread end, agent, representative figure, model, and member of the conversation: I hasten to repeat that these are simply five of the roles that individuals can play in our histories, and that others remain to be recognized and articulated. That said, it seems to me that these five cover a good deal of ground, which I will try to illustrate by my own writing-into-the-field of a heretofore forgotten figure, William McPhee.

WILLIAM N. MCPHEE

William McPhee is most widely remembered as the third author of *Voting*, but he was much more than that. Those who knew him know that much, but most of the rest of us don't. As I have begun to learn, McPhee conducted or had a hand in some of the more noteworthy media research that took place anywhere in the world from the 1940s through the 1960s. His life is fascinating in its own regard, and encounters with it left lasting impressions on people who knew him. Running like a thread through the fabrics of a formative era in the history of American media research, and the institutions that guided and funded it, McPhee's life also indexes fascinating but currently obscure research, including excellent work on radio, television, and mass culture. A pioneering figure in American opinion research, computer simulation, and mathematical sociology, McPhee was also a legend of sorts at the Bureau of Applied Social Research (BASR), where he worked from 1951 to 1960. As a senior research associate, he had a hand in some of the most important and historically significant media research done in the 1950s—from the controversial Near East study of media audiences in strategically significant countries (1951–1952), through potentially important but suppressed studies of the potentials of television (1953, 1957), to fascinating studies of radio disc jockeys and radio usage in the television age (1953–1956). Thereafter, as he moved into computer simulation models and mathematical sociology, he continued to write about media and culture and developed a style little known in U.S. media study—critical mathematical research.

I want to start by sketching contours of McPhee's life, which I do both to write him into greater collective awareness and to indicate ways that his life serves as a thread end for his era. He did varieties of media research for the U.S. Army, a small commercial survey research company, the BASR, the National Broadcasting Company (NBC), the Columbia Broadcasting System (CBS), the Ford Foundation, the National Science Foundation (NSF), and a series of other employers and funders. His life weaves through some of the dominant institutions of American mass media and organized research about it. As a researcher at the Bureau in the 1950s, McPhee was also agent and representative figure in the drama of postwar American media research, which I briefly develop in the last section of the chapter when I draw together threads of his life to fill out my five-part schema for the parts played by people in our histories. McPhee's work indexes both main directions and positions not pursued in American mass communications research of the 1950s and 1960s. It also, I will argue, deserves to be incorporated into the broader conversation and dialectical engagements of the field today, and can be taken in certain respects as a model.

William Norvell McPhee was born in Denver in 1921, the third boy in a family that would add one sister as well. Young "Will" had deep roots in the region. His mother, the former Sarah Eddy, was the daughter of John Eddy, who arrived in New Mexico in 1880, ranched, ran cattle drives, founded the town of Carlsbad ("Eddyville"), built the railroad from El Paso to Raton, New Mexico, and was said to have hosted Billy the Kid at his ranch in Eddy County. Will's father, Elmer McPhee, was the son of Charles McPhee, who in 1872 in Denver founded McPhee and McGinnity Lumber Company, which grew to be the largest lumber company in the Rocky Mountains. They logged both the old growth forests around Chama in the historic San Luis Valley on the western slopes of the Sangre de Cristo Mountains, and the forests around Dolores near Cortez in the southwestern corner of state, until they declared it "logged out" in 1923, and moved on. (McPhee Reservoir stands today near the site of the latter operation.) The family empire was wiped out in the crash of 1929, but a friend paid to send the bright young Will to Middlesex, the Concord, Massachusetts, boarding school, where he finished his secondary education among the Easterners and showed some independence of mind. While most of his class went to Harvard, McPhee went to Yale instead in the fall of 1940.[24]

When the Japanese bombed Pearl Harbor in December of his sophomore year, McPhee volunteered for the army, which gave him his first experience with media research, albeit an unusual variety. He was assigned one of the most dangerous missions in the world—flying "over the hump" of the Himalayas from Burma to China and back. The bulk of Allied planes were transport aircraft that carried provisions from the jungles of eastern India to Chiang Kai-Shek's Chinese forces beyond, but McPhee was assigned reconnaissance work and flew a Piper Cub instead. Unseen box canyons and fog-shrouded mountains contributed to high mortality rates, and McPhee himself crashed twice—not the last times he eluded death. The operation established the airlift as a large-scale military technique and veritable channel of mass communications, and McPhee was responsible for collecting information that made that communications possible.[25] It was survey research of a kind, conducted through a relatively new medium (the air-plane) experienced by few at the time.

After the war McPhee returned to Denver, where in 1945 he became cofounder of Research Services, the first survey research firm in the region. It is unclear where he learned about surveys. It may have been at Yale, as psychology classes of the era sometimes included instruction in the method.[26] More likely it was in Denver, where the National Opinion Research Center (NORC) had been founded by Harry Field in 1941, then housed at the University of Denver. Field (1900–1946) helped institutionalize the field of public opinion research in the United States, both through the formation of NORC (which later moved to the University of Chicago) and by energetically promoting opinion research as a means of pursuing peace and the public good after the war. He collaborated and hired research assistants locally, and almost surely came to know McPhee in some capacity. After organizing in the summer of 1946 a three-day public opinion conference in Central City, Colorado (which Jack Kerouac and his friends passed through soon after), Field died that fall in a plane crash, in Belgium. The follow-ing summer, the American Association for Public Opinion Research (AAPOR) was founded at the group's second conference.[27]

In his capacity at Research Services, Inc., McPhee was one node in the broader international development of public opinion research during the postwar period. His was one of one of the independent regional polling organizations that sprang up around the country immediately after the war. McPhee did work for the *Denver Post* and other clients in the area. He probably met Lazarsfeld (who supported and collaborated with Field) at the Central City conference in 1946.[28] In June 1950, the Fifth Annual Conference on Public Opinion Research was held at Lake Forest College, and on a Thursday afternoon, a group of academics and regional pollsters met to consider a cooperative research program for the con-gressional elections that would occur that fall. Two weeks later, a subcommittee was formed, consisting of Lazarsfeld, Stuart Dodd (University of Washington), Bernard Berelson (University of Chicago), and Sidney Goldish (The Minnesota Poll), and chaired by McPhee (Research Services, Inc., 1441 Welton Street, Denver 2 Colorado, in the address system of the day). A year later, Lazarsfeld brought McPhee to Columbia.[29]

Although McPhee had no undergraduate degree, he became research associ-ate at the Bureau and was admitted to Columbia's doctoral program in sociology. He appeared as the "native genius" Lazarsfeld had found out West. [30] Lazarsfeld thought enough of him to send him to speak in his place at Northwestern University's Centennial Conference on Communication in October, and Merton later called him the smartest graduate student he ever had.[31] McPhee quickly made an impact on the local scene at the Bureau. He was several years older than other core graduate students—Elihu Katz, James Coleman, Rolf Meyersohn, David Sills—who, with Philip Ennis, Ben Ringer, and others met as the informal "Traditions Group," so named because they had no traditions (but were perhaps interested in establishing some through the Bureau). McPhee was something of a ringleader for the group.

Sills describes McPhee as "a hick," the same designation he gave Jim Coleman, who grew up in small town southern Indiana. Meyersohn paints a subtler picture, with McPhee "self-consciously and perhaps disingenuously a Westerner, a poor country boy trying to get along. He was smart and sharp and modest." He was "in many ways larger than life," Meyersohn told me, "but without being intrusive or self-centered or domineering."[32] McPhee and Coleman developed a particularly close friendship, two harder-drinking gentiles from the provinces thrown among the Jews and Northeastern cosmopolitans who dominated Columbia's ranks. They often got together at the McPhees' rambling house at 24 Pinecrest Avenue in a modest neighborhood in the near suburb of Hastings. (Several years later, Charles Kadushin learned survey research analysis on Friday afternoons in McPhees' attic study, mediated by the martinis that McPhee made by the pitcher and poured out liberally.)[33] There was a small McPhee clan in Hastings—his wife, Minnow, and their three young kids, and then Minnow's sister and brother-in-law, who was a graduate student in another department at Columbia. McPhee was thirty years old when he arrived in 1951.

It was a heady time at Columbia, with huge numbers of graduate students crowding Merton's elegant lectures on social theory and Lazarsfeld's unprepared but brilliant seminars on methodology and empirical research. The Bureau was well funded, especially in comparison to its more precarious days a decade earlier, as Lazarsfeld the entrepreneur always managed to make contacts and find money to fund a welter of (ideally) dual-purpose projects that both satisfied the funding agency and advanced intellectual questions that the Bureau sociologists cared about.[34] There was excitement about the social science and (partially repressed) hope that mass media could be directed toward good in a world that had recently suffered enormous destruction. Women were present in the mix—as graduate students, staff researchers, and wives—but it was still very much a boys' club, especially at the higher echelons of status.[35]

Lazarsfeld's native genius was immediately put to work analyzing data for an ambitious research project recently begun with a grant from Leo Lowenthal, the Frankfurt School émigré who had worked for the Bureau from 1940 to 1949 before going to work for the State Department as head of research for the International Broadcasting Service and as one of the five directors of the Voice of America.[36] The study represented cold war politics worked out through media research. In 1948 and 1950, the Bureau had created surveys that were administered by native interviewers in strategic countries in Western Europe and the Near East. From 1951 on, Columbia researchers analyzed the data and wrote reports. McPhee and Meyersohn analyzed surveys administered in Lebanon and in Syria. Among other topics, the Lebanon study drew upon interviews designed to determine "the comparative role of the several mass media and face-to-face communications in the lives of various segments of the Lebanese population,"[37] a Bureau concern since at least the 1940 presidential voting study.[38] The Syria report was more menacing, and sought to gauge attitudes toward "Russia, America and the cold war" among citizens classified through a typology of political orientations—Revolutionary Left, Reform Left, Apolitical Center, Conservative Right, and Nationalistic Right.[39] The reports eventually contributed to Daniel Lerner's politically infamous cold war classic, *The Passing of Traditional Society* (1958).[40] McPhee had great talent analyzing tabulated survey data and qualitative responses alike—"as adept at locating the apercu as anyone I ever met," Meyersohn said of him. The Lebanon and Syria surveys gave him his first opportunities to exercise that talent at the Bureau.

McPhee was also put to work on a series of domestic voting and political research. Returning to the 1950 Congressional election study he helped organize, he gave the data gathered by the regional pollsters a preliminary analysis. He was included in an ambitious interdisciplinary grant application to the Ford Foundation for "analysis and systematization of theory and research in politics," which didn't come through,[41] and he set to work on pulling together a languishing Elmira study of the 1948 presidential elections, overseen by Lazarsfeld and Bernard Berelson. It was the last of Lazarsfeld's trilogy of big community-based studies in the United States—Sandusky (1940), Decatur (1945), and Elmira (1948), only the first of which had by 1952 been published. Berelson was by then working for the Ford Foundation, and Lazarsfeld was, as usual, busy with many other things. As Elihu Katz would do for the Decatur study in another year or so, McPhee brought the massive Elmira study to fruition as *Voting: A Study of Opinion Formation in a Presidential Campaign.*[42] As third author, McPhee hasn't gotten much credit for the book, which is still in print and among the most cited works in voting behavior research; but his role may have been decisive in its publication, and most likely included a combination of data analysis and manuscript writing.[43] Also in the political realm, McPhee conducted a study of community controversies over schools for the liberal Fund for the Republic that same year.[44]

Of greater interest to students of media, McPhee conducted a series of historically significant and often fascinating studies of American radio and television at the Bureau as well. In fact, from 1952 to 1957, McPhee directed or had a hand in some of the most noteworthy, though largely unpublished, radio and television research of the era—including the lead article in the important series of eleven reports produced in 1953 for the Ford Foundation and the New York State Television Commission, "New Strategies for Research on the Mass Media." (Part I of *Personal Influence* also originated as a report in this series by Katz.) [45] As David Morrison describes in this volume, that Ford Foundation project was cut short before the reports came to be published, a casualty it seems of the foundation's cold war anxieties about criticism of the mass media and other central institutions of American society. One can make the case that this episode, whose full details have yet to be unearthed, marks a crucial juncture in the history of the field as a social formation that might meaningfully shape television and other mass media in the United States.

Later that year, McPhee teamed with Meyersohn and Philip Ennis on a study of disc jockeys in America, funded by the musicians' group ASCAP (the American Society of Composers, Authors, and Performers), whose members were in the midst of suing the radio networks and their licensing agency Broadcast Music Incorporated (BMI). Sidney Kaye, with whom Lazarsfeld was friendly, was representing ASCAP and set the study in motion. To determine the factors that influence disc jockeys in choosing their music, and whether BMI played a part, McPhee, Ennis, and Meyersohn interviewed forty-two disc jockeys in twenty-four cities. Blending the sociology of work with broader analyses of the music and radio industries and their recent historical development, the authors produced a sophisticated and fascinating 154-page account of the contexts and actions that lead to the playing of particular records on radio stations. "The process of making decisions extends beyond the immediate experience of the individual making the choice," McPhee and his coauthors wrote.

> The criteria upon which he bases his decisions have roots both in his own past social background and in the network of economic and social relations in which he is currently enmeshed. The final act—whether it be selecting a record, pulling the lever in a voting booth, or selecting the TV set for the living room—is the culmination of a whole series of prior activities, many of which lie outside the person making the decision. [46]

That same year, McPhee also corresponded with the advertising director at *LIFE* magazine about the possibility of conducting a study of media flow into the classroom. [47]

In 1954, McPhee and Meyersohn again joined forces, for the first of two NBC-commissioned studies on radio's place in the television era. It issued in "Futures for Radio: An Interpretive Study of Radio Potentials Under Television Conditions," a 158-page report that had begun as an "exploratory study for the purpose of developing ideas" for radio programming and publicity (for NBC) and "a much-needed first step in viewing the mass media in the broader context of long range uses and gratification . . . rather than at short-range 'effects.'" The latter emphasis is worth pausing over, given the charge that the Bureau concentrated primarily on short-term behavioral effects of the mass media. For "Futures of Radio," the Bureau team designed a more open-ended project than they traditionally did, based on panel studies in Fort Wayne and Davenport and one-to-two-hour in-home interviews set up by telephone ahead of time. Interviewers were instructed to record their respondents' own words whenever possible, and to look around the house or apartment to see where televisions and radios were situated spatially.[48] The report offers us a snapshot of attitudes toward television and radio and a prescient view of persistent and new functions for radio in the television age.[49] It deserves a place among the classics of American radio studies, and of research done about older media supplanted by new ones.

A year or so later, McPhee and Francis Bourne conducted an additional study for NBC radio, positioned internally as an installment "in a Bureau series showing how technological change in the communication field has altered not only radio listening habits but the functions served by radio." NBC's new "Monitor" weekend service ran from 8 a.m. Saturday until midnight Sunday, and presented an array of news, feature stories, comedy, live and recorded music, and excerpts from NBC television shows as well. Interviews were conducted with listeners, ex-listeners, and nonlisteners of the service in the four heavy television markets of Baltimore, Hartford, Peoria, and San Francisco.[50] On a different communicative front, another Bureau researcher, Henry Lennard, undertook a study that same year on the effects of LSD on interpersonal communication.[51]

The last of McPhee's notable contributions to media research at the Bureau was perhaps his best, "Working Paper Suggesting Research on: The Potentialities of Television." He wrote it for CBS, which was looking to develop ideas for implementing proposals made by CBS President Frank Stanton in a 1955 speech, "The Role of Television in Society." Stanton had called the National Association of Radio and Television Broadcasters to visualize future possibilities for TV, and it was McPhee's job to offer specifics. He focused particularly on what kind of academic studies might contribute to the project, and criticized mass communications research for its "lack of inventiveness" and for operating "in the abstract scientific tradition" that was unlikely to produce "new aesthetic or entrepreneurial ideas" or to "stimulate the . . . creative imagination" of television writers or producers. As an alternative to what C. Wright Mills was about then calling abstracted empiricism, McPhee outlined seven kinds of projects that researchers might undertake to help television realize more of its potentialities. One involved organizing citizens' panels around the country to discuss possibilities for the new medium—"a months-long exploration of their needs and wants from television, as if in a jury or group psychotherapy," he said. He called the method "socio-analysis," which he said was "not an aloof inquiry but an attempt actively to create something like 'social insight' into group needs." He imagined other relevant research as well—the Environment-of-Viewing Study, which would try to understand viewing contexts through extensive family case studies (David Morley, where are you?); the Cultural Inventory of Television Resources, which would "attempt to map out some of the resources of the culture which television does not use, but might"; and the Decennial Survey, a large-scale once-a-decade survey about public behavior and attitudes toward television, with inquiries into popular criticisms and satisfactions with different aspects of the medium[52]—a study not unlike what Lazarsfeld and Harry Field had done for radio in 1946. McPhee's was a progressive and publicly oriented vision for mechanisms that might reform television and, potentially, other mass media as well. Thus an early Bureau emphasis, dating back to its days as the Office of Radio Research, was still alive.

After 1956, McPhee would pursue his reform-minded impulses through increasingly mathematical means. He and James Coleman shared this interest, bringing it together under the heading of research in "mass dynamics," a research program for which they submitted a grant to the Ford Foundation in 1957. They had in mind developing theories and models that would cover a range of mass phenomena, including "electorates, mass-communications audiences and consumers of popular-culture products."[53] (Coleman, who had the benefit of an undergraduate degree in engineering, would quickly outpace McPhee, a mathematical autodidact, in developing sophisticated models.) It was on the basis of this work that Berelson cited Coleman and McPhee as two "'new generation' sociologists" who offered fresh directions for a field, communication research, which in his view was "withering away."[54]

In 1958, the same year he was appointed as a lecturer in the graduate division of the Sociology Department at Columbia, McPhee took part in a Social Science Research Council–RAND training course on simulation, and helped found the Simulmatics Corporation, a high-powered group that offered its expertise and computer simulation models to health organizations, businesses, advertisers, and political organizations, including the Democratic National Committee.[55] Two years later he was writing a grant to the NSF for a project titled "Problems in the Analysis and Use of Computer Models for Social Processes." The kind of punch card sorting that Charles Kadushin had been doing with the congressional election data grew into plans for computer simulation of three kinds of social processes—voting, social mobility, and popular culture (e.g., television programs, movies, popular music). He had begun the voting model with two graduate assistants, Robert B. Smith and Jack Ferguson. With grants from IBM, NBC, and the Ford Foundation, they developed a computer simulation that used an IBM 650 machine accurately to predict the outcomes of the Wisconsin presidential primary and the general election that fall (an event whose historic significance the Smithsonian Institute later recognized in a display). Smith, fresh from engineering school, knew how to program the 650, and played a central role in the process.[56] McPhee was working on the popular culture model with Sam Becker, co-director of the Television Center at the University of Iowa, who had been in residence at Columbia in 1958–1959. Upon receiving a one-year trial grant for the project from the NSF for 1960–1961,[57] McPhee worked out a theory of what he called natural sampling and popular renown, developed through a mathematical model.[58] After teaching at the University of British Columbia's summer school on communication in 1960, McPhee set out for Stanford, where he spent the academic year at the Center for Advanced Study in the Behavioral Sciences (CASBS). In the spring of 1961, he applied for a grant to fund a project titled "Feasibility of a Computer Model of Alcoholic Drinking," a social phenomenon he was close to knowing first hand.[59]

McPhee wouldn't return to Columbia. He moved his family back to Colorado, and they took up residence in Colorado Springs, near his wife's brother and family, who returned west from Hastings. McPhee would take a job at the University of Colorado, in Boulder, where he commuted to work Tuesday till Thursday, returning home for the other four days. He taught in the Sociology Department and worked with the University's Institute for Behavioral Science, opened in 1959. He worked with Simulmatics and Research Services. He put in for another NSF grant.[60]

In his first two years at Boulder, McPhee published two significant books with the Free Press, *Public Opinion and Congressional Elections*,[61] and *Formal Theories of Mass Behavior*.[62] The first was the delayed fruition of the 1950 congressional election study conducted by the AAPOR group of regional opinion researchers. McPhee called it "one of the vanishing race of studies started by individual initiative and carried on by personal persistence," though he also thanked the variety of funders and collaborators who made it a snapshot of the institutional structure of U.S. public opinion research in the 1950s. McPhee and other Bureau alumni wrote most of the twelve papers, which addressed issues like election turnout and its prediction, the behavior of independent voters, the influence of voluntary associations, and the contextual dimensions of voting. In addition to writing the introduction, McPhee was lead author for three other important chapters, each of which drew upon mathematical models and made at least feints toward simulation, for which Smith (who inherited the Bureau's research program in computer simulations when McPhee left) had done the programming. Among the numbers are provocative insights and historical treasures, as in the chapter on "Political Immunization," a phenomenon he described as a "resistance to disturbances [that] is built up by disturbances (and a lack of resistance that arises from lack of exposure to them)." Illustrating his argument with reference to contemporary America and Weimar Germany, McPhee concludes that "moderation and indifference," not fanaticism, pose the greatest threat to create "overresponsive" political systems, which bow to the weight of a popularity that negates all opposition.[63] *Public Opinion and Congressional Elections* is an innovative and stimulating work, reviewed widely and mostly favorably at the time.[64]

Formal Theories of Mass Behavior[65] added culture, addiction, popularity, and informal social influence to the phenomena McPhee brought under mathematical consideration. Completed at the Center for Advanced Study at Stanford, it is a remarkable volume, another star in the galaxy of remarkable books about media written in the early 1960s—McLuhan's *Gutenberg Galaxy* (1962), Habermas's *Structural Transformation of the Public Sphere* (1962), Hugh Duncan's *Communication and Social Order* (1962), and Stuart Hall and Paddy Whannel's *The Popular Arts* (1964), among the more noteworthy.[66] I don't have space to discuss *Mass Behavior* here, nor am I suggesting it shines as brightly as Habermas's or McLuhan's books. Let me say, though, that the book is McPhee's break with Lazarsfeld and Columbia, even as he continued to build upon them.[67] In it the Columbia tradition of research on media, social influence, and culture gets refracted through mathematical modeling, dichotomous logic machines (he rejects the word "computer"),[68] and "aggregated dynamics of social significance." It also includes a paper on the campaign simulator he and Smith developed (later republished in the methods section of the second edition of Berelson and Janowitz's *Reader in Public Opinion and Communication*).[69] Of particular note for media research, however, are his theories of the survival of cultural products and his discussion of social exposure and popularity (chapters 1, 3). The mathematical lifting is in places heavy, but one can skip through these and can find bezels like "double jeopardy" (which explains the accumulated disadvantages of lesser-known or alternative cultural products) and see moments where McPhee is seeking to criticize and improve the culture industries by marshaling a blend of mathematics, systems theory, cultural discernment, and critical meliorism. It wasn't something many people were doing, then or at any time. Not everyone saw its brilliance, but some did.[70]

McPhee relaxed the pace of publication dramatically after his two books were published: I find only two articles and one book review, though I may have missed book chapters elsewhere. McPhee worked to develop his simulation models, and he occupied himself in other ways. He spent time with family. He skied. He drove his car weekly between Colorado Springs and Boulder, with the Rockies rising on one side and the Great Plains stretching out on the other. He drank and talked ideas with a couple of close friends. The pieces he published were noteworthy, however. "When Culture Becomes a Business," extends the mass culture analysis of *Mass Behavior* into something I might call a critical mathematical theory of commercial culture.[71] This is a remarkable genre indeed. I can think of no one in the Anglo-American traditions who works in it.

Two years later, McPhee reviewed a mathematics textbook for the *American Sociological Review*. Among his last published pieces, it gathered pedagogical thoughts of a self-taught social science mathematician, occasioned by a revised and enlarged second edition of *Introduction to Finite Mathematics*, or "the famous Kemeny, Snell and Thompson, as social scientists call their favorite pony in mathematics." The book was much expanded, McPhee noted, and the perfect alternative to the "engineering math" of calculus, offering a blend of logic, probability, linear algebra, and human applications thereof (e.g., sociometry, game theory, genetics, kinship, simulation, and learning). He used it as an opportunity to advocate a three-quarter sequence in mathematics for social scientists. This was February of 1968. The appeal to mathematics in the social sciences, never widely popular, would not stand much chance in the immediate post-1968 world.

A member of this unfolding world as well, in April 1968 McPhee joined 9,300 other faculty members from around the county who lent their names to an ad published in the *New York Times*, protesting the war and the indictments of critics of it like Benjamin Spock. Several years later, as mores loosened up on divorce, McPhee and his wife, Minnow, split up, as did many other middle-aged couples in America. He moved from Colorado Springs to Allenspark, a small town in the mountains above Boulder and a spot on the countercultural map of the era. He bought seven acres on a ridge, with a stream down below it, and built a cabin with his son and neighbors. He chopped wood and hiked and dropped fifty pounds, settling in at a solid 6'1", 200 lbs. He helped publish the local newspaper. He taught a course on sustainable communities in the Sociology Department at CU. He hung out with members of an acoustic folk band, Magic Music, which was popular at that time in the area. He tried a little pot, but mostly stuck with alcohol. He still smoked three packs a day and lived on junk food.

As Allenspark faded with the counterculture in the late 1970s, so did McPhee. What one former colleague called his "drinking program" got serious enough that he was pulled out of the classroom in the early 1980s. He lived alone, and winters depressed him. After briefly moving to Estes Park in the hopes of finding a less isolated environment, he determined to move south to Las Cruces, New Mexico, where the sun would help, and he could learn some new mathematical ideas from professors at New Mexico State. His move took him back to the region where his mother's family had built railroads and founded towns, and although he stayed for a time, the heat of southeast New Mexico wilted him, and he moved north to the mountains of Santa Fe, where his son youngest son, John, had settled after leaving Allenspark himself. In the last seven years of his life, Bill McPhee sobered up and wrote a book that developed his mathematical ideas further. He died in Santa Fe on April 1, 1998, at the age of seventy-seven. His final manuscript has not yet been published.

AGENT, CHARACTER, MODEL, CONVERSATIONALIST

The narrative threads of McPhee's life thus laid down, let me bring this figure together by returning to the rest of the scheme I developed in the chapter's first section—agent, representative figure, model, conversationalist. Within the history of media research, McPhee exercised agency of consequence on several occasions. Beginning with the formation of Research Services, Inc., in Denver in 1945, an important event in the social history of survey research in the region, and continuing through his work (mediated through a growing national network of academic and commercial opinion researchers) on the 1950 congressional election study, which would not have been analyzed or published without his personal involvement. Within the contexts of the Bureau, McPhee was one of the core players in the 1950s. His most influential accomplishment there involved bringing the Elmira study to fruition as *Voting*, which neither Berelson nor Lazarsfeld was in a position to finish, and which awaited someone like him to complete, as Decatur awaited someone like Elihu Katz. Perhaps his most noteworthy work revolved around developing computer simulations of voting and other social processes, for which he was nominated as a Columbian Ahead of His Time for the university's 250th anniversary.[72] However, contemporary media researchers would probably find his most interesting and relevant textual accomplishments to lie with his unpublished studies of radio and television in the 1950s, and with the wildly creative (if sometimes baffling) mathematical models of media and popular culture that he developed in the 1960s.

If McPhee was a midlevel mover in the postwar history of media research in America, he was also a larger-than-life character in several of its dramas. The man known to some as "Wild Bill" was a representative figure in the larger worldwide development of public opinion research in the immediate postwar era (1945–1951), at the BASR during its middle years (1951–1960), and in the dissemination and development of the Columbia lineage afterward (1961–1998). A grant recipient from central funding agencies of his day (the Ford Foundation, the NSF, and indirectly, the Voice of America), he was also among that small handful of media researchers who benefited from a year at Stanford's CASBS. He stands in too for the amphibious researchers who moved between commercial and academic worlds during the era. Like others who worked at the Bureau in the 1940s and 1950s, McPhee took advantage of relatively fluid institutional boundaries. He went from being a self-taught small businessman to an academic researcher—reversing the flow of Herta Herzog and Hans Zeisel, who went from the old Office of Radio Research to the McCann-Erickson advertising agency— but also stayed involved in the commercial realm through ongoing work with Simulmatics and Research Services. Finally, in his later work, McPhee represents those media researchers whose work falls outside the mainstream of their fields, and who drop from the sight of students and scholars. He is, in other words, a mostly forgotten figure.

That said, McPhee nevertheless stands available in multiple ways as a model. An innovating, think-beyond-the-field's-current-parameters kind of guy, he serves broadly as an intellectual model in this regard. His critically hued mathematical research offers a more particular model, little emulated by media researchers, but one that stands as an alternative point of light on the horizon of scholarly styles. It is a hybrid of sorts—part high-modernist turn-to-numbers, part democratic engagement with the culture industries—driven in his case by a distinctive kind of genius. His turn to simulation and computing technologies makes him a patron saint for those social research methodologies, while his return to Colorado from New York beckons as an example for all of those interested in leaving the faster track for the homelands. To put things in copious list-like form, I might say that McPhee is a model of sorts (positive or cautionary), for a number of subgroups in media research, existing and dormant, including: survey researchers, formal theorists, computer simulators, critical mathematical scholars, radio and television researchers, students of new and old technologies, hicks and apparent hicks, Coloradans, mountain people, intellectual oddballs, suburban-dwelling family people, commuters, cabin builders, drinkers, former drinkers, entrepreneurs, smokers, risk-takers, autodidacts, rugged individualists, late-blooming counter-culturalists, hikers, skiers, lifelong learners, and survivors.

McPhee is currently a minor conversationalist in the broader tradition of media research, but mostly at the margins, in neighboring or once-neighboring fields like election or political behavior studies, marketing research, and sociological theory and method. "Notes on a Campaign Simulator" does appear in Berelson and Janowitz's 1966 reader, and McPhee's books and articles are cited with semiregularity in the social sciences—approximately 230 references in all in the Social Science Citation index since the mid-1960s, including a few citations each year recently. He appears most often in the political science literature (in discussions of voting and the influence of small group processes), but also makes appearances in sociology (as a mathematical theorist and computer simulation methodologist), in addition to marketing and, very occasionally, media research. McPhee's idea of "double jeopardy" has entered the vocabularies of both of these last fields, sometimes with reference to his coinage and formulation, other times obliterated by incorporation. Use of the concept varies, but as developed in *Mass Behavior*, double jeopardy names the double disadvantage that relatively unknown products or program face in the marketplace, which in turn helps maintain the mass media status quo of the time.[73] The term is a minor commonplace in marketing, and broadcast researchers have sometimes used it as well to analyze patterns of television viewing and audience exposure to alternatives. For the most part, though, McPhee remains untouched and unremembered by contemporary media researchers and historians of the field. Within our field, he lives on mostly in the memories of those diminishing few people who knew the fleshly man, Bill McPhee, a striking figure in their worlds.

In conclusion, let me emphasize several points that I've tried to make in this piece. The first is that individuals matter, and we can productively draw upon them in all manners of history that we write. One kind of history focuses centrally upon individuals, and I have given a condensed example of it through the person of McPhee, whose significance I myself did not recognize until I researched and wrote this chapter, and whose story I tell incompletely. McPhee's life runs as a thread through a number of important projects and sites in the history of the field, and so following that thread functions as one productive means of discovery and invention. We learn not only about that life (often illuminating in its own right), but also about the institutions, dramas, and collective decisions with which it intersected.

Many other lives lie out there that are worth uncovering and engaging with from the perspectives of both history and contemporary media research. Two worth prioritizing are Helen McGill Hughes and Herta Herzog, significant intellectuals in their own right, through whom we can learn a great deal about the conditions for women in media research from the 1930s on.[74] We still know far more about the central men in the field than we do about the central women, or about the marginal figures male or female. In addition, Herzog (who moved among Austria, the United States, and Germany) and Hughes (who went between Canada and the United States) point us in a second direction we need to pursue: the writing of international histories of media research, peopled by those who made a difference and those unduly left behind as the field developed worldwide, among varying intellectual cross-fertilizations and enabling and constraining forces. That, it seems to me, is the next step we should take, as we focus on individuals and amplify other topics, in our efforts to write figures into the field, and to bring shape, detail, and vivacity to the greater tales we tell.[75]

NOTES

1. In the important Part I, "On Interpretation," of Burke, *Permanence and Change*, esp. 5–9, 37–49.
2. Berelson, Lazarsfeld, and McPhee, *Voting*.
3. "The parts played by people" is of course the elegant alliterative phrase composed by Elihu Katz and made famous in Katz and Lazarsfeld, *Personal Influence*, which I use here as both tribute and rhetorical continuation of the intellectual tradition of the field. As one small way of blending the history of media research with ongoing work in the field (be it theoretical, empirical, interpretive, or critical), I would advocate recycling well-known and overlooked phrases from the past ("rhetorical commonplaces of the field") in our discourses of today.
4. Berelson, "State of Communication Research."
5. Schramm, "Communication Research;" and Schramm, *Beginnings of Communication Study*.
6. Rogers, *History of Communication Study*.
7. Berelson, "State of Communication Research."
8. Schramm, "Communication Research."
9. Schramm, *Beginnings of Communication Study*.
10. Rogers, *History of Communication Study*.
11. Abbott, "Historicality of Individuals."
12. Ibid., 4.
13. Ibid., 10.
14. Ibid., 3. Abbott argues "that the sheer mass of the experience that individuals carry forward in time—what we might think of in demographic terms as the present residue of past cohort experience—is an immense social force," "a huge, recalcitrant weight of quite particular human material that severely limits what those large forces can accomplish." (Abbott, "Historicality of Individuals," 7.)
15. Quintilian, *Institutio Oratoria*, 1:2.
16. On threading things together in relation to "the investigation of things" (kê-wu) in eleventh-century neo-Confucian thinking, see Graham, *Two Chinese Philosophers,* 78–79.
17. Emerson, "Uses of Great Men," 3.
18. Ibid., 8.
19. Cooley, *Social Process*, 116.
20. Cooley, "Life Study Method."
21. Leff, "Hermeneutical Rhetoric," 199–204.
22. Burke, *Permanence and Change*.
23. Merton, *Shoulders of Giants*, 312; see also Sztompka, *Robert K. Merton*.
24. McPhee, "A Life Remembered;" and John McPhee, personal communications, November 13, 2006 and January, 6 2007.
25. The success of the massive (and harrowing) operation established the airlift as a tool for warfare, which the United States would apply in the first major skirmish of the cold war era, when an airlift kept West Berlin supplied after the Soviet Union cut off land routes to the city. The success McPhee enjoyed in staying alive persisted when he fortuitously missed a commercial flight in the 1950s that crashed and killed all but two of its passengers (the first Boeing 707 that crashed in New York City), walked away from an automobile crash in California's mountains when his sports car landed in a tree, and was discovered twice at his cabin on the verge of lapsing into a diabetic coma. (John McPhee, personal communication, November 13, 2006.)

26. Converse, "Changing Conceptions."

27. Field, an Englishman who came to the United States in the 1920s, learned survey research in the 1930s under George Gallup in Young and Rubicam's Research Department. He had dreams of using survey research not for marketing or predicting elections but rather to serve as a voice of the people, reflected in his short-lived pre-NORC People's Research Corporation, and continuing until his death. For further discussions of Field, see the three-part history on the NORC Web site, "NORC History"; as well as Converse, "Changing Conceptions"; and Simpson, *Science of Coercion*, 95.

28. Lazarsfeld and Field collaborated on *People Look at Radio*, a fascinating little book based on survey conducted by NORC, funded by the National Association of Broadcasters, and interpreted by the Bureau of Applied Social Research (BASR). Addressing the overall quality of programming and the problems with on-air commercials, the volume is an overlooked classic in media criticism, blending populism with client-based administrative research. For Lazarsfeld's history, see the hard-to-find but indispensable volume edited by Lautman and Lécuyer, *Paul Lazarsfeld*, as well as David Morrison's series of essays on him, listed on Jefferson Pooley's indispensable and still-growing History of Communication Research Bibliography Web site: url: http://www.history ofcommunicationresearch.org

29. Columbia University Special Collections, BASR Archives, Box 27, Folder B-0389.

30. "Lazarsfeld came back with McPhee, promoting his as a 'genius'—which he was," Elihu Katz remembered. "He was heralded as a 'native genius.'" (Elihu Katz, personal communication, November 30, 2006.)

31. The Northwestern conference is mentioned (in "University Parley Opens," *New York Times*). Merton identified him as such on a visit to Colorado and speaking with Richard Jessor, professor of psychology, who cofounded the University of Colorado's Institute for Behavioral Science, where McPhee was affiliated in the 1960s and 1970s. (Richard Jessor, personal communication, December 14, 2006.)

32. David Sills, personal communication, December 5, 2006; and Rolf Meyersohn, personal communication, November 22, 2006.

33. As a new graduate student at Columbia in 1956, Kadushin was assigned to McPhee as a Ford Foundation Intern, and was put to work running election punch card data through a counter sorter to test propositions made in the appendices of *Voting*. As Kadushin recounted, "I would run some data that I thought illustrated a proposition in *Voting* (a painful exercise at that time—the cards dropped into pockets and then each pocket had to be counted; a cross tab involved taking the contents of a pocket and running them on another column—each pocket represented one 'stub') and write up tables by hand and percentage them with my slide ruler. Every other Friday afternoon I would drive up to Hastings on the Hudson (near Bob Merton's house) and meet with Bill in his attic study. He was a big man and a serious drinker. He would pour from a large pitcher of martinis—one glass for him and one for me—and this went on and on. He yelled at me for my stupidity. 'Why in hell did you run that—what's that supposed to show,' etc. I learned the right way to do analysis in the haze of alcohol (remember, he was a big man who could drink most people under the table—and I was 5'9" and weighed 150 pounds)." The work would culminate in a chapter of McPhee and Glaser's *Public Opinion and Congressional Elections*. (Charles Kadushin, personal communication, October 26, 2006.)

34. We still lack a complete account of Lazarsfeld's media research, which ran from radio studies in the early 1930s through work on the telephone in 1970. David Morrison has covered a great deal of ground very well in a series of articles (Morrison, "Beginnings of Mass Communication"; Morrison, "Transfer of Experience;" and this volume), but the later years remain mostly unchart-ed. For a partial map of relevant unpublished work, see Pasanella, "Mind Traveller." The best

short introduction to Lazarsfeld's life is probably Sills, "Paul Lazarsfeld." For a fuller view, see the excellent edited volume by Lautman and Lécuyer, *Paul Lazarsfeld.*

35. For early and partial attempts to document gendered aspects of the Columbia scene, see the articles by Susan Douglas, John Summers, Paddy Scannell, and myself in Simonson, "Politics," as well as my introduction to a republication of Merton, *Mass Persuasion*, xl–xli.

36. As he described his position at the time, Lowenthal was "Director, Evaluation Staff, International Broadcasting Service, U.S. Department of State. In this capacity, is responsible for planning and supervising major international communications research projects, utilizing all methods of social research, and involving major research contracts with leading university and private research organizations, and for supervision of a staff of 50 to 100 sociologists, social psychologists, political scientists, etc. Also serves as one of the five directors of the Voice of America (1,500 employees), planning and formulating programming policy, propaganda strategy, etc, and acts as consultant to other U.S. agencies involved in psychological warfare. The position requires extensive liaison with U.S. Missions abroad, with foreign universities, and with policy officials of the Department of State" (Leo Lowenthal, curriculum vitae, 21 April 1954; in Robert K. Merton, personal files). This is an aspect of Lowenthal's career that isn't featured in more celebratory accounts, such as Hardt's "Conscience of Society," where the Frankfurt émigré is depicted as "the conscience of society." See Simpson, *Science of Coercion,* for an account of Lowenthal's role in linking communications research to psychological warfare during the cold war.

37. McPhee and Meyersohn, "Radio Audience."

38. Lazarsfeld, Berelson, and Gaudet, *People's Choice.*

39. McPhee and Meyersohn, "Syrian Attitudes," i.

40. As Simpson, *Science of Coercion,* 55–57, has shown, the studies were part of a broader institutional alignment between communications research and U.S. government funding agencies during the cold war era. Attitudes among researchers ranged from active and enthusiastic support for U.S. foreign policy to quieter complicity in its projects. For a fuller consideration of the study, see Samarajiva, "Murky Beginnings."

41. An extension of the Bureau's work on voting and political behavior, the proposal was written by Bureau Director Charles Glock and also included Columbia historian Richard Hofstadter, political scientist David Truman, social psychologist Herbert Hyman, and political sociologist Seymour Martin Lipset. Glock described the project to the Ford Foundation's Bernard Berelson (a Bureau alumnus himself) as an effort to "reanalyze the assertions that have been advanced about the political process, within systems of general propositions, from which can be derived a large number of empirically testable hypotheses" about politics and the individual, groups, and social structure. (Charles Glock to Bernard Berelson, May 16, 1952, BASR Archives, Columbia University, Box 104, Proposal 12.)

42. Berelson, Lazarsfeld, and McPhee, *Voting.*

43. "From conversation with Bill it was my impression that his major contributions to *Voting* were in the realm of the effects of the social surroundings of the individual on the probability of voting and the probability of voting in particular ways. I think Chapters 6 and 7 are almost wholly his. The Chart and note on page 146 were the program set for my work with him on the 1950 Congressional election. He first had me go through detailed historical voting records by county to try to work out the 'year off' pattern. It didn't quite work out." (Charles Kadushin, personal communication, January 30, 2007.)

44. McPhee, "Community Controversies."

45. Marjorie Fiske, yet another Bureau alum who took a job that controlled research funds, was in 1953 a member of the Implementation Committee on Television, and Glock approached her with

a proposal for funding "four background studies" for the committee. (Charles Glock to Marjorie Fiske, April 17, 1953, BASR Archives, Columbia University, Box 104, Proposal 37). In another sign of the small circles (or nepotism), Fiske married Leo Lowenthal, and served him for a time as research director of the Voice of America.

46. McPhee, Ennis, and Meyersohn, "Disc Jockey," II-1. Four decades later, Ennis included some of the findings (in *The Seventh Stream*). See also the related report prepared by Lowenthal in consultation with McPhee, "Trends in the Licensing of Popular Song Hits, 1940–1953," BASR, August 1953, a summary of best-selling records, most-played juke box titles, most-played disc jockey titles, most popular network songs, and sheet music sales for the period. Daniel Lowenthal was Leo Lowenthal's son from his first marriage.

47. Noting that "we are sensitive to occasional criticism from inside the Bureau about accepting projects which are 'merely' the supplying of fairly routine studies competitive with those of commercial organizations," McPhee proposed a study for *LIFE* that "falls into a general class of what we call 'flow' or 'transmission' studies" as well as "a class of what one might call 'use' or 'gratification' studies." The magazine was particularly interested in determining how *LIFE* was "intentionally being used in formal education today," but McPhee suggested something grander, which would involve classroom observation and follow-up interviews to determine something about media flow into classrooms (through both students and teachers) as well as the uses made and gratifications derived from media in those social and institutional settings. (William McPhee to Herb Breseman, Advertising Research Director at *LIFE*, 4 November 1953; BASR Archives, Columbia University, Box 104, Proposal 53.)

48. Charles Glock to Hugh M. Beville, Jr., NBC Director of Research, 27 January 1954; Memo from BASR to Members of the Projects Committee, Board of Governors, 27 January 1954. Both in BASR Archives, Columbia University Box 45, Folder B-0512.

49. "Humans originally adopt a device to perform some primary function and this is typically the same primary function the engineer had in mind," they wrote. "But then as the device is introduced into the homes of the people, or as it becomes woven into their lives in some way, it begins to take on additional functions for these people . . . [A] radio was once 'only a radio,' but after people have spent a generation weaving it into their lives, it is many things—an alarm clock to wake people up pleasantly, a kind of morning newspaper to bury one's thoughts in at breakfast, a traveling companion in the car, a day-long visitor to help pass the drearier hours of the day for a housewife, an education for the woman who learns about life from soap operas, game of suspense for the up-to-the-minute news follower or sports fan, a record player for teenagers, a partisan ritual for the avid follower of Fulton Lewis, Jr., a Muzak sound system for people whose moods respond to music, a prized personal possession for a child, and so on through many more. The uses to which people put a device even include contradictory ones, as for example when insomniacs use the same radio program to go to sleep as drousy drivers use to help keep awake!" (McPhee and Meyersohn, "Futures for Radio," 128–29.)

50. Bourne and McPhee, "Monitor."

51. BASR Folder B-0582; Lennard, Jarvik, and Abramson, "Lysergic Acid Diethylamide"; Lennard, Abramson, and Hewitt, "Drugs and Social Interaction"; and Lennard and Hewitt, "Study of Communications Processes."

52. McPhee, "Potentialities of Television."

53. Coleman and McPhee, "Proposal for a Research Development Program in 'Mass Dynamics,'" May, 1957, BASR Folder B-0553, BASR Archives, Columbia University, Box 49.

54. The end of Berelson's famous "requiem for a field" piece (where he sketches seven new directions for the future) is less remembered than the opening, which begins, "My theme is that, as for

communication research, the state is withering away." "*Mass* communication," Berelson wrote, noting that Coleman and McPhee had told him "that the first word needs more emphasis relative to the second. Their position is that the field is better seen as one of a variety of *mass* activities and that headway will be made by stressing similarities of such mass phenomena rather than the particularities attaching to a mass communication system." (Berelson, "State of Communication Research," 440, 445.)

55. The Simulmatics Research Division was headed by Ithiel de Sola Pool and included McPhee, Coleman, Yale psychology professor Robert Abelson, and Alex Bernstein, a former IBM mathematician. The company was run by Edward L. Greenfield, a New York businessman who saw the financial possibilities afforded by new computing technologies. (Freeman, "Advertising.") for more on Simulmatics, follow out the electronic paper trail in the *Times* historical database.

56. Robert B. Smith, personal communication, December 22, 2006. For more on the history of the voting simulation model at Columbia, see Smith, "Innovations at Columbia"; as well as McPhee, Ferguson, and Smith, "Model Simulating Voting."

57. McPhee submitted a grant proposal for "Problems in the Analysis of Computer Models of Social Processes," February 22, 1960. It said, among other things, "It is customary at the Bureau of Applied Social Research to encourage research with policy implications. One candidate for that possibility in our present work is the popular culture model, as it bears on the consequences of the audience ratings and sales polls; that is, the feedback devices controlling the prosecution of commercial culture." (BASR Archives, Columbia University, Box 106, Proposal 234.)

58. By natural sampling, McPhee meant the informal and uncontrolled "sampling" that people do on an everyday basis, which gives them a sense of their social worlds, to be able to answer questions about things like, for example, how Catholics vote ("A Theory of Natural Sampling and Popular Renown," [nd], BASR Archives, Columbia University, Box 62, B-0725). It was something like the social psychology Noelle-Neumann later developed, but McPhee folded the insight into a mathematical model of what he called "the processes of haphazard absorption of information and experience in large populations, appropriate for theories of mass phenomena such as public opinion and mass communications" (Ibid., 20).

59. "Feasibility of a Computer Model of Alcoholic Drinking," grant application made to the National Institutes of Health (a part of the US Department of Health, Education, and Welfare), May 31, 1961, BASR Archives, Columbia University, Box 106, Proposal 280.

60. Following up after receiving his one-year trial grant, McPhee submitted a proposal in early 1962 for a three-year project, "Formal Models of Mass Social Processes." Noting that he had written a paper, "Communication Effects" (a draft of which was set to appear in the Proceedings of the 1961 UCLA Conference on Simulation), McPhee set out to develop "a model of social addictions: 'pursuits,' 'passions,' and the like" in the service of developing "a larger simulation of the 'ecology' of motivation of contagious pursuits (those that depend on environment, e.g., narcotics or skiing, and feed on themselves, e.g., mathematical training in a discipline)" (10). He wanted to extend his model to what he called "ideological 'flare-ups,'" by which he meant both eras that came to swing strongly toward the right or left (the 1920s/1950s and the 1930s, respectively) and the "outbreaks" of those flareups at the community level, which he had examined in the school disputes of the mid-1950s (11–12). McPhee also sketched the model's application for what he called technological "replacement" theory, or the situation whereby new technologies partly supplant old ones. As McPhee and Meyersohn had discovered in their 1954 "Futures for Radio" study, new uses were emerging for the old technology, and new and old developed in mutual interrelation, creating "the seeming anomaly" that in the mid-1950s radio receivers were outselling TVs at the height of the TV boom. McPhee hoped to address these

phenomena and others with mathematical models. (BASR Archives, Columbia University, Box 107, Proposal 289.)

61. McPhee and Glaser, *Public Opinion.*

62. McPhee, *Formal Theories.*

63. McPhee and Glaser, *Public Opinion*, 171–79.

64. University of Michigan researchers in particular were taken by it, which is striking given the rivalry and divergence between Columbia and Michigan in the 1950s. Philip Converse noted that the book made him see that "the work of the Columbia and Michigan groups have converged upon one another," and observed that "McPhee's voting system model finds eloquent indorsement [*sic*] in all of the Michigan data." (Converse, Review of *Public Opinion.*) See the similarly warm review by Michigan's Lamb, review essay, 285–86.

65. McPhee, *Formal Theories.*

66. On the early 1960s and the particular importance of 1962 for mass communication, see Peters and Simonson, *Mass Communication*, 269–73. Peters and I missed McPhee when we were compiling that reader, which does better in representing the 1950s than it does the 1960s, a decade that is still coming into focus in the history of the field.

67. "If this book is any tradition, it is embodied by Robert Lazarsfeld, a contrary ten-year-old," McPhee wrote. "He has decided that in his time in sociology, the exciting uses of method will be in theory, and vice versa, the dramatic changes in theory will be in method. Yet his and mine is a strange 'revolution'; the local government keeps sending us supplies. I have not even asked Paul Lazarsfeld and Robert Merton to pretend even to the responsibility, however, of superficial reading of these examples completed when my absence prevented serious discussion." (McPhee, *Formal Theories*, vii.)

68. Ibid., 3.

69. Berelson and Janowitz, *Reader in Public Opinion.*

70. McPhee was spanked by Harrison White, a young MIT-trained theoretical physicist tuned Princeton-trained mathematical sociologist, who suggested that "the author's slipshod, chatty style is more appropriate for a pitchman's approach to a collection of mathematics amateurs than for a scholarly attempt to demonstrate the merits of a new scientific approach." As a doctor of scholarly learning, he intoned, "My diagnosis is that the technical work as such is too flawed by sheer incompetence to be worth the attention of the 'research fraternity.'" (White, Review of *Formal Theories*, 772.) The University of Michigan's Robert Somers (a friend of McPhee's) was more charitable, writing, "It takes a remarkably creative mind to (a) see the problematic character of everyday empirical observations, (b) translate the problematics into mathematical models that are simple enough to work with mathematically but not trivial in interpretation, and (c) deduce a variety of subtle applications from the formal properties of the models. This is not something one can expect from the typical mathematician, or the typical social scientist, or even the typical mathematical sociologist. This book is unique in the extent to which it has accomplished all three of these things." (Somers, Review of *Formal Theories*, 98–99.) Compare the passing reference to McPhee in Thomas Fararo's essay on the historical development of mathematical sociology. (Fararo, "Reflections on Mathematical Sociology," 80.)

71. "Most of our mass culture today is created for money or profit," the essay begins, and McPhee goes on to critique the idea that popularity indicates a kind of democratic expression of the vox populi. "Picture a system in which one is allowed to vote, here to register his cultural choice by buying something, not just once per person but as many times as he pleases. Specifically, suppose the number of times one votes is in proportion to how well he *likes* the existing regime, the existing fare in a culture. And conversely, in proportion that another person does *not* like the existing regime (the fare offered does not suit his cultural taste level), in that degree suppose he votes fewer

times. He is progressively *disenfranchised*, then, in proportion that he would otherwise register protesting votes. That would be a political joke, a George Orwell nightmare. It is not a joke, alas, but the real nightmare of the way we vote in commercial culture." (McPhee, "Culture Becomes a Business," 237.)

72. See the "Columbians Ahead of their Time" feature on the Columbia 250 Web site, and Robert Smith's nomination of McPhee.

73. McPhee, *Formal Theories*, 133–40.

74. For brief biographical starting points, see Perse, "Herta Herzog"; and Peters and Simonson, *Mass Communication*, 118, 139.

75. Thanks to the editors and Adrienne Stewart for offering insightful comments on earlier drafts, and to my informants—John McPhee, Robert Smith, Rolf Meyersohn, Charles Kadushin, Elihu Katz, David Sills, Dick Jessor, and Harriet Zuckerman—who kindly took the time to speak with me about McPhee. Archival research was supported with funds provided by the Department of Communication, University of Colorado, Boulder.

WORKS CITED

Abbott, Andrew. "The Historicality of Individuals." *Social Science History* 29 (2005): 1–16.

Berelson, Bernard. "The State of Communication Research." In *Mass Communication and American Social Thought,* edited by John Durham Peters and Peter Simonson, 440–45. Lanham, MD: Rowman & Littlefield, 2004.

Berelson, Bernard, and Morris Janowitz, eds. *Reader in Public Opinion and Communication.* 2nd ed. New York: Free Press, 1966.

Berelson, Bernard, Paul F. Lazarsfeld, and William N. McPhee. *Voting: A Study of Opinion Formation in a Presidential Campaign.* Chicago: University of Chicago Press, 1954.

Bourne, Frank, and William N. McPhee. "Monitor: An Evaluation of a New Experiment in Radio Programming." Report B-0541. Bureau of Applied Social Research, Columbia University: New York, 1956.

Bureau of Applied Social Research Archives. Columbia University Special Archives. Columbia University, New York.

Burke, Kenneth. *Permanence and Change: An Anatomy of Purpose.* New York: New Republic Press, 1935.

"Columbians Ahead of Their Time." Columbia 250. http://c250.columbia.edu/c250_celebrates/your_ columbians/william_mcphee.html (accessed May 2, 2007).

Converse, Philip E. "Changing Conceptions of Public Opinion in the Political Process." *Public Opinion Quarterly* 51 (1987): s12–s24.

———. Review of *Public Opinion and Congressional Elections,* edited by William N. McPhee and William A. Glaser. *American Journal of Sociology* 68 (1963): 604.

Cooley, Charles Horton. "The Life Study Method as Applied to Rural Social Research." In *Sociological Theory and Social Research,* edited by Charles Horton. Cooley, 331–39. New York: Henry Holt and Company, 1930.

———. *Social Process.* 1919. Reprint, Carbondale and Edwardsville: Southern Illinois University Press, 1966.

Douglas, Susan J. "*Personal Influence* and the Bracketing of Women's History." In "Politics, Social Networks, and the History of Mass Communication Research: Rereading *Personal Influence,*" edited by Peter Simonson, special issue. *Annals of the American Academy of Political and social science* 608 (2006): 41–50.

Emerson, Ralph Waldo. *Representative Men: Seven Lectures.* 1850. Reprint. Boston and New York: Houghton Mifflin, 1903.

———. "Uses of Great Men." In *Representative Men: Seven Lectures,* edited by Ralph Waldo Emerson, 3–39. Boston and New York: Houghton Mifflin, 1903.

Ennis, Philip H. *The Seventh Stream: The Emergence of Rocknroll in American Popular Music.* Hanover, NH: Wesleyan University Press, 1992.

Fararo, Thomas J. "Reflections on Mathematical Sociology." *Sociological Forum* 12 (1997): 73–101.

Freeman, William M. "Advertising: Life Is Imitated for Research." *New York Times* (August 27, 1961).

Graham, Angus C. *Two Chinese Philosophers: Ch'êng Ming-tao and Ch' êng Yi-ch'uan.* London: Lund Humphries, 1958.

Hardt, Hanno. "The Conscience of Society: Leo Lowenthal and Communication Research." *Journal of Communication* 41, no. 3 (1991): 65–85.

Katz, Elihu, and Paul F. Lazarsfeld. *Personal Influence: The Part Played by People in the Flow of Mass Communications.* Glencoe, IL: Free Press, 1955.

———. *Personal Influence: The Part Played by People in the Flow of Mass Communications,* with a new foreword by Elihu Katz. New Brunswick, NJ: Transaction, 2005.

Lamb, Karl. Review essay of *Rum Religion, and Votes: 1928 Re-Examined,* by Ruth C. Silva; *A Theory of Public Opinion,* by Francis Graham Wilson; and *Public Opinion and Congressional Elections,* by William N. McPhee and William A. Glaser. *Midwest Journal of Political Science* 7, no. 3 (August 1963): 283–86.

Lautman, Jacques, and Bernard-Pierre Lécuyer, eds. *Paul Lazarsfeld (1901–1976), Sociology from Vienna to New York.* Paris: L'Hartmattan, 1998.

Lazarsfeld, Paul F., and Harry Field. *The People Look at Radio.* Chapel Hill: University of North Carolina Press, 1946.

Lazarsfeld, Paul F., Bernard F. Berelson, and Hazel Gaudet. *The People's Choice.* New York: Duell, Sloan & Pearce, 1944.

Leff, Michael. "Hermeneutical Rhetoric." In *Rhetoric and Hermeneutics: A Reader,* edited by Walter Jost and Michael J. Hyde, 196–214. New Haven: Yale University Press, 1997.

Lennard, Henry, and Mollie P. Hewitt. "The Study of Communications Processes under LSD." In *Use of LSD in Psychotherapy,* edited by Harold A. Abramson, 199–240. New York: Josiah Macy, Jr. Foundation, 1958.

Lennard, Henry, Harold A. Abramson, and Mollie P. Hewitt. "Drugs and Social Interaction." In *Neuro-Psychopharmocology,* edited by P. B. Bradley, 625–30. Amsterdam, Elsevier Publishing, 1959.

Lennard, Henry, Murray E. Jarvik, and Harold A. Abramson. "Lysergic Acid Diethylamide (LSD-25); XII. A Preliminary Statement of Its Effects on Interpersonal Communication." *Journal of Psychology* 41 (1956): 185–98.

McPhee, John. "A Life Remembered: Dr. William McPhee." Family obituary. April 3, 1998.

McPhee, William N. "Community Controversies: Affecting Personal Liberties and Institutional Freedoms in Education." Report B-0533. Bureau of Applied Social Research, Columbia University: New York, 1954.

———. *Formal Theories of Mass Behavior.* Glencoe, IL: Free Press, 1963.

———. "The Potentialities of Television." Report B-0620. Bureau of Applied Social Research, Columbia University: New York, 1957.

———. "When Culture Becomes a Business." In *Sociological Theories in Progress,* edited by Joseph Berger, Morris Zelditch, Jr., and Bo Anderson, 227–43. New York: Houghton Mifflin, 1966.

McPhee, William N., and Rolf Meyersohn. "Futures for Radio: An Interpretive Study of Radio Potential Under Television Conditions." Report B-0470. Bureau of Applied Social Research, Columbia University: New York, 1955.

———. "The Radio Audience of Lebanon." Report B-0370-11. Bureau of Applied Social Research, Columbia University: New York, 1951.

———. "Syrian Attitudes Toward America and Russia." Report B-0370-15. Bureau of Applied Social Research, Columbia University: New York, 1952.

McPhee, William N., and William A. Glaser, eds. *Public Opinion and Congressional Elections*. Glencoe, IL: Free Press, 1962.

McPhee, William N., Jack Ferguson, and Robert B. Smith. "A Model Simulating Voting." In *Computer Simulation of Human Behavior*, edited by William Starbuck and John Dutton, 469–81. New York: John Wiley, 1971.

McPhee, William N., Phillip Ennis, and Rolf Meyersohn. "The Disk Jockey: A Study of the Emergence of a New Occupation and Its Influence on Popular Music in America." Report B-0470. Bureau of Applied Social Research, Columbia University: New York, 1953.

Merton, Robert K. *Mass Persuasion: The Social Psychology of a War Bond Drive*, with an introduction by Peter Simonson. New York: Howard Fertig Press, 2004.

———. *On the Shoulders of Giants: A Shandean Postscript*. Chicago: University of Chicago Press, 1991.

Morrison, David. "The Beginnings of Mass Communication Research." *European Journal of Sociology* 29 (1978): 347–59.

———. "The Transfer of Experience and the Impact of Ideas: Paul Lazarsfeld and Mass Communication Research." *Communication* 10 (1988): 185–209.

"National Opinion Research Center: NORC History." *NORC Online*. http://www.norc.org/Aboutus/history (accessed May 18, 2007).

Pasanella, Ann K. *The Mind Traveller: A Guide to Paul F. Lazarsfeld's Communication Research Papers*. New York: Freedom Forum 1994.

Perse, Elizabeth M. "Herta Herzog." In *Women in Communication: A Biographical Sourcebook*, edited by Nancy Signorielli, 202–11. Westport, CT: Greenwood Press, 1996.

Peters, John Durham, and Peter Simonson, eds. *Mass Communication and American Social Thought: Key Texts, 1919–1968*. Lanham, MD: Rowman & Littlefield, 2004.

Pooley, Jefferson. History of Communication Research Bibliography. http://www.historyofcommunicationresearch.org (accessed September 15, 2007).

Quintilian. *Institutio Oratoria*. 4 vols. Translated by H. E. Butler. Cambridge, MA: Harvard University Press, 1921

Rogers, Everett M. *A History of Communication Study: A Biographical Approach*. New York: Free Press, 1994.

Samarajiva, Rohan. "The Murky Beginnings of the Communication and Development Field: Voice of America and *The Passing of Traditional Society*." In *Rethinking Development Communication*, edited by Neville Jayaweera, Sarath Amunugama, and E.Tài Ariyaratna, 3–19. Singapore: Asian Mass Communication Research and Information Centre, 1987.

Scannell, Paddy. "*Personal Influence* and the End of the Masses." In "Politics, Social Networks, and the History of Mass Communications Research: Rereading *Personal Influence*," edited by Peter Simonson, special issue. *Annals of the American Academy of Political and Social Science* 608 (2006): 115-29.

Schramm, Wilbur. *The Beginnings of Communication Study in America: A Personal Memoir*, edited by Steven H. Chaffee and Everett Rogers. Thousand Oaks, CA: Sage, 1997.

Schramm, Wilbur. "Communication Research in the United States." In *The Science of Communication: New Directions and New Findings in Communication Research*, edited by Wilbur Schramm, 1–16. New York: Basic Books, 1963.

Sills, David. "Paul Lazarsfeld, 1901–1976." *Biographical Memoirs* 56 (1987): 251–82.

Simonson, Peter, ed. "Politics, Social Networks, and the History of Mass Communications Research: Re-Reading *Personal Influence*," special issue. *Annals of the American Academy of Political and Social Science* 608 (2006).

Simpson, Christopher. *Science of Coercion: Communication Research and Psychological Warfare, 1945–1960*. New York: Oxford University Press, 1994.

Smith, Robert B. "Innovations at Columbia." In *Paul Lazarsfeld (1901–1976), Sociology from Vienna to New York*, edited by Jacques Lautman and Bernard-Pierre Lécuyer, 353–60. Paris, L'Hartmattan, 1998.

Somers, Robert H. Review of *Formal Theories of Mass Behavior*, by William N. McPhee. *American Journal of Sociology* 70 (1964): 98–99.

Summers, John H. "Perpetual Revelations: C. Wright Mills and Paul Lazarsfeld." In "Politics, Social Networks, and the History of Mass Communications Research: Rereading *Personal Influence*," edited by Peter Simonson, special issue. *Annals of the American Academy of Political and Social Science* 608 (2006): 25–40.

Sztompka, Piotr. *Robert K. Merton: An Intellectual Profile*. New York: St. Martin's, 1986. "University Parley Opens." *New York Times*, (October 12), 1951.

White, Harrison. Review of *Formal Theories of Mass Behavior*, by William N. McPhee. *American Sociological Review* 29, no. 5 (October 1964): 771–73.

CHAPTER THIRTEEN

Media Research 1900–1945: Topics AND Conversations

JAMES A. ANDERSON AND JANET W. COLVIN

This chapter had its beginnings in work done to detail a 100-year history of media violence and aggression research.[1] We were struck then as we are now with the similarity of issues that attracted the attention of researchers across that period. We were also struck by how little the present day references the past. A contemporary journal article would barely dip a toe in the twentieth century. An advanced search might reach to the 1980s. That shortsightedness, of course, allows us to claim the old as new again, but it prevents us from understanding how we are joining a conversation that has been going on since the appearance of the penny-dreadfuls and the blood-running newspaper crime reports of the mid-nineteenth century.

METHOD

Given the opportunity here to more carefully examine the beginning products of social science media research, we developed a found sample of journal articles published in criminology, education, psychology, political science, and sociology journals from 1900 to 1945 (with excursions to 1947 to account for the impact of World War II on scholarship release). A found sample is a systematic convenience sample based on what is available to the researchers. In our case, it was the University of Utah library and access to full-text databases such as PsycINFO,

JSTOR, and CMMC. Two approaches were used: keyword searches were conducted, using broad terms such as newspapers, motion pictures, radio, and their variants. Physical and virtual searches of the journals were also conducted. In both of these searches, each available volume was examined for relevant articles. "Relevant" was defined by some topic involving mediated communication further addressed by empirical methodologies (experimental, survey, content analysis, case study, ethnographic), the development, or explication of theory, an educational/instructional application, or the critical analytical review of theory, methods, or results.

The method of textual analysis used here requires each case entry (article) to be in the form of a computer-readable, text file. That requirement meant that carbon-based articles (physical copy) had to be digitized. Our procedure was to use a digital camera and a copy stand to make a digital record of each page. The estimate is that we copied more than 3,500 pages from journal volumes. The digital copies and the downloaded PDF files had to be transformed from image files to readable text files through optical character recognition. Optical character recognition works nearly flawlessly with flat, high contrast, perfectly squared images, having no extraneous marks, smudges, or speckling. The greater distance from that standard, the more error appears and the more hand correction has to be done. Approximately twenty-two cases had to be abandoned because the work needed was prohibitive given the time constraints of this writing. We noted one systematic effect in these twenty-two cases: *School and Society*, a British education journal, routinely published correspondence on the use of media in the classroom. These were short comments but valuable to trace the impact of media on classroom instruction. Most of these were not processed because the typography caused a very high rate of processing errors. These comment articles accounted for sixteen of the twenty-two abandonments, although two exemplars are included in the sample.

Two hundred twenty-five articles from thirty-four different journals were collected and entered into the analysis. Table 13.1 presents the journals in the sample; the articles are listed in the Appendix.

The sample reported here, then, can make no claim that it represents all that is available from the literature of the period. Clearly, more journals could be searched, more databases examined, more articles selected, and more articles processed, and books and book chapters could be included. Nonetheless, the sample reported here may well be the only one of its kind, and it certainly represents the sort of work that was appearing in those nearly fifty years. Tables 13.2, 13.3, and 13.4 explore the sample by medium, article type, and overall content. In the tables, each case appears in only one category and all cases are coded. These tables are presented in the sections on the sample, coding, and conversations that follow.

Table 13.1. Journals appearing in the review

American Journal of Psychology	American Sociological Review	Annals of the American Academy of Political and Social Science
Archives of Psychology	California Journal of Elementary Education	Character and Personality
Clearing House	Educational Research Bulletin	Educational Review
English Journal	International Journal of Ethics	Journal of Abnormal and Social Psychology
Journal of American Institute of Criminal Law and Criminology	Journal of Applied Psychology	Journal of Applied Sociology
Journal of Educational Sociology	Journal of General Psychology	Journal of Pediatrics
Journal of Psychology	Journal of Social Psychology	Journalism Bulletin
Journalism Quarterly	Pedagogical Seminary	Proceedings of the American Sociological Society
Psychological Bulletin	Psychological Monographs	Psychological Review
Public Opinion Quarterly	Quarterly Journal of Speech	School Review
School and Society	Scientific Monthly	Social Forces
Studies in Sociology		

UNDERSTANDING THE SAMPLE

A few historical notes have to be provided to understand the character of the media research product of that era. These are drawn from both the history of media and the history of social science research.

Media Notes

Media enter the social science literature in three ways: professional practices such as journalism, advertising, and opinion polling; in their interface with the educational system in didactics, pedagogics, and accommodation; and in their effect on society. All three of the entrances require media to reach some threshold of market penetration, some critical convergence by which the majority of society is entailed. Newspapers reach that convergence by the 1880s and consequently that medium appears first in the sample. Motion pictures broke through between 1905 and 1910 and become the dominant form of entertainment for at least the next two decades. The cinema is the medium of the literature of the 1920s. For radio, although the technology was available in 1906, the business model of the medium did not develop until after World War I. In 1922, there were an estimated 440,000 receivers in the United States (for a population of 122 million). By 1933, there were more than 33 million sets in use.[2] The research literature of the 1930s is dominated by radio with regular appearances of the motion picture. Television shows its face in the late 1930s, mostly as a novelty, but it would have dominated the literature following its appearance at the 1939 World's Fair had it not been stopped by World War II. Table 13.2 presents the cases sorted by decade and medium of focus.

The doorway of effects appears to need additional cultural action for a medium to appear. Anderson[3] concludes his analysis of the effects literature by noting the conditions under which social consequences of media come into the purview of science:

> Historically, the actual process has begun when society has been required to accommodate a new, rapidly diffusing form of entertainment. That new form raises societal concerns in its own right and is also quickly connected to the on-going and often intractable problems of that society to the extent that political processes are engaged. Those political processes in turn further elevate the significance of what is now *the problem* and often provide the funding for the science that follows. That science has a mandate to solve *the problem* and *the problem* is a now a media problem.

As a result of this process, we first see newspapers with their "lurid and explicit" crime reporting held up as the source of decline in morality, the cause of juvenile delinquency, and as able participants in crime and violence. Newspapers were the videogame narratives of their day, read for and railed against their pleasures of violence. Newspapers hand off this dubious attention to the motion picture by 1920, although crime reporting is still an issue in criminology. Radio—particularly the radio crime drama—joins the motion picture in the 1930s to constitute the media of effects. We, of course, now know the rest of this history as

Table 13.2. Number of articles by medium and decade

Medium/Decade	1900s	1910s	1920s	1930s	1940s
Literature	2	1	0	2	1
Motion Pictures	0	7	7	47	4
Newspapers	6	1	12	11	6
Radio	0	0	3	32	20
Television	0	0	0	2	2
Not applicable	3	2	6	23	25

television supplants all other media from the end of the 1950s' comic book scare to the 1990s' introduction of the videogame. It remains to be seen how effects will be reconstituted in the current era of the consumer as producer, but we bet on online gaming.

Social Science Notes

Psychological science publication during the period from 1900 to 1920, and for some time following, exhibited the oligarchical characteristics of a relatively few major, tightly controlled outlets whose editors held their positions for a notable length of time. G. Stanley Hall held sway at the *American Journal of Psychology* (*AJP*) for more than twenty years. The ideological consequences of this arrangement become apparent by what was published. In psychology, behaviorism was the theory of choice until the mid-1920s. As a result, the topic of media is nearly absent in the mainline psychology journals. The first entry we found in *AJP*—a "limited effects" article (to use the common term for effects that happen through something else rather than directly from exposure)—is in 1923, a few years after Professor Hall's departure as editor and one year after the publication of Edward Tolman's[4] heresy of emotion in behaviorism that would soon become the revolutionary introduction of cognitivism.

Social psychology gained its own foothold in the pantheon of theory[5] and, along with cognitivism's attitudes and opinions, the demand for publication outlets following the cessation of worldwide hostilities broke the old oligarchy open. Consequently, it was the breakout journals of the 1930s, particularly the *Journal of Applied Psychology*, that provide most (20 items) of the psychology entries (61 total) in the sample.

We can see one other result of the liberalities introduced in the 1930s in the large increase of theoretical and analytical pieces that could not be attached to a particular medium (the N/A row of Table 13.2). These articles either addressed media in general—such as the processes of public opinion formation or advertising—or were concerned with the development of theory appropriate to the study of media and their effects or the methods of measurement.

Education (52 entries) and sociology (54 entries) provide the most continuous record of scholarly engagement of the media. The earliest records appear in the *American Journal of Sociology* and in *Educational Review*. The *Journal of Educational Sociology* carries literature on media through the 1940s. Education's interest has always been Janus-like, adopting new media for the classroom both in didactics and in media appreciation while expressing concern for the consequences of media on classroom performance and morality.[6] Sociology has focused on crime, propaganda, public opinion, and—not surprisingly—the consequences for society.

Public Opinion Quarterly, which describes itself as an interdisciplinary journal for the study of communication, begins to publish in 1936. It provides about 16 percent of the sample. Speech, at least one precursor of communication studies, also gets involved in the 1930s and 1940s. *Journalism Quarterly (JQ),* which begins publication as the *Bulletin* in 1925, offers a nearly continuous record of social science research. Its historical record is not yet archived, however, so only nine entries are represented here. About one-fourth of the sample (56), then, comes from communication-related publications, although another thirty or so articles remain untapped in *JQ.*

CODING AND ANALYSIS

About 85 percent of the 225 articles entered into the coding process dealt directly with media. The remaining 15 percent included articles that any media researcher would have been reading (e.g., Sherif and Cantril's 1945 article on attitudes).[7]

Coding

Coding was done by case, overall content, and content segment. Case coding records the global attributes of the case. Three case attributes (each with several values) were used: Decade of Publication, Focal Medium, and Article Type. Table 13.2 has presented the information on focal medium. Table 13.3 presents the article type by decade.

Overall content was one of eighteen topic values assigned to each case, according to our joint reading of each article. The overall content codes were drawn from the segment coding. Content segment coding used a grounded theory approach

Table 13.3. Article type by decade

Type/Decade	1900s	1910s	1920s	1930s	1940s
Clinical Study	0	2	1	3	1
Critical/Analytical Essay	9	4	20	49	33
Empirical Study	1	2	6	33	14
Media Appreciation/ Literacy	0	1	0	11	2
Media Pedagogics	0	2	0	5	2
Current Conditions	1	0	1	11	5
Review	0	0	0	5	1

in which codes are developed in the close reading of the articles. Nineteen content codes were developed using a "next unique paragraph" approach. The difference between the two coding schemes is that the overall content approach enters the introductory and concluding paragraphs of each case regardless of topical content into its typal analysis. It is coding based on structure. Segment coding responds to the content of segments (generally sentences and paragraphs) regardless of structural location. It is coding based on topic. Although this joint method does generate some duplicate text coding, it ensures that every case contributes to the analysis.

Analysis

The size of the database (more than 105MB of text) called for computer-assisted analysis. We used Galileo's Catpac II software. Catpac II produces concept clusters based on neuro-network analysis.[8] It looks at how words are used within word segments to extract the unique concepts (words) within the text and how they relate to one another. As with all clustering procedures, there is both a demand for intimate familiarity with the texts and a good deal of interpretation necessary to understand the meaning of the clusters in reference to those texts. Consequently, this is still an interpretive approach, but the reader can be assured that every word of the data has been evaluated.

To provide coherent textual entries for concept analysis, we adopted the notion of conversations in the scholarly record. The overall content codes were developed to represent those conversations. Other research has shown that scholarly interests follow an approximate ten-year cycle of growth and decay. Consequently, conversational topics that extended across time were examined for continuity, and new conversation categories were established as needed. Each major conversational unit had at least ten source elements (coded content from a case), and at least ten cases (articles) were represented. Two types of conversations are reported below: conversations that are represented by the cases on the basis of their overall topic, and conversations with the cases as identified in segment analysis.

THE CONVERSATIONS: CASE BY OVERALL CONTENT

Table 13.4 presents the overall content categories that were used to code the cases by decade. The table suggests that the conversations of sixty years ago would be familiar to us now, with the exception of radio listening and perhaps propaganda.

Table 13.4. Content topics by decade coded by case

Topic/Decade	1900s	1910s	1920s	1930s	1940s	Total
Advertising/Public Relations	1	1	1	7	2	12
Attitudes	0	0	0	6	4	10
Children	1	1	1	10	1	14
Content Analysis	0	0	1	2	3	6
Crime	0	0	5	0	0	5
Delinquency	0	2	1	2	0	5
Influences Thoughts/Actions	0	0	3	9	3	15
Radio Listening	0	0	1	5	2	8
Media Didactics/Pedagogics	0	3	2	11	1	17
Media Literacy/Appreciation	0	1	2	8	4	15
Media Training	1	0	0	7	0	8
Propaganda	0	0	0	12	4	16
Public Opinion	1	0	1	8	7	17
Public Policy	0	0	2	7	4	13
Research/Theory	1	1	4	8	17	31
Societal Implications	6	2	4	15	6	33

What is clearly missing is all the conversation that has been introduced by both feminist scholarship and cultural studies. The sensibilities of voice, power, race, and gender are simply absent, and their absence renders the discourse a bit quaint on some occasions and shockingly unreflective on others.

Ten topics met the initial criteria for further analysis:[9] Attitudes, Children, Influences, Media didactics/pedagogics, Media literacy/appreciation, Propaganda, Public opinion, Public policy, Research/theory, and Societal implications. In addition, the coded material of Crime and Delinquency was combined to test its coherence, and the categories of Research and Societal implications were broken by the topic shift that occurred at the end of the 1930s for Research and at the end of the 1920s for Societal implications.

Concept Analysis Results

Table 13.5 presents the key words and unanimity scores for each of the ten content codes. Key words are the concept descriptors associated with each code ordered roughly by the frequency of appearance and the number of cases in which they

Table 13.5. Case content codes by key words, complexity and unanimity scores

Code/Concept & Unanimity	Key Word	Key Word	Key Word	Key Word	Key Word	Unanimity
Attitudes	Social	Behavioral	Cultural	Biological	Acquired	0.44
Children	Motion Pictures	School	Radio	Effect	Parents	0.47
Crime/Delinquency	Crime News	Newspapers	Antisocial	Boys	Problem	0.47
Influences Thoughts/Actions	Public	Propaganda	Attitudes	Child	Political	0.39
Media Didactics/Pedagogics	Radio	Film	Educational	Classroom	Curriculum	0.48
Media Literacy/Appreciation	Radio	Film	People	Art	Reading	0.46
Propaganda	Radio	War	Nationalism	German	Press	0.43
Public Opinion	Public Will	Government	War	Political	Majority	0.55
Public Policy	Public Will	War	Neutrality	Belligerent	United	0.46
Research/Theory < 1940	News	Opinion	Psychology	Behaviorism	Phenomena	0.32
Research/Theory ≥ 1940	Radio	Operationalism	Communication	Scientific	Test	0.28
Societal Implications < 1930	Newspaper	Motion Picture	Modern	Changes	Moral	0.47
Societal Implications ≥ 1930	Public Will	Life	American	Culture	Mass	0.42

appeared. Teleological words (e.g., "opinion" in the Public opinion code) were not included. The software returns a value for the number of cases in which each unique descriptor appears. That value is used to construct a "unanimity" score. The unanimity score is the average proportion of cases in which the first twenty-five unique descriptors appeared. The higher that proportion the more likely the conversations across cases in that code were similar. Unanimity scores varied from a low of .28 for research of 1940s, indicating that more terms were used in single cases, to a high of .55 for public opinion, indicating a more common subject matter.

Text Exemplars

To complete this section, we offer a series of text exemplars taken from each of the ten codes. The exemplars were chosen from introductory or concluding sections of the articles to exemplify the key words in Table 13.5 and for their resonance to contemporary scholarship. They are presented by code without further comment.

Attitude. Every reader is familiar with the lively current controversy between "culturalists" and "individualists." The former declare that the roots of social attitudes lie in the folkways and in surrounding social patterns; personality, they say, is merely "the subjective aspect of culture." The individualists, on the other hand, find the root of attitudes in the native propensities of men, maintaining that personality develops primarily "from within."[10]

Children. Looking backward, radio appears as but the latest of cultural emergents to invade the putative privacy of the home. Each such invasion finds the parents unprepared, frightened, resentful, and helpless. Within comparatively short memory, the "movie," the automobile, the telephone, the sensational newspaper or magazine, the "funnies," and the cheap paper-back book have had similar effects upon the apprehensions and solicitudes of parents.[11]

Crime and Delinquency. The impression prevails that this alleged increase is due "to the war, " to a "break-down of criminal justice," to " immigration," to the " negro migration [sic]," to "break-down of parental authority," to "the newspapers," etc., and that the youth of the country are the chief offenders.[12]

Influence. How are attitudes measured? Since attitudes involve feelings, inclinations, or sentiments, we cannot measure them directly. We can, however, successfully measure attitudes through their verbalized expressions, namely, through opinions.[13]

Media Didactics/Pedagogics. It is obvious that the motion picture is well on the road to become an accepted educational asset. There is little doubt that its advent will be hailed with enthusiasm alike by progressive teachers and long-suffering pupils, because it will afford better pedagogical results at the same time that it makes memory an unconscious function instead of a labored effort involving all the agony of concentration on an uninteresting piece of work.[14]

Media Literacy. We do know definitely that the movie contains the most sexually suggestive elements. It shows lawlessness and crime in all their horror and brutality. It pictures drunkenness in its most licentious aspects. Home and family relations are made subjects of jest and ridicule us. The portraying of the sinister aspects of crime, drunkenness, and the rest, does not act as an object lesson to the adolescent. Quite the reverse. The movie serves to glorify indecency and immorality solely by its graphic presentation![15]

Propaganda. In no other war in history has propaganda played so important a part as in the world conflict of 1914–18. Although most of us know the extent to which the neutral countries were bombarded with propaganda from the civil authorities of the Allied and Central powers, few of us perhaps realize the importance placed upon propaganda as an instrument of warfare by the military authorities. The destruction of the enemy morale by the dissemination of defeatist, disheartening, and revolutionary leaflets, pamphlets, books and propaganda "news sheets," was recognized as an important part of the offensive against the enemy.[16]

Public Opinion. Government by Public Opinion is also a phrase in common use, Few will deny that public opinion does govern to a great extent in the United States and is at least a powerful governing force in all other civilized countries. But how this intangible, inexplicable force operates upon the machinery of governments; how it makes itself felt by rulers, and what are the sanctions which compel obedience to its mandates are problems little understood and subjects of much disagreement.[17]

Public Policy. Such organs of publicity as moving pictures, newspapers, magazines, advertising posters and the like, should not be allowed to contribute to the necessary burden of evil suggestion by the character of their productions. The purely commercial spirit should be tempered by a spirit of social welfare and education.[18]

Research before 1940. The obligation placed upon psychologists to re-define their contents in scientific terms first became obvious after the inadequacy of sensations, images, and feelings was revealed by the introspective studies of Binet, Woodworth, and 'the Wurzburg School'. It was this argument over the nature of conscious contents which more than anything else, I think, created a favorable moment for the appearance of behaviorism, and its denial of scientific validity to any and all these phenomena of consciousness.[19]

Research 1940 and Later. Paradoxical as it may seem, it is still true that the problem of meanings, which provided so much difficulty for the introspective psychologists, meets with a comparatively simple solution by the methods and materials of the

objective psychologist. ... By meanings we understand specific characteristic differential responses to particular stimulating objects in their appropriate settings.[20]
Societal Implications before 1930. If [newspapers] should neglect or fail to mend their ways, to remedy the serious defects justly complained of by so many intelligent and right-minded men and women, the penalty—unavoidable in the long run—which threatens them is moral decline, contempt, and a place among the forces of disorder and evil. There can be no justice, sanity, due process of law, decency in modern society unless the great and influential newspapers, with their smaller imitators and disciples, apply and respect these fundamental virtues in reporting events and holding the mirror up to life and human affair.[21]
Societal Implications 1930 and Later. Mass behavior, as I see it, is not confined to any stratum of society. People may participate in mass behavior regardless of class position, vocation, cultural attainment, or wealth. In war hysteria, the spread of fashion, migratory movements, "gold rushes" and land booms, social unrest, popular excitement over the kidnapping of a baby, the rise of interest in golf—the participants may come from all distinguishable strata of the population. The "mass," then (and shall use this term in preference to the term "masses") represents a population aggregate which cuts across the lines of class, vocational position, and cultural attainment.[22]

THE CONVERSATIONS: CASE BY SEGMENT CODING

The content within each article was coded across nineteen subject codes first developed by the "next unique subject matter" occurrence and then combined across similarities. Table 13.6 presents the segment codes and the number of content segments by decades in which they appear. In this table, the unit of analysis is a segment of content and the coding is mutually exclusive.

To avoid duplicating information already presented, we selected Advertising, Censorship, Morality, Motion Pictures, Newspapers, and Radio for analysis here. Table 13.7 presents the six codes' descriptive key words and unanimity scores.

The fact that children and/or school were key words in Censorship, Morality, Motion Pictures, and Radio connected to a finding by Anderson[23] that 51 percent of nearly 1,000 articles on media violence dealt with its relationship to children. Fear for our children, and fear of our children for what they might become, have been a continuing motivating force in media research.

Another recurring theme across journals was classroom utilization of the new media. As each new medium became more commonplace—newspapers in the earlier years of the century, movies starting around 1905, and radio starting in the late 1920s, with the latter two hitting a high point in the 1930s—educators

Table 13.6. Subject matter codes and the number of cases by decade in which they appear

Code/Decade	1900s	1910s	1920s	1930s	1940s	Total
Advertising	3	1	5	8	4	21
Appreciation	0	0	1	8	0	9
Attitudes	0	0	0	11	4	15
Censorship	0	2	3	10	0	15
Children	1	0	5	15	6	27
Criminals and Crime	0	1	6	3	2	12
Delinquency	1	2	2	3	1	9
Education	2	5	7	46	9	69
Effect on Society	0	2	6	11	3	22
Influence	0	0	7	15	9	31
Morality	4	4	6	9	0	23
Motion Pictures	0	2	6	28	5	41
Newspapers	8	0	13	9	4	34
Propaganda	0	0	2	23	12	37
Public Opinion	4	0	4	25	14	47
Radio	0	0	3	28	17	48
Research	0	0	1	13	9	23
Socioeconomic	0	1	1	5	2	9
Television	0	0	0	1	2	3

Table 13.7. Segment codes, their key word descriptors and unanimity scores

Code/Concepts & Unanimity	Key Word	Key Word	Key Word	Key Word	Key Word	Unanimity
Advertising	Belief	Business	Research	Radio	Magazine	.29
Censorship	Producers	Motion Pictures	News	War	Children	.38
Morality	Social Conduct	Standards	Children	Motion Pictures	School	.33
Motion Pictures	Children	School	Social Impact	Life view	Public	.43
Newspapers	Crime	Public will	Propaganda	Editorial Function	Influence	.53
Radio	Listening	Programs	Time	Audience	School	.49

started wondering how to use the various media in the classroom and what educational effects might occur. The earliest articles concerning each of the three media focused on how to obtain equipment and set it up, which classes could utilize such equipment, and what equipment was needed.

Several years after the initial "how to" articles came a focus on educational benefits—which classes could utilize the particular media the best, how teachers could save time, what students could learn, and whether there could be any negative effects. The potential for negative effects prompted many to consider the possibility of shaping the content available to schoolchildren to help them maximize educational benefits such as seeing a polar expedition instead of just reading about it, and minimize such deleterious effects as incitement to delinquent behavior. Bair describes this type of negative situation when he says, "it is desirable to protect children deliberately against the disintegrating impact of any sort of commercial movie, no matter how good—against an overdose, let us say, of love, sex and crime."[24] Children are viewed then, as even now, as an unknowing public that partakes full-heartedly without any sort of discrimination.

Censorship on behalf of children, specifically in educational settings, does not become an issue, however, until the late 1930s. Before this, censorship is not primarily focused on as a way to exclude certain information or prevent bad material from being broadcast to the public, but instead as a way of getting the media to monitor themselves (hence, the key word "Producers"). More than half of the articles we examined that deal with the issue of censorship highlight issues of media self-censorship or censorship from within. It is not until the 1930s that we begin to see suggestions that neither the public nor the media can monitor the content of media—that this must be done by an outside board of reviewers. Such impartial regulators begin to be mentioned at the same time articles begin appearing about the media, rather than censoring itself, providing whatever makes the most money and the public injudiciously partaking of whatever is provided regardless of its moral character or effect.

This idea of the public as an entity with an opinion that can be affected by media starts being discussed around the turn of the century. George W. Ochs says:

> Journalism as at presently developed is the teacher of the innumerable hosts of a self-conscious democracy; it is the moulder [sic] of public opinion. Without the newspaper, popular judgment would be dumb and formless. Unless this public opinion be kept sane, healthful, uncorrupted, our nation cannot endure.[25]

Ochs is one of the first to put into words an idea that dominates much of the earlier decades of the 1900s—that while people use the media, media also holds the power to move the public either for good or for bad.

Finally, it is interesting to trace the developing role of the researcher in industry as highlighted by the work in advertising and radio research. Advertising goes from the manly encouragement of Powers (see exemplar below) to quite sophisticated questions of effectiveness of size, placement, connections to content, relation to text, and so on. We see the effect of that change in the very low unanimity score of .29. In radio, an entire research approach from audience measurement to day-part programming had to be developed to advance the business model of an advertising supported, free-broadcast medium. That business model powered broadcast television (here yet to come) and powers much of the present-day Internet.

Text Exemplars

To provide further insight into the literature of these five decades we again provide a series of quotations chosen to connect to the key word analysis and current sensibilities.

Advertising. The secret of advertising success, then, is: first, have an article of high order that people really want; then sell it in an agreeable, tactful and honest way; advertise it in the same way. … It is as sure as anything in this world that the man who is manufacturing, selling, and advertising along these lines, and with thoroughness besides, will win for himself and his article a large measure of success, bounded only by the possibilities of time and the commercial area of the world.[26]

Not only do the advertisers check the news or at critical points influence the policy, so prostituting the public press to private or at least to conservative and stand pat commercial purposes, but also for obvious reasons the peculiar mentality of advertising with its lure and its stimulation spreads inevitably to the news and even to the editorial pages. A paper cannot be a great advertising medium on some pages and avoid, for something mentally better, a circulation-increasing pruriency and sensationalism on other pages.[27]

Censorship. The National Board [of Review of Motion Pictures] is opposed to legal censorship regarding all forms of the motion picture, as it is opposed to dictatorship of the screen by any arbitrary group, its own function being that of a nonpartisan, nonsectarian organization. It believes that far more constructive than censorship or repression is the method of selecting the better pictures, publishing descriptive, classified lists of them, and building up audiences and support for them through the work of community groups, thus speaking to the producer through the unmistakable terms of the box office.[28]

The imposition upon the public of a hard-and-fast partisan or other special view or preference is hardly an improvement upon the present situation which these very groups so deeply deplore. We gain nothing from such a censorship [of radio] by any group that has the power to exert special pressure. The negative approach is in the long run unproductive, although it is understandable as a manifestation of out-raged feelings.[29]

Morality. If by social conscience is meant the definition or content which society at any given time gives to the formal ethical norm which we call abstract right, then it is not public opinion, but law, using the word not in the lawyer's sense, but in its wider signification. Just as the individual gives a continually progressive definition to the abstract ethical norm—morality, so society, lagging behind the most advanced individuals, but far in advance of the morally backward classes, also gives a progressive definition to this ethical norm within the sphere of social relationships.[30]

The movies may well have deleterious as well as morally sound influences. But they are not all bad, nor are they the chief factor, perhaps, in stimulating antisocial conduct. Like the newspapers, so long blamed for crime waves and other antisocial conduct, the movies are but one of many influences at work in this period of rapid social change.[31]

Motion Pictures. We used to speak of three great institutions that set the pattern for American culture—the home, the school, and the church. Today we must add a fourth—the motion picture. More people attend motion picture theaters than attend all of our public schools and churches put together. About 29,000,000 are enrolled in our public schools. Our weekly church attendance—Catholic, Protestant, and Jewish—probably does not exceed 30,000,000. But every week 88,000,000 people flock into our theaters. Of these, 33,000,000 are under twenty-one years of age and 13,200,000 under fourteen years of age.[32]

Until very recent years many persons professed themselves to be shocked by the sensationalism and the cheapness of the film. Those persons were shocked, however, not so much by the film itself as by what the film showed them regarding the tastes, desires, and susceptibilities of a mass audience. The public was being served and served effectively by men who knew their business, perhaps in many instances more through instinct than reason. But they knew it, and they knew the public. The public got what it wanted. The motion picture at this point told much more about the public than it told to the public.[33]

Newspapers. While today the radio broadcast performs an important function with respect to foreign relations, and wields an incalculable influence over public attitudes, the broadcast still lacks the comprehensiveness and tangibility of the newspaper. Furthermore, the radio has made no attempt to duplicate the services of the great news agencies, and is largely dependent upon them for the content of news broadcasts.[34]

We are now in a position to consider the relationship of the newspapers to the four factors. The [Chicago Daily] News and The [Chicago] Tribune have their highest loadings in Factor I. This means that these papers have readers who are, above everything else, Republicans in national politics. The next highest loadings are found in Factor IV which indicates that the readers were strongly opposed to voting for the Republican candidates for state and local offices during the period under discussion.[35]

Radio. There are fundamental things in human nature that will prevent broadcasting from wielding any greater influence on us than the phonograph has. Radio does not make us congregate.—Radio does not satisfy man's desire to congregate with other persons in the mutual enjoyment of music, games, etc. Broadcasting cannot compete with other amusements.—Broadcasting does not encourage association or herding, and can, therefore, never compete injuriously with the theater, the concert, the church, or the motion picture. Radio's greatest benefit is to isolated persons.—It will, however, serve most effectively the sightless, the bedridden, the farmer, and the deaf [sic!].[36]

The distinction between the news commentator and the newscaster is an important one, not only for the radio listener and the sponsor, but also for the scientific observer of public opinion interested in analyzing the function of radio in the life of the listener. The wide appeal of both national and local commentators suggests that there are certain characteristics of their programs that serve unique psychological functions for their millions of followers.[37]

CONCLUDING THOUGHTS

From the rise of the popular newspaper in the mid-1800s with Benjamin Day's discovery that news stories sold more product than critical essays, media have been a lively source of scholarly attention. From that time to today, the attention has been focused on a fairly stable set of issues: the power and consequence of the media on the hearts and minds of citizens, on our children, in our schools, on the institutions of society, and on the constitution of culture. The lessons that can be drawn from the quotations are quite clear in that regard. With the change of a word here and there to reflect changing cultural sensibilities, the quotations could represent the contemporary scene as much as they do the past. Indeed, our questions or even the methodologies of our answers show little change—both of those have been in place across the more than a century of work. Certainly, the technology has changed, but that has not mattered much except to provide a new venue to ask the old questions. The historical tour teaches us that well.

What has changed remarkably has been us—the scholarly community. One can easily correlate the rise of media research with the rise of cognitivism and the social theories of psychology and sociology out of the iron grip of physiological behaviorism as well as the—albeit apparently temporary—recuperation of the critical from moralism. A modernist would call this progress; a postmodernist might see this as the good economic sense of job security.

The reason for this sense of deflation is that there can be no final answers to the fundamental set of questions concerning the media. The present day has a huge corpus of work, with certainty being claimed in particular communities of practice; but in fact there is little settled science across the vista of the mediascape. Today we are starting the conversation on interactive, online gaming. The *World of Warcraft* and *Second Life*, two immensely popular MMORPGs (massive multiplayer online role-playing game) and others like it will most likely provide the foundation for much of the media scholarship of the next decade. What are the consequences on the hearts and minds of citizens, on our children, in our schools, on the institutions of society, and on the constitution of culture? The initial dialogue—from attraction to addiction—reads the same as that surrounding the arrival of the motion picture, 100 years ago.

The senior author of this chapter has been involved in four major studies of media research that involved the cataloguing and coding of journal articles on media research from 1860 (earliest) to 2007.[38] In those studies, far more than 1,000 journal articles have been collected, catalogued, and content coded. These are not bibliographic studies but rather cultural analyses of the content of our field of work. Given that perspective, we would like to suggest that much of what we coded were examples of the symbolic uses of science. In this argument, we are following Edelman's conceptualization of symbolic politics.[39]

Edelman provides two markers for the symbolic uses of an activity. First, the discourse of the activity far exceeds what might be the expected consequences of the activity. In the politics of media content legislation, for example, we can review a continual parade of congressional hearings beginning at the turn of the twentieth century and continuing until today[40] with little effective legislation and certainly nothing that has been laid to rest. Second, the expressive function of the discourse serves the internal purposes of the agents of the discourse rather than the ostensible goals attributed to the activity. And so, in the politics of media content legislation, clearly there may be no practical solutions available, but something must be done and the doing is talking. The talking discharges the moral outrage, warns the perpetrators to cool it for a while (known as the contemporary political climate in business circles), and advances the stature of those conducting the hearings.[41]

How might we translate those ideas into the argument we are making here? The symbolic uses of science that implicate most social science and not just media are sharply delineated if we take the modernist goals of science—ahistoric, transcendental laws that formulate the last word of a knowledge claim. Certainly, we have achieved none of those, and further, we would argue that we currently do not have the theoretical or methodological means of achieving them. The lack of those means, however, does not preclude the trying, because we must do something, if we are to claim the mantle of science. The doing in this case is the scientific discourse of the field.

The issues we are charged with resolving—the power and consequence of the media on the hearts and minds of citizens, on our children, in our schools, on the institutions of society, and on the constitution of culture—are intractable problems. To begin with, nothing in the equation of technology, media, content, audience, social process, and culture has remained even remotely stable enough to properly ask the question, much less assay an answer. Consider the Sisyphean act of content analysis—one of the most common publications of the field. Whole careers have been built on counting violent acts, commercial messages, ad space, and the like. Yet not one of those studies tells us what content is today, and so we must do it again. The same is true of audience studies. Does it matter that in the 1940s farmers listened to fewer radio newscasts than townspeople?[42] (Did Schramm and Hurrer have any idea of what farmers did all day in the 1940s?) The question of what farmers listen to today in their air-conditioned combines equipped with satellite radio reception must be asked again and then tomorrow yet again.

Anderson and Colvin (in press), in their analysis of more than 300 media-violence-leads-to-aggression articles, were able to document several examples in which authors would first declare the relationship to be settled science and then proceed to report yet another study. If it is settled science, why do such studies continue to be funded and published? Our answer is, of course, that nothing is settled. Nothing is settled because nothing can be settled. We do not have the means to settle it.

However, we do have the means to do science and to create scientific discourse. Those means increased remarkably in the 1930s with the explosion of new journals and new theoretical approaches, but they have increased exponentially in the past two decades. The whole enterprise of science can now be "put in a box." From the "statement of the problem" to the "implications and conclusions," the activity can be conducted by responding to a few queries (or simply accepting the defaults). As the slogan of one formatting software puts it, "Just add words." We have not quite achieved the integration of, say, a PowerPoint or an InDesign, but that is just a matter of the market.

And we have the demand for it. That demand is both internal to our enterprise and external from those seeking to support their own enterprises. It is well beyond the scope of this chapter to talk about the latter, but consider the political processes for the various Surgeon General reports on media violence, consider why the cable industry would fund the National Television Violence Studies, or consider why the National Association of Broadcasters has a long-standing program of academic media research funding. Regardless of the findings, their purposes were and are being served.

We can spend just a bit more space on the internal demand for research. Even undergraduate research texts[43] are now pointing to the internal forces on the professoriate to produce research. Publish or perish is common knowledge; the institutional demand for grant monies may not be so common, but it has been with us from the beginning. Eric Allen, writing in 1929, calls us out:

> Money is being spent for research by the hundreds of thousand and even millions. Somebody told that a telephone company alone was spending sixteen millions for research. Everybody considers the press one of the greatest of modern problems, but I cannot discover that anybody is spending a cent in finding out about it through the academic channels or through schools of journalism .... The conclusion of this paper is to the effect that research in journalism will take money, and ought to take money and that money can probably be obtained for it.[44]

It makes the world go 'round.

The point of this argument is not to claim that our science is empty rhetoric driven by venal interests. It is to say that the symbolic value of our science by itself is of great worth to us and that its value surpasses the value that our conclusions hold for some transcendental archive of claim. For a modernist, this is an unfortunate state of affairs. It suggests that we have failed to achieve our putative goals. For a postmodernist, it has intrinsic value because it keeps the conversation going. Nothing in our history suggests that we are on the brink or even the pathway to solving the problems that arise in what is now being called the post-human era in which we are no longer individuals but nodes on the cellphone/SNS/Internet/ e-mail network. Consequently, it is the conversation and not the conclusion that matters.

As we look over the forty-five-plus year archive of that conversation analyzed here, we see that their conversations are our conversations; their voices are our voices. It is more than repetition, however. It is expressive of the manner by which we seek to control the reality we co-construct but cannot manage. In science, we "man up." We will penetrate the chaos through science even while creating more chaos in the process. Writing for NASA in 1993, Vernor Vinge advanced the prediction that "the creation of greater than human intelligence will occur during the next thirty years."[45] Should we survive the military applications of this greater-than-human intelligence, how will this conversation about media research change? The small effort of ours has opened the field to its past in ways not before available. What happens when the entire corpus of work is digitized and given over to the tireless investigation and limitless presentation of cyborg analysis? What will the conversation be then?

There is one other comment to make: improved and new computer hardware and methodologies now make the analysis of very large textual databases not only possible but also preferable to any other method of literature search or analysis. Further, these analyses can be conducted jointly with several researchers contributing data and insights all while protecting the primary database. In the future, articles like these will connect to a URL that will allow the reader to reconstruct and then advance the analysis at a click of the mouse. The only impediments at the present are a nostalgia for carbon-based publication forms and the labyrinth of licensing agreements that would have to be executed to have public access to the publications. If the music industry can resolve similar issues—so much so that the record album is a disappearing form—why not scholarship?

NOTES

1. Anderson and Colvin, "Is it Just Science?"
2. Miller, "Radio's Code of Self-Regulation."
3. Grimes, Anderson, and Bergen, *Media Violence*, 50.
4. Tolman, "A New Formula for Behaviorism."
5. Shepard, "Public Opinion."
6. Anderson, "Media Literacy."
7. Sherif and Cantril, "The Psychology of 'Attitudes.'"
8. Woelfel, "Artificial Neural Networks."
9. Advertising/public relations will be covered in the topics analysis that follows.
10. Allport and Schanck, "Are Attitudes Biological," 195.
11. Gruenberg, "Radio and the Child," 123.
12. Potter, "Spectacular Aspects," 1.
13. Rosenthal, "Change of Socio-Economic Attitudes," 5.
14. Blanchard, "The Motion Picture," 287.
15. Freeman, "The Movies in Education," 117.
16. Bruntz, "Allied Propaganda," 61.
17. Shepard, "Public Opinion," 32.
18. Poffenberger, "Motion Pictures and Crime," 339.
19. Ogden, "The Phenomenon of 'Meaning,'" 223.
20. Kantor, "An Objective Interpretation," 231.
21. "Is An Honest," 334.
22. Blumer, "Moulding of Mass Behavior," 115.
23. Anderson, "Production of Media Violence."
24. Bair, "Exploring the Educational Possibilities," 515.
25. Ochs, "Journalsim," 55.
26. Powers, "Advertising," 62.
27. Lloyd, "Newspaper Conscience," 201.
28. Barrett, "The National Board of Review," 177.

29. Gruenberg, "Radio and the Child," 127.
30. Shepard, "Public Opinion," 43.
31. Young, "Book Reviews," 255.
32. Eastman, "The Motion Picture," 44.
33. Quigley, "Public Opinion," 128.
34. Foster, "How America Became Belligerent," 465.
35. Gosnell and Schmidt, "Factorial Analysis," 383.
36. Beuick, "The Limited Social Effect."
37. Cantril, "The Role," 653–54.
38. Anderson, "Media Literacy."; Anderson, "The Production of Media Violence"; Anderson and Colvin, "Is It Just Science?": and this chapter.
39. Edelman, *The Symbolic Uses*; and Edelman, *Politics as Symbolic Action*.
40. Hoerrner, "Symbolic Politics."
41. Grimes, Anderson, and Bergen, *Media Violence*.
42. Schramm and Hurrer, "What Radio News Means."
43. Fry, Botan, and Kreps, *Investigating Communication*.
44. Allen, "Organization of Research," 16.
45. Vinge, "The Coming Technological Singularity."

WORKS CITED

Allen, Eric W. "Organization of Research." *Journalism Quarterly* 5, no. 4 (1929): 10–18.

Allport, G. W., and R. L. Schanck. "Are Attitudes Biological or Cultural in Origin." *Character and Personality: An International Psychological Quarterly* IV (1938): 195–205.

Anderson, James A. "Media Literacy: A Cultural Analysis." In *The Sage Handbook of Child Development, Multiculturalism and Media,* edited by Joy Keiko Asamen, Mesah L. Ellis and Gordon L. Berry. Thousand Oaks, CA: Sage, in press.

Anderson, James A. "The Production of Media Violence and Aggression Research: A Cultural Analysis." *American Behavioral Scientist,* in press.

Anderson, James A., and Janet W. Colvin. "Is It Just Science?" In *Media Violence and Aggression: Science and Ideology,* edited by Tom Grimes, James A. Anderson and Lori Bergen, 93–152. Thousand Oaks, CA: Sage, 2008.

Bair, Frederick H. "Exploring the Educational Possibilities of Motion Pictures." *Clearing House* 12, no. 9 (1938): 515–18.

Barrett, Wilton A. "The National Board of Review of Motion Pictures–How It Works." *Journal of Educational Sociology* 10, no. 3 (1936): 177–88.

Beuick, Marshall D. "The Limited Social Effect of Radio Broadcasting." *American Journal of Sociology* 32, no. 4 (1927): 615–22.

Blanchard, Phyllis. "The Motion Picture as an Educational Asset." *Pedagogical Seminary* XXVI, 3 (1919): 284–87.

Blumer, Herbert. "Moulding of Mass Behavior through the Motion Picture." *American Sociological Society* XXIX (August 1936): 115–27.

Bruntz, George G. "Allied Propaganda and the Collapse of German Morale in 1918." *Public Opinion Quarterly* 2, no. 1 (1938): 61–76.

Cantril, Hadley. "The Role of the Radio Commentator." *Public Opinion Quarterly* 3, no. 4 (1939): 654–62.

Eastman, Fred. "The Motion Picture and Its Public Responsibilities." *Public Opinion Quarterly* 2, no. 1 (1938): 44–46.

Edelman, Murray J. *Politics as Symbolic Action: Mass Arousal and Quiescence.* Chicago: Markham, 1971.

———. *The Symbolic Uses of Politics.* Urbana: University of Illinois Press, 1964.

Foster, H. Schuyler, Jr. "How America Became Belligerent: A Quantitative Study of War News, 1914–17." *American Journal of Sociology* 40, no. 4 (1935): 464–75.

Freeman, Bernadine. "The Movies in Education." *Educational Review* LXXII (June–December 1926): 115–18.

Fry, Lawrence R., Carl H. Botan, and Gary L. Kreps. *Investigating Communication: A Introduction to Research Methods.* Needham Heights, MA: Allyn & Bacon, 2000.

Gosnell, Harold F., and Margaret J. Schmidt. "Factorial Analysis of the Relation of the Press to Voting in Chicago." *Journal of Social Psychology* VII, no. 1 (1936): 375–85.

Grimes, Tom, James A. Anderson, and Lori Bergen. *Media Violence and Aggression: Science and Ideology.* Thousand Oaks, CA: Sage, 2008.

Gruenberg, Sidonie Matsner. "Radio and the Child." *Annals of the American Academy of Political and Social Science* 177 (1935): 123–28.

Hoerrner, K. L. (2000). "Symbolic Politics: Congressional Interest in Television Violence from 1950 to 1996." *Journalism and Mass Communication Quarterly* 76, no. 4 (2000): 684–98.

"Is an Honest and Sane Newspaper Press Possible?" *American Journal of Sociology* 15, no. 3 (1909): 321–34.

Kantor, J. R. "An Objective Interpretation of Meanings." *American Journal of Psychology* LIV, no. 1 (1941): 230–35.

Lloyd, Alfred H. "Newspaper Conscience–a Study in Half-Truths." *American Journal of Sociology* XXVII (July 1921): 197–210.

Miller, Neville. "Radio's Code of Self-Regulation." *Public Opinion Quarterly* 3, no. 4 (1939): 683–88.

Ochs, George W. "Journalism." *Annals of the American Academy of Political and Social Science* 28 (1906): 38–57.

Ogden, R. M. "The Phenomenon of 'Meaning.'" *American Journal of Psychology* XXXIV (1923): 223–30.

Poffenberger, A. T. "Motion Pictures and Crime." *Scientific Monthly* 12, no. 4 (1921): 336–39.

Potter, Ellen C. "Spectacular Aspects of Crime in Relation to the Crime Wave." *Annals of the American Academy of Political and Social Science* 125 (1926): 1–19.

Powers, John O. "Advertising." *Annals of the American Academy of Political and Social Science* 22 (1903): 58–62.

Quigley, Martin. "Public Opinion and the Motion Picture." *Public Opinion Quarterly* 1, no. 2 (1937): 129–33.

Rosenthal, Solomon P. "Change of Socio-Economic Attitudes under Radical Motion Picture Propaganda." *Archives of Psychology* 166 (1934): 1–46.

Schramm, Wilbur, and Ray Hurrer. "What Radio News Means to Middleville." *Journalism Quarterly* 23, 2 (1946): 173–81.

Shepard, Walter J. "Public Opinion." *American Journal of Sociology* 15, no. 1 (1909): 32–60.

Sherif, Muzafer, and Hadley Cantril. "The Psychology of 'Attitudes'–Part I." *Psychological Review* 52, no. 6 (1945): 295–319.

Tolman, Edward Chace. "A New Formula for Behaviorism." *Psychological Review* XXIX (1922): 44–53.

Vinge, Vernor. "The Coming Technological Singularity: How to Survive in the Post-human Era." http://adsabs.harvard.edu/abs/1993vise.nasa...11V (accessed April 13, 2007).

Woelfel, Joseph. "Artificial Neural Networks in Policy Research: A Current Assessment." *Journal of Communication* 43 (1993): 63–80.

Young, Kimball. "Book Reviews." *American Journal of Sociology* 41, no. 2 (1935): 249–55.

FROM Park TO Cressey: Chicago Sociology's Engagement WITH Media AND Mass Culture[1]

WILLIAM J. BUXTON

The Chicago School of Sociology is known not only for its contributions to the qualitative study of race relations, criminality, and urban life but also for the body of material that it produced on media, popular culture, and communications.[2] For the most part, however, commentators have largely assumed that the work on communications produced by the members of the Chicago School was homogeneous in terms of its scope of inquiry, methodology, and theoretical assumptions, and retained its basic orientation over time.[3] In this chapter, I argue that the approach to media studies taken by members of the Chicago School of Sociology was by no means uniform and unchanging.[4] The work of Paul G. Cressey and Frederic Thrasher on film in the 1930s, while continuous with certain features of Chicago sociology, also represented a decisive break from the views on media that were central to the School's work in this area. I will give particular attention to the work of the two authors for whom the point of reference was an initiative supported by the Payne Fund–sponsored Motion Picture Research Council during the 1930s. I argue that for Cressey and Thrasher (in line with the views of Park) media studies was not merely a subfield; their research on movies sought to produce knowledge that would both realign social practices and shift public opinion. I conclude by examining what Cressey (and to a lesser extent Thrasher) contributed to media studies, why they broke with the main tenets of Chicago sociology, and how their relocation from the institutional setting of sociology in Chicago to one of education at New York University (NYU) in New York City may have played a role in their changing views on film and mass communications. However, to situate their work on media in relation to that of the Chicago School (with which they always identified), it is necessary to examine the emergence of Chicago's media-research tradition, as centered in the contributions to the study of the press made by Robert Park.

The meaning of news and newspapers for Park can be better understood by briefly examining how his early intellectual trajectory provided the basis for his approach to the social sciences. Reflecting the influence of his teacher, John Dewey, and his engagement with Goethe's *Faust,* after his graduation from the University of Michigan in 1887, Park sought to experience "all the joys and sorrows of the world."[5] To this end, he worked for a number of years as a newspaperman in Minneapolis, Detroit, Denver, and New York. In 1892, just as he was on the verge of quitting the newspaper business to return to a position in his father's business, he was introduced by Dewey to a former editor named Franklin Ford, who wished to publish a new form of newspaper that would represent "a union between the newspaper and the increasing knowledge of the university scholar."[6] As Park would later write, his encounter with Ford led him to the view that a sociologist should report "the Big News. . .the long-term trends which recorded what is actually going on rather than what, on the surface of things, merely seems to be going on."[7] It was through the influence of Dewey and Ford, coupled with his own practical experience, that the "newspaper and news" became Park's research problem. Specifically, he sought to develop "a fundamental point of view from which [he] could describe the behavior of society, under the influence of news, in the precise and universal language of science."[8] This concern led Park to pursue graduate work in philosophy at Harvard, Berlin, Strasbourg, and finally Heidelberg, where he completed a dissertation under the direction of Wilhelm Windelband. His 1904 doctoral dissertation, however, did not directly address news and the newspaper per se; it dealt with what Park considered the two major emergent—and contesting—forms of social organization, namely the crowd and the public. Drawing on writers such as Le Bon and Sighele, Park claimed that the crowd was characterized by imitation, suggestibility, emotion, inhibition of individual impulses and interests, and a common drive.[9] He noted as well that "entrance to the crowd depends on the simplest conditions imaginable, namely, possessing the ability to feel and empathize."[10] In contrast, entrance to the public demands not only this ability, but also "the ability to think and reason with others."[11] Moreover, unlike the crowd, the public is characterized by "discussion among individuals who assume opposing positions."[12] This "critical behavior of various opposed individuals or groups," according to Park, results in "public opinion."[13] This discursive interaction was fuelled by "elements which must have the same meaning for all members of the group," namely, facts. Although Park does not elaborate on where these facts originate, he does obliquely suggest in his thesis that they are produced by modern journalism.[14]

The binary distinction between "crowd" and "public" informed Park's subsequent conceptualization of social change. This vision, in line with Park's long-standing commitment to practical reform, was predicated on the Manichean view that the public—as grounded in the emergence of public opinion, characterized by logic and reason, and mediated through factually based debate and discussion—was pitted against the crowd, as characterized by emotion, feelings, anarchy, and the lack of critical discourse. What made the formation of the public—as constituted by public opinion—possible, according to Park, was news. As a form of knowledge circulated by and through newspapers, news differed in degree but not in kind from the "body of special knowledge" that was produced as a form of "social advertising" by institutions that dealt with the public through the "medium of publicity." These included philanthropic organizations that produced texts, including "the Carnegie Report upon Medical Education, the Pittsburgh Survey," and "the Russel [sic] Sage Foundation Report on Comparative Costs of Public-School Education." These publications were, in Park's view, "something more than scientific reports. They are rather a high form of journalism, dealing with existing conditions critically, and seeing through the agency of publicity to bring about radical reforms."[15] Along the same lines, at a local level, bureaus of research (such as the Bureau of Municipal Research in New York), child-welfare exhibits, social surveys, and "propaganda in favor of public health" have all contributed to the formation of public opinion and thereby to the process of social control. Moreover, as Park stressed, among the agencies and devices that have taken in part in shaping public opinion were not only "the daily newspaper and other forms of current literature" but also "books classed as current."[16] By implication, then, academics—including sociologists—were involved in this process. It was not at all surprising, then, that Park and his colleagues sought to work closely with various social agencies within Chicago to produce knowledge that would have bearing on perceived social problems within the city.[17] "The Chicago School," as Taylor, Walton, and Young argue, "motivated by the journalist's campaigning and documentary concerns, was the example *par excellence* of determined and detailed empirical social research."[18] This took the form of identifying patterns of social disorganization, whether they be taxi-dance halls[19] or gangs,[20] and providing recommendations for how institutions and practices could be developed to assimilate pockets of social disorganization into the emergent public. Although newspapers were thought to help ensure social control within the public, other forms of communication, such as movies and radio (along with modes of transportation, such as the automobile) were assumed to contribute to social disorganization. In effect, the struggle between the crowd and the public that Park had originally identified in his doctoral dissertation[21] had moved to the civic realm of Chicago, where it pitted the news/knowledge-inflected sociologists, civic reformers, and journalists against the mass-media saturated denizens of urban disorder. Up until the late 1920s, it was rare to find any concerted attention paid to the forms of media that were thought to contribute to social disorganization.[22] This all changed with the massive Payne Fund–sponsored studies of the impact of movies on children.[23] To recruit researchers for the project, the program's director, W. W. Short, visited numerous social agencies, universities, and research institutes during the winter of 1928–1929. He was particularly interested in the participation of University of Chicago's Department of Sociology, because of its reputation as the leading center for urban research in the United States. To this end, he met with Robert Park, who enthusiastically agreed to participate; Park's young colleague, Herbert Blumer, served on the project as a junior scholar. Park's suggestion was that "he and Blumer could analyze case study data on juvenile delinquents already gathered by the Chicago courts and social welfare agencies to chart the impact of the movies on young people's identity and values formation."[24] Initially, Park and Blumer were slated to undertake two studies (of the fourteen planned) that explored these issues. Reflecting his binary view concerning the opposition between social disorganization and the public, Park's two studies were to compare "the conduct of middle-class high school and college students with those of juvenile delinquents and imprisoned criminals."[25] It was Park's view that the values to which children were exposed in dime novels and movies (through their membership in urban mass society) were at odds with those that were being taught in the central institutions of civic society, such as the school and the church. Hence, in negotiating their identities through role-playing, they could possibly develop antisocial values as part of their character structures. Indeed, Park speculated about whether two University of Chicago students who had murdered a young boy had been influenced by the movies that they had seen.[26]

Park, however, did not undertake these studies, as he had been given a fellowship to visit China and Japan in 1929; he turned the studies over to Herbert Blumer. Park's junior colleague, however, became absorbed in the study of middle-class high school students and college students to the neglect of that dealing with the juvenile delinquents. Nevertheless, by the spring of 1930, his analysis of student autobiographies was complete. Not surprisingly, in line with the assumptions of his mentor Robert Park, he concluded that Hollywood films provided standards for youth that could potentially have a harmful effect on their conduct.[27] He then turned his attention to the study of the relationship between movies, delinquency, and crime. To this end, he enlisted the services of graduate student Philip Hauser to serve as his co-author.[28] Even though Blumer and Hauser determined that movie-going could account for only 10 percent of criminal behavior among those that the studies examined, they nonetheless sought to make the case that movie watching worked in subtle and subconscious ways to engender criminal behavior.[29]

Robert Park—and his two protégés, Blumer and Hauser—were not the only University of Chicago sociologists who were involved in the Payne Fund Studies. Paul G. Cressey, a recent MA graduate of the department (and an assistant professor of sociology at Evansville College), joined the team of researchers in 1931.[30] He had been recruited by his former fellow graduate student (and future PhD advisor) Frederic Thrasher[31] to become the director of a motion-picture project that was part of a massive "Boys' Club Study" funded by the Bureau of Social Hygiene.[32] In this capacity, Cressey was to examine a tenement district in East Harlem (given the pseudonym of "Intervale"), where the Jefferson Park branch of the Boys' Club of New York was located.[33] This study involved more than fifteen hundred boys, many of whom were of Italian-American origin. The motion-picture project, while ostensibly still part of the "Boys' Club Study," for all intents and purposes evolved into a project largely subsidized by the Payne Fund.

While Cressey's study was arguably the most insightful work in the Payne Fund series, it was neither completed nor published.[34] Nevertheless, it is of interest for what it reveals about the encounter between Chicago sociology and other currents of thought and practices that were emerging during the Great Depression. Accordingly, I will examine Cressey's manuscript not so much in terms of the Payne Fund Studies per se, but rather how this work and his other writings from the same period[35] connect to the body of theory and research produced by Chicago sociologists.[36] In this regard, it is quite revealing to compare his study to those of Blumer and Hauser, given that all three of them were trained in sociology at the University of Chicago around the same time.

In reading Cressey's material, elliptical and fragmentary as it is, one has the distinct sense of coming upon a path that was well marked out, but was not taken up again for a half century. His work provides a fascinating glimpse into the world of youth, movies, and city life in a particular setting more than seventy years ago.[37] Like the other Payne Fund Studies researchers, Cressey began with the frame of reference established by Reverend William H. Short, the director of the Payne Fund–sponsored Motion Picture Research Council. Hence, Cressey examined the effect of movies on children, particularly in terms of how the movies engendered delinquent values, attitudes, morals, and behavior. As we have seen, Park enthusiastically endorsed this mode of analysis, which was faithfully adhered to by Blumer and Hauser in their studies. Cressey, however, found this framework to be much more problematic. In the fall of 1932, when his study was well underway, Cressey had an epiphany; he felt that he could no longer abide by the conceptual framework that was to guide all of the other studies. His new thesis was that "movies should not be linked to boys' delinquency, but must instead be viewed as a powerful source of 'informal education' that served boys in a far more direct and practical way than did schools or the Boys' Clubs."[38] As he explained to W. W. Charters, the director of the project, "I have been forced by the weight of evidence to see the motion picture not primarily in this relationship [to delinquency and misconduct], but in its varied functions in the lives of different groups and individuals within this community."[39] In effect, Cressey thus parted company with both Blumer and Hauser as well as with most of the other Payne Fund authors. Instead, with his focus on the films' role in "informal education," Cressey believed that Hollywood films were actually in a much better position to influence educational values then were schools. He noted that "the cinema is almost unique among the agencies in a community in that it presents what are interpreted as unified segments of life." This stood in contrast to the "traditional public school" that "from the point of view and experience of the typical school child," was "a disjunctive and a repressive agent."[40] Indeed, it was only able to maintain itself because it was backed up by public opinion and "the force of the truant officer."[41] In this sense, "the public school has often been able to continue and to gain strength because it was not forced to look to its own students for support."[42] By contrast, the cinema, "in order to survive commercially has been forced to adapt itself constantly to the immediate interests of patrons."[43] Cressey stressed, nevertheless, that the educational role of the commercial cinema could not be understood simply by examining its programming and advertising. It was only "by reference to the specific social backgrounds of each individual" and by seeing "the motion picture's impressions in terms of his own dominant interests and values, and his own *axeological world*" [emphasis Cressey's] that one could grasp "what any photoplay may mean for a spectator."[44]

Cressey's line of argument diverged considerably from that of Blumer and Hauser. It was also quite at odds with the Parkian premises upon which their studies were based. Rather than claiming that mainstream institutions such as the school system could integrate wayward youth into civic society, he maintained that they in fact served to marginalize and alienate disadvantaged young people. By the same token, Hollywood films did not bear the responsibility for antisocial behavior; they were important sources of informal education and could in fact help mold young people into upstanding citizens.[45]

Cressey's account also represented a significant departure from the ecological perspective on urban life developed by the Chicago School of Sociology. According to this approach, the city should be seen as a naturalistic world in which institutions and practices evolve in an organic fashion. "The structure of the city," as described by Park, "is just as much the product of the struggle and efforts of its people to live and work together collectively as are its local customs, traditions, social ritual, laws, public opinion, and the prevailing moral order."[46] Echoing these concerns, Cressey sought to examine the effect of movies on youth in what were called the "interstitial areas" that lay between the more highly institutionalized parts of the city. These urban sectors, owing to high rates of unemployment, a rapid increase in the volume and density of population, and the lack of stable communities, were thought to be characterized by "social disorganization."[47] Accordingly, since these areas were largely made up of individuals who were not bound together by reciprocal ties of loyalty and respect, they were thought to be prone to deviance, violence, and high crime rates. Cressey shared his Chicago colleagues' belief that interstitial urban areas could be examined in terms of social disorganization, or what he described as the lawlessness of a frontier area.[48] However, rather than viewing the urban community of East Harlem in naturalistic terms as a functioning organism, Cressey emphasized how this particular interstitial area was linked to a broader configuration of power and control. In effect, Cressey wished to demonstrate the extent to which the Hollywood film industry, through its penetration into American urban life—including its interstitial areas—was able to shape the lifeworlds of adolescent filmgoers.[49]

After having examined how the movie theaters in the local community were enmeshed in a broader web of production, circulation, exhibition, and publicity, Cressey turned his attention to the movie-going practices of the youth in question. He discovered that around 80 percent of the Boys' Club group and around 75 percent of the non-Boys' Club groups went to see motion pictures once or twice per week. This meant that for the majority of those surveyed, the amount of time they spent in movies was closely related to the amount of time they spent at home, at school, and on the streets. Yet at the same time, most of them indicated that they did not consider movie watching to be their favorite activity: the majority reported that sports and physical activity were their preferred pastimes. This might suggest that they went to films because they had nothing better to do, as they lacked the resources and facilities to engage in other activities such as sports, art, and reading.

This finding was also in line with another one of Cressey's major conclusions: that the movie theater was not just a place to view films. It had developed into a highly polyvalent institution within the urban community—a site where a bewildering variety of social practices were carried out. Cressey described the "special contacts, special activities, and certain practices" that are carried out in the "inexpensive motion picture theatre":

> Frequently it serves distraught mothers as a place to "check" their small children while they are busy about other cares. They can leave them at the cinema, confident that they will remain there, held by the spell of the picture. Sometime a local criminal

gang, when they are known by proprietor or attendants, may use attendance *en masse* at a motion picture theater as a means of establishing an alibi. During cold days of the winter the cheaper theaters may even be invaded by entire families who come in the morning carrying sandwiches and remain until evening when they can go to bed, thus saving a coal bill at the expense of the local cinema . . . a complete inventory of the activities in the local theaters in this community would have to include everything from the "spotting" and "planting" of victims for gangland bullets to clandestine sexual activity in the darkened movie house—and even to childbirth.[50]

Although Cressey placed a great deal of emphasis on the movie house as a community institution located within the matrix of various social practices, he by no means neglected the influence of movies and films on youth. To this end, he combined his elaboration of the ecological model with a variety of methodological approaches that were being deployed in the other Payne Fund–supported studies as well as the broader Boys' Club Study of which it formed a part. Indeed, he claimed that "at least twenty different methods or techniques have been used in the project," including delinquency records, "similar data regarding overt behavior," "statistical, ecological, and case-study techniques," written and dictated life histories, and the "controlled-interview technique" pioneered by Cressey's colleague, Frederic Thrasher.[51] He tried to develop an approach that examined the influence of moves on youth in a much more complex way than had been previously attempted.[52] In doing so, he stressed the importance of examining the response to films in terms of the specific characteristics of the youth involved, their place in the life cycle, and their educational and ethnic backgrounds. Drawing on aspects of Gestalt theory and the symbolic interactionism of such figures as Dewey, Mead, and Cooley, he sought to understand what movies meant to youth, and how they fit into their everyday life. "Instead of considering [the child] apart from the social world of which he is a part," Cressey reported,

> an attempt is being made to study him and his picture habits and attitudes as part of his normal social world. He is not scrutinized *in vacuo* but is seen as a dynamic personality interacting with the host of influences and social forces which constitute his normal social *milieu*.[53]

In this respect, Cressey's study of urban youth and the cinema was continuous with his earlier exploration of "The Taxi-Dance Hall as a Social World."[54] Consonant with contemporary theorizing, Cressey emphasized the active nature of moviewatching in the sense of how the meaning given to a film was constructed in relation to a boy's own needs, interests, and predilections. The cinema was conceived "not as a unilateral social 'force', but as a reciprocal interrelationship of screen and spectator, of screen patterns and values, and of social patterns and social values." In this respect, he stood in agreement with the views of another member of the Sociology Department, Ellsworth Faris, who maintained that cinema is able to modify behavior through a process of "imitation."[55] However, according to Cressey, even though the findings of Faris and others[56] found that the cinema has a *reflexive* influence on its viewers, "it is unwarranted to describe it as 'negative.' In a society in which there are many factors making both for disorganization and social amelioration, the cinema is an important social and educational force contributing directly and indirectly to both."[57]

Moreover, it was erroneous to view cinema patrons as "wholly passive agents who are merely 'played upon' through the arts and skills of cinematography."[58] This led him to the conclusion that films could best be seen not as causal agents exerting effects but as potentially powerful informal educators that could be actively drawn upon as resources for learning. With his emphasis on agency, Cressey, it can be argued, anticipated the emphasis on the role of the social actor as outlined by some of those who have come to be identified with the "second Chicago School," such as Herbert Blumer (in his mature phase), Everett Hughes, and Erving Goffman.[59]

The draft of Cressey's unpublished report—in conjunction with his other writings on film from the 1930s—has some intriguing implications for the history of media research. It underscores the fact that the body of work related to media and mass culture produced by those linked to Chicago sociology had a homogeneity that never in fact existed. Arguably, this state of affairs can be attributed to the presentist premises that have underpinned historical accounts of the emergence and development of Chicago sociology. It is assumed that those who were affiliated with this school of thought shared with contemporary sociologists the concern to produce "objective" material knowledge, if not entirely for its own sake, at least for the sake of building the discipline and strengthening its boundaries against other disciplinary fields. Accordingly, in examining the contributions of Chicago sociology, particular attention is given to its role in developing recognized subfields and methodological/theoretical orientations of the discipline, such as race relations, deviance, participant observation, and symbolic interactionism. However, media/mass culture was not simply just another (relatively minor) substantive field for Chicago sociologists—particularly those who came under the influence of Robert Park; these communication phenomena were inherently bound up with the production of knowledge that would allow one to identify and analyze instances of social disorganization, while providing the information that could have an effect on public opinion, thereby contributing to the formation of a public. Hence, for Park and his followers, sociology was inherently communicative in nature; it was through the practice of sociology that progress in areas such as race relations and deviance could be made. This meant that the work of Chicago sociologists on issues related to communication, the media, and public opinion

354 | THE HISTORY OF MEDIA AND COMMUNICATION RESEARCH

was often buried in their substantive findings on urban studies, race relations, and collective behavior, hence eluding the attention of media-research historians.[60]

The communicative concern in relation to the public/social disorganization was very much evident in the case of film. Unlike the newspaper, which had traditionally attracted attention for the role that it played in helping to constitute publics, film did not begin to arouse interest (in Chicago sociologists and others) until the late 1920s, when it had become a mass medium with a large public following. It is instructive that it was perceived by Park, Blumer, and Hauser as primarily a negative force that contributed to crime and social disorder, a position, as we have seen, that was contested by both Cressey and Thrasher.[61] This divergence of views demonstrates that Chicago sociology was by no means as monolithic as numerous commentators have suggested.

This raises some interesting questions about why this disparity of views emerged. It might be argued that for those who remained attached to the Department of Sociology at the University of Chicago during the 1930s, Parkian discourse was very much hegemonic and unchallenged. However, Cressey and Thrasher were not only removed spatially and institutionally from Chicago sociology but were also academically remote from the field of sociology as a whole. By virtue of practicing educational sociology within an education faculty, they were now subject to the currents that were sweeping this field during the 1930s. In particular, agencies such as the Progressive Education Association, the General Education Board, the National Board of Review, and the Motion Picture Producers and Distributors of America largely saw the new medium of film— whether commercial or instructional in origin—as a positive educational force.[62] Given their new institutional position, it would not have been surprising if both Cressey and Thrasher had modified the views on film that they had developed as sociology students at the University of Chicago.[63]

However, it may also have been the case that Cressey and Thrasher were never fully as beholden to Park's thinking as were other graduate students at the University of Chicago. For instance, that Cressey took it upon himself to draft his own statement on methodology[64] might be taken to mean that he was not totally enamored with the methodology (or lack thereof) that he had been taught. Moreover, the fact that he claimed the research perspective of the "anonymous stranger" was superior to that of the "sociological stranger" for gaining insights into the lives of members of subcultures might have been rooted in his disillusion with the role of the sociologist in relation to his/her subject matter—an implicit rejection of the stance that Park, Burgess, and others had so assiduously cultivated over the years.[65]

Finally, Cressey's work on film in the 1930s is of interest not only for the light that it sheds on Chicago sociology's engagement with media and mass culture, but for what it tells us about the historiography of the history of media studies. The fact that his writings on movies and youth—along with those of his advisor and co-worker, Frederic Thrasher—have largely escaped the notice of media studies historians should be a matter of concern.[66] It suggests that there may be legions of writers such as Cressey and Thrasher, whose important work has been ignored because their careers diverged from the academic mainstream. If we claim that we stand on the shoulders of giants, we ought to make sure that we do not inadvertently neglect colossal figures from the past that, for whatever reason, have fallen through the cracks of history.

NOTES

1. Earlier versions of this chapter were presented at a conference on "Cinema and Urban Remains," held in Montreal in April 2000, and at the annual meeting of the American Sociological Association, held in Montreal in August 2006. I wish to gratefully acknowledge the assistance of Christine Dancause, Larry Nichols, Andrew Abbott, and Dana Postan in preparing this chapter and its earlier versions.

2. Much of this commentary has centered on Robert Park's work on the press, such as *The Immigrant Press and its Control*; "Natural History of the Newspaper"; and "News as a Form of Knowledge." Although a number of the Sociology Department's master's and doctoral dissertations in their published form have become canonical works in urban studies (e.g., Anderson, *The Hobo*; Cressey, *The Taxi Dance Hall*; Thrasher, *The Gang*; and Zorbaugh, *The Gold Coast and the Slum*), only one study of newspapers would seem to fall into that category, namely Helen McGill Hughes's analysis of "the human interest story" (Hughes, *News and the Human Interest Story*) for which her supervisor, Robert E. Park, wrote the introduction (Park, "News and the Human Interest Story." To be sure, Frederick Detweiler's study of the Negro press in America (Detweiler, "The Negro Press in the United States") and Kisaburo Kawabe's study of the Japanese press (Kawabe, "The Japanese Newspaper") were both published by the University of Chicago Press (Kawabe, *The Press and Politics in Japan*; Detweiler, *The Negro Press*). Other master's theses and doctoral dissertations on newspapers written in the Department include the following: Clark, "News: A Sociological Study"; Elliot, "Negro News and News Agencies"; Jensen, "The Rise of Religious Journalism"; and Merrill, "The News and the Money Market." Most were likely written under Park's supervision.

3. Katz, Peters, Liebes, and Orloff, *Canonic Texts in Media Research*; Katz and Dayan, "The Audience is a Crowd"; Handelman, "Towards the Virtual Encounter"; Horton and Wohl, "Mass Communication;" and Rothenbuhler, "Community and Pluralism."

4. There is a good deal of ambiguity about who should be included as members of the Chicago School. Those working in the history of sociology tend to define the School narrowly, as centered in the Department of Sociology under the leadership of Robert E. Park and Ernest Burgess. (Bulmer, *The Chicago School of Sociology*; Carey, *Sociology and Public Affairs*). However, those working in the field of history of media have a tendency to view the Chicago School more broadly as including not only the members of the Department of Sociology during its heyday of the 1920s, but also thinkers who identified with issues of identify, culture, and interaction, such as George

Herbert Mead, John Dewey, and Charles Cooley (Carey, "The Chicago School"; Czitrom, *Media and the American Mind*; and Katz, Peters, Liebes, and Orloff, *Canonic Texts in Media Research*). In this chapter, the narrower, history-of-sociology definition will be used, as it allows one to examine tensions and developments during the 1930s with more clarity and precision than would be possible using the broader, more amorphous history-of-media-research definition.

5. Raushenbush, *Robert E. Park*, 15.

6. The newspaper, provisionally titled *Thought News*, never saw the light of day. Park and Dewey were among those who invested in it (Raushenbush, *Robert E. Park*, 20).

7. Ibid., 21.

8. Park, *The Collected Papers*, v–vi. Cited in Park, *The Crowd and the Public*, viii.

9. Ibid., 50.

10. Ibid., 80.

11. Ibid., 50.

12. Ibid., 56–57.

13. Ibid., 56–57.

14. Ibid., 57.

15. Park, Burgess, and McKenzie, *The City*, 37–38.

16. Ibid., 38.

17. Arguably, it was Park's contention that these social agencies on their own were incapable of adequately addressing these social problems. But aided and abetted by "objective" social-scientific knowledge, the point of reference for which was Park and Burgess's textbook (Park and Burgess, *Introduction to the Science of Sociology*), these organizations would be in a better position to alleviate social disorganization and incorporate wayward elements into the emergent civic realm.

18. Taylor, Walton, and Young, *The New Criminology*, 110.

19. Cressey, *The Taxi Dance Hall*.

20. Thrasher, *The Gang*.

21. Park, *Masse und Publikum*.

22. To be sure master's theses were written on motion pictures (Halley, "A Study of Motion Pictures in Chicago as a Medium of Communication") and radio (Woolbert, "The Social Effect of Radio").

23. Between 1933 and 1937, the monumental twelve-volume series, "Motion Pictures and Youth"—more commonly known as the Payne Fund Studies—was published under the auspices of the Motion Picture Research Council. This body of work still constitutes the most far-reaching evaluation that has ever been made of the role of motion pictures in American society—particularly with reference to their effect on children.

24. Jowett, Jarvie, and Fuller, *Children and the Movies*, 62.

25. Ibid., 70.

26. Ibid., 71.

27. Blumer, *Movies and Conduct*.

28. Hauser later completed a Master's thesis on criminals' film-watching behavior. (Hauser, "Motion Pictures.")

29. Blumer and Hauser, *Movies, Delinquency and Crime*; Jowett, Jarvie, and Fuller, *Children and the Movies*, 79.

30. Information on the life of Paul Goalby Cressey (1901–1955) is difficult to come by. He completed a BA from Oberlin College in 1922, and an MA in sociology from the University of Chicago in 1929, with a thesis titled "The Closed Dance Hall." A revised version of this work was published as *The Taxi Dance Hall*. In 1928, he accepted a position in the Sociology Department at Evansville

College in Evansville, Indiana. In 1931, he took a leave of absence from Evansville College (never to return) to take a position as an instructor in educational sociology and as associate director of the Motion Picture Study of the Boys' Club Study (1931–1934) directed by Frederic Thrasher in the School of Education at NYU. In addition to teaching at the University of Newark (1934–1937), he lectured in community organization at NYU from 1937 to 1942. On his arrival to NYU, Cressey began working toward a doctorate with a dissertation originally titled "The Social Role of Motion Pictures in an Interstitial Area." He evidently "received a PhD from NYU in 1942 for a dissertation titled 'The Role of the Motion Picture in an Interstitial Area,'" but no copy of it has ever been found in the library of NYU (Jowett, Jarvie, and Fuller, *Children and the Movies*, 30). (The completion of the degree was announced in the *American Sociological Review* for 1932, 1935, and again for 1939 [The title for the 1939 version of the thesis was "The Social Rise of the Motion Picture in an Interstitial Area: A Descriptive Study."].) After working as a public-opinion analyst in the Office of War Information (1942–1943), he served as executive director of the Social Welfare Council of the Oranges and Maplewood, New Jersey, from 1943 to 1950, giving particular attention to the support of community centers for children in slum areas. He was evidently working on a research project related to the subject of his dissertation until as late as 1946. In a "Census of Current Research Projects," listed in the *American Sociological Review* in 1945, he described his project (in the section on social psychology and social psychiatry) as "The Motion Picture Experience as Modified by Social Background and Personality;" he also listed a new research project titled "Community Organization and Process" (in the section on community and human ecology). For the 1946 Census, he listed a project bearing the title of his 1930s dissertation work, along with a project in the social problems and social work section, titled "The Utilization of Social Process in Community Organization." In 1950, Cressey was appointed professor of sociology at Ohio Wesleyan University. Reflecting his shifting interests, he taught a new course titled "Community Organization" that had the aim of "helping prepare students for effective citizenship in their own communities." He also taught "a preprofessional course in social work and courses in urban sociology and race relations." ("News and Notes," 492). Cressey died in Montclair, New Jersey (which seemed to have become his home base in the early 1930s) on July 7, 1955 at the age of fifty-four, survived by his wife, Germaine Poreau Cressey ("Obituary").

31. Frederic Milton Thrasher (1892–ca. 1970) was born in Shelbyville, Indiana in 1892. He obtained a BA in social psychology from DePauw University and completed an MA in sociology in 1918 from the University of Chicago with a thesis titled "The Boy Scout Movement as a Socializing Agency." He received a PhD in sociology from the University of Chicago in 1926. A revised version of his doctoral dissertation was published the following year as *The Gang*. In the late 1920s, he took up a position at the Steinhardt School of Education of NYU, eventually becoming professor of educational sociology. He served as associate editor of the *Journal of Educational Sociology* (founded in 1927), and, beginning in 1934, taught a pioneering course in film studies: "The Motion Picture: Its Artistic, Educational and Social Aspects." He also published research on the comic book, and served as a consultant for groups having an interest in motion pictures, crime, prison reform, and prevention of juvenile delinquency. He retired from NYU in 1959.

32. The final report of the Boys' Club Study (numbering 1150 pages) was submitted to the Bureau of Social Hygiene in 1934. It never was given wider circulation, however, very likely because it documented "the sexual practices of youth, prostitution and gang rape" and failed to substantiate the view that the Boys' Clubs helped to deter crime (Jowett, Jarvie, and Fuller, *Children and the Movies*, 131).

33. Ibid., 130.

34. Even though its forthcoming publication by Macmillan had been announced, the volume never materialized. Although Cressey's dire personal and financial circumstances undoubtedly contributed to his failure to complete the project, the main reason for his study's demise was likely the incommensurability between his emergent framework and that which was championed by William H. Short, the director of the Motion Picture Research Council, the body responsible for the Payne Fund series on movies and youth. Recently, however, an unfinished manuscript version of this book was discovered and published as "The Community—a Social Setting for the Motion Picture." It forms part of a recent volume containing commentary on the Payne Fund Studies, along with some of the early documents related to the project (Jowett, Jarvie, and Fuller, *Children and the Movies*).

35. Cressey, "A Comparison of the Roles"; Cressey, *The Taxi Dance Hall*; Cressey, "The Social Role of Motion Pictures"; Cressey, "The Motion Picture as Informal Education"; Cressey, "New York University Motion Picture Study"; Cressey, "The Community"; Cressey, "A Study in Practical Philosophy;" and Cressey, "The Motion Picture Experience."

36. As Lea Jacobs argues, given the working title for Cressey's project, namely, "Boys, Movies, and City Streets," the study was intended as "an homage" to a classical work of Chicago sociology written by Jane Addams more than two decades earlier. (Addams, *The Spirit of Youth and City Streets*.) In this work, "Addams had complained of the ways in which movies shaped the aspirations and expectations of male adolescents in the working-class communities she observed during her time at Hull House." (Jacobs, "Reformers and Spectators.")

37. The approach that he developed was very much inflected by his graduate training at the University of Chicago. His Master's thesis was on the women hired as dance partners in the slum-area dance halls of Chicago. (Cressey, "The Closed Dance Hall.")

38. Jowett, Jarvie, and Fuller, *Children and the Movies*, 86.

39. Ibid., 86–87.

40. Cressey, "The Motion Picture," 508–9.

41. Ibid., 508.

42. Ibid., 509

43. Ibid.

44. Ibid., 511.

45. That Cressey was not always consistent in the position that he took is suggested by a newspaper article on the hearings of the United States Senate sub-committee on crime control, held in New York in November, 1933. ("Movies Stir Row." *New York Times*, 32.) As a representative of the Motion Picture Research Council (responsible for the Payne Fund series), Cressey seemed to have provided testimony that was in the same vein as those of William Short and Henry James Forman, namely that films were linked to crime among youth and that children who regularly attended films were lacking in scholastic attainment and the spirit of cooperation. These claims were vigorously contested by defenders of the movie industry in attendance, such as Governor Carle E. Milliken of Maine (secretary of the Motion Pictures Distributors of America), Gabriel L. Hess (counsel to the same organization) and Anthony Muto (representing the Motion Picture Board of Trade).

46. Park, "Sociology," 29.

47. Carey, J. T., *Sociology and Public Affairs*, 99–104.

48. Cressey, "The Community," 159.

49. Ibid., 160–64.

50. Ibid., 171.

51. Cressey, "The Social Role of Motion Pictures," 239–41.

52. Cressey provides an insightful account of his methodological assumptions in a research note that was later published along with two commentaries. See Cressey, "A Comparison of the Rules"; Bulmer, "The Methodology of *The Taxi Dance Hall*," and Dubin, "The Moral Continuum of Deviancy Research."

53. Cressey, "The Social Role of Motion Pictures," 239.

54. Cressey, *The Taxi Dance Hall*, 31–53.

55. Faris, *The Nature of Human Nature*, 61–83.

56. These included Znaniecki, *Social Actions*, 304–11.

57. Cressey, "The Motion Picture Experience," 524.

58. Ibid., 522.

59. Colony and Brown, "Elaboration, Revision, and Progress," 17–81.

60. An exception to this tendency is Lal, *The Romance of Culture*. She convincingly demonstrates that Park's approach to the understanding of urban race and ethnicity was rooted in his perspective on culture, communication, and social control.

61. Park's views on film remained the same throughout the 1930s. Toward the end of the decade, he noted for instance that "the moving picture touches and enlivens men on a lower level of culture than it is possible to do through the medium of the printed page. (in Park, *The Crowd and the Public*, 116).

62. Jowett, Jarvie, and Fuller, *Children and the Movies*, 44; Jacobs, "Reformers and Spectators," 30–31; and Maltby, "The Production Code," 37–72.

63. For instance, at a symposium held at a conference marking the twenty-fifth anniversary of the National Board of Review of Motion Pictures held in New York in February 1934, Thrasher stressed the "value of motion pictures as an educative force and the need for exert technique in determining their effect on human character." ("Motives of Censors," 17.) Thrasher later wrote a book on film that was dedicated to the Warner Brothers and contained numerous references throughout the text to the contributions they had made to the movie industry. (Thrasher, *Okay for Sound.*)

64. Cressey, "A Comparison of the Roles."

65. By the same token, Dana Polan convincingly argues that Thrasher's interest in film studies was already evident in his study of gangs (Thrasher, *The Gang*), which conceived them in a strikingly visual manner. (Polan, *Scenes of Instruction.*)

66. By virtue of its detailed and nuanced examination of the Payne Fund Studies, *Children and the Movies* goes a long way toward drawing attention to the importance of Cressey and Thrasher as pioneering media scholars.

WORKS CITED

Addams, Jane. *The Spirit of Youth and City Streets*. New York: Macmillan, 1909.

Anderson, Nels. *The Hobo: The Sociology of the Homeless Man*. Chicago: University of Chicago Press, 1923.

Blumer, Herbert. *Movies and Conduct*. New York: Macmillan, 1933.

Blumer, Herbert, and Philip M. Hauser. *Movies, Delinquency and Crime*. New York: Macmillan, 1933.

Bulmer, Martin. *The Chicago School of Sociology: Institutionalization, Diversity, and the Rise of Sociological Research*. Chicago: University of Chicago Press, 1984.

———. "The Methodology of *The Taxi Dance Hall*: An Early Account of Chicago Ethnography from the 1920s." *Urban Studies* 12 (1983): 95–101.

Carey, James T. *Sociology and Public Affairs: The Chicago School*. Beverly Hills, CA: Sage, 1975.

Carey, James W. "The Chicago School and Mass Communication Research." In *American Communication Research: The Remembered History*, edited by Everette E. Dennis and Ellen Wartella, 21–38. Mahwah, NJ: Lawrence Erlbaum Associates, 1996.

Clark, Carroll Dewitt. "News: A Sociological Study." PhD diss., University of Chicago, 1931.

Colony, Paul, and J. David Brown. "Elaboration, Revision, and Progress in the Second Chicago School." In *A Second Chicago School? The Development of a Postwar American Sociology*, edited by Gary Alan Fine, 17–81. Chicago and London: University of Chicago Press, 1995.

Cressey, Paul G. "The Closed Dance Hall." Master's thesis, University of Chicago, 1929.

———. "The Community—A Social Setting for the Motion Picture." In *Children and the Movies: Media Influence and the Payne Fund Controversy*, edited by Garth S. Jowett, Ian C. Jarvie, and Karen H. Fuller, 133–216. Cambridge: Cambridge University Press, 1996. First written 1932.

———. "A Comparison of the Roles of the 'Sociological Stranger' and the 'Anonymous Stranger' in Field Research." *Urban Life* (1983): 102–20. First written 1932.

———. "The Motion Picture as Informal Education." *Journal of Educational Sociology* 8 (1934): 504–15.

———. "The Motion Picture Experience as Modified by Social Background and Personality." *American Journal of Sociology* 3 (1938): 516–25.

———. "New York University Motion Picture Study—Outline of Chapters." In *Children and the Movies: Media Influence and the Payne Fund Controversy*, edited by Garth S. Jowett, Ian C. Jarvie, and Karen H. Fuller, 220–36. Cambridge: Cambridge University Press, 1996. First written 1932.

———. "The Social Role of Motion Pictures in an Interstitial Area." *Journal of Educational Technology* 6 (1932): 238–243.

———. "A Study in Practical Philosophy: A Review of Mortimer J. Adler's 'Art and Prudence.'" *Journal of Higher Education* 9 (1938): 244–50.

———. *The Taxi Dance Hall: A Sociological Study in Commercialized Recreation and City Life*. Chicago: University of Chicago Press, 1932.

Czitrom, Daniel. *Media and the American Mind: From Morse to McLuhan*. Chapel Hill: University of North Carolina Press, 1984.

Detweiler, Frederick German. *The Negro Press in the United States*. Chicago: University of Chicago Press, 1922.

———. "The Negro Press in the United States." PhD diss., University of Chicago, 1922.

Dubin, Steven C. "The Moral Continuum of Deviancy Research." *Urban Life* 12 (1983): 75–94.

Elliot, Melissa Mae. "Negro News and News Agencies." Master's thesis, University of Chicago, 1931.

Faris, Ellsworth. *The Nature of Human Nature and Other Essays in Social Psychology*. New York, London: McGraw-Hill, 1936.

Halley, Lois Kate. "A Study of Motion Pictures in Chicago as a Medium of Communication." Master's thesis, University of Chicago, 1924.

Handelman, Don. "Towards the Virtual Encounter: Horton's and Wohl's 'Mass Communication and Para-social Interaction.'" In *Canonic Texts in Media Research. Are There Any? Should There be? How About These?* edited by Elihu Katz, John Durham Peters, Tamar Liebes, and Avril Orloff, 137–51. Cambridge: Polity, 2003.

Hauser, Philip M. "Motion Pictures in Penal and Correctional Institutions: A Study of the Reactions of Prisoners to Movies." Master's thesis, University of Chicago, 1933.

Horton, Donald, and R. Richard Wohl. "Mass Communication and Para-social Interaction: Observations on Intimacy at a Distance." *Psychiatry* 19, 3 (1956): 215–29.

Hughes, Helen McGill. *News and the Human Interest Story*. Chicago: University of Chicago Press, 1940.

Jacobs, Lea. "Reformers and Spectators: The Film Education Movement in the Thirties." *Camera Obscura* 22 (1990): 29–49.

Jensen, Howard Elkenberry. "The Rise of Religious Journalism in the United States." PhD diss., University of Chicago, 1920.

Jowett, Garth S., Ian C. Jarvie, and Karen H. Fuller, eds. *Children and the Movies: Media Influence and the Payne Fund Controversy*. Cambridge: Cambridge University Press, 1996.

Katz, Elihu, and Daniel Dayan. "The Audience is a Crowd, the Crowd is a Public: Latter–day Thoughts on Lang and Lang's 'MacArthur Day in Chicago.'" In *Canonic Texts in Media Research: Are There Any? Should There be? How About These?* edited by Elihu Katz, John Durham Peters, Tamara Liebes, and Avril Orloff, 121–36. Cambridge: Polity, 2003.

Katz, Elihu, John Durham Peters, Tamara Liebes, and Avril Orloff, eds. *Canonic Texts in Media Research: Are There Any? Should There be? How About These?* Cambridge: Polity, 2003.

Kawabe, Kisaburo. "The Japanese Newspaper and its Relation to the Political Development of Modern Japan." PhD diss., University of Chicago, 1919.

———. *The Press and Politics in Japan: A Study of the Relation between the Newspaper and the Political Development of Modern Japan*. Chicago: University of Chicago Press, 1921.

Lal, Barbara Ballis. *The Romance of Culture in an Urban Civilization: Robert E. Park on Race and Ethnic Relations in Cities*. London and New York: Routledge, 1990.

Maltby, Richard. "The Production Code and the Hays Office." In *Grand Design: Hollywood as a Modern Business Enterprise, 1930–1939*, edited by Tino Balio, 37–72. New York: Scribners, 1993.

Merrill, Francis Ellsworth. "The News and the Money Market." Master's thesis, University of Chicago, 1934.

"Motives of Censors are attacked by Rice." *New York Times*, February 10 (1934): 17.

"Movies Stir Row at Crime Hearing." *New York Times*, November 25 (1933): 32.

"News and Notes." *American Journal of Sociology* 56 (1951): 486–93.

"Obituary: Paul G. Cressey (1901–1955)." *American Sociological Review* 20 (1955): 579.

Park, Robert E. *The Crowd and the Public and Other Essays*. Edited with an introduction by Henry Elsner, Jr. Chicago: University of Chicago Press, 1972.

———. *The Immigrant Press and its Control*. New York: Harper, 1922.

———. *Masse und Publikum: Eine Methodologische und Soziologische Untersuchung (The Crowd and the Public: a Methodological and Sociological Investigation)*. Bern, Switzerland: Lack und Grunau, 1904.

———. "Natural History of the Newspaper." *American Journal of Sociology* 29 (1923): 80–98.

———. "News and the Human Interest Story." In *News and the Human Interest Story*, by Helen McGill Hughes, xi–xxiii. Chicago: University of Chicago Press, 1940.

———. "News as a Form of Knowledge: A Chapter in the Sociology of Knowledge." *American Journal of Sociology* 45 (1940): 669–86.

———. "Sociology." In *Research in the Social Sciences*, edited by Wilson Gee, 3–49. New York: Macmillan, 1929.

Park, Robert E., and Ernest W. Burgess. *Introduction to the Science of Sociology*. Chicago: University of Chicago Press, 1921.

Park, Robert E., Ernest W. Burgess, and Roderick D. McKenzie. *The City: Suggestions for the Investigation of Human Nature in the Urban Environment*. Chicago: University of Chicago Press, 1925.

Park, Robert E., Ernest W. Burgess, and Roderick D. McKenzie. (with an introduction by Morris Janowitz) *The City*. Chicago: University of Chicago Press, 1967.

Polan, Dana. *Scenes of Instruction: The Beginnings of the U.S. Study of Film*. Berkeley: University of California Press, 2007.

Raushenbush, Winifred. *Robert E. Park: Biography of a Sociologist*. Durham, NC: Duke University Press, 1979.

Rothenbuhler, Eric W. "Community and Pluralism in Wirth's 'Consensus and Mass Communication.'" In *Canonic Texts in Media Research: Are There Any? Should There be? How About These?* edited by Elihu Katz, John Durham Peters, Tamar Liebes, and Avril Orloff, 106–20. Cambridge: Polity, 2003.

Taylor, Ian, Paul Walton, and Jock Young. *The New Criminology: For a Social Theory of Deviance*. New York: Harper & Row, 1973.

Thrasher, Frederic. *The Gang: A Study of 1313 Gangs in Chicago*. Chicago: University of Chicago Press, 1927.

———, ed. *Okay for Sound: How the Screen Found its Voice*. New York: Duell, Sloan and Pearce, 1946.

Woolbert, Richard Latham. "The Social Effect of Radio." Master's thesis, University of Chicago, 1930.

Znaniecki, Florian. *Social Actions*. New York: Farrar & Rinehart, 1936.

Zorbaugh, Harvey. *The Gold Coast and the Slum*. Chicago: University of Chicago Press, 1929.

Appendix

ARTICLES ENTERED INTO ANALYSIS OF ANDERSON AND COLVIN,
"MEDIA RESEARCH, 1900—1945"

1900—1909

Adams, Myron E. "Children in American Street Trades." *Annals of the American Academy of Political and Social Science* 25 (1905): 23–44.

Anonymous. "Is an Honest and Sane Newspaper Press Possible?" *American Journal of Sociology* 15, no. 3 (1909): 321–34.

Fleming, Herbert E. "The Literary Interests of Chicago I and II." *American Journal of Sociology* 11, no. 3 (1905): 337–408.

———. "Literary Interests of Chicago III and IV." *American Journal of Sociology* 11, no. 4 (1906): 499–531.

Gilbert, Simeon. "The Newspaper as Judiciary." *American Journal of Sociology* 12, no. 3 (1906): 289–97.

Hysop, James H. "Newspaper Science." *Psychological Review* 1 (1900): 64–65.

Jastrow, Joseph. "Community and Association of Ideas: A Statistical Study." *Psychological Review* 1 (1900): 152–58.

Ochs, George W. "Journalism." *Annals of the American Academy of Political and Social Science* 28 (1906): 38–57.

Powers, John O. "Advertising." *Annals of the American Academy of Political and Social Science* 22 (1903): 58–62.

Shepard, Walter J. "Public Opinion." *American Journal of Sociology* 15, no. 1 (1909): 32–60.

Vincent, George E. "Laboratory Experiment in Journalism." *American Journal of Sociology* 11, no. 3 (1905): 297–311.

1910–1919

Anonymous. "School Children and Motion Pictures in England." *School and Society* VI, 134 (1917): 78–80.

Averill, Lawrence Augustus. "Educational Possibilities of the Motion Picture." *Educational Review* L (June–December 1915): 392–98.

Blanchard, Phyllis. "The Motion Picture as an Educational Asset." *Pedagogical Seminary* XXVI, no. 3 (1919): 284–87.

Du Breuil, Alice Jouveau. "The Moving Picture and the School: The First Organized Visual Instruction and What Preceded It." *Educational Review* XLIX (1915): 203–11.

Fleming, Pierce J. "Moving Pictures as a Factor in Education." *Pedagogical Seminary* XVII (1911): 336–52.

Howard, George Elliott. "Social Psychology of the Spectator." *American Journal of Sociology* 8, no. 1 (1912): 33–50.

Kellogg, Angie L. "Crime and Sociology." *Psychological Bulletin* XI (1914): 454–63.

Lickley, Ernest J. "Causes of Truancy among Boys Based on a Study of 1554 Cases." *Studies in Sociology* II, no. 2 (1917): 1–12.

Murray, Elva E. "Social Thought in the Current Short Story." *Studies in Sociology* II, no. 3 (1918): 1–12.

Strong, Edward K. "The Effect of Size of Advertisements and Frequency of Their Presentation." *Psychological Review* XXI (1914): 136–52.

Watson, Homer K. "Causes of Delinquency among Fifty Negro Boys." *Studies in Sociology* IV, no. 2 (1919): 1–12.

Woolbert, Charles H. "The Audience." *Psychological Monographs* XXI, no. 1 (1916): 37–54.

1920–1929

Allen, Eric W. "Journalism as an Applied Social Science." *Journalism Bulletin* 4, no. 1 (1927): 1–7.

———. "Organization of Research." *Journalism Quarterly* 5, no. 4 (1929): 10–18.

Beuick, Marshall D. "The Limited Social Effect of Radio Broadcasting." *American Journal of Sociology* 32, no. 4 (1927): 615–22.

———. "The Social Tendency in Newspaper Editorials. I. The Decline of the Political Editorial." *Social Forces* 4, no. 1 (1925): 156–62.

Dorland, W. A. Newman. "The Doctor's Prescription." *Journalism Quarterly* 5, no. 1 (1928): 1–12.

Eggen, J. B. "A Behavioristic Interpretation of Jazz." *Psychological Review* XXXIII (1926): 407–09.

Fowler, Rosalie Bunker. "A Study of Delinquency among School Girls." *Journal of Applied Sociology* VI, no. 2 (1921): 25–28.

Freeman, Bernadine. "The Movies in Education." *Educational Review* LXXII (June–December 1926): 115–18.

Gault, R. H. "The Standpoint of Social Psychology." *Journal of Abnormal Psychology and Social Psychology* XVI (1921–1922): 41–46.

Geiger, Roy. "The Effects of the Motion Picture on the Mind and Morals of the Young Joseph." *International Journal of Ethics* 34, no. 1 (1923): 69–83.

Higginbotham, Leslie. "Practice vs. Ph.D." *Journalism Bulletin* 1, no. 1 (1924): 10–12.

Highfill, Robert D. "The Effects of News of Crime and Scandal upon Public Opinion." *Journal of American Institute of Criminal Law and Criminology* 17, no. 1 (1926): 40–103.

Holmes, Joseph L. "Crime and the Press." *Journal of American Institute of Criminal Law and Criminology* 20, no. 1 (1929): 6–59.

———. "Crime and the Press (Concluded)." *Journal of American Institute of Criminal Law and Criminology* 20, no. 2 (1929): 246–93.

Lehman, Harvey C., and Paul A. Witty. "The Compensatory Function of the Movies." *Journal of Applied Psychology* XI, no. 1 (1927): 33–41.

———. "Newspaper vs. Teacher." *Educational Review* LXXIII (February 1927): 97–102.

Lloyd, Alfred H. "Newspaper Conscience—a Study in Half-Truths." *American Journal of Sociology* XXVII (July 1921): 197–210.

Lundberg, George A. "The Content of Radio Programs." *Social Forces* 7, no. 1 (1928): 58–60.

———. "The Newspaper and Public Opinion." *Social Forces* 4, no. 4 (1926): 709–15.

Ogden, R. M. "The Phenomenon of 'Meaning'." *American Journal of Psychology* XXXIV (1923): 223–30.

Poffenberger, A. T. "The Conditions of Belief in Advertising." *Journal of Applied Psychology* VII, no. 1 (1923): 1–9.

———. "Motion Pictures and Crime." *Scientific Monthly* 12, no. 4 (1921): 336–39.

Potter, Ellen C. "Spectacular Aspects of Crime in Relation to the Crime Wave." *Annals of the American Academy of Political and Social Science* 125 (1926): 1–19.

Stephens, Harmon B. "The Relation of the Motion Picture to Changing Moral Standards." *Annals of the American Academy of Political and Social Science* 128 (1926): 151–57.

Thurstone, L. L. "The Measurement of Opinion." *Journal of Abnormal Psychology and Social Psychology* XXII (1927–1928): 415–30.

Tolman, Edward Chace. "A New Formula for Behaviorism." *Psychological Review* XXIX (1922): 44–53.

Weiss, Benjamin S. "The Employment of Children in the Motion Picture Industry." *Journal of Applied Sociology* 6 (1921/1922): 11–18.

Willey, Malcolm M. "The Influence of Social Change on Newspaper Style." *Journal of Applied Sociology* 6 (1921/1922): 32–37.

1930–1939

Albig, William. "The Content of Radio Programs, 1925–1935." *Social Forces* 16, no. 3 (1938): 338–49.

Allport, G. W., and R. L. Schanck. "Are Attitudes Biological or Cultural in Origin." *Character and Personality: An International Psychological Quarterly* IV (1938): 195–205.

Alper, Benedict S. "Teen-Age Offenses and Offenders." *American Sociological Review* 4, no. 2 (1939): 167–72.

Anonymous. "Radio Educational Programs." *School and Society* 39, no. 1012 (1934): 640.

Arnspiger, V. C. "The Educational Talking Picture." *Journal of Education Sociology* 10, no. 3 (1936): 143–50.

Bain, Read. "Theory and Measurement of Attitudes and Opinions." *Psychological Bulletin* 27, no. 4 (1930): 357–79.

Bair, Frederick H. "Exploring the Educational Possibilities of Motion Pictures." *Clearing House* 12, no. 9 (1938): 515–18.

Baird, Thomas. "Civic Education and the Motion Picture." *Journal of Education Sociology* II, no. 3 (1937): 142–48.

Barrett, Wilton A. "The National Board of Review of Motion Pictures—How It Works." *Journal of Education Sociology* 10, no. 3 (1936): 177–88.

Benham, Albert. "War and Peace in the Movies." *Public Opinion Quarterly* 1, no. 4 (1937): 109–14.

Bernays, Edward L. "Molding Public Opinion." *Annals of the American Academy of Political and Social Science* 179 (1935): 82–87.

Beyle, Herman C. "Determining the Effect of Propaganda Campaigns." *Annals of the American Academy of Political and Social Science* 179 (1935): 106–13.

Bloodgood, Ruth. "Interest of the League of Nations in Motion Pictures in Relation to Child Welfare." *Journal of Educational Sociology* II, no. 3 (1937): 138–41.

Blumer, Herbert. "Moulding of Mass Behavior through the Motion Picture." *Proceedings of the American Sociological Society* (1935): 115–27.

Brown, Francis J. "Media of Propaganda." *Journal of Education Sociology* 10, no. 6 (1937): 323–30.

Bruel, Oluf. "A Moving Picture as a Psychopathogenic Factor: A Paper on Primary Psychotraumatic Neurosis." *Character and Personality: An International Psychological Quarterly* VII (1938–1939): 68–76.

Bruntz, George G. "Allied Propaganda and the Collapse of German Morale in 1918." *Public Opinion Quarterly* 2, no. 1 (1938): 61–76.

Cantril, Hadley. "The Effect of Modern Technology and Organization upon Social Behavior." *Social Forces* 15, no. 4 (1937): 493–95.

———. "The Role of the Radio Commentator." *Public Opinion Quarterly* 3, no. 4 (1939): 654–62.

Cantril, Hadley, and Hazel Gaudet. "Familiarity as a Factor in Determining the Selection and Enjoyment of Radio Programs." *Journal of Applied Psychology* XXIII, no. 1 (1939): 85–94.

Casey, Ralph D. "Republican Propaganda in the 1936 Campaign." *Public Opinion Quarterly* 1, no. 2 (1937): 27–44.

Church, George F. "The Socio-Psychological Nature of News." *Social Forces* 17, no. 2 (1938): 190–95.

Cooper, William John. "The Future of Radio in Education." *School and Society* 36, no. 916 (1932): 65–68.

Coutant, F.R. "Determining the Appeal of Special Features of a Radio Program." *Journal of Applied Psychology* XXIII, no. 1 (1939): 54–57.

Cressey, Paul G. "The Motion Picture as Informal Education." *Journal of Education Sociology* VII, no. 8 (1934): 504–15.

———. "The Motion Picture Experience as Modified by Social Background and Personality." *American Sociological Review* 3, no. 4 (1938): 516–25.

Crossley, Archibald M. "Straw Polls in 1936." *Public Opinion Quarterly* 1, no. 1 (1937): 24–35.

Curtis, Alberta. "The Reliability of a Report on Listening Habits." *Journal of Applied Psychology* XXIII, no. 1 (1939): 127–30.

Dale, Edgar. "The Motion Picture and Intergroup Relationships." *Public Opinion Quarterly* 3, no. 2 (1938): 39–42.

———. "Motion Picture Industry and Public Relations." *Public Opinion Quarterly* 3, no. 2 (1939): 251–62.

———. "Need for Study of the Newsreels." *Public Opinion Quarterly* 1, no. 3 (1937): 122–25.

Darvall, Frank. "The Film as an Agency of British-American Understanding." *Journal of Educational Sociology* II, no. 3 (1937): 129–37.

Dearborn, George Van Ness. "Children at the Movies." *School and Society* 40, no. 1022 (1934): 127–28.

DeWitt, S. A. "The Newspapers: Instruments of Propaganda or of Education." *Clearing House* VIII, no. 6 (1934): 368–69.

Doan, Edward N. "Chain Newspapers in the United States." *Journalism Quarterly* 9, no. 4 (1932): 229–338.

Doob, Leonard W., and Edward S. Robinson. "Psychology and Propaganda." *Annals of the American Academy of Political and Social Science* 179 (1935): 88–95.

Eastman, Fred. "The Motion Picture and Its Public Responsibilities." *Public Opinion Quarterly* 2, no. 1 (1938): 44–46.

Eldridge, Donald A. "Motion-Picture Appreciation in the New Haven Schools." *Journal of Education Sociology* II, no. 3 (1937): 175–83.

Elliott, Frank R. "Memory Effects from Poster, Radio and Television Modes of Advertising an Exhibit." *Journal of Applied Psychology* XXI (1937): 504–12.

Erskine, John. "The Future of Radio as a Cultural Medium." *Annals of the American Academy of Political and Social Science* 177 (1935): 214–19.

Ewbank, H. L. "Studies in the Techniques of Radio Speech." *Quarterly Journal of Speech* (1932): 560–71.

Foster, H. Schuyler Jr. "How America Became Belligerent: A Quantitative Study of War News, 1914–17." *American Journal of Sociology* 40, no. 4 (1935): 464–75.

Fowler, Rosalie B. "Motion Picture Shows and School Girls." *Journal of Applied Sociology* 3, no. 1 (1938): 76–83.

Freeman, Bernadine. "The Technique Used in the Study of the Effect of Motion Pictures on the Care of the Teeth." *Journal of Education Sociology* VI, no. 5 (1933): 309-11.

Gallup, George. "Testing Public Opinion." *Public Opinion Quarterly* 2, no. 1 (1938): 8–14.

Gaskill, Harold V. "Broadcasting Versus Lecturing in Psychology: Preliminary Investigation." *Journal of Applied Psychology* XVII (1933): 317–19.

Gaudet, Hazel. "The Favorite Radio Program." *Journal of Applied Psychology* XXIII, no. 1 (1939): 115–26.

Gosnell, Harold F., and Margaret J. Schmidt. "Factorial Analysis of the Relation of the Press to Voting in Chicago." *Journal of Social Psychology* VII, no. 1 (1936): 375–85.

Goudy, Elizabeth. "Pupil Broadcasts as Motivation." *Clearing House* 13, no. 1 (1938): 549–51.

Gruenberg, Sidonie Matsner. "Radio and the Child." *Annals of the American Academy of Political and Social Science* 177 (1935): 123–28.

Hard, William. "Radio and Public Opinion." *Annals of the American Academy of Political and Social Science* 177 (1935): 105–13.

Hauser, Philip M., Paul G. Cressey, and Edgar Dale. "How Do Motion Pictures Affect the Conduct of Children?: Methods Employed In 'Movies and Conduct' And 'Movies, Delinquency, and Crime'." *Journal of Educational Sociology* 6, no. 4 (1932): 231–50.

Hayakawa, S. I. "General Semantics and Propaganda." *Public Opinion Quarterly* 3, no. 2 (1939): 197–208.

Hearon, Fannin. "The Motion-Picture Program and Policy of the United States Government." *Journal of Educational Sociology* (1938): 147–62.

Inglis, Ruth A. "An Objective Approach to the Relationship between Fiction and Society." *American Sociological Review* 3, no. 4 (1938): 526–33.

Jester, Ralph. "Hollywood and Pedagogy." *Journal of Education Sociology* 12 (1938): 137–41.

Jones, Harold Ellis, and Herbert S. Conrad. "Rural Preferences in Motion Pictures." *Journal of Social Psychology* 1, no. 3 (1930): 419–23.

Jones, John Price. "Public Opinion, the Depression, and Fund-Raising." *Public Opinion Quarterly* 1, no. 1 (1937): 142–47.

Karol, John J. "Measuring Radio Audiences." *Public Opinion Quarterly* 1, no. 2 (1937): 92–96.

Katz, Elias. "Making Movies in the Classroom." *Clearing House* 11, no. 1 (1936): 153–56.

Keliher, Alice V. "The Motion Picture and Problems of Youth." *Public Opinion Quarterly* 2, no. 1 (1938): 53–54.

Kenney, Erle A. "Administrative Problems in the Use of Radio in the Classroom." *California Journal of Elementary Education* VII, no. 2 (1938): 101–08.

Landry, Robert J. "Radio and Government." *Public Opinion Quarterly* 2, no. 4 (1938): 557–69.

Lawton, Sherman P. "The Principles of Effective Radio Speaking." *Quarterly Journal of Speech* XVI, no. 3 (1930): 266–77.

Lazarsfeld, Paul F., and Marjorie Fiske. "The 'Panel' As a New Tool for Measuring Opinion." *Public Opinion Quarterly* 2, no. 4 (1938): 496–612.

Lucas, D. B. "The Impression Values of Fixed Advertising Locations in the Saturday Evening Post." *Journal of Applied Psychology* XXI (1937): 613–31.

Lumley, F. H. "Rates of Speech in Radio Speaking." *Quarterly Journal of Speech* (1933): 392–403.

Lundberg, George A. "Public Opinion from a Behavioristic Viewpoint." *American Journal of Sociology* 36, no. 3 (1930): 387–405.

Marvin, Kenneth R. "What Makes the Consumer Know Where to Buy?" *Journalism Quarterly* 15, no. 2 (1938): 185–90.

May, Mark A. "Educational Possibilities of Motion Pictures." *Journal of Education Sociology* II, no. 3 (1937): 149–60.

McMullan, Lois. "The Newspaper Goes to School." *Clearing House* 13, no. 1 (1938): 103–04.

Meyrowitz, Alvin, and Marjorie Fiske. "The Relative Preference of Low Income Groups for Small Stations." *Journal of Applied Psychology* XXIII, no. 1 (1939): 158–62.

Miller, Helen Rand. "Motion Pictures: A Social and Educational Force." *Journal of Education Sociology* II, no. 3 (1937): 164–65.

Miller, Neville. "Radio's Code of Self-Regulation." *Public Opinion Quarterly* 3, no. 4 (1939): 683–88.

Newcomb, Theodore. "Determinants of Opinion." *Public Opinion Quarterly* 1, no. 4 (1937): 71–78.

Noble, Lorraine. "Modernization, by Way of the Educational Film." *Journal of Education Sociology* 10, no. 3 (1936): 151–57.

Ollry, Francis, and Elias Smith. "An Index of 'Radio-Mindedness' and Some Applications." *Journal of Applied Psychology* XXIII, no. 1 (1939): 8–18.

Olson, Willard C. "Social Significance of the Cinema." *School Review* 42, no. 6 (1934): 466–70.

Orr, Ada E., and Francis J. Brown. "A Study of the Out-of-School Activities of High-School Girls." *Journal of Educational Sociology* 5, no. 5 (1932): 266–73.

Paley, William S. "Radio and the Humanities." *Annals of the American Academy of Political and Social Science* 177 (1935): 94–104.

Parsons, Talcott. "The Role of Ideas in Social Action." *American Sociological Review* 3, no. 5 (1938): 652–64.

Patton, Carl Safford. "Moving Pictures in the Church." *Journal of Applied Sociology* 3, no. 1 (1938): 59–64.

Pease, Kent. "Hamden's Course in Appreciation of Movies and Radio." *Clearing House* 12, no. 1 (1937): 39–43.

Peters, Charles C. "The Relation of Motion Pictures to Standards of Morality." *Journal of Education Sociology* VI, no. 5 (1933): 251–55.

———. "The Relation of Motion Pictures to Standards of Morality." *School and Society* 39, no. 1005 (1934): 414–18.

Phillips, Delight. "A Unit on the Use of Radio." *English Journal* 26, no. 1 (1937): 33–38.

Pinney, Jean B. "The Motion Picture and Social-Hygiene Education." *Journal of Education Sociology* 10, no. 3 (1936): 158–67.

Poole, De Witt Clinton. "Public Opinion and 'Value Judgments'." *Public Opinion Quarterly* 3, no. 3 (1939): 371–75.

Potter, Gary. "A Study of the Use of the Radio in a Group of California Schools." *California Journal of Elementary Education* V, no. 1 (1936): 59–64.

Quigley, Martin. "The Function of the Motion Picture." *Public Opinion Quarterly* 2, no. 1 (1938): 47–49.

———. "Public Opinion and the Motion Picture." *Public Opinion Quarterly* 1, no. 2 (1937): 129–33.

Reed, Thomas H. "Commercial Broadcasting and Civic Education." *Public Opinion Quarterly* 1, no. 3 (1937): 57–67.

Reitze, Arnold W. "Research Projects and Methods: The Motion Picture." *Journal of Educational Sociology* 12, no. 3 (1938): 177–81.

Remmers, H. H. "Propaganda in the Schools–Do the Effects Last?" *Public Opinion Quarterly* 2, no. 2 (1938): 197–210.

Riegel, O. W. "Nationalims in Press, Radio and Cinema." *American Sociological Review* 3, no. 1 (1938): 510–15.

———. "New Frontiers in Radio." *Public Opinion Quarterly* 1, no. 1 (1937): 136–41.

———. "Propaganda and the Press." *Annals of the American Academy of Political and Social Science* 179 (1935): 201–10.

Rosenthal, Solomon P. "Change of Socio-Economic Attitudes under Radical Motion Picture Propaganda." *Archives of Psychology* 166 (1934): 1–46.

Ruckmick, Christian A. "How Do Motion Pictures Affect the Attitudes and Emotions of Children?" *Journal of Education Sociology* VII, no. 8 (1934): 210–19.

Sartain, Geraldine. "The Cinema Explodes the Stork Myth." *Journal of Educational Sociology* (1938): 142–46.

Schank, R. L., and Charles Goodman. "Reactions to Propaganda on Both Sides of a Controversial Issue." *Public Opinion Quarterly* 3, no. 1 (1939): 107–12.

Schur, Abraham. "The Motion Picture as a Testing Device." *Clearing House* IX, 7 (1935): 504.

Sheridan, Marion C. "Rescuing Civilization through Motion Pictures." *Journal of Education Sociology* II, no. 3 (1937): 166–73.

Short, William H. "The Effect of Motion Pictures on the Social Attitudes of High-School Children." *Journal of Education Sociology* VII, no. 8 (1934): 220–28.

Stoddard, George D. "Measuring the Effect of Motion Pictures on the Intellectual Content of Children." *Journal of Education Sociology* VII, no. 8 (1934): 204–09.

Sumner, F. C. "Measurement of the Relevancy of Picture to Copy in Advertisements." *Journal of Psychology* 7 (1939): 399–406.

Talbert, E. L. "The Modern Novel and the Response of the Reader." *Journal of Abnormal and Social Psychology* 26, no. 4 (1932): 408–14.

Thomas, Frank W. "Radio—an Instrument in Progressive Education." *Clearing House* X, no. 5 (1936): 402–06.

Thrasher, Frederic M. "The Motion Picture: Its Nature and Scope." *Journal of Education Sociology* 10, no. 3 (1936): 129–42.

Thurstone, L. L. "Influence of Motion Pictures on Children's Attitudes." *Journal of Social Psychology* II, no. 1 (1931): 291–304.

Tyson, L. B. "The Radio Influences Speech." *Quarterly Journal of Speech* (1932): 219–24.

Wagner, Isabelle F. "Articulate and Inarticulate Replies to Questionnaires." *Journal of Applied Psychology* XXIII, no. 1 (1939): 104–15.

Western, John. "Television Girds for Battle." *Public Opinion Quarterly* 3, no. 4 (1939): 547–63.

Wiebe, Gerhart. "A Comparison of Various Rating Scales Used in Judging the Merits of Popular Songs." *Journal of Applied Psychology* XXIII, no. 1 (1939): 18–22.

Wilke, Walter H. "An Experimental Comparison of the Speech, the Radio, and the Printed Page as Propaganda Devices." *Archives of Psychology* 169 (1934): 1–32.

Wright, Quincy, and Carl J. Nelson. "American Attitudes toward Japan and China, 1937–38." *Public Opinion Quarterly* 3 (1939): 1.

Young, Kimball. "Book Reviews." *American Journal of Sociology* 41, no. 2 (1935): 249–55.

1940–1947

Allport, Floyd H. "Polls and the Science of Public Opinion." *Public Opinion Quarterly* 4, no. 2 (1940): 249–57.

Allport, Gordon W., and Janet M. Faden. "The Psychology of Newspapers: Five Tentative Laws." *Public Opinion Quarterly* 4, no. 4 (1940): 687–703.

Ammons, R. B. "Book Review: Propaganda, Communication, and Public Opinion; a Comprehensive Reference Guide." *Psychological Bulletin* 44, no. 1 (1947): 94–95.

Bartlett, Kenneth G. "Trends in Radio Programs." *Annals of the American Academy of Political and Social Science* 213 (1941): 15–25.

Boring, Edwin G. "The Use of Operational Definitions in Science." *Psychological Review* 52, no. 5 (1945): 243-245.

Bridgeman, P. W. "Some General Principles of Operational Analysis." *Psychological Review* 52, no. 5 (1945): 246–49.

Cantril, Hadley. "America Faces the War: A Study in Public Opinion." *Public Opinion Quarterly* 4, no. 3 (1940): 387–407.

———. "Introduction to Symposium." *Public Opinion Quarterly* 4, no. 2 (1940): 212–17.

Clark, Weston R. "Radio Listening Habits of Children." *Journal of Social Psychology* XII (1940): 131–49.

Crespi, Leo P. "Opinion-Attitude Methodology' and the Polls—a Rejoinder." *Psychological Bulletin* 43, no. 4 (1946): 562–69.

Crespi, Leo P., and Donald Rugg. "Poll Data and the Study of Opinion Determinants." *Public Opinion Quarterly* 4, no. 2 (1940): 273–76.

Daniel, Cuthbert. "Three Types of 'Like' Reactions in Judging Popular Songs." *Journal of Applied Psychology* XXIV, no. 6 (1940): 746–48.

Durant, Henry, and Ruth Durant. "Lord Haw-Haw of Hamburg: 2. His British Audience." *Public Opinion Quarterly* 4, no. 3 (1940): 443–50.

Edman, Marion. "Attendance of School Pupils and Adults at Moving Pictures." *School Review* December (1940): 753–63.

Erdelyi, Michael. "The Relation between 'Radio Plugs' and Sheet Sales of Popular Music." *Journal of Applied Psychology* XXIV, no. 6 (1940): 696–702.

Ewbank, H. L. "Trends in Research in Radio Speech." *Quarterly Journal of Speech* 26 (1940): 282–87.

Fay, Paul J., and Warren C. Middleton. "The Depauw Laboratory for Research on the Psychological Problems of Radio." *American Journal of Psychology* LIV, no. 1 (1941): 571–75.

Feigl, Herbert. "Operationism and Scientific Method." *Psychological Review* 52, no. 5 (1945): 250–59.

Fleischman, Earl E., and Richard Woellhaf. "The Motion Picture: A Neglected 'Liberal Art'." *Quarterly Journal of Speech* (1942): 182–84.

Fleiss, Marjorie. "The Panel as an Aid in Measuring Effects of Advertising." *Journal of Applied Psychology* XXIV, no. 6 (1940): 685–95.

Foulds, Graham. "The Child's Response to Fictional Characters and Its Relationship to Personality Traits." *Character and Personality: An International Psychological Quarterly* XI (1942): 64–75.

Franzen, Raymond. "An Examination of the Effect of Number of Advertisements in a Magazine upon the 'Visibility' of These Advertisements." *Journal of Applied Psychology* XXIV, no. 6 (1940): 791–801.

Gallup, George. "Is There a Bandwagon Vote?" *Public Opinion Quarterly* 4, no. 2 (1940): 244–49.

Graves, Harold N. Jr. "Lord Haw-Haw of Hamburg: The Campaign against Britain." *Public Opinion Quarterly* 4, no. 3 (1940): 429–42.

Gruenberg, Sidonie Matsner. "The Comics as a Social Force." *Journal of Educational Sociology* 18, no. 4 (1944): 204–13.

Guest, Lester. "Book Review: The People's Choice. How the Voter Makes up His Mind in a Presidential Campaign: Lazarsfeld, P.F., Berelson, B., Gaudet, H." *Psychological Bulletin* 43 (1946): 83–84.

Heider, Fritz. "Attitudes and Cognitive Organization." *Journal of Psychology* 21 (1946): 107–12.

Israel, Harold E. "Two Difficulties in Operational Thinking." *Psychological Review* 52, no. 5 (1945): 260–61.

Kantor, J. R. "An Objective Interpretation of Meanings." *American Journal of Psychology* LIV, no. 1 (1941): 230–35.

Katz, Daniel. "Three Criteria: Knowledge, Conviction, and Significance." *Public Opinion Quarterly* 4, no. 2 (1940): 277–84.

Kempf, Edward J. "The Law of Attitude." *Journal of General Psychology* 32 (1945): 81–102.

Lazarsfeld, Paul F. "The Use of Mail Questionnaires to Ascertain the Relative Popularity of Network Stations in Family Listening Surveys." *Journal of Applied Psychology* XXIV, no. 6 (1940): 802–16.

Miller, Neal E., Robert R. Sears, O. H. Mowrer, Leonard W. Doob, and John Dollard. "The Frustration-Aggression Hypothesis." *Psychological Review* 48, no. 4 (1941): 337–42.

Paley, William S. "Broadcasting and American Society." *Annals of the American Academy of Political and Social Science* 213 (1941): 62–68.

Pratt, Carroll C. "Operationism in Psychology." *Psychological Review* 52, no. 5 (1945): 262–69.

Preston, Mary I. "Children's Reactions to Movie Horrors and Radio Crime." *Journal of Pediatrics* 19, no. 2 (1941): 148–67.

Rachford, Helen Fox. "Developing Discrimination in Radio Listening." *English Journal* 33, no. 6 (1944): 315–17.

Reid, Seerley. "Reading, Writing, and Radio: A Study of Five School Broadcasts in Literature." *Journal of Applied Psychology* XXIV, no. 6 (1940): 703–13.

Riegel, O. W. "Press, Radio, Films." *Public Opinion Quarterly* 4, no. 1 (1940): 136–50.

Roper, Elmo. "Classifying Respondents by Economic Status." *Public Opinion Quarterly* 4, no. 2 (1940): 270–72.

Roslow, Sydney. "Measuring the Radio Audience by the Personal Interview Roster Method." *Journal of Applied Psychology* 27, no. 6 (1943): 526–34.

Ross, Jeanette. "The Speech Teacher Keeps Abreast of the Radio and the Motion Picture." *Quarterly Journal of Speech* (1940): 431–37.

Rowland, Howard. "Radio Crime Dramas." *Educational Research Bulletin* 23, no. 8 (1944): 210–17.

Ruch, Floyd L., and Kimball Young. "Penetration of Axis Propaganda." *Journal of Applied Psychology* XXVI (1942): 448–55.

Rugg, Donald, and Hadley Cantril. "The Wording of Questions in Public Opinion Polls." *Journal of Abnormal Psychology and Social Psychology* 37, no. 1 (1942): 469–95.

Sarnoff, David. "Possible Social Effects of Television." *Annals of the American Academy of Political and Social Science* 213 (1941): 145–52.

Sears, Robert R., Carl Iver Hovland, and Neal E. Miller. "Minor Studies of Aggression: I. Measurement of Aggressive Behavior." *Journal of Psychology* 9 (1940): 275–95.

Seldes, Gilbert. "The Nature of Television Programs." *Annals of the American Academy of Political and Social Science* 213 (1941): 138–44.

Sherif, Muzafer, and Hadley Cantril. "The Psychology of 'Attitudes'—Part I." *Psychological Review* 52, no. 6 (1945): 295–319.

———. "The Psychology of 'Attitudes' Part II." *Psychological Review* 53, no. 1 (1946): 1–24.

Smith, Mapheus. "Communicative Behavior." *Psychological Review* 53 (1946): 294–301.

Townsend, Howard W. "Factors of Influence in Radio Speech." *Quarterly Journal of Speech* (1944): 187–90.

———. "Psychological Aspects of Radio Speech." *Quarterly Journal of Speech* (1940): 579–85.

Turner, Ralph. "Culture, Change and Confusion." *Public Opinion Quarterly* 4, no. 4 (1940): 579–600.

Winslow, Charles N. "Sympathetic Pennies: A Radio Case Study." *Journal of Abnormal Psychology and Social Psychology* 39 (1944): 174–79.

Contributors

James A. Anderson (PhD, Iowa) is professor of communication and director of the Center for Communication and Community at the University of Utah. He regularly publishes on matters of communication epistemology and methodology. His latest book is *Media Violence and Aggression: Science and Ideology* (Sage, 2008), co-authored with Tom Grimes and Lori Bergen.

William J. Buxton is professor of communication studies at Concordia University in Montreal. He is the co-editor of *Harold Innis in the New Century: Reflections and Refractions* (McGill-Queens, 1999), and has published several articles on Innis. A collection he edited, *Patronizing the Public: The Impact of American Philanthropy on Communication, Culture, and the Humanities in the Twentieth Century* will be published by Lexington Books.

Janet W. Colvin is an assistant professor at Utah Valley State College. She received her doctorate from the University of Utah in 2005. Her interests include communication theory and instructional communication.

Hanno Hardt is professor of communication and media studies at the University of Ljubljana and professor emeritus of journalism and mass communication and communication studies at the University of Iowa. He is the author of scholarly books and articles, including, most recently, *Myths for the Masses* (Blackwell, 2004) and "Cruising on the Left: Notes on a Genealogy of 'Left' Communication Research in the United States" (*Fast Capitalism* 1999) (www.fastcapitalism.com)

Sue Curry Jansen received her PhD in sociology from the University at Buffalo (SUNY) and is professor of media and communication at Muhlenberg College in Allentown, Pennsylvania. She is the author of two books,

Censorship: The Knot that Binds Power and Knowledge (Oxford University Press, 1991) and *Critical Communication Theory: Power, Media, Gender, and Technology* (Rowman and Littlefield, 2002), and many chapters and journal articles.

Deborah Lubken is a PhD candidate at the University of Pennsylvania's Annenberg School for Communication. She is currently writing a dissertation on the cultural history of bells in the U.S.

David E. Morrison is currently professor of communications research at the University of Leeds, UK He has published widely on the history of communications research and the institutionalization of knowledge, journalists and war reporting, moral protest movements, audience responses to social issues, and methodological developments. His most recent book, *Media and Values: Intimate Transgressions in a Changing Moral and Cultural Landscape* (Intellect, 2007) examines moral incoherence in the contemporary world, and the way that this shows up in empirical research into individual attitudes/opinions/tastes/judgment.

Kaarle Nordenstreng is professor of journalism and mass communication at the University of Tampere, Finland. He has been in this position since 1971, before which he was head of research in the Finnish Broadcasting Company. He has published some thirty books and over 400 articles, mostly covering communication theory, international communication, and media ethics. He served as vice president of IAMCR from 1972–88 and president of IOJ from 1976–90. In Finland, he has served on several state committees on media and higher education, most recently as coordinator of social sciences for the national project of university degree reform.

David W. Park is Gustav E. Beerly, Jr. Assistant Professor in the Department of Communication at Lake Forest College. His research addresses intellectuals and the media, the history of the study of communication, and new media.

John Durham Peters is F. Wendell Miller Distinguished Professor in the Department of Communication Studies at the University of Iowa. He writes on media history, social theory, and sundry related and unrelated topics.

Veikko Pietilä is professor emeritus in the Department of Journalism and Mass Communication at the University of Tampere, Finland. His English publications on the development of mass communication studies include the monographs "On the Scientific Status and Position of Communication Research" (University of Tampere, 1978) and *On the Highway of Mass Communication Studies* (Hampton Press, 2005). He has also written about the development of science in general, the relationships between science and ideology, and the methodology of social sciences. He has moreover, reformulated literary narratological tools for news analysis and engaged lately in theorizing the relationships between the notions of "audience" and "public" in the framework of civic activity, public discussion, and the media.

Jefferson Pooley's research centers on the history of communication studies, as the field's emergence has intersected with the twentieth century rise of the other social sciences. His ongoing work in the history of communication research traces the emergence of a standard disciplinary memory, a storyline which helped to legitimate the infant social scientific field in the 1950s and 1960s. He is assistant professor of media and communication at Muhlenberg College.

Lana F. Rakow is a professor of communication and women studies at the University of North Dakota and director of the university's Center for Community Engagement. Rakow has a PhD from the Institute of Communications Research at the University of Illinois, Champaign-Urbana. She is the author or editor of four books on gender and communication, including *Feminist Communication Theory* (Sage, 2004), with Laura Wackwitz.

Wendy Worrall Redal, PhD., teaches media studies at the University of Colorado at Boulder. Her research interests include the history of mass communication theory and research, particularly the rise of cultural studies in Britain, and critical approaches to media coverage of the environment. Formerly with the University of Colorado's Center for Environmental Journalism, she now works as a freelance writer and editor on environment and health issues, in addition to teaching.

Peter Simonson is assistant professor in the Department of Communication at the University of Colorado at Boulder. He is editor of Politics, Social Networks *and the History of Mass Communications Research: Re-Reading* Personal Influence (sage, 2006), and, with John Durham Peters, co-editor of *Mass Communication and American Social Thought: Key Texts, 1919–1968* (Rowman & Littlefield, 2004). Among other work in the intellectual history of communication, he authored a new introduction to Robert K. Merton's *Mass Persuasion,* and served as executive producer of a documentary film, *The Long Road to Decatur: The Making of* Personal Influence.

J. Michael Sproule is professor of communication, and former dean of the College of Arts and Sciences, at Saint Louis University. He served in 2007 as president of the National Communication Association. Recognitions of his work in rhetoric, propaganda, and media history include a National Endowment for the Humanities Fellowship and the NCA Golden Anniversary Monograph Award.

Index